PROGRESS IN PHOTOSYNTHESIS RESEARCH

Progress in Photosynthesis Research

Volume 1

Proceedings of the VIIth International Congress on Photosynthesis
Providence, Rhode Island, USA, August 10–15, 1986

edited by

J. BIGGINS

Division of Biology and Medicine, Brown University
Providence, RI 02912, USA

1987 **MARTINUS NIJHOFF PUBLISHERS**
a member of the KLUWER ACADEMIC PUBLISHERS GROUP
DORDRECHT / BOSTON / LANCASTER

Distributors

for the United States and Canada: Kluwer Academic Publishers, P.O. Box 358, Accord Station, Hingham, MA 02018-0358, USA
for the UK and Ireland: Kluwer Academic Publishers, MTP Press Limited, Falcon House, Queen Square, Lancaster LA1 1RN, UK
for all other countries: Kluwer Academic Publishers Group, Distribution Center, P.O. Box 322, 3300 AH Dordrecht, The Netherlands

ISBN 90-247-3450-9 (vol. I)
ISBN 90-247-3451-7 (vol. II)
ISBN 90-247-3452-5 (vol. III)
ISBN 90-247-3453-3 (vol. IV)
ISBN 90-247-3449-5 (set)

GENERAL CONTENTS

VI

CONTENTS TO VOLUME I

1. Excitation Energy Transfer

3. Chlorophylls and Model Systems

4. Structure of Molecular Complexes: Crystallographic and Physical Studies

5. Oxygen Evolution

PREFACE

These Proceedings comprise the majority of the scientific contributions
that were presented at the VIIth International Congress on Photosynthesis.
The Congress was held August 10-15 1986 in Providence, Rhode Island, USA
on the campus of Brown University, and was the first in the series to be
held on the North American continent. Despite the greater average travel
distances involved the Congress was attended by over 1000 active
participants of whom 25% were registered students. This was gratifying
and indicated that photosynthesis will be well served by excellent young
scientists in the future.

As was the case for the VIth International Congress held in Brussels,
articles for these Proceedings were delivered camera ready to expedite
rapid publication. In editing the volumes it was interesting to reflect
on the impact that the recent advances in structure and molecular biology
had in this Congress. It is clear that cognizance of structure and
molecular genetics will be even more necessary in the design of
experiments and the direction of future research.

Shortly after the Brussels Congress in 1983 the photosynthesis community
was grieved to hear of the death of Professor Warren I Butler. Warren was
very enthusiastic about the prospect of holding the VIIth International
Congress in the USA and was anxious not only to participate in the
scientific program, but to welcome and host colleagues from overseas in
his country. A special issue of Photosynthesis Research will be published
shortly containing articles closely related to his field of study. Other
outstanding scientists who also passed away during this time period
include Drs. A Shlyk (USSR), E Roux (France), A Faludi-Dániel (Hungary)
and G Akoyunoglou (Greece). All were recognised during the Congress at
Brown University in symposia dedicated to their memory.

The development of the scientific program and planning of the VIIth
International Congress was the responsibility of the US Organising
Committee and the International Photosynthesis Committee, and their
assistance is gratefully acknowledged. At the local level I wish to thank
my colleagues at Brown University for their support and, in particular,
the outstanding effort provided by Professor Sam I Beale. Special thanks
are also due to Professor Frank Rothman, Dean of Biology, for extensive
logistical support and encouragement, and to Kathryn Holden, the Congress
Secretary. Finally it is a pleasure to acknowledge the long-term
assistance of Ir Adrian C Plaizier of Martinus-Nijhoff publishers for
providing guidance in the production of these Proceedings, and for
bringing the publication to fruition.

ACKNOWLEDGMENTS

The organizers and Congress participants wish to express their appreciation for the financial support received from the following agencies and companies:

United States National Science Foundation
United States Department of Energy
United States Department of Agriculture

VIth International Congress, residual fund

CIBA-GEIGY Corporation
Campbell Soup Company
*E.I. du Pont de Nemours and Company
Monsanto Company
*Pepsico Incorporated
Pfizer Incorporated
Philip Morris
Proctor and Gamble
*Rohm and Haas Company
Shell Development Company
Weyerhaeuser Company

* Benefactor contributor

Commercial products were gratefully received from the following companies:

R.C. Bigelow Incorporated
Coca Cola Bottling Company of Northampton
E.I. du Pont de Nemours and Company
R.T. French Company
Frito-Lay, Incorporated
J and J Corrugated Box Corporation
Nyman Manufacturing Company
Rhode Island Lithograph Company

INTERNATIONAL PHOTOSYNTHESIS COMMITTEE

J.M. Anderson (Australia)
C.J. Arntzen (U.S.A.)
M. Baltscheffsky (Sweden)
J. Barber (U.K.)
J. Biggins (U.S.A.), Chairman
R. Douce (France)
G. Forti (Italy)

H. Heldt (F.R.G.)
R. Malkin (U.S.A.)
N. Murata (Japan)
N. Nelson (Israel)
V.A. Shuvalov (U.S.S.R.)
C. Sybesma (Belgium)
R.H. Vallejos (Argentina)

US ORGANIZING COMMITTEE

C.J. Arntzen
J. Biggins (Chairman)
N.I. Bishop
L. Bogorad
A.L. Christy
A.R. Crofts
P.L. Dutton

M. Gibbs
G. Hind
A.T. Jagendorf
R.E. McCarty
K. Sauer
I. Zelitch

The committee is especially grateful to Dr. E. Romanoff of the National Science Foundation and Dr. R. Rabson of the Department of Energy for valuable advice and support during organization of the Congress.

CONGRESS SECRETARIAT: J. Biggins, Section of Biochemistry, Division of Biology and Medicine, Brown University, Providence, Rhode Island 02912

PICOSECOND ABSORPTION AND FLUORESCENCE SPECTROSCOPY OF ENERGY TRANSFER AND
TRAPPING IN PHOTOSYNTHETIC BACTERIA

R. VAN GRONDELLE, DEPARTMENT OF BIOPHYSICS, PHYSICS LABORATORY OF THE
FREE UNIVERSITY, DE BOELELAAN 1081, 1081 HV AMSTERDAM, THE NETHERLANDS

1. INTRODUCTION

The transfer of excitation energy in photosynthetic organisms is a
very efficient process. The characterization of the organization of the
light harvesting antenna and the physical principles involved is the
general subject of this work. In short, light absorbed by one of the
pigments is rapidly transferred down an energy gradient, until it reaches a
group of antenna pigments with the lowest excited state energy. In purple
bacteria this longwavelength antenna contains BChl 875, non covalently
linked to the pigment protein complex B875. B875 is assumed to form a network that at room temperature connects about 20 reaction centers (RCs). In
purple bacteria with all the RCs in an active ('open') state the lifetime
of the B875 excited state is less than 100 ps and during that time the
excitation has visited a large number of antenna pigment molecules [1-6].

2. THE TIME-RESOLVED TRANSFER PROCESS

In a photosynthetic bacterium such as Rhodopseudomonas (Rps)
sphaeroides excitations may be absorbed by for instance the carotenoids.
The carotenoid excited state is extremely short-lived ($\lesssim$ 1 ps [7,8]) and
within that time transfer must have taken place. Experimental estimates of
the in vivo carotenoid excited state lifetime are indeed close to 1 ps [8].
It is unknown what the mechanism of transfer is, but a close proximity of
the carotenoid and (bacterio)chlorophyll seems to be an absolute requirement [9]. However, it is well documented that in the antenna of photosynthetic purple bactera the carotenoid excited state can undergo a complicated sequence of reactions known as 'singlet fission' [10,11].
Especially in Rhodospirillum (Rs.) rubrum carotenoid triplet states are
formed very efficiently via carotenoid excited state singlet fission. For
this process to take place either two interacting carotenoid molecules must
be involved (homofission) or a carotenoid and a BChl molecule (hetero-
fission). The latter condition is exactly that required for excitation
energy transfer from carotenoid to BChl. On the other hand, intense
circular dichroism signals due to carotenoids are measured in Rps.
sphaeroides, indicating relatively strong excitonic interactions between
different carotenoid molecules.

In Rps. sphaeroides, besides B875, a second pigment protein complex
occurs, B800-850, that is found at the periphery of the B875 network [5]
and both complexes contain carotenoids. For the pigment organization in the
B800-B850 complex a minimum model was proposed that is shown in Fig. 1
[12]. In this model two types of carotenoid molecules can be distinguished:
one type transfers excitation energy to BChl 800, and is parallel to the
membrane plane, the other two transfer to BChl 850 and are perpendicular to
the membrane plane. Both transfer processes are highly efficient.

After transfer from carotenoid to BChl 800, or upon direct excitation

Biggins, J. (ed.), Progress in Photosynthesis Research, Vol. I. ISBN 90 247 3450 9
© *1987 Martinus Nijhoff Publishers, Dordrecht. Printed in the Netherlands.*

of BChl 800, the excited state bleaching due to BChl 800* can be observed
as a very short-lived absorption change ($\simeq$ 1 ps). This is shown in Fig. 2.
The lifetime corresponds to a distance of about 20 Å between BChl 800 and
one of the BChl 850's. The observed high polarization implies either that
the transfer rate among the two BChl 800 molecules is at most of the same
order as that from BChl 800 to BChl 850 or that the BChl 800 Q_y transition
moments are not perpendicular as suggested in the model in Fig. 1. Recent
experiments at 77 K support the latter possibility.

After transfer from BChl 800, or after transfer from the carotenoids
associated to BChl 850, or upon direct excitation, the BChl 850 excited
state is produced. At 800 nm excitation and detection the BChl 850 excited
state is strongly absorbing and the absorption change is completely depolarized (Fig. 2). This fast depolarization probably occurs due to three
different processes: (i) BChl 800 → BChl 800 transfer, (ii) BChl 800 →
BChl 850 transfer and (iii) BChl 850 → BChl 850 transfer. The model shown
in Fig. 1 does not predict complete depolarization. Therefore, this observation suggests a more random orientation of the 6 Q_y transition moments
than that shown in Fig. 1.

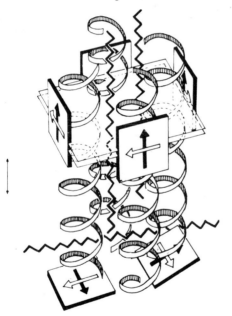

Figure 1. Schematic model of
the B800-850 complex. The upper
square boxes are the porphyrin
rings of BChl 850 and the lower
ones are those of BChl 800. Open
arrows, Q_y transition moments;
solid arrows, Q_x transition
moments. Zigzag lines represent
carotenoids and the spirals represent the α-helical transmembrane parts of the constituent
peptides. The plane of the membrane is horizontal; the vertical bar represents 5 Å (from
Kramer et al. [12]).

A pulse-probe experiment in which BChl 850 is directly excited is also
shown in Fig. 2. At 827 nm BChl 850* shows a strong absorption increase, at
860 nm an intense bleaching is observed (not shown). Both at 827 nm and at
860 nm the absorption signals are weakly polarized ($r(0) \simeq 0.1$). The fast
initial depolarization would correspond to extremely fast energy transfer
among the four BChl 850 molecules shown in the model in Fig. 1. It can be
estimated that a transfer rate $\gtrsim 3.10^{12}$ s^{-1} is required to generate the
observed anisotropy, i.e. a dipole-dipole distance of about 14 Å between
neighbouring BChl 850 molecules.

The decay of BChl 850* with all the traps closed is strongly biphasic.
A fast phase of about 35 ps is followed by a 200 ps decay. The first component represents the equilibration of the excitation density between

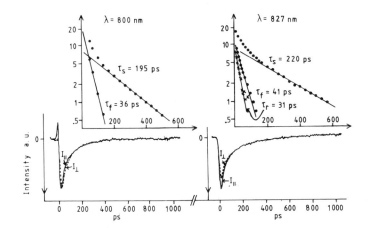

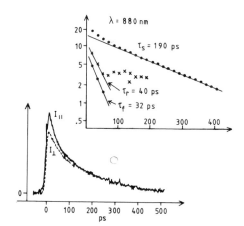

Figure 2. Parallel and perpendicular absorption changes in chromatophores of Rps. sphaeroides with all traps closed measured at 800, 827 and 880 nm. Plots of $\Delta I_{\parallel} + 2\Delta I_{\perp}$ are shown in the upper parts of each panel, including a two exponential analysis. In the panels for $\lambda = 827$ nm and $\lambda = 880$ nm the crosses (x —— x) indicate the anisotropy as a function of time (from Sundström et al. [6]).

B800-850 and B875. The second phase reflects the decay of the equilibrated excitation densities and the time constant (± 200 ps) agrees well with that extracted from the decay of the fluorescence with all the traps closed [3]. Essentially the same two lifetimes are found over the whole infrared absorption band (800 - 900 nm). The amplitude ratio of the fast 35 ps component to the slow 200 ps component is between 1 and 2 in the 800 - 850 nm region and decreases to 0.5 in the 870 - 900 nm region [6]. These decays are in agreement with the model proposed by Zankel [13] for the equilibration kinematics of the B800-850 and B875 excitation densities [6].

The transfer of excitation energy through the B800-850 antenna is associated to an additional depolarization. The r(t) curve shown in Fig. 2 for $\lambda = 827$ nm decreases from an initial value of about 0.1 to 0 with a time constant of the order of 30-50 ps. Assuming extremely fast transfer among the BChl 850 pigments in a minimum unit (4-8 BChl 850's), a slower transfer process of about 10 ps that represents the hopping of an excitation between different units may be associated to the additional slow anisotropy decay [6].

A B875 excited state may be produced through energy transfer from B800-850, via excitation of the B875 carotenoids or via direct BChl 875 excitation. Fig. 2 (lower) shows a pulse-probe experiment at 880 nm. With all the traps closed essentially the same two lifetimes are observed as at the other two wavelengths. The fast phase again represents the equilibration between B875 and B800-850. Note that in this case equilibration implies the uphill transfer of excitation energy.

As in the case of direct BChl 850 excitation, the initial polarization of the signal is low. Again a slow anisotropy decay is observed that is characterized by a time constant of about 30-50 ps. In analogy to the discussion of the B800-850 data, the fast depolarization is ascribed to fast ($\simeq 3.10^{12}$ s^{-1}) energy transfer among BChl 875 molecules in a minimum unit, while the slower phase is due to transfer between different units [6].

In Rhodospirillum (Rs.) rubrum with all the traps closed the decay of the B875 excited state is not monophasic, but can be resolved in a fast ($\simeq 50$ ps) and a slow ($\simeq 205$ ps) component. The fast phase reflects the equilibration of the excitation density between BChl 875 and a minor long wavelength component: BChl 896 [14]. Low temperature experiments confirm this interpretation [Bergström et al., unpublished observations]. The time-resolved depolarization in Rs. rubrum at 880 nm is very similar to that observed in Rps. sphaeroides [6].

With all the traps open the decay of the BChl excited states in Rps. sphaeroides remains multiphasic. At most wavelengths the decay is resolved in a 30 ps and a 110 ps component [6]. The former represents mainly equilibration, while the latter is dominated by the trapping of the equilibrated excitation density.

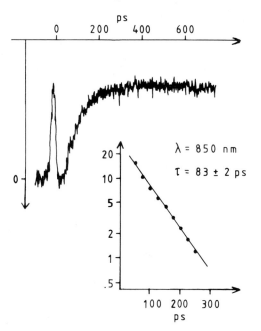

Figure 3. Isotropic absorption kinetics in Rps. sphaeroides chromatophores with all the traps open measured at 850 nm. From Sundström et al. [6].

A nice example of the type of trace that is observed under these conditions is given in Fig. 3, which shows a pulse-probe experiment at 850 nm. The initial increase in transmitted intensity represents excited BChl 850. The subsequent fast decrease is due to transfer to B875; the excited state of

B875 shows an absorption increase at this wavelength. After equilibration
a charge separation is formed, with a time constant of about 80 ps. The
state P^+Bpheo^- again leads to an absorption increase at this wavelength.

In Rs. rubrum with open traps the BChl excited decay is almost monophasic, $\tau \simeq 60$ ps at all wavelengths (830 - 900 nm). A very weak (less than
10% of the initial amplitude) slow phase, $\tau \simeq 250$ ps, is observed that may
represent electron transfer in the RC from Bpheo to Q.

3. TIME-RESOLVED DIFFERENCE SPECTRA

Time-resolved absorption difference spectra measured in Rs. rubrum
chromatophores upon excitation with a single 35 ps, 532 nm flash are shown
in Fig. 4 [15]. Both spectra were measured with overlapping excitation and
probing pulses. The spectrum represented by the solid circles (o———o) was
obtained under conditions that all the traps are closed and corresponds to
the B875 excited state. An intense bleaching is centered at about 890 nm,
while below 870 nm a strong increase in absorption is observed.

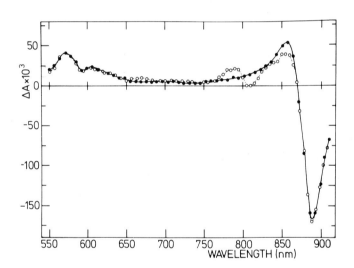

Figure 4. Absorbance difference spectra of chromatophores of Rs. rubrum
with coincident excitation and probe pulses with all the traps closed
(o———o) and with all the traps open (o———o) before the laser puls. A 35 ps,
532 nm flash was used as an excitation source. The flash excitation density
was 1.3 mJ/cm^2 (from Nuijs et al. [15]).

Apart from the fact that excited BChl 896 contributes to the red shift of
the difference spectrum, the main features are explained by assuming that
upon excitation of B875 the oscillator strength of about 6 BChl molecules
absorbing at 880 nm is lost and is replaced by the slightly (± 5 nm) blue
shifted absorption of 5 BChl molecules [8,25]. The net result is the
bleaching of the absorption of one BChl molecule.

The absorption increase in the 550 - 650 nm region represents the

carotenoid singlet excited state. A lifetime of about 1 ps can be calculated from these results [8,15]. As discussed above, the carotenoid excited
state decays in part via energy transfer to B875 and in part via singlet
fission.

The second spectrum shown in Fig. 4 (o---o) was measured with all the
traps open before the flash [15]. Again the strong bleaching at about 890
nm can be observed that represents the BChl 875 excited state. The bandshift around 800 nm indicates that a significant fraction (± 50%) of the
RCs was closed during the intense laser pulse. At relatively low flash
intensities the time required to obtain the full bleaching at 810 nm is
about 60 - 80 ps, in good agreement with the kinetic data discussed above.
On a longer time scale the transient bleaching at 810 nm shows a decay to
about 70% of the maximum level ($\tau \simeq 250$ ps) that reflects the electron
transfer process from reduced Bpheo to Q [15].

4. EXCITATION ANNIHILATION

Both the size of the domain of B875 that connects the RCs and the
dynamics of excitation transfer can be measured via the quenching of the
fluorescence yield by an intense picosecond laser pulse [1,2,16]. If such
an experiment is performed at room temperature the results indicate a rate
constant of excitation transfer between neighbouring BChl 875 molecules
that exceeds 10^{12} s^{-1}. Moreover, domain sized are obtained indicating a
1000 or more connected BChl 875 molecules [2,17]. If the temperature is
lowered to 4K the excitation annihilation curve changes dramatically. The
low-intensity fluorescence yield with open traps increases 4 - 5 fold, while
the total annihilation curve is shifted to much higher intensities. The
annihilation curves for Rs. rubrum chromatophores at 4K with all the traps
open (o——o) and all the traps closed (o——o) are shown in Fig. 5. In addition, the fluorescence yield measured by a weak xenon flash given 1.5 ms
after the intense laser pulse is shown both for the case that all the traps
are initially open (□——□) and closed (□——□). The simplest interpretation
of these data [17] indicates that the B875 domain size has decreased to
about 100 - 150 BChl molecules, that the 'average' rate of excitation transfer has dropped about a factor of 10, while in addition the rate constant
of trapping is a factor of 3 lower at 4K as compared to room temperature.
In Rps. sphaeroides chromatophores similar effects can be observed [17].
For Rps. sphaeroides the B800-850 domain size can be determined separately
by inspection of the B800-850 emission band at 880 nm and it is concluded
that this complex contains 30 - 40 connected BChl 850 molecules.

Small B875 domain sizes were also observed in Rps. sphaeroides cells
in a very early stage of pigment formation [18] and in Rps. sphaeroides
chromatophores fused with lipids [19]. All these observations suggest that
the minimum building block of the photosynthetic apparatus contains 3 - 4
RCs and 100 - 150 long wavelength BChl 875 pigments. Apparently, at
room temperature, these small systems can interact and form the large
domains that are observed experimentally. It may be worthwhile to study
these aggregation phenomena in more detail.

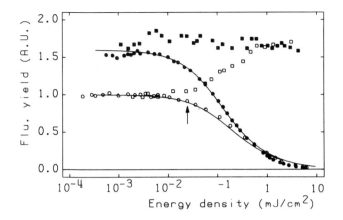

Figure 5. Annihilation curves of <u>Rs</u>. <u>rubrum</u> chromatophores at 4K, with RCs closed by continuous background illumination (o); RCs open before the arrival of the excitation flash (o). In addition the fluorescence yield of a weak Xenon flash given 1.5 ms after the laser flash is plotted against the laser intensity. Closed RCs (□); RCs open before the arrival of the actinic flash (□) (from Vos <u>et al</u>. [17]).

5. CONCLUSIONS AND PROSPECTIVES

Excitation energy transfer in a photosynthetic bacterium such as <u>Rps</u>. <u>sphaeroides</u> takes place according to the following reaction scheme:

$$BChl\ 800\ \rightarrow\ BChl\ 850\ \rightarrow\ BChl\ 875\ \rightarrow\ BChl\ 896 \rightarrow RC$$
$$1\ ps \qquad 35\ ps \qquad 30-50\ ps \qquad ?$$

A large part of the kinetic phenomena that are observed at room temperature indicate the equilibration of the excitation densities in the various pigment pools. The anisotropic absorption kinetics recorded at the various wavelengths indicate two types of depolarizing processes:

(i) Very fast energy transfer within a minimum B800-850 or B875 unit. The BChl molecules in such a unit interact rather strongly.

(ii) Slow energy transfer between different minimum units.

A preliminary analysis of the excitation annihilation data on the basis of such a model suggests that a reasonable fit is possible, although it seems that the average trapping times become somewhat too fast (30 - 40 ps). It should be kept in mind that in such an analysis it is assumed that the RC structurally acts as a 'normal' antenna complex. This is not necessarily so. Firstly, there is the putative BChl 896, about which we do not know how it is organized. Secondly, it seems possible that the excitations enter the RC in a specific way. The role of BChl 896 and the coupling of the RC to the light-harvesting antenna seem two fundamental questions that hopefully shall be answered in the near future.

References
1. Van Grondelle, R. (1985), Biochim. Biophys. Acta 811, 147-195
2. Bakker, J.G.C., Van Grondelle, R. and Den Hollander, W.Th.F. (1983), Biochim. Biophys. Acta 725, 508-518
3. Freiberg, A.M., Godik, V.I. and Timpman, K. (1984) in Advances in Photosynthesis Research (Sybesma, C. ed), Vol. I, pp. 45-48, Nijhoff/Junk Publishers, Dordrecht, The Netherlands

4. Vredenberg, W.J. and Duysens, L.N.M. (1963), Nature 197, 355-357
5. Monger, T.G. and Parson, W.W. (1977), Biochim. Biophys. Acta 460, 393-407
6. Sundström, V., Van Grondelle, R., Bergström, H., Åkesson, E. and Gillbro, T. (1986), Biochim. Biophys. Acta, in press
7. Dallinger, R.F., Woodruff, W.H. and Rodgers, M.A.J. (1981), Photochem. Photobiol. 33, 275-277
8. Nuijs, A.M., Van Grondelle, R., Joppe, H.L.P., Van Bochove, A.C. and Duysens, L.N.M. (1985), Biochim. Biophys. Acta 810, 94-105
9. Razi Naqvi, K. (1980), Photochem. Photobiol. 31, 523-524
10. Rademaker, H., Hoff, A.J., Van Grondelle, R. and Duysens, L.N.M. (1980), Biochim. Biophys. Acta 592, 240-257
11. Kingma, H., Van Grondelle, R. and Duysens, L.N.M. (1984), Biochim. Biophys. Acta 808, 383-399
12. Kramer, H.J.M., Van Grondelle, R., Hunter, C.N., Westerhuis, W.H.J. and Amesz, J. (1984), Biochim. Biophys. Acta 765, 156-165
13. Zankel, K.L. (1970) in The Photosynthetic Bacteria (Clayton, R.K. and Sistrom, W.R. eds.), pp. 341-347. Plenum Press, New York
14. Kramer, H.J.M., Pennoyer, J.D., Van Grondelle, R., Westerhuis, W.H.J., Niederman, R.A. and Amesz, J. (1984), Biochim. Biophys. Acta 767, 335-344
15. Nuijs, A.M., Van Grondelle, R., Joppe, H.L.P., Van Bochove, A.C. and Duysens, L.N.M. (1986), Biochim. Biophys. Acta 850, 286-293
16. Paillotin, G., Swenberg, C.E., Breton, J. and Geacintov, N.E. (1979), Biophys. J. 25, 513-533
17. Vos, M., Van Grondelle, R., Van der Kooy, F.W., Van de Poll, D., Amesz, J. and Duysens, L.N.M. (1986), Biochim. Biophys. Acta 850, 501-512
18. Hunter, C.N., Kramer, H.J.M. and Van Grondelle, R. (1985), Biochim. Biophys. Acta 807, 44-51
19. Westerhuis, W.H.J., Vos, M., Van Dorssen, R.J., Van Grondelle, R., Amesz, J. and Niederman, R.A. (1986), Biophys. J. 49, 488a

EXCITATION ENERGY TRANSPORT IN THE ANTENNA SYSTEMS OF PURPLE BACTERIA,
STUDIED BY LOW-INTENSITY PICOSECOND ABSORPTION SPECTROSCOPY.

V. Sundström,[a] R. van Grondelle,[b] H. Bergström,[a] E. Åkesson[a] and T. Gillbro[a]

a) Department of Physical Chemistry, University of Umeå, S-901 87 Umeå,
   Sweden;
b) Department of Biophysics, Physics Laboratory of the Free University, De
   Boelelaan 1081, 1081 HV, Amsterdam, The Netherlands.

1.   INTRODUCTION

The light harvesting system of the photosynthetic purple bacterium
Rhodopseudomonas sphaeroides contains two major pigment protein complexes,
B800-850 and B875. The latter is supposed to interconnect the reaction centers in a "lake"-like arrangement, consisting of at least 20 reaction centers and 1000 Bchl 875 molecules [1,2], while B800-850 is thought to constitute the peripheral part of the antenna. The long wavelength antenna of
Rs. rubrum is also organized like a lake [3], containing at least 10 and
probably not more than 20 reaction centers. The B880 complex of *Rs. rubrum*
shows several important spectroscopic properties similar to those of B875
of *Rps. sphaeroides* [4].
     Excited state lifetime measurements which have been quite numerous in
these systems suggested lifetimes ranging from 50 ps to 250 ps depending
on the state of the traps [5-8].
     In the present work we shall report on the excited state decay times
measured in chromatophores of *Rs. rubrum* and *Rps. sphaeroides*. We have, for
the first time, directly monitored the excited state decays by selective
infrared excitation and probing ($\lambda = 800-900$ nm) with low intensity picosecond pulses (5×10^{11}-5×10^{12} photons cm^{-2} pulse^{-1}). Moreover, we have recorded the picosecond time-resolved polarization changes associated with
these excited states.

2.   PROCEDURE

Picosecond kinetics were measured by means of absorption recovery,
using a mode-locked and cavity-dumped dye laser, synchronously pumped by
a mode-locked argon ion laser. The cavity dumper was operated in the 80 kHz
- 4 MHz range, typically giving 5-10 ps long pulses (fwhm) of 1-2 nJ energy
in the wavelength region 700-900 nm. The maximum excitation density used
in this work was 1×10^{13} photons cm^{-2} pulse^{-1} and by attenuating the excitation pulses with neutral density filters kinetics could be measured down
to ca 1×10^{12} photons cm^{-2} pulse^{-1}. These photon densities correspond to
maximum and minimum excitation degrees of $\approx 0.2\%$ and $\approx 0.02\%$, respectively.

Measurements with parallel ($I_{\parallel}$) and perpendicular ($I_{\perp}$) polarization
were used to follow the decay of induced anisotropy, $r(t) = (I_{\parallel}-I_{\perp})/(I_{\parallel} + 2I_{\perp})$, and measurements at the magic angle ($I_{54.7°}$) were used to obtain
the isotropic decay, free of orientational effects.

Biggens, J. (ed.), Progress in Photosynthesis Research, Vol. I. ISBN 90 247 3450 9

3. RESULTS AND DISCUSSION

The results of a representative kinetic measurement of *Rs. rubrum* and *Rps. sphaeroides* chromatophores are shown in the two figures below, for both open and closed reaction centers. A bleaching is represented as positive signal. Kinetic curves like those in the figure were measured at several wavelengths throughout the infrared absorption bands of *Rs. rubrum* and *Rps. sphaeroides* [10].

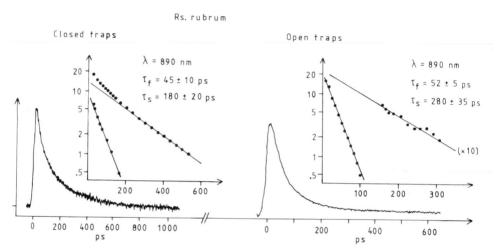

Figure 1.

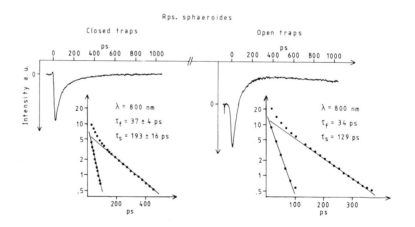

Figure 2.

The recorded absorption changes were seen to closely follow previously reported absorption difference spectra [11].The following lifetimes, averaged over all wavelengths, were obtained from these measurements. *Rs. rubrum*: Open traps, $\tau = 59 \pm 8$ ps; closed traps, $\tau_f = 47 \pm 4$ ps and $\tau_s = 200 \pm 20$ ps. *Rps. sphaeroides*: Open traps, $\tau_f = 32 \pm 6$ ps and $\tau_s = 112 \pm 12$ ps; closed traps, $\tau_f = 37 \pm 4$ ps and $\tau_s = 200 \pm 25$ ps. In addition to these lifetimes a very short lifetime $\tau < 1$ ps was observed for *Rps. sphaeroides* when exciting and probing in the 800 nm-band.

Picosecond absorption anisotropy measurements were also performed. Very low values of the anisotropy were generally observed (except at $\lambda \approx 800$ nm for *Rps. sphaeroides*).

The absorption anisotropy extrapolated to t=0 (r(0)) was found to be close to 0.1 in most cases, and decayed with a lifetime of approximately 40 ps to values very close to zero.

The results described above suggest the following schematic picture of the energy transfer and trapping dynamics. If in *Rs. rubrum* we take 6 Bchl 880's per B880 antenna complex then the excitation is delocalized among these six Bchl 880 molecules within 1 ps. Then, a random walk occurs among 8–9 of such complexes and assuming that trapping occurs at a rate of 3×10^{11} s^{-1} we can calculate, that the average jumping time from one B880 to another is of the order of 20 ps. This corresponds to a distance between the edges of the B880 complexes of about 20–25 Å.

The rapid delocalization within a B880 complex containing 6 Bchl 880 molecules implies that the transfer rate between pairs of Bchl 880 molecules in a complex is about 3×10^{12} s^{-1}. Assuming the transfer to be incoherent hopping a maximum distance of 14 Å between individual Bchl 880's within a complex is required. Very similar consideration apply for the B875 antenna of *Rps. sphaeroides*. In *Rps. sphaeroides* the effective transfer time from B800–850 to B875 is about 40 ps. The number of B800–850 complexes coupled together to transfer their excitation to B875 is unknown but recent excitation annihilation experiments at 4 K indicated that this number corresponds to about 30 connected Bchl 850's [Vos, M., personal communication]. Assuming these to be organized in B800–850 complexes, each containing 6 Bchl 800–850's, and assuming Bchl 875 to act as a perfect trap the transfer rate between two B800–850 complexes is about $1-2 \times 10^{11}$ s^{-1}.

We finally remark that the transfer time from Bchl 800 to Bchl 850 of about 1–2 ps corresponds to a maximum distance in the B800–850 complex of 22 Å between Bchl 800 and at least one of the Bchl 850's. From our experiment it is not clear to what extent rapid energy transfer between different Bchl 800's precedes the transfer to Bchl 850, but the high anisotropy of Bchl 800 suggest that no strongly depolarizing transfer process occurs.

Depending on the relative rates of energy transfer and trapping and efficiency of trapping, the energy transport process is described as diffusion limited or trap limited. The former notation refers to the case where the energy transport through the antenna molecules is the rate limiting process, while the latter is associated to the case where the trapping process at the reaction center is the rate limiting step. Our results suggest very fast transfer $k \approx 3 \times 10^{12}$ s^{-1} between individual bacteriochlorophyll molecules in a minimum unit of B880, B875 or B800–850, but slower transfer ($k \approx 5 \times 10^{10}$ s^{-1}) between such units. One might conclude that the transfer between the minimum units and between the different complexes is the rate limiting process.

REFERENCES

1. van Grondelle, R. (1985). Biochim.Biophys, Acta, 811, 147-195.
2. Clayton, R.K. (1980) in: Photosynthesis (Hutchinson, F., Fuller, W. and Mullins, L.J. eds.) Cambridge University Press, Cambridge.
3. Vredenberg, W.J. and Duysens, L.N.M. (1963) Nature 197, 355-357.
4. Kramer, H.J.M., Pennoyer, J.D., van Grondelle, R., Westerhuis, W.H.J., Niederman, R.A. and Amesz, J. (1984). Biochem. Biophys. Acta, 767, 335-344.
5. Borisov, A. Yu and Godik, V.I. (1972). J. Bioenerg. 3, 515-523.
6. Campillo, A.J., Hyer, R.C., Monger, T.G., Parson, W.W. and Shapiro, S.L. (1977). Proc. Natl, Acad. Sci., USA 74, 1997-2001.
7. Sebban, P., Jolchine, G. and Moya, I. (1984). Photchem. Photobiol. 39, 247-253.
8. Borisov, A. Yu, Freiberg, A.M., Godik, V.I., Rebane, K.K. and Timpman, K.E. (1985). Biochim. Biophys. Acta 807, 221-229.
9. Kingma, H. (1983) Doctoral Thesis, State University of the Netherlands.
10. Sundström, V., van Grondelle, R., Bergström, H., Åkesson, E. and Gillbro, T. (1986) Biochim. Biophys. Acta.
11. Nuys, A.M., van Grondelle, R., Joppe, H.L.P., van Bochove, A.C. and Duysens, L.N.M. (1985). Biochim. Biophys. Acta 810, 94-105.

THE ORGANIZATION OF THE LIGHT HARVESTING ANTENNA OF PURPLE BACTERIA

M. VOS[a], R.J. van DORSSEN[a], R. van GRONDELLE[b], C.N. HUNTER[c], J. AMESZ[a] and L.N.M. DUYSENS[a]

a) Department of Biophysics, Huygens Laboratory of the State University, P.O. Box 9504, 2300 RA Leiden, The Netherlands
b) Department of Biophysics, Physics Laboratory of the Free University, De Boelelaan 1081, 1081 HV Amsterdam, The Netherlands
c) Department of Pure and Applied Biology, Imperial College of Science and Technology, Prince Consort Road, London SW7 2BB, United Kingdom

1. INTRODUCTION.
    In photosynthetic systems, each RC is associated with a number of accessory pigment molecules which transfer their excitations to the RC. It has been known for a long time that several RCs share the same light-harvesting antenna. Such a large antenna plus RC is called a 'domain' [1,2].
    In Rhodobacter sphaeroides for each RC there are about 20 to 40 BChl 875 molecules, organized in the B875 complex, and between 50 and 400 BChl 800 + BChl 850 molecules, organized in the B800-850 complex. Both complexes have carotenoids.
    At room temperature a domain contains at least 20 RCs [1,3]. It has been assumed that these RCs are connected by B875, while the B800-850 complexes are arranged at the periphery [4].
    One technique to probe the functional structure of these domains is the measurement of the fluorescence yield induced by high intensity ps laser flashes. From the degree of quenching, resulting from singlet-singlet annihilation, the sizes of the domains can be determined, and information about the rates of energy transfer obtained [2,5]. For purple bacteria it was shown that in intact membranes the large domains that exist at room temperature are resolved into units with four RCs or less upon cooling to 4 K. Similarly, the B800-850 complexes seem to consist of relatively small units with about 50 BChl 850 molecules [3].
    This paper extends these observations to the separated antenna systems free from detergent artifacts by experiments on two carotenoid containing mutants of Rb. sphaeroides: a (B875 + RC)-less mutant (NF57) and a B800-850-less mutant (M21).

2. MATERIALS AND METHODS
    Low temperature emission and absorption spectra were recorded as described in [6]. Annihilation curves were measured with a 35 ps, 532 nm excitation flash as described in [3]. The properties of the two mutants of Rb. sphaeroides used in this work will be described elsewhere.

3. RESULTS
    Fig. 1 shows the 4 K absorption and emission spectra of M21 and NF57 membranes. From these spectra it can be concluded that M21 is almost devoid of B800-850. The RC bands at 760 nm and 800 nm are clearly observed. The RC : BChl ratio is about 1 : 40. The 4 K absorption spectrum of NF57 is typical for B800-850. However, the emission spectrum is clearly broadened, suggesting energy transfer to a small amount of long-wave emitting BChl, perhaps a remaining fraction of B875. NF57 contained less than 1 RC/10^4

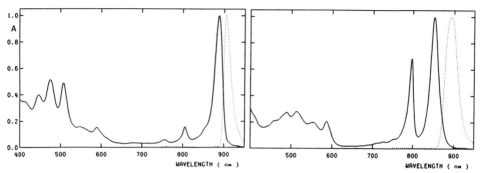

FIGURE 1. Absorption (——) and emission (—··—) spectra of Rb. sphaeroides
B800-850-less mutant M21 (left) and (B-875 + RC)-less mutant NF57
(right) at 4 K. Excitation at 590 nm.

BChl 850. In the wild type the B800-850 emission was only seen as a small
shoulder on the B875 emission. The emission yields ϕ_1^o (with open)
and ϕ_1^c (with closed RCs) of the two mutants are given in Table 1. NF57 can
be classified as a 'highly fluorescent' mutant; M21 has a normal low
fluorescence yield.

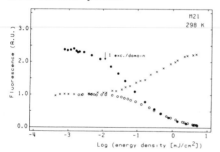

FIGURE 2. Integrated fluorescence
yield as a function of pulse energy
of the B800-850-less mutant M21.
Temperature 298 K, λ_{det} = 903 nm.
(o): all traps open before the
pulse; (●): all traps closed by
light; (x) fluorescence yield by a
weak xenon flash given 1.5 ms after
the laser flash.

Fig. 2 shows the time-integrated fluorescence yield as a function of
laser pulse energy of M21 at 293 K. The results of an analysis of this experiment are summarized in Table 1. At room temperature M21 clearly forms
large domains. The transfer (k_h) and trapping rate constants (k_t^o and k_t^c)
are similar to those obtained for other purple bacteria at room temperature
[1,3].

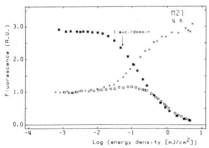

FIGURE 3. Same experiment as in Fig.
2, but at low temperature (4 K):
λ_{det} = 910 nm.

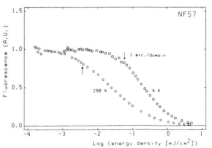

FIGURE 4. Annihilation curves in the
(B875 + RC)-less mutant N57 at room
temperature (o) (λ_{det} = 870 nm) and
at 4 K (□) (λ_{det} = 880 nm)

In Fig. 3 the same experiment but now at 4 K is shown. The results of the analysis of this experiment are again given in Table 1. Two important points must be noted: (i), the dramatic shift of the annihilation curve to higher laser intensities indicates that the domain size has decreased considerably upon cooling; (ii), at relatively low pulse intensities the fluorescence yield increases due to closing of the traps. At much higher pulse intensities annihilation takes over. Such an effect has been predicted [2], but has never been observed before. It is an indication that only a very small number of RCs are connected.

Fig. 4 shows the annihilation curves in mutant NF57 at room temperature and at 4 K and an analysis is given in Table 1. From these experiments it follows that in NF57 the B800-850 is strongly aggregated at room temperature. This mutant thus provides the first opportunity for a direct characterization of the B800-850 domain in membranes free from detergent artifacts. Upon cooling to 4 K the unit size decreases to about 30 BChl 850, a number remarkably similar to that obtained in intact Rb. sphaeroides at 4 K [3] and isolated LDS-B800-850 complexes [7]. From Table 1 it follows that the calculated rates of energy transfer are very low. This may reflect a subdivision into smaller units.

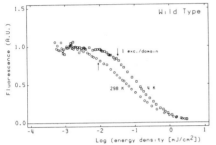

FIGURE 5. Annihilation curves in Rb. sphaeroides WT (B875 emission), with all traps closed, at room temperature (o) and at 4 K (□).

The annihilation curves obtained in Rb. sphaeroides WT, with all the traps closed at room temperature and at 4 K are shown in Fig. 5. The results of this experiment are very similar to those given in Ref. [3] (see Table 1).

4. CONCLUSIONS

The B875-RC complexes in mutant M21 and B800-850 in mutant NF57 are strongly aggregated at room temperature. At room temperature excitation energy transfer in both mutants proceeds as efficiently as in the wild type. Upon cooling to 4 K the rates of energy transfer among identical BChl molecules decreased drastically. In addition, at 4 K the antenna systems of both mutants and the wild type consist of isolated small units: about 100 BChls 875 in M21 and about 30 BChls 850 in NF57.

The number of connected RCs in M21 is sufficiently small to show, for the first time, an initial increase in the fluorescence yield upon illumination with ps laser pulses. A preliminary calculation of this effect suggests that we may have to assume some heterogeneity in the M21 domains to account for this effect.

The investigation was supported by the Netherlands Foundation for Biophysics, financed by the Netherlands Organization for the Advancement of Pure Research (ZWO). The authors wish to thank the European Molecular Biology Organization (EMBO) for a travel grant to CNH.

TABLE 1. Characterization of the energy transfer and trapping parameters in membranes of the B800-850-less mutant M21, the (B875 + RC)-less mutant NF57 and the wild type of <u>Rb</u>. sphaeroides

	M21		NF57		WT, B875		WT, B800-850
	298 K	4 K	298 K	4 K	298 K	4 K	4 K
ϕ_1^o	0.034	0.13			0.06	0.12	
ϕ_1^c	0.082	0.38	0.18	0.60	0.15	0.24	0.06
r	≥ 2	1	≥ 2	1	≥ 2	≥ 2	1
N_D	≥ 330	80-120	≥ 365	30	≥ 1000	≥ 100	45
N		40	40		25	25	23[a]
$k_h \times 10^{-10}$	≤ 100	14	≤ 53	1.7	≤ 200	≤ 8.7	26
$k_t^o \times 10^{-10}$	≥ 180	70			≥ 42	≥ 52	
$k_t^c \times 10^{-10}$	≥ 53	7.9			≥ 14	≥ 11	61[b]

r is the ratio of 'normal' decay and decay due to annihilation.
N_D is the number of BChl molecules per domain (domain size).
N is the number of BChl/RC (PSU size).
In all cases the lifetime of the BChl excited state in the absence of trapping processes is assumed to be 1 ns.

a) Two BChl 875 'traps' per B800-850 domain were assumed.
b) The rate of trapping in this case indicates the rate of excitation transfer from a BChl 850 molecule to a neighbouring BChl 875 trap.

REFERENCES
1 Bakker, J.G.C., van Grondelle, R. and den Hollander, W.T.F. (1983) Biochim. Biophys. Acta 725, 508-518
2 Den Hollander, W.T.F., Bakker, J.G.C. and van Grondelle, R. (1983) Biochim. Biophys. Acta 725, 492-507
3 Vos, M., van Grondelle,.R., van der Kooij, F.W., van de Poll, D., Amesz, J. and Duysens, L.N.M. (1986) Biochim. Biophys. Acta 850, 501-512
4 Monger, T.G. and Parson, W.W. (1977) Biochim. Biophys. Acta 460, 393-407
5 Paillotin, Swenberg, C.E., Breton, J. and Geacintov, N.E. (1979) Biophys. J.25, 513-534.
6 Kramer, H.J.M., Pennoyer, J.D., Van Grondelle, R., Westerhuis, W.H.J., Niedermann, R.A. and Amesz, J. (1984) Biochim. Biophys. Acta 767, 335-344
7 Van Grondelle, R., Hunter, C.N., Bakker, J.G.C. and Kramer, H.J.M. (1983) Biochim. Biophys. Acta 723, 30-36

PHOTOCHEMICAL AND NON-PHOTOCHEMICAL HOLEBURNING STUDIES

OF ENERGY AND ELECTRON TRANSFER IN

PHOTOSYNTHETIC REACTION CENTERS AND MODEL SYSTEMS

Steven G. Boxer, Thomas R. Middendorf,
David J. Lockhart, and David S. Gottfried
Department of Chemistry
Stanford University
Stanford, California 94305

1. INTRODUCTION

The initial processes in photosynthesis involve nearly degenerate energy transfer among chromophores in the antenna complex, energy transfer from the antenna to the reaction center (RC), and charge separation within the RC. Each of these processes is very rapid, typically occuring on a picosecond or faster timescale, and has been studied by time-domain (pulse-probe) spectroscopy. Recently we have begun to examine energy and electron transfer using frequency-domain methods, notably photochemical [1,2] and non-photochemical [3] holeburning spectroscopies. These methods can provide information on very rapid processes and on interactions between molecular excited states and lattice phonons.

The optical absorption spectra of chromophores in disordered systems are generally very broad due to the wide distribution of site energies in the sample. This inhomogeneous broadening is typically hundreds of wavenumbers and leads to featureless absorption spectra. This contrasts with the much smaller site energy variations in single molecular crystals or S'polskii matrices where the inhomogeneous linewidths are typically on the order of 1 cm^{-1}. In both cases the distribution of site energies masks the true homogeneous linewidth, which is simply related to the excited state lifetime (optical T_1) by the Uncertainty Principle in the absence of pure dephasing. Thus, the intrinsic homogeneous linewidth for a molecule such as chlorophyll is on the order of 10^{-3} cm^{-1}, corresponding to a 5 ns excited state lifetime. The relationship between excited state lifetime and homogeneous linewidth is presented in Table 1.

Table 1. Conversion between lifetime (T_1) and homogeneous linewidth (Γ_o) in the absence of pure dephasing.

T_1	50fs	500fs	5ps	50ps	500ps	5ns
Γ_o (cm^{-1})	100	10	1	.1	.01	.001
Γ_o (GHz)	3000	300	30	3	0.3	0.03

In addition to the minimum linewidth determined by the excited state lifetime, rapid fluctuations in the site energy during the excited state lifetime contribute to the homogeneous linewidth. Such pure dephasing processes, characterized by time T_2^*, generally have a strong temperature dependence which is characteristic of excited state-phonon coupling. For example, the temperature dependence of the pure dephasing contribution to the homogeneous linewidth is very different in single crystal vs. glass host matrix. In general, the homogenous linewidth (FWHM = Γ_h) is given by:

$$\Gamma_h = \frac{1}{\pi T_2} = \frac{1}{\pi T_2^*} + \frac{1}{2\pi T_1} \cdot$$

Biggens, J. (ed.), Progress in Photosynthesis Research, Vol. I. ISBN 90 247 3450 9
© *1987 Martinus Nijhoff Publishers, Dordrecht. Printed in the Netherlands.*

The homogenous linewidth can be measured by a photon echo or a holeburning experiment. A variation on the photon echo method has been applied to RCs by Meech et al. [4]. For holeburning experiments the measured holewidth (FWHM) is equal to twice the homogeneous linewidth when the hole is narrow compared to the inhomogeneous line and the laser linewidth is small compared to the holewidth.

In photochemical holeburning, the excited state undergoes a photochemical reaction leading to products whose absorption bands are removed from those of the reactant. If a subset of reactants with a certain transition energy is selectively removed from the ground state with a narrow bandwidth light source, then the photochemistry produces a hole in the reactant's inhomogeneous absorption band which can be detected as long as the products do not reform reactants. This situation arises in bacterial RCs, where the absorption of the electron donor (P870 in R. sphaeroides, P960 in R. viridis) is bleached upon excitation. Electron transfer to the initial electron acceptor, a bacteriopheophytin (BPheo), and subsequent reduction of the first quinone acceptor preserves the bleach for many ms, allowing observation of the hole. Transient absorption measurements show that reduction of the BPheo takes place with a 1/e time of 2-4ps [5-7], therefore the minimum possible homogeneous linewidth is about 1 cm^{-1}.

In non-photochemical holeburning the excited state does not undergo conversion to products. Rather a slight configurational change of the lattice in the vicinity of the excited chromophore changes the site energy of the chromophore. So long as the site energy is changed more than several times the homogeneous linewidth, a hole can be observed in the absorption spectrum at the burn frequency. We have used this type of holeburning to measure the homogeneous linewidth of chlorophyllide in apo-myoglobin (Mb) [8] and of chlorophyllide-substituted hemoglobin [9,10]. In the latter, efficient energy transfer occurs among the chromophores at room temperature.

2.   PROCEDURE

The method for photochemical holeburning in RC samples has been described in detail elsewhere [1]. Briefly, RCs in polyvinylalchohol (PVA) films are immersed in pumped liquid He in an optical cryostat. The sample is excited with pulses from a Nd:YAG pumped dye laser whose output is Raman shifted to provide near infrared excitation (laser linewidth 1-2 cm^{-1}). The change in absorption (hole spectrum) is probed with a weak probe beam passed through a monochrometer (ultimate resolution ~0.3cm^{-1}). The non-photochemical holeburning experiments are performed using an ultra-high resolution dye laser (Coherent 699-29, linewidth about 1MHz). The holes are burned at very low power (<20μW/cm^2) for about 1min., and the absorption is then probed as the fluorescence excitation spectrum by scanning the same laser at about 1/10 of the burn power. The burn power and burn time dependence of the holewidth were measured at each temperature in order to avoid the "pitfalls" [11] of power broadening. The temperature was varied between 1.35 and 2.5K by adjusting the pressure above the liquid He (accuracy ±.02K).

RCs were prepared by standard methods. The film containing R. viridis RCs was a generous gift from Professor Parson; the film containing R. sphaeroides RCs treated with NaBH$_4$ to remove one BChl monomer [12] was a generous gift from Professor Holten. Mg- and Zn-pyrochlorophyllide a-substituted sperm whale myoglobin (Mg and ZnPChlaMb), ZnPChla-substituted human apo-Mb (prepared by recombinant DNA methods [13]), and human hemoglobin [(ZnPChla)₄Hb] were prepared by methods described earlier [8,9] and were embedded in PVA films. Zn-pyrochlorophyllide a esterified with 3-(3'-hydroxypropyl)pyridine [14] was embedded in a polystyrene film. The concentration of chromophores was ~10⁻⁴ M.

3. RESULTS

3.1 Reversible Photochemical Holeburning in RCs

The changes in absorption following photoexcitation at various frequencies within the lowest energy electronic absorption band of the primary electron donor of both species of RCs are compared in Fig. 1. The absence of any dependence of the holewidth on burn power or temperature within the sensitivity of the experiment is shown in Fig. 2 for R. sphaeroides RCs. Parallel experiments on R. viridis RCs gave the same results. Hole spectra for NaBH₄-treated R. sphaeroides RCs were indistinguishable from those of normal RCs, including the holewidth and peak position variations with burn frequency, and the absence of a temperature or power dependence (data not shown).

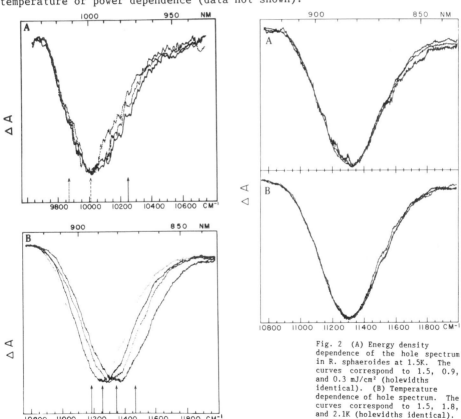

Fig. 1 (A) Photochemical holes observed in R. viridis RCs at 1.4K. Holes were burned at wavelengths indicated by vertical arrows. The FWHM was about 400 cm⁻¹. (B) Photochemical holes observed in R. sphaeroides RCs at 1.5K. The burn wavelengths and holewidths (all in cm⁻¹) are 11470 (440); 11345 (420); 11253 (410); and 11180 (390). All hole depths are scaled to the same magnitude for comparison.

Fig. 2 (A) Energy density dependence of the hole spectrum in R. sphaeroides at 1.5K. The curves correspond to 1.5, 0.9, and 0.3 mJ/cm² (holewidths identical). (B) Temperature dependence of hole spectrum. The curves correspond to 1.5, 1.8, and 2.1K (holewidths identical). All hole depths are scaled to the same magnitude for comparison.

3.2 Non-photochemical Holeburning in Chlorophyll-Protein Complexes

A typical example of a non-photochemical hole observed in ZnPChlaMb is shown in Fig. 3. Identical holewidths were observed at other burn frequencies within the inhomogeneous absorption band. Comparable results were observed for MgPChlaMb and for ZnPChla in human apo-Mb. The temperature dependence of the holewidth was studied between 1.35 and 2.5K. The homogeneous linewidth was obtained from the best fit to a Lorentzian lineshape (Fig. 3). The data were fit to an expression of the form: $\Gamma_h = \Gamma_0 + bT^\alpha$, where Γ_0 was assumed to be 40MHz, corresponding to the room temperature singlet lifetime of the chromophore [8]. A convenient representation of the data is the log-log plot shown in Fig. 4 from which one obtains the slope (α) equal to 1.26±.074. A parallel study of Zn-pyrochlorophyllide a esterified with 3-(3'-hydroxypropyl)pyridine in polystyrene gave very similar results, though the quality of the data obtained to date was not as high as for ZnPChlaMb. The slope (α) of the temperature dependence was 1.32±.21. In a preliminary study, the holewidth of (ZnPChla)$_4$Hb was examined. The holewidth at 1.4K was 2.1GHz, considerably broader than for monomeric ZnPChlaMb, and there was very little increase in the holewidth as the temperature was increased up to 2.1K, in striking contrast with monomeric ZnPChlaMb. Due to the temporary demise of the laser, this experiment has not been repeated as of this writing.

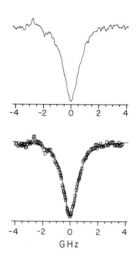

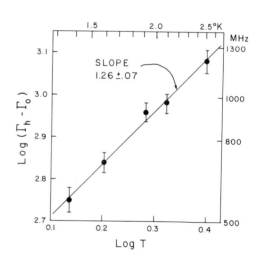

Fig. 3 Non-photochemical hole in ZnPChlaMb at 1.35K and the best-fit Lorentzian. The burn and probe powers were 20µW/cm² and 1.2µW/cm², respectively. The burn frequency and burn time were 15050.0cm^{-1} and 1min, respectively. The holewidth is 1.3±0.1GHz.

Fig. 4 Log-log plot of the temperature dependence of the pure dephasing contribution to the homogeneous linewidth for ZnPChlaMb.

4. DISCUSSION

4.1 Reversible Photochemical Holeburning in RCs

Four striking results emerge from the holeburning studies of RCs.

(i) The holes are very broad -- on the order of 400 cm^{-1}, which is comparable to the inhomogeneous linewidth. (ii) The results for the two species are very similar. (iii) Removal of one monomer BChl in R. sphaeroides RCs has no effect on the hole spectrum. (iv) The holewidths show no variation with temperature between 1.3 and 2.1K.

We believe that it is reasonable to neglect pure dephasing contributions to the photochemical holewidths. The pure dephasing contribution in crystals and glasses is generally less than 1 cm^{-1} at the temperatures of these experiments [11,15]. It is clear from the data shown in Fig. 4, that a protein matrix (at least Mb) is not fundamentally different from a glass in this regard (see below). The complete absence of a temperature dependence also argues against pure dephasing as the dominant contribution to the holewidth.

Using the conversion in Table 1, the roughly 400 cm^{-1} holewidth translates into an excited state lifetime of ~20fs. This interpretation of the holewidth has been favored by Meech et al. [4] and has been discussed by us [1,2]. Although little is known about the vibronic structure associated with the P870 or P960 absorption bands, it is reasonable to assume that it is not unlike that associated with other chlorophyll or porphyrin molecules. For this reason it is more appropriate to consider the minimum excited state lifetime which would broaden the vibronic structure sufficiently to conceal any evidence of substructure in the observed holes. Since vibronic spacings on the order of 50 cm^{-1} are likely, a comparable minimum linewidth is required. This leads to an excited state lifetime on the order of 200 fs, still much shorter than what is observed in transient absorption spectroscopy [5-7].

Although a simple excited state lifetime analysis such as that presented above is appealing, it is difficult to reconcile with recent transient absorption measurements [5-7], so we have argued that an alternative viewpoint may be more appropriate [1,2]. If photoexcitation of the primary electron donor involves a substantial change in equilibrium nuclear coordinates, this could lead to complete suppression of the zero-phonon line leaving a broad (featureless) band.

The holeburning measurements to date do not allow one to distinguish between a physical model involving strong excited state-phonon coupling or strong coupling to vibronic states of the reactive chromophore complex. Nonetheless, a considerable body of data and analysis in the literature can be collected together to propose a reasonable and consistent physical model as follows. The precise origin of the red shift of the lowest energy absorption band of the primary electron donor has been a subject of much speculation. Strong interchromophore interactions within the special pair and a substantial degree of charge-transfer character have been suggested by the analysis of Parson and co-workers [16]. Charge transfer character in P870 is also suggested by the observed Stark effect which is considerably greater for P870 than other absorption bands in the RC [17]. EPR analysis of the triplet state of P870 in RC single crystals by Norris and co-workers [18] demonstrates that none of the principal magnetic axes is coincident with the reaction center C_2 axis [19], though one must be by symmetry. This suggests an environmentally induced asymmetry in the triplet excited state of P870, as could be the case for a state with charge transfer character. Mixing of charge transfer character may also explain the reduction in zero-field splittings for the triplet state of P870 relative to a monomeric BChl [20].

Thus, the physical picture we favor involves direct excitation to a state with substantial charge transfer character. This picture is consistent with ps and fs transient absorption and stimulated emission results [5-7], with the holeburning results, and with the spectroscopic data discussed above. Given the highly symmetric arrangement of the chromophores in the R. viridis x-ray structure [19], any reduction in symmetry must be imposed by the protein. An attractive hypothesis is that the highly directional transfer of charge towards the L-branch of the RC is favored by hydrogen bonding to the ring E keto carbonyl group on the L-side of the special pair, possibly as follows:

Special Pair
L-Side

Additional stabilization by electron withdrawing amino acid side chains which are more prevalent on the L than the M side is also likely. Although we tend to favor a charge transfer state localized on the special pair and stabilized by specific interactions with the protein, it is possible that strong coupling to charge transfer states involving the monomer BChl and BPheo is important. This view has been put forward recently by Won and Friesner to explain our results [21]. Since removal of one BChl monomer has no effect on the holeburning results or any aspect of the initial electron transfer kinetics [22], charge transfer states involving this BChl monomer must not be important. Again, at the level of the chromophores there is no distinction between the monomer BChls or BPheos on the M- and L-side of the RC in R. viridis. Assuming that the BChl monomer which is removed is from the M-side, any difference in energy between $P^+BChl_M^-$ and $P^+BChl_L^-$ must be the result of interactions with the protein matrix.

4.2 Non-photochemical Holeburning in Chlorophyll-Protein Complexes

The non-photochemical holeburning experiments reported here are part of a continuing investigation of excited state-phonon coupling in protein matrices and the effects of energy and electron transfer on homogeneous linewidths in ordered biological systems. Several noteworthy conclusions emerge from the results presented in section 3.2. (i) Non-photochemical holeburning occurs in protein matrices as it does in simpler polymeric or organic glasses. (ii) The homogeneous linewidths obtained at very low burn powers and pumped He temperatures are comparable for a chromophore in a protein matrix and in other glassy matrices. (iii) The $T^{1.3}$ temperature dependence of the dephasing contribution to the homogeneous linewidth for a chromophore in a protein matrix is very similar to that of the same chromophore in a polymeric glassy matrix.

Comparison of the holewidth in Fig. 3 and the conversion in Table 1 demonstrates that the homogeneous linewidth in ZnPChlaMb, even at temperatures as low as 1.35K, is dominated by pure dephasing. This is what is found in other polymeric glasses, where it has been shown that extremely low temperatures (often <<0.3K) are required in order to eliminate pure dephasing [15,23]. Although there is considerable controversy about whether holeburning measurements accurately reflect the true homogeneous linewidth, in the most careful studies the holewidths extrapolate to the known lifetime-limited value and the differences between results of holeburning and photon echo experiments appear to be small [15].

A variety of models has been introduced to rationalize the temperature dependence of the homogeneous linewidth in glasses. Most are related to the model of a distribution of nearly isoenergetic two-level systems of the glassy matrix coupled to the excited state of the chromophore. Such models were first introduced by Anderson and co-workers [24] and Phillips [25] to explain the anomolous low-temperature specific heat of glasses, and were introduced by Small and co-workers [26] to explain the phenomenon of non-photochemical holeburning. Although it is not at all obvious that a $T^{1.3}$ temperature dependence follows from these models, most treatments lead to an exponent between 1 and 2. Most importantly, an exponent of 1.3 ± 0.1 has been repeatedly observed in a wide variety of glasses both by holeburning and by photon echo methods, leading to the qualitative conclusion that such a temperature dependence is a fundamental characteristic of coupling to disordered host matrices [15]. Thus, the nearly identical temperature dependence of the dephasing contribution to the homogeneous linewidth in ZnPChlaMb and that of ZnPChla and many other chromophores in glasses suggests that the coupling between the ZnPChla excited state and the apo-Mb matrix is comparable with that in any disordered matrix, with no evidence for unusual modes in the protein [3].

The preliminary results on the homogeneous linewidth for ZnPChla in tetrameric hemoglobin are quite different from those in the monomeric ZnPChlaMb complex. In the Hb complex, chromophores on the α_1 and β_2 chains are 25Å (center-to-center) apart. The Q_y transition moments are oriented such that the interchromophore dipole-dipole interaction is quite large (about 6-8cm^{-1}) leading to very rapid energy transfer at room temperature [9]. It seems reasonable that energy transfer could contribute to the homogeneous linewidth if its rate is comparable to the other deactivation rates and the value of $1/T_2^*$ which is characteristic of the chromophore in its matrix in the absence of energy transfer. The temperature dependence of this contribution to the homogeneous linewidth would be that of energy transfer, not necessarily the $T^{1.3}$ dependence that we find in monomeric ZnPChlaMb. We are not aware of experimental studies of the temperature dependence of energy transfer in systems where the three dimensional structure is known, however, this problem has been considered in detail theoretically [27,28]. Although energy transfer is not likely to be responsible for the holeburning results in RCs, the temperature dependence of energy transfer and its effect on the homogeneous linewidth in a variety of chlorophyll dimers and antenna complexes is currently being studied in our lab.

5. SUMMARY

In conclusion, we note that the strong temperature dependence for the homogeneous linewidth in our monomeric ZnPChla-protein complexes is in striking contrast to the absence of a temperature dependence for the holewidth in RCs. Furthermore the 4-5 orders of magnitude difference in holewidth makes it clear that we are dealing with fundamentally different phenomena in model compounds and the RC. Although novel dephasing mechanisms in strongly coupled chromophore arrays such as the RC can not be ruled out, we are confident that a pure dephasing mechanism does not play a role in determining the holewidth in RCs. Instead we believe that the RC result is best explained by excitation to a state with substantial charge transfer character, possibly accompanied by other changes in the protein matrix which stabilize this nascent charge-separated state.

ACKNOWLEDGEMENTS Portions of this work were supported by NSF Grants

PCM8303776, PCM8352149 and DMB-8607799, and the Dow Chemical Corporation. The Coherent 699-29 was borrowed from the San Francisco Laser Center, supported by NSF and NIH Grants GHE79-16250 and P41 RR01613-02, respectively. D.S.G. is an NSF Pre-doctoral Fellow.

REFERENCES

1. Boxer, S.G., Lockhart, D.J. and Middendorf, T.R. (1986) Chem. Phys. Lett. 123, 476-482
2. Boxer, S.G., Middendorf, T.R. and Lockhart, D.J. (1986) FEBS Lett. 200, 237-241
3. Boxer, S.G., Gottfried, D.S., Middendorf, T.R., and Lockhart, D.J. (1986) J. Chem. Phys., submitted for publication
4. Meech, S.R., Hoff, A.J. and Wiersma, D.A. (1985) Chem. Phys. Lett. 121, 287-292
5. Woodbury, N.W., Becker, M., Middendorf, D. and Parson, W.W. (1985) Biochem. 24, 7516-7521
6. Martin, J.L, Breton, J., Hoff, A.J., Migus, A. and Antonetti, A. (1986) Proc. Natl. Acad. Sci. USA 83, 957-961
7. Breton, J., Martin, J.-L., Migus, A., Antonetti, A. and Orszag, A. (1986) Proc. Natl. Acad. Sci. USA 83, 5121-5125
8. Wright, K.A. and Boxer, S.G. (1981) Biochem. 20, 7546-7556
9. Kuki, A. and Boxer, S.G. (1983) Biochem. 22, 2923-2933
10. Moog,R.S., Kuki,A., Fayer,M.D. and Boxer,S.G. (1984) Biochem. 23, 1564-1571
11. Thijssen, H.P.H. and Völker, S. (1985) Chem. Phys. Lett. 120, 496-502
12. Ditson, S.L., Davis, R.C. and Pearlstein, R.M (1984) Biochem. Biophys. Acta 766, 623-629
13. Varadarajan, R., Szabo, A. and Boxer, S.G. (1985) Proc. Natl. Acad. Sci. U.S.A. 82, 5681-5684
14. Kuki, A., Ph.D. Thesis, Stanford University (1985)
15. Thijssen, H.P.H., van den Berg, R. and Völker, S (1985) Chem. Phys. Lett. 120, 503-508
16. Parson, W.W., Sherz, A. and Warshel, A. (1985) in Antennas and Reaction Centers of Photosynthetic Bacteria (Michel-Beyerle, M.E. ed.), Springer Series in Chemical Physics, Vol. 42, pp.122-130, Springer, Berlin
17. DeLeeuv,D., Malley,M., Butterman,G., Okamura,M.Y. and Feher,G. (1982)Biophys. J.37,111a
18. Norris, J.R., Budil, D.E., Crespi, H.L., Bowman, M.K., Gast, P., Lin, C.P., Chang, C.H. and Schiffer, M. (1985) in Antennas and Reaction Centers of Photosynthetic Bacteria (Michel-Beyerle, M.E., ed.), Springer Series in Chemical Physics, Vol. 42, pp.147-149
19. Deisenhofer, J., Epp., O., Miki, K, Huber, R. and Michel, H. (1984) J. Mol. Biol 180, 385-398; (1985) Nature 318, 618-624
20. Levanon, H. and Norris, J.R. (1978) Chem. Rev. 78, 185-198
21. Won, Y. and Friesner, R.A., Proc. Natl. Acad. Sci., USA, in press.
22. Maroti, P., Kirmaier, C. Wraight, C., Holten, D. and Pearlstein, R.M. (1985) Biochim. Biophys. Acta 810, 132-139
23. Gorokhovskii,A., Korrovits,V., Palm,V. and Trummal,M.(1986)Chem. Phys. Lett. 125, 355-359
24. Anderson, P.W., Walperin, B.I. and Varma, C.M. (1972) Phil. Mag. 25, 1
25. Phillips, W.A. (1972) J. Low Temp. Phys. 7, 351
26. Hayes, J.M., Fearey, B.L., Carter, T.P. and Small, G.J. (1986) Intl. Rev. Phys. Chem. 5, 175-184
27. West, B.J. and Lindenberg, K.(1985) J.Chem.Phys.83,4118-4135
28. Brown,D.W., Lindenberg,K. and West, B.J.(1985)J.Chem.Phys.83,4136-4143

THE TEMPERATURE DEPENDENCE OF ELECTRON BACK-TRANSFER FROM THE PRIMARY RADICAL PAIR OF BACTERIAL PHOTOSYNTHESIS

David E. Budil, Stephen V. Kolaczkowski, and James R. Norris, Chemistry Division, Argonne National Laboratory, Argonne IL, 10439

Introduction

The evolution of the primary radical pair in RC's where secondary electron transfer is blocked by removal or reduction of the endogenous quinone (Q) may be represented by the simple scheme

$$
^1P^* \, I \rightarrow \quad ^1(P^+ \, I^-) \xleftrightarrow{\ \omega\ } \, ^3(P^+ \, I^-) \tag{1}
$$
$$
\downarrow k_s \qquad\qquad \downarrow k_t
$$
$$
P \ I \qquad\qquad\quad ^3P \ I
$$

where P and I represent the primary donor and acceptor, $^1(P^+ \, I^-)$ and $^3(P^+ \, I^-)$ the singlet- and triplet- correlated radical pair, ω the rate of radical pair intersystem crossing, and k_s and k_t the rate constants for charge recombination to form the ground and triplet states of the donor P. Since both static magnetic fields [1-3] and resonant microwaves [4-5] modulate the intersystem crossing rate ω, they also change the yield of triplet product ^3P and the overall decay rate of $(P^+ \, I^-)$ [6].

Haberkorn and Michel-Beyerle [1] have given a simple relationship between the lifetime τ of $(P^+ \, I^-)$ and the absolute triplet yield Φ_T formed by radical pair intersystem crossing:

$$
\tau = \tau_s + (\tau_t - \tau_s)\Phi_T \tag{2}
$$

where $\tau_{s,t} = 1/k_{s,t}$. This expression approximates $(P^+ \, I^-)$ decays with a single exponential, and applies to the τ measured at a wavelength where only $(P^+ \, I^-)$ is observed.

Since it is most practical to measure some combination of $(P^+ \, I^-)$ decay and ^3P rise, we have derived an expression for the τ measured at a wavelength with an arbitrary ratio η of the extinction coefficients of $(P^+ \, I^-)$ and ^3P. This is achieved by treating radical pair intersystem crossing as a classical reaction interconverting $^1(P^+ \, I^-)$ and $^3(P^+ \, I^-)$, and solving the resultant coupled differential equations for the time evolution of these two states. We then write down the time dependence of the function $[(P^+ \, I^-)] + \eta[^3P]$, subtract the asymptote, and normalize to unity at t=0 before integrating to give the effective lifetime

$$
\tau = \tau_s + (\tau_t - \tau_s)\Phi_T - \tau_t \left[\frac{(1 - \Phi_T)}{(\omega\tau_t + 1)} \frac{\eta\Phi_T}{(1 - \eta\Phi_T)} \right] \tag{3}
$$

This derivation assumes $\omega \ll (k_t - k_s)$, in which limit intersystem crossing may be replaced by classical reactions, and the observed optical transient may be approximated by a single exponential. Equation (3) reduces to equation (2) for $\eta=0$ as expected; for $\eta=1$, it gives the $\tau(\Phi_T)$ for a wavelength where only the conversion of $(P^+ \, I^-)$ to ground state is observed, e.g., 870 nm.

Figure 1 shows the calculated $\tau(\Phi_T)$ at 870 nm with varying ratios of k_t/k_s. A pronounced curvature is apparent in the $\tau(\Phi_T)$ function when the approximation $\omega \ll (k_t - k_s)$ breaks down; however, for sufficiently slow intersystem crossing, $\tau(\Phi_T)$ approaches the line $\tau = \tau_s - \tau_s\Phi_T$ (dashed line), which has an x intercept of unity. This may be verified using equation (3) with $\eta = 1$ in the limit $\omega\tau_t \ll 1$. It is also apparent

Biggens, J. (ed.), Progress in Photosynthesis Research, Vol. I. ISBN 90 247 3450 9
© *1987 Martinus Nijhoff Publishers, Dordrecht. Printed in the Netherlands.*

from figure 1 that $\tau(\Phi_T)$ is a line extrapolating to $\tau = \tau_s$ at $\Phi_T = 0$, in the limit $\omega\tau_t \ll 1$. This conclusion is independent of wavelength, and is consistent with one's intuition that $(P^+ I^-)$ would decay entirely via the singlet pathway with lifetime τ_s if no triplet were formed.

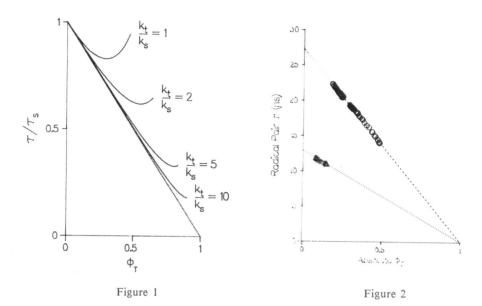

Figure 1   Figure 2

Experimental

The apparatus used for optical measurement of $(P^+ I^-)$ kinetics under the influence of magnetic fields and resonant microwaves has been described [7]. The basic experiment is to modulate τ and Φ_T with the applied fields and plot the resultant (Φ_T,τ) pairs. The lifetime τ is obtained by a least-squares fit of a single exponential to the observed $(P^+ I^-)$ decay measured on a ns timescale; the asymptote of this fit is taken to be an optical density difference (ΔOD) proportional to Φ_T.

Figure 2 gives representative experimental $\tau(\Phi_T)$ lines measured in RC's with Q removed (dashed line) or reduced by dithionite (dotted line) obtained using 0–300 G fields and high power microwaves (upper curve). To a good approximation, $\tau(\Phi_T)$ is linear in both sample types even near zero field; thus, the limit $\omega\tau_t \ll 1$ applies reasonably well to the $(P^+ I^-)$ system at the fields used. Absolute Φ_T is obtained from these curves by normalizing the ΔOD_{870} axis to an intercept of 1 (zero field corresponds to the highest Φ_T plotted); τ_s is measured by extrapolating the lines to $\Phi_T = 0$. A series of extrapolations similar to Figure 2 gives $\Phi_T = 0.47 \pm 0.04$, $\tau_s = 26 \pm 2$ ns for Q-depleted RC's, and $\Phi_T = 0.15 \pm 0.04$, $\tau_s = 13 \pm 2$ ns for Q-reduced RC's at zero field and room temperature.

Figure 3 shows a preliminary temperature study of the electron transfer rate from $^1(P^+ I^-)$ in Q-depleted (circles and triangles) and Q-reduced (squares) RC's. The k_s values were determined using both low fields (circles) and high power microwaves (triangles). Charge recombination from $^1(P^+ I^-)$ in Q-depleted RC's exhibits a weak negative dependence down to near 120 K with an apparent activation energy of .040 $\pm$ .010 eV; below this temperature, k_s is temperature independent. The very preliminary study of Q-reduced RC's suggests that k_s has a shallower temperature dependence than in Q-depleted samples, but is slower at low temperatures.

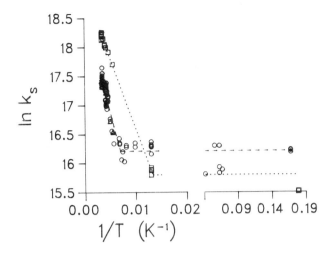

Figure 3

Discussion

The fast k_s found in Q-reduced RC's is sufficient to explain the shorter $(P^+ I^-)$ lifetime and lower Φ_T in this sample type, assuming the rates of electron transfer to ground state and ^3P are relatively unchanged. Such a situation is possible if the rate of activated electron transfer back to $^1P^*$ is increased by an upward shift in the $(P^+ I^-)$ energy owing to coulombic repulsion with Q^-. To account for the observed τ_s in Q^- RC's, this mechanism would require about 50% of the $^1(P^+ I^-)$ population to decay via $^1P^*$. If such a large fraction of $^1(P^+ I^-)$ decays via this activated route, the temperature dependence of k_s should be much steeper than that shown in figure 3. This might be rationalized by correcting the apparent activation energy for the red shift of the $^1P^*-P$ energy at low temperatures; however, the hypothesis of significant $^1(P^+ I^-)$ decay via P^* also contradicts recent measurements of the delayed fluorescence yield from $(P^+ I^-)$ [8].

The temperature dependence of electron transfer from $^1(P^+ I^-)$ to ground state is remarkably similar to that of $(P^+ Q^-)$ ion pair recently reported by Gunner $et\ al.$ [9]. As these authors note, a weak temperature dependence is expected for ET coupled to more than one high frequency vibrational mode of the reactants or protein, especially for such an exothermic reaction. The smaller k_s in Q^- RC's at low temperature is noteworthy: since the exothermicity of the reaction to ground state is presumably in the Marcus "inverted region", this result would also indicate that $(P^+ I^-)$ is further from the ground state in the presence of Q^-.

References

1. Haberkorn, R. and Michel-Beyerle, M. E., *Biophys. J.* **26**:489-498 (1979)
2. Ogrodnik, A., Kruger, H. W., Orthuber, H., Haberkorn, R. and Michel-Beyerle, M. *Biophys. J.* **39**:91-99 (1982)
3. Boxer, S. G., Chidsey, C. E. D. and Roelofs, M. G., *Ann. Rev. Phys. Chem.* **34**:389-417 (1983)
4. Norris, J. R., Bowman, M. K., Budil, D. E., Tang, J., Wraight, C. A. and Closs, G. L., *Proc. Natl. Acad. Sci. USA* **79**:5532-5536 (1982)
5. Lersch, W. and Michel-Beyerle, M. E., *Chem. Phys.* **78**:115-126 (1982).
6. M. R. Wasielewski, C. H. Bock, M. K. Bowman and J. R. Norris, *J. Am. Chem. Soc.* **105**:2903-2904 (1983)
7. Wasielewski, M. R., Bowman, M. K., and Norris, J. R., *Farad. Discuss. Chem. Soc.* **78**:279-288 (1984)
8. Woodbury, N. T. W. and Parson, W. W., *Biochim. Biophys. Acta* **767**:345-361 (1984).
9. Gunner, M. R., Robertson, D. E., and Dutton, P. L., *J. Phys. Chem.*, in press (1986)

SUPRAMOLECULAR ORGANIZATION OF LIGHT-HARVESTING PIGMENT-PROTEIN COMPLEXES
OF RHODOBACTER SPHAEROIDES STUDIED BY EXCITATION ENERGY TRANSFER AND
SINGLET-SINGLET ANNIHILATION AT LOW TEMPERATURE IN PHOSPHOLIPID-ENRICHED
MEMBRANES

WILLEM H. J. WESTERHUIS, MARCEL VOS*, ROB J. VAN DORSSEN*, RIENK VAN
GRONDELLE[+], JAN AMESZ* AND ROBERT A. NIEDERMAN

Department of Biochemistry, Rutgers University, PO Box 1059, Piscataway,
NJ 08854, USA, *Department of Biophysics, Huygens Laboratory of the State
University, PO Box 9504, 2300RA Leiden, and [+]Department of Biophysics,
Physics Laboratory of the Free University, De Boelelaan 1081, 1081HV
Amsterdam, The Netherlands

1. INTRODUCTION

The intracytoplasmic membranes of the facultative photoheterotrophic
bacterium Rhodobacter (Rhodopseudomonas) sphaeroides contain peripheral
and core light-harvesting pigment-protein complexes designated as B800-850
and B875, respectively, based upon near-IR absorbance maxima (1). Radiant
energy harvested by B800-850 is transferred to B875 which surrounds,
interconnects and funnels excitation energy to the reaction centers.

In a previous study of the structural organization of the light-
harvesting system, the phospholipid content of the membrane was increased
by fusion of chromatophores to liposomes (2). Excitation energy transfer
measurements in these preparations suggested that a portion of the B800-850
had dissociated from the rest of the photosynthetic unit (2). Here, a more
gentle fusion technique (3) was employed which preserved the B800 absorption band and permitted a detailed study of the temperature dependence of
fluorescence emission from the photosynthetic units. Singlet-singlet
annihilation studies were consistent with the possibility that in these
fused preparations, B800-850 multimers were dissociated into smaller units.

2. EXPERIMENTAL PROCEDURE

The freeze-thaw procedure of Casadio et al. (3) was used to fuse chromatophores to small unilamellar liposomes. Chromatophores (0.95 mg
BChl/ml) prepared as in (2) were mixed with liposomes obtained by sonica-
tion of egg yolk phosphatidylcholine (75 mg/ml) in a 6:4 (vol/vol) ratio,
respectively, and frozen quickly by immersion in dry ice-ethanol. The
sample was thawed at room temperature, sonicated at 0°C for 30 s, layered
onto a discontinuous sucrose density gradient and centrifuged for 18 h at
96,000 x g. Three bands were visible after centrifugation, two of which
sedimented at the same rates as native chromatophores (fraction 1) and
liposomes (fraction 4), respectively, with the third one (fraction 3) at
an intermediate rate.

Absorbance, fluorescence excitation and emission spectra were obtained
as described previously (2). The absorbance spectra showed that only
small losses of B800 BChl occurred upon fusion. For measurements of
excitation energy transfer efficiencies, fractional absorption spectra and
excitation spectra of 920-nm emission from B875 were equalized at 885 nm.
Singlet-singlet annihilation was measured from the quenching of integrated
fluorescence yield as a function of the intensity of a picosecond laser
pulse with instrumentation described elsewhere (4).

Biggens, J. (ed.), Progress in Photosynthesis Research, Vol. I. ISBN 90 247 3450 9
© *1987 Martinus Nijhoff Publishers, Dordrecht. Printed in the Netherlands.*

3. RESULTS AND DISCUSSION

Fluorescence excitation spectra at 4 K showed that ∿5-fold phospholipid enrichment resulted in decreased energy transfer of ∿35% from B850 to the B875 core complex; that from B800 to B850 and from carotenoids to BChl were essentially unaffected (Table 1). This confirms the selective phospholipid-induced effect on inter-complex excitation energy transfer observed previously in preparations obtained by acidic-pH and polyethylene glycol fusion procedures (2). Decreased energy transfer to B875 is also reflected in the enhanced emission from B850 BChl of up to ∿10-fold relative to that from B875 BChl observed at 4 K (Fig. 1).

TABLE 1. Efficiency of excitation energy transfer in fused preparations at 4 K as determined from fluorescence excitation spectra

Fraction	Phospholipid/ protein (-fold increase)	BChl B850 → B875	BChl B800 → B850	BChl Q_x^c → B875	Carotenoid 512 nm → Q_x
1	1.0	1.0	0.85	0.91	0.92
2[a]	1.1	0.94	0.86	0.78	0.96
3	5.3	0.64	0.86	0.69	0.86
4	_b	0.54	0.80	0.75	0.73

[a]Derived from upper portion of chromatophore band.
[b]Phospholipid content not determined since free liposomes also present.
[c]Q_x band measured at 590 nm; contains contributions from B800, B850 and B875 BChls.

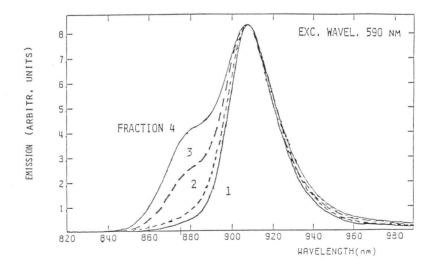

FIGURE 1. Fluorescence emission spectra of fused preparations at 4 K. The spectra (not corrected for response of measuring system) were normalized at their emission maxima.

With emission spectra obtained in the temperature range 200–300 K, the relative fluorescence yield from B850 and B875 ($\emptyset870/\emptyset900$) was plotted as a function of temperature (Fig. 2). Assuming efficient energy transfer and equal fluorescence rate constants for B850 and B875, the energy difference (ΔE) between their excited states was calculated using $\emptyset870/\emptyset900 = [N850/N875]\exp(-\Delta E/kT)$, where N = number of connected BChls, k = Boltzmann's constant and T = absolute temperature (5). This yielded $\Delta E = 0.035$ eV, which is in reasonable agreement with the value calculated from the absorption spectrum ($\Delta E_{850-875\ nm} = 0.042$ eV). The fused preparations showed little or no dependence of $\emptyset870/\emptyset900$ on temperature (Fig. 2). This indicates that after membrane fusion, thermal equilibrium no longer exists between B850 and B875 because the B800–850 complex becomes disconnected and the energy distribution is shifted toward B850.

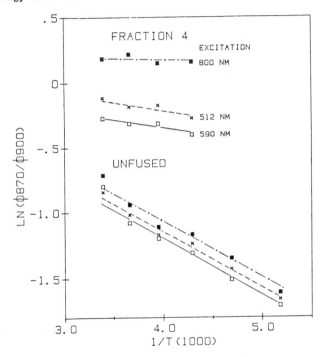

FIGURE 2. Temperature dependence of ratio of B850 ($\emptyset870$) and B875 ($\emptyset900$) fluorescence emission yields. See text for details.

Singlet-singlet annihilation was examined to measure the domain size (N_D), i.e. the cluster of connected antenna BChls among which excitations are transferred freely. In the unfused chromatophores, analysis of B875 BChl emission was consistent with an N_D value ≫550 at ~300 K which was reduced ~8-fold at 4 K (Table 2, fraction 1). This can be interpreted as a decreased connectivity of B875 units at low temperature and lack of back transfer to B850 (4). In the fused preparations, N_D values of 75–90 were calculated from the B875 BChl emission at both ~300 and 4 K. The temperature independence of the domain size observed in the fused preparations suggests that phospholipid enrichment decreased B875 connectivity to that of a minimal unit in which 3–4 reaction centers are embedded (4).

TABLE 2. Domain sizes calculated from singlet-singlet annihilation data[a]

| Fraction | B875 BChl emission | | | | B850 BChl emission | |
| | 300 K(0897)[b] | | 4 K(0907) | | 4 K(0875) | |
	r[c]	N_D	r	N_D	r	N_D
1	5	>550	1	65	1	45[d]
2	1	75	2	80	0	3
4	2	90	2	85	1	12

[a]Domain sizes (N_D) estimated from curves of integrated fluorescence yield vs energy density of 532-nm laser pulse excitation as described in ref. (4).
[b]Wavelength (nm) of fluorescence measurements.
[c]Parameter reflecting efficiency of annihilation relative to mono-excitation decay processes (4); high r values correspond to low annihilation efficiencies.
[d]Values for R. sphaeroides strain 2.4.1. obtained from 880-nm emission (ref. 4). Emission from B850 BChl in chromatophores of strain NCIB8253 used here was insufficient for these measurements.

Such minimal units have also been observed in the developing photosynthetic membranes of R. sphaeroides (6) and are of essentially the same size as the B875 clusters determined for the unfused chromatophore preparations at 4 K.

In unfused chromatophores, an N_D value of 45 was calculated with B850 BChl emission data obtained at 4K (ref. 4); after fusion, domains of only ~3 to 12 connected B850 BChls were observed (Table 2). This is consistent with a phospholipid-induced dissociation of B800-850 multimers into much smaller units.

ACKNOWLEDGEMENTS

Supported by U.S. National Science Foundation grant DMB85-12587 and Netherlands Foundation for Biophysics.

References

1. Cogdell, R. J., Zuber, H., Thornber, J. P., Drews, G., Gingras, G., Niederman, R. A., Parson, W. W. and Feher, G. (1985) Biochim. Biophys. Acta 806, 185-186
2. Pennoyer, J. D., Kramer, H. J. M., van Grondelle, R., Westerhuis, W. H. J., Amesz, J. and Niederman, R. A. (1985) FEBS Lett. 182, 145-150
3. Casadio, R., Venturoli, G., Di Gioia, A., Castellani, P., Leonardi, L. and Melandri, B. A. (1984) J. Biol. Chem. 259, 9149-9157.
4. Vos, M., van Grondelle, R., van der Kooij, F. W., van de Poll, D., Amesz, J. and Duysens, L. N. M. (1986) Biochim. Biophys. Acta 850, 501-512
5. van Grondelle, R., Kramer, H. J. M. and Rijgersberg, C. P. (1982) Biochim. Biophys. Acta 682, 208-215
6. Hunter, C. N., Kramer, H. J. M. and van Grondelle, R. (1985) Biochim. Biophys. Acta 807, 44-51

CORRELATION BETWEEN THE EFFICIENCY OF ENERGY TRANSFER AND THE POLYENE
CHAIN STRUCTURE OF CAROTENOIDS IN PURPLE PHOTOSYNTHETIC BACTERIA

H. HAYASHI, K. IWATA, T. NOGUCHI, and M. TASUMI

DEPARTMENT OF CHEMISTRY, FACULTY OF SCIENCE THE UNIVERSITY OF TOKYO,
BUNKYO-KU, TOKYO 113, JAPAN

1. INTRODUCTION

Previously we reported the resonance Raman spectra of carotenoids in
light harvesting (LH) systems of several purple photosynthetic bacteria
[1].   The relative intensity of a Raman band at about 960 cm^{-1} of
carotenoids in some species of bacteria *in vivo* is higher than that of the
corresponding band of extracted carotenoids.   This band is assigned to the
CH out-of-plane wagging, and we have concluded that the polyene chains of
the all-*trans* carotenoids contained in those bacteria are distorted from
planarity. On the other hand, for some other bacteria, the relative
intensity of the 960 cm^{-1} Raman band is as weak as that of extracted
carotenoids.   In the latter bacteria most of carotenoids probably have the
planar all-*trans* polyene chain. Thus, the Raman band at about 960 cm^{-1}
gives information on the polyene chain structure of *in vivo* carotenoids.

It is known that the efficiency of the energy transfer from
carotenoids to bacteriochlorophyll (Bchl) in purple photosynthetic bacteria
is varied from one species to another [2].   Boucher et al. [3] have
suggested that the efficiency is specific to the carotenoid species. On
the other hand, Cogdell et al. [4] have concluded that it is independent of
the carotenoid species and determined by the interaction between
carotenoids and proteins which are specific to each bacterial strain.
Thus, the factors which affect the efficiency of the energy transfer are
not yet clearly understood.   In order to gain an insight into this
problem, we studied the correlation between the efficiency of the energy
transfer and the intensity of the 960 cm^{-1} Raman band of carotenoids
contained in photosynthetic bacteria.   These measurements were carried out
not only for intrinsic carotenoids contained naturally in the bacteria but
also for an extrinsic carotenoid artificially bound to the carotenoid-
deficient LH protein complexes.

2. MATERIALS AND METHODS

Efficiency of the energy transfer and resonance Raman spectra of
carotenoids were measured for chromatophores and LH Bchl protein complexes
from *Rhodopseudomonas sphaeroides*, its green mutant, *Rp. capsulata*, *Rp.
gelatinosa*, *Chromatium vinosum*, *Rhodospirillum rubrum*, and *Rp. palustris*.
The carotenoid-deficient LH complex was isolated from *Ch. vinosum* cultured
in a medium containing 12 mg/l of diphenylamine (DPA).   Spheroidene
isolated from *Rp. sphaeroides* was bound to this complex using a method
similar to that reported by Boucher et al. [3].   The isolated LH complex
(absorbance at 800 nm ca. 30) was mixed with spheroidene (Bchl:spheroidene
= 1:30) and sonicated for 5 min in the presence of 2.0 % sodium
deoxycholate.   After dialyzing this mixture against Tris buffer overnight,
excess and loosely bound spheroidene was separated from the LH complex with
SDS polyacrylamide gel electrophoresis.   This step was repeated until no
free spheroidene appeared.

Resonance Raman spectra of carotenoids were measured in the manner
reported previously [1].   The 488.0, 501.7, and 514.5 nm lines of an Ar+
laser were used for Raman excitation.   Efficiency of the energy transfer
from carotenoids to Bchl was estimated by comparing the excitation spectrum

Biggens, J. (ed.), Progress in Photosynthesis Research, Vol. I. ISBN 90 247 3450 9

of Bchl fluorescence with the absorption spectrum in the region of 350-650 nm. The intensities of the two kinds of spectra were normalized at 590 nm.

3. RESULTS AND DISCUSSION

In Fig. 1 are shown the absorption and excitation spectra in the 350-650 nm region as well as the resonance Raman spectra in the 1300-900 cm^{-1} region of chromatophores from *Rp. sphaeroides* and those from *Ch. vinosum*. The efficiency of the energy transfer from carotenoids in *Rp. sphaeroides* is higher than 90 % and that of carotenoids in *Ch. vinosum* lower than 40 %. The Raman band at about 960 cm^{-1} of carotenoids is weak for *Rp. sphaeroides* and relatively high for *Ch. vinosum*. Results for several bacteria are summarized in Table 1. For *Rp. palustris* and *Ch. vinosum*, the measurements were carried out for both chromatophores and the isolated LH complex, and similar results were obtained.

The bacterial strains used in this study can be classified into two groups according to the carotenoid composition which is determined by the pathway of carotenoid synthesis [5]. One of them is a group of bacteria containing spheroidene series carotenoids (spheroidene, spheroidenone, neurosporene, methoxyneurosporene, and chloroxanthin), the other is that of bacteria containing spirilloxanthin series carotenoids (spirilloxanthin, rhodopin, rhodovibrin, and lycopene). From the results summarized in Table 1, it is obvious that the bacteria containing sheroidene series carotenoids exhibit high efficiencies and relatively weak 960 cm^{-1} bands.

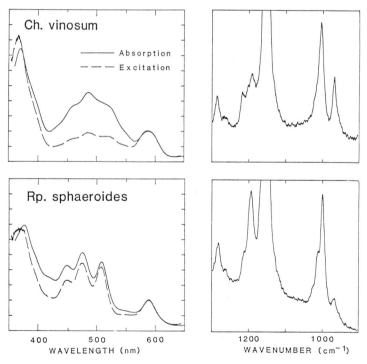

Figure 1. Absorption and excitation spectra (left), and resonance Raman spectra (right) of carotenoids in *Rp. sphaeroides* (upper) and *Ch. vinosum* (bottom).

Table 1. Efficiency of the energy transfer and the intensity
of the 960 cm^{-1} Raman band

Sample	960 cm^{-1} Band*	Efficiency (%)	Carotenoids**
Ch. vinosum	S	30	A
Rp. palustris	S	30	A
Rs. rubrum	S	40	A
Rp. sphaeroides	W	90	B
Rp. sphaeroides Gm.	W	80	B
Rp. capsulata	W	80	B
Rp. gelatinosa	W	80	B
Ch. vinosum DPA cul.	S	50	B
Ch. vinosum +carotenoid	S	40	B

*S: strong (distorted), W: weak (planar)
**A: rhodopin, spirilloxanthin
  B: neurosporene, chloroxanthin, spheroidene, spheroidenone

On the other hand, those containing spirilloxanthin series carotenoids exhibit low efficiencies and relatively strong 960 cm^{-1} bands. Thus, it seems that carotenoids species, the efficiency of the energy transfer, and the intensity of 960 cm^{-1} Raman bands are correlated.

To make the above correlation clearer, the efficiency and Raman spectra were measured for an extrinsic carotenoid component contained in the LH protein of *Ch. vinosum*. The LH complex from cells cultured with DPA for 2-3 days contained a very small amount of carotenoids (about 1/30 of the normal contents). To this carotenoid-deficient LH complex, about 0.25 mole of spheroidene per 1 mole of Bchl could be bound. The excitation spectrum of the spheroidene-binding LH complex is shown in Fig. 2 together with the absorption spectrum. The efficiency of the energy transfer was estimated to be about 40 %. The resonance Raman spectrum of carotenoids contained in the spheroidene-binding LH complex is shown in Fig. 2. Only the bound spheroidene is responsible for this Raman spectrum since the 501.7 nm line of Ar+ laser was used for excitation and at this

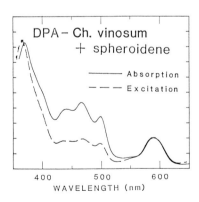
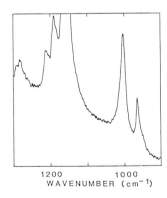

Figure 2. Absorption and excitation spectra (left), and resonance Raman spectra (right) of spheroidene-binding LH complex of *Ch. vinosum* cultured with DPA.

wavelength the intrinsic carotenoids remaining in the carotenoid-deficient LH complex show a very weak absorption. It is noticeable that relative intensity of the 960 cm^{-1} band is much higher for spheroidene in the LH complex of *Ch. vinosum* than for spheroidene in *Rp. sphaeroides* (Fig. 1).

The LH complex from *Ch. vinosum* cultured with DPA for more than 5 days contained more carotenoids than the carotenoid-deficient LH complex (about 1/10 of the normal contents). Major carotenoid components of this partially carotenoid-deficient LH complex were chloroxanthin and OH-ζ-carotene. This result is reasonable in view of the action of DPA to the carotenoid synthesis. The efficiency of the energy transfer from chloroxanthin to Bchl was estimated to be below 50 %. The resonance Raman spectrum which mainly arises from chloroxanthin shows a relatively intense 960 cm^{-1} Raman band.

There are definite differences in chemical structure between spheroidene series and spirilloxanthin series carotenoids, apart from the peripheral substituents of individual carotenoids. The former carotenoids have nine or ten conjugated C=C bonds and their conjugated system is asymmetrical with respect to the central $C_{15}=C_{15'}$ bond. On the other hand, the latter carotenoids have the symmetrical conjugated systems with eleven or thirteen conjugated C=C bonds. However, these differences of the conjugated polyene parts have no effects to the intensity of the 960 cm^{-1} Raman band, because spheroidene and chloroxanthin contained in the LH complex of *Ch. vinosum* show the relatively intense 960 cm^{-1} Raman band, similarly to spirilloxanthin series carotenoids which are typical components of *Ch. vinosum*. This means that the distortion of the polyene chain is primarily due to the interaction between carotenoids and proteins, independently of the carotenoid species. It is also clear that the low intensities of the 960 cm^{-1} band of spheroidene series carotenoids in *Rp. sphaeroides*, *Rp. capsulata*, and *Rp. gelatinosa* reflect the planar structure of the conjugated polyene chain.

From the results obtained for not only the intrinsic components but also the extrinsic components of carotenoids, it is clear that the efficiency of the energy transfer is specific to the bacterial strain but not to the carotenoid species. The planarity of the polyene chain is also specific to the bacterial strain but not to the carotenoid species. The planarity of the carotenoid polyene chain seems to be an important factor for ensuring the efficient energy transfer from carotenoids to Bchl in purple photosynthetic bacteria. The planarity of polyene chain probably depends on the protein moiety of the LH complex of the individual bacterial strain. As a result, the efficiency of the energy transfer from carotenoids to Bchl is specific to the bacterial strain.

4. REFERENCES

1. Iwata, K., Hayashi, H., and Tasumi, M. (1985) Biochim. Biophys. Acta 810, 269-273
2. Goedheer, J.C. (1959) Biochim. Biophys. Acta 35, 1-8
3. Boucher, F., Van der Rest, M., and Gingras, G. (1977) Biochim. Biophys. Acta 461, 339-357
4. Cogdell, R.J., Hipkins, M.F., MacDonald, W., and Truscott, T.G. (1981) Biochim. Biophys. Acta 634, 191-202
5. Schmidt K. (1978) in "The Photosynthetic Bacteria" ed by Clayton, R.K. and Sistrom W.R., Plenum Press, New York, 729-750

TRIPLET ENERGY TRANSFER BETWEEN PHOTOSYNTHETIC PIGMENTS: AN ESR STUDY OF
B800-850 LIGHT-HARVESTING COMPLEXES AND SYNTHETIC CAROTENOPORPHYRIN
MOLECULES

HARRY A. FRANK, BARRY W. CHADWICK, CHAOYING ZHANG AND JUNG JIN OH,
DEPARTMENT OF CHEMISTRY, 215 GLENBROOK ROAD, UNIVERSITY OF CONNECTICUT,
STORRS, CT 06268 USA; DEVENS GUST, PAUL A. LIDDELL AND THOMAS A. MOORE,
DEPARTMENT OF CHEMISTRY, ARIZONA STATE UNIVERSITY, TEMPE, AZ 85281 USA;
RICHARD J. COGDELL DEPARTMENT OF BOTANY, UNIVERSITY OF GLASGOW, GLASGOW G12
8QQ UK

1. INTRODUCTION

The antenna systems of purple photosynthetic bacteria are comprised of
discrete pigment-protein complexes in which non-covalently bound
bacteriochlorophyll (BChl) and carotenoid molecules are capable of
transferring excitation energy to the reaction center [1]. One such
complex, denoted B800-850 for its approximate wavelengths of maximum
absorption in the near infrared spectral region, makes up a large part of
the antenna system in several photosynthetic bacteria [1]. Excitation
transfer between the pigments in the B800-850 complex has been studied
extensively by optical spectroscopic methods [2]. One approach to
examining the molecular features which control both singlet and triplet
excitation transfer between carotenoids and chlorophylls is to study
covalently-linked carotenoporphyrin molecules [3]. These molecules consist
of carotenoid polyenes linked to porphyrin derivatives and are capable of
mimicking both the light-harvesting and photoprotective functions of
carotenoids. Recently, electron spin resonance (ESR) techniques have been
used to probe the structures, geometries and dynamics of carotenoids in
several strains of photosynthetic bacteria [4]. In this paper we present a
parallel investigation of the triplet state ESR properties of a series of
B800-850 complexes and synthetic carotenoporphyrin molecules.

2. MATERIALS AND METHODS

The B800-850 complexes were isolated from the photosynthetic bacteria
essentially as previously described [5]. The carotenoporphyrins were
prepared by dissolving the molecules in 2-methyltetrahydrofuran which had
been purified by distillation over sodium metal. Triplet state ESR
spectroscopy was carried out as previously described [4].

3. RESULTS AND DISCUSSION

The triplet state ESR spectra of the B800-850 light harvesting complexes
are shown in Fig. 1. Figure 2 shows the structures of four different model
compounds analyzed here. The triplet state ESR spectra from the model
compounds are shown in Fig. 3. All of these ESR spectra display the
polarization pattern eae aea. In addition to the free-base
carotenoporphyrin systems the zinc-substituted porphyrin analogs of
carotenoporphyrin a and b were analyzed by triplet state ESR. The zinc-
substituted carotenoporphyrin a displayed precisely the same spectrum as
the free-base carotenoporphyrin a shown in Fig. 3. The zinc-substituted

Biggens, J. (ed.), Progress in Photosynthesis Research, Vol. I. ISBN 90 247 3450 9
© *1987 Martinus Nijhoff Publishers, Dordrecht. Printed in the Netherlands.*

carotenoporphyrin b, however, displayed an ESR spectrum that was markedly different from the free-base compound b. (See Fig. 4.) The lineshape arising from the zinc-substituted carotenoporphyrin b is consistent with two triplet species giving rise to the spectrum. The first of these triplets has precisely the same zero-field splitting parameters as carotenoporphyrin a. The lineshape of the second triplet species resembles the spectrum of zinc-tetraphenylprophyrin reported previously [6]. The triplet state zero-field splitting parameters are summarized in Table I.

The triplet state ESR spectra which are observed in the B800-850 light-harvesting complexes of photosynthetic bacteria (Fig. 2) are due to carotenoids. This assignment is supported by the following arguments:

I. The zero-field splitting parameters which characterize the triplet state spectra correlate with the structures of the various carotenoids present in the different B800-850 complexes. Rb. sphaeroides GA contains neurosporene chromophores. Anaerobically and aerobically grown Rb. sphaeroides wild type contain spheroidene and spheroidenone, respectively, as their major carotenoid pigments. Rps. acidophila 7750 contains rhodopin. The differences in the zero-field splitting parameters presented in Table I may be understood in terms of variations in the extent of dipole-dipole interaction arising from different amounts of π-electron conjugation in the various carotenoids. The $|D|$ values follow the order neurosporene (nine carbon-carbon double bonds) > spheroidene (ten carbon-carbon double bonds) > spheroidenone (ten carbon-carbon double bonds plus one carbon-oxygen double bond) > rhodopin (eleven carbon-carbon double bonds).

II. There is a correlation between the trends in the wavelength maxima observed in the optical spectroscopic experiments carried out previously on B800-850 complexes and the trends in the zero-field splitting parameters presented here [7]. The triplet-triplet absorptions have been shown to blue shift as the length of the π-electron conjugation is shorted. This is consistent with the ESR data which show that the zero-field splitting parameters increase as the extent of π-electron conjugation decreases.

III. The ESR of the spectra of the B800-850 complexes shown in Fig. 2 bear a marked similarity in lineshape to the ESR spectra observed from the carotenoids in synthetic carotenoporphyrin molecules (Fig. 3). When the carotenoid is absent (e.g. in tetratoluoporphyrin) or when it is linked by a long chain (i.e. carotenoporphyrin c), one observes only the porphyrin triplet state spectrum either optically or by ESR. As the link between the porphyrin and the carotenoid is shortened the rate and efficiency of energy transfer from the porphyrin to the carotenoid increases. If the rate of energy transfer is comparable to the decay of the carotenoid triplet to the ground state (~10μsec), both the porphyrin and carotenoid triplets can be observed simultaneously. (See Fig. 3.) If the carotenoid is linked in very close proximity to the porphyrin, such as provided by the amide link in carotenoporphyrin a, the triplet transfer rate is extremely fast (~30ns), the transfer efficiency is very high (~100%) and one observes only the carotenoid triplet state.

A comparison of the triplet state ESR spectra of the synthetic compounds with those of the isolated light-harvesting complexes indicates that a very close geometric proximity is achieved between the BChl and the carotenoid in the B800-850 complexes. The optical and ESR results from the synthetic compounds suggest that this close proximity is an essential structural feature for efficient porphyrin-to-carotenoid triplet-triplet energy transfer because the porphyrin triplet signal intensities become more pronounced relative to the carotenoid signals as the length of the link is increased [3].

The most obvious feature of the data presented here is the fact that all of the triplet state ESR spectra attributable to carotenoids exhibit the spin polarization pattern (eae aea). The present data show that the carotenoid triplet ESR lineshapes are not affected by changes in the structure of the donor (porphyrin, zinc-porphyrin or BChl), are insensitive (apart from the spectral line splittings) to the extent of conjugation of the carotenoid acceptor, and are not dependent on the relative orientation of the pair (i.e. the ortho, meta and para analogs of carotenoporphyrin a gave rise to the same ESR spectra). The molecular basis for this spectral uniformity is the subject of an ongoing investigation.

TABLE I

ZERO-FIELD SPLITTING PARAMETERS OF THE OBSERVED TRIPLET STATES

$|D|$ and $|E|$ are the triplet state zero-field splitting parameters (in cm^{-1}) which characterize the ESR spectra. The parameters were obtained from computer simulations of the experimental triplet state spectra. The errors in the numbers give the range of parameters for which the simulations fell within the reproducibility of the experimental spectra.

sample	$|D|$	$|E|$	triplet assignment
B800-850 complexes			
Rb. sphaeroides GA	0.0365 ± 0.0002	0.0035 ± 0.0002	neurosporene
Rb. sphaeroides wild type (anaerobic)	0.0324 ± 0.0002	0.0036 ± 0.0002	spheroidene
Rb. sphaeroides wild type (aerobic)	0.0318 ± 0.0002	0.0032 ± 0.0002	spheroidenone
Rps. acidophila 7750	0.0279 ± 0.0003	0.0029 ± 0.0003	rhodopin
Model compounds			
carotenoporphyrin a	0.0356 ± 0.0002	0.0036 ± 0.0002	carotenoid
Zn-carotenoporphyrin a	0.0356 ± 0.0002	0.0036 ± 0.0002	carotenoid
carotenoporphyrin b			
Triplet 1	0.0356 ± 0.0002	0.0036 ± 0.0002	carotenoid
Triplet 2	0.0378 ± 0.0002	0.0080 ± 0.0002	tetratoluoporphyrin
Zn-carotenoporphyrin b			
Triplet 1	0.0356 ± 0.0002	0.0036 ± 0.0002	carotenoid
Triplet 2	0.0308 ± 0.0005	0.0103 ± 0.0005	Zn-tetratoluoporphyrin
carotenoporphyrin c	0.0398 ± 0.0002	0.0078 ± 0.0002	tripyridylporphyrin
tetratoluoporphyrin	0.0378 ± 0.0002	0.0080 ± 0.0002	tetratoluoporphyrin

REFERENCES

1. Thornber, J. P., Cogdell, R. J., Pierson, B. K. and Seftor, R. E. B. (1983) J. Cell. Biochem. 23, 159-169
2. Van Grondelle, R. (1985) Biochim. Biophys. Acta 811, 147-195
3. Gust, D., Moore, T. A., Bensasson, R. V., Mathis, P., Land, E. J., Chachaty, C., Moore, A. L., Liddell, P. A. and Nemeth, G. A. (1985) J. Amer. Chem. Soc. 107, 3631-3640
4. McGann, W. J. and Frank, H. A. (1985) Biochim. Biophys. Acta 807, 101-109
5. Cogdell, R. J., Durant, I., Valentine J., Lindsay, J. G. and Schmidt, K. (1983) Biochim. Biophys. Acta 772, 427-435
6. Van Willigen, H. and Chandrashekar, T. K. (1983) J. Chem. Phys. 78, 7093-7098
7. Kingma, H., R. van Grondelle and L. N. M. Duysens (1985) Biochim. Biophys. Acta 808, 383-399

ACKNOWLEDGEMENTS

This work was supported by grants from the National Science Foundation (PCM-8408201) and the University of Connecticut Research Foundation.

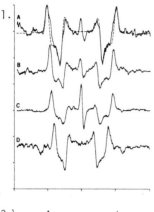

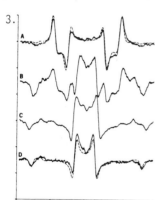

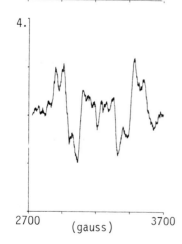

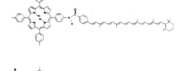

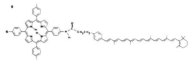

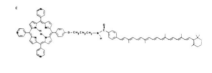

FIGURE CAPTIONS

Figure 1. Triplet state ESR spectra of B800-850 complexes. (a) Rb. sphaeroides GA; (b) Rb. sphaeroides wild type (anaerobically grown); (c) Rb. sphaeroides wild type (aerobically grown); (d) Rps. acidophila 7750.

Figure 2. Structures of the carotenoporphyrins used in the present study.

Figure 3. Triplet state ESR spectra of the carotenoporphyrin molecules shown in Fig. 2.

Figure 4. Triplet state ESR spectrum of zinc-carotenoporphyrin b.

2700 (gauss) 3700

PICOSECOND EXCITATION ENERGY TRANSFER BETWEEN DIFFERENT LIGHT-HARVESTING
COMPLEXES AND REACTION CENTRES IN PURPLE BACTERIA

V.I.GODIK, A.FREIBERG[*], K.TIMPMANN[*], A.Yu.BORISOV, and K.K.REBANE[*]

A.N.BELOZERSKY LABORATORY OF MOSCOW STATE UNIVERSITY, MOSCOW 119899 and
[*]INSTITUTE OF PHYSICS, ESTONIAN SSR ACADEMY OF SCIENCES, TARTU 202400,
U.S.S.R.

1. INTRODUCTION

This paper represents the results of the first direct measurements of
heterogeneous picosecond excitation energy transfer rates between different
bacteriochlorophyll (BChl) spectral forms of light-harvesting antenna and
reaction centres of purple photosynthetic bacteria. Fluorescence decay kinetics of B800, B850 and B875 (B890) spectral forms of non-sulfur bacteria
Rhodopseudomonas sphaeroides and sulfur one *Chromatium minutissimum* were
studied as a function of the reaction centre state at room temperature and
77 K. On the basis of the obtained data, BChl excited state dynamics in the
photosynthetic membranes of purple bacteria was described and a lateral to-
pographly of different antenna complexes was proposed. In particular, strong
evidence was obtained which favours a heterogeneity of the B800-850 complexes *in vivo*. A part of these complexes seems to be directly coupled to
the reaction centre protein, instead of being at the periphery of the photosynthetic units, as it was suggested by Monger and Parson [1].

2. MATERIALS AND METHODS

2.1. Objects: Cells of *Rps.sphaeroides* and *Chr.minutissimum* (wild types,
   Moscow University Collection) were grown and chromatophores were isolated as described [2]. Chromatophore suspensions with optical density of 150-200 at 880 nm were stored at 0°C and diluted by tris-HCl buffer, pH 7.5 (or by a mixture of the buffer and glycerol, 1 : 3 (v/v) in
   low temperature experiments) to a final concentration of about $5 \cdot 10^{16}$
   molecules/ml just before measurements.

2.2. Methods: BChl fluorescence lifetimes were measured with a picosecond
   spectrochronograph [2,3]. Excitation light of 373, 772 and 796 nm from
   a "Spectra - Physics" mode locked cw oxazine 1 dye laser, synchronously
   pumped with 82 MHz by krypton-ion laser, was employed. Fluorescence,
   thus excited, came through a subtractive dispersion double-grating
   monochromator (spectral bandwidth, 4 nm) and was detected by a synchro-
   scan streak camera. Time resolution of the apparatus was 3 ps. Experimental decay curves were simulated by a sum of exponential components
   taking into account the apparatus response function.

3. RESULTS AND DISCUSSION

On exciting the chromatophore suspensions of either *Rps.sphaeroides* or
Chr.minutissimum by light of 772 or 796 nm, absorbed mainly by B800 of the
B800-850 antenna complex, fluorescence decay kinetics were found to change
substantially with the wavelength of observation. A part of the data concerning *Rps.sphaeroides* has been published elsewhere [4]. In the spectral

range of 790–820 nm a weak short-lived fluorescence from B800, judging from the spectral distribution of the amplitudes of the decay curves (Fig.1), was observed.

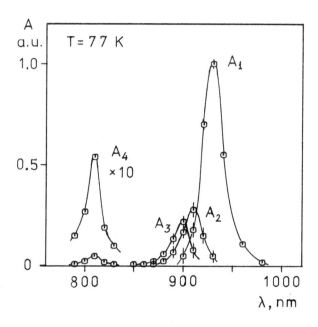

FIGURE 1. Time resolved fluorescence spectra of individual decay components of *Chr.minutissimum* chromatophores without additions at 77 K. Absolute amplitudes of the components were plotted as a function of the observation wavelength: A_1 – for the 190 ps component; A_2 – for the 60 ps component; A_3 – for the 8 ps component; A_4 – for the ≤5 ps component. Excitation light was at 772 nm and its density was 0.4 W/cm². The monochromator bandwidth, 4 nm.

Fluorescence decay curves may be simulated in this spectral range by one dominant component with a lifetime shorter than 3 – 5 ps and at least one minor more long-lived ($\tau \simeq 60$ ps) component both for *Rps.sphaeroides* and *Chr. minutissimum*. In the 840 – 920 nm range at least three lifetime components were necessary to suggest for deconvolution of experimental decay curves, viz. $\tau_1 \simeq 8$ ps; $\tau_2 \simeq 15$ ps; $\tau_3 \simeq 70$ ps for open reaction centres and $\tau_1 \simeq 8$ ps; $\tau_2 \simeq 60$ ps; $\tau_3 \simeq 190$ ps for closed photooxidized reaction centres (Fig.2). The ratio of the amplitudes of these components changed with the registration wavelength in a way that may be cleared up from the dependences shown in Fig.1. It was somewhat different in different cultures and depended on the conditions of growth in the same culture. It may be inferred from the data of Fig.1 that two of the three decay components, the short-lived and the intermediate one, belong to B850 fluorescence and that the most long-lived component arises from B875 (B890). Thus, B850 emission is heterogeneous: it contains two kinetically different components whose picosecond fluorescence spectra are slightly (∿10 nm) shifted one relative the other.

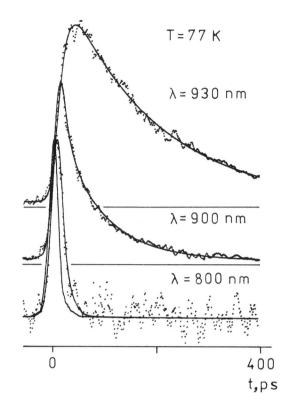

T = 77 K

λ = 930 nm

λ = 900 nm

λ = 800 nm

0   400
                     t,ps

A part of B850 emission (that with the intermediate lifetime) was found to be variable: the value of the lifetime increased from ∿15 ps to ∿60 ps when reaction centres turned from open to closed photooxidized state.[*)]

Such situation that the B850 emission is sensitive to the reaction centre state may be realized if a part of B800–850 pigment-protein complexes is directly coupled to the reaction centre protein. This conclusion is supported by the observation of Peters et al. [5] that there is a chemical cross-linking between the polypeptides of B800–850 complex and those of the reaction centre H subunit. The remaining part of B800–850 complexes, fluorescing with the lifetime of about 8 ps, apparently irrespective of the reaction centre state, seems to be in a direct contact with B875 (B890) complexes, since the lifetime of excitation decay in these complexes was within the experimental uncertainty equal to the risetime of B875 (B890) fluorescence. However, the alternative possibility of the 60 ps component being independent of the reaction centre state cannot be fully rejected on the basis of the above

FIGURE 2. The picosecond kinetics of *Chr. minutissimum* fluorescence at 800, 900 and 930 nm (dotted curves). Continuous curves are model computer simulations of the experimental data using the single exponential decay with $\tau = 5$ ps at $\lambda = 800$ nm, three exponential decay with $\tau_1 = 185$ ps, $\tau_2 = 60$ ps and $\tau_3 = 8$ ps and amplitude ratio $A_1 : A_2 : A_3 = 1 : 3 : 4$ at $\lambda = 900$ nm; the single exponential decay with $\tau = 185$ ps and a risetime of 9 ps at $\lambda = 930$ nm. Conditions of measurements an in Fig.1, except that the monochromator spectral bandwidth was 8 nm. The apparatus response is shown.

[*)] At the used exciting light density of the order of 0.1 W/cm², all the reaction centres were in the photooxidized state. Open reaction centre state was achieved by addition of 10^{-3} M $Na_2S_2O_4$, which turned the reaction centres to PQ^- state. As shown [2], excitation trapping in the picosecond time range by the reaction centres in PQ^- state occurs very much like that with open reaction centres.

data, since the signal to noise ratio was not sufficient to make an unambiguous simulation of the experimental data by a sum of exponentials, having
comparable lifetime values (as it would be the case of $Na_2S_2O_4$ reduced
chromatophores if the lifetimes of the long-lived (the A_1 spectrum in Fig.1)
and the intermediate (the A_2 spectrum in Fig.1) components have similar
values).

Fluorescence decay at $\lambda > 900$ nm (at 77 K 10 - 30 nm at longer wavelengths)
was well approximated by a single exponential with the lifetime equal to
that of the most long-lived component, observed at shorter wavelengths,
viz. 60 ps for open and 190 ps for closed photooxidized reaction centres.
Spectral distribution of this component (Fig.1) shows that it belongs to
B875 (B890) fluorescence. Thus, all of the B875 (B890) picosecond fluorescence is variable, which is in agreement with earlier data [2,4,6].

Lowering the temperature made the spectral dependence of fluorescence
kinetics much more pronounced, especially in *Rps.sphaeroides*, since thermal
mixing of different components, which is very effective in this bacterium
at room temperature, became much less operative at 77 K [4].

The obtained data are summarized in the following hypothetical scheme
of BChl excited state kinetics in the photosynthetic membranes of purple
bacteria, containing both B800-850 and B875 (B890) light-harvesting complexes, after picosecond pulse excitation into the B800 spectral form. The
scheme supposes a peculiar mutual arrangement of the antenna and the reaction centre complexes:

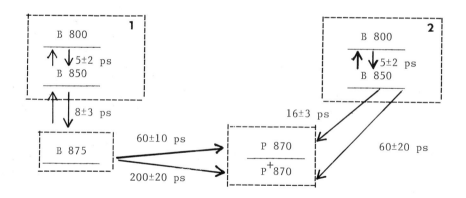

REFERENCES
1 Monger, T.G. and Parson, W.W. (1977) Biochim. Biophys. Acta 460, 393-407
2 Borisov, A.Yu., Freiberg, A.M., Godik, V.I., Rebane, K.K. and Timpmann,
  K.E. (1985) Biochim. Biophys. Acta 807, 221-229
3 Freiberg, A. and Saari, P. (1983) IEEE. J. Quantum Electron., QE-19,
  622-630
4 Godik, V.I., Timpmann, K.E., Freiberg, A.M., Borisov, A.Yu. and Rebane,
  K.K. (1986) Dokl. Acad. Nauk SSSR 289, No 3, 714-717
5 Peters, I., Takemoto, I.Y. and Drews, G. (1983) Biochemistry 22, 5660-5667
6 Zankel, K.H. and Clayton, R.K. (1969) Photochem. Photobiol. 9, 7-15

SPECTRAL DEPENDENCE OF THE FLUORESCENCE LIFETIME OF *RHODOSPIRILLUM RUBRUM*.
EVIDENCE FOR INHOMOGENEITY OF B880 ABSORPTION BAND

A.FREIBERG, V.I.GODIK[*)] and K.TIMPMANN

INSTITUTE OF PHYSICS, ESTONIAN SSR ACADEMY OF SCI., TARTU 202400 and
[*)]A.N.BELOZERSKY LABORATORY OF MOSCOW UNIVERSITY, MOSCOW 119899, U.S.S.R.

1. INTRODUCTION
   Light-harvesting antenna of *Rhodospirillum rubrum* consists of the single
pigment-protein complex B880. Lately, there have been several publications
where, together with the main B880 spectral form, existence of an additional minor form, absorbing 10-25 nm to the red of the main band, was suggested
[1,2]. In this work, spectral dependence of picosecond fluorescence lifetime was measured for *R.rubrum* chromatophores both at room temperature and
77 K. The data obtained show that B880 band is spectrally inhomogeneous and
this inhomogeneity favours an increase in excitation density at the long-
wavelength bacteriochlorophyll (BChl) molecules, as compared to the molecules absorbing at shorter wavelengths of B880 band, due to directed energy
transfer. Hence, the organization of the B880 antenna complex provides the
conditions of a directed flow of excitations to the reaction centres, which,
according to the model calculations [3], may shorten considerably the time
interval needed for an excitation to be trapped by a reaction centre and,
thereby, increase the quantum yield of photosynthetic process.

2. MATERIALS AND METHODS
2.1. Objects: Cells of *R.rubrum* (wild type, Collection of Moscow University)
     were grown and chromatophores were isolated as described [4]. Chromato-
     phore suspensions of high optical density were stored at $0^{o}C$ in anaer-
     obic conditions and diluted before measurements by 50 mM tris-HCl buf-
     fer (pH 7.5), containing 220 mM sucrose and 2 mM $MgSO_4$, to the final
     concentration of about $5 \cdot 10^{16} - 10^{17}$ BChl molecules per ml. For low tem-
     perature measurements, glycerol was added to the chromatophore suspen-
     sions in a 3 : 1 ratio, v/v.

2.2. Methods: Fluorescence lifetime measurements were performed with a pico-
     second fluorescent spectrochronograph [5]. "Spectra-Physics" mode
     locked cw oxazine 1 dye laser (tuning range 685-800 nm, 345-400 nm
     with frequency doubling, pulse duration 3 ps, average power up to 200
     mW) synchronously pumped with 82 MHz by a krypton-ion laser, served as
     an excitation source. Light of three different wavelengths, 373, 746,
     and 796 nm,was employed in our experiments. Fluorescence was analyzed
     through a subtractive dispersion double-grating monochromator (spectral
     bandwidth, 4 nm) by a synchroscan streak camera, operating in a mode
     of continuous streaking in synchronism with the dye laser operation.
     Time resolution of the apparatus was 3 ps. All experimental decay
     curves were treated as a sum of exponentials, the apparatus response
     function being taken into account.

Biggens, J. (ed.), Progress in Photosynthesis Research, Vol. I. ISBN 90 247 3450 9
© *1987 Martinus Nijhoff Publishers, Dordrecht. Printed in the Netherlands.*

3. RESULTS AND DISCUSSION

The kinetics of fluorescence decay of either cells or chromatophores of *R.rubrum* at room temperature are well approximated by a single exponential with the lifetime of 60 ps for open and 160 - 200 ps for closed reaction centres independent of registration wavelength, in agreement with our previous data [6], both for blue and red excitation light. When temperature was lowered to 77 K, the decay curves were found to change considerably with the wavelength of observation (Fig.1). They could be simulated now by a single exponential with the lifetime approximately equal to that at room temperature only at the long wavelength part of the fluorescence spectra. At $\lambda \leq$ ≤ 917 nm (the wavelength of the fluorescence maximum at 77 K, Fig.2) the kinetics was evidently non-exponential suggesting the presence of at least two components with wavelength dependent amplitudes. In this spectral range the kinetics were treated as a sum of two exponentials, one of which having the lifetime equal to that of the long-wavelength fluorescence, the lifetime of the other component and their relative amplitudes being varied. As a result, an additional short-lived decay component with the lifetime equal to 16 ± 4 ps was revealed at $\lambda \leq 917$ nm. The contribution of this component to the overall emission increased greatly to the short-wavelength part of the fluorescence spectrum (Fig.2). When exciting light of 796 and, especially, 746 nm was employed, the most short-wavelength part of the B880 absorption band was excited (together with the reaction centre P800 and bacteriopheophytin bands). Under these conditions the above spectral dependence of fluorescence decay was much more pronounced than for blue exciting light, which exites B880 band uniformly.

Two conclusions may

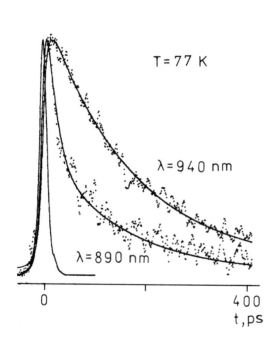

FIGURE 1. The picosecond kinetics of *R.rubrum* (chromatophores without additions) fluorescence decay at 890 and 940 nm (dotted curves). Continuous curves are model computer simulations of the experimental data using the single exponential decay with $\tau = 166$ ps at $\lambda = 940$ nm and two exponential one with $\tau_1 = 166$ ps and $\tau_2 = 16$ ps, amplitude ratio $A_1 : A_2 =$ $= 1 : 1.7$ at $\lambda = 890$ nm. Excitation light density was 0.3 W/cm² at 796 nm. Temperature of 77 K. The apparatus response function is shown.

be derived from these observations: first, that B880 absorption band of *R.rubrum* is spectrally inhomogeneous and, second, that a directed energy transfer across this inhomogeneous band occurs. As a result of this transfer, a partial localization of excitations at BChl molecules, absorbing at

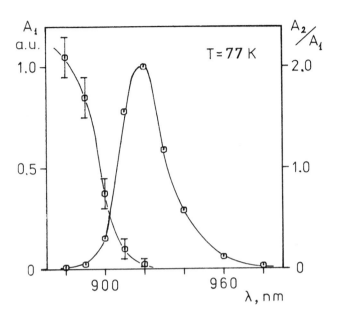

FIGURE 2. The amplitude of the long-lived component (A_1) of *R.rubrum* fluorescence decay and the ratio of amplitudes of the short-lived and long-lived components (A_2/A_1) as a function of the observation wavelength. Conditions of measurements as in Fig.1.

the red edge of the B880 band, takes place at 77 K. The measurements of polarized fluorescence of B880 at 4 K [2] have shown that these BChl molecules are nearly perfectly oriented with their Q_y transition dipoles parallel to the membrane plane. If these BChl molecules are directly coupled to the reaction centres, forming the "proximal antenna" [6], then the time interval, necessary for an excitation, produced somewhere in the photosynthetic unit, to reach a reaction centre and to be trapped, should be much shorter than that in the case of spectrally homogeneous BChls forming B880 band (see model calculations in Ref. [3]). At room temperature, unlike the case at low one, the localization of excitation at the long-wavelength BChl molecules does not occur, judging from the absence of both the increase in fluorescence polarization at the long-wavelength part of B880 band [2] and the short-lived component(s) of fluorescence. Nevertheless, when a directed energy transfer takes place across spectrally inhomogeneous absorption band, equal excitation distribution density is never achieved. It means, that even at room temperature, excitation density will be higher at the long-wavelength part of the Q_y BChl band. We suppose that the BChl molecules of antenna, absorbing at this spectral region, are close neighbours of the reaction centres. Both of the two interrelated factors: a directed, rather than random excitation flow, and an increased density of excitations in close vicinity of the reaction centres, are essential for shortening the lifetime of excitation trapping and, thereby, fluorescence lifetime as compared with the case of homogeneous and isotropic migration.

REFERENCES

1 Borisov, A.Yu., Gadonas, R.A., Danielius, R.V., Piskarskas, A.S. and Razjivin, A.P. (1982) FEBS Lett. 138, 25-28
2 Kramer, H.J.M., Pennoyer, J.D., Van Grondelle, R., Westerhuis, W.H.J., Niederman, R.A. and Amesz, J. (1984) Biochim. Biophys. Acta 767, 735-744
3 Fetisova, L.G., Borisov, A.Yu. and Fok, M.V. (1985) J. Theor. Biol. 112, 41-75
4 Nazarenko, A.V., Samuilov, V.D. and Skulachev, V.P. (1971) Biokhimija 36, 780-782
5 Freiberg, A. and Saari, P. (1983) IEEE J. Quantum Electron., QE-19, 622-630
6 Godik, V., Freiberg, A. and Timpmann K. (1984) Proc. Acad. Sci. of Estonian SSR, Physics, Mathematics 33, 211-219

ACKNOWLEDGEMENTS

   The authors are indebted to Profs. K.K.Rebane and A.Yu.Borisov for stimulating interest and support.

PROTEIN PHOSPHORYLATION: A MECHANISM FOR CONTROL OF EXCITATION ENERGY
DISTRIBUTION IN PURPLE PHOTOSYNTHETIC BACTERIA

Nigel G. Holmes and John F. Allen, Department of Plant Sciences, University
of Leeds, Leeds, LS2 9JT, UK.

1. INTRODUCTION
    In higher plant chloroplasts, the regulation of energy transfer within
the pigment matrix is known to involve the redox-controlled phosphorylation
of LHC-II (1-3). Evidence from Loach and co-workers (4) and from Holuigue
et al (5) indicates that phosphorylation of light-harvesting polypeptides
also occurs in purple photosynthetic bacteria and that this plays a part
in regulating energy transfer between photosynthetic units. We have
recently (6) suggested a general role for protein phosphorylation in the
regulation of energy transfer in both prokaryotic and eukaryotic systems.
We have proposed a model in which phosphorylation of polypeptides of
pigment-protein complexes gives rise to a mutual electrostatic repulsion
of neighbouring complexes parallel to the plane of the membrane. Figure
1 shows how this model would apply to an organism such as Rhodospirillum
rubrum with one type of reaction centre and one type of light-harvesting
complex. We have identified phosphorylated polypeptides in both
cyanobacteria (7) and purple photosynthetic bacteria (8) as possible energy
transfer regulation sites between pigment-protein complexes. The work
discussed below extends these observations to Rds. rubrum.

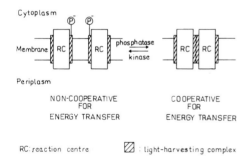

Figure 1.
Model for the
regulation of
energy transfer
between photo-
synthetic units
in Rds. rubrum

2. PROCEDURE
2.1 Materials and methods
    Cells of Rds. rubrum were grown and chromatophores isolated as in
(9,10). Conditions for incubation of cells with photosynthetic units
cooperating or non-cooperating in energy transfer were similar to those
developed by Loach et al (4), ie 20mM KPi, pH 6.8, 2 hours incubation
in the dark for non-cooperative cells and 20mM KPi, 20mM MgSO$_4$, pH 6.8, 2
hours incubation in the light for cooperative cells. For whole cell
labelling experiments, cells were grown overnight in a 7 ml bottle with
1 mCi ^{32}P-Pi. The cells were pelleted in an Eppendorf centrifuge and

Biggens, J. (ed.), Progress in Photosynthesis Research, Vol. I. ISBN 90 247 3450 9
© *1987 Martinus Nijhoff Publishers, Dordrecht. Printed in the Netherlands.*

resuspended under the conditions given, in the presence of a further 250 µCi ^{32}P-Pi per ml of original culture. Samples were precipitated in 5% trichloroacetic acid and SDS-PAGE performed, using an 11.5% to 16.5% gel (8).

In vivo bacteriochlorophyll fluorescence was measured, at right angles, using a fluorimeter constructed in this department. Filter combinations of Corning 4-96 with Calflex C and Scott 891 nm interference filter with Wratten 88A were used on the exciting light and the photodiode respectively.

3. RESULTS AND DISCUSSION

3.1 We have used the transient rise of the in vivo bacteriochlorophyll fluorescence yield, on a millisecond time scale, to monitor the extent to which energy is able to pass between photosynthetic units (11); as the probability of transfer increases, the curve of the transient becomes more sigmoidal in appearance (12). Figure 2 compares the transients in whole cells of Rds. rubrum incubated as above. Under conditions of light+Mg^{2+} the transient is clearly more sigmoidal than under conditions of dark-Mg^{2+}.

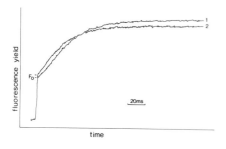

Figure 2.
Fluorescence yield transients
Rds. rubrum cells
1) Light+Mg^{2+}
2) Dark-Mg^{2+}

3.2 Cells were also incubated, in the presence of ^{32}P-Pi, under conditions giving different degrees of cooperativity. Analysis by SDS-PAGE and autoradiography (Fig. 3) shows that several pronase sensitive bands are phosphorylated with M_r 51 kDa, 20.5 kDa, 17 kDa, 13 kDa and 10.5 kDa. Of these labelled polypeptides, the 17 kDa and 13 kDa were phosphorylated only in the cooperative cells, whereas the 10.5 kDa polypeptide was phosphorylated strongly in non-cooperative cells but only weakly in cooperative cells. In contrast, the labelled bands of M_r 51 kDa and 20.5 kDa were phosphorylated to a similar extent under both incubation conditions. The 13 kDa and 10.5 kDa bands appearing on the autoradiograph correspond with bands staining with Coomassie Blue on the gel.

3.3 Chromatophores of Rds. rubrum were incubated with [γ^{32}P]-ATP and the polypeptides analysed by SDS-PAGE and autoradiography as for whole cells (Figure 4). Several components were phosphorylated, including polypeptides of M_r 13 kDa and 10.5 kDa which were strongly labelled after illumination for 15 minutes, but only weakly labelled in the dark. In the presence of 2 mM potassium ferricyanide only the 13 kDa polypeptide was phosphorylated. 5 mM sodium dithionite inhibits labelling of most low molecular weight components as was previously reported in chromatophores of Rhodopseudomonas sphaeroides (8). This effect of dithionite may be a direct effect on the kinase(s), possibly a direct

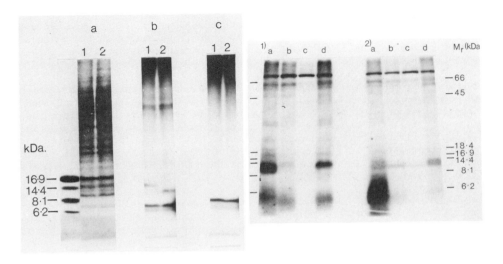

Figure 3. a) gel stained with Coomassie Blue; b) autoradiograph of (a); c) autoradiograph of pronase-treated samples of cells of Rds. rubrum incubated: 1) Light+Mg^{2+}; 2) Dark-Mg^{2+}.

Figure 4. Autoradiograph of chromatophores of Rds. rubrum incubated for 15 minutes with [γ^{32}]-ATP, with (2) or without (1) 144,000 g supernatant fraction, as follows, a) light; b) dark; c) dark + 5mM sodium dithionite; d) dark + 2mM potassium ferricyanide. Chromatophores were made and incubated in 20 mM KPi, 20 mM $MgSO_4$, pH 6.8.

reduction of the enzyme. Adding back the supernatant gives strong phosphorylation of a polypeptide of M_r 10.5 kDa; the phosphorylation of this polypeptide is not inhibited by the addition of dithionite. Phosphorylation of 13 kDa and 11 kDa components in Rds. rubrum chromatophores has also been observed by Holuigue et al (5). However, in their experiments, they observed that phosphorylation of the 13 kDa polypeptide required the presence of both a membrane fraction and a soluble fraction, in contrast to the results reported here.

3.4. In conclusion we have confirmed that cooperative behaviour is associated with phosphorylation of a 13 kDa polypeptide which we identify as B880-α (as has been previously suggested by Loach et al (4)) and that non-cooperative behaviour is associated with phosphorylation of a 10.5 kDa polypeptide, which we provisionally identify as B880-β. In terms of the model presented above, mutual electrostatic repulsion between phosphorylated B880-β would, by increasing the distance between photosynthetic units, then provide a plausible mechanism for inhibiting energy transfer between those units. Phosphorylation of B880-α, which may occur on the periplasmic side of the membrane (4), could then reinforce the effect of the dephosphorylation of B880-β. This reinforcement could possibly involve neutralising an existing positive charge on the molecule on the opposite side of the membrane to the phosphorylation side of B880-β on the cytoplasmic site.

The chromatophore studies reported above confirm that phosphorylation of components of Mr 13 kDa and 10.5 kDa is light activated. Phosphorylation of the 13 kDa polypeptide is responsive to ferricyanide, suggesting that the kinase responsible for phosphorylating this polypeptide may be activated by oxidation of an electron transport component. The factors regulating the phosphorylation of the 10.5 kDa component remain unclear, although a component present in the soluble fraction appears to be involved

ACKNOWLEDGEMENTS
We are grateful to the Royal Society, SERC and Nuffield Foundation for financial support. NGH is a University of Leeds Research Fellow.

REFERENCES
(1) Allen, J.F., Bennett, J., Steinback, K.E. and Arntzen, C.J. (1981) Nature 291, 25-29
(2) Bennett, J. (1983) Biochem. J. 212, 1-13
(3) Barber, J. (1982) Annu. Rev. Plant Physiol. 33, 261-295
(4) Loach, P.A., Parkes, P.S. and Bustamante, P. (1984) in: Advances in Photosynthesis Research Vol. II (Sybesma, C. ed.) pp. 189-197 Martinus Nijhoff/Dr W. Junk. The Hague. The Netherlands.
(5) Holuigue, L., Lucero, H.A. and Vallejos, R.H. (1985) FEBS Lett. 181, 103-108.
(6) Allen, J.F. and Holmes, N.G. (1986) FEBS Lett. 202, 175-181.
(7) Allen, J.F., Sanders, C.E. and Holmes, N.G. (1985) FEBS Lett. 193, 271-275.
(8) Holmes, N.G., Sanders, C.E. and Allen, J.F. (1986) Biochem. Soc. Trans. 14, 67-68.
(9) Sistrom, W.R. (1960) J. Gen. Microbiol. 22, 778-785.
(10) Holmes, N.G. and Allen, J.F. (1986) FEBS Lett. 200, 144-148.
(11) Joliot, A. and Joliot, P. (1964) C.R. Acad. Sci. Paris 258, 4622-4625.
(12) Pradel, J., Lavergne, J. and Moya, I. (1978) Biochim. Biophys. Acta. 502, 169-182.

A MODEL FOR THE FUNCTIONAL ANTENNA ORGANIZATION AND ENERGY DISTRIBUTION IN THE PHOTOSYNTHETIC APPARATUS OF HIGHER PLANTS AND GREEN ALGAE

Alfred R. Holzwarth

Max-Planck-Institut für Strahlenchemie, D-433 Mülheim/Ruhr, West Germany

1. Introduction

Elucidation of the fate of light energy absorbed by photosynthetic antenna pigments and the accurate determination of the kinetic laws governing the exciton decay are of fundamental importance for an understanding of the energy transfer processes and the functional organization of the photosynthetic apparatus. The remarkably complex organization of the antenna systems and reaction centers presents an outstanding challenge to researchers trying to unravel the multitude of parallel and sequential processes related to exciton migration, exciton trapping and charge separation. The data obtained from time-resolved fluorescence spectroscopy may, in principle, provide detailed information on parameters such as the energy transfer kinetics and pathways, the absorption and emission spectra of the connected antenna, the relative absorption cross-sections of the different photosystems, variation in communication between photosynthetic units, spill-over, the redox state of the reaction centers, etc. Despite a large amount of work devoted to picosecond fluorescence studies during the last few years, no general agreement has been achieved between the various research groups as far as the interpretation of the data is concerned, however. In an attempt to clarify the situation this paper will be focussed on the following questions:

i) Which guidelines can be followed when analyzing multicomponent fluorescence kinetic data from photosynthetic antenna systems?

ii) Using these guidelines, how many kinetic components can be identified at the extreme states of fully open (F_o) and fully closed (F_{max}) PSII reaction centers?

iii) Is there evidence from time-resolved fluorescence to support the concept of PSII heterogeneity?

iv) What is the functional organization of the photosynthetic apparatus in terms of energy transfer processes?

2. Results and Discussion

2.1 Analysis and assignment of decay components

It has been demonstrated recently that the Chl fluorescence kinetics from higher plant chloroplasts and green algae generally contain at least 3 or even 4 exponentials (for a recent review see [1]). The conventional approach has been to separately analyze each individual fluorescence decay which had been recorded by the single-photon counting technique under particular conditions of excitation and emission wavelength, light intensity, inhibitor and salt concentration etc. using various least squares methods [1]. The decay law generally assumed is a sum of exponentials

$$I(t, \lambda_{exc}, \lambda_{em}) = \sum_{i=1}^{n} A_i(t, \lambda_{exc}, \lambda_{em}) \cdot exp(-t/\tau_i) \qquad (1)$$

where $I(t, \lambda_{exc}, \lambda_{em})$ is the experimental fluorescence intensity as a function of time (t) and wavelength ($\lambda_{exc}, \lambda_{em}$), $A_i(t, \lambda_{exc}, \lambda_{em})$ is the preexponential factor or amplitude of a decay component, and τ_i is its lifetime. Both A_i and τ_i are free parameters which have to be determined by the fitting procedure, using the criterion of minimum deviation between the experimental and theoretical curves. Under the

Biggens, J. (ed.), Progress in Photosynthesis Research, Vol. I. ISBN 90 247 3450 9
© *1987 Martinus Nijhoff Publishers, Dordrecht. Printed in the Netherlands.*

conditions that prevail in higher plant chloroplasts and green algae one usually finds that a model function of three exponentials fits an individual decay very well when judged on the basis of statistical criteria. A good fit to an individual decay function using n exponentials does not prove that one has found the correct decay law, however. The reason for this discrepancy consists in the fact that the result of data fitting to multicomponent kinetics often is not unique. Other criteria must therefore be used in addition for analyzing complex fluorescence kinetics.

It has been shown that conventional three-exponential analysis of fluorescence kinetic data of green algae, recorded as a function of the emission wavelength, yields sets of substanially different lifetime parameters at each wavelength [2]. This finding is in contrast to what one would expect from a spectroscopic point of view. The expectation would generally be that the lifetimes should be invariant over a substantial wavelength range, while the corresponding amplitudes should change with wavelength in a specific manner. In fact, in many cases one would expect the lifetimes to stay constant over the entire emission wavelength range. Changes in experimentally determined lifetimes across the emission spectrum therefore generally indicate that the decay law used for fitting the data is in fact inadequate. Then a decay law more complex than what can actually be handled using the conventional single decay analysis techniques must be applied. In such cases the use of the most powerful data analysis techniques becomes mandatory. We have recently applied a data analysis technique which fits simultaneously a whole set of decay curves which have been recorded as a function of emission and/or excitation wavelength [4,5]. Thus the whole three-dimensional I(t,λ) data surface is fit to a particular decay law. This procedure is known also as the "global analysis technique". Fitting one set of parameters to the entire data surface provides a number of advantages which are essential for analyzing complex kinetic data sets:

i) As compared to the individual analysis the number of total parameters is reduced, which makes the procedure numerically more stable and often drastically improves the precision of each parameter.

ii) The capability of multicomponent resolution is substantially increased, and

iii) in its simplest form, the global analysis will ensure that the lifetimes are constant over the entire data set.

iv) Any inadequacy of the chosen model function will show up more pronouncedly as compared to single decay analysis procedures.

A parameter of extreme importance which has often been ignored when analyzing and assigning fluorescence decays is the amplitude of a kinetic component as a function of wavelength. For a single decay component the amplitude A is given by

$$A(\lambda_{exc}, \lambda_{em}) = C \cdot N_{Chl} \cdot \epsilon(\lambda_{exc}) \cdot k_{rad} \cdot F(\lambda_{em}) \qquad (2)$$

where $A(\lambda_{exc}, \lambda_{em})$ is the same as in eqn. (1), C is a proportionality constant, N_{Chl} is the number of Chls connected to the antenna pool giving rise to that decay component, ϵ is the absorption coefficient at the excitation wavelength, k_{rad} is the radiative lifetime, and F(λ_{em}) is the normalized emission spectrum. Thus the amplitude is proportional to the total absorption cross-section $N_{Chl} \cdot \epsilon(\lambda_{exc})$ of that particular antenna pool. This relationship provides one of the most important handles for assigning decay components at, e.g., F_o- (PSII reaction centers open) and F_{max}- (PS II reaction centers closed) conditions. We do not expect that any of the parameters in eqn. (2) would change when the primary quinone acceptor Q_A of PSII is reduced (F_{max}). Thus, when assigning kinetic phases one should test for such components which have about equal amplitude at F_o and F_{max}, but different lifetimes. The functions $\epsilon(\lambda_{exc})$ and F(λ_{em}) describe the excitation and emission wavelength dependence, respectively, of the amplitudes, i.e. they describe the time-resolved excitation and emission spectra (TRES). A TRES, which is a plot of the amplitudes vs. wavelength, thus in most cases should be identical to the fluorescence and excitation spectra that would be obtained if that particular antenna pool would have been isolated. It follows from the above that TRES will provide basic information on the Chl composition and size of the connected antenna pool(s). TRES are extremely useful therefore for assigning the origin of the various decay components as well as for analyzing the effects on the functional organization of the thylakoid membrane brought about by processes such as, e.g., state transitions, thylakoid phosphorylation, changes in salt concentration etc.

For the reasons described above, the amplitudes $A_i(\lambda_{exc},\lambda_{em})$ in a time-resolved experiment and their wavelength dependence in a way describe the static properties of the antenna system in a given situation. In contrast to that the lifetimes τ_i describe the energy transfer and trapping processes, i.e. the kinetic properties of the system. It is important to consider the static and kinetic parameters obtainable from a time-resolved experiment separately, rather than mixing these different kinds of information. In the case of exponential kinetics there exists a simple relationship which describes the fluorescence yield Φ_i of a kinetic component such that

$$\Phi_i = \int_0^\infty A_i \cdot exp(-t/\tau_i)dt = A_i\tau_i \qquad (3)$$

It follows from eqn. (3) that the yield Φ_i, due to its composite nature, provides less information than do the individual amplitude A_i and lifetime τ_i parameters. Thus the component yields Φ_i should only be used when a comparison is made to steady state fluorescence experiments.

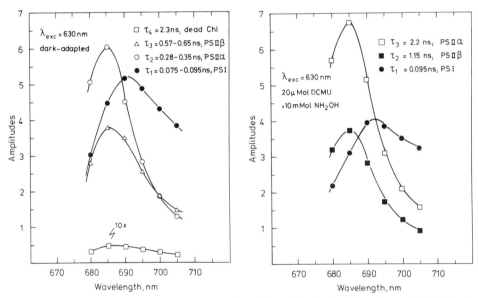

Fig. 1: (left) Time-resolved fluorescence spectra for dark-adapted cells of *Scenedesmus* at the F_o-level of fluorescence.

Fig. 2: (right) Time-resolved fluorescence spectra for dark-adapted cells of *Scenedesmus* at the F_{max}-level of fluorescence. The cells were excited at 630 nm and the data have been analyzed using the "global data analysis" technique.

2.2 Number and assignment of decay components

Using the guidelines and principles described above, the global analysis of the fluorescence kinetics and TRES from both green algae systems and higher plant chloroplasts so far revealed four components at F_o-conditions and three components at F_{max} conditions over the emission range of about 670 to 740 nm. It has been shown, however, that the fourth, long-lived, component at F_o has a very small amplitude. It represents some sort of an "impurity" and can therefore be ignored as far as the assignment of the essential components is concerned [5,6].

An example of emission TRES data sets is given in Figs. 1 and 2 for the green alga *Scenedesmus* in the F_o and F_{max}-states [5]. Very similar data were obtained for *Chlorella*. A set of excitation TRES data analyzed by the global analysis procedure is shown in Fig. 3 [1,2]. The information obtained from a large variety of such lifetime experiments is compiled in Table 1. Three spectral and kinetic components can be distinguished from TRES. Based on its spectral properties and its insensitivity to the state of the PSII reaction centers, the shortest-lived, bathochromically shifted, component has been assigned to PSI units (Component I in Table 1). Comparison of the TRES of green algae (Figs. 1-4) and pea chloroplasts (Figs. 5 and 6) reveals that this component has different spectral properties in the two systems. In pea chloroplasts the TRES of the PSI component shows a long-wavelength (720 nm) band in addition to the shorter wavelength maximum (~695 nm) which is present in both green algae and peas [6].

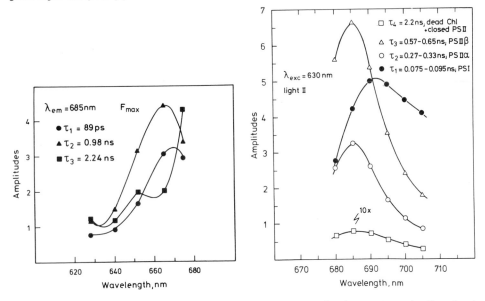

Fig. 3: (left) Time-resolved fluorescence excitation spectra for cells of *Chlorella* at the F_{max}-level of fluorescence. The detection wavelength was 685 nm. PS II reaction centers had been closed using DCMU and hydroxylamine. The data have been analyzed using the "global data analysis" technique.

Fig. 4: (right) Time-resolved fluorescence spectra for light II-adapted cells of *Scenedesmus* (state II) at the F_o-level of fluorescence. Data from the same sample as shown in Fig. 1.

The other two kinetic and spectral components, i.e. compounds II and III, are clearly related to PSII. Their emission spectra are virtually identical, but their excitation spectra show distinct differences [2]. Only component II shows a significant portion of Chl b pigments in the antenna, i.e. it is associated with most of the LHCP. Using the amplitude criterion for assigning the corresponding components at F_o- and F_{max}-conditions clearly indicates that the 200-300 ps component at F_o and the 2.2-2.4 nsec component at F_{max} correspond to each other. Accordingly the 500-600 ps and the 1.0-1.4 nsec components derive from the same PSII units at F_o and F_{max}, respectively. We also find that in the F_o and F_{max} states the TRES of the corresponding components have the same shape and about the same amplitudes within the error limits. Our findings are therefore in contrast to the recent interpretation and assignments of Hodges and Moya [3].

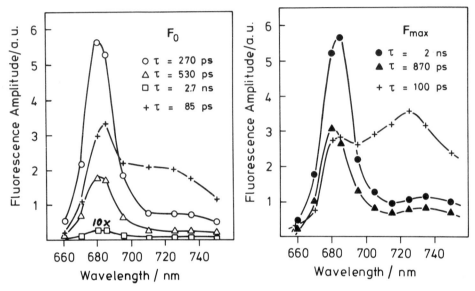

Fig. 5: (left) Time-resolved fluorescence spectra for dark-adapted pea chloroplasts at the F_o-level of fluorescence.

Fig. 6: (right) Time-resolved fluorescence spectra for pea chloroplasts at the F_{max}-level of fluorescence.

2.3 Are the PSII units heterogeneous?

In order to study the effects of state transitions, the time-resolved emission spectra have been studied for the green alga *Scenedesmus* as a function of the adaptation to different light conditions. The example of a TRES obtained from a sample in state II (after adaptation to light II of ~650 nm) is given in Fig. 4. This should be compared to the TRES of the same, but dark-adapted, sample shown in Fig. 1. The striking difference between the two conditions is the change in the relative amplitudes of the PSII decay components. The changes in the lifetimes of the two components are relatively minor only. In state II the amplitude of the 600 ps decay component is high while the amplitude of the 300 ps component is low. In state I their relative amplitudes are reversed, the 300 ps component now being more prominent than the 600 ps component. It is important to note that the sum of the amplitudes of the two components is independent of the light adaptation. Both the lifetime and amplitude of the fast

(PSI) component do not change under the different conditions of light adaptation. Bearing in mind that the amplitude of a kinetic component directly reflects the total size of the connected antenna pool, we conclude from these results that a state transition is characterized by a redistribution in the relative antenna sizes of two different pools of PSII units. These data show furthermore that the direct antenna size of PSI is unafffected by a state transition.

Our data as a whole can not be reconciled with a homogeneous pool of PSII units but rather require a heterogeneous model. The two types of PSII units differ not only in their antenna properties like, e.g., their pigment-protein complexes, their spectra, their antenna sizes, their F_{max}/F_o ratios etc. Our data rather suggest that most probably there also exist distinct differences in their reaction centers. Of course it is not possible to analyze the details of these differences in the reaction centers by fluorescence methods.

Table 1. Properties of fluorescence lifetime components from data obtained for the green algae *Scenedesmus* and *Chlorella* as well as for pea chloroplasts

	kinetic components		
	I	II	III
lifetime τ, ps at F_o	70-100	180- 300	500- 600
lifetime τ, ps at F_{max}	70-l00	2000-2400	1000-1400
emission maximum, nm	~695[+]	685	685
excitation maximum*, nm	680-685	~675,652	~665
amplitude in TRES for excitation at 695 nm*	dominant ~90%	minor ~5%	minor ~5%
major Chl b contribution in excitation TRES[‡]	No	Yes	No
does lifetime depend on the state of PSII reaction center(s)	No	Yes	Yes
F_{max}/F_o	1	~ 8-10	~ 2-3
amplitude change upon State I→II transition	~constant	down	up
amplitude at 8th hour of cell cycle in synchronized cultures[o]	intermediate	high	low
amplitude at 16th hour of cell cycle	intermediate	low	high

[+] in pea chloroplasts this component has two maxima at ~690 and 720 nm
*data for *Chlorella* only
[‡] in dark-adapted state
[o] data for *Scenedesmus*

Fluorescence induction measurements on these samples show that a transition to state II is related to an increase in the antenna size and/or number of those PSII units which give rise to the slow fluorescence induction component. This component is generally attributed to PSIIβ centers [7,8]. The parallel increase upon transition to state II in both the amplitude of the 600 ps fluorescence component (at F_o) and the antenna size of β-units, as indicated by fluorescence induction, strongly suggest that both phenomena may have the same origin. Furthermore, the agreement of a number of other properties of the two PSII fluorescence components with those reported previously for PSII

α- and β units [7,8] makes it likely that the heterogeneous PSII units found in the lifetime experiments reflect the same kind of PSII heterogeneity. Large changes in the ratio of α- and β-units, similar to the effects of state transitions, have also been observed during the life cycle of synchronized cells of *Scenedesmus*. These changes are also accompanied by corresponding changes in the fluorescence induction curves [12].

2.4 Model for the functional organization of the antenna

Recently Hodges and Moya [3] invoked a homogeneous tripartite model in order to interpret their time-resolved fluorescence data of green algae and thylakoid fragments of green plants. Thus the fast (200-300 ps) PSII component of F_o has been assigned to the energy transfer from LHCP to the PSII core antenna. The corresponding process at F_{max} was identified by these authors to give rise to the 1.0-1.4 nsec component. Accordingly it was suggested that the 600 psec and the 2.2 nsec components reflect the trapping time of excitons in the Chl_{aII} core complexes with open and closed PSII reaction centers, respectively. Their model is clearly in conflict with the criteria developped in Chapter 2.1 for the assignment of corresponding components based on the amplitude analysis both on their own data as well as ours. Their assignment would also require the LHCP to be associated primarily with the 1.2 nsec component, quite in contrast to our time-resolved excitation spectra which show the LHCP to be associated with the 2.2 nsec component. The homogeneous tripartite model furthermore is also in conflict with our data on state transitions, which suggest a redistribution of antenna between two different types of PSII units. The requirement of a heterogeneous PSII population for the interpretation of time-resolved fluorescence experiments has also been noted by Berens et al. [9].

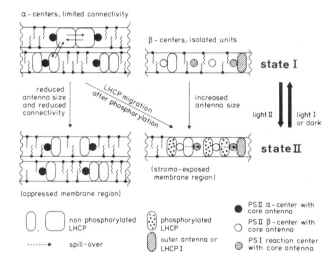

Fig. 7: Proposed schematic model of the PSII antennae organization in states I and II. The regulation mechanism involves the reversible migration of phosphorylated LHCP from α-centers to β-centers. The model indicates that up to 30% of the total Chl in PSII may be mobile.

In view of the inadequacy of the homogenous tripartite model to account for the full body of available kinetic data we have proposed a model of the functional antenna organization which is outlined in Fig. 7. Our model involves PSII α-units which are assumed to be located preferentially in the stacked regions of the thylakoids. They carry most of the LHCP antenna complexes (in state I). The α-units give rise to the 200-300 psec and the 2.2 nsec components at F_o and F_{max}, respectively. The β-units, located in the vicinity of the PSI units, give rise to the 600 psec and 1.2 nsec components, respectively. We presently do not see a need to invoke a new or different kind of PSII heterogeneity in order to explain our data. We rather use the α,β-heterogeneity as a working model, bearing in mind that at the present stage the difference between α- and β-reaction centers is hardly understood. It has been argued that the only difference between the PSII units giving rise to the fast and slow induction phases consists in their different binding capacity do inhibitors. This view ist not supported by our data. Rather the PSII heterogeneity expresses itself in our data also in the absence of inhibitors like, e.g., DCMU.

It has been shown by several methods that phosphorylation of the LHCP causes this pigment complex to migrate from stacked regions of the thylakoids to the stroma-exposed regions [11]. It is also generally believed that the latter mechanism is responsible for *in vivo* state transitions. These ideas can be easily incorporated into our model. We thus propose that the phosphorylated LHCP attaches preferentially to PSII β-centers. As a consequence the antenna size of the efficient α-centers would be reduced upon a transition to state II while that of the β-units increases. This model is also consistent with a large body of spectroscopic, biochemical and fine structural data from other groups.

Based mostly on circumstantial evidence it has been proposed that the phosphorylated LHCP should attach to PSI. Since the amplitude of the time-resolved spectrum is a direct measure of the antenna size, we can exclude this possibility at least for the green algae which we have studied. Instead our data indicate that any increase in the antenna size of PSI in state II is negligible as compared to the effects on the relative antenna sizes of the two different PSII units upon a state transition. We have some preliminary evidence, however, that the increase in the antenna size of β-units in state II might give rise to an increased spill-over efficiency from PSII β-centers to PSI, as indicated in Fig. 7. Such a mechanism could then account for an increase in the *effective* cross-section of PSI units, while leaving their *direct* antenna size unchanged.

References

1. Holzwarth A.R. (1986) Photochem. Photobiol. **43**, 707-726.
2. Holzwarth A.R., Wendler J. and Haehnel W. (1985) Biochim. Biophys. Acta **807**, 155-167.
3. Hodges M. and Moya I. (1986) Biochim. Biophys. Acta **849**, 193,202.
4. Holzwarth A.R., Wendler J. and Suter G.W. (1986) Biophys. J. in print.
5. Wendler J. and Holzwarth A.R. (1986) Biophys. J. submitted.
6. Schatz G. and Holzwarth A.R. (1986) Poster contribution this Conference.
7. Melis A. and Homann P.H. (1975) Photochem. Photobiol. **21**, 431-437.
8. Melis A. and Anderson J.M. (1983) Biochim. Biophys. Acta **724**, 473-484.
9. Berens S.J., Scheele J., Butler W.L. and Magde D. (1985) Photochem. Photobiol. **42**, 59-68.
10. Hodges M. and Barber J. (1986) Biochim. Biophys. Acta **848**, 239-246.
11. Kyle D.J., Staehelin L.A. and Arntzen C.J. (1983) Arch. Biochem. Biophys. **222**, 527-541.
12. Bittersmann E., Senger H. and Holzwarth A.R. (1986) Poster contribution this Conference.

PICOSECOND TRANSIENT ABSORBANCE SPECTRA AND FLUORESCENCE DECAY KINETICS IN
PHOTOSYSTEM II PARTICLES.

A.R. HOLZWARTH, H. BROCK AND G.H. SCHATZ
Max-Planck-Institut für Strahlenchemie
Stiftstraße 34-36, D-4330 Mülheim a.d. Ruhr 1, F.R. Germany

1. INTRODUCTION

The processes of excitation trapping, charge separation and primary charge
stabilization in photosynthetic reaction centers (RC) can be monitored by
picosecond absorbance spectroscopy, as proved in many studies on bacterial RC
preparations. In PS II-particles, however, the number of antenna pigments
still is large as compared to that of bacterial RC-preparations. This implies
a higher probability of singlet annihilation processes, which in fact were
the main problem in earlier studies (1). We have performed experiments at
extremely low excitation energy densities (<10 $\mu J/cm^2$ pulse) in order to
avoid (artificial) contributions by singlet annihilation. We used oxygen-
evolving PS II particles with about 80 Chla/PS II to measure absorbance transients and fluorescence decay kinetics with picosecond resolution under
almost identical conditions.

2. MATERIALS AND METHODS

Fluorescence decay kinetics were measured with a single-photon-timing system
as described in (2) after excitation at 675 nm with an energy density E $<$
1 $\mu J/cm^2$ pulse. Absorbance transients were determined by the pump-probe-
technique. The "pump" pulse (675 nm, 10 $\mu J/cm^2$ pulse, if not otherwise
stated) and the "probe" pulse (620 - 700 nm, 0.1 $\mu J/cm^2$ pulse) were provided
by two synchronized cavity-dumped (400 kHz) dye lasers. The time delay
between the pulses was variable from -300 ps to 1500 ps. Noise reduction was
accomplished by modulation of the pump-pulse train and lock-in amplification
of the probe signal. Intensity fluctuations of the "probe" beam were
monitored in a reference channel. This allowed reliable measurements of
signals as small as 10^{-5} in $\Delta I/I$. The sample was rapidly (500 ml/min) pumped
through a flow cuvette (d=1 mm) to prevent the PS II from being closed by the
laser pulses. The PS II particles were prepared from a thermophilic cyanobacterium Synechococcus sp. as described in (3), (4). They have been characterized in (4), (5) to contain about 80 Chl per PS II-RC. The PS II particles
were suspended in 50 ml of buffer A (500 mM sucrose, 10 mM $MgCl_2$, 2 mM
K_2HPO_4, 20 mM MES-NaOH, pH=6.0) at a chlorophyll concentration of 20-30 μM
for absorbance or at 10 μM for fluorescence measurements. F_o-conditions
(open PS II centers) were established by addition of 2 mM $K_3Fe(CN)_6$, F_{max}-
conditions (closed PS II centers) by addition of 2 mM dithionite or 20 μM
DCMU, respectively.

3. RESULTS AND DISCUSSION

3.1 Fluorescence decay kinetics

Typical fluorescence decay kinetics of PS II particles are shown in Fig. 1
for open PS II centers. Comparison of the weighted error residuals (top)
shows that 3 exponential decay components are necessary to fit the experimental data. This is also found for closed PS II (not shown). Characteristic
results are given in table I. They show:
1.) Fluorescence decay kinetics at F_o are dominated by a 60-80 ps component.

Biggens, J. (ed.), Progress in Photosynthesis Research, Vol. I. ISBN 90 247 3450 9
© *1987 Martinus Nijhoff Publishers, Dordrecht. Printed in the Netherlands.*

The main component is by a factor 3-4 shorter lived as compared to unfractionated PS II in thylakoids with a 3-4 fold larger antenna size (see
proceeding contributions by A.R. Holzwarth and G.H. Schatz + A.R. Holzwarth).
This strongly supports the idea that the exciton decay is trap limited (6).
2.) Under both conditions the minor contribution ($\leqslant 10\%$) by very long lived
components are attributed to impurities and/or to loosely bound chlorophyll.
Hence, the fluorescence decay kinetics of coupled chlorophyll are assumed to
be biphasic.
3.) Closing the RC of the PS II particles by reduction of Q_A results in an
increase of the lifetimes by approximately a factor of 3 and in a clear
decrease of the amplitude ratio A_1/A_2. The major contribution is then by a
1.3-1.4 ns emission.

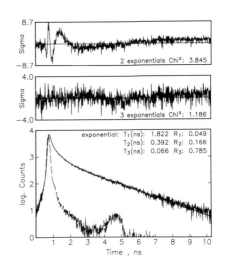

FIGURE 1: Fluorescence decay at 685 nm
of PS II particles plus 2 mM $K_3Fe(CN)_6$
after excitation at 675 nm (bottom) and
weighted error residuals between the
experimental curve and an exponential
fit with 2 and 3 components, resp.(top)

TABLE I: Characteristic lifetimes τ
(and relative amplitudes A) of the
three exponential decay components
in PS II particles at F_o and F_{max}.

F_o	F_{max}
60- 80 ps (75-80%)	240-260 ps (40%)
400-450 ps (15-20%)	1.3-1.4 ns (50%)
1.7-1.8 ns (5%)	3.5-4.5 ns (10%)

It is noteworthy that fluorescence decay measurements are the method of
choice to precisely characterize the (multiple) kinetics of Chl*. In absorbance transients the decay of Chl* will be monitored together with the consecutive processes of radical pair formation and charge stabilization.

3.2 Absorbance changes in open PS II centers.
In order to restrict artificial contributions by Chl*-Chl*-annihilation in
the PS II antenna we studied absorbance changes at 685 nm in dependence on
the energy density (E) of the pump-pulse.
The resulting kinetics were analyzed in terms of
3 exponentials. Their corresponding amplitudes
are shown in a double reciprocal plot (Fig. 2)
and indicate two types of kinetics: a) kinetics
obeying linearity of 1/ A versus 1/E with almost
the same half saturation value (after linear
extrapolation) and b) a rapid (40 ps) decay with

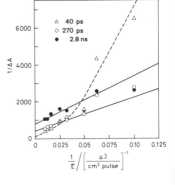

FIGURE 2: Double reciprocal plot of the amplitudes of three components of absorbance changes
at 685 nm versus the energy density of the pump
pulse. PS II particles plus 2 mM $K_3Fe(CN)_6$,
30 µM Chl; optical pathlength 1 mm.

a distinct saturation behaviour. The latter kinetic was dominant at flash energy densities $\geq$50 µJ/cm^2·pulse and markedly decreased at values of $\leq$16 µJ/cm^2·pulse. It is attributed to artifact processes (stimulated emission and singlet-singlet annihilation). Therefore we chose a flash energy density of $\leq$10 µJ/cm^2·pulse (corresponding to $\langle 3\cdot 10^{13}$ photons/cm^2·pulse) for all subsequent experiments to minimize such artifacts. We note that kinetics measured at high energy densities cannot be fitted to exponential decays similar to those obtained from the more sensitive fluorescence experiments.

FIGURE 3: Kinetics of absorbance transients in PS II particles plus 2 mM $K_3Fe(CN)_6$. Energy density of pump-pulse: 10 µJ/cm^2 / pulse; 30 µM Chl; optical pathlength 1 mm.

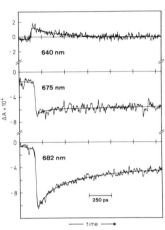

Absorbance transients were measured in the wavelength range from 620 to 700 nm. Typical kinetic traces with open RC are shown in fig. 3. They were subjected to a global analysis which simultaneously fits the kinetics at all wavelength to a 3 exponential function. Free running fits result in the following lifetimes: τ_1=80-100 ps, τ_2=460-500 ps, $\tau_3\gg$10 ns (beyond our resolution). The corresponding component absorbance difference spectra are shown in fig. 4. At a first glance the difference spectra of (A) the long- and (B) the middle-lived component show the most typical features of the (P680$^+$-P680) and (Pheo$^-$-Pheo) difference spectra, respectively, which are well known ((7) and references there). The extrapolated absolute extinction coefficients at saturating intensity would be in the order of 50.000 M^{-1} cm^{-1}

We conclude that the long-lived kinetic (A) results from P680$^+$-reduction by the intact donor side. From earlier studies on the same material these reduction kinetics are known to proceed within 20 ns $\leq$ t(1/2) $\leq$ 300 ns (5). The reoxidation of the intermediate acceptor pheophytin (I) (8) appears to proceed with τ_2 = 480 ps (corresponding to t(1/2) = 330 ps). This is slower than the corresponding kinetic in bacterial reaction centers (7). The fastest kinetic component (fig. 4C) in these experiments has a lifetime which is in agreement with that of the short-lived fluorescence decay component suggesting that it reflects the processes of trapping and charge separation. This phase is associated with an absorbance difference spectrum which can be explained as composite of several positive and negative absorbance changes. This can be rationalized by a qualitative reasoning: as long as the rate constants of the forward reactions in the consecutive reaction sequence (see fig. 6) are k$_1$ $\gg$ k$_2$ $>$ k$_{-1}$ $\gg$ k$_3$ the absorbance experiment will yield almost pure component spectra with lifetimes corresponding to 1/k$_i$. The consecutive steps can be separated as follows:

τ_3 $\gg$ 10 ns	P$^+$ - P	fig. 4A
τ_2 = 450-500 ps	I$^-$ - I	fig. 4B
τ_1 = 80-100 ps	Chl*PI - ChlP$^+$I$^-$	fig. 4C
at t = 0	Chl* - Chl	fig. 4A + 4B + 4C

Therefore, the sum of all decay component spectra is expected to result in the $(Chl^*(S_1) - Chl(S_0))$ difference spectrum. Our result shown in fig. 4D is very similar to the (Chl^*-Chl) spectrum in (1).

3.3 Absorbance changes in closed PS II centers

After closing the PS II RC also the absorbance decay kinetics change. Their analysis in terms of 3 exponentials (one lifetime fixed to τ_2 = 1300 ps) results in τ_1 = 110 - 130 ps, τ_2 = 1300 ps and $\tau_3 \gg$ 10 ns with component spectra shown in fig. 5.

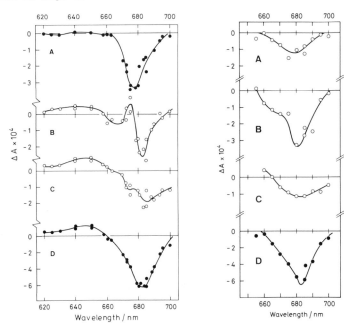

FIGURE 4 (left) and FIGURE 5 (right): Absorbance difference spectra of the kinetic components in PS II particles, in presence of 2 mM $K_3Fe(CN)_6$ (open RC, FIG.4) and in presence of 2 mM sodium dithionite (closed RC, FIG.5), respectively. The components are associated with the following lifetimes: (4A and 5A) $\tau \gg 10$ ns, (4B) $\tau=450-500$ ps; (5B) $\tau=1300$ ps; (4C) $\tau=80-100$ ps; and (5C) $\tau=110-130$ ps. In both figures (D) shows the sum of A,B and C (corresponding to the initial amplitude). From several experiments normalized to 20 µM Chl.

It is evident from fig. 4D and fig. 5D that the initial concentration of Chl^* is the same in both cases. However, the decay pathways are altered: exciton trapping and charge separation (fig. 5C) and P680$^+$ decay (fig. 5A) have become less important processes. Their amplitudes are reduced to about 1/2 to 1/3 as compared to open PS II. The absorbance decay component associated with the largest amplitude is that with τ_2 = 1300 ps (fig. 5B) and shows a difference spectrum with features of (Chl^*-Chl) as well as (I^--I). This suggests a slower exciton decay from Chl^* in agreement with our results from fluorescence kinetics. We consider these findings as evidence for the following effects of Q_A reduction in PS II particles:

1.) a decrease in the yield of radical pair $(P^+ I^-)$ formation.

2.),*a shift of the Chl*PI/ChlP^+I^- ratio towards a higher concentrations of Chl* PI, i.e. longer lifetimes for an exciton within the antenna. Charge separation in closed PS II centers is no longer possible at the high yield as in open centers. Furthermore, the smaller fraction of P^+I_+ formed does obviously not decay by recombination since in that case P^+ and I^- should dissapear by the same kinetic. Either P^+ formed in closed PS II decays slower than I^- or the radical pair rapidly converts to an intermediate (e.g.triplet) with similar optical properties as P_{680}.

FIGURE 6: Kinetic model for the
initial processes in PS II.

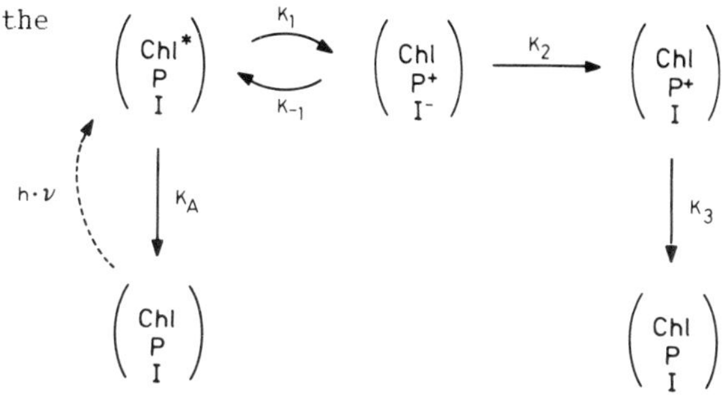

3.4 Kinetic model

Our experimental data can be used for a kinetic model describing trap limited charge separation (k_1) and charge stabilization processes (k_2 and k_3) as shown in fig. 6. The analytical solution of the corresponding rate equations* of this scheme (unpublished) predicts 2 exponential decay kinetics for Chl* and 3 exponential kinetics for the radical pair.

Model calculations then corroborate the conclusion from our qualitative reasoning: the changes in fluorescence (by Chl*) and absorbance (by Chl*, P^+, I^-) kinetics upon closing PS II by reducing Q_A are only compatible with a pronounced decrease in the value of k_1; i.e. a reduction of the rate constant of trapping and/or charge separation. Similar conclusions are drawn from a kinetic analysis of fluorescence data in algae and chloroplasts (see 9).

4. REFERENCES

1 Nuijs,A.M., van Gorkom,H.J., Plijter,J.J., and Duysens,L.N.M.,
   (1986) Biochim. Biophys. Acta 848, 167-175
2 Holzwarth,A.R., Wendler,J., and Haehnel,W., (1985) Biochim.
   Biophys. Acta 807, 155-167
3 Schatz,G.H. and Witt,H.T.,(1984) Photobiochem. Photobiophys. 7, 1-14
4 Schatz,G.H. and van Gorkom,H.J., (1985) Biochim.Biophys.Acta 810, 283-294
5 Schlodder,E., Brettel,K., Schatz,G.H., and Witt,H.T., (1984) Biochim.
   Biophys. Acta 765, 178-185
6 Pearlstein,R.M., (1982) Photochem. Photobiol. 35, 835-844
7 Parson,W.W. and Ke,B., (1982) in Photosynthesis, (Govindjee, ed.),
   Vol. 1, pp. 331-385, ed.: Govindjee, Academic Press, New York
8 Klimov,V.V. and Krasnovsky,A.A., (1982) Biophys. 27, 186-198
9 Schatz, G.H. and Holzwarth,A.R., (1986) Photosynth.Res. in press

PICOSECOND TIME RESOLVED CHLOROPHYLL FLUORESCENCE SPECTRA FROM PEA CHLOROPLAST THYLAKOIDS.

G.H. SCHATZ AND A.R. HOLZWARTH
Max-Planck-Institut für Strahlenchemie, Stiftstr. 34-36
D-4300 Mülheim/Ruhr, F.R.Germany

1. INTRODUCTION

The picosecond decay kinetics of chlorophyll fluorescence measured (at 685-±10 nm) in green algae and higher plants at ambient temperatures have commonly been analyzed in terms of three exponential components: a fast ($\tau(1/e)$=50-150 ps), a middle ($\tau(1/e)$=450-750 ps), and a slow ($\tau(1/e)$=1400-2300 ps) one (1,2). Recently, the spectral resolution in the range from 665 nm to 725 nm of chlorophyll fluorescence in green algae (3,4) revealed a separate, very short-lived (80 ps) component with a red shifted emission maximum (at 695 nm). This component was attributed to chlorophyll associated with photosystem I (PSI).
In this work, we present the spectral and kinetic resolution of chlorophyll-fluorescence from pea chloroplasts with completely open or completely closed PS II reaction centers. By a simultaneous kinetic analysis which fits all the decays recorded at different wavelengths to a multiexponential function we very accurately obtain the amplitude spectra of PS I and PS II components. These amplitudes can be taken as a direct measure of the size of antenna pools connected with the respective photo-system (cf. A.R. Holzwarth, these proceedings) and provide the basis for a model of the energy partition between chlorophyll-excitons and the radical pair [P 680$^+$ Pheo]. The adaptation of this model to our experimental findings shows that the long-lived (2 ns) fluorescence decay from closed PS II centers is prompt emission from the antenna rather than delayed emission after charge recombination.

2. MATERIALS AND METHODS

Chloroplast thylakoids were prepared from peas (12 to 14 days old) as described in (5). They were resuspended in 1,5 1 SNMT-buffer (100 mM Sucrose, 10 mM NaCl, 5 mM MgCl$_2$ and 20 mM Tricine / NaOH pH = 7.8) giving a chlorophyll concentration of about 8 µg Chl/ml. Temperature was kept at 15°C and the sample was rapidly (550 ml/min) pumped through the cuvette (1.5 x 1.5 mm^2 cross section). Thus, renewal of the sample was as rapid as the repetition rate of the exciting laser pulses (400 kHz) providing conditions for completely dark adapted PS II centers (F$_o$ condition). Closure of the PS II centers(F$_{max}$ conditions) was accomplished by adding either 2 mM sodium dithionite or 20 µM DCMU plus 10 mM hydroxylamine and illumination of the sample before entering the cuvette. The laser pulses (FMHW ≈ 15 ps, λ = 630 nm, energy density < 50 nJ / pulse · cm^2) were provided by a cavity dumped dye laser pumped by a mode locked Ar ion laser. Fluorescence was passed through a monochromator (4 nm optical bandwidth) and was monitored by the single photon timing method. Data analysis was performed by a global deconvolution procedure as described in (6).

Biggens, J. (ed.), Progress in Photosynthesis Research, Vol. I. ISBN 90 247 3450 9
© 1987 Martinus Nijhoff Publishers, Dordrecht. Printed in the Netherlands.

3. RESULTS AND DISCUSSION

Fig. 1 shows the time resolved fluorescence decay component spectra from pea chloroplasts with open (F_o-condition, left) and closed (F_{max} condition, right) PS II reaction centers.

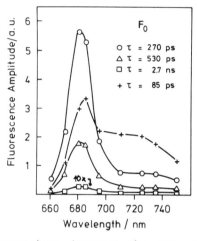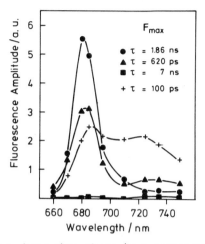

A comparison between the two states shows that there is a component of almost invariant lifetime ($\tau = 90 \pm 10$ ps) and with almost unchanged spectra. The associated spectra have relative maxima at 688 um and 725 nm. From the independence of this fluorescence decay component on the redox state of Q_A we conclude that it is due to Chl coupled to PS I. Also in green algae a similarly invariable and short-lived component was found (3), (7) and attributed to PS I. However in green algae this emission has only one maximum which is at 695 nm. These differences in the PS I emission spectra are in agreement with results of an analysis of CW measurements in (8).

All the other decay components show one type of spectrum with an emission maximum at about 680 nm. In open centers their lifetimes are typically 250 – 300 ps and 500 – 600 ps. The dominant amplitude is associated with the 250 – 300 ps phase. Closing the PS II centers by addition of dithionite or of DCMU plus hydroxylamine results in the following changes: the lifetime of the component with the largest amplitude is about 7 fold increased to 1.8 – 2 ns; that of the other component increases only slightly from 500 – 600 to 600 – 800 ps. Interestingly, the amplitude of the 250 – 300 ps phase in open centers is the same as that of the 1.8 ns – 2 ns phase in closed centers. Minor contributions (< 3% of total amplitude) are associated with long lifetimes of 2.7 ns in open and of 7 ns in dithionite-treated PS II centers. They are attributed to uncoupled chlorophyll, released from the protein matrix. The integration of PS I and PS II emission spectra in the range of 660 to 760 nm gives a measure of the relative antenna sizes of the photosystems. Results from our spectra are: Chl (PS I) : Chl (PS II) = 0.9 $\pm$ 0.1 in agreement with values reported in (9), (10).

We conclude from these findings:

a) PS I fluorescence in chloroplasts can be separated from the PS II emissions by its spectral and kinetic features.

b) After separation of the PS I component there is definitively no other fast fluorescence decay at F_{max} conditions.

c) At both extreme state, at F_o and F_{max}, the PS II fluorescence decays with 2 exponential terms.

d) The similar spectral shape of the PS II components suggests an origin from a common chlorophyll-pool.

e) Charge separation in PS II is reflected by the 250-300 ps phase.

We emphasize on the complementary relationship between the amplitudes of the fast (250-300 ps) and the slow (2 ns) PS II -component. This is an aspect decisive for the interpretation of the slow phase: The hypothesis that it might be delayed luminescence after charge recombination (11), (1) which necessarily would require a preceding charge-separation (fast phase) is in clear conflict with the disappearance of the fast phase at F_{max}. This qualitative argument is corroborated by a kinetic model which takes into account charge separation and charge recombination. Model-assumptions:

a) Chlorophylls of the core and of the LHC form a tightly coupled domain.

b) Trapping of an exciton by P_{680} from Chl is reversible (P_{680} is a shallow trap) and the equilibrium is reached within a time short compared to the overall exciton decay time. Thus, the probability of P_{680}^{*}/P_{680} becomes inversely proportional to the antenna size and is small (about 1/200). Under such conditions the apparent rate of primary charge separation is trap-limited (12). This assumption is strongly supported by our findings (A.R.Holzwarth, H.Brock and G.H.Schatz, these proceedings) that in isolated PS II particles with an antenna of about 80 Chl/PS II the major exciton decay phase in open centers is shortened to 60-80 ps. Thus, as compared to chloroplasts a 3-4 fold decrease in antenna size is correlated with a 3-4 fold decrease of the exciton lifetime as predicted by theory (12).

This model is represented by the scheme. We have shown (13) with plausible values for the different rate constants that the experimental results from fluorescence kinetics at F_o and F_{max} are compatible with the following mechanism:

Reduction of Q_a will decrease the value of k_1 (by about a factor of 10), and thus decrease the efficiency of charge separation; therefore, the excitation energy will reside in the antenna much longer than at F_o. Consequently, the observed 2 ns lifetime at F_{max} is prompt fluorescence rather than delayed luminescence after charge recombination.

4. REFERENCES

1 Karukstis,K.K. and Sauer,K., (1983) J. Cell. Biochem. 23, 131-158
2 Holzwarth,A.R., (1986) Photochem. Photobiol. 43, 707-725
3 Holzwarth,A.R., Wendler,J., and Haehnel,W., (1985) Biochim. Biophys. Acta 807, 155-167
4 Wendler,J. and Holzwarth,A.R., (1986) Biophys. J. in press
5 Polle,A. and Junge,W., (1986) Biochim. Biophys. Acta 848, 257-264
6 Holzwarth,A.R. and Suter,G.W., (1986) Biophys. J. in press
7 Holzwarth,A.R., (1986) in Topics in Photosynthesis (J.Barber, ed.)
8 Marchiarullo,M.A. and Ross,R.T., (1985) Biochim.Biophys.Acta 807, 52-63
9 Andersson,B. and Haehnel,W., (1982) FEBS Lett. 146, 13-17
10 Whitmarsh,J. and Ort,D.R., (1984) Arch. Biochem. Biophys. 231, 378-389
11 Klimov,V.V. and Krasnovsky,A.A., (1982) Biophys. 27, 186-198
12 Pearlstein,R.M., (1982) Photochem. Photobiol. 35, 835-844
13 Schatz,G.H. and Holzwarth,A.R., (1986) Photosynth. Res. in press

PICOSECOND FLUORESCENCE SPECTRA OF SYNCHRONOUS CULTURES OF THE GREEN ALGA SCENEDESMUS OBLIQUUS.

E. Bittersmann, H. Senger,[+] and A. R. Holzwarth

Max-Planck Institut für Strahlenchemie, D-4330 Mülheim/Ruhr, and [+]Fachbereich Biologie-Botanik der Philipps-Universität, D-3550 Marburg, West Germany

1. Introduction

Time-resolved chlorophyll (chl) fluorescence spectra provide very helpful means of studying the behaviour of photosynthetic systems (1,2). In this work we investigated changes in the time-resolved spectra and fluorescence lifetimes of synchronous cultures of the green alga Scenedesmus obliquus as a function of their life cycle. In addition we measured the fluorescence induction kinetics of the same samples and correlated these results with our findings on the fluorescence decay measurements. The aim of these studies was to test our proposed assignment of the kinetic lifetime components to α- and β-units (3).

2. Materials and Methods

Wild type Scenedesmus obliquus was grown as described (4). The algae were synchronized under a light : dark regime of 14:10 hours. At the beginning of each life cycle the cultures were diluted automatically. The degree of synchronization was checked by microscopical inspection to be at least 95%. Cell division took place between the sixteenth and seventeenth hour of the life cycle.

The fluorescence decays were recorded as a function of emission wavelength using the single photon timing method with picosecond time resolution. The excitation source was a synchronously-pumped and cavity-dumped dye laser system as described previously (5). The excitation wavelength was 630 nm. Data analysis was accomplished by a global deconvolution procedure resulting in time-resolved fluorescence spectra (6,7). The amplitudes of these spectra give a direct measure of the number of chl´s in the antenna pool . During the fluorescence decay measurements the algae were pumped through a 1.5 x 1.5 mm flow cuvette at a rate of 700 ml/min to prevent closing of the reaction centers.

The fluorescence induction kinetics were recorded in a home-built apparatus with excitation also at 630 nm. For the measurements the algae were dark adapted for 20 minutes. After this adaptation the samples were incubated with 20µM DCMU two minutes before the induction curves were recorded. The rise kinetics was analyzed in terms of two exponentials with a conventional fitting procedure.

Biggens, J. (ed.), Progress in Photosynthesis Research, Vol. I. ISBN 90 247 3450 9
© *1987 Martinus Nijhoff Publishers, Dordrecht. Printed in the Netherlands.*

3. Results and Discussion

Fluorescence measurements at room temperature were carried out during the eighth and sixteenth hour of the algal life cycle. These stages correspond to the maximum and minimum of the photosynthetic capacity, respectively.

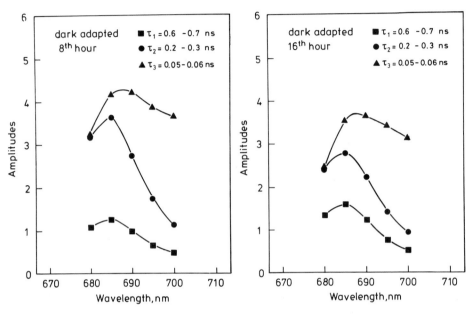

Figure 1: Time-resolved emission spectra of synchronous cultures of <u>Scenedesmus obliquus</u> at λ_{exc}=630 nm during the eighth and sixteenth hour of the life cycle. The spectra are normalized to the same quantum yield at both stages.

As can be seen from figure 1 the time-resolved spectra consist of three decay components, i.e. a long-lived component with a time constant of $\tau_1 \simeq 600 - 700$ ps, a medium-lived with $\tau_2 \simeq 200 - 300$ ps and a short-lived with $\tau_3 \simeq 50 - 60$ ps. Sometimes a fourth very long-lived ($\approx$2ns) component with very low amplitude is also needed to obtain a good fit. It has its origin in either functionally decoupled chl or in partially closed reaction centers. The short-lived component has a red-shifted spectrum (λ_{max}^{em} = 690 nm) and is assigned to photosystem (PS) I emission according to our previous results (5). The amplitude of this spectrum is almost constant in relation to the other two components which are attributed to PS II and have an emission maximum at λ_{max}^{em} = 685 nm. During the eighth hour the amplitude of the short-lived (200 – 300 ps) PS II component is large as compared to the amplitude of the longer-lived (600 – 700 ps) PS II component. During the sixteenth hour, the amplitude of the short-lived (τ_2) PS II component decreases while that of the longer-lived (τ_1) component shows an increase relative to the spectra of the eighth hour. These results are summarized in table 1.

TABLE 1: Lifetimes (τ_i) and relative amplitudes (α_i) of the two PS II components.

	8th hour		16th hour	
τ_i (ns)	0.2 - 0.3	0.6 - 0.7	0.2 - 0.3	0.6 - 0.7
α_i	0.745	0.255	0.635	0.365

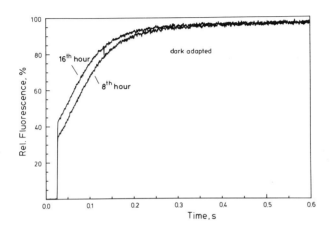

Figure 2: Fluorescence induction curves of dark adapted
Scenedesmus obliquus cells during the eighth and sixteenth
hour of synchronous growth.

Figure 2 shows the fluorescence induction curves of
Scenedesmus obliquus cells during the eighth and sixteenth
hour of their life cycle. Although the kinetics were recorded
for at least four seconds only the first part of the curves is
shown in the picture.

TABLE 2: F_m/F_o ratio and results from the exponential fit of
the induction kinetics, k_β is the rate constant of
the slow exponential induction phase.

	8th hour	16th hour
F_m/F_o	3.40	2.44
k_β (s^{-1})	2.25	4.02

Table 2 summarizes the results of the fluorescence
induction experiments. There is a decrease in the F_m/F_o (the
F_m/F_o ratio gives a measure of the maximum efficiency of
PS II) ratio during the sixteenth hour of the life cycle as
compared to that of the eighth hour. These values are in
excellent agreement with the results of Butko et.al. (8) who
performed measurements on Chlorella cells. The same trend can
also be seen in the data of Krupinska et.al. (9)

for synchronized Scenedesmus obliquus cells in the presence and absence of Mg^{2+}-ions. In addition to the F_m/F_o ratio the parameters of an exponential fit to the rise kinetics are given. Only the slow exponential β-phase (10) is considered for our interpretation. The rate constants k_β are a direct measure of the antenna size because the larger the antenna the shorter will be the time needed to close all reaction centers. During the eighth hour the rate constant k_β is smaller by a factor of two than during the sixteenth hour. This indicates an increased antenna size of the units giving rise to the slow induction phase during the sixteenth hour relative to the eighth hour.

From a comparison of the fluorescence lifetime and induction measurements we conclude that an increase in the unit size of the β-centers corresponds to an increase in the amplitude of the long-lived (600 - 700 ps) lifetime spectrum. These findings strongly support our former interpretation that the longer-lived (600 - 700 ps) PS II component arises from β-centers and the faster (200 - 300 ps) PS II component from α-centers. The differences in the fluorescence parameters between the eighth and sixteenth hour of synchronous growth of the algae are thus similar but not necessarily identical to state I/II transitions (7).

References
1 Karukstis,K.K. and Sauer,K.,
   (1983) J. Cell. Biochem. 23, 131-158.
2 Holzwarth,A.R., (1986) in Encyclopedia of
   Plant Physiology: Photosynthesis III New
   Series 19, 299-309, Springer, Berlin.
3 Wendler,J. and Holzwarth,A.R.,
   (1986) Biochim. Biophys. Acta submitted.
4 Bishop,N.I. and Senger,H., (1971) in Methods
   in Enzymology 23, 53-66 Academic Press, New York.
5 Holzwarth,A.R., Wendler,J., and Haehnel,W.,
   (1985) Biochim. Biophys. Acta 807, 155-167.
6 Suter,G.W., Holzwarth,A.R., Klein-Bölting,P.,
   Bittersmann,E., and Stempfle,W., (1986) submitted.
7 Holzwarth,A.R., (1986) Symposium contribution
   this conference.
8 Butko,P. and Szalay,L.,
   (1985) Photobiochem. Photobiophys. 10, 93-103.
9 Krupinska,K., Akoyunoglou,G., and Senger,H.,
   (1985) Photochem. Photobiol. 41, 159-164.
10 Melis,A. and Homann,P.H.,
   (1976) Photochem. Photobiol. 23, 343-350.

MEASUREMENTS AND KINETIC MODELING OF PICOSECOND TIME-RESOLVED FLUORESCENCE
FROM PHOTOSYSTEM I AND CHLOROPLASTS

BRUCE P. WITTMERSHAUS, DEPARTMENT OF PHYSICS AND ASTRONOMY, UNIVERSITY OF
ROCHESTER, ROCHESTER, NY 14627, (present address, BIOPHYSICS DEPARTMENT)

1. INTRODUCTION
    Photosystem I (PSI) is a network of antenna chlorophyll-proteins of
different sizes [1] and different spectral forms of chlorophyll (Chl) [1,2,3,4]
which absorb light and efficiently transfer excitation energy to its primary
electron donor P700 [5,6]. The association of absorption spectra deconvolution
components with particular fluorescence bands has illustrated the heterogeneity
of Chl spectral forms in PSI [2,3,4]. Simplifying these results, approximately
80% of the Chl in PSI can be assigned to a component absorbing maximally at
680 nm (C680), reflecting the 678 to 682-nm absorption peak of isolated
PSI. Intact PSI preparations (Chl/P700$\geq$100) contain small amounts of Chl
absorbing at longer wavelengths, the two most prominent being C697 and C705
[2,3,4,5]. At 295 K, fluorescence from isolated PSI particles is characterized
by a maximum at 685-690 nm and a weak shoulder from 700-740 nm [2]. The
former is attributed to the large antenna-Chl pool C680(F690). The latter
has contributions from C697(F720) and C705(F735) [2,3,4] and the vibrational tail of fluorescence from C680 Chls [7,8].
    Decreasing the temperature of PSI to 77 K, the fluorescence yield of
C680(F690) is observed to remain virtually constant while the fluorescence
from C697(F720) and C705(F735) increase dramatically [2,6,9,10]. For intact
PSI particles at 77 K from higher plants, fluorescence is dominated by a broad
emission band centered at 735 nm with an asymmetry towards shorter wavelengths,
indicating a 720-nm component and a very weak fluorescence component at 690
nm [2,3,6,9]. PSI from many algae, including P. luridum, have a 720-nm
fluorescence maximum similar to that from the PSI-65 preparation from higher
plants [2]. Tapie et al. [3] have concluded that algae exhibiting a 720-nm
fluorescence maximum at 77 K, and not at 735 nm as in higher plants, do not
contain the C705(F735) component. They also demonstrated that the C680 and
C697 components make up a "core" of PSI common to higher plants and algae.
Since the discovery of spectral forms of protein-bound Chls with absorption maxima at wavelengths > 690 nm in PSI [7], it has been proposed that
these Chls act as traps for excitations [7,11,12]. Fluorescence spectra,
excitation spectra and yield results from PSI at 295 and 77 K support this
idea [6,7,13]. In this role, C697 and C705 might act as part of a funneling
system for channeling excitations to P700.
    A model proposed by Mullet et al. [2] forms a basis for the models
examined in this work. The Chl-proteins of PSI are divided into three
complexes. Spectral components are assigned to each complex based on spectral
measurements from preparations with successively lower Chl/P700 values
[2]. The core complex contains P700 and $\sim$ 40 Chls and is defined by the
proteins which remain most persistent in their bonding to the P700-protein
complex. The internal complex consists of $\sim$ 20-25 Chls and contain the
C697(F720) component. The peripheral complex is made up of 40-45 Chls and
contains C705(F735) in higher plants [14]. All three complexes contain

Biggens, J. (ed.), Progress in Photosynthesis Research, Vol. I. ISBN 90 247 3449 5
© *1987 Martinus Nijhoff Publishers, Dordrecht. Printed in the Netherlands.*

antenna Chls which are part of the major spectral form C680(F690) [2].

Multi-exponential analyses of photon-counting fluorescence measurements from higher plants and algae have revealed a component with a decay time ranging from 40 to 130 ps attributed to PSI [15,16]. Early investigations of PSI-enriched fractions reported single-exponential decays from 10 to 130 ps [17,18]. Current studies have resolved multi-exponential decays with a prominent fast component in the 45 ps [15,16] to 20 ps [9,19] range. Generally, the fast decay component is attributed to fluorescence from C680(F690) Chls [9,15,16,19] and reflects the efficient quenching and low fluorescence yield ($\sim$0.004 [6]) of excitations in the antenna-Chl proteins of PSI at 295 K. Earlier time-resolved fluorescence studies of PSI at 77 K have either not resolved a C680(F690) decay component [9,18] or have observed a 3-ns lifetime for 690-nm emission [20] indicative of Chl-proteins isolated from a quenching source (non-coupled) [16].

This study reports time-resolved fluorescence data from isolated, intact PSI preparations from the blue-green alga _Phormidium luridum_ and from pea chloroplasts. To explain the fluorescence rise, decay and yield measurements, a model describing the rates of excitation transfer between distinct groups of Chl-proteins is proposed. The kinetic equations defined by this model are solved numerically and their parameters selected to fit the experimental data. The kinetic modeling results for isolated PSI particles are applied to measurements from chloroplasts [21,22] to develop a consistent model characterizing the network of heterogeneous antenna Chl-proteins associated with PSI. An important aspect of the model is the inclusion of small numbers of Chls with lower-than-average excited-state energy levels as intermediate traps for directing a majority of the excitations in the Chl-protein network of PSI to P700.

2. PROCEDURE

Detergent-free PSI (df-PSI) samples were prepared and obtained courtesy of Dr. Donald S. Berns and Dr. Cinnia Huang, Wadsworth Center for Laboratories and Research, New York State Department of Health, Albany, New York. Isolation of a P700-enriched complex was performed without detergents as previously described [10]. The samples were suspended in a 50 mM Tris buffer solution (pH 8.0 at 22°C). The Chl$\underline{a}$/P700 ratio of df-PSI was $\sim$100 [10]. The fluorescence maximum of df-PSI particles were 687 nm at 295 K and 720 nm at 77 K. Samples of the photosystem I-110 preparation (PSI-110) were obtained courtesy of Jacques Breton, Departement de Biologie, CEN-Saclay, Gif-sur-Yvette, France. PSI-110 samples were isolated from pea chloroplasts according to [2], relying on minimal use of Triton X-100 detergent. PSI-110 particles had a Chl$\underline{a}$/P700 ratio of 110±10 [2]. At 293 K the fluorescence spectrum had a maximum at 690 nm with a broad shoulder at 710-740 nm. The fluorescence at 77 K was dominated by a broad emission peaking at 736 nm [2]. Spinach chloroplasts were prepared with closed PSII as previously described [21] in a buffer solution of 0.4 M sucrose, 20 mM KCl and 20 mM Tris buffer with no Mg^{2+} added (pH 8 at 4°C).

The experimental apparatus and data acquisition procedures have been decribed in detail in [21-23]. Samples were excited with single, 30-ps, 532-nm laser pulses from a frequency-doubled mode-locked Nd^{3+}:YAG laser at a 0.5 Hz repetition rate. Fluorescence was time-resolved using a low-jitter ($\sim$2) streak camera and detected by a optical multi-channel analyzer. Samples were prepared with optical densities < 0.15 at 532 nm in a 65-70% glycerol-buffer solution. Fluorescence was collected through long-pass cut-off filters and 685(14) nm and 735(30) nm interference filters (bandwidth).

Two techniques were used for analysis of time-resolved fluorescence measurements. The first was to compare experimental data with theoretical curves generated by integrating kinetic equations based on a specific model as described in [21,22]. The kinetic parameters were varied using estimates for their range of reasonable values until a good fit to the data was obtained or a model was determined unfeasible. A sum-of-least-squares fitting routine was also developed for fitting data to a sum of exponential-decay components by iterative convolution [22,23]. Each component was defined by its lifetime τ and its percentage of the total fluorescent population (%N1, %N2, etc.) assuming equal absorbtion cross-sections for the components.

TABLE 1. Average parameters using two exponential-decay components to fit 685-nm time-resolved fluorescence from PSI at 295 and 77 K.*

Temperature(K)	Sample	τ_1 (ps)	τ_2 (ps)	%N1	%N2
295	df-PSI	15.7(3.5)	106(43)	86.8(5.0)	13.2(5.0)
	PSI-110	11.4(1.9)	89(8)	74.6(3.8)	25.4(3.8)
77	df-PSI	13.7(2.7)	189(29)	89.6(1.0)	10.4(1.0)
	PSI-110	12.1(0.7)	305(65)	90.3(3.1)	9.7(3.1)

* Values in parentheses are standard deviations.

3. RESULTS

The average parameters of fits using two exponential-decay components to describe time-resolved fluorescence from df-PSI and PSI-110 at 295 and 77 K are given in Table 1. The fluorescence decays from both preparations are similar and are characterized by a dominant, fast decay component and a small, long-lived population. At 295 K, no significant changes in the fluorescence decays from df-PSI and PSI-110 were observed when using a 685-nm interference filter or a 590, 665 or 695-nm cutoff filter [23]. Distinct fluorescence decay components associated with the long-wavelength spectral forms of Chl were not clearly resolved at 295 K. The fluorescence decays measured were independent of excitation fluences from 2×10^{12} to 4×10^{16} photons/cm^2 for df-PSI and from 1×10^{13} to 3×10^{14} photons/cm^2 for PSI-110 [23]. Lowering the temperature from 295 to 77 K did not dramatically alter the 685-nm fluorescence decay from df-PSI or PSI-110 (Table 1), although a tendency for τ_2 to increase at 77 K was noted. The 735-nm fluorescence from df-PSI and PSI-110 at 77 K (Fig. 1) has a delayed risetime indicating excitation transfer from some source to that population emitting at 735 nm. The kinetic model chosen to fit the data (Fig. 2, Table 2) is characterized by a large pool of antenna Chl-proteins, C680(F690), transferring their excitations to small pools of lower energy Chl-proteins, P700 and C697(F720) for df-PSI (P. luridum) and P700, C697(F720) and C705(F735) for PSI-110 (pea). These energy transfer relationships manifest themselves in the fast decay component ($\sim$ 12 ps) observed at 685 nm and attributed to C680(F690) for df-PSI and PSI-110 (Table 1) and the correlated delayed rise of 735-nm fluorescence from C697(F720) (Fig. 2). The long-lived 685-nm component is ignored except in the yield calculations and the fluorescence lifetime representing C680(F690) in the kinetic modeling is only the fast decay (12 ps). P700 is assumed to be a constant trapping source [7,19]. The fluorescence decay curves measured using the 735-nm, 30-nm bandwidth interference filter are predominantly characteristic of fluorescence from C697(F720). This is true of df-PSI by default since the C705(F735) component is not present. For PSI-110, the 310-ps lifetime observed for the 735-nm data (Fig. 1) had to be reconciled with the 2 to 3 ns lifetime reported for isolated PSI [9,18,20] and in chloroplasts [16,17] at 77 K from higher plants. The similarity of 735-nm results from df-PSI, PSI-110 (Fig. 1) and a reported 330-ps lifetime for 720-nm fluorescence from PSI of pea [9] suggested that the 735-nm fluorescence decay observed from PSI-110 was not from C705(F735) but from C697(F720). The PSI-110 samples may have developed emission similar to PSI-65 [2] but the exact cause has not been established.

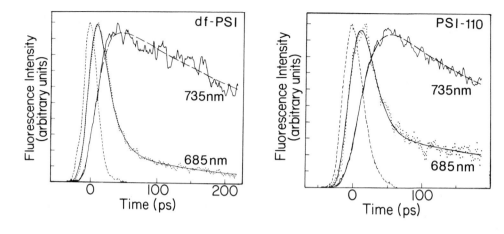

FIGURE 1. Time-resolved fluorescence at 685 and 735 nm from df-PSI and PSI-110 at 77 K. Emission at 685 nm (·····) has average fit parameters listed in Table 1. The kinetic equations defined by the model in Fig. 2 and the parameters in Table 2 are used to calculate the fit (— —) to the rise and decay of 735-nm fluorescence (———) from C697(F720). The curves are normalized and each is the sum of 200 shots with an excitation pulse (-----) fluence of 4.0×10^{14} photons/cm^2 for the df-PSI data and 3.2×10^{13} photons/cm^2 for PSI-110 data.

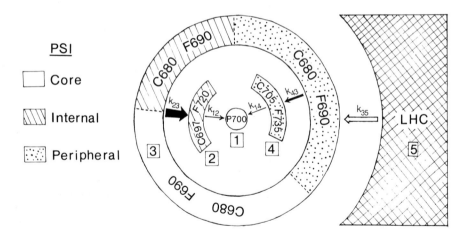

FIGURE 2. Model for the excitation transfer network of PSI-related antenna Chl-proteins. Areas are approximately proportionate to the relative amount of Chl in each component. C697 in df-PSI and C697 and C705 in PSI-110 act as intermediate traps for all excitations to P700 and are coupled to a common pool of C680(F690) Chls in the core, internal, and peripheral antenna complexes. For fitting data from isolated PSI (Fig. 1), LHC is not included. To model the data from chloroplasts (Fig. 3), LHC transfers excitation to the entire C680(F690) component of PSI. The theoretical fluorescence decays for each population at 77 K are calculated from the parameters in Table 2.

To narrow the range of values for the kinetic parameters, constraints based on this study's results and from the work of other researchers were applied. The absorption cross-sections at 532 nm for C697 and C705 were estimated to be between one to two times the value for C680. The relative numbers of Chl for each spectral component (Fig. 2, Table 2) were estimated from deconvoluted absorption spectra and absorption changes in PSI particles of various Chl/P700 values [2,3]. Relative fluorescence yields (ϕ) were determined using deconvoluted fluorescence spectra giving for df-PSI [10], $\phi(F690):\phi(F720) = 1:11$ and for PSI-110 [2,0], $\phi(F690):\phi(F720):\phi(F735) = 1:11:32$. The fluorescence lifetimes of non-coupled antenna Chl-proteins are assumed to be from 1.2 to 3 ns [16]. These constraints were used as guidelines in attempting to model the 685-nm fast decay and the rise and decay of 735-nm fluorescence from df-PSI and PSI-110 at 77 K.

TABLE 2. Parameters for the proposed model of excitation transfer in PSI.

Sample	Excitation Transfer Rates*				Population Parameters§					
	$N_{10}k_{12}$	$N_{20}k_{23}$	$N_{40}k_{43}$	$N_{30}k_{35}$	i=	1	2	3	4	5
PSI-110	367^{-1}	17^{-1}	44^{-1}	80^{-1}	N_{i0}	2.0	5.9	80	5.9	80
& chloro-					σ_i	2	2	1	2	1
plasts					τ_i	1	2000	1250	1500	1250
df-PSI	395^{-1}	12.1^{-1}	-	-	N_{i0}	2.0	5.9	80	-	-
P.					σ_i	2	2	1	-	-
luridum					τ_i	1	2000	1250	-	-

* $N_{i0}k_{ij}$ is rate from j to i in $(ps)^{-1}$. Rates not listed are assumed zero. In particular, $N_{40}k_{14}=0$ at 77 K.
§ Chromophore densities (N_{i0}) are in units of $\times 10^{18} cm^{-3}$ and reflect the relative number of Chls. Ground-state absorption cross-sections (σ_i) are in units of $\times 10^{-17} cm^2$. Fluorescence lifetimes (τ_i) are in ps.

The constraints discussed were to a large degree satisfied in the choice of fitting parameters (Table 2). The most difficult constraint to reconcile in attempting to fit the PSI data was $\phi(F690):\phi(F720) = 1:11$. The yield and lifetime measurements are the most important tests of the model. To reach the yield constraint, the model required large percentages of the excitations in C680(F690) to be trapped initially by C697(F720) (Fig. 2, Table 2). Assuming the excitation pathways involving C705(F735) (population 4 in Fig. 2) do not exist for df-PSI, the parameters in Table 2 for df-PSI give ϕ(F690): $\phi(F720) = 1:11.3$ just reaching the constraint value when all the excitations in C680 are transferred to C697. For PSI from higher plants, calculations based on the results of Moya et al. [9] suggest that $\sim$ 72% of the sum of excitations in C697 and C705 are in C697(F720). The fit to the fluorescence from C697(F720) in PSI-110 (Fig. 1) has 72% of the excitations in C680(F690) going to C697(F720) (Table 2, Fig. 2). Allowing for those trapped by C705 (F735), this suggests that nearly all the excitations created in the C680(F690) antenna-Chls of PSI-110 pass through either C697 or C705 Chls before being transferred to P700. Even in this apparently optimal case the value of $\phi(F690):\phi(F720) = 1:5.9$ (Table 2). Unfortunately, the small amount of long-lived peripheral antenna-Chls in PSI-110 at 77 K (Table 1) contributes $\sim$ 70% of the 690-nm fluorescence. The variability of its lifetime causes uncertainties in the 690-nm yield making reliable comparisons between the yield measurements on one sample to lifetime results for another difficult [23]. This may be the reason the yield constraint could not be satisfied for PSI-110.

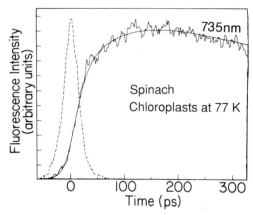

FIGURE 3. Kinetic model fit to time-resolved 735-nm fluorescence from spinach chloroplasts at 77 K. The experimental data (——) is fit (smooth ——) using the model in Fig. 2 and parameters from Table 2. The curve is the sum of 300 shots with an excitation pulse (-----) fluence of 2.2x10^13 photons/cm².

Fig. 3 is a measurement of 735-nm fluorescence from spinach chloroplasts at 77 K previously reported [21]. The delayed, biphasic rise of the time-resolved fluorescence was originally attributed to the presence of at least two different sources transferring excitations to the C705(F735) component of PSI [21]. By adding the LHC component to the kinetic model defined by the data from isolated PSI particles (Fig. 2), the 735-nm fluorescence from PSI in chloroplasts can be successfully described (Fig. 3, Table 2). The biphasic rise of 735-nm emission has an initial fast component from excitation original-ly created in C680(F690) being transferred to C705, reflecting the 12-ps lifetime of C680(F690) fluorescence observed in PSI-110 (Fig. 1, Table 1). The slower rise component is the transfer of excitations in LHC to C705 indirectly through the C680 antennae (Fig. 2) with a rate of $\sim (80 \text{ ps})^{-1}$ (Table 2). The transfer ratios are such that 72% of C680's excitations are trapped by C697(F720), 28% by C705(F735) and none by P700 directly (Table 2).

4. DISCUSSION

PSI from algae and higher plants do not share all the spectral forms of Chl-proteins. Most notably, C705(F735) is not present in many algae [3,5] including P. luridum. Despite these differences the time-resolved fluorescence measurements for df-PSI and PSI-110 at 295 and 77 K are very similar for emission from C680(F690) and C697(F720). The kinetic modeling results of Table 2 and Fig. 2 reflect the conclusion that the core and internal Chl-protein complexes are approximately the same in df-PSI and PSI-110 as suggested by Tapie et al. [3].

The largest antenna-Chl component, C680(F690), encompasses $\geq$ 80% of the Chl in PSI and dominates the absorption spectra of most PSI preparations [2], reflecting its predominance in the core, internal, and peripheral antenna complexes [2]. Despite its small lifetime (12-16 ps), C680's large number of Chls and the presence of a small amount of poorly coupled peripheral antenna-Chls with a lifetime of 100-300 ps permits fluorescence from C680(F690) to dominate the 295 K fluorescence spectra of PSI. The long-lived C680(F690) component plays a minor role in the initial time-resolved fluorescence but makes up 50-75% of the yield of fluorescence from C680(F690). Native PSI in the photosynthetic membranes is not expected to have these uncoupled peripheral Chl-proteins. The observed 12-16 ps lifetime for $\sim$ 90% of C680 (F690) is the result of rapid energy transfer from the C680 component to P700 via C697 and C705 causing small fluorescence yields from C680(F690) in intact PSI. The temperature-invariance of the yield [6,8,20,24,25] and lifetime of the fast decay component of C680(F690) from 295 to 77 K implies that the structural organization of the antenna-Chl components and the excitation transfer rates which define their relationships to the C680 proteins do not change. The fast decay component is also insensitive to the redox state of P700 and only slightly affected by annihilation at high fluences [22,23]. These observations provide an important link between the conclusions based on measurements from PSI at 77 and 295 K.

The antenna-Chl spectral form C697(F720) of the internal complex of PSI and C705(F735) found in the peripheral complex of PSI from higher plants are small amounts of lower energy-level, protein-bound Chl which act as intermediate traps for funneling excitation from C680(F690) to P700. This greatly increases the overall efficiency of PSI [11]. The correlation of the fast decay component of C680(F690) ($\sim$ 12 ps) to the delayed rise of emission from C697(F720) at 77 K is direct evidence for C697's role as such an intermediate trap of C680's excitations. The observation of a delayed rise of C705(F735)'s fluorescence in higher plants [16,20] implies that C705 has a similar role. C697 is the major trap of C680's excitations. Therefore, C697 must be a trap for excitations at 295 K for C680(F690)'s fast decay component to remain unchanged from 295 to 77 K, supporting the proposal that these long-wavelength antenna-Chl traps exist at 295 K.

In higher plants C705(F735)'s temperature-dependent fluorescence yield has been directly correlated with its lifetime [13]. An increase in C697 (F720)'s fluorescence lifetime with decreasing temperature has also been observed [9], although a thorough comparison of lifetime and yield versus temperature has not been reported. It is proposed that C697 and C705 are closely associated with P700 and their lower energy levels result in a temperature-dependent excitation back transfer rate similar to that proposed [13] and modeled [8] for C705(F735) in higher plants. This is consistent with the above observations and with the decrease in lifetimes observed at 77 K as the antenna-Chl component's energy level approaches that of P700, C697(F720)=330 ps [9], and C705(F735)=2-3 ns [16,17]. The effect of the relative energy level is also evident in comparing the yield versus temperature results for C705(F735) [8,12,24] and C697(F720) [25] where their fluorescence yields become constant below $\sim$ 80 K and $\sim$ 10 K respectively. The only excitation transfer relationships which change in PSI from 295 to 77 K appear to be those between the long-wavelength antenna Chl-proteins and P700. The decrease in transfer of excitations from C697(F720) and C705(F735) to P700 with decreasing temperature is responsible for the increases in fluorescence lifetime and yields of these components at 77 K [8,12,24,25] and the resulting dramatic temperature-dependent changes in PSI fluorescence spectra.

The kinetic modeling of the rise, decay, and relative yield of fluorescence from C697(F720) in df-PSI and PSI-110 at 77 K required that greater than 72% the excitations in the C680(F690) Chl-proteins be transferred to C697(F720), making it the major Chl-protein intermediate trap for excitations in both df-PSI and PSI-110. The long-wavelength antenna-Chl components appear to act in parallel instead of in series in channeling excitation to P700 (Fig. 2). No 330-ps rise component has been observed in the fluorescence from C705(F735) [16,21]. The similarity of the lifetime of C697(F720) emission in df-PSI where no C705 is present and the 330-ps observed for C697(F720) in PSI-150 [9] and PSI-110 is indirect evidence that C697 does not transfer significant amounts of excitations to C705 in PSI of higher plants.

The proposed model illustrates that the coupling of LHC to PSI is important in determining the rise kinetics of 735-nm fluorescence from chloroplasts at 77 K. The biphasic rise of 735 nm emission was explained as a fast rise component ($\sim$12 ps) from transfer from C680(F690) to C697(F720) and C705(F735) and a slow rise component ($\sim$75-100 ps) attributed to excitations transferred from LHC. The results of time-resolved fluorescence measurements from isolated PSI particles and PSI in chloroplasts were brought together to form a consistent kinetic model explaining in detail the excitation transfer relationships of the network of antenna Chl-proteins related to PSI *in* vivo.

5. ACKNOWLEDGEMENTS
This research was supported in part by USDA grant 82-CRCR-1-1128 and by NSF grant PCM-83-03004, and by the sponsors of the Laser Fusion Feasibility Project of the University of Rochester. My thanks to D.S. Berns., C. Huang, J. Breton, N.E. Geacintov, and R.S. Knox for their help and encouragement.

6. REFERENCES
1) Melis, A. and Anderson, J.M. (1983) Biochim. Biophys. Acta 724, 473-484.
2) Mullet, J.E., Burke, J.J.and Arntzen, C.J. (1980) Plant Physiol. 65, 814-822.
3) Tapie, P., Choquet, Y., Breton, J., Delepelaire, P. and Wollman, F.-A. (1984) Biochim. Biophys. Acta 767, 57-69.
4) Wollman, F.-A. and Bennoun, P. (1982) Biochim. Biophys. Acta 680, 352-360.
5) Bose, S. (1982) Photochem. Photobio. 36, 725-731.
6) Boardman, N.K., Thorne, S.W. and Anderson, J.M. (1966) Proc. Nat'l. Acad. Sci. 54, 586-593.
7) Butler, W.L., (1978) Ann. Rev. Plant Physiol. 29, 345-378.
8) Tusov, V.B., Korvatovskii, B.N., Pashchenko, V.Z. and Rubin, L.B. (1980) Doklady - Biophysics [English trans.] 252, 112-115.
9) Moya, I., Mullet, J.E., Briantais, J.-M. and Garcia, R. (1981) in Proc. 5th Int'l. Congr. on Photosynthesis (Akoyunoglou, G., ed.),pp.163-172, Balaban Int'l. Sci. Services, Philadelphia, PA.
10) Huang, C. and Berns, D.S. (1983) Arch. of Biochem. Biophys. 220, 145-154.
11) Seely, G.R., (1973) J. Theor. Biol. 40, 173-199.
12) Butler, W.L., Tredwell,C.J., Malkin, R. and Barber, J. (1979) Biochim. Biophys. Acta 545, 309-315.
13) Satoh, K. and Butler, W.L. (1978) Biochim. Biophys. Acta 502, 103-110.
14) Lam, E., Oritz, W., Mayfield, S. and Malkin, R. (1984) Plant Physiol. 74, 650-655.
15) Holzwarth, A.R. (1985) in Encyclopedia of Plant Physiology (Staehlin, L.A. and Arntzen, C.J., eds.), Vol. 19, pp.299-309, Springer-Verlag, Berlin.
16) Karukstis, K.K. and Sauer, K. (1983) J. of Cellular Biochemistry 23, 131-158.
17) Breton, J. and Geacintov, N.E. (1980) Biochim. Biophys. Acta 594, 1-31.
18) Searle, G.F.W., Barber, J., Harris, L., Porter, G. and Tredwell, C.J. (1977) Biochim. Biophys. Acta 459, 390-401.
19) Il'ina, M.D., Krasauskas, V.V., Rotomskis, R.J. and Borisov, A.Y. (1984) Biochim. Biophys. Acta 767, 501-506.
20) Avarmaa, R.A., Kochubey, S.M. and Tamkivi, R.P. (1979) FEBS Lett. 102, 139-142.
21) Wittmershaus, B.P., Nordlund,T.M., Knox, W.H., Knox, R.S., Geacintov, N.E. and Breton, J. (1985) Biochim. Biophys. Acta 806, 93-106.
22) Wittmershaus, B.P., (1986) Ph.D. thesis, University of Rochester.
23) Wittmershaus, B.P., Berns, D.S. and Huang, C. (Submitted to Biophys. J., 1986).
24) Rijgersberg, C.P., Melis, A., Amesz, J. and Swager, J.A. (1979) Ciba Foundation Symposium, Excerpta Medica, 305-322.
25) Rijgersberg, C.P. and Amesz, J. (1980) Biochim. Biophys. Acta 593, 261-271.

TIME-RESOLVED FLUORESCENCE DECAY KINETICS IN PHOTOSYSTEM I. EXPERIMENTAL
ESTIMATES OF CHARGE SEPARATION AND ENERGY TRANSFER RATES.

T.G. Owens, S.P. Webb, D.D. Eads, R.S. Alberte, L. Mets and G.R. Fleming
Dept. of Chemistry and Dept. of Molecular Genetics and Cell Biology, The
University of Chicago

The analysis of fluorescence decay kinetics in photosynthetic systems
has provided valuable information on excitation migration, trapping
dynamics and structural organization of the photosynthetic unit. Analysis
of chloroplasts and green algae indicates that a minimum of four
exponential components are required to adequately describe in vivo
fluorescence decay (1,2). It is generally agreed that a fast, 50-100 ps
component should be assigned to the lifetime of excitations in the antenna
pigments of PS I (1-3). We have approached the analysis of fluorescence
decay in PS I using isolated P700 Chl a-protein complexes and mutants of
Chlamydomonas reinhardii that lack the PS II reaction center/core antenna
complex. Mutant B1 contains a deletion in the chloroplast psbA gene
encoding the 32 kDa thylakoid membrane protein of PS II. Strain A4d
contains the B1 mutation and a nuclear mutation that reduces the chl a/b-
protein content by 50-80%. Using time-correlated single photon counting, we
have measured the spectral and kinetic properties of the pigments
contributing to each decay component in PS I. Application of the data to
lattice model equations provides experimental estimates of charge
separation and energy transfer rates in PS I.

MATERIALS AND METHODS
The P700 Chl a-protein was isolated from barley and Synechococcus sp.
by the procedure of Shiozawa et al (4). Preparations contained no
detectable chl b and had chl a/P700 ratios between 23 and 55. C. reinhardii
strains A4d and B1 were grown mixotrophically at 10 μE m^{-2} s^{-1} (1).
Fluorescence decay kinetics were measured using time-correlated single
photon counting (1). Excitation pulses (8-10 ps FWHM, 590-710 nm) were
produced by a synchronously pumped DCM dye laser which was acousto-
optically dumped at 75 kHz. Maximum excitation intensities of 3 x 10^{11}
photons/cm^2/pulse resulted in < 0.05 photons/reaction center/pulse.
Fluorescence was detected at 90° using a monochrometer (bp = 4 nm) and a
red-sensitive microchannel plate/photomultiplier. The instrument response
function for scattered excitation pulses had a FWHM of 60-80 ps allowing a
time resolution of ~10 ps. Decay curves were fit to a sum of exponentials
convoluted with the instrument response function. Fluorescence excitation
and emission spectra of the individual decay components were computed using
the procedure of Holzwarth et al (2).

RESULTS AND DISCUSSION
Table 1 presents the time-resolved fluorescence decay measured at 690
nm for the P700 Chl a-protein (excitation: 665 nm) and C. reinhardii
strains A4d and B1 (excitation: 652 nm). The decays are fit to the sum of
two or three exponentials with X^2 parameters < 1.10. All three samples
exhibit a fast decay component with 1/e lifetimes of 20-40 ps.
Figure 1 shows the time-resolved fluorescence emission spectra of the
fast (20 ps) and slow (6 ns) decay components in the P700 Chl a-protein.

Biggens, J. (ed.), Progress in Photosynthesis Research, Vol. I. ISBN 90 247 3450 9
© 1987 Martinus Nijhoff Publishers, Dordrecht. Printed in the Netherlands.

Table 1. Summary of fluorescence decay kinetics in the P700 Chl a-protein and PS II mutants of C. reinhardii. Lifetimes have units of ps.

	A_1	T_1	A_2	T_2	A_3	T_3	x^2
P700 Chl a	.892	20			.108	6370	1.09
A4d	.906	42	.082	232	.010	2375	1.02
B1	.328	42	.136	372	.563	2090	1.04

The spectra are normalized to a value of one at their respective emission maxima. Both spectra are characterized by emission peaks at 680-690 nm and a rapid decline in emission at wavelengths > 700 nm. The fast decay component is assigned to the lifetime of excitations in the PS I core antenna. The short lifetime is due to efficient photochemical quenching by P700. Treatment of the P700 Chl a-protein with 0.2 to 1.0% LDS induces a parallel loss of P700 and the fast decay component without removing pigments from the complex (data not shown).

The lifetime of excitations contributing to the slow decay component in the P700 Chl a-protein is the same as that of free chl a in solution. However, the fluorescence emission spectrum of the slow component is red-shifted by 10 nm compared with detergent-solubilized chl a (Figure 1). Accordingly, the slow decay component is assigned to a small population of PS I core antenna complexes which lack a functional PS I trap. Time integration of the individual decay components shows that the steady-state fluorescence emission of the P700 Chl a-protein is derived almost entirely (>95%) from the 6 ns component.

The decay kinetics in C. reinhardii strains A4d and B1 are best fit to the sum of three exponentials with lifetimes of 40, 300 and 2200 ps (Table 1). The fluorescence emission spectra of the fast and intermediate lifetime components are very similar, showing maxima at 685 and 695 nm (Figure 2).

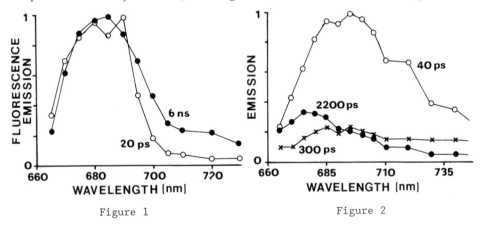

Figure 1   Figure 2

Figure 1. Fluorescence emission spectra of the two decay components in the P700 Chl a-protein (excitation: 650 nm). Spectra are normalized at their respective maxima. The relative maximum of the 6ns component is 5-15% of the 20 ps component.

Figure 2. Relative fluorescence emission spectra of the three decay components in C. reinhardii strain A4d (excitation: 652 nm).

These peaks are shifted by ~ 5 nm to the red compared with the fast
component in the P700 Chl a-protein. The fast component in A4d and B1 is
assigned to the PS I core antenna while the intermediate lifetime component
is attributed to core antenna pigments with reduced efficiencies of energy
transfer to P700.

The lifetime of the slow component in A4d and B1 is similar to the
isolated chl a/b-protein and C. reinhardii mutant C2, which lacks both the
PS I and PS II reaction center/core antenna complexes (1). The fluorescence
emission spectrum of the slow component is dominated by short wavelength
emission characteristic of the chl a/b-protein. This slow component is
assigned to a population of light-harvesting complexes that are incapable
of energy transfer to the PS I core antenna. The size of the uncoupled pool
is much smaller in A4d than in B1, correlating with the 50-80% reduction in
chl a/b-protein content in A4d.

The fluorescence excitation spectra for the fast decay component in
strains A4d and B1 show a large contribution of pigments absorbing at
wavelengths < 680 nm (data not shown). The lifetime of the fast component
in the PS I core antenna is the same for excitations that originate on chl
b in the peripheral light-harvesting antenna (652 nm) or direct absorption
by the core antenna (680 nm). This suggests that the time required for the
excitation to migrate through the light-harvesting to the core antenna does
not contribute significantly to the measured lifetime. This is supported by
the time-dependent evolution of fluorescence emission from all pigments
(Figure 3). Immediately following a δ-function excitation, the fluorescence
emission spectrum shows only emission from the core antenna. During the
next 30 ps, the spectrum shifts to the red by only 1-2 nm. At times that
are long compared with the lifetimes of excitations in the core antenna,
the spectrum shifts to wavelengths characteristic of the chl a/b-protein
This indicates that energy transfer from coupled chl a/b-proteins to the
core antenna is rapid (<10 ps) and confirms that the slow decay component
results from uncoupled light-harvesting complexes.

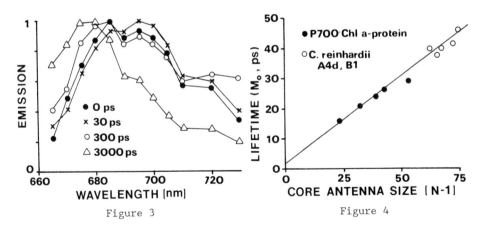

Figure 3   Figure 4

Figure 3. Calculated time-dependent evolution of total fluorescence
emission in C. reinhardii strain A4d following a δ-function excitation
(spectra are normalized at their respective maxima).

Figure 4. Dependence of the lifetime of excitations in the PS I core
antenna on antenna size.

The lifetime of excitations in the PS I core antenna differs by about a factor of two between the P700 Chl a-protein (13-27 ps) and mutants A4d and B1 (38-45 ps). These variations in lifetime can be attributed to differences in the core antenna size between the samples using the mathematical expressions of Pearlstein (5). These expressions are derived from model equations of excitation migration and trapping in a regular lattice of photosynthetic pigments. In this treatment, the mean lifetime, M_o, is given by:

$$M_o = [(1/qF_T) - (1/qF_A)][(N-1)^2/N] + \alpha N/F_A + (F_D/F_T)(N-1)(1/k_p) + 1/k_p$$

where F_T, F_D and F_A are the Förster rate constants for trapping, detrapping and transfer among lattice pigments, k_p is the photochemical rate constant and N is the antenna size (q and α are lattice constants).

Assuming that $(N-1)^2/N = N-1$, this equation predicts that M_o will be a linear function of N-1 and that the y-axis intercept ($= 1/k_p + \alpha/F_A$) provides an estimate of k_p in PS I. Using data derived from isolated P700 Chl a-proteins and core antenna sizes calculated for C. reinhardii strains A4d and B1 (6), a plot of M_o against N-1 (Figure 4) confirms the expected linear dependence. Estimation of the ratios F_D/F_T and F_T/F_A from the Förster overlap integrals permits calculation of k_p, the average Förster rate constants (F_A, F_T and F_D) and the single-step transfer time (SSTT). Given an average PS I core antenna size of 68 chl a (6), values may also be calculated for the in vivo first-passage time (τ_{FPT}), detrapped excitation lifetime (τ_{DT}) and the average number of visits an excitation makes to the trap before photoconversion (N_v). Data are calculated for monomeric and dimeric P700 (Table 2). Estimates of the photochemical lifetime ($1/k_p = 1.9\pm.6$ ps) are about twice that measured directly in reaction centers of Rhodopseudomonas sphaeroides ($2.8\pm.2$ ps)(7). The value of N_v ($1.9\pm.2$) indicates that excitation migration in PS I is nearly diffusion limited.

Table 2. Parameters describing excitation migration and trapping in PS I. Förster rate constants have units of 10^{12} s^{-1}, lifetime paramenters have units of ps.

P700	F_A	F_T	F_D	$1/k_p$	SSTT	τ_{FPT}	τ_{DT}	N_v
monomer	1.29	0.90	0.08	1.67	0.19	28.4	10.1	1.89
dimer	0.95	1.32	0.12	2.12	0.26	25.3	12.8	1.85

This research was sponsored by National Science Foundation grants PCM 8409014 and DMB 8509590.

REFERENCES
(1) Gulotty, R.J., L. Mets, R.S. Alberte and G.R. Fleming (1985) Photochem. Photobiol. 41,487-496
(2) Holzwarth, A.R. J. Wendler and W. Haehnel (1985) Biochim. Biophys. Acta 807,155-167
(3) Karukstis, K. and K. Sauer (1985) Biochim. Biophys. Acta 806,374-383
(4) Shiozawa, J.A. R.S. Alberte and J.P. Thornber (1974) Arch. Biochem. Biophys. 165,388-397
(5) Pearlstein, R.M. (1982) Photochem. Photobiol. 35,835-844
(6) Owens, T.G., S.P. Webb, D.D. Eads, L. Mets, R.S. Alberte and G.R. Fleming (in prep)
(7) Martin, J.-L., J. Breton, A.J. Hoff, A. Migus, and A. Antonelli (1986) Proc. Nat. Acad. Sci. USA 83,957-961

SPECTRAL PROPERTIES OF PHOTOSYSTEM I FLUORESCENCE AT LOW
TEMPERATURES

J. WACHTVEITL & H. KRAUSE, UNIVERSITY OF REGENSBURG, INSTITUT
PHYSIK II, UNIVERSITÄTSSTRASSE 31, 8400 REGENSBURG, F.R.G.

1. INTRODUCTION

Isolated PSI complexes [1] exhibit a Chl a fluorescence which depends on temperature,redox changes, and the excitation wavelength. Room temperature PSI-
emission spectra show a maximum at 685 nm. They have in general a very low quantum yield. At low temperatures a long wavelength emission peak becomes dominant.
Light induced absorbance changes lead to the suggestion that the PSI-primary
electron donor is probably a Chla-Dimer [2]. Measurements of excitation spectra,
obtained by selective excitation of fluorescence bands are relevant to the study of
pigment compositions.

In the present work fluorescence spectra and absorption difference spectra of PSI
and RCI complexes from **spinach** and **Anabaena variabilis** were measured between
r.t. and liquid He temperatures. The obtained results support the suggested model
of PS I and give further information about details of the system.

2. MATERIALS AND METHODS

Preparation

PS I and RC I complexes from spinach (Chl a/P 700 = 95) and the cyanobacterium
Anabaena variabilis (Chl a/P 700 = 80) were isolated according to ref. [3]. RC I-
particles are SDS-treated PS I-preparations, highly enriched in the large subunit.

Instrumentation

Absorption spectra were measured with an Aminco DW-2 spectrophotometer containing a low temperature equipment. For fluorescence measurements, the samples
were placed in a He-bath-cryosystem (temperature range 1.3 K-room temperature)
and illuminated with a pulsed dye laser system, an Ar-Ion-Laser or a Xe-spectral-
lamp with monochromator. The spectra were corrected for the sensitivity of the
Hamamatsu R 1463 photomultiplier.

3. RESULTS

Chla-fluorescence depends on the redox state of the acceptor X as reported earlier
[4,5]. When dithionite is added the RC I preparation of **anabaena** shows a slow
increase in fluorescence yield which depends on the light intensity (fig. 1). This
effect is due to the subsequent photoreduction of X which consequently looses
its function as a quencher of fluorescence. Illumination drives the following reactions:

$$P\ 700\ X A \xrightarrow{h\nu} P\ 700^* X A \longrightarrow P\ 700^+ X \bar{A} \quad \text{(Niveau 1)}$$

Addition of dithionite reduces the FeS-acceptor A, accordingly X becomes photoreduced:

$$P\ 700\ X A^- \xrightarrow{h\nu} P\ 700^* X A^- \longrightarrow P\ 700^+ X^- A^- \xrightarrow{e^-(ED)} P\ 700\ X^- A^-$$

$$\text{(Niveau 2)}$$

Biggens, J. (ed.), Progress in Photosynthesis Research, Vol. I. ISBN 90 247 3450 9
© *1987 Martinus Nijhoff Publishers, Dordrecht. Printed in the Netherlands.*

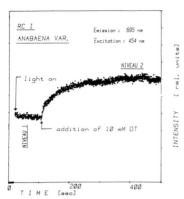
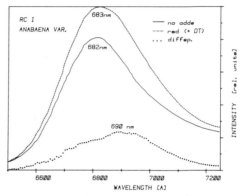

Fig. 1: Variable fluorescence of RC I   Fig. 2: Emission spectra of RC I

This result provides evidence for the fact, that the RC I-core-protein is the binding site for the primary acceptor X.

Emission spectra (fig. 2) of reduced RC I exhibit a red shift more drastically seen in the difference spectrum. This indicates that the variable part of the fluorescence is mainly caused by energetically lowered Chl a-molecules close to the charge separation site.

Absorption

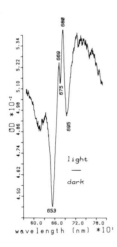

Fig. 3: Absorption difference spectrum light minus dark of PS I from "Anabaena" at low temperature

Fig. 4: Absorption difference spectrum light minus dark of PS I + DT of Anabaena at low temperature

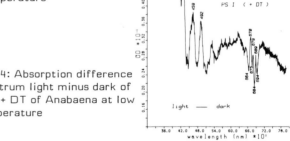

Fig. 3 shows the low temperature absorbance difference spectrum of PS I from **Anabaena** upon illumination. The strong apparent bleaching around 653 nm is caused by a band shift in the illuminated sample towards slightly longer wavelengths. The signal that is due to photooxidation of P 700 shows a positive (680 nm) and two negative (675 and 695 nm) components.

The light minus dark spectrum of PS I frozen in the presence of dithionite shows an additional band with a minimum at 684 nm, which should be related to photo-reduction of a Chl a-type acceptor in the PS I reaction center (fig. 4). The spectra can be explained by the assumption that RC I involves two Chl a molecules. In agreement with ref. [2] we suggest, that the peaks are caused by disappearance of excitonic interaction which follows the oxidation of one Chl a molecule of the dimer. An additional new absorption of the unoxidised Chl a appears.

Fluorescence spectroscopy

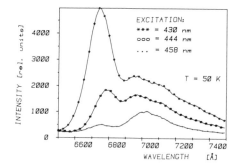

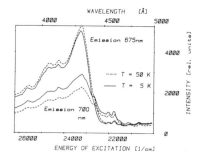

Fig. 5: Emission spectra of PS I temperature 50 K and different excitation

Fig. 6: Excitation spectra of PS I measured at 675 nm and 700 nm and temperatures 5 K and 50 K

The low temperature emission spectra of PS I **Anabaena** show a strong dependence on excitation wavelength. The antenna Chl a fluorescence is excited by a band around 430 nm. Excitation with longer wavelengths results in a drastic decrease of the antenna fluorescence around 675 nm. The 700 nm core-antenna emission band can be excited selectively at 458 nm (fig. 5). The red shift of the core Chl a molecules compared to the antenna pigments in the blue region of the spectrum is detectable in absorption and in fluorescence excitation spectra. Cooling from 50 K to 5 K increases the efficiency of the 700 nm emission, but no spectral shifts or changes in relative intensities occur (fig. 6). Temperature effects arise from changes of transfer rates to the different PS I pigment systems.

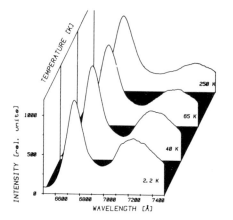

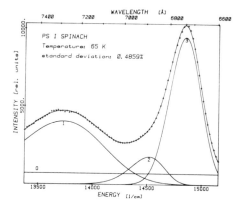

Fig. 7: Temperature dependence of the emission spectrum of PS I

Fig. 8: Emission spectrum of PS I with three components

The PS I from spinach shows a long wavelength emission band around 720 – 730 nm.
When treated with Triton this fluorescence is relatively weak compared to fluorescence effects seen by other groups [6]. Fig. 7 shows the temperature dependence
of this band. The emission characteristics can be explained by the assumption of
three different PS I Chl a-antenna forms and a decreasing transfer efficiency of
excitation between the different domains. The standard deviation of the fit curves
is below0.8%at all temperatures (fig. 8) . The presented results lead to the model of
the PS I structure as shown in fig. 9.

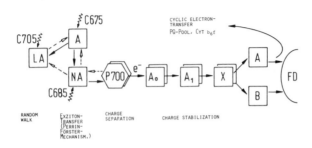

Fig. 9: Model for PS I structure

4. CONCLUSIONS

- The RC I-core-protein is the binding site of the primary acceptor X.
- The absorption difference spectra provide evidence for the fact that P 700 is
  a Chl a dimer.
- Fluorescence excitation and absorption spectra show energetically lowered
  states of core Chlorophyll molecules in the Soret-band.
- A three domain model of PS I explains the temperature effects sufficiently.

References

1   Ort, D.R. (1986) Photosystem I Complex, in: Enc. Plant Physiol. 19, chapt. 4,
    168–172
2   Philipson, K.D., Sato, V.L., Sauer, K. (1972) Biochem. 11, No 24, 4591–4595
3   Schoeder, H.U., Lockau, W. (1986) FEBS Lett., 199, 23–26
4   Telfer, A., Barber, P., Heathcote, P., Evans, M.C.W. (1978) BBA, 504, 153–164
5   Tripathy, B.C., Draheim,J.E., Anderson, D.P., Gross, E.L. (1984) Arch. Biochem.
    Biophys. 235, 449–360
6   Mullet, J.E., Burke, J.J., Arntzen, J. (1980) Plant Physiology 65, 814–822

ANALYSIS OF PIGMENT SYSTEM I CHL a FLUORESCENCE AT ROOM TEMPERATURE BY THE STEADY STATE SPECTRUM AND THE TIME RESOLVED-SPECTRUM IN PICOSECOND TIME RANGE

Mamoru MIMURO, Iwao YAMAZAKI[*], Naoto TAMAI[*], Tomoko YAMAZAKI[*] and Yoshihiko FUJITA
National Institute for Basic Biology, Myodaiji, Okazaki, Aichi 444, Japan and [*]Institute for Molecular Science, Myodaiji, Okazaki, Aichi 444, Japan.

Introduction
    In the analysis of the energy transfer in photosynthetic pigment systems, fluorescence has been used as an essential index. Photosystem I (PS I) chl a has been known to show a temperature-dependent fluorescence yield; at room temperature, it is low, whereas at low temperature, it becomes high. Due to this character, the analysis of the fluorescence from PS I chl a has been done at very low, thus non-physiological tempe-rature. However, the correlation between the fluorescence components observed at physiological and non-physiological temperatures has not yet been determined. The effect of detergent used for isolation of PS I particles on the chl a forms has posed another problem [1]. The present study was undertaken to make the fluorescence analysis of PS I chl a excluding the problems described above. We measured the fluorescence spectrum of PS I chl a at room temperature by the two methods with PS I particles prepared without detergent. One is the measurement of fluo-rescence spectrum at steady state and the other, the time-resolved fluo-rescence spectrum in picosecond (ps) time range. Both measurements gave essentially identical results on the composition of the fluorescence band.

Materials and method
    Spinach chloroplasts were isolated by using 50 mM tricine-NaOH (pH 7.5) containing 0.33 M sucrose, 2 mM $MgCl_2$ and 10 mM NaCl. Chloro-plast subparticles were isolated by the combination of mechanical breakage and differential centrifugation (cf. 2). The purity of PS I particles was more than 90 %, based on the fluorescence spectrum at -196°C.
    Fluorescence spectrum was measured with a Hitachi 850 spectro-fluorometer. The spectral sensitivity of the apparatus was numerically corrected by a microcomputer (an HP 216). Time-resolved fluorescence spectrum was measured with the apparatus described earlier [3, 4]. The excitation pulse was obtained from the Ar^+-pumped dye laser (630 nm, pulse width 6 ps (FWHM), and the intensity, 10^8 to 10^9 photons/cm^2). The fluorescence was monitored by a time-correlated single photon counting system. The spectral sensitivity was not corrected.

Results and discussion
    At steady state, the fluorescence maximum of the PS I particles was observed at 684 nm (Fig. 1a), whereas that of chloroplasts or PS II parti-cles was located at 683 nm (Fig. 1d). The bandwidth of the 684 nm fluo-rescence was wider than that of the 683 nm component, indicating a pre-sence of an additional band.

    A new fluorescence band in PS I chl a was detected by three methods.

Biggens, J. (ed.), Progress in Photosynthesis Research, Vol. I. ISBN 90 247 3450 9
© *1987 Martinus Nijhoff Publishers, Dordrecht. Printed in the Netherlands.*

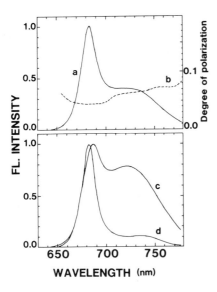

Fig. 1. Fluorescence properties of PS I particles at steady state. Fluorescence (a) and fluorescence polarization (b) spectra. (c) Calculated spectrum of PS I by subtraction of PS II spectrum (d). Excitation, 435 nm.

(1) The second derivative spectrum indicates presence of the band at 688 nm, in addition to the main 683 nm fluorescence (data not shown). (2) The fluorescence polarization spectrum (Fig. 1b) was not monotonous, suggesting presence of plural components. The lowest degree of polarization was observed around 690 nm, and it increased sharply around 705 nm and gradually increased toward longer wavelength. This spectrum indicates presence of at least three fluorescence bands; around 690, 700 and longer than 720 nm. (3) The intensity of the 688 nm fluorescence at -196°C specifically increased when the acceptor(s) of reaction center I (RC I) was reduced by sodium dithionite (data not shown). These results clearly indicate an occurrence of the 688 nm fluorescence emitted from the intrinsic component of PS I chl a.

The fluorescence spectrum of PS I particles was contaminated by the fluorescence from PS II. The subtraction of PS II fluorescence from the spectrum of PS I particles gives the intrinsic spectrum of the fluorescence from PS I chl a. Subtraction was made to give the fluorescence maximum at 688 nm. The spectrum thus obtained is shown in Fig. 1c. Besides the 688 nm maximum, the second maximum was observed around 722 nm. Since the intensity at 688 nm varies depending on the redox state of RC I, it seems to be closely related to RC I. The fluorescence yield of the PS I chl a is about 6 % of that of PS II chl a.

The time-resolved fluorescence spectrum in ps time range (Fig. 2) clearly showed the presence of plural components in PS I chl a. When the PS I particles were excited by a 630 nm light, the fluorescence maximum was observed at 688 nm in a time range close to the excitation pulse, and the maximum shifted with time to a shorter wavelength. About 60 ps after flash, the maximum was observed at 683 nm, the same wavelength for the maximum of the PS II chl a fluorescence. An additional fluorescence band was observed around 700 nm throughout the time. Since this component was not observed with PS II particles (data not shown), it is assigned as the fluorescence band specific to PS I chl a, which probably corresponds to the component observed by the polarization spectrum (cf. Fig. 1b). When we compared the spectrum with the spectrum around 1 ns after laser flash (shown by dotted lines in Fig. 2), the band 700 nm was clear. The difference spectrum gives the maximum at 703 nm and a clear shoulder around 720 nm, which might correspond to the maximum observed in the steady state spectrum (cf. Fig. 1c).

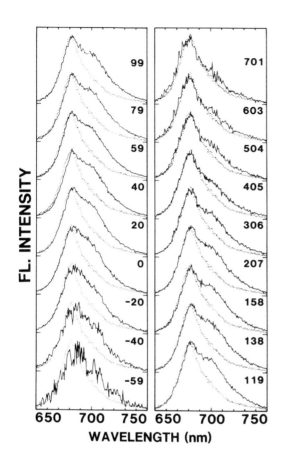

FL. INTENSITY

99
79
59
40
20
0
-20
-40
-59

701
603
504
405
306
207
158
138
119

650 700 750 650 700 750

WAVELENGTH (nm)

Fig. 2. Time-resolved fluorescence spectra of PS I particles. Figures indicate the times in ps after excitation pulse. Dotted lines show the spectta at about 1 ns after excitation pulse.

The decay kinetics of the fluorescence from PS I particles were monitored at various wavelengths (Fig. 3). A fast decay component was clearly observed, and the fraction of this component increased at longer wavelength. At longer than 700 nm, the decay kinetics were almost identical irrespective of the monitoring wavelength (Fig. 3d and 3e). When compared with the decay kinetics in PS II particles monitored at 683 nm (Fig. 3a), the slowest decay component observed in PS I particles has almost the same lifetime as in PS II particles. This slow component, therefore, most probably originates from PS II particles contaminating in PS I particles. The fast decay component is characteristics of PS I, indicating a low fluorescence yield of PS I chl a. As clearly shown in Fig. 3, there are several decay components in PS I fluorescence. The fastest decay component has a lifetime about 40 ps in the case of four component convolution.

The presence of the fast decay component around 690 nm strongly indicates that the fluorescence band at 688 nm of PS I chl a is not an artificial form. In the time-resolved fluorescence spectrum of PS I particles at -196°C, the 688 nm component was observed in the early time range (data not shown). Therefore, the 688 nm component is most probably

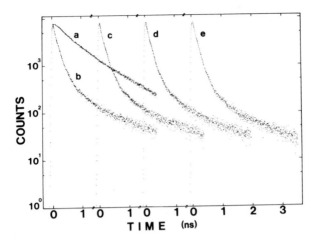

Fig. 3. Decay kinetics of PS I particles at various wavelengths. (b) At 683 nm, (c) 690 nm, (d) 700 nm and (e) 720 nm. (a) The decay kinetics of PS II particles at 683 nm.

the intrinsic fluorescence band of PS I chl a. Our data are not consistent with the proposal by Nechushtai et al. [1], that is, the fluorescence band around 680 to 690 nm at -196°C is artifact.

The degree of polarization at longer than 700 nm was higher than that around 690 nm (Fig. 1b). This suggests that the chl a for the 688 nm component is not the energy donor for chl a's for the bands located at the wavelengths longer than 700 nm. The absorption cross section of the components located at longer than 700 nm might be small. Thus, the antenna pigments might be present, which can transfer the energy both to the 688 nm component and to the components longer than 700 nm.

Summary
    The fluorescence spectrum of PS I chl a at room temperature was analyzed by the measurement of the steady state fluorescence and by the time-resolved fluorescence spectrum. At least three fluorescence components were detected by both methods at 688, 703 and around 720 nm. The 688 nm component is closely related to the RC I. The energy flow among these components is not necessarily straightforward to the components of lower energy levels.

References
1. Nechushtai, R., Nourizadeh, S.D., and Thornber, J.P. (1986) Biochim. Biophys. Acta, 848, 193-200.
2. Sane, P.V., Goodchild, D.J. and Park, R.B. (1970) Biochim. Biophys. Acta, 216, 162-178.
3. Yamazaki, I., Mimuro, M., Murao, T., Yamazaki, T., Yoshihara, K., and Fujita, Y. (1984) Photochem. Photobiol., 39, 233-240.
4. Mimuro, M., Yamazaki, I., Tamai, N., Yamazaki, T. and Fujita, Y. (1985) Photochem. Photobiol., 41, 597-603.

SPECTRAL SHIFTS IN PICOSECOND TRANSIENT ABSORPTION SPECTRA DUE TO
STIMULATED EMISSION FROM CHLOROPHYLL *IN VITRO* AND IN PROTEIN COMPLEXES.

D.R. Klug, B.L. Gore, L.B. Giorgi and G. Porter
Davy Faraday Research Laboratory, The Royal Institution, 21 Albemarle
Street, London W1X 4BS, U.K.

1. INTRODUCTION

Picosecond transient absorption spectroscopy has had significant success
in investigating the widely varying chromophore-protein complexes of bacteria and algae. The potential presence of stimulated emission in such data
has tended to be overlooked even though it can be a major spectral feature.
Here we demonstrate the presence and usefulness of stimulated emission in
data from chlorophyll-b *in vitro*, higher plant light harvesting complex and
Photosystem 1 (PS1) reaction centres.

2. MATERIALS AND METHODS

Chlorophyll-b was purchased from Sigma Chemical Co. The chlorophyll-b concentration used was $\sim 10^{-5}$M in a solvent consisting of ether plus 1% pyridine.
Light-harvesting chlorophyll-protein complex (LHC2) was extracted from pea
chloroplasts (1) and resolubilised according to previously determined conditions (2). LHC2 samples contained 10µg chlorophyll/ml. PS1 reaction centres were isolated from pea chloroplasts (see ref.3) with a chlorophyll/
P700 ratio of 50.

The dual beam picosecond transient absorption spectrometer is of a common
design. The 600nm excitation pulses are generated by a synchronously pumped
dye laser and passed through a series of four dye amplifiers pumped by the
530nm second harmonic of a 10Hz Q-switched Nd:YAG laser. The resulting
pulses contain 2.5mJ and produce an autocorrelation trace of 10ps (full
width half maximum). These pulses are then split, with 10% of the pulse
used as a pump after passing down a variable delay line, the remainder
being used to generate a white light continuum in a cell containing a
mixture of D_2O and H_2O. The spectrometer has a reference arm by means of
which the continuum intensity is monitored, and this data is simultaneously
collected along with the signal from the sample by a dual track vidicon
system. Samples were continuously stirred in a 10mm pathlength cell and
maintained at $\sim 20^{\circ}$C.

3. RESULTS AND DISCUSSION

It is straightforward to show that the optical density difference due to an
excited singlet state population can be written:

$$\Delta A(\nu) = h\nu N_A l C_1 (B_{12} - B_{10} - B_{01})/2.303 \cdot 4000 \qquad (1)$$

where N_A is the Avogadro number, C_1 is an excited state (S_1) concentration
(Molar) and B_{ij} is an Einstein coefficient for the photon stimulated process

Biggens, J. (ed.), Progress in Photosynthesis Research, Vol. I. ISBN 90 247 3450 9
© *1987 Martinus Nijhoff Publishers, Dordrecht. Printed in the Netherlands.*

$S_i \rightarrow S_j$. Molecules (such as chlorophyll) which show a high degree of mirror symmetry between their emission and concomitant absorption bands, must have similar vibrational wavefunctions in their ground and excited-electronic-vibrationally-relaxed states; and as such will be expected to have similar B-coefficients as shown below:

$$B_{0'1} = K \left| <\phi_{0'} | \underline{\mu}_{0'1} | \phi_1> \right|^2 \qquad (2)$$

$$B_{1'0} = K \left| <\phi_{1'} | \underline{\mu}_{1'0} | \phi_0> \right|^2 \qquad (3)$$

$$\text{if } \phi_{1'} \overset{\sim}{\sim} \phi_{0'} \text{ then } B_{01} \overset{\sim}{\sim} B_{1'0} \qquad (4)$$

where K is a constant, ϕ_i is a vibrational wavefunction, $\underline{\mu}_{ij}$ is a transition dipole moment and the prime signifies a vibrationally relaxed state. From (1) and (4) we would expect the strength of the stimulated emission pseudo-bleach to equal that of the ground state bleach itself, however, the red-shift of the emission centroid, combined with the reflection of any absorption sidebands (if they emit) distinguishes them. Absence of stimulated emission is thus a simple diagnostic for non-radiatively coupled states such as the T_1 state of an S_0 ground state molecule which can be formed by processes such as intersystem crossing.

Here we present data from four quite different chlorophyll systems in which stimulated emission is a useful interpretive phenomenon (figures 1-4).

Fig.1 shows transient spectra of chlorophyll-b in ether plus 1% pyridine. The two spectra correspond to an entirely singlet populated excited state (10ps) and a largely triplet populated excited state (12ns). The shift of the bleach centroid of 3nm is half the Stokes shift of 6nm usually observed which is what would be expected when bleach and pseudo-bleach centroids are not resolved. (N.B. the similarity in relative magnitudes of the two spectra shown are due to a difference in pumping intensity).

Fig.2 shows transient spectra of solubilized LHC2 trimers.

Fig.3 shows transient spectra of PS1 reaction centres with P700 chemically oxidized (3). Under global analysis these data yield a 15ps lifetime with a background consisting of the triplet (270ps) spectrum. The triplet bleach disappears in 1-2ns, presumably due to quenching by carotenoid(s). It can be seen that the triplet yield of these particles is high 30-50%.

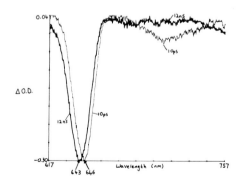

Fig.1 Transient spectra of chlorophyll-b in ether plus 1% pyridine.

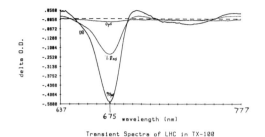

Transient Spectra of LHC in TX-100

Fig.2 Transient spectra of
solubilized LHC2 trimers.

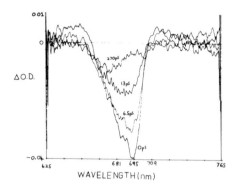

Fig.3 Transient spectra of
PS1 reaction centres with
P700 chemically oxidized.

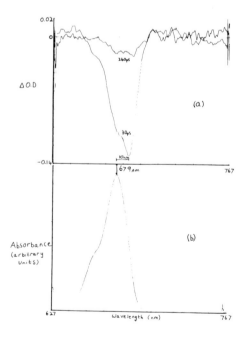

Fig.4(a) Transient spectra of
LHC2 aggregates.
Fig.4(b) Ground state absorption spectrum of LHC2 aggregates
taken on the picosecond
spectrometer

Fig.4(a) shows transient spectra of LHC2 aggregates. LHC2 aggregates form quenching traps *in vitro* (2) and thus shorten the excited state lifetime. The aggregation redshift is seen to be greater for emission than for absorption, hence the separation of the negative centroid into bleach (shoulder) and pseudo-bleach (emission peak). It can be seen from this data that the Einstein B-coefficient for emission is actually ∿10% greater in strength than that for absorption in this case. Fig.4(b) shows the ground state absorption spectrum of LHC2 aggregates taken on the picosecond spectrometer. This serves to highlight the presence and position of stimulated emission.

4. ADDENDUM

Since these experiments were performed, stimulated emission has been used by three separate groups to help determine the kinetics of charge separation in the bacteriochlorophyll special pair of reaction centres from R.Viridis and R. Sphaeroides (4-6).

ACKNOWLEDGEMENTS

We would like to thank the SERC and British Petroleum plc for financial support. The LHC2 was prepared by W. Kuhlbrandt at the Department of Pure and Applied Biology, Imperial College, London.

REFERENCES

1 Kuhlbrandt, W. et al. (1983) J. Cell Biol. 96, 1414-1424
2 Ide, J.P. et al. (1986) Proceedings of Vllth International Congress on Photosynthesis
3 Giorgi, L.B. et al. (1986) Proceedings of Vllth International Congress on Photosynthesis
4 Martin, J.L. et al. (1986) Proc. Natl. Acad. Sci. USA 83, 957-961
5 Woodbury, N.W. et al. (1985) Biochemistry 24, 7516-7521
6 Shuvalov, V.A. and Duysens, L.N.M. (1986) Proc. Natl. Acad. Sci. USA 83, 1690-1694

FAST FLUORESCENCE AND ABSORPTION MEASUREMENTS OF PHOTOSYSTEM 1 FROM A
CYANOBACTERIUM

E. Hilary, Evans, Raymond Sparrow, Robert G. Brown*
Schools of Applied Biology and Chemistry*, Lancashire Polytechnic,
Preston, Lancashire, U.K.,
David Shaw
SERC Daresbury Laboratory, Warrington, Cheshire, U.K.
John Barr, Martin Smith and William Toner
SERC Laser Facility, Rutherford-Appleton Laboratory, Didcot, Oxon,
U.K.

1. INTRODUCTION
Recently, attempts have been made to probe the energy transfer and
electron transfer of chlorophyll on Photosystem 1 (PSI) by using fast
absorbance changes on a picosecond time scale (1-4). Il'ina et al (2)
identified a reversible photobleaching at 704 nm under continuous
saturating light conditions in the presence of TMPD which they
attributed to a light harvesting chlorophyll a. Niujs et al, using a
laser of 35 ps. half-band width, saw no re-reduction of P700 under
conditions where the iron sulphur acceptors were oxidised. More
recently Giorgi et al (3,4) have measured time resolved spectra of
PSI excited by a 10 ps. pulse which indicated two absorbing
components, one at 690 nm and the other at 700 nm, bleaching of the
former occurring before the latter under conditions where P700 was
reduced before the flash. All these experiments have been performed
using PSI preparations from lettuce chloroplasts, and no fluorescence
data measured over the same time-scale for the same samples is
available. In this paper we present fast absorption and fluorescence
measurements made on a PSI preparation from the cyanobacterium
Chlorogloea fritschii.

2. MATERIALS AND METHODS
C. fritschii was maintained, grown and PSI preparations extracted as
previously described (5). Time resolved fluorescence spectroscopy was
performed using the single photon counting technique with the
synchroton Radiation Source at Daresbury as the excitation source, as
previously described (6). Absorption changes were measured following
a 4 ps. pump pulse at 585 nm produced by a mode locked Nd:YAG laser
synchronously pumping a dye laser and amplified by a pulsed dye
amplifier pumped by a Q-switched frequency doubled Nd:YAG laser. A
probe pulse was generated by splitting the 585 nm pulse and
generating a continuum which was monochromated using interference
filters. Absorption changes were detected as previously described
(7).

3. RESULTS AND DISCUSSION
Figure 1 shows a typical fluorescence decay profile exhibited by C.
fritschii PSI preparations excited at 430 nm and measured at 700 nm.
All the fluorescence decay profiles measured required three

Biggens, J. (ed.), Progress in Photosynthesis Research, Vol. I. ISBN 90 247 3450 9
© *1987 Martinus Nijhoff Publishers, Dordrecht. Printed in the Netherlands.*

FIGURE 1. FLUORESCENCE DECAY PROFILE OF PSI PARTICLES
EXCITED AT 430 NM AND MEASURED AT 700 NM
PREPARATION SUSPENDED TO 50 µG CM^{-3} IN
0.06 M TRIS HCL, 0.03 M EDTA PH 7.8

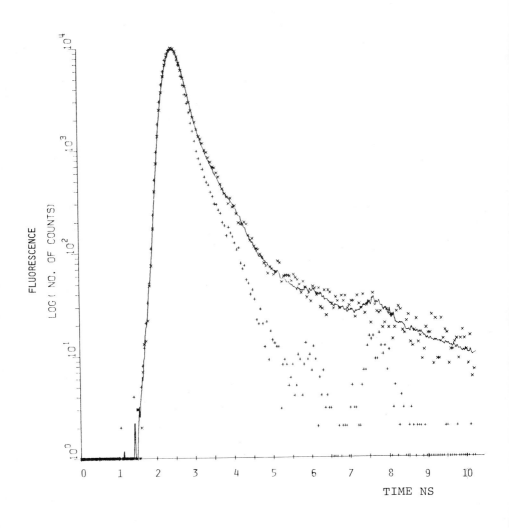

FIGURE 2. TRANSIENT ABSORBANCE CHANGE OF PSI PREPARATIONS
PREPARATION SUSPENDED TO 26 µG CM^{-3} IN 0.06 M TRIS HCL 0.03 M EDTA PH 7.8

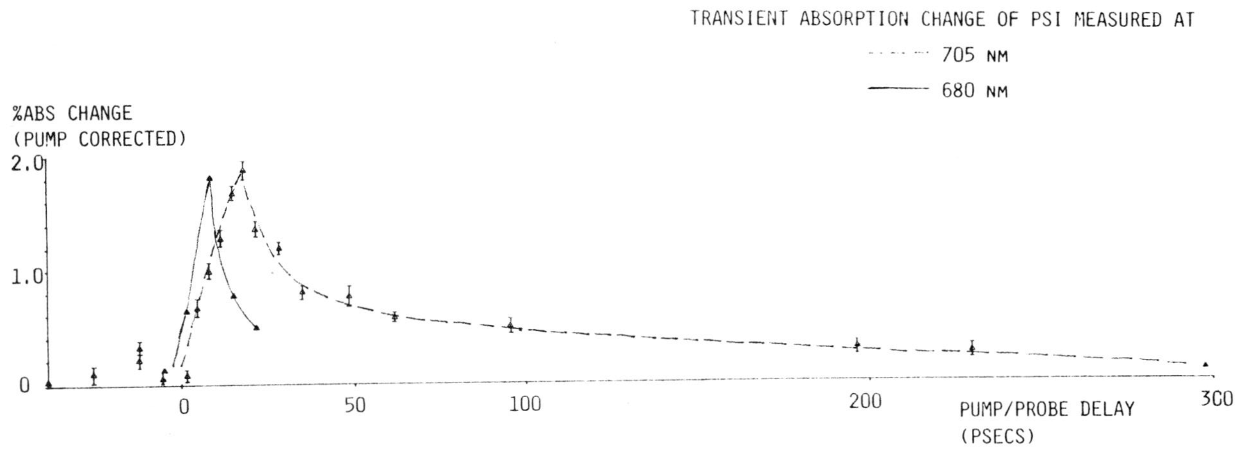

exponential components to achieve an acceptable fit ($\chi^2 < 1.2$ and a random distribution of residuals). The three decay components thus obtained had τ values of τ_1 16-33 ps, τ_2 400-600 ps. and τ_3 3-4 μs. The largest component is τ_1, typically 60-80% of the total decay, τ_2 is constant at about 12% and τ_3 between 5-30%. The variability of τ_1 and τ_3 depend on the redox state of the reaction centre, τ_1 decreasing as a proportion of the total and τ_3 increasing in the presence of either sodium dithionite or sodium ascorbate/DCPIP as reductants. This variation is wavelength dependant, the greatest effect being at 680 nm, with a negligible effect at 720 nm. The peak of fluorescence intensity and of the fast component (τ_1) is at 710 nm. The peak of the τ_3 contribution is at 680 nm. These preparations have a low steady state fluorescence emission, with maximum emission at 710 nm, unlike preparations from chloroplasts (8).

Figure 2 shows the kinetics of the absorption change exhibited by the same preparations following a laser pulse. Measurements are presented at 680 nm and 705 nm. The bleaching at 680 nm is very rapid and is reversible within the time scale that the bleaching at 705 nm occurred. The bleaching at 705 nm is also reversible in these particles in the absence of reductant, unlike lettuce chloroplast PSI (2,4). The rise of the bleaching at 705 nm is complete by 20 ps., but we cannot yet resolve the kinetics of the rise. The 680 nm bleaching is reversed by 50% at 15 ps. The spectrum of the change shows peaks at 690 nm and at 705 nm. These results are in general agreement with those of Giorgi et al (4), and could be explained as a bleaching due to the excitation of light harvesting chlorophyll at 680 nm (but possibly with maximal absorbance at 690 nm) which transfers the energy to a component absorbing at 705 nm (P700) within 20 ps. It could be proposed that the fast fluorescence decay component (τ_1 16-33 ps) is associated with this light harvesting chlorophyll. The recovery of this bleaching at 705 nm cannot, however, be associated with any fluorescence decay component and must therefore be assigned to a non-radiative deactivation process or redox reactions of P700 and/or the proposed Chla acceptor (9).

REFERENCES
1 Nuijs, A.M., Vasmel, H., Duysens, L.N.M. and Amesz, S. (1986) Biochim. Biophys. Acta, 848, 167-175
2 Il´ina, M.D., Krasauskas, V.V., Rotomskis, R.R. and Borison, Y.Y. (1984) Biochim. Biophys. Acta, 767, 501-506
3 Giorgi, L.B., Gore, B.L., Klug, D.R., Ide J.P., Barber, J. and Porter, G. (1986) Biochem. Soc. Trans. 14, 47-48
4 Giorgi, L.B., Gore, B.L., Klug, D.R., Ide, J.P., Barber, J. and Porter, G., J. Chem. Soc. Faraday Trans., in press
5 Pullin, C.A. and Evans, E.H. (1981) Biochem. J. 196, 489-493.
6 Sparrow, R., Brown, R.G., Evans, E.H. and Shaw, D., J. Chem. Soc. Faraday Trans., in press
7 Smith, M.J.C. and Toner, W.T. (1986) SERC Central Laser Facility, Annual Report to the Laser Facility Committee 1986, Rutherford Appleton Laboratory, Didcot, Oxon., U.K., p. B4.7-B4.11
8 Pullin, C.A., Brown, R.G. and Evans, E.H. (1979) FEBS Letts. 101, 110-112
9 Bonnerjee, J. and Evans, M.C.W. (1982) FEBS Letts. 148, 313-316

ANOMALOUS FLUORESCENCE INDUCTION ON SUBNANOSECOND TIME SCALES AND EXCITON-EXCITON ANNIHILATIONS IN PSII

A.DOBEK*°, J.DEPPEZ*[+], M.E.GEACINTOV[#] and J.BRETON*
*Service de Biophysique, Département de Biologie, Centre d'Etudes Nucléaires de Saclay, 91191 Gif-sur-Yvette Cedex, France
+Université Paris Sud, I.U.T., 94230 Cachan, France
#Chemistry Department, New York University, New York, NY 10003, U.S.A.
°On leave from: Quantum Electronics Laboratory, Institute of Physics, A. Mickiewicz University, Poznań, Poland.

INTRODUCTION.

The photosystem II reaction center consists essentially of the Chlorophyll a donor molecule(s) (P680), situated close to a pheophytin (Phe) molecule, which acts as an intermediate acceptor of charge. The negative charge is subsequently transfered to a quinone molecule (Q) on which it is stabilized. The time required for the stabilization of the charge on Q is unknown (but is probably of the order of 200-300 ps [1]). An unidentified donor D reduces P680$^+$ on the time scale of ns [2,3], while the oxidation of Q$^-$ by a secondary acceptor occurs on time scales of some hundreds of μs [4].

When the reaction centers are open (state P680 Phe Q), the fluorescence yield of the antenna-reaction center system is low. When the reaction centers are closed (state P680 Phe Q$^-$) either by illumination or by chemical reduction of Q, their is an increase by a factor 3-4 in the fluorescence yield. This classic fluorescence induction process take place in the tens of nanoseconds range [3].

We have utilized either single, or double pulses to study the dependence, on the subnanosecond time scale, of the fluorescence yield on the energy of the laser pulse and on the state of the reaction centers. In this method, the first laser pulse P1 is followed by a second pulse P2 a time interval Δt later. In this manner, it is shown that the evolution of the state of the reaction centers during the initial stages of primary photochemistry can be studied with a resolution of approximately 25 ps [3,5].

MATERIALS and METHODS.

Room temperature and 77 K PSII fluorescence measurements were carried out at 685 nm on suspensions of spinach chloroplasts prepared according to methods outlined earlier [3] and using an apparatus previously described [5,6]. The 77 K PSI fluorescence was detected at 735 nm. A Neodynium YAG mode-locked laser provided 30 ps excitation pulses at 530 nm at a rate of 5 Hz. The laser output was divided into two pulses P1 and P2 spaced apart in time in the range of 0-5 ns. The fluorescence signal was measured on dark adapted chloroplasts or in the presence of a saturating He-Ne background illumination. With Δt values lower than 5 ns, it was not possible, as in classical pump-probe fluorescence measurements [3], to discriminate the fluorescence produced by the two pulses P1 and P2. Then, three different types of measurements were performed : (i) the fluorescence yield F1 generated by P1 alone , (ii) the fluorescence yield F2 generated by P2 alone and (iii) the overall fluorescence yield F1+2 created by both P1 and P2 pulses.

Biggens, J. (ed.), Progress in Photosynthesis Research, Vol. I. ISBN 90 247 3450 9
© *1987 Martinus Nijhoff Publishers, Dordrecht. Printed in the Netherlands.*

The energy E of the two pulses is ajusted to be the same (within ± 10%) and F1 = F2 = F. The ratio:

$$\Delta F = (\; F1+2 \; - \; \mathbf{F}) \; / \quad \mathbf{F}$$

is a measure of the degree of correlation of the effects generated by the two pulses P1 and P2.

RESULTS and DISCUSSION.

Fluorescence yields measured using single pulses.

Both the yields F and FHeNe obtained in the presence of open and closed reaction centers (figure 1) decrease at the higher pulse intensities. This effect is attributed to exciton-exciton annihilations [3,7]. In the case of open reaction centers, there is an increase in F at relatively low pulse intensities. A similar result has been reported by Hirsh et al. [8]. These results suggest that a new type of fluorescence induction phenomenon may be responsible for the rise in F observed in the range of laser pulse energies corresponding to 5×10^{12} to 5×10^{13} photons.cm^{-2} pulse^{-1}. This rise cannot be attributed to the usual induction mechanism because only a negligible fraction of the reaction centers are transformed to the state P680 Phe Q$^-$ during the lifetime of the excitons [3,9].

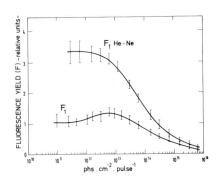

FIGURE 1: Dependence of the fluorescence yields F (open RC) and FHeNe (closed RC) on the intensity E of a single picosecond excitation pulse.

PSII fluorescence yields measured with the double pulse technique.

The values of ΔFHeNe measured experimentally are equal to zero when the energy of the two pulses is lower than 10^{12} photons.cm^{-2}, regardless of the value of Δt. However, for the case of open reaction centers, the values of ΔF are positive below energies of $\approx 10^{13}$ photons.cm^{-2}, and the magnitudes depend on Δt. Variations of ΔF as a function of Δt obtained at fixed energies of the two pulses ($\approx 10^{12}$ photons.cm^{-2}) are shown in figure 2a. An attempt to fit the experimental curve suggests that the 1/e risetime and decaytime of the correlation between the P1 and P2 pulses have values of **60±20 ps** and **150±50 ps** respectively. These effects are unambiguously related to the existence of open reaction centers, since analogous rise and decay times are not observed when the PS II reaction centers are closed.

The values of ΔF and ΔFHeNe, measured under conditions corresponding to energies of the two pulses equal to $\approx 10^{13}$ photons.cm^{-2} or more, are negative regardless of the values of Δt. Typical values of these two quantities measured as a function of Δt at relatively high excitation energies E are shown in figures 2b and 3. Assuming an exponential decay of the correlation between the two pulses, characteristic decay times of **150±50 ps** and **400±100 ps** can be estimated for the case of open and closed reaction centers respectively. At high energies, ΔF and ΔFHeNe represent the annihilations caused by singlet excitons produced by P1 and which survive for time intervals Δt.

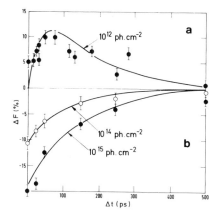

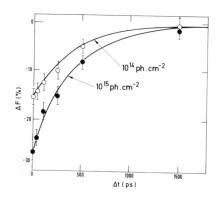

FIGURE 2: Amplitude of the fluorescence yield change ΔF as a function of the delay time between the two pulses P1 and P2 for **dark-adapted** chloroplasts at three fixed energies. Continuous curve in (a) is calculated from ΔF=K[exp(-Δt/150)-exp(-Δt/60)]. Continous curves in (b) indicate a 150 ps decay time constant.

FIGURE 3: Amplitude of the fluorescence yield change ΔFHeNe as a function of the delay time between the two pulses P1 and P2 for **light-adapted** chloroplasts at two fixed energies. Continous curves indicate a 400 ps decay time constant.

Anomalous fluorescence induction phenomena: fluorescence properties of intermediate states of PSII reaction centers.

In the case of open reaction centers subjected to low energies of excitation ($< 10^{13}$ photons.cm^{-2}), our results can be interpreted in terms of the appearance of an intermediate state, characterized by a lower fluorescence quenching efficiency than that of the open state. We propose that the risetime of this state corresponds to the rise in ΔF observed in figure 2 and is $\approx 60 \pm 20$ ps. We further suggest that this new state of the reaction center corresponds to P680$^+$Phe$^-$Q. If we assume that the reaction P680* ---> P680$^+$ is rapid (< 10 ps), the ≈ 60 ps risetime could correspond to the exciton trapping time (antenna* + P680 --> P680* + antenna). The 150 ± 50 ps decay phase could correspond to the charge stabilization step P680$^+$Phe$^-$Q --> P680$^+$PheQ$^-$.

In the case of closed reaction centers, excitons created by pulse P1 generate the state P680$^+$Phe$^-$Q$^-$. The charge recombination mechanism in the P680$^+$Phe$^-$ radical pair [10] provides a source of excitons which are capable of interacting with excitons generated by pulse P2 as long as Δt is shorter than the lifetime of this radical pair state (400 ± 100 ps). An analogous mechanism in open reaction centers can also be invoked to explain the 150 ± 50 ps decay phase [11].

Excitation annihilations effect in PSI and PSII at 77K.

The values of ΔF measured at 735 and 685 nm at 77K are equal to zero, regardless of the value of Δt, for energies of the two pulses lower than 10^{12} photons.cm^{-2}. Only negative values of ΔF are detected for both PSI and PSII at higher energies.

For each values of Δt, the absolute value of ΔF is higher for 735 nm than for 685 nm, corresponding to a higher singlet-singlet quenching efficiency in the PSI than in the PSII.

An identical heterogeneity was observed in [12] concerning singlet-triplet annihilations after μs pulse excitation, but was not observed in the ps excitation range.

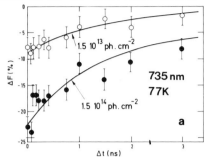 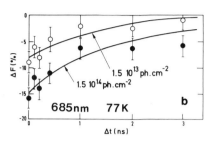

FIGURE 4: Low temperature fluorescence yield change ΔF in PSI (a) and PSII (b) as a function of the delay time between the two pulses. Continuous curves indicate 2.0 ns and 1.5 ns decay time constants respectively.

At 735 and 685 nm, the time constants corresponding to a monoexponential approximation of ΔF are respectively equal to **2±0.5 ns** and **1.5±0.5 ns**. These values are similar in magnitude to the mean decay time constants of the fluorescence measured at 77K in PSI and PSII [13, 14].

ACKNOWLEDGEMENTS
This work was in part supported by the National Science Foundation, Grant PCM 83-08190, to N.E.G.

REFERENCES
[1] Nuijs A.M., van Gorkom, H.J., Plijter, J.J. and Duysens, L.N.M. (1986), Biochim, Biophys. Acta, 848, 167-175.
[2] Van Best,J.A. and Mathis,P. (1978) Biochim. Biophys. Acta 503,178-188.
[3] Deprez, J., Dobek, A., Geacintov, N.E., Paillotin, G. and Breton, J. (1983), Biochim. Biophys. Acta 725,444-454 .
[4] Mathis, P. and Paillotin, G. (1981) in The Biochemistry of plants (Hatch,M.D. and Boardman,N.K. Eds.),Vol.8,pp.97-161,Acad.Press, New York.
[5] Dobek,A., Deprez,J., Geacintov,N.E., Paillotin, G. and Breton, J. (1985), Biochim. Biophys. Acta 806, 81-92.
[6] Deprez,J., Dobek,A., Geacintov,N.E., Paillotin,G. and Breton,J. (1985),Proceedings of the 5th Int. Sem. on Energy Transfer in Condensed Matter (Pancoska,P. and Pantoflicek,J. Eds.), Prague, 204-210.
[7] Breton, J. and Geacintov,N.E. (1980) Biochim. Biophys. Acta 594,1-32.
[8] Hirsh, I., Neef, E. and Fink, F. (1982) Biochim. Biophys. Acta 681,15-20.
[9] Geacintov,N.E., Paillotin,G., Deprez,J., Dobek,A. and Breton,J. (1984) in: Advances in Photosynthesis Resarch (Sybesma, C. Ed.), Vol.1, pp 37-40, Martinus Nijhoff/Dr.W.Junk publishers, The Hague.
[10] Klimov, V.V. , Klevanik, A.V. , Shuvalov, V.A. , and Krasnovski, A.A. (1977), FEBS Lett. 82, 183-186.
[11] Breton, J. (1983) FEBS Lett. 159,1-5.
[12] Geacintov, N.E., Breton, J., Swenberg, C., Campillo, A.J., Hyer, R.C., and Shapiro, S.L. (1977), Biochim. Biophys. Acta 461, 306-312.
[13] Reisberg, P., Nairn, J.A. and Sauer, K. (1982), Photochem. Photobiol. 36, 657-661.
[14] Moya, I. and Garcia, R. (1983), Biochim. Biophys. Acta 722, 480-491.

LASER FLASH–INDUCED NON–SIGMOIDAL FLUORESCENCE INDUCTION CURVES IN CHLO-ROPLASTS

NICHOLAS E. GEACINTOV, JACQUES BRETON[*], LEE FRANCE, JEAN DEPREZ[*+], and ANDRZEJ DOBEK[*++], CHEMISTRY DEPARTMENT, NEW YORK UNIVERSITY, NEW YORK, NY 10003, [*]C.E.N. de SACLAY, 91191 GIF sur YVETTE, FRANCE, [+]I.U.T. CACHAN, UNIVERSITY of PARIS–SUD, CACHAN, FRANCE, and [++]INSTITUTE of PHYSICS, A. MICKIEWICZ UNIVERSITY, 60780 POZNAN, POLAND.

INTRODUCTION

The fluorescence yield of Photosystem II (PS II) in green plants depends on the photochemical state of the PS II reaction centers [1]. If P680 is denoted as the primary donor, I (pheophytin) the intermediate electron acceptor, and Q the primary electron acceptor, variations in the fluorescence yield can be interpreted in terms of the following scheme:

$$\text{Exciton} + \text{Z-P680-I-Q} \rightleftharpoons \text{Z-P680}^+\text{-I}^-\text{-Q} \longrightarrow \text{Z}^+\text{-P680-I-Q}^- \qquad (1)$$

where Z is a primary electron donor which, in intact chloroplasts, reduces $P680^+$ with a rate constant of about 30 ns [2,3]. In the state Z-P680-I-Q the reaction centers are said to be in the open state and the fluorescence yield is denoted by F_o, while in the state Z^+-P680-I-Q$^-$ the reaction centers are said to be closed, while the fluorescence yield is denoted by F_{max}. The ratio F_{max}/F_o is usually in the range of 3–5. As the reaction centers are progressively closed by illumination, the fluorescence yield increases from a value of F_o to a value of F_{max}, and the resulting traces are called fluorescence induction curves.

CLASSICAL FLUORESCENCE INDUCTION CURVES

The fluorescence induction curves are normally measured under conditions of steady state illumination on time scales of 100 ms to about 5 s. The shapes of these induction curves have been a subject of considerable interest [1,4]. The variable fluorescence at a given time t, $F_v(t)$, is defined by $[F(t) - F_o]/[F_{max} - F_o]$; this quantity depends on q, the fraction of closed reaction centers, according to the following equation derived by Joliot and Joliot [5]:

$$F_v = \frac{(1 - p)q}{1 - pq} \qquad (2)$$

where p is the connectivity parameter, and is defined as the probability of transfer from a photosynthetic unit containing a closed reaction center to another one which contains an open reaction center. The fluorescence induction curves are sigmoidal in shape. However, under certain conditions, e.g. low temperatures, low magnesium ion concentration, and low redox potentials, the induction curves become less sigmoidal and more exponential in shape [reviewed in 1 and 4]. The sigmoidicity is explained in terms of the connectivity between photosynthetic units (PSU) [5], where a PSU is visualized as a PS II reaction center with its complement of chlorophyll antenna molecules.

Biggens, J. (ed.), Progress in Photosynthesis Research, Vol. I. ISBN 90 247 3450 9
© *1987 Martinus Nijhoff Publishers, Dordrecht. Printed in the Netherlands.*

If there is no connectivity between photosynthetic units, $p = 0$, $F_v = q$, and the shape of the fluorescence induction curves is exponential [1,4].

Generally, in the presence of herbicides which are used to prevent the secondary electron transfer in PS II reaction centers, the fluorescence induction curves at room temperature can be resolved into a fast, sigmoidal phase and a slower exponential one [6]. The biphasic nature of the induction curves is generally interpreted in terms of the PS II$_\alpha$ and PS II$_\beta$ heterogeneity of photosystem II. The fast sigmoidal phase is attributed to connected, relatively large PS II$_\alpha$ units in the stacked membranes in the granal regions of the chloroplasts. The slower exponential phase is attributed to the smaller PSII$_\beta$ units in non-appressed end margins and stromal lamellae, with no connectivity between units.

LASER FLASH INDUCED FLUORESCENCE INDUCTION CURVES

The double pulse method for measuring fluorescence induction was introduced by Mauzerall [3]. In this method, an actinic laser pulse $P_1(I_1)$ of variable energy I_1 closes a fraction q of the reaction centers, while a second, weak probe pulse P_2, which follows the actinic flash after a suitable time interval Δt, probes the fluorescence yield of the system. In selecting the time interval Δt between the two pulses, three considerations are important: (1) the state P680$^+$ is as good a fluorescence quencher as the state P680 (when Q is not yet reduced) [2]; thus, about 100 ns must be allowed for a full reduction of P680$^+$ by Z according to Eq.1. (2) The actinic pulse also creates triplet quenchers [2], and about 10 μs should be allowed for the complete decay of these quenchers. (3) In the absence of the herbicide DCMU, oxidation of Q$^-$ by the secondary electron acceptors becomes important. For these reasons, a value of $\Delta t = 30$ μs [7] to about 100 μs (this work) is found to be most suitable.

Classical induction curves are measured by turning on the illumination of constant fluence rate $I_c(t)$ (energy cm^{-2} s^{-1}) and monitoring the fluorescence yield as a function of time. After an illumination time t, the fluence (or total energy incident/unit area) is just equal to the integral $\int I_c(t)dt = I_c t$. In laser flash experiments, the fluence rate $I_1(t)$ is time-dependent; however, integration over the width of the pulse yields the fluence I_1 of the pulse P_1. If q is only a function of the number of absorbed photons, then the shapes of the F_v induction curves should be independent of the mode of excitation, i.e. steady-state or pulsed.

In the case of separated units, or when the exciton lifetime does not change when a PS II reaction center has already been visited by another exciton (e.g. at time intervals Δt between the two pulses of the order of 30 ns or less [2], q depends on I_1 in the following manner [2,4,7-9]:

$$q = (1 - \exp[- \sigma I_1]) \qquad [3]$$

where σ is the absorption cross section in units of cm^{-2}. On the other hand when the units are connected, and there is a change in lifetime when the excitons visit a reaction center which has previously captured an exciton, the dependence between q and I_1 is more complex and depends on the integration of the following equation:

$$-dq(t) = \frac{I_1(t)[\ 1 - q(t)\ ]}{1 - pq(t)} \, dt \qquad [4]$$

and q no longer depends exponentially on the integral $I_1 = \int I_1(t)dt$.

Even in the case of connected units, q depends on I_1 exponentially if the P_1 laser pulse width is of the order of 30 ns, or so. On these time scales, reduction of $P680^+$ has not yet occured, and thus Eq. 3 is applicable. If the actinic pulse width is >> 30 ns, the relationship between q and I_1 is expected to be non-exponential, unless exciton-exciton annihilation prevents the expected increase in the lifetimes of the excitations due to the closure of the PS II reaction centers; however, this appears to be a minor effect since exciton-exciton annihilation in the range of fluences relevant to the filling of PS II reaction centers does not appear to be important [8].

Regardless of the relationship between q and I_1, which depends on the properties of the actinic P_1 pulse, the relationship between the fluorescence yield F_2 and q, as determined by the probe flash P_2, is expected to be sigmoidal since the steady state experiments suggest that $F_v \neq q$.

We have utilized the two-pulse method to measure the shapes of the laser-flash induced fluorescence induction curves in suspensions of spinach chloroplasts [2] without herbicides. Under these conditions, the fluorescence induction curves measured under steady-state excitation with a He-Ne laser are strongly sigmoidal (data not shown). The actinic pulses were either (1) 530 nm, 30 ps wide Nd:Yag laser, (2) 337 nm, 200 ps wide N_2 laser, and (3) 650 nm, 700 ns flash lamp-pumped dye laser pulses. The probe pulse was usually a weak xenon flash incident on the same spot of the spinach chloroplast samples 50-100 μs after the actinic pulse. A typical induction curve utilizing 200 ps wide N_2 laser pulses is shown in the Figure. The fluorescence induction curves are exponen-

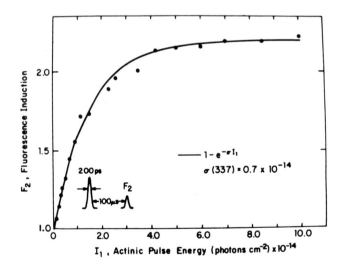

tial rather than sigmoidal, in contrast to the results of the steady-state ilumination experiments in which the induction curves are sigmoidal. Similar exponential induction curves are obtained when either the 30 ps or 700 ns pulse-width lasers are utilized [4,8].

CONCLUSIONS

The exponential shape of the laser flash-induced and xenon pulse-detected fluorescence induction curves indicates that (1) the relationship between q and I_1 is exponential, (2) that the variable fluorescence F_v measured with a 1 μs wide xenon flash is proportional to q, the fraction of closed PS II reaction centers, and that (3) the degree of sigmoidicity of fluorescence induction curves may depend on the time scales of illumination in these experiments. On longer time scales, a more complex set of photochemical states of PS II, each possibly characterized by somewhat different fluorescence yields, may play a role in determining the shapes of the classical, steady-state illumination induction curves.

Analogous results on laser flash-induced fluorescence induction curves have been recently published by Ley and Mauzerall [7,9], who concluded that the effective cross section of PS II reaction centers does not increase as the reaction centers are progressively closed (determined by monitoring both F_v and the oxygen flash yields).

Because the double-flash fluorescence induction curves are exponential in shape, the α, β heterogeneity of PS II [6] deduced from the steady-state classical sigmoidal induction curves, is not apparent in the laser flash induced fluorescence induction experiments. While there are other experiments which suggest that there is indeed connectivity between PSU's and a heterogeneity of the photosynthetic apparatus of green plants [4], it is by no means certain that such conclusions can be based on the shapes of fluorescence induction curves alone. The reasons for the differences in the shapes of the induction curves determined by the laser flash double-pulse techniques and classical steady-state illumination experiments should be investigated.

ACKNOWLEDGEMENTS

This work was supported by the National Science Foundation, Grant PCM 83-08190.

REFERENCES

1. Lavorel, J. and Etienne, A.L. (1977) in: Primary Processes of Photosynthesis, Vol. 2, Barber, J., ed., Elsevier, Amsterdam, 203-268.
2. Deprez, J., Dobek, A., Geacintov, N.E., Paillotin, G. and Breton, J. (1983) Biochim. Biophys. Acta 725, 444-454.
3. Mauzerall, D. (1972) Proc. Natl. Acad. Sci. U.S.A. 63, 1358-1362.
4. Geacintov, N.E. and Breton, J. (1987) C.R.C. Crit. Rev. Plant Sciences, Conger, B.V., ed., in press.
5. Joliot, A. and Joliot, P.(1964) C.R. Acad. Sci. Paris 258, 4622-4625.
6. Melis, A. and Homann, P.H. (1976) Photochem. Photobiol. 23, 343-350.
7. Ley, A.C. and Mauzerall, D.(1982) Biochim. Biophys. Acta 680,174-180.
8. Geacintov, N.E., G. Paillotin, Deprez,J., Dobek, A. and Breton, J. (1984) in: Adv. Photosynth. Res., Vol. 1, Sybesma, C., ed., 37-40.
9. Ley, A.C. and Mauzerall, D.(1986) Biochim.Biophys.Acta 808, 192-200.

IS VARIABLE FLUORESCENCE DUE TO CHARGE RECOMBINATION ?

I.MOYA, M. HODGES AND J-M. BRIANTAIS

Laboratoire de Photosynthèse, C.N.R.S., BP 1, 91190 Gif sur Yvette, France.

1. INTRODUCTION

Early fluorescence lifetime measurements (1-3) have shown the constancy of the ratio $\tau_{mean}/\varnothing_T$ during the fluorescence induction rise in algae (where τ_{mean}=average lifetime, $\varnothing_T$=fluorescence yield). This observation has been considered as evidence for energy transfer between PS2 units and means that the major part of the fluorescence at Fo is homogeneous to the variable part (Fv). Other measurements including the shape of the relationship between the steady-state rate of O_2 emission and the number of open RC2 (4), as well as the sigmoidal shape of the Chl fluorescence induction rise in the presence of DCMU (5) agree with such a conclusion. This question was reinvestigated using the single photon counting technique which allowed the decomposition of the fluorescence decay into, at least, three exponential components. It was concluded that Fv arose mainly from the increase in the yield of a component whose lifetime (1-2.2ns) exhibited only minor changes upon closing RC2's (6). This finding seemed to be consistent with the hypothesis that Fv was a fast luminescence arising from charge recombination between P680$^+$ and I$^-$, when Q_A was reduced (7). However this proposed delayed emission does not have the 4.3 ns lifetime reported by Klimov et al. (7) and is still observed in a mutant which lacks the reaction center proteins of PS2 (8). Recent ps decay measurements with increased time resolution (9) have shown larger changes in component lifetime, and the need of 4 decay components (10). Closing RC2 produced two variable lifetime components (0.25-1.4ns and 0.45-2.6ns) the yield of which paralleled the change in their lifetimes, and 2 constant components (50 and 220 ps). In this report we question the origin of Fv by comparing the change in the lifetimes produced by either photochemistry or the addition of m,dinitrobenzene (DNB), which acts as an external quencher of Chl fluorescence, in wild type and PS2 lacking C. reinhardtii cells.

2. MATERIALS AND METHODS

C. reinhardtii wild type and F139 mutant were grown as described in (11). All samples were diluted in the growth medium to 20 µg Chl/ml. Fluorescence decay measurements were carried out using an improved ps fluorimeter having an instrumental response function of 60 ps (FWHM) as described in (9). The apparatus and deconvolution program were checked using the dye oxazine which gave a single exponential decay with a lifetime of 785 ps and a Chi^2 of 1 (Fig.1). Fo was generated by a flow method. F_M was generated under stationary conditions in the presence of DCMU. Intermediate fluorescence levels were produced by a saturating preillumination (to reduce all Q_A) and by varying the dark time allowed to reoxidize Q_A.

3. RESULTS
3.1. DECAY KINETICS OF THE WILD-TYPE

The fluorescence decays of C.reinhardtii monitored at 685nm with 625nm excitation have been recorded at several different fluorescence states between Fo and F_M. It has already been shown in (10), from time resolved emission spectra that, at least, 4 components are required to obtain a constant lifetime versus emission wavelength relationship. However, new results suggest the presence of even more components (Hodges _et al._ these proceedings). Fig.2 shows the relationship between the component lifetime and τ_{mean} (as τ_{mean} is proportionnal to ϕ_T, not shown). Two striking observations emerge; the lack of long lived components at Fo and the almost proportional increase in the lifetimes of two components with τ_{mean}. The lifetimes of the two remaining components (50 and 250) stayed almost insensitive to PS2 trap closure and their total weight contribution at Fo was 22%. As for the corresponding lifetime changes, only the yield of the two slow components are variable and their changes mirror the changes in their lifetime.

3.2 DECAY KINETICS IN A MUTANT LACKING PS2

F139 is a mutant of _C.reinhardtii_ having no PS2 activity and in which the 45, 42 and 33 Kda proteins associated with PS2 are absent (12). It exhibits no Fv but the PS1 and LHC2 seems unaltered. The decomposition of the fluorescence decay into 4 components yields pratically the same lifetime parameters as those seen in the wild-type (Table 1). It is obvious that the slow components cannot be attribued to the recombination luminescence mechanism. However they may have origins different from those seen in the wild-type. In order to check this point, we have investigated, in both the mutant and wild type cells, the effect of DNB addition.

3.3 EFFECT OF DNB ON THE DECAY KINETICS

It has been shown that, at concentrations $<10^{-4}$M, DNB has a specific quenching action on Fv, whereas Fo is unaltered (13). We observe, in the wild-type case, a parallel decrease in the lifetime and yield of the two slow variable components (Fig.2A). Lifetimes and yields of the two constant components are unaffected by the treatment. The effect produced by the addition of low DNB concentrations on lifetime and yield appears to be identical to the effect obtained by partial quenching by open RC2.

DNB addition to the PS2 mutant, in the same concentration range, produces a specific quenching of the two slow components, as in the wild-type case (fig.2B), infering that the origin of the fluorescence emissions are the same in both cases.

4. DISCUSSION

Two types of experiments are invoked to assign Fv to a nanosecond recombination luminescence mechanism. i/ The finding, using TSF2 particles, of a 4.3 ns lifetime component which paralleled the disappearance of an absorption signal in the range 400-600nm, when Q_A is reduced by the addition of dithionite (7). We have repeated the same type of experiment using BBY-type PS2 particles without finding such long lifetime component (unpublished results). ii/ Earlier decompositions of the fluorescence decay into a sum of 3 exponential components yielded a slow component (1-2.2 ns) whose lifetime exhibited a small change when RC2 became closed but whose yield incresed by 20-60 fold (6). Such properties were judged to aggree with the recombination luminescence hypothesis. As a consequence of the increased time resolution of our ps fluorimeter and the precautions taken to analyze Fo, no slow component is usually observed at Fo with intact algae. This result is independent of any deconvolution problems (9). Fv is generated by the almost parallel increase in both lifetime and yield of two components. Even more clear results are found with "BBY" or "Murata" particles. In such case only three lifetime components are required to best fit the decay, as a consequence of the absence of PS1 components (10 and Hodges _et al._ these proceedings). Thus we conclude that under normal conditions, closing RC2 leads to a continuous increase in both the lifetime and yield of two components that are pre-existent at Fo and represent 70-80% of this fluorescence level. A recent report in this field (14)

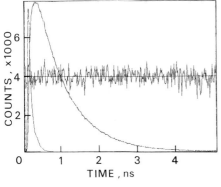

Fig.1. Excitation profile and
fluorescence decay of oxazine
in methanol measured at 685
nm, fitted by one exponential
giving a lifetime of 785 ps
with a Chi^2 of .92. The weighted
differences between the measured
and fitted curves are shown
in the centre of the plots
(scale +10 to -10).

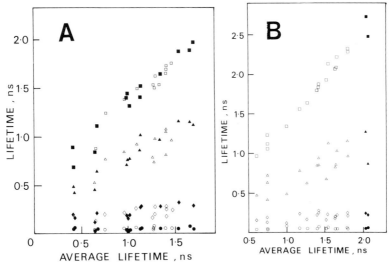

Fig.2. Variations in component lifetimes brought about either i/ closing
RC2 (closed symbols) or ii/ increasing additions of DNB (open symbols)
A. C.reinhardtii Wild type   B. F139 mutant lacking RC2.

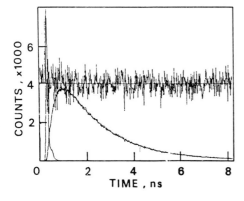

Fig.3. An excitation profil and
a fluorescence decay of $Fv=F_M-$
Fo. The best fit of 4 components
analysis, is superimposed on the
actual decay :1.92ns (109%),
1.11ns (14%), .41ns (-21%), .15ns
(-2%). Compare these values with
those of table 1 for F_M. The
obvious lag, which is generated
by the substraction of the Fo
decay from the F_M decay, is fitted
by the 2 negative amplitude
components.

TABLE 1. Lifetime and yield values of the fluorescence decays from the wild-type and PS2-lacking mutant of <u>C.reinhardtii</u> measured at F_M

Wild-type	τ (ns)	.052	.32	1.2	2.0	τ_{mean}=1.69
	ø_ (%)	2	4	26	68	
Mutant	τ (ns)	.042	.23	1.28	2.64	τ_{mean}=2.03
	ø (%)	4	7	25	64	

claimed that a lag (<u>i.e</u> a component having a negative amplitude) is seen by analysing Fv in <u>Chlorella</u> cells. This observation was found to be consistent with the slow component arising from recombination luminescence. Such a negative amplitude component has, in fact, a trivial origin by the fact that Fv (=F_M-Fo) was directly deconvoluted. As the substraction removes mainly the non-variable components, common to Fo and F_M, there remains four exponential components with different lifetimes, two of them having negative amplitudes. Therefore, the conclusion of (14) are unfounded.

The finding that the interaction between open RC2 and excitons in the antenna of PS2 appears as a dynamic quenching means than on average, the energy is localised in the antenna; consequently if charge recombination occurs in closed RC2 this mechanism introduces a very small delay (<100ps) otherwise a non zero extrapolation of the fluorescence lifetime when $ø_T$->0 would be expected. However, recombination luminescence does not occur in the mutant lacking RC2. DNB was therfore used to see if the long lived components of the mutant had similar origins to the decay components of the wild-type. In both cases the quenching by DNB mimics the quenching by open RC2 (Figs.2A & B), infering that the origin of the two slow components is probably the same in both cases. (See Hodges <u>et al.</u> these proceedings for the origin of the fluorescence decay components).

5. REFERENCES

1. Tumerman, L.A. and Sorokin, E.M. (1967) Mol. Biol. 1,628-638
2. Briantais, J.M., Merkelo, H. and Govindjee (1972) Photosynthetica 6,133-141
3. Moya, I., Govindjee, Vernotte, C. and Briantais, J.M. (1977) FEBS Lett. 75,13-18
4. Joliot, P. and Joliot, A. (1964) C.R. Acad. Sci. 258,4622-4625
5. Delosme, R. (1967) Biochim. Biophys. Acta. 143,108-128
6. Haehnel, W., Nairn, J.A., Reisberg, P. and Sauer, K. (1982) Biochim. Biophys. Acta. 680,161-173
7. Klimov, V.V., Allakhverdiev, S.I. and Paschenco, V.Z. (1978) Dokl. Akad. Nauk. 242,1204-1207
8. Green, B.R., Karukstis, K.K. and Sauer, K. (1984) Biochim. Byophys. Acta. 767,574-581
9. Moya, I., Hodges, M. and Barbet, J.C. (1986) FEBS Lett. 198,256-261
10. Hodges, M. and Moya, I. (1986) Biochim. Biophys. Acta. 849,193-202
11. Maroc, J. and Garnier, J. (1981) Biochim. Biophys. Acta. 637,473-480
12. Maroc, J., Guyon, D. and Garnier, J. (1983) Plant and Cell Physiol. 24,1217- 1230
13. Etienne, A.L., Lemasson, C. and Lavorel, J. (1974) Biochim. Biophys. Acta. 333,288-300
14. Mauzerall, D.C. (1985) Biochim. Biophys. Acta. 809,11-16

AKNOWLEDGEMENTS. We are indebted to D. Guyon and J. Garnier for the <u>C.reinhardtii</u> wild-type and PS2-lacking mutant. M.H. wishes to thank the Royal Society (London,U.K.) for financial support.

TIME RESOLVED CHLOROPHYLL FLUORESCENCE STUDIES OF PHOTOSYNTHETIC PIGMENT
PROTEIN COMPLEXES: CHARACTERISATION OF FIVE KINETIC COMPONENTS

M. HODGES, I. MOYA, J-M. BRIANTAIS AND R. REMY

Laboratoire de Photosynthèse, C.N.R.S., BP 1, 91190 Gif sur Yvette, France.

1. INTRODUCTION

Studies concerning the lifetime of chlorophyll fluorescence after ps excitation of LHC-containing photosynthetic organisms have revealed multi-
exponential decay kinetics. The overall decay can usually be statistically
well defined by three components which undergo complex changes in lifetime
and yield as PS2 reaction centres (PS2 RC) are closed. However, recent
studies (1-3) suggest that this is an oversimplistic model and that the
decay contains, at least, four components with τ's of approx. 50, 250, 250-
1400 and 450-2600 ps (3). It is not surprising that the observed decay is so
complex when considering the numerous different pigment-proteins located in
the thylakoid membrane. For a mechanistic description of the decay, in terms
of structure, antenna organisation and energy transfer processes, it is necessary to assign each individual component to deactivation processes of
functional constituents of the photosynthetic membrane. A popular current
concept involves the assignment of the decays to PS1 and a heterogeneous
pool of PS2 (termed α and β centres) (1). To better understand the *in vivo*
decay, which is probably too complex to fully describe in terms of each
individual component, it is necessary to characterise simplified systems (eg
isolated pigment-protein complexes or photosynthetic mutants). The results
of such an approach are presented in this work in which the overall decay is
described in terms of the two photosystems and their associated peripheral
antenna systems.

2. MATERIALS AND METHODS

'Murata-type' PS2-enriched membranes were prepared according to (4). A PS1
fraction (including LHC1) was isolated using Triton X100 as described in
(5). LHC2 was isolated, in an aggregated form, by density gradient centrifugation as in (6). Electrophoretic separation of chlorophyll protein
complexes was performed as in (7). The *C. reinhardtii* wild-type and double
mutant ($Fl39Pg28$) were grown as described in (8). All samples were diluted
to 20µg Chl/ml and fluorescence measurements were carried out and analysed
using the apparatus and criteria described in (3). The measurements on the
CP1, CP1a and LHC2 monomer were carried out directly on gel slices.

3. RESULTS

Fig. 1a shows the change in the component's lifetimes, exhibited by PS2-
enriched membranes, as PS2 RC are closed to photochemistry, as indicated by
the change in average lifetime (see 3). A proportional increase in the lifetime of each of the three components necessary to best fit the overall decay
is observed (by a factor of approx. 6) on going from F_0 (all Q_A oxidised) to
F_M (all Q_A reduced). Fig. 1b shows the change in relative fluorescence yield
of each kinetic component, exhibited by the same particles, in response to

Biggens, J. (ed.), Progress in Photosynthesis Research, Vol. I. ISBN 90 247 3450 9
© *1987 Martinus Nijhoff Publishers, Dordrecht. Printed in the Netherlands.*

PS2 RC closure. It can be seen that the yield of each component increases so as to parallel the modification in their respective lifetimes. It has been previously shown (3) that the three components have their fluorescence emission maxima at 681 nm which is indicative of PS2 origins.

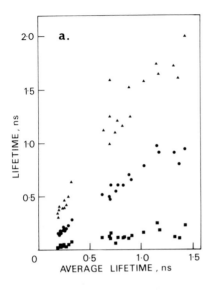 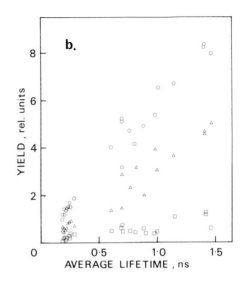

Fig. 1 The effect of PS2 trap closure on a) the lifetimes and b) the relative yields of the individual decay components exhibited by PS2 enriched membranes. The minimum fluorescence level was achieved in the presence of 100 µM potassium ferricyanide while the maximum level was attained in the absence of DCMU.

Table 1 gives the lifetime and preexponential factor (amplitude, Aexp) values of the kinetic components exhibited, at F_M, by photosynthetic material varying in degree of complexity. The 'intact' system (*C. reinhardtii*, in this case) produces two, variable long-lived and two, constant rapid lifetime decays of which only the 41ps component appears to arise from PS1 pigments (as judged from time-resolved emission spectral analyses) (3). At F_M, the PS2 particles lack this fast, red-shifted component and give rise only to the three PS2- associated components. Table 1 indicates that isolated LHC2 requires, at least, two exponential decays to fit the overall decay. The exact lifetimes of these two components vary with the aggregation state of the LHC2, however the difference in the two lifetimes are very similar to that of the two long lived PS2-associated components and are in close agreement with those already published (2,9). It might be that the two long-lived decays observed in the 'intact' system and the PS2-enriched membranes originate from LHC2. Isolated PS1, whether in solution (Table 1) or associated with electrophoresis gel, gives rise to several decay components of which the two rapid decays (τ's <220ps) produce 99% of the relative total amplitude and have red-shifted emission maxima (not shown).

TABLE 1 LIFETIME AND AMPLITUDE (Aexp) VALUES OF THE INDIVIDUAL DECAY
COMPONENTS EXHIBITED BY *C. REINHARDTII* (WILD-TYPE AND MUTANT *Fl39Pg28*
LACKING PS2 AND LHC2), PS2-ENRICHED PARTICLES, LHC2 AND PS1

						τ_{mean}
C. reinhardtii	τ	2.171	1.193	0.267	0.041	1.810
(wild-type)	Aexp	0.30	0.19	0.13	0.39	
PS2 particles	τ	2.006	0.937	0.244		1.449
(type 'Murata')	Aexp	0.27	0.47	0.26		
LHC2	τ	0.886	0.419			0.816
(aggregated)	Aexp	0.73	0.27			
LHC2	τ	3.460	0.988			3.118
(monomeric)	Aexp	0.64	0.36			
PS1	τ	2.766	0.549	0.109	0.027	0.743
	Aexp	0.004	0.006	0.22	0.77	
C. reinhardtii	τ	2.210	0.546	0.148	0.034	0.490
(*Fl39Pg28*)	Aexp	0.005	0.05	0.17	0.775	

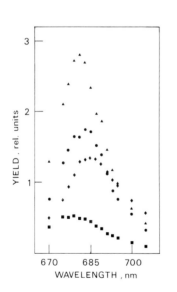

YIELD, rel. units

670   685   700
WAVELENGTH , nm

This trend is also seen in a double mutant
of *C. reinhardtii* (*Fl39Pg28*) which
contains only PS1-associated pigment-
protein complexes. Fig. 3 shows the
time-resolved emission spectra of the
individual decay components exhibited by
this mutant. The 148 ps decay produces a
maximum at 686 nm while the 35 ps decay
emits maximally at 688 nm. The two minor
amplitude components (τ=3.144 and 0.577
ns) have their emissions at 678 nm and
682 nm respectively and probably arise
from the presence of a small amount of PS2
associated pigment undetected by gel
electrophoresis. It is possible that the
two kinetic components are controlled by
the deactivation processes associated with
the LHC1 and the PS1 core proteins. This
is apparently confirmed by the difference
observed in the ratio of the rapid decay
component's amplitudes in isolated CP1 and
CP1a (which differ in the quantity of
associated LHC1 where the CP1a contains
more LHC1). The CP1a produced a ratio
Aexp217/Aexp22 of 0.1 while the CP1 gave
a ratio Aexp165/Aexp24 of 0.065 (where the
subscripts refer to the component τ in ps)

Fig. 3 Time-resolved emission spectra of the τ components exhibited by the
Fl39Pg28 mutant at F_M (3.144ns;   577ps;   116ps;   28ps)

4. DISCUSSION

It can be seen that the overall decay of 'intact' thylakoids is too complex to beable to define all of the individual decay components. There appears to be three PS2-associated lifetime components which are sensitive to PS2 RC closure and which give rise to the so-called variable fluorescence. The proportional increase in both lifetime and yield of these three components suggest a system of well connected PS2 RC favouring excitation energy transfer between centres. The trends seen in Fig. 1 can not be explained by a slow (ns) charge recombination mechanism giving rise to the slow decay components and suggest that variable chlorophyll fluorescence is indeed prompt fluorescence and not luminescence as previously proposed (10). To produce the changes observed in Fig. 1, based on charge recombination taking place between P^+ and I^- to reexcite the antenna chlorophylls, then it must be a rapid (<100 ps) process. It is possible that the two longer lived emissions originate from the deactivation processes governed by LHC2 (see Table 1 and references 2,9). The need to model the overall LHC2 decay by two components might suggest the presence of an heterogeneous pool of LHC2. There are, at least, two constant PS1-associated emissions which produce lifetimes in the order of 100-200ps, possibly associated with LHC1, and 20-30ps probably arising from chlorophyll containing proteins close to the PS1 RC. With respect to the 'intact' system, which has been shown to give rise to four kinetic components (3), it is probable that it was not possible to define the three rapid decays shown to exist in this work. Furthermore, it seems that the overall decay exhibited by 'intact' LHC-containing thylakoid membranes does not require the presence of PS2 α/β heterogeneity to explain the observed decay components.

5. REFERENCES

1. Holzwarth, A.R., Wendler, J. and Haehnel, W. (1985) Biochim. Biophys. Acta 807, 155-167.
2. Gulotty, R.J., Mets, L., Alberte, R.S. and Fleming, G.R. (1985) Photochem. Photobiol. 41, 487-496.
3. Hodges, M. and Moya, I. (1986) Biochim. Biophys. Acta 849, 193-202.
4. Kuwabara, T. and Murata, M. (1982) Plant and Cell Physiol. 65, 533-539.
5. Mullet, J.E., Burke, J.J. and Arntzen, C.J. (1980) Plant Physiol. 65, 814-822
6. Burke, J.J., Ditto, C.L. and Arntzen, C.J. (1980) Arch. Biochem. Biophys. 187, 252-263.
7. Remy, R. and Ambard-Bretteville, F. (1984) FEBS Lett. 170, 174-180.
8. Maroc, J. Guyon, D. and Garnier, J. (1983) Plant and Cell Physiol. 24, 1217-1230.
9. Lotshaw, W.T., Alberte, R.S. and Fleming, G.R. (1982) Biochim. Biophys. Acta 682, 75-85.
10. Klimov, V.V., Allakhverdiev, S.I. and Paschenco, V.Z. (1978) Dokl. Akad. Nauk. SSSR 242, 1204-1207.

ACKNOWLEDGEMENTS

M.H. wishes to thank the Royal Society (London,U.K.) for financial support. We are indebted to D. Guyon and J. Garnier for the *C. reinhardtii* wild-type and double mutant. This work was also financed by grant No. SAV 7738 CNRS-CEA.

MULTIVARIATE ANALYSIS OF PHOTOSYSTEM II CHLOROPHYLL FLUORESCENCE QUENCHING BY QUINONES

K. K. KARUKSTIS, S. C. BOEGEMAN, S. M. GRUBER, C. R. MONELL, J. A. FRUETEL, and M. H. TERRIS

Department of Chemistry, Harvey Mudd College, Claremont, CA 91711 USA

INTRODUCTION

Photosynthetic efficiency in higher plants is dependent upon the transfer of absorbed light to photochemical reaction centers and subsequent electron transport along a specialized pathway of electron donors and acceptors. One indirect measure of photosynthetic efficiency is the amount of excitation that is lost from the photochemical path in the form of chlorophyll fluorescence. Certain substituted quinones, when added to chloroplast suspending media, quench the level of in vivo room temperature chlorophyll fluorescence [1-3]. The mode of action of these substituted quinones is not clearly known, but studies suggest that they function to dissipate excitation energy by interaction with Photosystem II light-harvesting chlorophyll-protein complexes [1,4-6]. We have used Stern-Volmer analysis methods [7,8] to determine those physicochemical parameters of substituted quinones that enhance interaction with photosynthetic membranes as evidenced by chlorophyll fluorescence quenching in barley chloroplasts. Both classical and modified Stern-Volmer equations enable the determination of fluorescence quenching constants and corresponding fractions of chlorophyll fluorescence accessible to extrinsic quinones.

MATERIALS AND METHODS

Chloroplasts were isolated from freshly-harvested growth-chamber barley as previously described [9]. For fluorescence measurements the chloroplast suspension was diluted with 0.1 M sucrose/50 mM HEPES (pH 7.5)/5 mM NaCl buffer to a concentration of 10 μg Chl per ml. Substituted quinones were purchased from Aldrich Chemical Company, Eastman Organic Chemicals, and Alfa Products and, as necessary, were further purified by recrystallization or sublimation. Quinones were dissolved in ethanol or DMSO to give final constant solvent concentrations of 1% (v/v) in chloroplast samples. Both solvents showed no quenching effects of chlorophyll fluorescence at this concentration. Quinone concentrations ranged from 0 to 200 μM.

Room-temperature fluorescence emission spectra were measured with a Perkin-Elmer LS-5 fluorescence spectrophotometer, excitation λ = 620 nm, emission λ = 650-750 nm. All measurements were made for chloroplasts in the F_{max} state, i.e., with saturating light intensity to close Photosystem II reaction centers and induce maximal chlorophyll fluorescence levels.

For experiments involving the quenching effect of quinones on chlorophyll in organic solution, chlorophyll was extracted from isolated

Biggens, J. (ed.), Progress in Photosynthesis Research, Vol. I. ISBN 90 247 3450 9

pea chloroplasts using an 80% (v/v) acetone–water mixture. Fluorescence measurements were made with chlorophyll samples of 5 µg/ml and quinone concentrations ranging from 0 to 1600 µM, using the excitation and emission parameters employed above.

Analysis of chlorophyll fluorescence quenching involved application of the Stern–Volmer equation,

$$I_o/I = 1 + K_{SV}[Q] \qquad (1)$$

where I_o and I are chlorophyll fluorescence intensities in the absence and presence of quinone quencher, respectively, K_{SV} is the Stern–Volmer quenching constant, and $[Q]$ is the concentration of quinone quencher. A linear Stern–Volmer plot of I_o/I vs. $[Q]$ is generally indicative of a single class of chlorophyll fluorophores with equal accessibility to quencher [7,8]. A Stern–Volmer plot exhibiting downward curvature indicates the existence of a chlorophyll population which is inaccessible to quinone quencher. The quenching is then described by a modified Stern–Volmer equation [7,8]:

$$I_o/\Delta I = 1/(f_a \cdot K_{SV} \cdot [Q]) + 1/f_a \qquad (2)$$

where ΔI is the magnitude of the fluorescence intensity decrease observed in the presence of quinone at concentration $[Q]$ and f_a is the fraction of the initial fluorescence which is accessible to quencher.

RESULTS AND DISCUSSION

The benzoquinones examined included 1,4–benzoquinone and the following substituted 1,4–benzoquinones: 2,5–dihydroxy; 2,3,5,6–tetrachloro; 2,3–dichloro–5,6–dicyano; 2,5–dichloro–3,6–dihydroxy; 2,3,5,6–tetramethyl; 2,5–dibromo–3–methyl–6–isopropyl (DBMIB); and tetrahydroxy–1,4–quinone hydrate. Fluorescence quenching activity was observed for benzoquinones with hydrophobic substituents or hydrogen atoms at positions 2, 3, 5, and 6. The presence of hydrophilic groups (e.g., –OH, –CN) at one or more of these positions resulted in no measurable fluorescence quenching, presumably as a consequence of the absence of an interaction of quinone with the receptor site on the thylakoid membrane. Benzoquinones with only electron–withdrawing hydrophobic substituents had similar K_{SV} values to that of 1,4–benzoquinone (4.2 x 10^4 M^{-1}), while those with electron–releasing hydrophobic substituents exhibited lower K_{SV} values (e.g., 2,3,5,6–tetramethyl–1,4–benzoquinone, 3.5 x 10^3 M^{-1}). The presence of both electron–releasing and electron–withdrawing hydrophobic groups increased K_{SV} above the value for 1,4–benzoquinone (e.g., DBMIB, 1.4 x 10^5 M^{-1}).

The naphthoquinones examined included 1,4–naphthoquinone and the following substituted 1,4–naphthoquinones: 2–hydroxy; 2–methoxy; 2–methyl; 2–amino; 2–bromo; 2,3–dichloro; 2–chloro–3–morpholino; 2–amino–3–chloro; 2–chloro–3–pyrrolidino; 2,3,5,6–tetrahydroxy; 5–hydroxy; and 2,3–epoxy–2,3–dihydro–1,4–naphthoquinone. The lipophilic character of the substituents on the α–ring of 1,4–naphthoquinone (i.e., at positions 2 and 3) determined the fraction of chlorophyll fluorescence accessible to naphthoquinone. Hydrophobic substituents increased f_a to a maximum value of 1, while hydrophilic substituents lowered f_a to a minimum value of 0. In the case of 2,3–substitution of a hydrophilic and a hydrophobic substituent, the effects on f_a were "additive" and therefore often canceled each other. An exception to these generalizations on f_a effects was observed for 2-amino-1,4-naphthoquinone, with a linear Stern-Volmer plot (f_a = 1.00). The electronic character of the substituents at positions 2 and 3 alters the magnitude of the observed Stern-Volmer quenching constant. The presence of one or two electron-releasing substituents at these positions lowers K_{SV} below the 1,4-naphthoquinone value (e.g., 2-methyl-1,4-naphthoquinone, 7.5 x 10^3 M^{-1}), while the presence of one or two electron-withdrawing substituents increases K_{SV} (e.g., 2,3-dichloro-1,4-naphthoquinone, 2.7 x 10^5 M^{-1}). The K_{SV} value in the presence of one electron-withdrawing and one electron-releasing group remains approximately at the 1,4-naphthoquinone value (1.8 x 10^4 M^{-1}).

A number of hydroxy-substituted 9,10-anthraquinones were examined including: 1-hydroxy; 1,2-dihydroxy; 1,4-dihydroxy; 1,8-dihydroxy; 1,2,4-trihydroxy; and 1,2,5,8-tetrahydroxy. In general, hydroxy substitution dramatically increased f_a above the value observed for 9,10-anthraquinone. Hydroxy substitution at positions 1, 4, 5, and/or 8 increased the K_{SV} value above that observed for 9,10-anthraquinone (8.9 x 10^4 M^{-1}), while hydroxy substitution at position 2 lowered the K_{SV} value.

The effects of a number of substituted benzoquinones and naphthoquinones on the fluorescence of chlorophyll in acetone solution were measured. Each quinone induced a quenching of chlorophyll emission to give a linear Stern-Volmer relation with comparable K_{SV} values (average K_{SV} = 9.9 x 10^1 M^{-1}). The extent of quenching in acetone solution was much weaker than the quenching effects observed _in vivo_, as previously noted [1].

The ability of various thiols to reverse quinone-induced fluorescence quenching when added to quinone-treated chloroplasts was examined. In particular, measurements were made of the amount of thiol needed to reverse chlorophyll fluorescence quenching and the extent to which reversal occurred. The thiols dithiothreitol, 2-hydroxy-mercaptoethanol, and glutathione were tested for reversal effects on 2,3-dichloro-1,4-naphthoquinone; DBMIB; 2-bromo-1,4-naphthoquinone; 1,4-anthraquinone; and 1,8-dihydroxy-9,10-anthraquinone. For all quinones examined, the thiol concentrations for reversal were in the order [dithiothreitol] < [2-mercaptoethanol] < [glutathione]. 1,4-anthraquinone required the lowest thiol concentrations for reversal: 20 μM dithiothreitol, 60 μM 2-hydroxy-mercaptoethanol, and 110 μM glutathione. 2,3-dichloro-1,4-naphthoquinone required the highest thiol concentrations for reversal: 100 μM dithiothreitol, 400 μM 2-hydroxy-mercaptoethanol, and 700 μM glutathione. No thiol was able to reverse the fluorescence quenching induced by 1,8-dihydroxy-9,10-anthraquinone. The extent of quenching reversal varied with the quinone studied and the thiol used, from a minimum of no reversal for 1,8-dihydroxy-9,10-anthraquinone (for all thiols) to a maximum reversal of DBMIB effects by dithiothreitol to a fluorescence level equal to 150% of the fluorescence intensity in the absence of quinone.

In conclusion, we have used Stern-Volmer analyses of chlorophyll fluorescence quenching in barley chloroplasts by added quinones to assess those physicochemical factors of quinones that govern the quenching process. For substituted benzoquinones and naphthoquinones, hydrophobic moieties promote the partitioning of quinones between the lipophilic thylakoid membrane and the surrounding aqueous environment. Furthermore, the activity of a bound quinone at a receptor site is sensitive to the charge distribution on the quinone ring. Hydroxy substituents enhance fluorescence quenching of 9,10-anthraquinone by increasing the fraction of chlorophyll fluorescence accessible to quinone. Significant increases in Stern-Volmer quenching constants are also observed for hydroxy substitution at the 1, 4, 5, and/or 8 positions. It should be noted that those quinone substituents that promote fluorescence quenching do not necessarily also enhance electron transport inhibition [4,10–12]. A multiparameter approach to structure-activity relationships is currently in progress to gain additional insight into the mechanism of chlorophyll fluorescence quenching by extrinsic quinones in chloroplasts.

ACKNOWLEDGMENTS

This research was supported by an Exxon Education Foundation Grant of Research Corporation, a grant from the California Foundation for Biochemical Research, a Harvey Mudd College Faculty Research Award, and a grant from the Harvey Mudd College Arnold and Mabel Beckman Research Fund. The authors also gratefully acknowledge support from the Chemistry Department of Harvey Mudd College.

REFERENCES

1. Amesz, J. and Fork, D.C. (1967) Biochim. Biophys. Acta **143**, 97–107.
2. Arnon, D.I., Tsujimoto, H.Y., and McSwain, B.D. (1965) Proc. Natl. Acad. Sci. U.S. **54**, 927–934.
3. Thomas, J.B., Voskuil, W., Olsman, H., and De Boois, H.M. (1962) Biochim. Biophys. Acta **59**, 224–226.
4. Pfister, K., Lichtenhaler, H.K., Burger, G., Musso, H., and Zahn, M. (1981) Z. Naturforsch. **36c**, 645–655 (1981).
5. Kitajima, M. and Butler, W.L. (1975) Biochim. Biophys. Acta **376**, 105–115.
6. Berens, S.J., Scheele, J., Butler, W.L., and Magde, D. (1985) Photochem. Photobiol. **42**, 51–57.
7. Lakowicz, J.R., Principles of Fluorescence Spectroscopy, Plenum Press, New York, 1984.
8. Lehrer, S.S. (1971) Biochemistry **10**, 3254–3263.
9. Karukstis, K.K. and Gruber, S.M. (1986) Biochim. Biophys. Acta (in press).
10. Trebst, A., Harth, E., and Draber, W. (1970) Z. Naturforsch. **25b**, 1157–1159.
11. Oettmeier, W., Reimer, S., and Link, K. (1978) Z. Naturforsch. **33c**, 695–703.
12. Sarojini, G. and Daniell, H. (1981) Z. Naturforsch. **36c**, 656–661.

ENERGY TRANSFER IN CHLOROPHYLL ANTENNAE OF ISOLATED PSII PARTICLES

Tomas Gillbro, Åke Sandström and Villy Sundström
Department of Physical Chemistry, University of Umeå, S-901 87 Umeå, Sweden

Michael Spangfort and Bertil Andersson
Department of Biochemistry, University of Lund, S-211 00 Lund, Sweden

Göran Lagenfelt
Department of Biochemistry, University of Göteborg, S-412 96 Göteborg, Sweden

Introduction

In this work we have investigated how the size of the antenna pigment system influences the rate of excitation energy transfer to the reaction center of photosystem 2 (PSII). We have for this purpose studied the rate of energy transfer in isolated PSII particles of spinach and the blue-green bacterium Synechococcus 6301. The antenna system of these PSII particles consists of about 240 and 60 chlorophylls, respectively. The difference is due to the lack of Chl a/b proteins in blue-green bacteria. We assume that the antenna core is the same in both particles and that also the transfer rate from the chlorophyll a core to the reaction center is the same. Finally a model for the transfer of excitation energy within the chlorophyll antennae will be presented.

Experimental

The PSII particles were prepared according to published procedures [1,2] and the picosecond pump-probe experiments were made on samples containing 2.5 mM ferricyanide and 0.45 mM PPBQ, in order to keep the reaction centers in the open state. We also added 0.3% triton X-100 to reduce light scattering. In the pump-probe picosecond technique the sample (OD≈0.5) is excited with a ≈10 ps laser pulse at 685 nm. The recovery of the induced signal is probed at the same wavelength by a pulse which is delayed relative the excitation pulse. This time delay can be varied between 0 ps and 2 ns. Since the repetition rate of the system is 80 kHz, the sample has to be quickly removed between two excitation pulses in order to avoid artifacts from closed reaction centers. This is achieved by using a rotating sample cell. We have also used low light intensity, i.e. < 10^{13} photons cm^{-2} $pulse^{-1}$, which means that annihilation effects can be neglected in our measurements.

Results and Discussion

In Fig. 1 we show the kinetics of the absorption recovery at 685 nm of PSII (S 6301) kept in the open state. The observed signal can best be fitted to a sum of three exponential decays with the lifetimes 20, 134 and 2200 ps, respectively. The fast 20 ps component contributes 71% of the total intensity, while the other two components contribute only 11 and 18%, respectively. We interpret the fast decay to the transfer of excitation energy from the chlorophyll a core complex to the reaction center

Biggens, J. (ed.), Progress in Photosynthesis Research, Vol. I. ISBN 90 247 3450 9
© *1987 Martinus Nijhoff Publishers, Dordrecht. Printed in the Netherlands.*

of PSII. The second component is probably due to the charge separation $P^+I^-Q \rightarrow P^+IQ^-$, which then should have a rate constant of about 7.5×10^9 at room temperature. The long lifetime of 2.2 ns is supposed to be the reduction of P^+. In this report we will only discuss the fast component, i.e. the energy transfer from the chlorophyll a antennae to the reaction center.

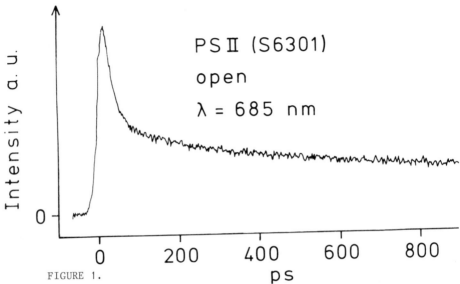

FIGURE 1.

In Fig. 2 we have performed a picosecond absorption experiment on PSII from BBY particles under the same experimental conditions as for PSII of S 6301 in Fig. 1. In this case (Fig. 2) there is a marked increase of the short lifetime. A fit with three exponentials gives the lifetimes 70 and 307 ps for the fast component and 59 ns for the long lived component.

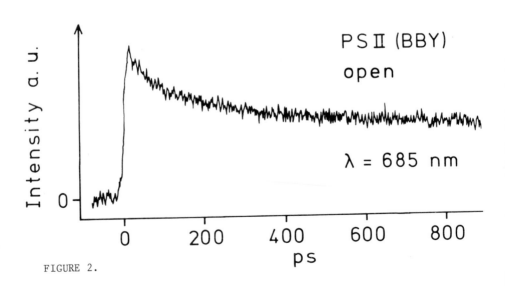

FIGURE 2.

Independent of the way of analyzing the data we obtain a longer life-time for the overall excitation energy transfer from the chlorophyll an-tennae to the reaction center in PSII of BBY particles as compared to S 6301. In order to analyze our results we have used the simplified model of PSII below.

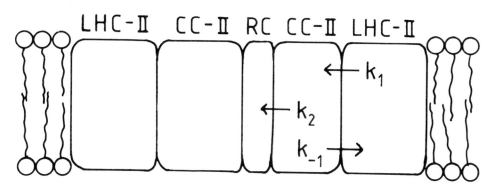

We have assumed that the antennae chlorophylls are in the light har-vesting complex (LHC-II) and the core complex (CC-II) and that the energy can be transferred between these complexes with the rate constants k_1 and k_{-1}. CC-II can transfer energy to the reaction center (RC) with the rate constant k_2. We then obtain the following rate equation:

$$k_{a,b} = \frac{1}{2}\left\{(k_1 + k_{-1} + k_2) \pm \left[(k_1 - k_{-1} - k_2)^2 + 4k_1 \cdot k_{-1}\right]^{\frac{1}{2}}\right\}$$

where k_a can be thought of as the rate for equilibration of excitation energy between LHC-II, and k_b is the rate for energy transfer to the reac-tion center.

Now assuming k_2 has the same value in PSII of BBY and S 6301, then $k_2 = 5 \times 10^{10}$ s^{-1} (from Fig. 1). Assuming the same value for k_1 and $k_{-1} = 2.5 \times 10^{10}$ s^{-1} [since there is only a small energy difference between the excited chlorophylls of LHC-II and CC-II] , we obtain $k_a = 8 \times 10^{10}$ s^{-1} and $k_b = 1.5 \times 10^{10}$ s^{-1}, i.e. $\tau_a = 12$ ps and $\tau_b = 68$ ps. The fast (12 ps) com-ponent should have a low intensity and would be difficult to detect in an absorption experiment, since the different chlorophyll a pigments have similar absorption spectra. The calculated lifetime of $\overline{68}$ ps for the trans-fer of excitation energy to the RC is in good agreement with our experimen-tal data (Fig. 2). Our simple model thus gives a reasonable explanation of the 70 ps lifetime observed in BBY particles.

Acknowledgements

This work was supported by the Swedish Natural Science Research Council.

I.1.126

References

1. Lagenfelt, G., Hansson, Ö. and Andréasson, L.-E. (1986)
   Acta Chem. Scand. Ser. B. in press.
2. Berthold, D.A., Babcock, G.T. and Yocom, C,F, (1981)
   FEBS Letters 134, 231-234
   modified by
   Ford, R.C. and Evans, M.C.W. (1983)
   FEBS Letters 160, 159-164

POLARIZED SPECTRA OF PS2 PARTICLES IN PVA FILMS

D. Frackowiak[a,b], W. Hendrich[c], M. Romanowski[a,b], A. Szczepaniak[c] and
R.M. Leblanc[b]

a) Poznan Technical University, 60-965 Poznan, Poland; b) Centre de
recherche en photobiophysique, U.Q.T.R., Trois-Rivières, Québec, G9A 5H7,
Canada; c) University of Wroclaw, 50-137 Wroclaw, Poland.

1. INTRODUCTION

The thermal deactivation (TD) of excitation, the emission of fluorescence (F) and their transfer of excitation energy (ET) to other molecule
are competative processes occuring in excited antenna chromophores.

Therefore when the yield of ET of given type of chromophores is changed, as a result of their reorientation, also their yield of fluorescence
and efficiency of thermal deactivation are affected. The polarized photoacoustic, fluorescence and absorption spectra of photosystem 2 (PS2) particles embedded in isotropic and stretched polyvinyl alcohol (PVA) film
were measured.

The aim of this paper is to establish the relation between the orientation of PS2 chromophores and the processes of excitation deactivation
(ET, TD and F).

2. MATERIAL AND METHODS

PS2 RC complexes were isolated from pea leaves (Ghanotakis, Yocum
(1)). Methods of PVA film preparation and stretching were described previously (2). PAS were measured with single beam photoacoustic spectrophotometer constructed in Trois-Rivières (3). The methods of measuring the
polarized photoacoustic, fluorescence and absorption spectra have been
published (4, 5). It was shown previously (5) on the basis of polarized
fluorescence spectra, that PS2 particles exhibit some planar orientation in unstretched PVA films.

As a result of the film stretching in addition to this planar orientation, the uniaxial orientation around the axis located in a film plane is
formed. Orientation in now investigated preparations is similar, therefore all presented fluorescence result were obtained at the same geometry
of experiments, such that $F_{\parallel}$ refers to contributions from transition
moments (TM) located parallel to the film stretching axis, $F_{\perp}$ from TM
located perpendicular to this direction, but both being almost parallel to
PVA plane (5).

Spectra $A_{\parallel}$, $A_{\perp}$, A_{o}; $F_{\parallel}$; $F_{\perp}$; F_{o} and $PAS_{\parallel}$; $PAS_{\perp}$; PAS_{o} refer to the
absorption, fluorescence, and photoacoustic data taken at light polarized
parallel ($\parallel$) or perpendicular ($\perp$) to the film stretching axis, and with
unpolarized light for unstretched films respectively (o).

3. RESULTS AND DISCUSSION

Unpolarized and polarized absorption spectra (A_o, $A_{\parallel}$ and $A_{\perp}$, Fig. 1)
have different shapes. Especially the ratio of maximum at 435 nm to that
at 422 nm is changed with film stretching and light polarization. Linear

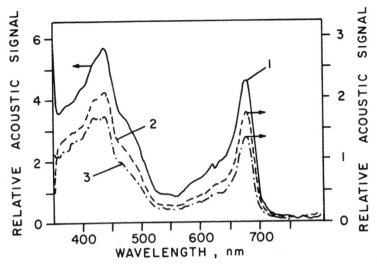

Figure 3: Photoacoustic spectra 1 - PAS_0; 2 - $PAS_{\parallel}$; 3 - $PAS_{\perp}$.

Fig. 4 shows fluorescence spectra of PS2 in PVA. Previously (6) for PS2 enriched particles, containing small contamination from other parts of photosynthetic apparatus (Chl a - Chl b core complexes of PS1; CF_1 coupling factor α and β), besides the maximum at 683–685 nm, characteristic for PS2, additional maximum at 730 nm region was observed (6). Now, this emission is absent, what shows, that it was related with PS1 contamination. Also decay of fluorescence (Table II) which previously was composed of three exponential component, is now only two exponential. Third one, long living component was related, as it was supposed previously (6), with 730 nm emission.

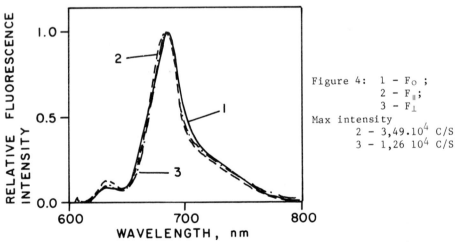

Figure 4: 1 - F_0 ;
          2 - $F_{\parallel}$;
          3 - $F_{\perp}$
Max intensity
          2 - $3,49.10^4$ C/S
          3 - $1,26\ 10^4$ C/S

Lifetime change with film stretching (Table II) is complex phenomenon including the change in mutual pigments interactions and in pigment interactions with environment (6). Observed $\tau_{\parallel} < \tau_{\perp}$ (Table II) is in the agreement with $TD_{\parallel} > TD$ (Table I), because at higher TD yield of fluo-

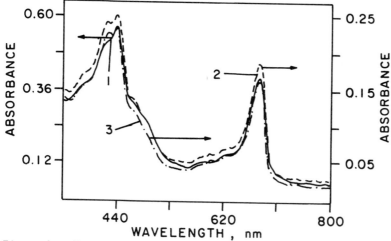

Figure 1: Absorption spectra 1 - A_o; 2 - $A_{\parallel}$; 3 - $A_{\perp}$

dichroism (LD) of absorption (Fig. 2) has higher value at 435 nm than at 422 nm. Anisotropy of absorption $\left(\frac{A_{\parallel} - A_{\perp}}{A_{\parallel} + 2A_{\perp}}\right)$ is higher ($\sim 0,04$) in red band than in the Soret band ($\sim 0,02$), what is related with predominantly "y character" of red band and, "mixed" (x and y) polarization of Soret band.

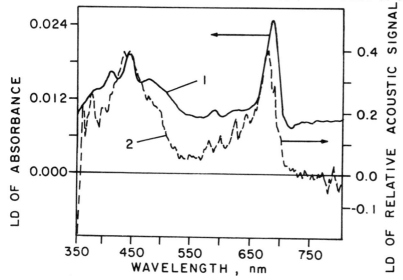

Figure 2: Linear dichroism of absorption (1) and LD of PAS (2).

Shape of PAS (Fig. 3) and LD of PAS (Fig. 2) are different of respective absorption spectra (Fig. 1 and 2), what shows that thermal deactivation (TD) of excitation in various types of PS2 chromophores has different efficinency. TD (in arbitraty units) is obtained as a ratio of PAS to absorption (Table I). TD in a red band region is higher than that for the Soret band; it decreases as a result of film stretching and is higher for parallel (TD$_{\parallel}$) than for perpendicular (TD$_{\perp}$) component.

<table>
<tr><td colspan="4">Table I.
Thermal deactivation (a.u.)</td></tr>
</table>

[nm]	420	435	676
TD$_0$	9,74	9,95	11,38
TD$_{\parallel}$	8,24	8,28	9,31
TD$_{\perp}$	7,42	7,35	8,35

Table II
Fluorescence lifetime (τ)
analysis

	τ_1 [ns]	Int.$_1$ %	τ_2 [ns]	Int.$_2$ %
F$_0$	0,364	64	4,742	36
F$_{\parallel}$	0,513	55	5,386	45
F$_{\perp}$	0,530	66	5,558	34

rescence and lifetime have to be lower. Fluorescence at 683-685 nm region is polarized positively ($F_{\parallel} > F_{\perp}$), as it was found previously (6).

Anisotropy of emission at main maximum is: $\dfrac{F_{\parallel} - F_{\perp}}{F_{\parallel} + 2F_{\perp}} = 0,37$; much higher than anisotropy of absorption at excitation region (430 nm), or in red band. Because of used geometry of fluorescence experiment, to our spectra contribute predominantly TM located in PVA plane (at planar orientation of particles in membrane plane). Tapie et al. (7) have investigated polarized absorption and fluorescence in squeezed gel, comparing contributions from out of membrane plane oriented TR with that given by TM located in this plane. Observed by us emission at about 685 nm is according Tapie et al. (7) related with TM forming small angles with membrane and is related with absorption dichroism maximum at 676 nm. Our polarized absorption (Fig. 1) shows exactly the maximum at 676 nm and the emission at 683-685 region (Fig. 3). It is therefore the same type of complexes, as reported by Tapie et al. (7). Observed by us emission seems to be predominantly related with PS2. Detail discussion of the orientation and reorientation of various complexes in our samples and their TD, based also on polarized fluorescence excitation spectra, will be a subject of the following publication.

ACKNOWLEDGMENT

Two of us wish to thank the Université du Québec à Trois-Rivières for visiting scientist position (Danuta Frackowiak) and the exchange visitor grant (Marek Romanowski). This work was supported by the NSRC at Canada (Roger M. Leblanc) and Polish grant (RP II 11, 4.2) (W.H., A. Sz., D.F., M.R.).

REFERENCES

1. Ghanotakis, D.F., Yocum, C.F. (1986) FEBS Lett. 197, 244-248.
2. Fiksinski, K. and Frackowiak, D. (1980) Spectroscopy Lett. 13, 873-899.
3. Ducharme, D., Tessier, A. and Leblanc, R.M. (1979) Rev. Sci. Inst. 50, 42-43.
4. Frackowiak, D., Szych, B. and Leblanc, R.M. (1986) Acta Phys. Pol. A69, 121-133.
5. Frackowiak, D., Hotchandani, J. and Leblanc, R.M. (1983) Photobiochem. Photobiophys. 6, 339-350.
6. Frackowiak, D., Hendrich, W., Kieleczawa, J., Szczepaniak, A., Szurkowski, J., Romanowski, M. and Leblanc, R.M. Biochim. Biophys. Acta (sended).
7. Tapie, P., Choquet, Y., Wollman, F.-A., Diner, B. and Breton, J. (1986) Biochim. Biophys. Acta 850, 156-161.

THE DEPENDENCE OF THE ENERGY TRANSFER KINETICS OF THE HIGHER PLANT LIGHT
HARVESTING CHLOROPHYLL-PROTEIN COMPLEX ON CHLOROPHYLL/DETERGENT RESOLUBIL-
ISATION RATIOS.

J.P. Ide, D.R. Klug, B. Crystall, B.L. Gore, L.B. Giorgi, W. Kuhlbrandt*,
J. Barber* and G. Porter
Davy Faraday Research Laboratory, The Royal Institution, 21 Albemarle
Street, London W1X 4BS, U.K.; *Department of Pure & Applied Biology,
Imperial College, Prince Consort Road, London SW7 2BB, U.K.

1. INTRODUCTION

The main functional role of the LHC *in vivo* is to transfer energy from sun-
light to the reaction centres of the thylakoid membrane. The mechanism by
which this occurs is thought to be a Forster like process, each transfer
step between chlorophylls being of the order of a picosecond. In this work
we investigate the effect of changing the chlorophyll/detergent molar ratio
on the resolubilised LHC energy transfer kinetics. Steady-state and time
resolved fluorescence techniques have been used to observe changes in
kinetics; the state of the protein has been deduced from circular dichroism
spectra which have been interpreted in light of the known geometry of the
crystallised LHC (1).

2. EXPERIMENTAL

Excited singlet state lifetimes were measured using the technique of time
correlated single photon counting (2). The excitation source consisted of
an Argon-ion laser, mode-locked at 514.5nm, synchronously pumping a tunable
Rhodamine 6-G dye laser system (3). This provided pulses of ca.15ps duration (FWHM) in the range 590-630nm. Our time resolution was estimated from
the measured instrument response function to be approximately 120ps. Steady-
state fluorescence (SSF) spectra were recorded on an MPF4 Perkin-Elmer
spectrafluorimeter and circular dichroism (CD) spectra on a Jasco J40CS
spectrapolarimeter. The LHC was extracted from pea thylakoid membranes (4)
and resolubilised in low (0.2×10^{-3}:1), medium (0.46×10^{-3}:1) and high
(0.025:1) chlorophyll/n-octylglucoside (detergent) molar ratio conditions.
The corresponding conditions for Triton X-100 detergent were found to be
low (1.4×10^{-3}:1) and high (0.24:1). All samples contained 10µg chloro-
phyll-a/b per ml of solution.

3. RESULTS AND DISCUSSION

Figure 1 shows steady-state and time resolved fluorescence data together
with the circular dichroism spectrum for the medium molar ratio case. The
data is qualitatively the same for both detergents under the same chloro-
phyll/detergent molar ratio conditions.

Under low molar ratio conditions (data not shown) we observe: (1) Two peaks
in the SSF at the chlorophyll-b and chlorophyll-a emission maxima; the shape
of the SSF is time and excitation wavelength dependent. (2) Complex decay
kinetics; main component $\approx$5.7ns.

Biggens, J. (ed.), Progress in Photosynthesis Research, Vol. I. ISBN 90 247 3450 9
© *1987 Martinus Nijhoff Publishers, Dordrecht. Printed in the Netherlands.*

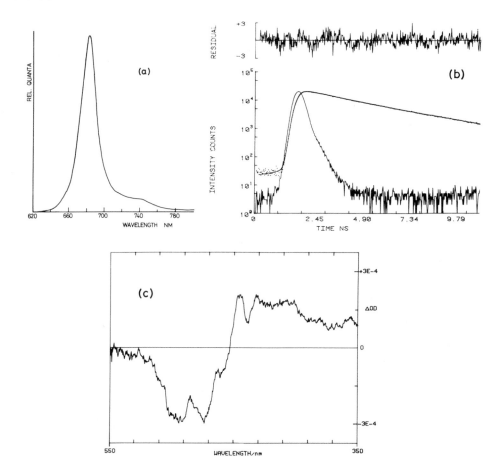

Fig.1. Spectroscopic data for the LHC under medium molar ratio resolubilisation conditions:
a)   steady state fluorescence spectrum; excitation wavelength 470nm
b)   single photon counting data; reduced chi-square = 1.19, biexponential
     fit with lifetimes of 3.6 and 1.2ns and pre-exponential factors in a
     ratio of $\sim$ 4:1 respectively; excitation and emission wavelengths 605
     and 675nm respectively
c)   Soret circular dichroism spectrum

Under medium molar ratio conditions (see Fig.1) we observe: (1) Single
chlorophyll-a emission peak in the SSF; shape of SSF spectrum is excitation
wavelength and time invariant. (2) Biexponential decay kinetics; lifetimes
of 3.53ns ($\pm$0.04) and 1.12ns ($\pm$0.12) and pre-exponential factors in a ratio
of $\sim$ 4:1 respectively. (3) Soret CD shows a large amplitude chlorophyll-b
excitonic feature; a smaller chlorophyll-b feature with reversed chirality
with respect to the previous chlorophyll-b excitonic feature and a chloro-
phyll-a/b exciton; all are conservative. (4) Excitation at the chlorophyll-
b Q_y absorption maximum leads to greater depolarisation of the chlorophyll-a
SSF emission than excitation at the chlorophyll-a Q_y absorption maximum.

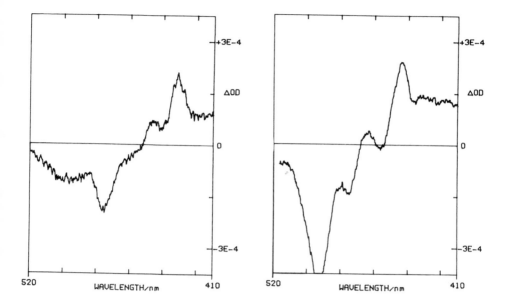

Fig.2. CD spectrum showing loss of chlorophyll-b exciton.

Fig.3. CD spectrum showing loss of chlorophyll-a/b exciton.

Under high molar ratio conditions (data not shown) we observe: (1) A single peak in the SSF at chlorophyll-a emission maximum. The integrated area under this peak is considerably less than for the medium molar ratio condition despite identical chlorophyll concentrations. (2) Complex subnanosecond decay data. (3) Identical CD features to the medium molar ratio condition except that the "small" chlorophyll-b feature has acquired considerable amplitude; we refer to this as the quenching exciton. (4) A very close similarity exists between the data obtained under these conditions and that obtained for the suspensions of crystalline LHC.

Other detergent dependent effects: Two other effects have also been found. Firstly, a moderately low chlorophyll/detergent molar ratio causes a gradual collapse of chlorophyll-b to chlorophyll-a energy transfer over a period of a few hours. However, the C_3 symmetric chlorophyll-b exciton is largely lost within 5 minutes (see Fig.2), followed by a gradual decline in the magnitude of the chlorophyll-a/b exciton, which roughly follows the course of the collapse of energy transfer. Secondly, resolubilisation of the LHC2 with sodium dodecyl sulphate (SDS) causes an apparently immediate loss of the chlorophyll-a/b exciton (Fig.3), whereas both energy transfer and the C_3 symmetric chlorophyll-b exciton are retained over a period of hours. The soret CD spectra in the presence of SDS are qualitatively identical to those previously obtained by other workers (5,6).

From the results we deduce the following: (1) Under high molar ratio conditions extensive aggregation of protein trimers has occurred in a qualitatively similar way to salt induced crystallisation. The quenching chlorophyll-b CD feature results from a new chlorophyll-b interaction arising through the aggregation process, leading to quenching of both chlorophyll-a fluorescence lifetime(s) and quantum yield. (2) Under medium molar ratio conditions we have predominantly non-aggregated trimers which are associated with the long decay component (3.53ns). The short decay component (1.12ns) and the small amplitude chlorophyll-b exciton feature are associated with the presence *in vitro* of a small amount of aggregated trimers. Forster energy-transfer appears efficient and the fluorescence depolarisation /CD implies a C_3 symmetric chlorophyll-b interaction together with a dimeric chlorophyll-b/a interaction. It seems probable that the C_3 symmetry of the protein trimers is a pre-requisite for the C_3 symmetric chlorophyll-b interaction/excitonic feature. (3) Under low molar ratio conditions the C_3 symmetric chlorophyll-b exciton feature is lost suggesting that the C_3 symmetry of the protein trimers has also been lost. The presence of a chlorophyll-a/b exciton feature suggests that the protein monomers are initially intact, however at longer times the protein appears to denature resulting in the detachment of chromophores from their binding sites on the protein. The decay data supports this assertion on the basis that free chlorophyll-a in detergent micelle has a lifetime of $\approx$ 5.7ns (7). (4) The detailed action of SDS remains unclear. It is probable that the loss of the chlorophyll-a/b exciton is due to chlorophyll-a molecules being stripped from the protein. It is possible, however, that this is due to the use of too high an SDS concentration. We did not carry out a full study of the effect of the SDS/chlorophyll ratio as was done for TX-100 and octylglucoside.

In summary, the high molar ratio condition leads to ordered aggregation of trimer functional units, the medium molar ratio leads to predominantly trimer functional units with some aggregate present, while the low molar ratio condition leads to loss of trimer geometry and eventual denaturing of the protein monomers.

ACKNOWLEDGEMENTS

We would like to thank Dr A. Drake of University College London for obtaining the CD spectra and for his advice and assistance in their interpretation.

J.P. Ide wishes to thank the U.S. Army and D.R. Klug the SERC for financial support during this research.

REFERENCES

1 Kuhlbrandt, W. (1984) Nature 307, 478-480
2 Knight, A.E.W. and Selinger, B.K. (1973) Australian J. Chem. 26, 1
3 O'Connor, D.V. and Phillips, D. (1984) in 'Time-Correlated Single-Photon Counting', Academic Press, London
4 Kuhlbrandt, W., Thaler, T. and Wehrli, E. (1983) J. Cell Biol. 96, 1414-1424
5 Van Metter, R.L. (1977) Biochim. Biophys. Acta 462, 642-658
6 Shepanski, J.F. and Knox, R.S. (1981) Israel J. Chem. 21, 325-331
7 Ide, J.P., Klug, D.R., Kuhlbrandt, W., Giorgi, L.B., Porter, G., Gore, B., Doust, T. and Barber, J. (1986) Biochem. Soc. Trans. 14, 34

CHARACTERIZATION OF THE FLUORESCENCE DECAYS OF THE CHLOROPHYLL A/B
PROTEIN

D. D. Eads, S.P. Webb*, T. G. Owens, L. Mets, R. S. Alberte, G. R.
Fleming*
Dept. of Molecular Genetics and Cell BIology, *Dept of Chemistry,
Univ. of Chicago, Chicago IL 60637

INTRODUCTION

Characterization of chlorophyll excited state lifetimes
provides a sensitive means for investigating the excitation
migration and trapping kinetics in the photosynthetic apparatus.
The fluorescence decay of intact photosynthetic membranes is complex
and the interpretation of these decays is nontrivial. We have been
using biochemical purification of photosynthetic subcomponents and
genetic manipulation of whole cells to simplify analysis and resolve
features that may be hidden in the more complex in vivo decays.

The Chl a/b protein was examined by using both of these
approaches in order to characterize more precisely this ubiquitous
protein. In the first stage of experiments the protein was isolated
from barley and characterized under a variety of ionic and
denaturing conditions in vitro. Secondly, we studied the protein in
vivo using the Chlamydomonas mutant, C2, which does not has neither
reaction center I nor II. We have found that 1) the in vitro
protein has an ionic strength dependent absorption spectrum that
does not seem to affect the fluorescence decay; 2) the time resolved
fluorescence of the in vitro protein is a biexponetial decay of the
form $f(t) = 15\exp(-t/0.7 \pm .15 \text{ ns}) + 85\exp(-t/3.2 \pm 0.2 \text{ ns})$
demonstrating that >90% of the steady state fluorescence is due to
the 3.2 ns component; and 3) the in vivo decay of the C2 mutant is a
single exponetial with a lifetime of 2.4 ±0.1 ns.

MATERIALS AND METHODS

The Chl a/b protein was isolated from barley by the method of
Lotshaw et.al (1) with the following modifications: 1) the washed
membranes were homogenized in a solution of 1% sodium dodecyl
sulfate in 20 mM tricine (pH 7.2) (1 ml/mg chlorophyll) for 2.5
minutes and immediately centrifuged to remove insoluble material;
and 2) the green eluate of the hydroxylapatite column was collected
in fractions, each fraction was brought to 0.31 mM Triton X-100,
those fractions which had a minimal chlorophyll b emission were
combined, and the combined fractions were desalted by passage
through a Sephadex G-25 column equilibrated with 1.8 mM SDS, 0.31 mM
Triton X-100, 20 mM Tricine (pH 7.4).   The C. reinhardii strain
C2 was grown in partial light to a typical cell density of 1x 10^5
cells/ml in an acetate containing medium. The C2 strain has

Biggens, J. (ed.), Progress in Photosynthesis Research, Vol. I. ISBN 90 247 3450 9
© 1987 Martinus Nijhoff Publishers, Dordrecht. Printed in the Netherlands.

acquired a secondary mutation heritable starch granule   that is
characterized by an extremely large starch granule encasing the
pyrenoid (E. Beasley and L. Mets personal communication). C2
cultures in late log phase growth display the starch accumulation so
extreme that normal cell division is blocked and aggregation occurs.

Steady state absorption measurements were performed on an
Aminco DW 2 and the steady state fluorescence measurements were
performed with a corrected Aminco SPF 500. Fluorescence decays were
determined by time correlated single photon counting (2).

Results and Discussion

The Chlamydomonas mutant C2 is a PSI⁻ PSII⁻ strain that has no
detectable P700 activity, does not show fluorescence induction, and
lacks the reaction center proteins which are normally detectable
upon gel electrophoresis (L. Mets unpublished). The C2 strain can
thus be used to establish the behavior of the Chl a/b protein in the
thylakoid membrane. The absorption spectrum of C2 is
identical to that of the in vitro Chl a/b protein.   The
fluorescence decay of the C2 mutant fits well to a single exponetial
decay with a lifetime of 2.4 ±0.07 ns. The lifetime of 2.4 ns is
independent of excitation and emission wavelengths at the
experimental excitation wavelengths of 652 and 665 nm and the
emission range of 680 to 720 nm.   We attribute the very slight
nonexponetiality observed in our previous study (2) to cultures
which contained nonphysiological cells (see Materials and Methods).

Figure 1 shows the absorption spectra of the isolated Chl a/b
protein from barley at 0.24 M and 0.12 M Na_2HPO_4 and in salt free
buffer (pH 7.2). As the salt concentration is raised the Chl a and
Chl b absorption peaks broaden, but do not shift in wavelength. The
two simplest explanations for this salt effect are that either the
number of peptides per micelle increases with increasing ionic
strength or that specific ionic interactions exist between the
protein and the ions. Since we can reproduce the same effects with
$MgCl_2$, NaCl, or Na_2HPO_4 and since the same broadening occurs at
nearly equal ionic strengths of the solutions, the existance of
either a specific cationic or anionic interaction with the protein
seems unlikely. To understand the characteristics of a salt-
dependent aggregation of the protein in the micelle, the details of
the micelle-protein environment must be considered.

Little is known about the macrochemistry of mixed detergent
solutions and even less on mixed detergent-protein solutions.
Though, the micelles that are formed in these experiments are a
mixture of both SDS and Triton X-100 (3). The Triton X-100
concentration of the protein solution is 0.31 mM which is in excess
of the critical micelle concentration (cmc) of Triton X-100 (.24 mM,
3).   Furthermore, since the ionic strength of a solution has an
inverse effect on the cmc (the number of detergent monomers per
micelle), all of the detergent-salt buffers have detergent in excess
of the cmc. One possible effect of an increased ionic strength is
to favor the formation of larger micelles. Thus, these larger

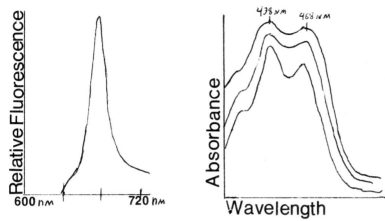

438 NM   468 NM

Relative Fluorescence

600 nм   720 nM

Absorbance

Wavelength

Figure 1   Fluorescence emission spectrum of Chl a/b protein;
Figure 2   Absorption spectra of Chl a/b protein;   salt free,
            0.12 M Na$_2$HPO$_4$,   0.24 M Na$_2$HPO$_4$.

micelles could contain more peptides. Another possibility is that
salt-induced aggregation of the Chl a/b protein could be occuring.
Aggregation of the protein is known to occur at high salt
concentrations and is exploited in the precipatation of the protein
as crystals (4).   Whether the number of peptides present in a
micelle increases under higher ionic strengths is presently being
investigated.

     The fluorescence spectra of the isolated Chl a/b protein and the
C2 mutant are shown in Figure 2.   The fluorescence spectrum is a
sensitive measure of the status of the in vitro protein.   When the
protein denatures the efficient energy transfer from Chl b to Chl a
no longer occurs and Chl b fluorescence is observed.   The C2 mutant
(Fig. 2) shows that the Chl a/b protein in its native environment
has no Chl b emission, whereas our best isolates of the Chl a/b
protein have a small shoulder at 652 nm that is attributed to Chl b
emission.   The emission spectra of the high salt and salt free
protein solutions are indistinguishable.

     The fluorescence decay of the isolated Chl a/b protein fits
well to a biexponetial function of the form f(t)= 15exp(-t/0.7 + .15
ns) + 85exp(-t/3.2 + .2 ns).   Over 97% of the steady state
fluorescence is accounted for by the 3.2 ns component in our best
preparations. Further, the integrity of the samples can be checked
by time resolved fluorescence measurements.   Depending on the degree
of denaturation of the protein via time or heat, the fits to the
fluorescence decays change.   The long component increases in
lifetime while the short component decreases and the inclusion of a
third component becomes necessary to achieve a good reproduction of
the data.   As the protein denatures two possible effects occur which
disrupt the Forster transfer of energy from one chromophore to
another: 1) the chlorophylls may dissociate from the protein and
become free in the core region of the micelles; and/or 2) the

chlorophylls may remain associated with the denatured protein but assume new positions relative to each other that would lead to either a higher probability of nonradiative quenching (decreased separation) or a lower probability of both energy transfer and nonradiative quenching (increased separation). The changes on denaturation are very similar to those observed by Ide et. al (1986), where the authors isolated the Chl a/b protein from pea seedlings and maintained the protein in a n-octyglucoside detergent solution. We propose that the biexponetial fluorescence behavior of the in vitro Chl a/b protein is most likely due to a small amount of denaturing of protein inherent in the isolation procedure.

The disparity between the C2 mutant lifetime of 2.4 ns and the lifetime of in vitro protein from barley of 3.2 ns is not fully understood. Two possiblities exist which are presently being investigated. The first possibility is that the Chl a/b protein from barley and Chlamydomonas are inherently different. An experiment is underway to answer this question by isolating the protein from C2 and measuring its fluorescence decay. A second possibility is that the macroscopic environment of the protein in the micelle causes the protein to have a slightly different structure, aggregation states, or boundary interactions and these effects lead to the increased lifetime. To test this possibility a technique was developed that of reconstitutes the protein into mixed lipid vesicles containing the lipids sulfoquinovosyldiacylglycerol, monogalactosyldiacylglycerol and digalactosyldiacylglycerol. These vesicles, when formed with the proper ratio of the lipids, retain the steady state fluorescence properties of the native sample (Figure 3). By determining the fluorescence decays of these Chl a/b protein vesicles we will be able to establish if the protein must be in a thylakoid lipid environment to emulate in vivo conditions.

REFERENCES

(1) Lotshaw, W.T., R.S. Alberte, G.R. Fleming (1982) Biochim. Biophys. Acta 682, 75-82

(2) Gulotty, R.J., L. Mets, R.S. Alberte and G.R. Fleming (1985) Photchem. Photobiol. 41, 487-496

(3) Tanford, C. The Hydrophobic Effect: Formation of Micelles and Biological Membranes 2nd ed., John Wiley & Sons, N.Y. (1980)

(4) Kuhlbrandt, W. Nature 307, 478-480 (1984)

(5) Ide, J.J., B. Crystall, D.R. Klug, B.L. Gore, L.B. Giorgi, W. Kuhlbrandt, J. Barber, and G. Porter; VII International Congress on Photosyntesis, 1986.

FLUORESCENCE DECAY AND DEPOLARIZATION KINETICS CALCULATED USING FÖRSTER
INDUCTIVE RESONANCE AND THE MOLECULAR COORDINATES FOR C-PHYCOCYANIN

KENNETH SAUER, CHEMISTRY DEPARTMENT, UNIVERSITY OF CALIFORNIA, BERKELEY,
CALIFORNIA, 94720, USA; AND HUGO SCHEER, BOTANISCHES INSTITUT DER
UNIVERSITÄT MÜNCHEN, MENZINGERSTR. 67, D-8000 MÜNCHEN 19, FRG

1. INTRODUCTION

Excitation energy transfer among photosynthetic antenna pigments is an
ubiquitous process in nature for producing efficient photon capture. The
transfer of excitation to the photosynthetic reaction centers occurs
rapidly, within a few hundred picoseconds, and with minimal loss of energy
to fluorescence or heat. It has been proposed that this transfer occurs
through a combination of exciton interactions among closely coupled
chromophores and inductive resonance interactions among weakly coupled
chromophores of pigment protein complexes. Detailed descriptions of the
nature of this process have been lacking because, until recently, the
molecular structures of the pigment proteins have not been known. The
first publications of such structures with sufficient resolution to
determine both the positions and orientations of the intrinsic chromophores were presented by Schirmer, et al. for C-phycocyanins (C-PC) from
two different organisms (1,2). Examination of the structures supports the
conclusion from spectroscopic measurements that exciton interactions
probably do not play an important role in C-PC (3).

We have applied the inductive resonance mechanism of Förster to C-PC
to calculate the rate constants for successive pairwise excitation
transfer steps among the chromophores. Because C-PC consists of two
distinct subunits--the α-subunit with one covalently attached phycobilin
chromophore and the β-subunit with two, and because these can be prepared
in $(\alpha\beta)$-monomeric, $(\alpha\beta)_3$-trimeric and $(\alpha\beta)_6$-hexameric forms (3), the
calculations have been carried out for aggregates of different sizes.
Experimental data are available for the kinetics of fluorescence intensity
decay and for the rate of depolarization of fluorescence on the picosecond
to nanosecond time scale for a variety of these preparations (4-9).
Comparison of the calculated results with the best time resolved data
indicates that the Förster transfer mechanism is sufficient to account for
essentially all of the excitation energy transfer in C-PC.

2. PROCEDURE

Details of the method of calculation are described in a related
publication (10). Briefly, the molecular structure information for C-PC
from Agmenellum quadruplicatum was used to specify the interchromophore
separations and the transition dipole moment orientations (2). The
spectral overlap integrals were determined using absorption and emission
spectral components obtained from a deconvolution of the experimental
spectra of the α- and β-subunits and of the $\alpha\beta$-monomer. Values of the
extinction coefficients and fluorescence yields were taken from Mimuro et
al (11). Fluorescence lifetimes of the chromophores were based on
measurements of the α-subunit by Switalski and Sauer (12) and by Hefferle
et al. (5). Chromophore assignments were α-84 = intermediate (λ,max 616
nm), β-84 = fluorescer (λ,max 622) and β-155 = sensitizer (λ,max 598)(13).

Biggens, J. (ed.), Progress in Photosynthesis Research, Vol. I. ISBN 90 247 3449 5
© *1987 Martinus Nijhoff Publishers, Dordrecht. Printed in the Netherlands.*

A set of simultaneous differential equations was formulated to describe the kinetics of excited state populations for each chromophore in the pigment protein. Because of the extensive symmetry of the larger aggregates, the calculational procedure could be appreciably simplified. Where desired, specification of the excitation and emission wavelengths was accounted for in terms of weighting parameters associated with the different chromophores. The differential equations were solved by an iterative (reverse Euler) method using a personal computer. (The authors wish to acknowledge the contribution of Peter Sauer, Carleton College, Northfield, MN in writing most of the programs used in the calculations). Programs were developed to calculate the time evolution of the excited state lifetimes and of the fluorescence emission expected. These results were then analyzed in terms of two or three exponentials, as required by the symmetry of the molecular species.

Depolarization kinetics were calculated using the data describing the time evolution of the excited state population of each chromophore in the complex. Excitation was assumed to initiate on one of the three chromophore types, and fluorescence was assumed to be depolarized by a factor of $(3 \cos^2 \theta - 1)/2$, where θ is the angle between the chromophore initially excited and the fluorescing chromophore (14). The sum of contributions from each chromophore in the complex, normalized by the corresponding excited state population, produced a time-dependent depolarization profile.

3. RESULTS AND DISCUSSION

Relaxation kinetics to compare with experimental fluorescence (isotropic) lifetime measurements were calculated for the separate β-subunit, the $\alpha\beta$-monomer, the $(\alpha\beta)_3$-trimer and the $(\alpha\beta)_6$-hexamer. In the kinetic model the excitation was localized initially on one chromophore in the complex, and the transfer to other chromophores was calculated using the transfer rate constants obtained from the Förster formulation. Radiative plus non-radiative decay processes led to a loss of excited state population with a rate constant $k_F = 0.67$ ns^{-1}, corresponding to the experimental longest lifetimes of approximately 1.5 ns. This entire calculation was then repeated for excitation present initially on a chromophore of a second class and, (except for the β-subunit), then on a chromophore of the third class. The results of these two or three calculations were multiplied by weighting parameters appropriate to the particular excitation and emission wavelengths. From these results describing the time evolution of the excited state population of all chromophores present, those for the individual chromophore classes or for the total excited state population were obtained by appropriate summations. The symmetry requires that the calculated decays are the sums of no more than three exponential components (two in the case of the β-subunit), one of which is the excited state decay with a 1.5 ns lifetime (15). The other shorter lifetimes are presented along with some recent results of time-resolved fluorescence relaxation in Table I. Relative amplitudes of the individual decay components are not included; they will depend on the particular excitation and emission wavelengths. Holzwarth (personal communication) has solved the sets of differential equations by inverting the matrix of coefficients (rate constants) and obtained identical values for the lifetimes and amplitudes of the exponential components.

The data are compared to experimental results on C-PC of different aggregate sizes isolated from the cyanobacteria, Synechococcus 6301 (4,15)

and <u>Mastigocladus</u> <u>laminosus</u> (5,7,16). The known structure of the latter
PC (1) is similar to that of <u>A</u>. <u>quadruplicatum</u> (2), and a similar arrangement is also believed for the former on the basis of extensive sequence
homologies (17). The experimental decay profiles have been resolved into
a sum of 2-4 exponentials. General agreement is seen between the calculated and observed lifetimes, especially in the decreasing value of the
short lifetime with increasing extent of aggregation. The intermediate
decay component is not well matched by the experimental data; however,
there is a fourth component resolved experimentally in several studies and
that is not predicted by our model. This fourth component may be related
to the second one seen for the isolated α-subunit, where only one
chromophore is present and only one decay component is expected (5,12).
Resolution of these problems must await further study.

<u>Table 1</u>. Comparison of calculated with measured lifetimes. Each
calculated relaxation curve is the sum of three exponentials (except, only
two for the β-subunit). The slowest (τ_F = 1500 ps) reflects the overall
excited state decay. Comparison is with wavelength-resolved experimental
measurements. All lifetimes are given in psec.

	These Calculations	Synechococcus 6301 (4,15)	M. laminosus (7)	M. laminosus (16)
β-subunit	48, 1500			
αβ - Monomer	45	47		
	700	200		
	1500	675, 1320		
(αβ)$_3$- Trimer	16	20	36	45-61
	27	122	203	
	1500	600, 1300	807, 1420	1130-1640
(αβ)$_6$- Hexamer	15	10		
	18	40-50		
	1500	1800		

The anisotropic fluorescence (depolarization) relaxation kinetics was
extracted from the excited state populations by taking into account the
relative orientations of the chromophore that initially absorbed the
radiation and the one(s) that finally emit it (factorization with
$(3 \cos^2\theta-1)/2$). Again, symmetry may be invoked to decrease the number of
initially excited chromophores to one member of each class, but now each
emitting chromophore needs to be treated separately with respect to the
orientation factor prior to the final summation. It was possible to
"resolve" the time dependence of anisotropic fluorescence decay empiri-
cally into three exponential components. In this case none of the decays
corresponds to the excited state lifetime, and there is no reason to
expect that the deconvolution is exact. The results of this analysis are
presented in Table II along with the available experimental data. The
agreement is less satisfactory than for the isotropic fluorescence decay;
however, more extensive experimental results are needed, especially with
picosecond time resolution, before definite conclusions can be drawn. It
is significant to compare the fastest (5-15 ps) depolarization calculated
for the (αβ)$_6$-hexamer with the experimental values of 10 ps for phycobilisomes from <u>Synechococcus</u> (8). The fastest depolarization steps were
attributed to excitation transfer in C-PC in the phycobilisomes.

Table II. Comparison of calculated with measured depolarization lifetimes.
Each calculated depolarization relaxation curve is resolved into 1 to 3
exponentials. Values obtained assuming (1) equal initial excitation and
(2) equal emission intensity for each chromophore class. All values in
psec.

	These Calculations	M. laminosus (5)	M. laminosus (16)
β-subunit	50	403	
$\alpha\beta$-Monomer	46, 1300	580	
$(\alpha\beta)_3$-Trimer	8, 24, 46	70	36-57, 800-1150
$(\alpha\beta)_6$-Hexamer	5, 15, 31		

ACKOWLEDGEMENTS We wish to thank Peter Sauer for writing the programs
used in solving the simultaneous differential equations. This work was
supported by the Alexander-von-Humboldt Stiftung, Bonn (award to K.S.) and
by the Deutsche Forschungsgemeinschaft (SB 143 and the CIP botany computer
facilities, H.S.).

REFERENCES
1 Schirmer, T., Bode, W., Huber, R., Sidler, W. and Zuber, H (1985) J.
   Mol. Biol. 184, 257-277
2 Schirmer, T., Huber, R., Schneider, M., Bode, W., Miller, M. and
   Hackert, M.L. (1986) J. Mol. Biol. 188, 651-676
3 Scheer, H. (1982) in Light Reaction Path of Photosynthesis (Fong, F.K.
   ed.) pp. 7-45, Springer-Verlag, Berlin
4 Holzwarth, A.R. (1985) in Antennas and Reaction Centers of
   Photosynthetic Bacteria (Michel-Beyerle, M.E. ed.), pp. 45-52,
   Springer-Verlag, Berlin
5 Hefferle, P., Geiselhart, P., Mindl, T., Schneider, S., John, W. and
   Scheer, H. (1984) Z. Naturforsch. 39c, 606-616
6 Hefferle, P., John, W., Scheer, H. and Schneider, S. (1984) Photochem.
   Photobiol. 39, 221-232
7 Wendler, J., John, W., Scheer, H. and Holzwarth, A.R. (1986) Photochem.
   Photobiol. 44, 79-86
8 Gillbro, T., Sandström, Å., Sundström, V., Wendler, J. and Holzwarth,
   A.R. (1985) Biochim. Biophys. Acta 808, 52-65
9 Earlier and related work is reviewed in: Holzwarth, A.R. (1986)
   Photochem. Photobiol. 43, 707-725
10 Sauer, K., Scheer, H. and Sauer, P. (1986) Photochem. Photobiol.
   Submitted for publication
11 Mimuro, M., Füglistaller, P., Rümbeli, R. and Zuber, H. (1986) Biochim.
   Biophys. Acta 848, 155-166
12 Switalski, S.C. and Sauer, K. (1984) Photochem. Photobiol. 40, 423-427
13 Siebzehnrübl, S., Fischer, R. and Scheer, H. (1986) Z. Naturforsch. C.
   Submitted for publication, see also: Scheer, H., these proceedings
14 Dale, R.E. and Eisinger, J. (1975) in Biochemical Fluorescence:
   Concepts (Chen, R.F. and Edelhoch, H., eds.) Vol 1, pp 115-284, Marcel
   Dekker, NY
15 Holzwarth, A.R. (1986) Biophys. J. In press
16 Schneider, S., Geiselhart, T., Scharnagl, C., Schirmer, T., Bode, W.,
   Sidler, W. and Zuber, H. (1985) Book, ref. 4, pp. 36-44
17 Zuber, H. (1985) Book, ref 4, pp. 2-14

PHOTOCHEMISTRY AND PHOTOPHYSICS OF C-PHYCOCYANIN

Hugo Scheer, Botanisches Institut der Universität, Menzinger Str.67
D-8000 München 19, Federal Republic of Germany

Introduction

Phycocyanin (**PC**) belongs to a group of pigments functional for light-harvesting in cyanobacteria, red and cryptophyte algae. In the former two classes of organisms, it is a major constituent of phycobilisomes, the light-harvesting complexes located at the outer surface of the photosynthetic membrane. There, it absorbs light energy in the spectral range between 580 and 640 nm, and transfers it via a second biliprotein, allophycocyanin (**APC**), to the chlorophyll within the membrane. In many species, PC also accepts energy from a third type of biliprotein, e.g. phycoerythrin (**PE**), thus acting as an intermediate carrier in the energy transfer from the latter to **APC**.

The simplest **PC**, which is found in cyanobacteria (**C-PC**), contains three chromophores of the dihydrobilindion type, each of them being attached covalently to the apoprotein via a single thioether bond to cysteine. The same chromophore is present in APC, and a chromophore differing only in one of the β-pyrrolic substituents is found in the plant photomorphogenetic pigment, **phytochrome**. According to the different functions of these three pigments, the properties of the chromophores in each of them are quite different from each other, and they all differ considerably from the properties of free pigments bearing this type chromophore (1). The factors responsible for the different adaptations of these structurally so similar chromophores are still only partly understood. From reversible denaturation studies in **C-PC**, it appears that they are mostly due to non-covalent protein-chromophore interactions.

The recent elucidation of the x-ray structures (2,3) of **C-PC**s from two different organisms, has greatly advanced our knowledge of these pigments. It has for the first time in any photosynthetic antenna system become possible to 'look' at the native chromophore structures on a molecular level, and to obtain direct information on their conformations and relative orientations. This renders it possible to test the viability of theoretical models applied in the calculation of the their spectral properties, of the energy transfer pathways, the kinetics among them, etc., by using the structural data as input parameters.

This report summarizes recent work carried out along these lines in München. It contains data on the photochemistry of **C-PC** from the cyanobacterium, Mastigocladus (M.) laminosus, which are compared to the respective properties of the photochromic plant photoreceptor, phytochrome, as well as theoretical and experimental results on the energy transfer in aggregates of different sizes from this chromoprotein.

Biggens, J. (ed.), Progress in Photosynthesis Research, Vol. I. ISBN 90 247 3450 9

Chromophore structure

The three chromophores of **C-PC** are bound to the protein via single thioether bonds to cysteine 84 on the α-subunit (cys α-84), and to cys β-84 and cys β-155 on the β-subunit. A linkage to the C-3 ethyl-substituent of the hydrogenated ring A had been established for chromophores of all type of plant biliproteins including phytochrome (1,4). A more complex binding pattern has recently been proposed, which involves a different linkage and structure for the chromophore β-155 (5). Instead of being hydrogenated at ring A, this chromophore is hydrogenated at ring D and bound via the 'exo' 18-thioethyl substituent.

All chromophores are present in more or less extended conformations (2,3), which account for the observed absorption increase of the red bands, and a concomitant decrease of the near-uv bands (1). It is also likely, that the chromophores have a reduced conformational mobility, which accounts for their high fluorescence quantum yields, and for their inertness to a variety of chemical reagents (metal ions, reducing agents) known to react readily with the free chromophores (see 1 for leading references). The confirmation of this general structural principle in the crystal structure of C-PCs from two different organisms (<u>M. laminosus</u> and <u>Agmenellum (A.) quadruplicatum</u>), makes it likely that the spectrally similar chromophores of **APC** and phytochrome have similar native structures as well.

Chromophore assignment

While the x-ray results supported this general structure principle, they show on the other hand pronounced differences among the details of the chromophore conformations, of the binding sites and of likely interactions with the apoprotein. Such differences account for a variety well documented spectroscopic and chemical results indicating the presence of a set of distinct chromophores in almost any phycobiliprotein (1). In **C-PC**, these allowed the definition of three distinct chromophores, e.g. α-1, β-1 and β-2.

In the case of **C-PC** from <u>M. laminosus</u>, the following data were combined for the spectral resolution: The integral pigment can be separated into two subunits, the α-subunit bearing only one chromophore, and the β-subunit bearing two chromophores. Since the absorption spectra of the two subunits -weighted properly according to the subunit stoichiometry- add up to the spectrum of the monomeric pigment, it is likely that the states of the chromophores remain unchanged during subunit separation, and that strong inter-subunit chromophore-chromophore interactions are absent. This yields directly the required absorption and fluorescence spectra of the α-subunit, and reduces the problem to the resolution of the β-subunit spectrum. The presence of two spectrally distinct chromophores is this subunit is derived from several lines of evidence:

Reversible photochemistry: Native **PC** has a high fluorescence quantum yield. Its photochemistry is characterized by an irreversible bleaching, which has a low quantum yield (0.4%) and proceeds probably via the triplet state because it is slowed down in the presence of oxygen. This irreversible reaction occurs with a similar quantum yield also in phytochrome (Scheer, unpublished results). Addition of urea to **PC** causes a gradual, reversible unfolding of the protein and a concomitant loss of its interactions with the chromophore. At 8 M urea, the protein is completely denatured, and the chromophores then attain the properties characteristic of free bile pigments (1). Here, the fluorescence is greatly reduced ($\phi < 10^{-3}$). The photochemical reactivity is increased, but it is again irreversible, leading to a variety of tri- and tetrapyrroles absorbing at shorter wave lengths. The onset of unfolding at intermediate urea concentrations is characterized by a reduction of fluorescence and the concomitant occurrence of a reversible photochemical reaction, which is maximum at about 5 M urea. Similar reactions have been observed as well with other denaturants at moderate concentrations, and they have been related to the reversible Z,E-isomerization of the phytochrome chromophore, and to the primary reactions of the less well understood phycochromes (see 1). If the isolated subunits of **C-PC** from M. laminosus are titrated with urea this reaction is negligibly in the α-subunit, but much more pronounced in the β-subunit. The absorption difference spectrum of the latter shows a single negative band in the visible spectral range peaking at 624 nm, which is considerably to the red of the absorption maximum at 606 nm, and its shape is similar to that of a typical bile pigment (6). This suggests, that only one of the two chromophores on the β-subunit is susceptible to this reaction, and that this chromophore absorbs at longer wavelengths than the second, inactive one.

Fluorescence polarization: The fluorescence polarization spectra of nearly all phycobiliproteins show distinct discontinuities, and the anisotropy rises in discrete steps towards longer wavelengths (13, 14, see also 1). Since the red absorption band of bile pigments corresponds to a single electronic transition, this has been interpreted as the result of several distinct chromophores being present, with different orientations and different absorption spectra, among which energy transfer occurs. The fluorescence polarization spectrum of the β-subunit of **C-PC** from M. laminous shows two distinct regions of anisotropy (7). Below 600 nm, it is nearly wavelength independent about 0.2, and then rises sharply to 0.4. There are, therefore, at least two chromophores present absorbing below and above this threshold wavelength. Assuming a similar Stokes' shift for the fluorescing chromophore of the β-subunit ($\lambda_{max}^{f\,luor} = 643$ nm) and the one of the α-subunit ($\lambda_{max}^{a\,bs} = 616$, $\lambda_{max}^{f\,luor} = 641$ nm), an absorption around 620 nm can be estimated for the former.

Circular dichroism: The cd spectrum of the α-subunit shows a single positive band in the visible spectral region peaking close to its absorption maximum. The cd-spectrum of the β-subunit also shows a single positive peak. Its intensity is decreased by 40% on a molar basis, increased by 20% on a chromophore basis, and centered well to the blue (590nm) of the absorption maximum. At longer wavelengths, the band trails slightly and indicates the presence of a smaller, much less intense band. This result, which has been reported independently by Mimuro et al. (7), is again, best interpreted as to arise from two different chromophores, one of them is strongly optically active and absorbs around 595 nm, the second one is much less active and absorbs above 610 nm.

Curve resolution of absorption spectrum: To better define the absorption bands of these chromophores, the spectrum of the β-subunit was resolved by computer analysis. It was assumed, that the shape and width of the bands were identical to that of the alpha chromophore, and the starting wavelengths for the analysis were estimated from the aforementioned data. The absorption band was fit best with two bands peaking at 598 and 622 nm, with molar absorptivities of 92 and 60%, respectively, of that of the α-subunit. For an estimation of the individual fluorescence spectra, similar Stokes' shifts and fluorescence lifetime was furthermore assumed. The resulting spectral data are given in table 1 and compared to a similar analysis by Mimuro et al. (7).

Chemical reactivity and assignment to binding sites

Whereas the aforementioned results allowed the distinction of two spectrally defined chromophores on the β-subunit, the correlation between the different chromophores defined above (α-1, β-1 and β-1), and the ones defined by their binding sites (α-84, β-84, β-155), respectively, was still lacking. Following a suggestion by Schirmer, Bode and Huber (2,3), we have been able to make this assignment by treatment of **C-PC** with organic mercurials (8). There is only a single free cysteine present in **C-PC** located at position β-111 (9), which is very close to the chromophore β-84, but more than 22A from chromophores β-155 and β-84 (2,3). This cysteine is the only site to which mercurials were bound in heavy-atom derivatives of **C-PC** crystals, and it was expected that a binding of the bulky reagent at this position would have a discernible effect on the spectral properties of the neighboring chromophore, and only on this one. In solution, titration of **C-PC** from M. laminosus leads to a decrease of absorption at about 620nm, and a concomitant, albeit smaller, increase around 650nm, and a general absorption increase in the near-uv spectral region (Fig. 1). In the β-subunit, the same spectral changes are observed, but the relative amplitudes are increased. Titration experiments showed, that the reaction is complete after the addition of 1 ± 0.2 moles of the mercurial (p-chloro-mercuribenzenesulfonate, PCMS) per mole of **C-PC** or β-subunit , respectively, and that the reaction can be reversed to more than 80% by addition of thiols. In the α-subunit, the reaction is negligible (Fig.1).

These findings are interpreted in the following way:

1: The reaction site is the single free cys-111 on the β-subunit, similar to the situation in the crystal.
2: No irreversible reaction occurs between mercurials and the chromophores, because the α-subunit is inert.
3: Since the difference absorption maximum is on the red side of the absorption maximum in the β-subunit, it must be related to a spectral change of the long-wavelength absorbing chromophore, e.g. the one defined above spectroscopically as β-1.
4: Due to the spatial relationships, this chromophore is the one bound to cys-β-84, e.g. chromophore β-84 is identical with chromophore β-1.
5: The indirect effect of the mercurial binding on the absorption of chromophore β-84 involves probably a conformational change from the native, extended to a more denatured, cyclic-helical conformation, as indicated by the overall decrease of absorption in the red, and an increase in the near-uv spectral region (1).
6: The resulting spectral data for the three chromophores are summarized in table 1.
7: Due to the inertness of the chromophores to a direct reaction with mercurials, these reagents are suitable to test the accessibility of cys-111 in higher aggregates, and at the same time to identify the absorption of the β-84 chromophores.

	α-84	β-1 = β-84	β-2 = β-155
This work	616(120)	622(72)	598(106)
Mimuro et al. (7)	618(108)	624(103)	594(113)

Table 1: Absorption maxima [λ_{max} (ϵ x 10^{-3})] of the individual chromophores of C-phycocyanin from Mastigocladus laminosus.

The assignment (β-1 = β-84, β-2 = β-155) agrees with the ad-hoc assignment of by Mimuro et al. (7), and there is also a reasonable agreement of the spectral data of the individual chromophores. The most pronounced differences are the position of the β-155 absorption, which is displaced to the blue by appx. 4nm by these authors, and a lower absorptivity of the β-84 absorption in our calculations.

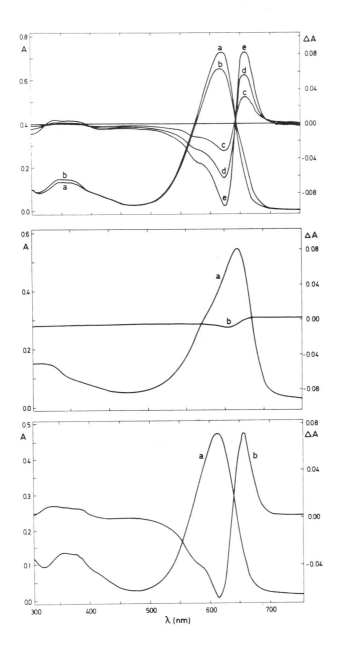

Fig.1: Reaction of trimeric C-phycocyanin from <u>Mastigocladus laminosus</u> (A), its α- (B) and β-subunit (C). Figure (A) shows both the absorption spectra before (a) and after (b) reaction with **PCMS** (1 mole / mole **C-PC**), and the difference spectra during the titration (c,d,e). The other figures show the absorption spectrum before addition of **PCMS**, and the absorption difference spectrum after its addition (1 mole / mole subunit). All reactions in potassium phosphate buffer (50mM, pH 7.5).

Kinetics of energy transfer

Energy transfer in phycobiliproteins has been suggested to occur mainly via the induced dipole or Förster type mechanism. The main arguments to this have came originally from the absence in most biliproteins, including **C-PC** from M. laminosus, of strong, s-shaped signals indicative of strong exciton couplings, and from polarization spectroscopy (see 1). Over the past 5 years, the amount of kinetic data on energy transfer has tremendously increased. It has been found in particular, that the fastest energy transfer processes take place on a time scale in the order of tens of picoseconds or even less (10,11). With all the necessary information at hand, it was then intriguing to test if such fast kinetics could be matched by theory on the basis of a pure Förster transfer mechanism.

The rate constant for energy transfer is described by:

$$k \simeq \frac{\kappa^2}{R^6} \int F_{Donor} \cdot A_{Acceptor}$$

The major variables in this equation are the distance R between the transition dipoles of the donor and acceptor chromophores, the relative orientations factors κ of the two, and the overlap of the fluorescence of the donor (F_{Donor}) with the absorption of the acceptor ($A_{Acceptor}$). The only other variable in the equation, e.g., the refractive index of the medium, is hidden in the proportionality constant. A value of 1.56 has been used throughout. The first two pieces of information have been taken from the x-ray data of Schirmer et al. for **C-PC** from A. quadruplicatum (3). This pigment has a very similar chromophore arrangement as **C-PC** from M. laminosus (2), but it has been resolved to higher accuracy. Moreover, it crystallizes as stacks of hexamers rather than trimers, so that orientations and distances between chromophores on different trimers and hexamers are available. These authors have tabulated the distances of the centers of gravity of the π-systems for all chromophores, as well as the relative orientations of chromophores as defined by the masses of the atoms present in the chromophore π-systems projected on a straight line. These distances and orientations do, therefore, not strictly correspond to the transition dipoles, but rather to their reduced masses. In view of the elongated structure of the chromophores, it is likely, however, that the deviations are reasonable. The overlap integral was finally calculated from the individual absorption and fluorescence bands of the three different chromophores as given in the top row of table 1.

The details of these calculations, which are the result of a continuing cooperation between K. Sauer and our group, are being published elsewhere (12), and a summary is presented in the poster abstracts of this conference. There is a good agreement with most of the currently available experimental data. The calculations show an increased transfer rate with increasing aggregate size. They support, in particular, a preferential energy transfer along the rods of trimer-stacks, as compared to energy transfer within trimers. Such a preferential transfer would greatly facilitate the funelling of energy towards the reaction centers. According to these results, the energy transfer in **C-PC** can be accounted for well by the Förster mechanism, and the flow of excitation energy in these moderately complex aggregates can be analyzed on a molecular basis. Since the data can be transformed readily to mimic a variety of experimental conditions (different excitation and emission wavelengths, static and dynamic depolarization), a comparison with new data and/or assistance in the choice of experimental conditions are expected to further evaluate this conclusion critically.

Acknowledgements

This work was supported by the Deutsche Forschungsgemeinschaft, Bonn (SFB 143 and Forschergruppe "Pflanzliche Tetrapyrrole"). The cooperation with K. Sauer, Berkeley, was rendered possible by an Alexander-von-Humboldt Award. The following coworkers were involved in this work as detailed in the references: R. Fischer, W. John, G. Schmidt, G. Schoy, S. Siebzehnrübl (all Botanisches Institut der Universität München) and P.Sauer (Carlton College).

References

1. Scheer, H. (1982) in Light Reaction Path of Photosynthesis (Fong, F.K., ed.) pp. 7-45, Springer-Verlag, Berlin.
2. Schirmer, T., Bode, W., Huber, R., Sidler, W. and Zuber, H. (1985) J. Mol. Biol. 184, 257-277.
3. Schirmer, T., Huber, R., Schneider, M., Bode, W., Miller, M. and Hackert, M. L. (1986) J. Mol. Biol. 188, 651-676.
4. Rüdiger, W. and Scheer, H. (1983) in Handbook of Plant Physiology, Vol. 16 (Shropshire, W. and Mohr, H., eds), pp. 119-151, Springer Verlag, Berlin
5. Bishop, J. E., Lagarias, J. C., Nagy, J. O., Schoenleber, R. W., Rapoport, H., Klotz, A. V. and Glazer, A. N. (1986) J. Biol. Chem. 261, 6790-6796.
6. Siebzehnrübl, S., Fischer, R. and Scheer, H. (1985) Proc. Int. Symp. Energy Transfer, Charles University Press, Praha.
7. Mimuro, M., Füglistaller, P., Rübeli, R. and Zuber, H. (1986) Biochim. Biophys. Acta 848, 155-166.
8. Siebzehnrübl, S., Fischer, R. and Scheer, H. (1986) FEBS Lett., submitted.
9. Zuber, H. (1985) in Antennas and Reaction Centers of Photosynthetic Bacteria (Michel-Beyerle, M. E., ed.) pp. 3-25, Springer Verlag, Berlin.
10. Scheer, H. (1986) in Handbook of Plant Physiology, Vol. 19 (Staehelin, L. A. and Arntzen, C. J., eds.) pp. 327-337, Springer Verlag, Berlin.
11. Earlier and related work has been reviewed in: Holzwarth, A. R. (1986) Photochem. Photobiol. 43, yearly review.
12. Sauer, K. and Scheer. H., this volume, p.
13. Dale, R. E. and Teale, F. J. W. (1970) Photochem. Photobiol. 12, 99-117
14. Grabowski, J. and Gantt, E. (1978) Photochem. Photobiol. 28, 39-46 and 47-56

PRIMARY REACTIONS OF PHOTOSYNTHESIS : DISCUSSION OF CURRENT ISSUES

Paul MATHIS

Département de Biologie, Service de Biophysique, CEN Saclay
91191 Gif-sur-Yvette cedex, France

1. INTRODUCTION

A few years ago, primary reactions of photosynthesis were envisioned essentially from a photochemical point of view : how can excited chlorophyll react with a neighbouring species to induce a stable charge separation ? It now appears that understanding the primary reactions requires the cooperation of widely different approaches ranging from molecular biology to the most sophisticated physical techniques. In this review I shall not attempt to present a complete description of primary reactions (these will be understood as the reactions taking place in reaction centers), but rather to discuss a few of the problems which are now the object of particularly active research, and of controversy, in the field of plant photosystems.

The recent determination (1) of the three-dimensional structure of the reaction center from the purple bacterium Rps. viridis had and continues to have a profound impact on our field of research. Firstly it indeed permits a "calibration" of various spectroscopic methods in which many scientists had not a fully quantitative confidence. Secondly, knowledge of the structure of the reaction center in one species of bacteria is the basis for working hypotheses concerning other types of reaction centers. These hypotheses are founded on comparative spectroscopic and biochemical studies, and also on a reasonable belief in the unity of biological processes.

2. THE PS-II REACTION CENTER

The comparative approach is especially worthwhile for PS-II since the sequence of electron acceptors in that center has properties strikingly similar to that of purple bacteria. I shall briefly recall these resemblances (2-5) :

- presence of two quinones Q_A and Q_B, which function in series. Q_A is firmly bound, is a one-electron carrier, has a Em of about $- 200$ mV. Q_B is bound weakly in the neutral forms Q and QH_2, but more strongly in the form Q^-. Q_B is a two-electron carrier.

- an iron atom interacts with both Q_A and Q_B. The iron is normally Fe^{2+}. The interaction is well observed with the radical anions (Q_A^-, Q_B^-) which both have a characteristically broadened EPR spectrum. Depending on the pH, this spectrum has two forms in PS-II, which are also found in photosynthetic bacteria (6).

- the primary acceptor is a (bacterio)pheophytin, which then reduces Q_A in a few hundred ps. When Q_A is chemically reduced, the pheophytin can be stably reduced by continuous illumination. The state (Pheo$^-$, Q_A^-, Fe^{2+}) displays a characteristic EPR spectrum, often named "split signal", resulting from the interactions between the three chemical species.

Biggens, J. (ed.), Progress in Photosynthesis Research, Vol. I. ISBN 90 247 3450 9
© *1987 Martinus Nijhoff Publishers, Dordrecht. Printed in the Netherlands.*

– electron transfer from Q_A to Q_B is efficiently inhibited by
chemicals (including important herbicides) which act by nearly competitive
displacement of Q_B.
– the plane of the (bacterio)pheophytin is perpendicular to the
membrane plane.

These similarities do not fully extend into the details. Discrepancies
can thus be noticed in the pattern of proton uptake associated to electron
transfer, in the relative efficiency of some inhibitors, in the properties
of the iron atom (which can participate in electron transfer, under some
conditions, in PS-II (7,8)). I estimate nevertheless that the properties of
the sequence (Pheo, Q_A, Fe, Q_B) are sufficiently similar in purple bacteria
and PS-II to predict that the spatial relationships between these species
are nearly as indicated schematically in Fig. 1. The possible role of the
second bacteriopheophytin in bacterial reaction center remains totally
puzzling. That an analogous pheophytin is present in PS-II is indicated by
recent quantitative determinations which have shown that two pheophytin
molecules are present per PS-II reaction center (9).
The structural properties of the set of redox centers are actually
determined by their binding to the polypeptides of the reaction center. It
is thus quite coherent that the determination of amino-acids sequences has
shown rather strong homologies between the subunits L and M of purple
bacteria reaction centers and peptides of 32 and 34 kD of PS-II (10-12). In
a recent work (12) it was shown that the amino-acid sequence of the 32 and
34 kD proteins (also named D_1, D_2, respectively) follows a pattern of
hydrophobicity compatible with the formation of 5 transmembrane helices, a
structure already found in L and M. The predicted structure also
incorporates two important features :

– the localization at appropriate positions on D_1 and D_2 of four
conserved histidine residues which are involved as ligands to the Fe^{2+} in
Rps. viridis ;
– the localization in the same region, close to the external face of
the membrane, of four amino-acids residues which had been shown to be
substituted in four herbicide-resistant mutants. This region contains the
putative site for Q_B binding, which is also the site for herbicide binding.

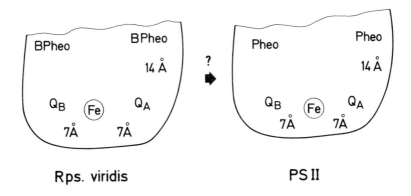

Rps. viridis   PS II

Fig. 1. A plane projection of the redox centers of the Rps. viridis reaction center (according to ref.1) and its hypothetical counterpart in PS-II.

Such comparative speculations have the main value of providing new working hypotheses. In ref. 12, a reasonably detailed model has been proposed for the binding sites of Q_A, Q_B and Fe^{2+} in PS-II. This model is very attractive, although a limitation may reside in the great symmetry between the proposed sites for Q_A and Q_B, whereas these two quinones have widely different functional properties. It is quite possible, however, that large differences in the binding of Q_A and Q_B originate in rather subtle differences in their environment.

The organization of polypeptides in the PS-II reaction center is still a matter of debate. Six polypeptides have been evidenced, with approximate MW of 47,43,34,32,10 and 4 kDa (these are apparent values for plants ; slightly different values were reported in algae). The two smallest ones are the apoproteins of cytochrome b_{559}, which is always found in PS-II although its role is still not known (13). The peptide of 43 kDa binds chlorophyll a, making the pigment-protein complex CP-43, which presumably plays a role in light-harvesting or energy transfer, but not in electron transfer. There remain three peptides (47,34,32 kDa) to possibly hold all the redox centers (14,15) : electron acceptors (as already discussed) and electron donors : the primary donor P-680 (Chl a, presumably in a dimeric state), the secondary donor Z (presumably a plastohydroquinone, in a highly acidic pocket), the accessory donor D, and also the site for binding the Mn cluster involved in oxygen evolution. It is now clear that this last function can be performed in PS-II particles which contain no other peptides than those indicated before, although a few more peptides are implicated under physiological conditions. These protein data are in agreement with previous functional data showing a close interaction between the oxygen-evolving system and primary photochemistry (re-reduction of $P-680^+$, photoreduction of C-550), from what I have advocated a proximity of the sites of oxygen evolution and of primary electron transfer. At the moment there are no informations on which of the reaction center peptides carry Z, D and the site for the Mn complex (many models grant these roles to the 47 kDa peptide, but this is a simple surmise).

On which polypeptides are located the redox centers participating in primary photochemistry, P-680 and Pheo ? Two widely different points of view have been supported (Figs. 2 and 3). In a first hypothesis, P-680 and Pheo are held by the 47 kDa polypeptide (this peptide has often been isolated as a pigment-protein complex complex named CP-47, a name which will be used below). A second hypothesis considers that the couple D_1, D_2 (32, 34 kDa peptides) holds P-680 and Pheo, in addition to Q_A and Q_B ; in that case, CP-47 could be a specific reaction center antenna and/or also be involved for the electron donors to P-680. Both hypotheses have experimental support, but a definitive conclusion cannot yet be reached.

In support of the CP-47 hypothesis, it appears that CP-47 is present in all photoactive sub-fractions of PS-II (see e.g. refs 16, 17 ; a remarkable exception has been reported recently in PS-II particles partially digested with trypsin, ref. 18). A fluorescence at 695 nm, and light-induced absorption changes were also attributed to the primary reactants (see ref. 14 for a review). The major objection towards these experimental data is that the absence of D_1 and D_2 has not always been established. These polypeptides are difficult to visualize and may have been present in all of the reported photoactive preparations. In a fraction of the PS-II reaction center from a thermophilic bacterium it has been reported (17) that primary photochemistry takes place with CP-47 as the only pigmented complex.

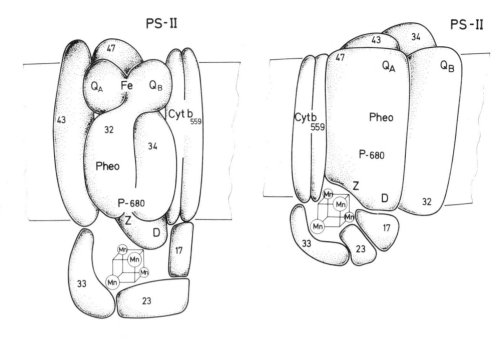

Fig. 2. Schematic representation of the PS-II reaction center (plus the polypeptides involved in O_2 evolution) in the hypothesis where polypeptides D_1 and D_2 hold the primary reactants (adapted from ref. 5).
Fig. 3. Same as Fig. 2, but supposing that the 47 kD polypeptide holds the primary reactants.

The alternate hypothesis (D_1, D_2 hypothesis) is based on the analogy between the D_1, D_2 couple in PS-II and the L, M couple in purple bacteria (11,18) in which the association of L and M polypeptides holds all of the redox centers from the primary donor to Q_B. The same situation could occur in PS-II. The strength of this hypothesis resides in its good predicting value for the ensemble (Q_A, Fe, Q_B), as discussed before. The extremely good analogy between Pheo in PS-II and BPheo in purple bacteria permits to think that Pheo is also held by the couple D_1, D_2 in PS-II. The most extreme proposal (P-680, Pheo, Q_A, Fe, Q_B held by D_1 and D_2) suffers from a few weak points :

- D_1 and D_2 have not been found to bind pigments. However this may result from the harsh detergent treatments required to isolate these polypeptides. Pigments could be solubilized during the isolation (ref. 19 provides an indication in that direction).
- the polypeptide D_1 is rapidly degraded and replaced under high light conditions. How is it possible to imagine this turn-over of a single peptide in the integrated structure of the reaction center ?
- the analogy between PS-II and purple bacteria ceases at the level of primary donors. P-680 seems to be dimeric as P-870 is, but its redox properties (very high oxidation potential) and its orientation (20) are

different from those of the bacterial primary donor.

A conclusion is difficult to reach with the presently available data. It will certainly be worthwhile to develop careful and quantitative work on all polypeptides of PS-II. A number of ambiguities still relate to the identification of polypeptides of about 32-34 kD in PS-II and to their coding by already sequenced genes. The present uncertain situation is largely due to technical problems : proteins which are difficult to stain ; lack of good assay for P-680, for the biradical ($P-680^+$, $Pheo^-$) and for Z ; etc.

3. THE PS-I REACTION CENTER

The PS-I reaction center differs widely from the reaction center of purple bacteria and of PS-II. In some respects it resembles more to the reaction center of anaerobic green bacteria. The singularities of PS-I reside both in the chemical nature of redox centers and in the polypeptidic composition (1,5,21,22).

The redox centers of PS-I comprise the primary donor P-700 and a sequence of electron acceptors : A_0, A_1, F_X, (F_A, F_B). P-700 shares many properties in common with the primary donor of purple bacteria. Recent controversies have concerned the chemical nature of its constituants (Chl a or related derivatives such as Chl a' or Chl-RC I) and its monomeric or dimeric nature. A final answer is not yet achieved, although a dimeric nature seems to be more probable (5,22). The primary acceptor A_0 is presumably a specifically positioned molecule of Chl a, as shown by recent picosecond absorption data (23). The secondary acceptor A_1 is presumably a quinone, vitamin K_1, as shown by recent flash absorption experiments (24, and Brettel et al, these Proceedings). This is in agreement with previous analytic data showing that vitamin K_1 is present in PS-I particles (25,26). F_X is probably an iron-sulfur center, and F_A and F_B are certainly iron-sulfur centers. This sequence of electron acceptors is thus apparently well understood. In fact many questions remain unanswered (5), mainly because none of the steps of forward electron transfer has ever been resolved. The actual sequence of electron carriers, under physiological conditions, is still largely hypothetical. The great specificity of PS-I is undoubtedly the low redox potential of its electron acceptors, the study of which is thus greatly hampered. The Em of F_X is around -0.73 V and the rapid forward electron transfer requires that A_1 and A_0 have Em around -0.9 and -1.1 V, respectively. This raises a serious question for A_1, since the Em of vitamin K_1, for the couple Q/Q^-, is of $\simeq$ -0.5 V in an aprotic solvent (27). What can bring its Em as low as -0.9 V ? It is well known that a large energy is required to force an electric charge in a medium of low dielectric constant. It can thus be envisioned that A_1 is located in a highly apolar pocket, rendering it difficult to reduce and rendering A_1^- a highly reducing species, capable of reducing F_X. It should also be realized that F_X, supposed to be an iron-sulfur center, is much more reducing than the other known iron-sulfur centers.

How can we define the PS-I reaction center ? There are several ways of isolating PS-I particles which have about 100 Chl a per P-700 and which contain all the membranous redox centers mentioned before (2,28). It is then possible to specifically remove pigment-protein complexes amounting to 60 Chl a per P-700 : this treatment leaves small PS-I particles, with 40-50 Chl a and with the redox centers, carried by several polypeptides. I think that these particles deserve to be named "PS-I reaction centers". It is possible to further simplify these particles by treatment with SDS, which practically does not remove any pigment, but which removes a few small polypeptides and the iron-sulfur centers. The remaining particle, often named CP-1, contains two big polypeptides, together with 40-50 Chl a and P-700. The primary biradical (P-700$^+$, A_0^-) is still photoinduced in these particles (see refs. 5,22), which thus contain A_0 in an active configuration (A_1 may still be present, but essentially inactive). There has been some controversy as to the stoechiometry between polypeptides and P-700 in CP-1, and as to the significance of these two polypeptides : in some cases they are not separated on electrophoretic gels and it could have been that one of them were a degraded form in cases where they appear separated. Recently it has been shown that two genes with slightly different sequences were coding for the big PS-I polypeptides (29). I consider as the most probable structure that one copy of each of the two big polypeptides (with MW of $\approx$ 62 and 65 kD) holds the primary reactants P-700 and A_0, as shown in Fig. 1. Vitamin K_1 is also present in CP-1, and thus also held by the same polypeptides. A recent work (30) permits to think that F_X is also held by the same polypeptides, although this iron-sulfur center is absent from CP-1. About 40-50 molecules of Chl a are thus held by the couple of polypeptides, so that there is probably no hope for a further selective decrease in the number of chlorophylls associated with the PS-I center. This hypothesis implies that both of the big polypeptides are required for holding the primary photoreactants. This putative requirements is simply based on the structure of the reaction center from Rps. viridis in which the primary reactants are at the interface between L and M subunits. The situation could be different in PS-I (for example P-700 could be held entirely by one of the subunits).

Electrophoretic analysis revealed that the PS-I reaction center also contains a few smaller polypeptides (MW : 8-10 kDa and 16-24 kDa), the functions of which are not elucidated. Small peptides (8-10 kDa) are involved as apoproteins of the iron-sulfur centers F_A and F_B (31). It may happen that some of the other ones are involved in secondary electron transfer and are thus not intrinsic components of the reaction center.

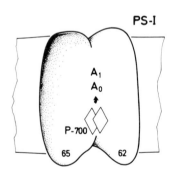

Fig. 4 : Hypothetic structure of the core of the PS-I reaction center

4. EXCITED STATES AND FLUORESCENCE

In isolated bacterial reaction centers it is clear that the primary biradical is formed in a very short time of a few picoseconds after a flash (see e.g. 32). Such a rapid reaction can be initiated only by a singlet

excited state, since intersystem crossing is much too slow to permit a
significant triplet state formation within that time. The reactive state is
usually considered as the singlet excited state of P, the primary donor
("special pair"). The precise nature of that state is not simple to define,
as shown by several lines of evidence :

i) interpretation of the absorption spectrum of the reaction center
requires the admixture of a very large number of states (exciton and charge
transfer states) (33) ;

ii) a consideration of both absorption properties and electron
transfer has led to propose (34) a strong interaction between P and other
molecules. The structural evidence that a bacteriochlorophyll is located
between P and BPheo, but does not appear as a well-defined electron
carrier, is in good agreement with such a model.

iii) hole-burning experiments (35,36) indicate that the initial
excited state of P lasts much less that the few picoseconds measured by
flash absorption. This indicates that some kind of relaxation (such as a
charge separation within the dissymmetric dimer P) takes place before
electron transfer to BPheo.

The primary biradical is formed as a singlet state and normally
evolves by rapid electron transfer to the secondary electron acceptor Q_A.
When this secondary step is inhibited, however, the primary biradical lasts
for nanoseconds and has sufficient time to evolve into a triplet state.
Eventually the biradical recombines and produces the localized triplet
state of the primay donor ^{3}P. It is remarkable that ^{3}P has been found in
all the studied reaction centers (purple and green bacteria, PS-I, PS-II).
A detailed reaction scheme has been established only in purple bacteria,
but the properties of ^{3}P are sufficiently unique, particularly its
polarized EPR spectrum at low temperature, to conclude to a common scheme
of formation in all reaction centers. This property differentiates the
photochemical behaviour in photosynthetic reaction centers from artificial
model systems in which a polarized triplet state has not been formed as a
result of charge recombination.

Electron transfer appears to be a mechanism of quenching of the
singlet excited state of P : it should thus appear as a route of quenching
of (bacterio)chlorophyll fluorescence. A phenomenon of that type has been
observed in chloroplasts a long time ago and interpreted as a competition
between fluorescence and electron transfer for the decay of excited
chlorophyll (37). With good experimental support, it is usually assumed
that in chloroplasts the redox state of Q_A modulates the fluorescence
yield : the yield is low (F_O) when Q_A is oxidized and high (F_M) when Q_A is
reduced. When Q_A was thought to be the primary electron acceptor in PS-II,
it was reasonable to assume that F_O and F_M were the fluorescence of PS-II
chlorophyll (antenna and perhaps, P, i.e. P-680) in the presence or absence
of photochemistry, respectively. Such a view cannot be accepted as such,
since Q_A is not the primary acceptor. Following his proposal that
pheophytin was the primary acceptor of PS-II, Klimov (38) then proposed
that the variable fluorescence ($F_v = F_M-F_O$) is a luminescence resulting
from a back-reaction : $P-I-Q_A^- \xrightarrow{\text{(light)}} P^+I^-Q_A^- \longrightarrow P-I-Q_A^- +$ luminescence.
The origins of F_O, F_M and F_v are in fact the object of much controversy and
confusion, and I will thus attempt to briefly describe this problem which
is of interest because of the developments in studies of primary
photochemistry (rapid flash absorption, fluorescence decay measurements)
and because fluorescence is so much used a a tool in studies of PS-II.

The relationships between the antenna and the PS-II reaction center,
as leading to photochemical electron transfer and to fluorescence, can be
considered according to three models (the antenna will be considered as a

single ensemble although it is known to be non-homogeneous). In the first model we consider that the reaction center is always a good trap for excitation, be Q_A oxidized or reduced : the primary biradical (P^+, Pheo$^-$) is formed and disappears by forward electron transfer (Q_A oxidized) or by back-reaction with emission of luminescence (Q_A reduced). If electron transfer from excited P to Pheo takes place in a few picoseconds, then it follows that F_O is the fluorescence in the antenna before the excitation reaches the reaction center (its lifetime reflects the transfer time in the antenna) or the luminescence which can take place before electron transfer from Pheo$^-$ to Q_A (its lifetime reflects that electron transfer step). F_v (F_M in the second alternative) is the luminescence resulting from charge recombination and is emitted in the reaction center. Its lifetime is that of the back-reaction of the primary biradical.

In the second model we consider that the reaction center is a good trap only when Q_A is oxidized ; when Q_A is reduced, for some reason (for example because of electrostatic repulsion) the primary biradical (P^+, Pheo$^-$) is formed with a low yield. According to this model the fluorescence is always emitted in the antenna ; its lifetime is that for trapping by the reaction center (Q_A oxidized) or by other permanent traps such as PS-I, chlorophyll aggregates, etc. (when Q_A reduced).

The third model supposes that an equilibrium is rapidly reached between the three states considered :

$$(Chl^*) \text{ antenna} \rightleftharpoons P^*PheoQ_A^- \rightleftharpoons P^+Pheo^-Q_A^-$$

$$\downarrow \qquad\qquad\qquad \downarrow$$

$$\text{fluorescence} \qquad \text{fluorescence}$$

The antenna, however, has a much larger size than the reaction center, and the equilibria are greatly displaced to the left. Thus the fluorescence emission always practically originates in the antenna. Its lifetime is determined by the electron transfer from Pheo$^-$ to Q_A (Q_A oxidized) or by the back-reaction of the primary biradical (Q_A reduced).

These three models can be used to discuss experimental data and to make some predictions :

a) Published data on energy transfer between PS-II units or from PS-II to PS-I, when Q_A gets reduced, are not consistent with model 1.

b) Model 1 is the only one which assumes that light emission originates mainly from reaction center constituants, as proposed on the basis of fluorescence polarization measurements (39).

c) Each model can be supported by some of the fluorescence lifetime data. However, most recent data (40,41) are in better agreement with models 2 and 3.

d) Recent absorption data (42) were interpreted as indicating that a picosecond flash induces a large (P^+, Pheo$^-$) charge separation when Q_A is either oxidized (duration 250 ps) or reduced (duration 2 ns). These data are consistent only with model 1, as were also older absorption data (43). Models 2 and 3 would require to completely change the interpretation of the data. Absorption changes attributed to the biradical (P^+, Pheo$^-$) could be due to the singlet excited state of chlorophyll a.

e) In model 3, the equilibrium will vary with the size of the antenna associated with the PS-II reaction center, as proposed by P. Joliot (personal communication). A preferential population of the reaction center states should become observable with smaller PS-II particles.

f) The work on bacterial reaction centers has shown that the primary biradical is formed nearly as efficiently when Q_A is reduced or oxidized. If the analogy developed earlier holds at that level, it follows that model 2 is not justified.

g) In conclusion, it appears that a critical reexamination of published data, as well as new experiments, are required before we can reach a good understanding of the PS-II fluorescence.

Other quenching states have been observed under special conditions. The triplet state of carotenoids quenches chlorophyll fluorescence in different sub-sets of the chloroplast membrane (44). Oxidized P-680 is also known as a fluorescence quencher for which I have proposed a mechanism based on the existence of low energy levels in chlorophyll radical ions (45) ; this mechanism would also explain a quenching by the reduced pheophytin in the reaction center. It should be noted that a quenching by $P-680^+$ has not to be invoked if we accept model 1 : the primary biradical cannot be formed any more. The quenching by $Pheo^-$ would explain the low fluorescence reached under light, at low redox potential (38). A good use of fluorescence for monitoring PS-II properties should take into account several additional properties : membrane potential, Q_{400}, membrane stacking, etc.

5. CONCLUSION

In conclusion to this rapid overview, it should be stressed that reaction centers are an intimate and ordered association of pigments, redox centers and proteins. These proteins have an obvious role of maintaining the other components at well defined positions and in suitable environment (acidity and polarity are important parameters). Do they have other functions ? It becomes more and more probable that proteins are more directly involved in electron transfer reactions. This can be realized in two ways :
- firstly, the distance separating two redox centers cannot be considered as a homogeneous medium. It is occupied by amino-acids or their side groups, which may provide favorable paths for electron transfer ;
- secondly, electron transfer requires a precise energy tuning between the initial and final states. This can be provided by a coupling with vibrational modes in the proteins.

REFERENCES

1. Deisenhofer, J., Epp, O., Miki, K., Huber, R., Michel, H. (1984) J. Mol. Biol. 180, 385-398
2. Okamura, M.Y., Feher, G. and Nelson, N. (1982) in Photosynthesis : Energy Conversion by Plants and Bacteria (Govindjee, ed.) Vol. 1, 195-272, Academic Press, New York
3. Parson, W.W., Ke, B. (1982) as ref. 2, pp. 331-335
4. Mathis, P. (1985) Photosynth. Res. 8, 97-111
5. Mathis, P. and Rutherford, A.W. (1987) in New Comparative Biochemistry: Photosynthesis (Amesz, J., ed.) Elsevier, Amsterdam (in press)
6. Rutherford, A.W., Zimmermann, J.-L. (1984) Biochim. Biophys. Acta 767, 168-175
7. Petrouleas, V. and Diner, B.A. (1986) Biochim. Biophys. Acta 849, 264-275
8. Zimmermann, J.-L. and Rutherford, A.W. (1986) Biochim. Biophys. Acta (in press)
9. Murata, N., Araki, S., Fujita, Y., Suzuki, K., Kuwabara, T. and Mathis, P. (1986) Photosynth. Res. 9, 63-70
10. Youvan, D.C., Bylina, E.J., Alberti, M., Begush, H. and Hearst, J.E. (1984) Cell 37, 949-957
11. Michel, H. and Deisenhofer, J. (1986) in Encyclopedia of Plant Physiology, vol. 19, Photosynthesis III (L.A. Staehelin and C.J.

Arntzen, eds.) pp. 371-381, Springer Verlag, Berlin
12. Trebst, A. (1986) Z. Naturforsch. 41c, 240-245
13. Widger, W.R., Cramer, W.A., Hermodson, M. and Herrmann, R.G. (1985) FEBS Lett. 191, 186-190
14. Satoh, K. (1985) Photochem. Photobiol. 42, 845-853
15. Diner, B. (1986) as ref. 11, pp. 422-436
16. Pakrasi, H.B., Riethman, H.C. and Sherman, L.A. (1985) Proc. Nat. Acad. Sci. USA 82, 6903-6907
17. Yamagishi, A. and Katoh, S. (1985) Biochim. Biophys. Acta 807, 74-80
18. Trebst, A. and Depka, B. (1985) in Antennas and Reaction Centers of Photosynthetic Bacteria (M.E. Michel-Beyerle, ed.) pp. 216-224, Springer-Verlag, Berlin
19. Machold, O. (1986) FEBS Lett., in press
20. Rutherford, A.W. (1985) Biochim. Biophys. Acta 807, 189-201
21. Rutherford, A.W. and Heathcote, P. (1985) Photosynth. Res. 6, 295-316
22. Setif, P. and Mathis, P. (1986) as ref. 11, pp. 295-316
23. Nuijs, A.M., Shuvalov, V.A., van Gorkom, H.J., Plijter, J.J. and Duysens, L.N.M. (1986) Biochim. Biophys. Acta 850, 319-323
24. Brettel, K., Setif, P. and Mathis, P. (1986) FEBS Lett. in press
25. Interschick-Niebler, E. and Lichtenthaler, H.K. (1981) Z. Naturforschg. 36c, 276-283
26. Schoeder, H.U. and Lockau, W. (1986) FEBS Lett. 199, 23-27
27. Prince, R.C., Dutton, P.L. and Bruce, J.M. (1983) FEBS Lett. 160, 273-276
28. Wollman, F.A. (1986) cf. ref. 11, pp. 487-495
29. Fish, L.E., Kück, U. and Bogorad, L. (1985) J. Biol. Chem. 260, 1413-1421
30. Golbeck, J.H. and Cornelius, J.M. (1986) Biochim. Biophys. Acta 849, 16-24
31. Lagoutte, B., Sétif, P. and Duranton, J. (1984) FEBS Lett. 174, 24-29
32. Martin, J.L., Breton, J., Hoff, A.J., Migus, A. and Antonetti, A. (1986) Proc. Natl. Acad. Sci. USA 83, 957-961
33. Parson, W.W., Scherz, A. and Wharshel, A. (1985) as ref. 122-130
34. Friesner, R. and Wertheimer, R. (1982) Proc. Nat. Acad. Sci. USA 79, 2138-2142
35. Meech, S.R., Hoff, A.J. and Wiersma, D.A. (1985) Chem. Phys. Lett. 121, 287-292
36. Boxer, S.G., Lockhart, D.J. and Middendorf, T.R. (1986) Chem. Phys. Lett. 123, 476-482
37. Duysens, L.N.M. and Sweers, H.E. (1963) in Studies on Microalgae and Photosynthetic Bacteria, pp. 353-373, University of Tokyo Press, Tokyo
38. Klimov, V.V., Allakhverdie, S.I. and Pashchenko, V.Z. (1978) Dokl. Akad. Nauk. SSSR 242, 1204-1207
39. Breton, J. (1982) FEBS Lett. 147, 16-20
40. Moya, I., Hodges, M., Briantais, J.M. and Hervo, G. (1986) Photosynth. Res., in press
41. Schatz, G.H. and Holzwarth, A.R. (1986) Photosynth. Res., in press
42. Nuijs, A.M., van Gorkom, H.J., Plijter, J.J. and Duysens, L.N.M. (1986) Biochim. Biophys. Acta 848, 167-175
43. Shuvalov, V.A., Klimov, V.V., Dolan, E., Parson, W.W. and Ke, B. (1980) FEBS Lett. 118, 279-282
44. Mathis, P., Butler, W.L. and Satoh. K. (1979) Photochem. Photobiol. 30, 603-614
45. Mathis, P. (1981) in Photosynthesis III. Structure and Molecular Organization of the Photosynthetic Apparatus (G. Akoyunoglou, ed.) pp. 827-837, Balaban I.S.S., Philadelphia

SELECTIVE REDUCTION AND MODIFICATION OF BACTERIO-
CHLOROPHYLLS AND BACTERIOPHEOPHYTINS IN REACTION
CENTERS FROM RHODOPSEUDOMONAS VIRIDIS

V.A.SHUVALOV, A.Ya.SHKUROPATOV AND M.A.ISMAILOV,
INSTITUTE OF SOIL SCIENCE AND PHOTOSYNTHESIS, USSR
ACADEMY OF SCIENCES, PUSHCHINO

1. INTRODUCTION

Picosecond and subpicosecond measurements of absorbance
changes (Δ A) in reaction centers (Rc's) from purple bacteria
[1-3] give the information about the primaty charge separation between pigment molecules (4 molecules of bacteriochlorophyll (BChl)
and 2 molecules of bacteriopheophytin (Bph)) and quinone (Qa) in
Rc's. The interpretation of these data should be based on the
correct assignment of the optical transitions in Rc's to the absorption of the pigments and their interaction complexes. At present
three-dimention structure of the protein and chromophores is known
for Rhodopseudomonas viridis Rc's [4]. The chromophores model
[4,5] shows two pigment chains: P-Ba-Ha-Qa and P-Bb-Hb-Qb
situated symmetrically in L and M protein subunits, respectively
(where P is the primary electron donor, BChl dimer; B is the
"monomeric" BChl; H is Bph). The picosecond electron transfer has
been suggested to occur in photoactive chain P-Ba-Ha-Qa [4,6],
however, the relative contribution of Ba and Ha in the electron
transfer from P* to Qa is not completely clear (see also [1-3])
since in picosecond measurements the bleaching of BChl band at
830 nm is only observed in near infrared region [7,8].
  Another method giving the information about optical properties
of acceptor molecules is the method of photoaccumulation of these
molecules in the reduced state [9-11]. Combination of this method
and picosecond data can clarify the sequence of the electron
transfer in Rc's. At low redox potentials when Qa is reduced, the
photoreduction of pigments giving absorption at 830, 810 and 790 nm
at room temperature [9] and at 810 nm at low temperature [10-11],
is observed due to the electron transfer from cytochrome which is
present in R.viridis Rc's.
  In this work the conditions for the selective reduction of Hb
(absorbing mainly at 790 nm), Ha (810 nm) and Ba (830 nm)
have been found. The reduction of both latter pigments is accompanied by the bleaching of the band at 545 nm which is usually
attributed to the Bph absorption only [5,7,12]. This result has been
used for the interpretation of picosecond data. It was found also
that the modification of Rc's with $NaBH_4$ leads to the decrease of
absorbance at 820 nm and to the decrease of Δ A at 810 nm in
the spectrum of P^+ formation at 77K. This shows that the oand at
820 nm belongs to the absorption of second BChl monomer (Bb) by
analogy with data obtained for Rhodopseudomonas spaeroides Rc's
[6].

2. MATERIALS AND METHODS

Rc's from R.viridis were isolated as described earlier [11]. For measurements at low temperature Rc's were frozen in the presence of 60% glycerol, 10 mM ascorbate and 1 mg/ml dithionite. Under these conditions Qa and cytochrome were completely reduced. Modification of R.viridis Rc's with $NaBH_4$ was carried out as described earlier for R.sphaeroides Rc's [6].

3. RESULTS

Fig.1A shows the absorption spectrum (solid curve) of R.viridis Rc's frozen in the dark at 77K and 250 mV. In near infrared region the six transitions are observed at 790, 810, 820, 830, 850 and 1000 nm in agreement with results described earlier [10,11].

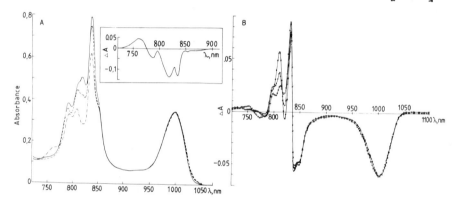

Fig. 1 (A) The absorption spectra of R.viridis Rc's at 250 mV and 77K for two different preparations (solid and dashed curves) and after modification with $NaBH_4$ (curve with points). Inset shows the spectral changes in Rc's induced by the modification with $NaBH_4$. (B) Difference (light–dark) absorption spectra of P^+ formation for two preparations of Rc's (open and closed points) and for modified Rc's (crosses). The amplitude of ΔA at 810 nm corresponds to the amplitude of the 820–nm band in absorption spectra.

The amplitude of the 820–nm band is different for different preparations (Fig.1A, solid and dashed curves; [10,11]). Modification of Rc's with $NaBH_4$ induces remarkable decrease of the 820–nm band (see curve with points and inset), the decrease of 790– and 833–nm bands and the absorption increase at 760 nm. Difference (light–dark) absorption spectra of the P^+ formation for native Rc's (Fig.1B, curves with points) correspond to those described earlier [5,10]. The decrease of the absorption at 820 nm in preparation corresponds to the decrease of the development at 810 nm in P^+ spectra. The decrease of the 820–nm band induced by the modification with $NaBH_4$ of Rc's is accompanied by the remarkable decrease of the development at 810 nm and of the trough at 820 nm with respect to the bleaching at 1000 nm (Fig.1B, curve with crosses). One can also see a new blue shift of new 760-nm band. Thus the spectral features near 810-820 nm in P$^+$ spectrum are determined by the blue shift of the 820-nm band.

Fig.2 demonstrates the absorption spectrum (solid curve) of native R.viridis Rc's frozen in the dark at 77K and -400 mV. The illumination of Rc's at 77K for 1 hour induces the following absorbance changes (see dashed absorprion curve and solid curve of Δ A): i) the bleaching of pigment bands at 545 and 810 nm; ii) the blue shift of the 830-nm and 1000-nm bands; iii) the bleaching of cytochrome bands at 547 and 552 nm and iv) the red shift of 610-nm band.

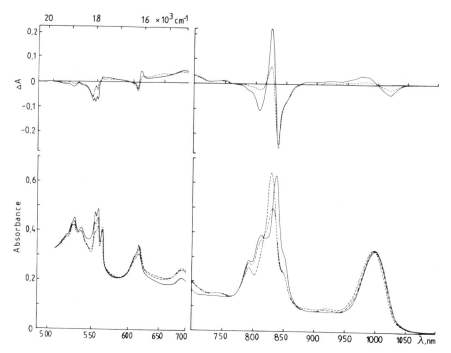

Fig. 2 The absorption spectra (lower curves) of R.viridis Rc's at -400 mV and 77K frozen in the dark (solid, 1), illuminated for 1 hour at 77K (dashed, 2) and warmed after illumination in dark up to 220K for several minutes (curve with points, 3). Upper spectra show the difference between the absorption spectra: 2-1 (solid); 3-1 (dashed).

The spectral curve with points in Fig.2 shows the absorption spectrum at 77K of Rc's at -400 mV illuminated at 77K and warmed up to 220K for several minutes. One can see that the 810-nm band is recovered but the 830-nm band is partially bleached together with a shoulder at 850 nm (see also dashed curve of Δ A). In visible region the additional bleaching at 610 nm and the recovery of cytochrome absorption are observed. The recovery of the 545-nm band is not observed.

If Rc's at -400 mV are illuminated for 1-2 minutes at 293K and frozen under illumination at 77K, the main absorption decreases are observed for the 830- and 545-nm bands (Fig.3). The bands at 810 and 790 nm are slightly decreased also. The illumination of such Rc's at 77K leads to additional bleaching at 830 nm and to remark-able bleaching at 790 nm. The band at 810 nm is recovered almost completely. In near infrared region the small narrow band at 1060 nm appears. The bleaching at 545 nm is not changed, but the additional bleaching at 535 nm is observed.

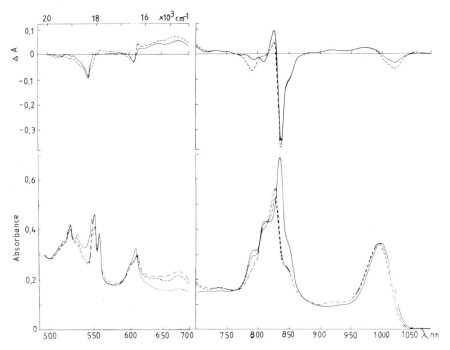

Fig.3 The absorption spectra (lower curves) of R.viridis Rc's at
−400 mV and 77K frozen in the dark (solid, 1), frozen after 2
minutes of illumination at 293K (dashed, 2) and additionally
illuminated at 77K for 1 hour (curve with points, 3). Upper
spectra show the difference between the absorption spectra:
2−1 (solid); 3−1 (dashes).

Fig.4 presents the data obtained in ref.7 using the selective picosecond excitation at 960 nm of R.viridis Rc's at 293K. At −38 ps the bleaching at 960 nm is accompanied by the bleaching of the 850-nm band. Then during 12 ps the bleaching of the 830-nm band is observed without any bleachings or troughs near 790 and 810 nm.

4. DISCUSSION

The methods of photoselection [5,10] and linear dichroism [12] applied to R.viridis Rc's allow to clarify the orientation of vectors of electric dipole transition moments of pigments with respect to dipole transition moment of P. The obtained results should be compaired with X-ray analysis data for the same Rc's [4] which show that Qy

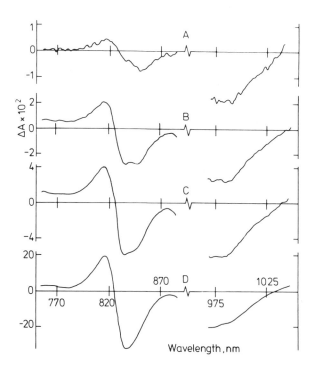

Fig.4 Difference absorption spectra of R.viridis Rc's at 293K measured upon selective picosecond (33 ps duration) excitation at 960 nm at −38 ps (A), −30 ps (B), −10 ps (C) and 70 ps (D). Quinone acceptors were reduced. Data are taken from ref. 7.

(N1–N3) transitions of "monomeric" BChl and Bph are oriented approximately parallel and perpendicular to Qy transition of P, respectively. Optical measurements show that the transitions at 790 and 810 nm are oriented approx.perpendicular and at 820 and 830 nm parallel to the transition at 1000nm, respectively [5,10,12]. It means that the transitions at 790 and 810 nm belong to Bph mainly, while those at 820 and 830 nm to BChl. The contribution of all pigments, however, to each transition in Rc's should be taken into account [5,13]. The selective bleachings of the bands at 790, 810, 820 and 830 nm induced by the reduction or modification show that one pigment is mainly responsible for one transition in the region of 780–830 nm in Rc's of R.viridis (see below about the 850–nm transition).

Data described here show that the illumination of Rc's at low or room temperature as well as changes of temperature after or during the illumination lead to the selective and reversible reduction at low redox potential of the following pigments: i) Bph with the absorption at 810 nm (Bph–810); ii) BChl–830 and iii) Bph–790. The modification of Rc's with NaBH$_4$ selectively induces the decrease of the absorption of BChl–820.

In Rc's frozen in the dark the complete photoreduction of Bph–810 is observed with the bleachings at 810 and 545 nm and with the blue shift of the 830–nm band (Fig.2). In Rc's with the reduced Bph–810 the changes of the temperature from 77K up to 220K and down to 77K lead to the electron transfer from Bph–810$\bar{}$ to BChl–830. It is accompanied by the bleachings at 830 and 610 nm and by the recovery of the band at 810 nm. The recovery of the band at 545 nm is not observed. Thus the 545-nm band is bleached when the reduction of Bph-810 as well as BChl is observed. This implies that the transition at 545 nm does not only belong to one pigment.

The illumination of Rc's during the freezing induces the more complete reduction of BChl-830 and the partial reduction of Bph-810 and Bph-790 (Fig.3). The additional illumination of such Rc's at 77K induces almost complete reduction of Bph-790 and recovery of 810-nm band without remarkable changes of BChl-830. In this state the bleaching of the band at 545 nm with a shoulder at 535 nm is observed. Thus in Rc's in which BChl-830 is reduced the photoreduction of Bph-810 is not observed but an electron is transfered to Bph-790. Futhemore the illumination at 77K induces the recovery of the 810-nm band, i.e. partially reduced Bph-810 is photooxidized.

Modification of R.viridis Rc's with $NaBH_4$ mainly decreases the band at 820 nm (Fig.1). This is accompanied by the decrease of the trough at 820 nm and of the development at 810 nm in the spectrum of P^+ formation. Thus the development at 810 nm in the P^+ spectrum is due to the blue shift of the 820-nm band induced by the field of P+Qa- and is not due to the band of P^+ itself as suggested earlier [5]. Since ΔA of the P^+ spectrum in the region of 810-820 nm have the positive polarization [5,10], the band at 820 nm probably belongs to BChl-820.

By analogy with R.sphaeroides Rc's in which the treatment with $NaBH_4$ leads to the modification of BChl-800 situated in non-active pigment chain P-Bb-Hb [9], one can suggest that in R.viridis Rc's BChl-820 is also situated in non-active chain P-B820-Hb (where Hb is probably Bph-790; see below). This pigment is possibly modified during the isolation of Rc's from the cells also. Therefore its contain in Rc's is not stable (Fig.1, [5,10,12]). The remarkable shift of the 820-nm band induced by P^+ is probably related to small distance between P and BChl-820.

Another band at 830 nm shifted by the field of P^+ (as well as of BChl-810) belongs to BChl-830. This shift is not changed in the P^+ spectrum after the treatment with $NaBH_4$. The band at 830 nm is also bleached in the primary process of the electron transfer in R.viridis Rc's (Fig.4, [7,8]). Therefore one may suggest that BChl-830 is situated in the active chain P-BChl-830-Ha (where Ha is Bph-810, see below).

The subtraction of the P^+ spectrum from the P+I- spectrum [8] (where I is the complex of Ba and Ha) is not probably correct way to obtain the I- spectrum in the region of 810 nm since the field of P+I- can differ from that of P+Qa- near BChl-820. Therefore in the picosecond measurements at room temperature the bleachings at 835 and 545 nm in the absence of the bleaching at 790-810 nm can indicate the formation of $P^+BChl-830^-$ rather than P^+Bph^- since the bands at 830 and 545 nm are bleached when BChl-830 is reduced (see above). Then an electron is transfered to Qa, probably, with the participation of Bph-810. The localization of an electron density on Bph-810 seems to be small because of the energy of $P^+Bph-810^-$ is possibly higher than that of $P^+BChl-830^-$ at room temperature.

The suggestion about the location of Bph-810 in the active pigment chain P-BChl-830-Bph-810 is supported by the following facts: i) Bph-810 (in contrast to Bph-790) is an single electron trap at low redox potentials and low temperature (Fig.2) and ii) the field of Bph–810$^-$shifts the band of BChl–830. The electron transfer in the dark from Bph–810$^-$ to BChl–830 at 220K probably indicates the medium reorganization and the decrease of the energy of PBChl830$^-$relatively to that of PBph810$^-$. This agrees with the pico-second measurements [7] in which P$^+$BChl830$^-$ is registrated as a most stable state at room temperature (Fig.4).

It is unlikely that the simultaneous reduction of both pigments BChl830 and Bph810 in the active chain is possible because of the close location of these pigments. Therefore the photoreduction of Bph–810 at low temperature is not observed when BChl830 is already reduced. In this case an electron is probably transfered to the non–active chain with the reduction of Bph–790. In spite of the complete block of the electron transfer to Bph–810 the complete bleaching of the band at 830 nm is not observed even in Rc's from which BChl–820 was removed. The absorption spectrum of such Rc's in the region of 820–860 nm is, however, similar to that of BChl **b** in solution [14].

It should be noted that the transition at 850 nm bleached or shifted when BChl–830 is reduced or P is oxidized, is probably due to the interaction between P and BChl–830 (see also [6]). This band does not belong to P itself since the formation of triplet state of P only shifts the 850–nm band [15,16]. The modification of Rc's which induces the decrease of the 820–nm band, does not change the 850–nm band (Fig.1). Therefore one may suggest that BChl–820 does not participate in the transition at 850 nm in Rc's.

The following scheme shows the relative energy positions of the states in R.viridis Rc'c discussed above.

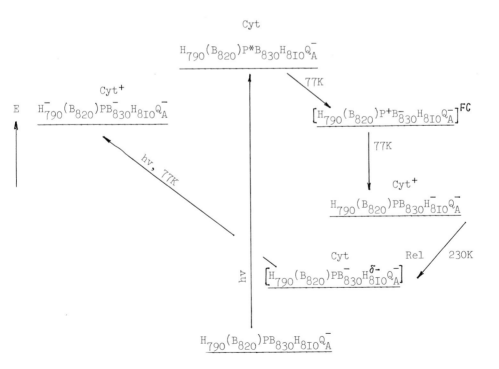

REFERENCES
1 Shuvalov, V.A. and Duysens, L.N.M. (1986) Proc.Natl.Acad.Sci. USA 83, 1690-1694
2 Woodbury, N.W.T., Becker, M., Middendorf, D. and Parson, W.W. (1985) Biochemistry 24, 7516-7521
3 Martin, J.-L., Breton, J., Hoff, A.J., Miqus, A. and Antonetty, A. (1986) Proc.Natl.Acad.Sci.USA, in the press
4 Deisenhofer, J., Epp, O. ,Miki, K., Huber, R. and Michel, H. (1985) Nature 318, 618-624
5 Shuvalov, V.A. and Asadov, A.A. (1979) Biochim.Biophys.Acta 545, 296-308
6 Shuvalov, V.A., Shkuropatov, A.Ya., Kulakova, S.M., Ismailov, M.A. and Shkuropatova, V.A. (1986) Biochim.Biophys.Acta 849, 337-346
7 Shuvalov, V.A., Amesz, J. and Duysens, L.N.M. (1986) Biochim. Biophys.Acta, in the press
8 Kirmaier, C., Holten, D. and Parson, W.W. (1983) Biochim. Biophys. Acta 725, 190-202
9 Shuvalov, V.A., Krakhmaleva, I.N. and Klimov, V.V. (1976) Biochim. Biophys.Acta 449, 597-601
10 Vermeglio, A. and Paillotin, G. (1982) Biochim.Biophys.Acta 681, 32-40
11 Maslov, V.G., Klevanik, A.V., Ismailov, M.A. and Shuvalov, V.A. (1983) Dokl.AN SSSR 269, 1217-1221
12 Breton, J. (1985) Biochim.Biophys.Acta 810, 235-245
13 Knapp, E.W., Fisher, S.F., Zinth, W., Sander, M., Kaiser, W., Deisenhofer, J. and Michell, H. (1985) Proc.Natl.Acad.Sci. USA 82, 8463-8467
14 Fajer, J., Davis, M.S., Brune, D.C., Spaulding, L.D., Borg, D.C. and Forman, A. (1976) In Brookhaven Symposia in Biology, N.28, pp. 74-104
15 Shuvalov, V.A. and Parson, W.W. (1981) Biochim.Biophys.Acta 638, 50-59
16 Den Blanken, H.J. and Hoff, A.J. (1982) Biochim.Biophys.Acta 681, 365-374

SPECTROSCOPIC AND PRIMARY PHOTOCHEMICAL PROPERTIES OF MODIFIED
RHODOPSEUDOMONAS SPHAEROIDES REACTION CENTERS

Dewey Holten, Christine Kirmaier and Leanna Levine

1. INTRODUCTION
    Reaction centers (RCs) from the photosynthetic bacterium Rhodo-
pseudomonas sphaeroides are a pigment protein complex which contain four
molecules of bacteriochlorophyll (BChl), two molecules of bacteriopheophytin (BPh), two quinones (Q) and a non-heme high-spin Fe^{2+} atom
[1,2]. Of the six bacteriochlorin molecules present, only three have been
shown to date to participate in the initial photochemistry. Two BChls
form a dimer, P, which initially absorbs a photon [3,4]. Within about 4
ps an electron arrives on a BPh [5-7], yielding the state P^+BPh^- (also
called P^+I^-) [8-10]. P^+BPh^- subsequently transfers an electron to one of
the quinones (Q_A) with a time constant of about 200 ps at room temperature
[10-13], resulting in formation of state $P^+Q_A^-$. An electron moves from
Q_A^- to the second quinone (Q_B) in about 150 μs [10].
    It has been possible for many years to remove the native quinones from
RCs, and to substitute other quinones in their place [16,17]. Studies on
quinone-depleted and quinone-substituted RCs have provided much valuable
information [2,10,16-18]. In a similar way, one approach to better understanding the roles (if any) of the two BChls which are not part of P, the
second BPh, and the Fe^{2+} atom, would be to remove a particular component
and examine the effect its removal has on the primary photochemistry. It
has become possible to do this for two of these moities. Ditson et al.
[19] first reported that a BChl (most likely one of the two not part of P)
could be chemically altered by treatment with $NaBH_4$. A subsequent study
showed that the modified pigment could be removed, and that the resulting
RC exhibits photochemistry and electron transfer properties which are
indistinguishable from those of native RCs [20]. Debus et al. [21]
recently developed a new procedure by which the Fe^{2+} could be removed.
These metal-free RCs have been shown to display impaired primary photochemistry [22]. Here, we will summarize these findings on both of these
two kinds of modified RCs (including some new results on $NaBH_4$-treated
samples), and discuss what can be learned from them in regard to the
spectroscopy of the RC and the mechanism of charge separation.

2. $NaBH_4$-TREATED REACTION CENTERS
    The two BChl molecules which are not part of P are thought to be
largely responsible for the 800 nm absorption band in the ground state
spectrum of the Rps. sphaeroides RC. As discovered by Ditson et al. [19],
when RCs are treated for a period of several hours with $NaBH_4$, this band
decreases in size by 20-40%, while the other absorption bands in the
spectrum are largely unaffected. At the same time a large new absorption
band near 715 nm appears, which is assigned to the product of the reaction
between a BChl and $NaBH_4$. Maroti et al. [20], succeeded in removing this
product from the samples leaving, presumably, some large fraction of RCs
with one less BChl. These $NaBH_4$-treated RCs were shown to retain full
photochemical activity [20]; the yield of initial formation of P^+BPh^-, the
200 ps time constant for electron transfer from BPh^- to Q_A, the charge

Biggens, J. (ed.), Progress in Photosynthesis Research, Vol. I. ISBN 90 247 3450 9
© *1987 Martinus Nijhoff Publishers, Dordrecht. Printed in the Netherlands.*

recombination times of $P^+Q_A^-$ and $P^+Q_B^-$, the Q_A/Q_B two-electron gate function, and the quantum yield of cytochrome photooxidation, are all the same
in $NaBH_4$-treated and native RCs. An example of these similarities is
shown in Fig. 1, which compares transient-state difference spectra for
native and $NaBH_4$-treated RCs taken at 33 ps for state $P^+BPh^-(P^+I^-)$ and at
1.6 ns for state $P^+Q_A^-$. These spectra were acquired with samples that had
the same ground-state absorption in P's band at 865 nm, so the magnitudes
of the absorption changes for the two samples are directly comparable.

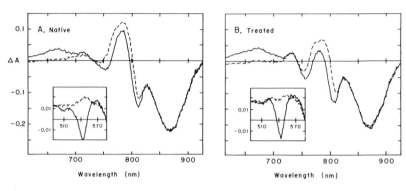

Fig. 1.   Transient-state difference spectra for Rps. sphaeroides RCs taken
          33 ps (——) and 1.6 ns (---) after the center of a 30-ps, 600-nm
          excitation flash: (A) native (B) $NaBH_4$-treated. From [20].

It can be seen in Fig. 1 that the $NaBH_4$-treated and native samples have
the same size bleaching at 865 nm at both delay times. The 33-ps (P^+BPh^-)
spectra for both samples show the same size bleaching at 542 nm (due to
bleaching of the BPh ground-state absorption) and the same broad absorption band centered at 665 nm due to the presence of BPh^-.
     The conclusion drawn by Maroti et al. is that one of the two "mono-
meric" BChl molecules plays no essential role in the photochemistry of the
RC. This has important implications for understanding the detailed nature
of the molecular states involved in the earliest photochemical events.
For example, it now seems unlikely that it is necessary for the initial
excitation on P^*, or the hole on P^+ to be delocalized over all four neighboring BChls, since RCs with some amount of BChl removed perform essentially identical photochemistry as native RCs [20].
     An area to which these $NaBH_4$-treated samples may contribute a more
direct understanding is the basic spectroscopy of the RC. With the
knowledge of the Rps. viridis crystal structure in hand [23], it is
currently of interest to attempt to calculate how the pigments and the
interactions among them give rise to the ground-state absorption spectrum.
The simplest view is that the two BChls which comprise P are responsible
for the 865-nm absorption band, that the two "monomeric" BChls are largely
responsible for the 800-nm absorption and that the two BPh molecules give
rise to the band near 765 nm. However, the situation probably is more
complex [24-28]; interactions among the pigments and between the pigments
and the protein may be important in determining the overall absorption
spectrum. If we can selectively remove one of the BChl molecules, then we
can hope to understand something about these interactions, and the assignment of the absorption bands in not only the ground-state spectra, but
also the transient-state spectra.

What is required for this analysis, however, is a better understanding of the nature of the sample one is left with after treatment with NaBH$_4$. There seems little doubt that the two BChls which comprise P are not affected directly by the NaBH$_4$, but of the other two is one and only one BChl selectively removed, and if so which one is it? Recently we have been performing pigment extraction analyses on NaBH$_4$ treated RCS, and low temperature spectroscopy to see if one can learn something more definite about the composition of the samples, and their spectra. The ratio of the ground-state absorption bands A$_{800}$/A$_{865}$ in native RCs is typically 2.2. Our recent experience with the NaBH$_4$ treatment is that this ratio comes down to about 1.4 to 1.6 over a period of about 10 hours, and seems to have a natural tendency to stop there, in that more prolonged exposure to NaBH$_4$ results in the degradation of the absorption bands at 865 and 765 nm, as well as further reduction of the 800-nm band. To date, our pigment extraction analyses of RCS with A$_{800}$/A$_{865}$ of 1.5 ± 0.2, give a total pigment content of 4.8 ± 0.3, compared to 5.8 ± 0.3 we found on (control) native RCs (L. Levine, C. Kirmaier, C. Wraight, D. Holten, unpublished). (The procedures we followed were essentially as outlined in Straley et al [1].) These results indicate that one pigment (net) has been removed by the treatment. Again, the spectral changes indicate that BChl is predominantly affected. On this basis, it seems reasonable to suggest that modification/removal of one BChl does not result in a decrease by half of the 800-nm band. In other words, the two "monomeric" BChls do not appear to contribute equally to the 800-nm band. This could arise, for example, if the local environments of the two molecules are not the same and/or if there is a difference in the interactions between the monomeric BChls and those that constitute P. Our observation could be reconciled with equal contribution of the two monomeric BChls to the 800-nm band, however, if the higher-energy excitonic component of P contributes significantly to the absorption in this region. Although this does not appear to be the case, there is still debate on this point [24-28].

The contributions of the monomeric BChls to the near-infrared region can be examined further from the low temperature spectrum of treated RCs. Figure 2 compares ground state spectra in the near-infrared region of native (dashed) and NaBH$_4$-treated (solid) RCs in polyvinyl alcohol films

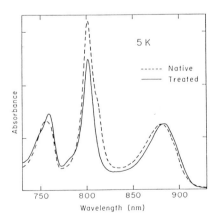

Fig. 2. 5 K ground-state absorption spectra of native (---) and NaBH$_4$-treated (——) <u>Rps. sphaeroides</u> RCs in PVA films.

at 5 K. In the native RC spectrum one should note the shoulder near 812 nm on the 800-nm absorption band. This shoulder has been reported previously [28,29] and we suggested [28] that it could reflect a resolution of the absorption of one BChl from that of the other. The absorption spectrum of the NaBH$_4$ treated sample would seem to confirm this, as we note that the shoulder is completely absent. Another interesting point can be made from the 5-K spectrum of native RCs. If one simply takes the ratio of heights of the peak at 802 nm and the shoulder at 812 nm from Fig. 2 as a naive measure of the contributions of the two monomeric BChl molecules, one finds a ratio $A_{802}:A_{812}$ of about 2:1. Assuming that the two molecules have the same temperature dependence of their spectra, this ratio implies that one BChl contributes roughly twice as much as the other to the overall 800-nm band at room temperature. Carrying this further, if the BChl with the smaller contribution were removed from the spectrum, then one would predict that the ratio of A_{865}/A_{800} would be reduced from 2.2 (in native) to 1.47. This is consistent with the results of our pigment analyses, and the tendency of the NaBH$_4$ reaction to seemingly reach a completion near this point, as described above.

The transient-state difference spectra of NaBH$_4$ treated RCs can be expected to exhibit different features reflecting their altered pigment content. (This can be seen, for example, in the 780-820 nm region of the room temperature spectra of Fig. 1B). Previously we reported [28] a splitting at 800 and 810-nm in the transient-state difference spectra of native RCs at 5 K which is not seen at room temperature (see Fig. 1A). On the basis of the above discussion, one could predict that in the low-temperature transient-state spectra of NaBH$_4$-treated RCs, the 810-nm bleaching would be gone. This is exactly what one sees. Figure 3 compares 5-K spectra of native and NaBH$_4$-treated RCs, where it can be seen that the NaBH$_4$-treated samples show only the bleaching at 800 nm, and not the additional trough observed at 810 nm in native RCs. Notice also that the band-shift-like feature with a peak near 780 nm and a trough near 800 nm is very symmetrical in the NaBH$_4$-treated (but not in the native) spectrum. Such a symmetrical feature is consistent with a blue-shift of the absorption band of a single BChl.

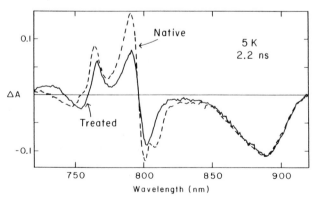

Fig. 3. 5-K difference spectra for state P$^+$Q$_A{}^-$ taken at 2.2 ns. Other conditions as in Fig. 1.

Thus we have a relatively consistent picture of an RC that appears to have three instead of the usual four BChl molecules. (At present we cannot eliminate the possibility that the BPh content also may be affected, but certainly to a much lesser degree than BChl.) The NaBH$_4$-modified RC performs all of the photochemistry associated with native RCs. Via both ground-state and transient-state spectra it appears that the removed BChl has its low-temperature absorption maximum near 810 nm (in PVA films) while the remaining monomeric BChl absorbs near 800 nm. We previously suggested [28] that the 810-nm BChl is likely on the "M side" (the apparently non-photochemically active side) of the RC [23]; this seems the most reasonable choice in the absence of any proof one way or the other. Further studies comparing additional properties of these modified RCs with those of native may help us understand how the pigments interact with one another (and their protein host) and give rise to the special properties of the RC.

3. Fe-DEPLETED AND Zn-RECONSTITUTED REACTION CENTERS

Most of the RCs from photosynthetic bacteria that have been analyzed to date contain an Fe^{2+} atom; some, such as Chloroflexus, contain Mn^{2+}. Rps. sphaeroides contains Fe^{2+}, and recently Debus et al. succeeded in developing a new procedure to remove it. Further, they were able to replace it with several other divalent first-row transition-metal ions [21]. Debus et al. found that in metal-free RCs the rate of electron transfer from Q_A to Q_B was affected but only by a factor of two (slower) compared to native RCs, and that reconstitution with Mn^{2+}, Co^{2+}, Cu^{2+}, Ni^{2+}, Zn^{2+}, or Fe^{2+} restored the 150 μs rate found in native RCs. They also showed that the yield of $P^+Q_A^-$ formation in metal-free RCs after a saturating 400-ns flash was significantly reduced from the 100% seen in native samples. Again, reconstitution with the above mentioned metals restored the native photochemistry. Recently measurements employing both 30-ps and 10-ns excitation flashes were performed to further investigate the primary photochemistry of both metal-free and Zn-reconstituted RCs [22]. We will briefly review some of the results.

Figure 4 compares transient-state difference spectra taken on native, Fe-depleted and Zn-reconstituted RCs. (The samples used in this series of experiments all had the same RC concentration, so the sizes of the absorption changes are directly comparable.) The 35-ps (solid) spectra for all three types of RCs are the same within experimental error. The 35-ps spectra can be ascribed to state P^+BPh^- and are characterized by bleaching of P's band at 865 nm, and the broad transient absorption band centered near 665 nm which is characteristic of BPh^-. [The bleaching of the BPh 545-nm absorption band also has the same magnitude in all three samples (not shown).] These observations suggest that the initial formation of P^+BPh^- is not affected by the removal of the Fe^{2+} atom or its replacement with Zn^{2+}.

For both native and Zn-reconstituted samples the magnitude of P's bleaching at 865 nm is the same at 35 ps and at 12 ns, at which time $P^+Q_A^-$ has formed. This observation reflects the 100% yield of conversion of state P^+BPh^- to state $P^+Q_A^-$ and supports the conclusion that the photochemistry of the Zn-reconstituted RCs is the same as native RCs. However, the 12-ns spectrum is different in the Fe-depleted sample. The bleaching of P's band has recovered by about 28% at 12 ns (Fig. 4B), indicating that as P^+BPh^- decays in metal-free RCs, some fraction do so by returning to the ground state.

Figure 5 compares the lifetime of state P^+BPh^- in all three types of RCs, as measured by the decay of the 665-nm absorption band of BPh^-. Native and Zn-reconstituted RCs show similar fast kinetics (205 and 255 ps, respectively). Since the $P^+Q_A^-$ yield is 100% in both samples, these

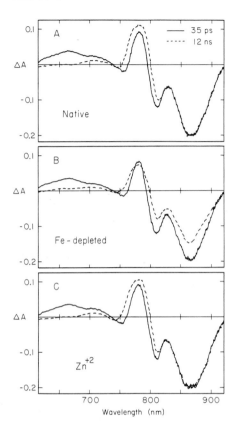

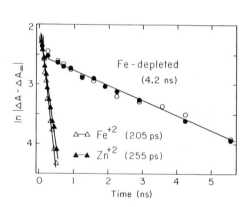

Fig. 4 (Above left.) Transient-state difference spectra for native (A), Fe-depleted (B), and Zn-reconsituted (C) RCs taken 35 ps (——) and 12 ns (---) after the center of a 30-ps 600-nm excitation flash. From [22].

Fig. 5 (Above right.) Log plots of the kinetics measured at 665 nm in native (△) Zn-reconstituted (▲) and Fe-depleted (o and •) RCs. From [22].

time constants reflect the rate of electron transfer from BPh⁻ to Q_A. In the Fe-depleted sample, however, the lifetime of P^+BPh^- is seen to be longer by a factor of about twenty. The 4.2-ns time constant is sufficiently short that at 12 ns the recovery by 28% of P's bleaching (Fig. 4B) is a good measure of the fraction of P^+BPh^- decay which regenerates the ground state. The fate of the remaining 72% of the RCs was determined by using 10-ns flashes and probing the absorption changes and kinetics at longer times. It was determined that there was about a 25% absolute yield of P^R, the triplet state of P, and only a 47% yield of formation of $P^+Q_A^-$.

The overall photochemistry of Fe-depleted RCs is summarized in Fig. 6. This model, however, is simplified, neglecting the complicated spin dynamics of state P^+BPh^-. Therefore it is not straightforward to calculate the rate of electron transfer from BPh⁻ to Q_A in Fe-depleted RCs, but it was estimated that the rate is reduced by a factor of 20-50 compared to

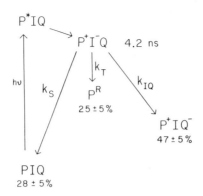

Fig. 6. Simplified scheme of the decay pathways of P^+I^- in metal free RCs. From [22].

native (and Zn-reconsitituted) RCs [22].

Thus, the metal ion does have a pronounced effect on the primary photochemistry, and one can speculate on the possible reasons [22]. One possibility that can be eliminated is some type of effect due to the spin or magnetic field of the metal, since RCs containing the diamagnetic Zn^{2+} show the same photochemistry as those which contain the native Fe^{2+}. There are still three other equally likely possibilities. One is that a metal is needed to preserve the structural integrity of the RC complex, notably the distance and orientation between BPh and Q_A. A second is that the divalent metal ion may help to poise the redox potential of Q_A, which in turn means that it is important in determining the free energy change accompanying the electron transfer from BPh^- to Q_A. Or, third, the metal may affect the reorganization energy of the electron transfer reaction via some type of vibronic coupling mechanism. Some combination of these effects may be important, and determining which if any is the dominant one will require further investigation.

REFERENCES
1. Straley, S.C., Parson, W.W., Mauzerall, D.C. and Clayton, R.K. (1973) Biochim. Biophys. Acta 305, 597-609.
2. Okamura, M.Y., Feher, G. and Nelson, N., in Govindjee (ed.) (1982) "Photosynthesis: Energy Conversion by Plants and Bacteria", Academic Press, NY, 195-272.
3. Norris, J.R., Uphaus, R.A., Crespi, H.L. and Katz, J.J. (1971) Proc. Natl. Acad. Sci. USA 71, 4897-4900.
4. Feher, G., Hoff, A.J. Isaacson, R.A. and Ackerson, L.C. (1975) Ann. N.Y. Acad. Sci. 244, 239-259.
5. Holten, D., Hoganson, C., Windsor, M.W., Schenck, C.C., Parson, W.W., Migus, A., Fork, R.L. and Shank, C.V. (1980) Biochim. Biophys. Acta 592, 461-477.
6. Martin, L.L., Breton, J., Hoff, A.J., Migus, A. and Antonetti, A. (1986) Proc. Natl. Acad. Sci. 83, 957-961.
7. Woodbury, N.W., Becker, M., Middendorf, D. and Parson, W.W. (1986) Biochem. 24, 7516-7521.
8. Fajer, J., Bruce, D.C., Davis, M.S., Forman, A. and Spaulding, L.D. (1975) Proc. Natl. Acad. Sci. USA 72, 4956-4960.

9. Tiede, D.M., Prince, R.C. and Dutton, P.L. (1976) Biochim. Biophys. Acta 449, 447-469.
10. Parson, W.W. and Ke, B., in Govindjee (ed.) (1982) "Photosynthesis: Energy Conversion by Plants and Bacteria", Academic Press, NY, pp. 331-385.
11. Rockley, M.G., Windsor, M.W., Cogdell, R.J. and Parson, W.W. (1975) Proc. Natl. Acad. Sci. USA 72, 2251-2255.
12. Kaufmann, K.J., Dutton, P.L., Netzel, T.A., Leigh, J.S. and Rentzepis, P.M. (1975) Science 188, 1301-1304.
13. Kirmaier, C., Holten, D. and Parson, W.W. (1985) Biochim. Biophys. Acta 810, 33-48.
14. Kleinfeld, D., Okamura, M.Y. and Feher, G. (1984) Biochim. Biophys. Acta 766, 126-140.
15. Debus, R.J., Feher, G. and Okamura, M.Y. (1986) Biochem. 25, 2276-2287.
16. Okamura, M.Y., Isaacson, R.A. and Fehrer, G. (1975) Proc. Natl. Acad. Sci. USA 72, 3491-3495.
17. Gunner, M.R., Tiede, D.M., Prince, R.C. and Dutton, P.L. (1982) in "Function of Quinones in Energy Conserving Systems", (B.L. Trumpower, Ed.), p 271-276.
18. Boxer, S.G., Chidsey, C.E.D. and Roelofs, M.G. (1983) Ann. Rev. Phys. Chem. 34, 389-417.
19. Ditson, S.L., Davis, R.C. and Pearlstein, R.M. (1984) Biochem. Biophys. Acta 766, 623-629.
20. Maroti, P., Kirmaier, C., Wraight, C., Holten, D. and Pearlstein, R.M. (1985) Biochim. Biophys. Acta 810, 132-139.
21. Debus, R.J., Feher, G. and Okamura, M.Y. (1986) Biochem. 25, 2276-2287.
22. Kirmaier, C., Holten, D., Debus, R.J., Feher, G. and Okamura, M.Y. Proc. Natl. Acad. Sci. (in press).
23. Deisenhofer, J., Epp, O., Miki, K., Huber, R. and Michel, H. (1984) J. Mol. Biol. 180, 385-398.
24. Vermeglio, A., Breton, J., Paillotin, G. and Cogdell, R. (1978) Biochim. Biophys. Acta 501, 514-530.
25. Shuvalov, V.A. and Asadov, A.A. (1979) Biochim. Biophys. Acta 545, 296-308.
26. Parson, W.W., Scherz, A. and Warshel, A. (1985) in "Antennas and Reaction Centers of Photosynthetic Bacteria" (Michel-Beyerle, ed.), pp 120-130, Springer, Berlin.
27. Zinth, W., Knapp, E.W., Fischer, S.F., Kaiser, W., Deisenhofer, J. and Michel, H. (1985) Chem. Phys. Lett. 119, 1-4.
28. Kirmaier, C., Holten, D. and Parson, W.W. (1985) Biochim. Biophys. Acta 810, 49-61.
29. Feher, G. (1971) Photochem. Photobiol. 14, 373-387.

Authors' Address: Department of Chemistry, Washington University, St. Louis, MO 63130.

FOURIER TRANSFORM INFRARED (FTIR) SPECTROSCOPIC INVESTIGATIONS OF THE PRIMARY REACTIONS IN PURPLE PHOTOSYNTHETIC BACTERIA

E. NABEDRYK[a], B.A. TAVITIAN[a], W. MÄNTELE[b], W. KREUTZ[b] and J. BRETON[a]

[a]Service de Biophysique, CEN Saclay 91191 Gif-sur-Yvette cedex
[b]Institut für Biophysik und Strahlenbiologie der Universität D7800 Freiburg

1. INTRODUCTION

During the primary photochemical act in bacterial photosynthesis, an electron is transferred from the excited primary donor P, a dimer of BChl a (Pa) or of BChl b (Pb) to an intermediary acceptor H, (a BPh a (Ha) or BPh b (Hb) in about 3 ps and subsequently to a primary quinone (Q) in about 200 ps. If electron transfer is blocked between H and Q, i.e. when Q is either prereduced chemically or absent, the generated P^+H^- state has a 12 ns half-life (1). In BChl a - and BChl b -containing species having a cytochrome (Cyt) tightly bound to the reaction center (RC) - e.g. Ch. vinosum and Rps. viridis respectively - reduced Cyt competes with H^- in transferring electrons to P^+. Depending on the temperature and the illumination conditions, an equilibrium can be reached where P^+ is reduced rather by Cyt than by H^- and the RCs are photochemically trapped in a long-lived PH^- state.

Here, we use Fourier transform infrared (FTIR) difference spectroscopy to characterize molecular changes associated with the light-induced i) primary electron donor oxidation in chromatophores of Rps. sphaeroides, Rps. capsulata, Rsp. rubrum, Ch. vinosum and Rps. viridis and ii) intermediary acceptor reduction in RCs of Rps. viridis and quantasomes of Ch. vinosum. As IR spectroscopy is able to monitor all bonds of the BChl molecules - conjugated or side groups - as well as polypeptides and lipids, it allows molecular changes in pigment-protein interactions occuring during the primary reactions to be detected. Furthermore, a comparison between the observed changes associated with the primary donor (Pa or Pb) oxidation or the intermediary acceptor (H_a or H_b) reduction and the ones observed in the FTIR difference spectra of the electrochemically generated radicals of isolated pigments (W. Mäntele et al, these proceedings) will lead to a better understanding of the mechanisms of the primary photosynthesis steps on a molecular basis.

2. PROCEDURE

FTIR difference spectra were recorded before, during and after continuous illumination in situ by actinic light (715 nm $< \lambda <$ 1000 nm) on a Nicolet 60 SX spectrophotometer. Samples of chromatophore membranes, quantasomes and RCs were air-dried on CaF_2 windows. For the investigation of P oxidation, films were mounted in a closed hydration cell (2). For the investigation of H reduction, films of RCs of Rps. viridis and quantasomes of Ch. vinosum (prepared according to 3) were prereduced with sodium dithionite (4). Samples (O.D. between 0.7 and 1.1 absorbance units in the amide I region) were thermostatically controlled down to 100 K. Further experimental details can be found in (2) and (4).

Biggens, J. (ed.), Progress in Photosynthesis Research, Vol. I. ISBN 90 247 3450 9

3. RESULTS AND DISCUSSION

3.1 Primary electron donor photooxidation

Fig. 1 compares the light-induced FTIR absorbance difference spectra at 250 K of chromatophores from Rps. sphaeroides, Rps. capsulata, Rsp. rubrum, Ch. vinosum (all containing BChl a) and Rps. viridis (BChl b). These difference spectra were obtained from a spectrum recorded with additional bleaching light (state P^+Q^-) and a spectrum without (state PQ). They will further on be referred to as P^+ spectra. Absorbance changes as small as 5×10^{-4} O.D. units can be detected. The P^+ spectra from chromatophores of Rps. sphaeroides, Rps. capsulata and Rsp. rubrum, compare fairly well showing identical band frequencies notably in the carbonyl region at 1750, 1706, 1684-1690, 1655 cm^{-1} and also at 1552, 1476, 1400, 1284 and 1058 cm^{-1}. In general, spectral band features reflect absorbance changes of specific chemical groups in the photoinduced state. They may be either due to an intrinsic change in oscillator strength upon radical formation or to a change of environment, e.g. hydrogen bonding. As already argued for RCs of Rps. sphaeroides and Rps. capsulata (2), the sharp positive band at 1750 cm^{-1} might arise from the carbomethoxy or propionic ester C=O bond of the BChl molecules. It might correspond to a localized C=O bond in P^+ compared to an extensive hydrogen bonding in P. The differential feature at 1706/1684-1690 cm^{-1} could be interpreted in terms of a shift - from 1684-1690 cm^{-1} in the P state to 1706 cm^{-1} in P^+- of the 9-keto C=O vibration on ring V of BChl a.

In contrast to the previous spectra which are all very similar (Fig. 1 and ref. 2), the P^+a spectrum from chromatophores of Ch. vinosum shows several different features. In particular, the ester C=O signals detected at 1743 and 1737 cm^{-1} are very weak while the magnitude of a differential band at 1663/1653 cm^{-1} is strong. It suggests a surprisingly small involvement of the BChl a ester C=O groups in the P^+a formation of Ch. vinosum. The differential keto C=O band at 1706/1684-1690 cm^{-1} common to Rps. sphaeroides, Rps. capsulata and Rsp. rubrum appears shifted to 1712/1695 cm^{-1} in Ch. vinosum indicating a slightly different keto environment. The sharp dispersive band at 1663/1653 cm^{-1} located at the position of the strongly absorbing amide I could possibly reflect some change in the protein - RC or Cyt - during the P^+a formation in Ch. vinosum; however, from its magnitude, it would only involve one or two peptide bonds. Moreover, only a very weak negative band at 1663 cm^{-1} can be detected in the P^+a spectrum of the quantasomes while a positive band at 1653 cm^{-1} is still observed (data not shown).

In general, light-induced difference spectra of chromatophores, quantasomes and RCs closely correspond to each other but with possible differences in the relative intensity of the individual bands. This property also extends to chromatophores (Fig. 1) and RCs of Rps. viridis (2) although the P^+b spectra are different from the ones of BChl a-containing species. Upon lowering the temperature down to 100 K, no significant temperature dependence in the P^+ spectra has been observed for both chromatophores of Rps. viridis and isolated RCs of Rps. sphaeroides. Furthermore, extensive deuteration of RCs of Rps. sphaeroides for three days did not shift the P^+ bands position, excluding absorption changes due to accessible protonated groups from protein side chains.

The difference in position between the P^+a and P^+b spectra would then rather point to a slightly different type of bonding to the protein than to the difference between BChl a and BChl b itself. This interpretation is strengthened by the observation of FTIR difference spectra of the

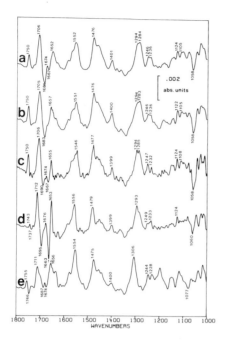

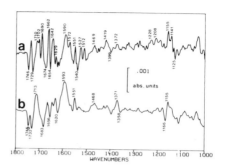

FIG. 2 : Light-induced FTIR difference spectra of a) <u>Ch. vinosum</u> quantasomes b) <u>Rps. viridis</u> RCs covered with dithionite redox buffer in the microcell.
Resolution : 4 cm^{-1}, T : 230 K

FIG. 1 : Light-induced FTIR difference spectra of a) <u>Rps. sphaeroides</u>, b) <u>Rps. capsulata</u>, c) <u>Rsp. rubrum</u>, d) <u>Ch. vinosum</u> and e) <u>Rps. viridis</u> chromatophores films. Resolution : 4 cm^{-1}, T : 250 K

electrochemically generated radicals - BChl a$^+$ and BChl b$^+$ - which clearly show remarkable similarities in the carbonyl frequency region both between themselves and with the P$^+$ spectra of chromatophores and RCs (W. Mäntele et al, these proceedings).

3.2. Intermediary acceptor photoreduction

In order to trap photochemically the state PH$^-$, chromatophores and RCs of <u>Rps. viridis</u> as well as chromatophores and quantasomes of <u>Ch. vinosum</u>, all containing tightly bound Cyt, were prereduced with dithionite and illuminated at 230 K. Fig. 2 compares the light-induced FTIR absorbance difference H$^-$ spectra of <u>Ch. vinosum</u> quantasomes and <u>Rps. viridis</u> RCs. The noise level is seen in the 1750-1800 cm^{-1} region. The <u>Ch. vinosum</u> H$^-$a spectrum displays an abundance of difference bands. All the reproducible signals are labelled. It appears that several band features are common to both H$^-$a and H$^-$b spectra. In particular, in the carbonyl region, the two negative signals at 1746 and 1732 cm^{-1} in <u>Rps. viridis</u> which have been tentatively attributed to a decrease in absorption of both 7c propionic and 10a carbomethoxy ester C=O vibrations of the Hb molecule after photoreduction (4), are also detected in <u>Ch. vinosum</u> (at 1746 and 1729 cm^{-1} respectively). We suggest that these two negative bands can well be considered as characteristic features of the H photoreduction in vivo as they are also observed in photosystem II particles after photoreduction of the pheophytin a acceptor (5, B. Tavitian et al., these proceedings). Moreover, upon reduction of the isolated BPh a, a decrease of absorption is also found for the C=O ester vibration (W. Mäntele et al, these proceedings).

In the Ch. vinosum H^-a spectrum, three highly reproducible positive signals at 1710, 1702, 1690 cm^{-1} (instead of one at 1712 cm^{-1} in the P^+a spectrum) appear in the frequency range where keto carbonyl groups are expected. Indeed, in Rps. viridis, the positive band at 1713 cm^{-1} together with a negative one at 1683 cm^{-1} were interpreted in terms of a shift of a 9 keto C=O vibration upon Hb photoreduction (4). According to resonance Raman data (6) and model compound studies (7), the bands at 1662, 1654, 1642 and 1629 cm^{-1} could arise from a change of environment of acetyl carbonyl groups. However, as in P^+, a possible contribution of peptide groups at 1662-1654 cm^{-1} cannot be excluded. In both H^- spectra, further strong absorbance change at 1590-1593 cm^{-1} could be due to C-C bridge of the tetrapyrrole rings. Among the small intensity bands between 1550 and 1000 cm^{-1}, some of them can be tentatively ascribed to shifts of C-C (1419/1399 cm^{-1}) and C-N (1372 cm^{-1}, 1166/1155 cm^{-1}) vibrations.

4. CONCLUSIONS

The FTIR data presented above provide evidence for specific molecular changes associated with the oxidation of the Pa or Pb primary electron donor as well as with the reduction of the Ha or Hb intermediary acceptor. The largest difference bands are seen in the C=O stretching frequency region. Their intensities are in the order of single bond absorption. Some of these bands are comparable to the ones observed in the FTIR difference spectra of the electrochemically generated radicals of isolated BChls and Bphs (W. Mäntele et al., these proceedings). In all species, for both P^+ and H^- states, only small signals are observed in the amide I and amide II regions which excludes the eventuality of large protein conformational changes during the primary reactions in purple photosynthetic bacteria.

ACKNOWLEDGEMENTS

We would like to thank S. Andrianambinintsoa for preparing the chromatophore, quantasome and RC samples. Part of this work was funded by A.F.M.E. grant (No. 4.320.2285).

REFERENCES

1. Parson, W.W. (1982) Annu. Rev. Biophys. Bioeng. 11, 57-80
2. Mäntele, W., Nabedryk, E., Tavitian, B.A., Kreutz, W. and Breton, J. (1985) FEBS Lett. 187, 227-232
3. Andrianambinintsoa, S., Bardin, A.M., Berger, G., Bourdet, A., Breton, J., Hervo, G. and Nabedryk, E. Colloque de Photosynthèse, Saclay 24-25 April 1986
4. Nabedryk, E., Mäntele, W., Tavitian, B.A. and Breton, J. (1986) Photochem. Photobiol. 43, 461-465
5. Tavitian, B.A., Nabedryk, E., Mäntele, W. and Breton, J. (1986) FEBS Lett. 201, 151-157
6. Lutz, M. (1984) in : Advances in Infrared and Raman Spectroscopy (Clark, R.J.H. and Hester, R.E., eds) vol. 11, pp. 211-300, Wiley-Heyden
7. Ballschmitter, K. and Katz, J.J. (1969) J. Am. Chem. Soc. 91, 2661-2677

PICOSECOND CARACTERIZATION OF PRIMARY EVENTS IN RHODOPSEUDOMONAS VIRIDIS WHOLE CELLS BY TRANSMEMBRANE POTENTIAL MEASUREMENTS.

J. DEPREZ*°, H.-W. TRISSL# and J. BRETON*

*Service de Biophysique, Département de Biologie, Centre d'Etudes Nucléaires de Saclay, 91191 Gif-sur-Yvette Cedex, France.
°Université Paris Sud, I.U.T., 94230 Cachan, France.
#Schwerpunkt Biophysik, Universität Osnabrück, Barbarast. 11, D4500 Osnabrück, FRG.

INTRODUCTION.

The first event of the photosynthetic process consists in the migration of the singlet excitation energy resulting from photon absorption by an antenna pigment toward the reaction centers (RC), where it creates the excited state of the primary donor, P*. In the case of the photosynthetic bacterium Rps. viridis, the rate of this trapping by the RC is not known. However, in Rps. sphaeroides and in R.rubrum, a time constant of 50 to 100 ps has been deduced from fluorescence decay measurements [1-3]. Informations about the further charge separations in the RC of Rps.viridis have been provided by picosecond and subpicosecond absorption measurements on isolated RCs. After P* formation, the charge separated state between P and the bacteriopheophytin electron acceptor H, (P^+H^-), appears in 2.8 ± 0.2 ps [4] and the subsequent electron transfer to a quinone, Q, requires 175 ± 10 ps [5]. The electron transfer from Q to a secondary quinone takes place in some 100 µs [6], whereas the reduction of P^+ by the associated cytochrome c occurs with a 270 ns relaxation time [7].

As determined by X-ray cristallography of Rps.viridis RC [8], H is located half-way between P and Q. Thus, during the sequential electron transfer from P to H and further to Q, an increase of the spatial separation of electric charges and therefore an increase of the strength of the transmembrane electric field must occur.

In this study, the measurements in the picosecond time scale, with a light gradient technique [9], of the photovoltage induced by the charge separations in the RC after excitation of whole cells of Rps.viridis by a picosecond excitation has been used as an alternative approach to the fluorescence lifetime or absorption changes techniques for the caracterization of the first events of photosynthesis

MATERIALS AND METHODS.

The electric measurements of the light gradient type were carried out essentially as described [9,10]. the measuring cell was built in a SMA-type micro-coaxial connector of 3 mm diameter. The bottom electrode was a Pt disk and the top electrode was a Pt mesh (wire diameter 60µm). The spacing between the two electrodes was 0.1 mm.

The photovoltage signals were recorded as single sweep on a 4 GHz oscilloscope cascaded with two 20 dB, 6 GHz amplifiers. The trace was digitized by a vidicon camera and subsequentely stored in a computer for signal analysis. The 10% to 90% response time of the whole detection device was found to be 100 ps using a pulse generator with a rise time < 25 ps. The excitation source was a mode-loked Nd-YAG laser delivering single 30 ps wide pulses at 1064 nm or 532 nm. For experiments where saturating preflash were needed, a dye laser delivering 500 ns pulses at 600 nm was available.

Biggens, J. (ed.), Progress in Photosynthesis Research, Vol. I. ISBN 90 247 3450 9
© *1987 Martinus Nijhoff Publishers, Dordrecht. Printed in the Netherlands.*

    The experimental traces were compared with the results of a convolution calculation assuming mono- or biexponential charge displacement current and taking into account the response time of the experimental set-up , the time distribution of the laser pulse and the time constant of the discharge of the capacitance measuring cell into the input impedance of the preamplifier.

    Rps.viridis whole cells were suspended in 20 mM Tris-HCl buffer, containing 10μM phenazine methosulfate, at an optical density of 50 cm^{-1} at 532 nm. When needed, 100 mM Na dithionite was added just prior the measurements.

RESULTS AND DISCUSSION.

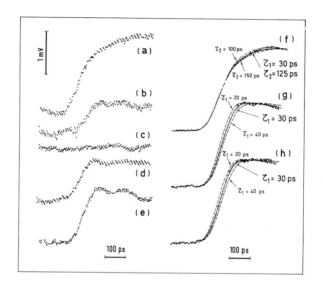

    FIGURE 1. Traces a-d: single-sweep records of the photovoltage elicited by a 30 ps, 532 nm laser flash at an energy of 2.10^{14} photons.cm^{-2} on Rps. viridis whole cells in the presence of 10 μM phenazine methosulfate. Cells were either dark-adapted for 20 s (trace a), dark-adapted with the addition of 0.1 M Na dithionite (trace b), continuously illuminated in the presence of 0.1 M Na dithionite (trace c), or exposed to a saturating preillumination flash (600 nm, 500 ns wide) 10 us before excitation. Trace e: single sweep record of the photovoltage of purple membranes electrically preoriented (500 V/cm for 5 s). A baseline signal, recorded in the dark, was substracted from all signal traces. The normalized traces e, f and g depict the kinetic traces obtained by averaging 15 individual curves corresponding to the conditions used to record the traces shown in a, b and d respectively. The continuous and broken curves are the calculated responses of the measuring equipment to assumed mono- or biexponential charge separations with the indicated time constants.

    A typical set of data is shown in figure 1. The photovoltage observed on dark adapted Rps.viridis whole cells is biphasic (trace a).

The slower phase only occurs when the quinone acceptor is oxidized (trace a) and disappears when it is reduced, either chemically in the presence of Na dithionite (trace b) or photochemically when a saturating preflash is given 10 μs prior the measuring ps-flash (trace d). We therefore assign this phase to the electron transfer from H to Q. The fast phase, close to the instrumental limitation, is independant of the redox state of the quinone and caracterizes the rise of the state P^+H^-Q of the RCs. It disappears when the sample is illuminated continuously in the presence of Na dithionite so that the state PH^-Q^- accumulates (trace c).

The 10% to 90% rise time observed in dark adapted <u>Rps. viridis</u> cells (Fig.1, trace a) was found to be 180±15 ps over the excitation range 5.10^{13} to 2.10^{15} photons.cm^{-2}. According to our deconvolution procedure, two electrogenic phases, with time constant **125±25 ps** and **30±10 ps**, contribute each about **50%** of the total amplitude of the photovoltage (traces f, g and h).

No significant differences between the kinetic traces recorded with ps-excitation pulses at 532 nm or 1064 nm were observed, meaning that our measurements obtained upon 532 nm excitation were not affected by partial direct absorption by the RC pigments.

As mentionned in the introduction, the state PHQ^-, which is present approximately 1 μs after a preflash, converts in some 100 μs to the state PHQ. Thus measurements of the photovoltage at various delay Δt between the preflash and the measuring ps-pulse probe the different ratios of theses states offering the possibility to assess the consistency of the assignment of the slow phase to electron transfer from H to Q. As depicted in figure 2, both amplitude and rise time increase when increasing Δt with a t1/2 equal to ≈ 100 μs. The values obtained at the shortest Δt of 5 us are identical to the values obtained in the presence of Na dithionite.

<u>FIGURE 2</u>: Dependence of the amplitude (a) and of the 10%-90% rise time (b) of the photovoltage measured after preillumination of <u>Rps. viridis</u> cells on the delay Δt between the preillumunation flash (600 nm, 500 ns) and the 30 ps, 532 nm measuring pulse. The amplitude is normalized to the amplitude measured in the same excitation condition from dark-adapted cells. The two horizontal broken lines indicate the mean value obtained in the dark-adapted state in the presence (lower line) or in tne absence (upper line) of 0.1 M Na dithionite.

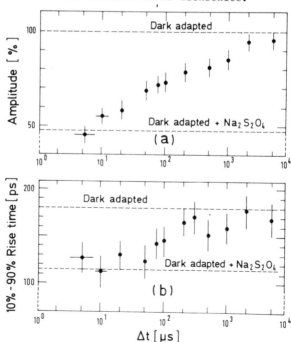

The 125±50 ps time constant of the slow phase is of the same order of magnitude as the value 175±10 ps reported for the electron transfer from H to Q in isolated RC [5].

The rapidly rising phase, which is not time resolved, results from the appearance of the state P^+H^-Q. Two kinetics are involved in this process: one is the migration of the excitation energy in the antenna pigments toward the RC where trapping occurs, the other is the charge separation between P and H. Subpicosecond spectroscopic measurements have shown that the state P^+H^-Q is created in less than 3 ps in isolated RC of Rps. viridis [4]. Therefore the 30±10 ps time constant in our measurements should be assigned to the migration of energy in the antenna. This is the fastest reported trapping time in an intact photosynthetic organism.

The location of H in the middle between P and H deduced from X-Ray diffraction on the crystallized RC of Rps.viridis [8] and the equal amplitudes of the two rising phases assigned to the successive electron transfer from P to H and from H to Q indicate a homogeneous dielectric between theses primary electron carriers in the photosynthetic membrane.

CONCLUSION.

The improved time-resolution of photoelectric measurements down to 100 ps in this study has made possible an investigation of the dynamics of exciton migration and trapping by the electrogenicity connected with the first steps of primary photosynthetic charge separation. The results obtained on whole cells of Rps.viridis are in agreement with the information deduced from other techniques on isolated RCs. Thus, for the study of primary events in intact photosynthetic membranes, fast photoelectric measurements turn out to be an alternative to measurements of absorption changes and fluorescence lifetimes.

REFERENCES.

[1] Campillo, A.J., Hyer, R.C., Monger, T.G., Parson, W.W. and Shapiro, S.L.(1977), Proc. Natl. Acad. Sci. USA 74, 1997, 2001.
[2] Sebban, P. and Moya, I. (1983), Biochim. Biophys. Acta 722, 436–442.
[3] Borisov, A.Yu., Freiberg, A.M., Godik, V.I., Rebane, K.K. and Timpman, K.E.(1985), Biochim. Biophys. Acta 807, 221–229
[4] Breton, J., Martin, J.L., Migus, A., Antonetti, A. and Orszag, A. (1986),Proc. Natl.Acad. Sci. USA 83, in press.
[5] Kirmaier, C., Holten, D. and Parson W.W. (1986), Biophys.J. 49, 586a.
[6] Carithers, R.P. and Parson, W.W. (1975), Biochim. Biophys. Acta 387,194–211.
[7] Holten, D., Windsor, M.W., Parson, W.W. and Thornber, J.P. (1978), Biochim. Biophys. Acta 501, 112–126.
[8] Deisenhofer, J., Epp, O., Miki, K., Huber, R. and Michel, H. (1984), J. Mol. Biol. 180, 385–398.
[9] Trissl, H.,-W. and Kunze, U. (1985), Biochim. Biophys. Acta 806, 136–144.
[10] Deprez, J., Trissl, H.,-W. and Breton, J. (1986), Proc. Natl. Acad. Sci. USA 83,1699–1703.

EXCITATION OF ANTENNA PIGMENTS AND ELECTRON TRANSFER UPON PICOSECOND FLASH
ILLUMINATION OF MEMBRANES OF <u>CHLOROFLEXUS</u> <u>AURANTIACUS</u>

A.M. NUIJS, H. VASMEL, L.N.M. DUYSENS and J. AMESZ

Department of Biophysics, Huygens Laboratory of the State University,
P.O. Box 9504, 2300 RA Leiden, The Netherlands

1. INTRODUCTION

<u>Chloroflexus</u> <u>aurantiacus</u> has been classified as a green photosynthetic
bacterium on the basis of its pigment composition and structure. However,
the antenna structure of the cytoplasmic membrane, which contains bacteriochlorophyll (BChl) <u>a</u>, organized in the B808-866 complex, bears more resem-
blance to that of purple bacteria than of green sulfur bacteria [1,2] and
studies with isolated reaction centers have shown that the same is true for
its primary photochemistry [3]. The present communication reports a study
by means of picosecond absorption difference spectroscopy of membranes of
<u>C. aurantiacus</u>. The results provide information on the lifetimes of the
excited antenna BChl <u>a</u> and on the rate of energy transfer from short-wave
to long-wave absorbing BChl <u>a</u>. Data are also reported on primary and secondary electron transport in the reaction centers in situ.

2. RESULTS AND DISCUSSION
2.1. <u>Oxidized reaction centers</u>

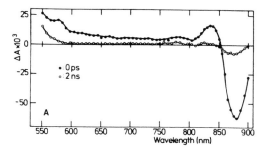 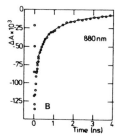

FIGURE 1. (A) Absorbance difference spectra of membrane fragments of <u>C.</u>
<u>aurantiacus</u> at 0 ps (o) and 2 ns (o) after the excitation pulse
in the presence of 2 mM $K_3Fe(CN)_6$. Continuous background
illumination was given at 528 nm. The flash energy density was
0.9 mJ/cm². The absorbance of the sample was 0.45 at 532 nm in a
2 mm cell. (B) Kinetics of the absorbance changes at 880 nm.

Fig. 1A shows the absorbance difference spectra of membrane fragments
of <u>C. aurantiacus</u> with coincident 35 ps excitation and probe pulses (0
ps) and at 2 ns after the 532 nm excitation pulse. The primary donor
P-865 was kept in the oxidized state by ferricyanide and continuous
background illumination. The prominent bleaching around 880 nm can be
ascribed to formation of singlet excited antenna bacteriochlorophyll

Biggens, J. (ed.), Progress in Photosynthesis Research, Vol. I. ISBN 90 247 3450 9
© 1987 Martinus Nijhoff Publishers, Dordrecht. Printed in the Netherlands.

(BChl*). The excitations are mainly localized on BChl 866; the absorbance increase between about 750 and 850 nm can be ascribed to a concomitant blue shift of at least 5 neighboring BChl molecules, in a similar way as was done for <u>Rhodospirillum</u> <u>rubrum</u> [4]. The absorbance changes below 650 nm are probably due to carotenoid excited singlet and triplet states.

The kinetics of absorbance changes at 880 nm are shown in Fig. 1B. The rapid decay during the first 200 ps can be explained by singlet-singlet annihilation. From 200 ps after the flash onwards the decay could be fitted by a bi-exponential decay with lifetimes of 200 ps and 1.2 ns, and a constant component (solid line); the relative amplitude of the 200 ps phase was more than 80 % of the total.

2.2. Open reaction centers
In the presence of N-methyl-phenazonium methosulfate (PMS) and ascorbate both the antenna excited states and the charge separation in the reaction centers were observed. Fig. 2 shows the difference spectra at 150 ps and at 2 ns after the pulse.

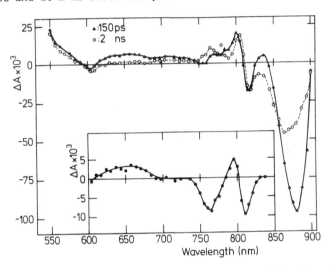

FIGURE 2. Difference spectra at 150 ps (Δ) and 2 ns (o) after the flash in the presence of 10 mM ascorbate and 20 µM PMS. Inset: difference between the 2 ns and 150 ps spectra corrected for absorbance changes in the antenna. The spectrum is attributed to reduction of I.

Above 600 nm the 2 ns spectrum mainly represents the formation of the state P-865$^+$Q$^-$, where Q denotes the secondary acceptor (menaquinone). In the near-infrared region the spectrum is similar to that obtained with isolated reaction centers [5]. The oxidation of P-865 causes the bleaching around 865 and 605 nm, while the maximum and minimum at 800 and 815 nm are probably due to a blue shift of a BChl <u>a</u> band around 813 nm. Measurement of the extent of P-865$^+$ formation as a function of intensity of the exciting flashes indicated an efficiency of the charge separation upon excitation of antenna BChl of at least 80 %. A BChl <u>a</u> to reaction center ratio of about 33 was obtained, in agreement

with the value of 30 - 35 reported by Vasmel et al. [2]. The absorbance increase at 760 nm can be ascribed to bacteriopheophytin a (BPh
a) and arises from both the oxidation of P-865 and the reduction of Q
[5]. The negative shoulder at 880 nm is probably caused by some remaining excited antenna BChl a (cf. Fig. 1). The 150 ps spectrum shows
a much larger bleaching around 880 nm and, in the region 600 - 830 nm,
changes that are related to the reduction of the primary electron acceptor I. Subtraction of the 2 ns spectrum from that measured at 150
ps and correction for the contribution by BChl* yielded the difference
spectrum shown in the inset of Fig. 2. This spectrum, with bleachings
around 765 and 810 nm and an increase around 795 nm, should mainly depict the absorbance difference between I$^-$ and I. Similar changes have
been observed by Kirmaier et al. in isolated reaction centers [5] and
were interpreted to result from the reduction of a complex of BPh a
and BChl a. The absorbance increase around 650 nm observed by us is
probably due to the absorption of the anionic species.

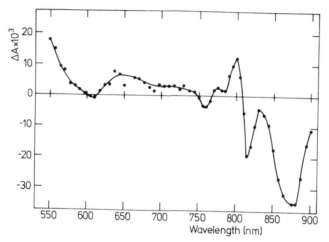

FIGURE 3. Difference spectrum at 2 ns after the flash in the presence of
dithionite, signifying the formation of P-865$^+$I$^-$.

The difference spectrum at 2 ns after the flash in the presence of
dithionite is shown in Fig. 3. The bleaching around 875 nm can be
ascribed to both P-865$^+$ and BChl*866. The broad increase between 610
and 700 nm and the bleaching around 760 nm indicate the reduction of I
while the asymmetry of the band shift around 810 nm may be explained
by the same bleaching around 810 nm that is observed in the spectrum
in the inset of Fig. 2. These results show that upon chemical reduction of Q the lifetime of I$^-$ is increased.

Fig. 4 displays the kinetics of the absorbance changes at various
wavelengths in the presence of either PMS and ascorbate (o) or dithionite (o). The decay of the bleaching at 880 nm (A) up to 2 ns showed
a time constant of about 200 ps due to BChl*866. One might expect this
phase to be replaced by a faster decay in the presence of open reaction centers. However, at the rather high energy density applied most
or all of the reaction centers already became closed during the flash,
and so the effect of trapping on the lifetime of BChl*866 was not ob-

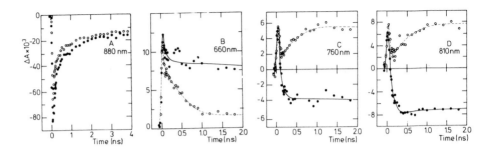

FIGURE 4. Kinetics at various wavelengths in the presence of ascorbate and PMS (o), or of dithionite (●).

served. The residual bleaching at 880 nm is mainly caused by P-865$^+$. From the kinetics with dithionite it can be observed that the lifetime of P-865$^+$ when the reaction center is in the state P-865$^+$I$^-$Q$^-$ is probably more than 10 ns. At 660 nm (B), 760 nm (C) and 810 nm (D) the first 200 ps again mainly show the decay of antenna excited singlet states. Subsequently, in the presence of PMS and ascorbate, a 350 – 450 ps phase can be observed at all three wavelengths, which is presumably due to reoxidation of I$^-$, probably by Q. This phase is lacking in the presence of dithionite, which again indicates that the lifetime of I$^-$ is increased upon chemical reduction of Q.

This investigation was supported by the Netherlands Foundations for Biophysics and for Chemical Research (SON), financed by the Netherlands Organization for the Advancement of Pure Research (ZWO).

REFERENCES

1 Feick, R.G. and Fuller, R.C. (1984) Biochemistry 23, 3693-3700
2 Vasmel, H., van Dorssen, R.J., de Vos, G.J. and Amesz, J. (1986) Photosynth. Res. 7, 281-294
3 Blankenship, R.E. (1984) Photochem. Photobiol. 40, 801-806
4 Nuijs, A.M., van Grondelle, R., Joppe, H.L.P., van Bochove, A.C. and Duysens, L.N.M. (1985) Biochim. Biophys. Acta 810, 94-105
5 Kirmaier, C., Holten, D., Mancino, L.J. and Blankenship, R.E. (1984) Biochim. Biophys. Acta 765, 138-146

ELECTRON TRANSPORT IN HELIOBACTERIUM CHLORUM

H.W.J. SMIT, J. AMESZ, M.F.R. van der HOEVEN and L.N.M. DUYSENS
Department of Biophyscs, Huygens Laboratory of the State University,
P.O. Box 9504, 2300 RA Leiden, The Netherlands

1. INTRODUCTION
    The recently discovered strictly anaerobic photosynthetic bacterium
Heliobacterium chlorum contains BChl g as major antenna pigment [1]. The
primary reaction consists of the transfer of an electron from P-798 (presumably a BChl g dimer) to a BChl c-like acceptor molecule [2]. Only frag-
mentary information is available on the secondary electron transport in
this species.
    This communication describes a study by absorbance difference spectro-
scopy of electron transport in membranes of H. chlorum.

2. RESULTS AND DISCUSSION
    The kinetics of flash-induced absorbance changes in membranes of H.
chlorum are shown in Fig. 1. At 800 and 450 nm rapid absorbance changes
were observed, together with a minor contribution by a much slower component which did not reverse during the first 100 ms after the flash. The
rapid decay could be fitted with two exponential components of 4 and 25 ms
with combined amplitudes of about 90 % of the total absorbance change. The
kinetics at 425 and 553 nm showed only the 25 ms component, together with a
much larger irreversible phase.

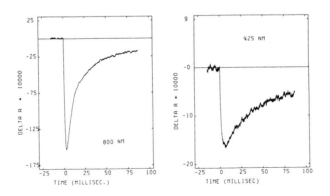

FIGURE 1. Kinetics of absorbance changes induced by a saturating 13 μs
          xenon flash in membranes of H. chlorum at 800 (A) and at 425 nm
          (B). The membranes were suspended in 10 mM Tris, 10 mM ascorbate,
          pH = 8.0, under anaerobic conditions.

Biggens, J. (ed.), Progress in Photosynthesis Research, Vol. I. ISBN 90 247 3450 9
© *1987 Martinus Nijhoff Publishers, Dordrecht. Printed in the Netherlands.*

Fig. 2 (circles) shows the difference spectrum, by plotting the total absorbance changes, together with the spectrum of the irreversible component. In the red and near-infrared region the first spectrum is very similar to that obtained by Nuijs et al. [2] upon illumination with continuous light, which spectrum has been attributed to photooxidation of the primary electron donor P-798 [3]. In the Q_x region negative bands are observed at 595 and 570 nm; in the Soret region a negative band at 400 and a positive band at 450 nm are seen. The spectrum of the irreversible component has negative bands at 553 and 420 nm, suggesting that it is due to photooxidation of cytochrome c_{553} [2,4], but its oxidation is clearly much faster than earlier reported [4]. The amount of oxidized cytochrome was quite small. If an oxidized-minus-reduced extinction difference coefficient equal to that of isolated cytochrome c_{552} from the green bacterium Chlorobium limicola f. thiosulfatophilum ($\Delta\varepsilon_{553} - \Delta\varepsilon_{540} = 20$ mM^{-1} cm^{-1}) and $\Delta\varepsilon_{799} = 100$ mM^{-1} cm^{-1} for P-798 are assumed, then it can be concluded that only in about 10 % of the reaction centers cytochrome c oxidation induced by xenon flashes could be detected. Our results thus indicate that in our preparation about 80 % of the P-798$^+$ formed in a flash decays by a back reaction with one or more reduced acceptors. About 10 % reacts in a rapid reaction with cytochrome c_{553} and the remaining 10 % is rereduced very slowly. No evidence was found for a photoreduction with a time constant of about 6 ms coupled to cytochrome c_{553} oxidation as reported by Prince et al. [4]. This indicates that the absorbance changes in the spectrum of Fig. 2 (circles) in principle must contain contributions by the photooxidation of P-798 as well by the photoreduction of (an) acceptor(s) X.

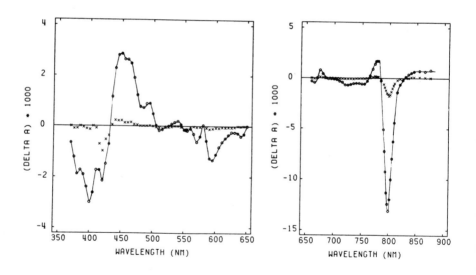

FIGURE 2. Absorption difference spectrum of the absorbance changes shown in Fig. 1. (o) Total absorbance change; (x) the irreversible component.

In the presence of PMS (40 µM) and of 10 mM ascorbate the 25 ms compo-
nent disappeared, and a mono-exponential decay of 4 - 6 ms was observed.
However, the amplitude of the absorbance change at 800 nm was not affected.
The difference spectrum showed only minor deviations from that obtained
without PMS. This indicates that the contribution by X^- to the spectrum of
Fig. 2 is only small.

In the presence of PMS and dithionite (E_h = -450 mV) to keep P-798 reduced, we observed the photoaccumulation in continuous light of a reduced
electron acceptor. The difference spectrum thus obtained is shown in Fig.
3. In the blue region the spectrum shows a broad absorbance decrease with
minima at 395, 425 and 450 nm, which suggests that it is due to reduction
of an iron-sulfur center, as was also observed in membranes of green sulfur
bacteria [5]. An ESR signal attributed to reduction of an iron-sulfur center at low temperature has been reported by Brok et al. [6]. If the reduction of one iron-sulfur center per reaction center is assumed then a differential extinction ccefficient of about 7.5 mM^{-1} cm^{-1} is calculated at
450 nm. This value is somewhat larger than that obtained for green sulfur
bacteria [5], indicating that at least one, but possibly two iron-sulfur
centers were reduced per reaction center in our preparation. Between 320
and 380 nm the absorbance changes were very small. This indicates that reduced quinones were not accumulated [6]. The absorbance changes around 670
nm and at 775 - 825 nm may be ascribed to electrochromic band shifts of the
pigment absorbing at 670 nm and of BChl $\underline{g}$, respectively.

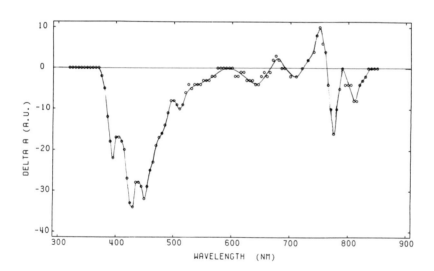

FIGURE 3. Spectrum obtained with continuous light at E_h = -450mV (see
text).

Under strongly reducing conditions (in the presence of 15 mM dithionite
at pH = 9,5, E_h = -560 mV) the formation of a 35 µs decaying component occurred. The spectrum and kinetics of this component are shown in Fig. 4.

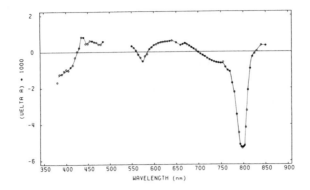 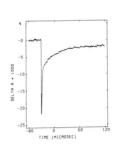

FIGURE 4. Spectrum of the absorbance changes induced by a 15 ns, 532 nm laser flash at E_h = -560 mV (see text). Inset: kinetics at 798 nm.

The spectrum shows a bleaching centered at 798 nm. We conclude that the spectrum is due to formation of the triplet of P-798. The bleaching band is significantly broader than that of P-798$^+$, which can be explained by the absence of the band shift that causes the maximum at 776 nm in the P-798$^+$ spectrum (Fig. 2). A similar broadening was observed in the triplet spectrum of P-700 in PS I particles [7]. It is interesting to note that in the Q_x region the spectrum, in contrast to that of P-798$^+$ (Fig. 2), shows a single bleaching at 575 nm, approximately at the same wavelength as the antenna BChl g Q_x absorption. This indicates that the larger bleaching at 595 nm of P-798$^+$ does not represent a disappearance of the Q_x band but may be due to a change in interaction with neighbouring pigments. The amplitude of the band at 798 nm indicates that the triplet was formed with a yield of 30 % in a flash.

The investigation was supported by the Netherlands Foundation for Biophysics, financed by the Netherlands Organization for the Advancement of Pure Research (ZWO).

REFERENCES
1 Gest, H., Favinger, J.L. (1983) Arch. Microbiol. 136, 11-16
2 Nuijs, A.M., van Dorssen, R.J., Duysens, L.N.M. and Amesz, J. (1985) Proc. Natl. Acad. Sci. USA 82, 6865-6868
3 Fuller, R.C., Sprague, S.G., Gest, H. and Blankenship, R.E. (1985) FEBS Lett. 182, 345-349
4 Prince, R.C., Gest, H. and Blankenship, R.E. (1985) Biochim. Biophys. Acta 810, 377-384
5 Swarthoff, T., Gast, P., Hoff, A.J. and Amesz, J. (1981) FEBS Lett. 130, 93-98
6 Brok, M., Vasmel, H., Horikx, J.T.G. and Hoff, A.J. (1986) FEBS Lett. 194, 322-326
7 Shuvalov, V.A., Nuijs, A.M., van Gorkom, H.J., Smit, H.W.J. and Duysens, L.N.M. (1986) Biochim. Biophys. Acta 850, 319-323

A POSSIBLE MECHANISM FOR ELECTRON TRANSFER IN THE DIQUINONE ACCEPTOR
COMPLEX OF PHOTOSYNTHETIC REACTION CENTERS[1]

S.K. Buchanan, K. Ferris and G. C. Dismukes, Princeton University,
Department of Chemistry, Princeton, New Jersey, USA 08544

1. INTRODUCTION

The function of the non-heme Fe(II) ion found in the "ferroquinone"
acceptor complex of bacterial and Photosystem II reaction centers has
been investigated using EPR spectroscopy of reaction center protein
complexes isolated from wild type *Rb. sphaeroides* strain Y which are
biosynthetically replaced with Cu(II) (Buchanan and Dismukes, submitted).
The local ligand structure of the Cu(II) has been determined and compared
to that observed in both the Fe-extracted/Cu-reconstituted complex (1)
and to the iron site in the *Rps. viridis* reaction center determined by
x-ray diffraction (2,3). We also propose a mechanism, supported by
electronic structure calculations, for electron transfer between the
quinones in which the metal acts as an electrostatic force to activate
proton transfer and to stabilize a negatively charged transition state.

2. PROCEDURE

The isolated "Cu" reaction centers (4,5) had the following
properties: A(280nm)/A(800nm)=1.3-1.6; reaction center L, M and H bands
and no detectable LHC2 bands on SDS-gels; 100% of the control yield of
light-induced P865$^+$; no detectable g 1.8 signal for Fe(II)Q$^-$; 0.25 atom
Fe/RC, 0.85 Cu/RC, <0.02 (Mn+Zn)/RC. Semi-empirical SCF-MO calculations
were carried out using the MOPAC program. (6) Individual molecules
included in the interacting triad were imidazole, benzoquinone and 2+
charge (7).

3. RESULTS AND DISCUSSION

The EPR spectrum of the Cu(II) center in dark adapted reaction
centers is shown in Fig. 1. The derived spectroscopic parameters are
given in Table I. The Cu site has a very high degree of conformational
uniformity. It is tetragonal in symmetry with coordination to 4
equivalent ^{14}N atoms, derived most probably from the N_ϵ atoms of the 4
histidines seen coordinated to iron in the X=ray structure from *Rps.
viridis* (2,3). This is derived from the well resolved 9-line hyperfine
structure due to ^{14}N seen on four of the Cu hyperfine peaks. Comparison with a number of structurally characterized Cu complexes, including
Cu(ImH)$_4$$^{2+}$ (Table I), reveals that the 4 N_ϵ atoms have tetragonal
symmetry with a departure from coplanarity by a dihedral angle no greater
than α=15°, in agreement with the low resolution x-ray structure of the
viridis reaction center (iron;3). This strong constraint will enable a
closer comparison to be made once the resolution of the latter can be
improved. This should be interesting because these two metals adopt
different ligation symmetries in unrestrained complexes (Fe(II)
tetrahedral; while Cu(II) square planar). Upon illumination at 300K to
trap the state Cu(II)Q$^-$, the dark signal is replaced by a featureless
axial resonance at 6K with $g_{\parallel}$=2.03 and $g_\perp$= 2.01, attributed to weak spin
exchange coupling between Cu(II) and Q$^-$, analogous to the interaction
seen with Fe(II). The geometry of the Cu(II) site derived from this work
and its coordination to protein ligands (2,3) is given in Fig. 2.

[1] Supported by grants from the Exxon Educ. Found. and the N.J.Comm. Sci.
Tech. (86-240020-11). GCD thanks the A.P. Sloan Foundation for an award.

Table I. EPR and structural parameters for Cu(II).

Sample	$g_{\parallel}$	$g_{\perp}$	$A_{\parallel}$	A_N	N lig.	geometry	α	Ref
Cu RC,WT biosynthetic	2.18	2.01	190G	15.3G	4	tetragonal	<15°	This work
^{65}Cu RC,R26 reconstituted	2.31	2.07	143	10.9	3	unknown	-	1
^{65}Cu(IMH)$_4$HPO$_4$	2.25	2.0	196	14.2	4	sq. planar	0	13

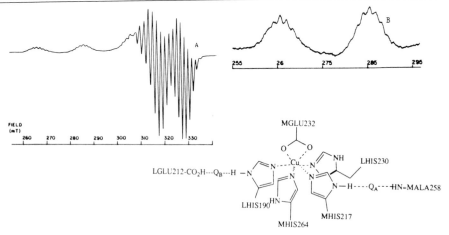

Fig. 1.A. EPR spectrum of the biosynthetically introduced Cu(II) site site at 85K. B. Expansion of the Cu M_I=-3/2 and -1/2 peaks.

Fig. 2. Structure and coordination (3) of the Cu(II) site.

   There is a very large difference between the Cu site in reaction centers obtained by biosynthetic substitution in wild type cells (strain Y) and that obtained by reconstitution of Cu to Fe-extracted reaction centers in the carotenoid-free R26 mutant (1). The EPR parameters of the latter (Table I) show coordination of only 3 N atoms to Cu(II), and the increased $g_{\parallel}$ and the decreased A_N and $A_{\parallel}$(Cu) values show a large departure from tetragonal symmetry. Despite these gross structural differences, the reconstituted Cu(II) reaction center exhibits normal kinetics and yield of electron transfer (1,8). At present it is not known if the low symmetry 3-coordinate Cu site, converts to the tetragonal 4-coordinate form after light-induced electron turn-over (M. Okamura, private commun.). We don't yet understand the origin of the structural differences in the metal ligation. It could be that we are looking at two distinct conformations that form during normal functioning of the reaction center.
   We believe that it is now possible to propose a specific role for the metal based upon the clear structural picture that is available from the X-ray work,and from the clear lack of a direct redox involvement (4,8). The central location of the metal ion coordinated to the two imidazoles which bind to Q_A and Q_B via H-bonds from the N_γ atoms points to an electrostatic function for the metal in enhancing the acidity of the A site imidazole through stabilization of the imidazolate anion. A related example, where a carboxylate anion which is H-bonded to imidazole serves

Figure 3.

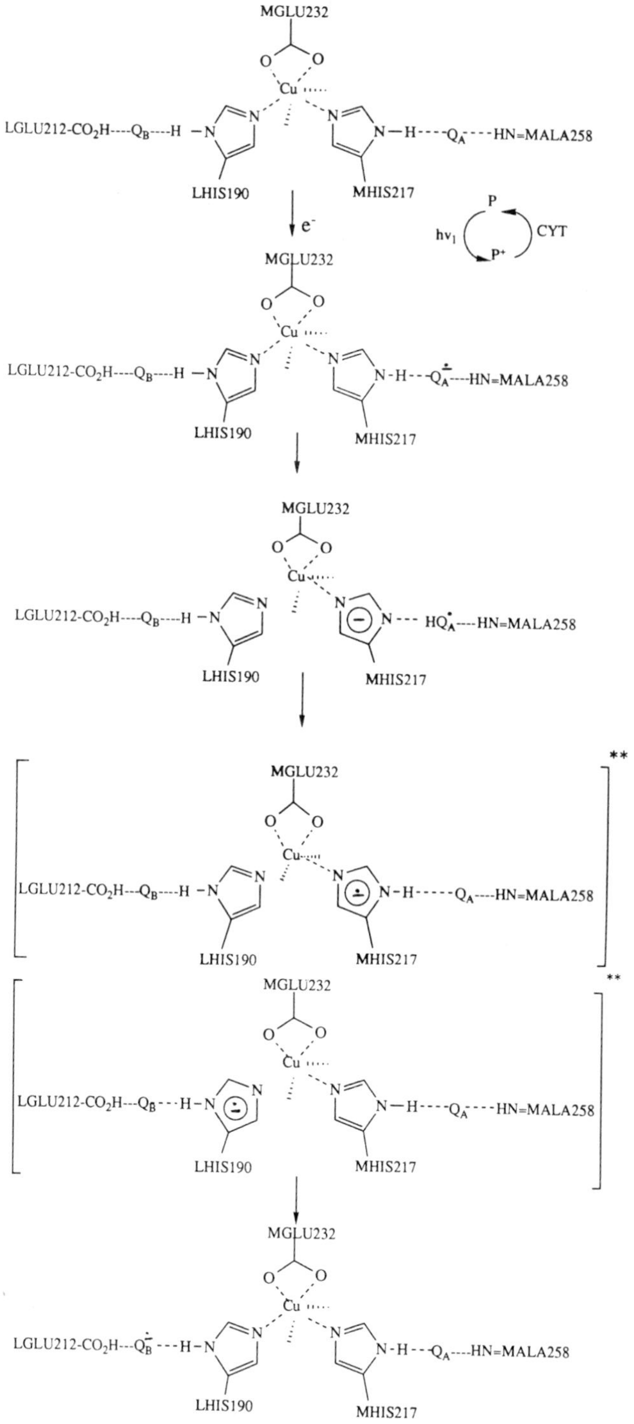

to enhance the basicity of the latter group, is well documented in the activation of the catalytic serine residue in the serine proteases (9).

We propose the mechanism given in Fig. 3 in which the imidazoles serve two functions. First, ImH_A transfers its N_γ proton to the initially formed π anion radical of $Q_A^{\bar{}}$, thereby creating the neutral semiquinone and the $Im_{\bar{A}}$ anion. Semi-empirical SCF-MO calculations show that this process is favorable energetically by about -60 kcal, of which -50 kcal is due to the dication-Im$^-$ interaction (7). Second, both ImH_A and ImH_B ultimately provide the orbital pathway through which the electron is transfered via formation of the imidazolate π anion radical in the transition state(**). This is proposed to be facilitated by the dissociation of ImH_B from the metal upon deprotonation of ImH_A. An uncoordinated ImH_B is capable of restricted migration nearer to $Im_A^{\bar{}}$ that may contribute to electron transfer in the transition state: $Im_A^{\bar{}}ImH_B \rightarrow Im_A ImH_B^{\bar{}}$. Dissociation of ImH_B from the metal upon formation of the anion $Im_{\bar{A}}$ is reasonable to propose because it is known that for metal complexes the bonding of ligands coordinated *trans* to Im$^-$ is considerably weaker than it is for the neutral ImH ligand(10). This proposed dissociation of ImH_B may also account for the functional 3-coordinate Cu site observed in the metal reconstituted reaction centers (1).

In this mechanism the initial protonation of $Q_A^{\bar{}}$ to form $QH_A\cdot$ is important because the recombination rate with the oxidized donor chlorophyll, P^+, should be considerably slowed owing to the activation energy that must be overcome in reforming the π anion radical prior to recombination (7, 8,11). This could contribute significantly to the high efficiency for charge separation in the reaction center. The purpose of the metal is to lower the pK_A of ImH_A and to orient the two imidazoles, otherwise it has no direct redox involvement. The curious observation that the metal extracted reaction center functions nearly as well as the native complex may be understandable if protons replace the metal. This possibility is supported by SCF-MO calculations showing that the acidity of the N_γ proton is also increased by coordination of a proton in place of the metal at the N_ϵ nitrogen(7).

REFERENCES
1. Feher, G.,Isaacson, R.A., Debus, R.J., Okamura,M.Y. (1986) *Biophys.J.* 49,585a.
2. Deisenhofer, J., Epp, O., Miki, K., Huber, R., Michel, H. (1985) *Nature, Lond.*, 318, 618.
3. Michel, H., Epp, O.,Deisenhofer, J. (1986) *Eur.Mol.Biol.J.* in press.
4. Nam, H. K., Austin, R.A., Dismukes, G. C. (1984) *Biochim.Biophys.Acta*,765,301.
5. Jolchine, G., Reiss-Husson, F. (1974) *FEBS Lett.*,40,5.
6. Dewar, M.J.S. QCPE Progr 455 version 1, MOPAC, Indiana University.
7. Ferris, K. F. (1986) Ph.D. Thesis, Princeton University.
8. Debus, R. J., Feher, G., Okamura, M.Y. (1986) *Biochemistry*, 25,2276.
9. Blow, D. M., Birktoft, J.J., Hartley, B.S. (1969) *Nature,Lond.*,225,811.
10. Quinn, R., Nappa, M., Valentine, J.S. (1982) *J.Amer.Chem.Soc.*,104,2588.
11. Ferris, K. F., Petrouleas, V., Dismukes, G.C. (1983) Proc. 186th Nat. Meeting Amer. Che. Soc. Inorg. Div. Abstra. 364, Wash. D.C.
12. VanCamp, H.L., Sands, R.H., Fee, J.A. (1981) *J. Chem. Phys.*, 75,2098.

TRIPLET-MINUS-SINGLET ABSORPTION DIFFERENCE SPECTRA OF SOME BACTERIAL
PHOTOSYNTHETIC REACTION CENTERS WITH AND WITHOUT CAROTENOIDS RECORDED BY
MAGNETO-OPTICAL DIFFERENCE SPECTROSCOPY (MODS) AT 290 AND 20 K

E.J. LOUS and A.J. HOFF
Department of Biophysics, Huygens Laboratory of the State University,
P.O. Box 9504, 2300 RA Leiden, The Netherlands

1. INTRODUCTION
    Bacterial photosynthetic reaction centers (RC) contain 6 or 7 pigments:
4 bacteriochlorophylls (BChl), 2 bacteriopheophytins (BPh) and for some RC
also 1 carotenoid (Car). At the same time at least one ubi- or menaquinone
is present. Following the photosynthetic electron transport chain we meet
the primary donor, P (BChl-dimer), the accessory BChl, B, the intermediary
acceptor I (BPh) and the first acceptor Q (quinone), respectively. By removing Q out of the RC or by prereducing it the formation of the stable radical pair P^+Q^- is blocked, and the pair P^+I^- decays by recombination. One
of the recombination products is the triplet state of P, P^T, which at higher temperatures may transfer its excitation energy to Car to form Car^T
(Fig. 1). This step is temperature dependent; below 30 K the triplet transfer is inhibited [1].
    The formation of P^T and/or Car^T can be followed by monitoring the
triplet-minus-singlet (T-S) absorption difference spectrum of the RC.
Techniques that have been used in the past comprise (Fig. 1):
1. CFS: conventional flash spectroscopy [2]. The P^T, Car^T triplets are
    created by excitation of Car^* or P^* via an intense laserflash, while

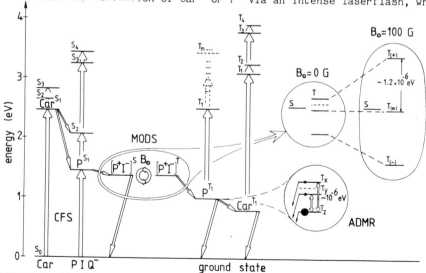

FIGURE 1. Electron transport scheme and energy level diagram of the primary
          photosynthetic reactants in which the centers of action of the
          different (T-S) spectroscopic techniques are indicated. The en-
          larged parts are not mutually scaled.

Biggens, J. (ed.), Progress in Photosynthesis Research, Vol. I. ISBN 90 247 3450 9
© *1987 Martinus Nijhoff Publishers, Dordrecht. Printed in the Netherlands.*

monitoring the difference spectrum by a wide band absorption spectrometer.
2. ADMR: absorption detected magnetic resonance [3]. P^T is populated by a
    stable, continuous light source. The concentration of P^T is varied by
    the application of microwaves resonant between two of the three triplet
    sublevels, while the (T-S) changes are monitored with a narrow band ab-
    sorption spectrometer (Fig. 1). ADMR, however, is restricted to liquid
    helium temperatures, as spin-lattice relaxation between the triplet
    sublevels has to be inhibited.
3. MODS: magneto-optical difference spectroscopy [4]. In the RC the P^T
    yield can be decreased by lowering the efficiency of the singlet-
    triplet (S-T) mixing of the radical pair P^+I^- by the application of a
    steady-state (or slowly modulated) magnetic field B_o. This change is
    again monitored by narrow band absorption spectroscopy.
By the use of modulated microwaves c.q. magnetic field and phase sensitive
detection a sensivity of $\Delta A \sim 10^{-6}$ can be reached for ADMR and MODS, that
of CFS is routinely $\Delta A \sim 10^{-4}$. Although in CFS a higher P^T concentration
can be created, the signal-to-noise ratio of the MODS-recorded (T-S) spectra is at least comparable or 10 times better than that of CFS in the whole
temperature range of 20 to 290 K. Therefore, and because in MODS the (T-S)
spectrum is continuously scanned at a rate of 0.3-3 s/nm, this method appears to be the method of choice for investigating the temperature dependence of photosynthetic reaction centers.
    In this work we present MODS recorded (T-S) spectra of RC of four
photosynthetic bacteria at 290 and 20 K.

2. MATERIALS AND METHODS
    RC of Rhodopseudomonas (Rps.) viridis were prepared as described in [3]
by 5 % (v/v) LDAO (lauryl dimethylamine oxide) incubation. RC of Rhodospirillum (R.) rubrum S1 and FR1-VI were isolated by SDS (sodium dodecyl
sulphate) incubation [5] and RC of Rhodobacter (Rb.) sphaeroides 2.4.1 by
LDAO incubation following [6]. The RC were photochemically prereduced by
adding 10 mM sodium ascorbate and diluted with ethylene glycol (67 % v/v)
to avoid cracking the sample during cooling. The experimental set-up is
described in [4].

3. RESULTS AND DISCUSSION
    (T-S) spectra recorded at 20 and 290 K by the MODS spectrophotometer
are shown in Fig. 2. Remarkable features are:
1. The BChls are excitonically coupled by transition dipole-dipole inter-
    actions. This coupling is disturbed by the triplet state on P, which
    besides a bleaching of P at 860/890 (960/1007) nm (figures in paren-
    theses for Rps. viridis) causes sharp absorption changes around 800
    (830) nm. Hardly any absorption change is observed around 750 - 790 nm
    where the BPhs absorb. Apparently, these pigments are only weakly
    coupled to P.
2. At 20 K the P^T concentration is much larger than the Car^T concentra-
    tion. This situation is reversed at 290 K. Note, however, that at 290 K
    the transfer $P^T \rightarrow Car^T$ is not 100 % effective, as there is still an ab-
    sorption change at 860 (960) nm.
3. The absence of a shift of the BChl and BPh bands (especially the P
    band) on Car^T formation indicates that Car only weakly couples to P.
4. The assignment of the features in the blue region of the (T-S) spectra
    at 290 K of the Car containing RC of R. rubrum S1 and Rb. sphaeroides
    2.4.1 is given in Table 1. Although the number of features seems to be

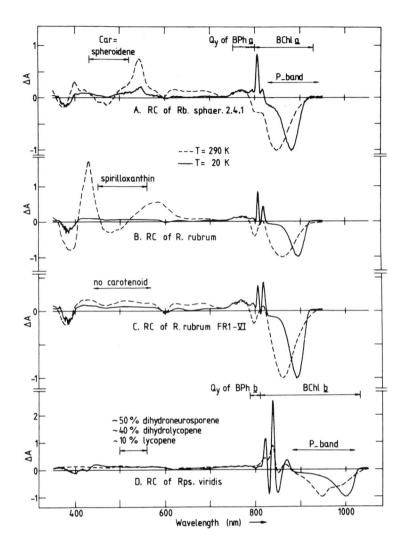

FIGURE 2. (T-S) spectra of RC of various bacteria recorded by MODS, normalized at the top of the long wavelength P band. The main area of
absorption of the RC pigments are indicated. Experimental conditions: scan rate: 3 s/nm, resolution: 2.2 nm.

the same for the two organisms, large differences in peak ratios are
observed. By taking the differential extinction coefficient of spheroi-
dene at 540 nm $\Delta\varepsilon$ (Car-Car^T) = 29 mM^{-1} cm^{-1} [7] and that of spirilloxanthin at 580 nm $\Delta\varepsilon$ (Car-Car^T) = 43 mM^{-1} cm^{-1} [7], and assuming $\Delta\varepsilon$
(P-P^T) = 100 mM^{-1} cm^{-1} at 860 nm [5], the concentration ratio of
Car^T/P^T at 290 K is 1.30 for RC of R. rubrum S1 and 2.60 for RC of Rb.
sphaeroides 2.4.1, and at 20 K 0.00 and 0.69, respectively.
5.   In the (T-S) spectra of Rps. viridis RC only very small absorption
changes are observed around 500 nm. This indicates that in this orga-

TABLE 1. Assignment of the absorbance changes at 290 K in the Car area of Fig. 2a,b (sh, shoulder).

	R. rubrum S1	Rb. sphaeroides 2.4.1
BChl bleaching	388 nm	375 nm
triplet-triplet absorption of Car^T	417 (sh)	399
	430	416 (sh)
Car bleaching	478	456 (sh)
	500 (sh)	472
triplet-triplet absorption of P^T partly		
set off by Car bleaching	540 (sh)	508 (sh)
triplet-triplet absorption of Car^T [2,1,7]	580	544

nism the transfer $P^T \rightarrow Car^T$ is negligible. The differences in the $P^T \rightarrow Car^T$ transfer efficiency of the three Car containing RCs may be caused by different distances between P and Car in these RC. Moreover, in Rps. viridis the Car may be detached out of the RC complex during isolation [8], which could explain its low transfer efficiency of $P^T \rightarrow Car^T$.

6. For RC of Rps. viridis the temperature induced absorption change around 830 nm is smaller than that for the other bacterial RC in the corresponding wavelength region around 800 nm. This may be (partly) caused by the lesser extent of spectral overlap with the long wavelength band of P in Rps. viridis.

This investigation was supported by the Netherlands Foundation for Chemical Research (SON), financed by the Netherlands Organization for the Advancement of Pure Research (ZWO).

REFERENCES
1 Schenck, C.C., Mathis, P. and Lutz, M. (1984) Photochem. Photobiol. 39, 407-417
2 Cogdell, R.J., Monger, T.G. and Parson, W.W. (1975) Biochim. Biophys. Acta 408, 189-199
3 Den Blanken, H.J. and Hoff, A.J. (1982) Biochim. Biophys. Acta 681, 365-374
4 Lous, E.J. and Hoff, A.J. (1986) Photosynth. Res. 9, 89-101
5 Slooten, L. (1973) Doctoral thesis, University of Leiden
6 Vadeboncoeur, C., Mamet-Bratley, M. and Gingras, G. (1979) Biochemistry 18, 4308-4314
7 Bensasson, R., Land, E.J. and Maudinas, B. (1976) Photochem. Photobiol. 23, 118-193
8 Thornber, J.P., Cogdell, R.J., Seftor, R.E.B. and Webster, G.B. (1980) Biochim. Biophys. Acta 593, 60-75

AN E.P.R. SIGNAL ARISING FROM Q_B^-Fe IN CHROMATIUM VINOSUM STRAIN D.

P. HEATHCOTE[a] AND A.W. RUTHERFORD[b]
a. SCHOOL OF BIOLOGICAL SCIENCES, QUEEN MARY COLLEGE, UNIVERSITY OF LONDON, MILE END ROAD, LONDON, El 4NS, U.K.
b. SERVICE BIOPHYSIQUE, DEPARTEMENT DE BIOLOGIE, CENTRE D'ETUDES NUCLEAIRES DE SACLAY, GIF-SUR-YVETTE, CEDEX, FRANCE.

1.   INTRODUCTION

Previous studies (1) have failed to detect an e.p.r. signal arising from the bound semiquinone-iron electron acceptor Q_B^- Fe in photochemical reaction centres of C. vinosum strain D. In this study a new e.p.r. signal has been observed and characterized in C. vinosum chromatophores, and it is concluded that it arises from Q_B^-Fe, analogous to the signal arising from Q_B^-Fe in Rhodopseudomonas viridis (2). Preliminary observations indicate that the characteristic e.p.r. signal g_z = 2.93 arising from the oxidised low - potential cytochrome C_{553}^+(3) shifts to g = 2.70 when C. vinosum chromatophores are subjected to a pH >9.0.

2.   MATERIALS AND METHODS

Chromatophores were isolated from cells of C. vinosum in a buffer containing either 50 mM MOPS pH 7.0, 50mM TRIS-HCl pH 9.0 or 50 mM glycine - KOH pH 10.0 as indicated. The buffer also contained 100 mM KCl and 1 mM Mg Cl_2. E.p.r. spectra were obtained using a Bruker ER - 200 T - X - band spectrometer fitted with an Oxford Instruments liquid helium cryostat and temperature control system. Samples were illuminated in the cryostat at 5 K using an 800 W projector and perspex light guide. Flashes at room temperature were provided by a Quantel YAG laser (80 m J, 15 ns pulse, 530 nm). Samples were frozen approximately 1 s after flash excitation.

3.   RESULTS AND DISCUSSION

Figure 1a shows that a broad semiquinone - iron type e.p.r. signal is present in unpoised C. vinosum chromatophores at pH 9.0 frozen in the dark. This signal is similar to that previously observed from Q_A^-Fe in C. vinosum (4) but is distinguished by its narrower line width of 240 gauss (measured from the peak maximum around g = 1.84 to the high field trough at g = 1.73) compared to 500 gauss for Q_A^-Fe (high field trough at g = 1.63). In this respect this signal of resembles that of Q_B^-Fe in Rhodopseudomonas viridis (2). Monitoring of the redox state of the reaction centre by observations of the photoinduced triplet from P_{860} and the split Bph^- signal photoinduced at 200 K indicated that this new e.p.r. signal could be attributed to Q_B^-Fe. Illumination of this sample at 5 K results in irreversible photoreduction of Q_A^-Fe and photooxidation of the low-potential cytochrome C_{553}^+ with characteristic (3) gz = 2.93 and gy = 2.24 peaks (Figure 1a and 1c). If dithionite is added to C. vinosum chromatophores at pH 9.0 Q_A^-Fe is reduced

Biggens, J. (ed.), Progress in Photosynthesis Research, Vol. I. ISBN 90 247 3450 9

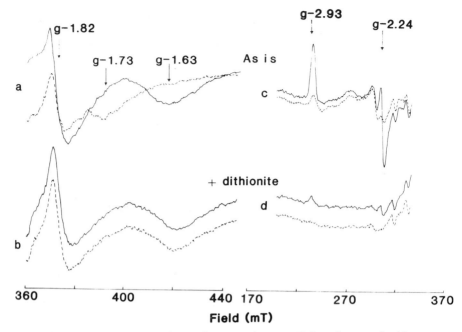

FIGURE 1. E.p.r. spectra from chromatophores of C. vinosum in the
presence and absence of dithionite, before (---) and after (-)
illumination at 5K. The chromatophores were in pH 9.0 buffer. The sample
in spectra a and c was incubated for 20 minutes at room temperature
in the dark and frozen in the dark. The sample in spectra b and
d was incubated with 0.4% w/v sodium dithionite for 10 minutes in
the dark at room temperature and frozen in the dark. The E.P.R. spectrometer settings for spectra a and b were modulation amplitude 20G,
microwave power 8dB down from 200 mW ($\simeq$ 35 mW), temperature 4.6
K. For spectra c and d the modulation amplitude was 32 G, microwave
power 17 dB down from 200 mW, temperature 15 K.

in the dark (Figure 1b), and illumination of the sample at 5 K
causes little further photochemistry, with only a slight amount
of cyt C_{553}^{+} being oxidised (Figure 1d).
The signal attributed to Q_B^-Fe could also be formed in equilibrium
redox titrations, showing that this semiquinone is thermodynamically
stable. Figure 2 presents the results of titrations of the g
= 1.82 signal (Q_A^-Fe, Q_B^-Fe) at pH 7.0 and pH 9.0. The stability
of the semiquinone Q_B^-Fe is less than that observed previously
for Q_B^-Fe in Rps. viridis (5), and Rps. sphaeroides (6) with
only 30% of the semiquinone being reduced in equilibrium redox
conditions. This is reflected in the mid-point potentials determined
for Q_B / Q_B^-(H^+) and Q_B^-(H^+) / QBH_2 couples. (Em 7.0 Q_B / Q_B^-(H^+) $\simeq$ + 95 mv, Q_B^-(H^+)/ Q_B H_2 $\simeq$ + 90, Em 9.0 Q_B / Q_B^-(H^+)
$\simeq$ 5 mv, Q_B^-(H^+) / QBH_2 $\simeq$ - 10 mv). Thus only a small proportion
of the Q_B^-Fe signal will be observed under equilibrium redox
conditions explaining the previous failure to detect this signal
(1). The estimated pK of Q_B / Q_B^-(H^+) is 8.5.

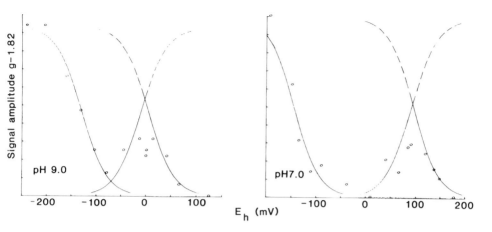

FIGURE 2. Redox titrations at pH 7.0 and 9.0 of the extent
of the g = 1.82 e.p.r signal in chromatophores of C. vinosum.
Points were obtained from titrations done in both oxidising and
reducing directions. The curves fitted are drawn for 1 e⁻ acceptors.
The extent of the g = 1.82 signal was measured in dark samples
with the following EPR spectrometer settings : modulation amplitude
20G, microwave power 8dB down from 200 mW, temperature 4.6 K.

If C. vinosum chromatophores in pH 7.0 buffer are washed with
2 mM potassium ferricyanide the dark e.p.r. spectrum demonstrates that
Q_B ⁻Fe is oxidised. (Figure 3). If this sample is illuminated at
room temperature with a single turnover flash followed by rapid freezing
the e.p.r. spectrum shows that Q_B ⁻Fe has been formed. If orthophenanthroline is added to the sample to prevent electron transfer from
Q_A ⁻Fe to Q_B ⁻Fe, a single flash at room temperature followed by
rapid freezing causes the reduction of Q_A ⁻Fe (Figure 5).

A previously unobserved e.p.r. signal at g = 2.70 appeared in C.
vinsoum chromatophores at pH 9.0 upon illumination at 200 K. Figure
4 indicates that this is due to a reversible pH-induced shift in
the g_z of at least one of the cytochrome $C_{553}+$ from g = 2.93 to g
= 2.70. Chromatophores poised at pH 10.0 with cyt C $_{553}+$ oxidised
have the feature at g = 2.70, but when returned to pH 7.0 this signal
has dissapeared and the g = 2.93 feature has increased considerably
in size.

ACKNOWLEDGEMENTS
Peter Heathcote thanks the Royal Society and Central Research
Fund for financial support. A.W.R. is supported in part by the
C.N.R.S. We would like to thank L. Richardson and J. Antoniew
for their expert technical assistance.
REFERENCES
1. Rutherford, A.W. (1980) Ph.D. thesis, University of London.
2. Rutherford, A.W. and Evans, M.C.W., (1979) FEBS Lett. 104,
227- 230.
3. Dutton, P.L. and Leigh, J.S. (1973) Biochim. Biophys. Acta
314, 178-190.

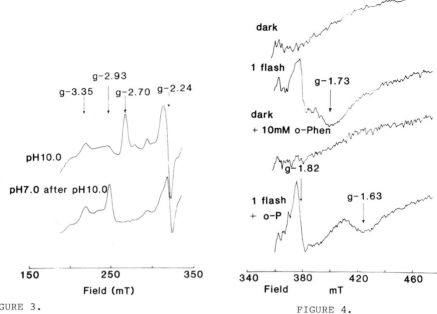

FIGURE 3.   FIGURE 4.

FIGURE 3

EPR spectra of chromatophores of C̲. vinosum before and after illumination
at room temperature with a single flash, in the presence and absence
of orthophenanthroline. The chromatophores were prepared in pH
7.0 buffer and washed once in buffer containing 2 mM potassium
ferricyanide. E.P.R. spectrometer instrument settings were modulation
amplitude 32 G, microwave power 8dB down from 200 mW, temperature
4.8 K.

FIGURE 4. E.p.r. spectra of cytochromes in chromatophores of
C̲. vinosum at pH 7.0 and pH 10.0. Chromatophores were prepared
in pH 10.0 buffer, and washed once with 2 mM (potassium ferricyanide)and
then a fraction taken and washed and resuspended in pH 7.0 buffer
as indicated. E.P.R. spectrometer instrument settings were modulation
amplitude 32 G, microwave power 15dB down from 200 mW, temperature
15 K.

4.   Leigh, J.S. and Dutton, P.L. (1972) Biochem. Biophys. Res.
Commun. 46, 414-422.
5.   Rutherford, A.W., Heathcote, P. and Evans, M.C.W. (1979) Biochem.
J. 182, 515-523.
6.   Rutherford, A.W. and Evans, M.C.W. (1980) FEBS Lett. 110,
257- 261.

PHOTOCHEMICAL REDUCTION OF EITHER OF THE TWO BACTERIOPHEOPHYTINS IN BACTERIAL PHOTOSYNTHETIC REACTION CENTERS

SANDRA FLORIN AND DAVID M. TIEDE
CHEMISTRY DIVISION D-200, ARGONNE NATIONAL LABORATORY, ARGONNE, IL 60439

1. INTRODUCTION

X-ray crystallographic studies on bacterial photosynthetic reaction centers from Rps viridis (1) and Rps sphaeroides (2) have shown that the pigments are arranged with C_2 symmetry. Two bacteriochlorophylls, BChl, lie closest to the primary electron donor, a bacteriochlorophyll dimer, B_2. Next to each BChl is a bacteriopheophytin, BPh, which is in turn followed by a quinone. The structural symmetry suggests that there may be two pathways for electron transfer.

The two bacteriopheophytins can be distinguished in low temperature optical spectra. In Rps sphaeroides the Q_x absorption band splits into peaks at 542 nm and 530 nm (3). In Rps viridis the bacteriopheophytins are distinguished by a splitting of the Q_y bands, with maxima at 790 nm and 805 nm (4). Picosecond transient absorption spectra of the first charge separated state ($B_2^+BPh^-$), show that it is predominately only the red shifted of the two bacteriopheophytins which functions as the electron acceptor below 100 K (5,6). A determination of the extent which each bacteriopheophytin contributes to the transient spectra at room temperature can not be made due to the broadness of the overlapping bands. However, using a photochemical "trapping" technique, Robert et. al. (7) found that either of the BPhs in Rps sphaeroides reaction center could be reduced at room temperature.

In the photochemical trapping experiment, the quinone is reduced, and electron donors are used to re-reduce the B_2^+ in light induced $B_2^+BPh^-$ (P^F) states: $c[B_2BPh] + h\nu \rightleftharpoons c[B_2^+BPh^-] \longrightarrow c^+[B_2BPh^-]$ (7). With Rps sphaeroides reaction centers exogenous electron donors must be added. With Rps viridis, the reaction center associated c-cytochromes serve as electron donors down to 4 K.

In this paper, we have used low temperature optical and EPR spectroscopies to further characterize the B_2BPh^- redox states generated at room temperature in Rps sphaeroides reaction centers, using methyl viologen or cytochrome c_2 as electron donors. In addition, we have compared the B_2BPh^- redox states generated in Rps viridis reaction centers at 100 K and 295 K. We find that only the red shifted BPh (BPh-542, $Q_{x,y}$= 542,762 nm in Rps sphaeroides; BPh-805, $Q_{x,y}$=542,805nm in Rps viridis) functions as an electron acceptor below 100 K. At room temperature in both Rps viridis and sphaeroides reaction centers, we find that either of the BPhs can be reduced. The extent to which each is reduced is variable, and may be related to the redox or protonation state of the quinone. These experiments show that while only one pathway is predominately functional at low temperature, either BPh possibly serves as the primary electron acceptor at room temperature.

2. PROCEDURE

Reaction centers (3 μM) in 0.1% triton x-100, 60% glycerol, 10 mM tris pH 8, were poised at E_h -460 mV in a sealed, argon purged redox vessel with either 50 μM methyl viologen, or sodium dithionite alone. Samples were transferred to a sealed, argon purged cuvette and either cooled to 80 K for spectroscopic assay, or first illuminated at 295 K with red light (590 nm visible cutoff) to accomplish the photochemical trapping. Illumination times were 15 min. (at 295 K) for Rps sphaeroides, and 2 min. (295 K) and 15

Biggens, J. (ed.), Progress in Photosynthesis Research, Vol. I. ISBN 90 247 3450 9
© *1987 Martinus Nijhoff Publishers, Dordrecht. Printed in the Netherlands.*

min (80 K) for Rps viridis reaction centers. Electron donors to the Rps sphaeroides reaction center were either 50 μM methyl viologen, or 25 μM cytochrome c$_2$.

Optical and EPR spectra were recorded with a Cary 14 and Varian E-9 spectrometers respectively.

3. RESULTS AND DISCUSSION
3.1. Rps sphaeroides Reaction centers
3.1.1. Optical Spectra

Visible optical absorption spectra for Rps sphaeroides reaction centers at 80 K are shown in figure 1. The solid line, larger trace shows the spectrum for reaction centers frozen in the dark, and illustrates the splitting seen for the bacteriopheophytin Q$_x$ bands. The dotted line trace was obtained following photochemical reduction of the bacteriopheophytin acceptors at room temperature, using methyl viologen as an electron donor. The blue shifted Bph (530 nm max) is seen to be nearly completely reduced, while the red shifted BPh (542 nm max) is only partially (40 %) reduced. The corresponding absorption spectra in the near infrared are shown in figure 2. A bleaching on the blue side of the bacteriopheophytin Q$_y$ peak shows that the blue shifted BPh has Q$_{x,y}$ maxima at 530 nm and 752 nm. The concomitant bleaching of the 600 nm and 810 nm bands shows that the blue shifted Bph is electronically coupled to one of the accessory BChl. This can be distinguished form the bleaching seen at 802 nm which accompanies the red shifted BPh reduction (4,7). These spectra demonstrate that the 810 nm band is associated with the second accessory BChl, and is not an exciton component of the B$_2$ spectrum.

Photochemical trapping at room temperature with samples poised with methyl viologen at pH 8 consistently reduces nearly all the BPh-530, but the BPh-542 is variably reduced in the range 10-50%. In samples poised without methyl viologen, but having cytochrome c$_2$ as an electron donor, both BPh-530 or BPh-542 are reduced to a variable extent. This suggests that the ability for either BPh to function as an acceptor is variable, and may be linked to the protonation or redox state of the quinone.

3.1.2. EPR Assays of Photochemical Activity

Light-induced electron transfer between the B$_2$ and the BPhs can be measured by formation of the B$_2$ triplet EPR signal. In quinone reduced reaction centers, a spin polarized triplet is formed on the B$_2$, which is diagnostic of charge recombination from a biradical, charge separated state: B$_2^+$BPh$^-$ → ^{T}B$_2$BPh (7-9). The amplitude of this signal is proportional to the extent of primary charge separation.

B$_2$ triplet signals are seen in reaction center samples in which BPh-530 is fully reduced. The amplitude of the signal is directly proportional to the extent which BPh-542 remains unreduced. This suggests that reduction of BPh-530 does not alter electron

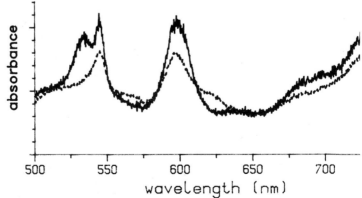

FIGURE 1. Absorption spectra of Rps sphaeroides reaction centers at 80 K, E$_h$-460 mV.

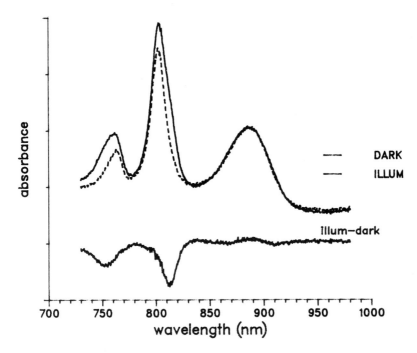

FIGURE 2. Absorption spectra of <u>Rps sphaeroides</u> reaction centers, 80 K and E_h-460 mV.

transfer to BPh-542. Further, the linewidths and peak positions of the B_2 triplet EPR signal are unaffected by reduction of BPh-530, indicating that BPh-530 reduction does not significantly perturb the electronic structure of the B_2 triplet state.

3.2. Rps viridis Reaction Centers
3.2.1. Optical Spectra

In <u>Rps viridis</u> reaction centers the bacteriopheophytin photochemically reduced at 80 K was compared to that reduced at 295 K. Figure 3 shows illuminated-dark difference spectra measured at 80 K for reaction centers poised at E_h -460 mV, and either cooled to 80 K and then illuminated, or illuminated at 295 K and then cooled in the dark to 80 K.

For the sample illuminated at 80 K, only the red shifted, BPh-805, is seen to be reduced. The bleaching of the BPh 805 nm band is seen to be accompanied by a blue shift of the accessory BChl Q_y transition at 833 nm. This trapping experiment suggests, in agreement with the ps transient measurements, that only the red shifted BPh functions as an acceptor at low temperature. In contrast, the lower spectrum in figure 3 shows that photochemical trapping at 295 K reduces the blue shifted BPh-790, which implies that this BPh may additionally function as an acceptor at room temperature.

3.2.2. EPR Assays of Photochemical Activity

Assay of the B_2 triplet formation at 10 K with the 80 K illuminated sample shown in figure 3 exhibited an 80% reduction of the triplet signal compared to the unilluminated sample. This attenuation is consistent with the notion that only the BPh-805 serves as an electron acceptor at low temperature. The triplet generated in approx. 20% of the reactions may be due to a population in which the BPh-805 was not reduced, or may represent a partial functioning of the BPh-790 as an electron acceptor. This latter possibility is supported by the small shoulder at 790 nm seen in the 80 K difference spectrum (figure 3), and by the observation that further illumination neither appreciably reduced the triplet EPR signal nor produced further optical absorbance changes.

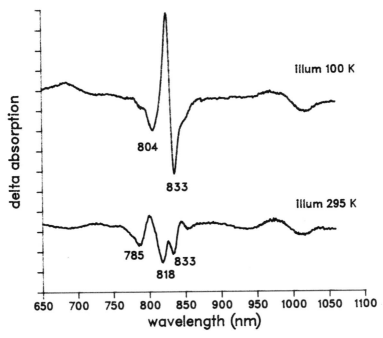

FIGURE 3. 80 K difference spectra due to BPh reduction at 80 K and 295 K in Rps viridis.

4. CONCLUSIONS

1. The photochemical trapping experiments suggest that at temperatures below 100 K, the red shifted BPh predominately functions as an electron acceptor in reaction centers of Rps sphaeroides and viridis, independent of the redox poise. This conclusion agrees with the findings from ps transient measurements. However, the trapping experiments further suggest that either BPh may function as an electron acceptor at room temperature. The efficacy of the blue shifted BPh acting as an acceptor may be determined by the quinone protonation or redox state.

2. The 810 nm absorption band in Rps sphaeroides reaction centers is assigned to the accessory BChl coupled to the blue shifted BPh.

This work was supported by the Office of Basic Energy Sciences, Department of Chemical Sciences, U.S. Department of Energy, under Contract No. W-31-109-ENG-38.

REFERENCES
1 Deisenhofer, J., Epp, O., Miki, K., Huber, R. and Michel, H. (1985) Nature 318, 614-624.
2 Chang, C.-H., Tiede, D.M., Smith, U., Norris, J. and Schiffer, M. (1986) FEBS Lett. in press.
3 Dutton, P.L., Prince, R.C. and Tiede, D.M. (1978) Photochem. Photobiol. 28, 939-949.
4 Vermeglio, A. and Paillotin, G. (1982) Biochim. Biophys. Acta 681, 23-40.
5 Kirmaier, C., Holten, D. and Parson, W.W. (1985) Biochim. Biophys. Acta 810, 49-61.
6 Wasielewski, M.R. and Tiede, D.M., unpublished.
7 Robert, B., Lutz, M. and Tiede, D.M. (1985) FEBS Lett. 188, 326-330.
8 Tiede, D.M., Prince, R.C. and Dutton, P.L. (1976) Biochim. Biophys. Acta 449, 447-467.
9 Thurnauer, M.C., Katz, J.J. and Norris, J.R. (1975) Proc. Natl. Acad. Sci., USA 72, 3270-3274.

RECONSTITUTION OF REACTION CENTERS IN PLANAR BILAYER LIPID MEMBRANES (BLM)

H. TI TIEN, MEMBRANE BIOPHYSICS LABORATORY, DEPARTMENT OF PHYSIOLOGY, MICHIGAN STATE UNIVERSITY, EAST LANSING, MI 48824 (USA)

1. INTRODUCTION

On the basis of evidence from electron microscopy, the structure of a thylakoid membrane is that of a lipid bilayer with embedded pigment-protein complexes. To explain the light-driven redox reactions from water to NADP, the so-called Z-scheme, with two photosystems (PS-I and PS-II) was proposed in the 1960s (1-3). A hypothesis for accounting electron transfer and photophosphorylation via an ion--impermeable lipid membrane, PI-I and PS-II of the Z-scheme have been incorporated into the thylakoid membrane in the late 1970s (1-5). Concurrently with afore-mentioned developments, the BLM system was developed in 1962 with the observation of photoelectric effects in chlorophyll-containing BLM in 1968 (6-8). These past accomplishments have enhanced our understanding of light transduction processes in the thylakoid membrane and set the stage for future experiments. In this communication, a brief review of the recent literature of photoeffects in BLM is presented. Additional-ly, I would like to discuss a redox membrane electrode model, based on our findings and others with pigmented BLM, in relation to the thylakoid membrane of photosynthesis (10).

2. SUMMARY OF RECENT EXPERIMENTS

BLM containing chlorophylls and related compounds have been studied (6-9). Techniques and setups have been described (11,12). To elicit appreciable photoelectric effects, asymmetrical conditions across the pigmented BLM are necessary. For electron transfer and charge separation studies, BLM containing Chl or porphyrins with and without added modi-fiers has been investigated (13). A close proximity between a donor species and an electron acceptor molecule is a prerequisite. To test this hypothesis, covalently linked porphyrin-quinone and porphyrin-carotene complexes have been synthesized (14,15). Incorporation of these newly available compounds, as a model for the initial photophysico-chemical event in reaction centers of photosynthesis, into reconstituted BLM has been carried out (16). The photopotentials obtained on some of these new BLM systems are indeed more interesting than those of chlorophyll-containing BLMs. Rich and Brody (17) found that Chl-BLM in the presence of dihydroxy-carotenoids give rise to much greater photo-currents than either the simple carotenes or the diketo- carotenoids. The negligible photocurrent in pheophytin might be due to the mis-matching of redox potential between pheophytin and ferricyanide. Vacek et al. (18) observed a decrease of photopotential of Chl-beta carotene BLM upon repeated flash excitation, which might be owing to lipid oxidation. The kinetics of photopotential was studied in more detail (19). The conclusion is that the pigment cation does not transverse

Biggens, J. (ed.), Progress in Photosynthesis Research, Vol. I. ISBN 90 247 3450 9
© *1987 Martinus Nijhoff Publishers, Dordrecht. Printed in the Netherlands.*

the BLM in less that 10 msec. To obtain the so-called "action" spectra, photoelectrospectrometry has been developed (20,21). Sielewiesiuk (22) reported the photooxidation of Chl in BLM on the basis of a decrease in photocurrent. Kadoshnikov and Stolovitsky (23) reported that about 25 Chl molecules were incorporated in the aggregates. The BLM, with incorporated Chl and pheophytin, generated photopotentials of opposite signs. Koyama et al. (24) obtained a spectrum of carotenoid-BLM by resonance Raman spectroscopy. Brasseur et al. (25) have developed a new procedure for conformation analysis to define the position of chlorophyll in BLM. Hattenbach et al. (26) raised the question, "does phytochrome interact with lipid bilayers?" Mueller and his colleagues (26) have found no detectable change in membrane conductance during illumination with red light (660 nm).

A unique feature of the BLM system is that a coupled photosensitized redox reaction may be independently activated at the two interfaces of the BLM, mimicking the Z-scheme of photosynthesis (3-6). One way to accomplish this is to use ZnTPP (or MgTPP) in conjunction with tris($2,2'$-bipyridine) ruthenium ion, $Ru(bpy)_3^{2+}$ (10). Photoexcitation of $Ru(bpy)_3^{2+}$ and/or ZnTPP leads to viologen reduction, as evidenced by the polarity of photopotentials. Since acceptor (MV^{2+} or Fe^{3+}) and donor (EDTA) are physically immobilized, some transmembrane redox reactions must take place. The most direct interpretation of the data is that reduction of $Ru(bpy)_3^{2+}$ occurs on the side containing EDTA with transmembrane electron tunneling facilitated by ZnTPP and VK_3. The drop in photocurrent may be caused by the diffusion of reduced MV^+, which could become more soluble in the membrane phase.

Reconstitution with entire reaction centers has also yielded a photoelectric effect (27). Upon illumination, the PS-I reaction center (RC) containing BLM generated transient and steady-state light-induced voltages and currents. The wavelength dependence of the photoresponse matched the absorption spectrum of RC. The main features of the photoresponse are explained by the redox membrane electrode model as discussed below. This RC-containing BLM provides direct evidence that the RC is an intrinsic membrane protein that span the lipid bilayer. In connection with reconstitution experiments, mention should also be made of pigmented liposomes, reported by Chen and Li (28).

3. THE REDOX MEMBRANE ELECTRODE MODEL

In the first report describing the BLM photoelectric effect, three possible mechanisms were proposed, one of which is the redox electrode mechanism; electronic charges are generated as a result of light excitation which cause one side of the BLM-solution interface to be reducing and the other side oxidizing (29). The number of pigment molecules in a BLM is estimated to be about 3×10^{13} molecules per cm^2 with average distance of about 10 ± 5 Å between photoactive molecules. Therefore, energy transfer is possible from a donor (D) to an acceptor (A), if the bands of respective D and A are overlapped.

Evidence favors a redox electrode mechanism in which the excited pigment molecule is directly involved. The excited pigment molecule (P) ejects an electron to an acceptor (A). This is followed by accepting an electron from a donor (D) by the oxidized pigment molecule (P^+). The separated charges are prevented from recombining by the presence of the lipid bilayer core of the BLM. Thus, an electric field should appear shortly after the absorption of photons by the pigments in the BLM.

Mechanistically, two modes of operation are possible, i.e., via a one-photon or two-photon process (10). In the former case, the ion-radicals (P^- and P^+) generated are reduced or oxidized, respectively, by D and A. In the two-photon process, the reaction of P with A on the acceptor side (left) is to produce A^- and P^+. Being in the membrane phase, this P^+ is highly likely to be reduced by an adjacent P^* on the donor side (right) of the membrane. The resulting P on the donor side is, in turn, reduced by D. In the two-photon scheme, it is envisioned that transmembrane diffusion of P^* (or P^+) is unlikely. Rather, the electron translocation across the membrane is explained in terms of a tunneling mechanism between P^+ and P^* (see the figure below).

It is worth restating that one of the aims of pigmented BLM stu-dies is to elucidate the mechanism of quantum conversion of photosyn-tic thylakoid membranes. In natural photosynthesis, the uniqueness of having an ultrathin lipid membrane, besides providing an appropriate environment for pigments and redox proteins, serves as a barrier to suppress the back reaction of the photoproducts. With the photosystems located in different parts of the membrane, these make possible for the upgrading of the energy of photons. For two photons absorbed, only one electron and one hole are involved in the ensuing redox reactions.

In concluding, it should be mentioned that a pigmented BLM/aqueous interface has been likened to a double-Schottky barrier (or p-n type) cell, with the aqueous solution playing the role of metal. Bearing this in mind, one side of the membrane acts as a photocathode (p-type junc-tion) and the other acts as a photoanode (n-type junction). The photovoltages observed are influenced by the redox compounds (electron donors and acceptors) present in the bathing solution and/or externally applied voltage.

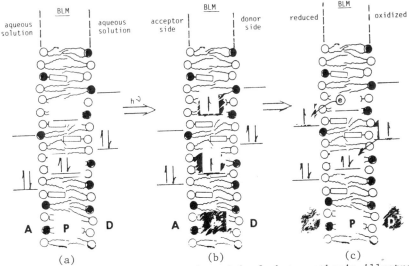

Figure. The redox membrane electrode model of photosynthesis illustrating the mechanism of light-induced charge separation and electron transfer processes in a pigmented BLM separating two aqueous solutions. (a) the system at the ground state, (b) after absorption of photons by the pigment resulting in charge separation, (c) Electron transfer processes (see Nature, 219, 272, 1968; 227, 1232, 1970).

REFERENCES

1. J.M. Olson and G. Hind (ed.) Chlorophyll-Proteins, Reaction Centers and Photosynthetic Membranes, Brookhaven Symposia Biology, 28, 103-131 (1976).

2. H. Metzner (ed.), Photosynthetic Oxygen Evolution, Academic Press, NY, (1978) pp. 411-438.

3. G. Akoyunoglou, (ed.). Fifth International Congress on Photo-synthesis,, Balaban International Science Services, Rehovot, Israel, 1, 254-262, (1981).

4. J. Barber (ed.) Photosynthesis in Relation to Model Systems, Topics in Photosynthesis, 3, Elsevier/North-Holland, NY, 1979, pp. 115-173.

5. J.R. Bolton and D.O. Hall, Ann. R. Energ., 4, 353 (1979).

6. (a) H.T. Tien, Bilayer Lipid Membranes (BLM): Theory & Practice, Dekker, Inc., NY, 1974, pp. 655. (b) Progress in Surface Science, (S.G. Davison,ed.) 19,(3) 169-274 (1985).

7. H.T. Tien (ed.). Photoelectric BLMs, Photochem. Photobiol., 24, 95-207, (1976); H.T. Tien and B. Karvaly in Solar Energy and Fuels (J.R. Bolton, ed.), Academic Press, New York, 1977, pp. 167-225.

8. R. Antolini, A. Gliozzi and A. Gorio (ed.) Transport in Membranes: Model Systems and Reconstitution, Raven Press, NY, 1982, pp. 57-75.

9.   A. Gliozzi and R. Rolandi, in Membranes and Sensory Transduction, (G. Colombetti and F. Lenci, ed.) Plenum Press, NY, 1984, pp. 1-69.

10. H.T. Tien and N.B. Joshi, Photobiochem. Photobiophys., in press (1986).

11. J.S. Huebner, Photochem. Photobiol., 30, 233 (1979).

12. P. Tancrede, P. Paquin, A. Houle and R.M. Leblanc, J. Biochem. Biophys. Meth., 7, 299 (1983).

13. Wang, C.-B., et al. (1982). Photobiochem. Photobiophys., 4, 177.

14.   Tabushi, I., Koga, N., and Yakagita, M. (1979). Tetrahedron Lett, 3, 257.

15. Moore, A.L., Dirks, G., Gust, D., and Moore, T.Z. (1980). 32, 691.

16. N.B. Joshi, J.R. Lopez and H.T. Tien, J. Photochem., 20, 139 (1982).

17. M. Rich and S.S. Brody, FEBS Lett., 143(1), 45 (1982).

18. K. Vacek, O. Valent, and A. Skuta, Gen. Physl. B., 1, 135 (1982).

19. T.M. Liu and D. Mauzerall, Biopohys. J., 48, 1 (1985).

20. J.R. Lopez and H.T. Tien, Biochim. Biophys. Acta, 597, 433 (1980).

21. A.V. Putvinskii, Biotizika, 22, 725 (1977).

22. J. Sielewiesiuk, Stud. Biophys., 96, 117 (1983).

23. S.I. Kadoshnikov and Yu.M. Stolovisky, Bioelectrochem., Bioenerg., 9, 79 (1982).

24. Y. Koyama, M. Komori and K. Shiomi, J. Coll. Int. Sci., 90, 293 (1982).

25.   R. Brasseur, J.D. Meutter and J.-M. Ruysschaert, Biochim. Biophys. Acta, 764, 295 (1984).

26. A. Hattenbach, J. Gundel, G. Hermann, D. Haroske and E. Mueller, Biochem. Physiol. Pflanzen., 177, 611 (1982).

27. J.R. Lopez and H.T. Tien, Photobiochem. Photobiophys., 7, 25 (1984).

28. Chen, Q.-S., and Li, S.-J. (1986). Acta Phytophysiol. Sinica, 12, 9.

29. H.T. Tien, Nature, 219, 272 (1968).

[Funding of this work was provided by a DHHS-NIH grant GM-14971]

^{3}(P$^+$I$^-$) LIFETIME AS MEASURED BY B$_1$ FIELD DEPENDENT RYDMR TRIPLET YIELD

Stephen Kolaczkowski, David Budil and James R. Norris, Chemistry Division, Argonne National Laboratory, Argonne, Illinois, 60439

INTRODUCTION

The observation of a triplet state of spheroidene, ^{3}C, the carotenoid in reaction centers, RCs, of Rhodopseudomonas sphaeroides Wild Type naturally led to the suggestion that a triplet state of (P$^+$I$^-$) was being quenched by spheroidene (1). The original observation of ^{3}C by Cogdell, Monger and Parson (1) indicated that the radical pair state (P$^+$I$^-$), labeled P^F (2), could be quenched directly by spheroidene. A subsequent experiment by Parson and Monger (3) concluded on the basis of light saturation curves of (P$^+$I$^-$) and ^{3}C formation that the ^{3}C quantum yield was well below that of (P$^+$I$^-$). They then suggested that a recombination product of the radical pair state, ^{3}P, was the intermediate being quenched by spheroidene (3). As a result, experiments concerning the nature of ^{3}C formation have followed the premise that ^{3}P, or P^R, is being quenched in the formation of ^{3}C (4,5). Close examination of the absorbance-change transients of Parson and Monger (3) and Schenck et al. (5) indicated that an early formation of ^{3}C is not consistent with the ^{3}P as the sole source of ^{3}C formation. In the course of our investigations concerning the mechanism of triplet energy transfer between ^{3}P and spheriodene, it is clear that reevaluation of ^{3}(P$^+$I$^-$) quenching by spheroidene was necessary.

Optically detected magnetic resonance (ODMR) experiments on RCs from the carotenoidless mutant strain R. sphaeroides R-26 in this laboratory have placed limits on the lifetimes of the two spin states of the radical pair, ^{1}(P$^+$I$^-$) and ^{3}(P$^+$I$^-$) at 27 and 2.0 nsec, respectively. By performing similar experiments on wild type RCs we measured the effect of spheroidene on ^{3}(P$^+$I$^-$) lifetime in order to determine if ^{3}(P$^+$I$^-$) is quenched by spheroidene. If the lifetime of ^{3}(P$^+$I$^-$) lifetime were shortened in wild type RCs, while no change in ^{3}(P$^+$I$^-$) lifetime indicates that ^{3}(P$^+$I$^-$) is not quenched by spheroidene, evidence would be gained for spheroidenes quenching of ^{3}(P$^+$I$^-$).

MATERIALS AND METHODS

RCs from wild type and R-26 were isolated from the respective bacterial strains and the quinone removed by modified procedures of Wraight (6) and Okamura et al. (7). The transient spectra of (P$^+$I$^-$), ^{3}P and ^{3}C and ODMR of (P$^+$I$^-$) were taken on instrumentation described elsewhere (8).

RESULTS

Figure 1 shows the transient absorbtion spectra of (P$^+$I$^-$) and ^{3}P recorded at 1.5 and 135 nsec. after the excitation laser flash in RCs from R. sphaeroides R-26, also shown is the spectrum of ^{3}C taken under identical conditions in RCs from R.sphaeroides wild type, 2.9uM, 1cm cell path length. The ODMR spectra were obtained using 420 nm for ^{3}P detection and 545 nm for ^{3}C detection. Figure 2 shows the increase in relative triplet yield as a function of resonant field for the induced transitions from the S-T$_0$ mixed sublevels to the T$_+$ and T$_-$ spin sublevels of (P$^+$I$^-$), B$_1$=17 gauss. The resonant signal at 3250 gauss is similar in amplitude and linewidth, 30 gauss, for both ^{3}P and ^{3}C detected magnetic resonance of (P$^+$I$^-$).

Biggens, J. (ed.), Progress in Photosynthesis Research, Vol. I. ISBN 90 247 3450 9
© 1987 Martinus Nijhoff Publishers, Dordrecht. Printed in the Netherlands.

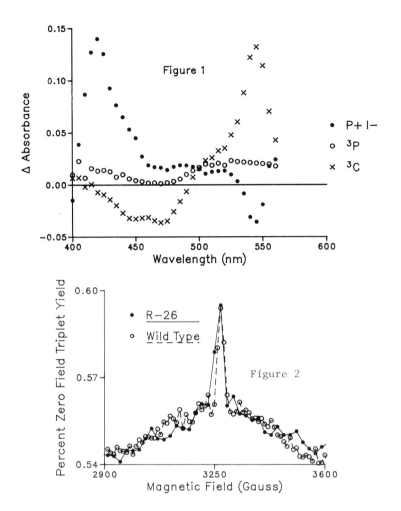

Figure 1

Figure 2

Figure 3 shows the "state locked" spectra of (P$^+$I$^-$) taken under conditions where the B$_1$ field, 45 gauss, is large enough to prevent the hyperfine induced dephasing of the two spins in the radical pair. In this situation the triplet yield decreases as the radical pair is locked into its initial singlet state and cannot form triplet as readily. Figure 4 shows the effect of varying the B$_1$ field, by changing the incident microwave power in the ODMR experiment, on the relative triplet yields, the data curves maxima correspond to ^{3}(P$^+$I$^-$) lifetimes of 2.0 and 2.3 nsec for the 12-26 and Wild type RCs, respectively.

DISCUSSION

From the linewidths of the low power ODMR spectra we can gain insight about the lifetime of ^{3}(P$^+$I$^-$). For homogeneously broadened spectral lines the linewidth provides lifetime information via the uncertainty principle:

$$\Delta E \times \Delta t \geq \hbar/2$$

The linewidth of the signal <u>increases</u> with a <u>decrease</u> in the lifetime of the spin state and is generally known as uncertainty broadening. The linewidths from figure 2 indicate that

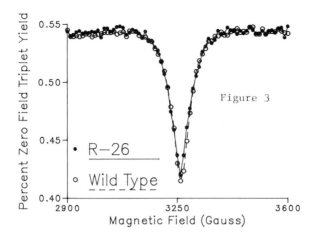

Figure 3

• R-26

○ Wild Type

very little or no uncertainty broadening occurs when spheroidene is present in the RC. This indicates that spheroidene in wild type RCs doesn't shorten the lifetime of $^3(P^+I^-)$.

However the error in the spectra of figure 2 are of the order that small changes in the lifetime of $^3(P^+I^-)$ might go undetected. A much more sensitive method of detecting a change in $^3(P^+I^-)$ lifetime is by comparing the relative triplet yields of ^3P and ^3C by varying the magnetic field strength in the rotating frame, or B_1 field. The B_1 field effect closely mimics the low field magnetic field effect or MARY experiment (9). In the low field magnetic field effect as the field is increased the T_+ and T_- sublevels of $^3(P^+I^-)$ split away in energy from the S and T_0 sublevels. When the T_+ level crosses the S level a maximum in triplet yield occurs because the two states are degenerate and crossing from singlet to triplet is more allowed. When the magnetic field increases beyond the crossing point triplet formation is decreased since the only triplet level readily accessible to S is T_0.

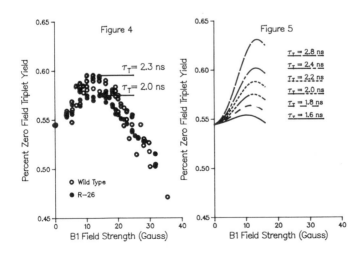

The B_1 effect on triplet yield is given in equation 1 (9).

$$\frac{\Phi_T(B_1)}{\Phi_T(B_1=0)} = \sum_{\lambda=\pm 1} \frac{(2J)^2 + [1/2(k_T-k_S)]^2}{(2J+\lambda B_1)^2 + [1/2(k_T-k_S)]^2} \tag{1}$$

As in the low field magnetic field effect as the B_1 field is increased the triplet yield increases as the B_1 field more closely matches 2J, the splitting between the S and T_0 levels, when B_1 matches 2J the triplet yield is maximized. When the B_1 strength surpasses 2J the triplet yield decreases due to overpowering effect B_1 has over the hyperfine interaction to cause state locking and prevention of triplet formation. The lifetime of $^3(P^+I^-)$ is the reciprocal of k_T and figures directly in the triplet yield, as k_T decreases the lifetime of $^3(P^+I^-)$ increases and the triplet yield increases. Small changes in lifetime, on the order of 0.5 nsec increase the maximum triplet yield by 0.05%, and would easily be observed in the B_1 field dependence experiments. Figure 5 demonstrates the effect of increasing the lifetime from 2 to 4 nsec on the same scale as the data presented in figure 4. By fitting the Triplet Yield vs B_1 Field curves with the parameters of 2J = 7 gauss and k_S = 27 nsec (13) k_T may be determined. For R-26 and Wild type RCs lifetimes of 2.0 ± 0.1 and 2.3 ± 0.1 nsec are found for $^3(P^+I^-)$ respectively.

CONCLUSIONS

From both the linewidth measurements and the triplet yield dependence on the lifetime of $^3(P^+I^-)$ we conclude that $^3(P^+I^-)$ is not being quenched by spheroidene in the course of forming spheroidenes first excited triplet state. We agree with the conclusion of Parson and Monger (3) that ^3C is not formed from (P^+I^-) but are not certain that ^3P is the only triplet state in normal photosynthesis in the RC that is quenched by spheroidene. The slight increase in $^3(P^+I^-)$ lifetime cannot be explained in terms of any reaction that would be occurring between $^3(P^+I^-)$ and spheroidene. It is possible that the presence of spheroidene in the RC slightly changes the geometry of P and or I so that $^3(P^+I^-)$ recombines at a slightly slower rate.

This work was supported by the Division of Chemical Sciences, Office of Basic Energy Science, U.S. Department of Energy under contract W-31-109-ENG-38.

REFERENCES

1. Cogdell, R. J., Monger, T. G. and Parson, W. W. (1975) Biochim. Biophys. Acta 408, 189-199.
2. Parson, w. W., Clayton, R. K. and Cogdell, R. J. (1975) Biochim. Biophys. Acta 387, 265-278.
3. Parson, W. W. and Monger, T. G. (1976), Brookhaven Symp. Biol. 28, 195-212.
4. Frank, H. A., Machnicki, J. and Friesner, R. (1983) Photochem. Photobiol. 38, 451-455.
5. Schenck, C. C., Mathis, P. and Lutz, M. (1984) Photochem. Photobiol. 39, 407-417.
6. Wraight, C. A. (1976) Biochim. Biophys. 548, 309-327.
7. Okamura, M. Y., Issacson, R. A. and Feher, g. (1975) Proc. Natl. Acad. Sci. USA 72, 3491-3495.
8. Wasielewski, M. R., Norris, J. R. and Bowman, M. K. (1984) Faraday Disc. Chem. Soc. 279-288.
9. Lersch, W. and Michel-Beyerle, M. E. (1983) Chem. Phys. 78, 115-126.
10. Norris, J. R., Bowman, M. K., Budil, D. E., Tang, J., Wraight, C. A. and Closs, G. L. (1982) Proc. Natl. Acad. Sci. USA, 79, 5532-5536.

Electron Transfer in Reaction Center Protein from *R. sphaeroides*: Generation of a Spin Polarized Bacterio-chlorophyll Dimer EPR Signal Whose Formation is Modulated by the Electron Transfer Rate From Bacteriopheophytin to Q_A

Gunner, M.R., Robertson, D.E., LoBrutto, R.L., McLaughlin, A.C. and Dutton, P.L., Dept. of Biochemistry and Biophysics, Univ. of Pennsylvania, Philadelphia, Pa 19104

INTRODUCTION

The early light activated events in the reaction center protein (RC) of photosynthetic bacteria involve electron transfer-mediated charge separation. In simplified form (presented with approximate rate constants (s^{-1}) at 300K) these virtually temperature independent charge separation and charge recombination processes can be written:

$$
\begin{array}{llll}
& (BChl)_2 & BPh & Q_A \\
& (BChl)_2^* & BPh & Q_A \\
10^1 & (BChl)_2^+ & BPh^- & Q_A \\
& (BChl)_2^+ & BPh & Q_A^-
\end{array}
$$

$h\nu$ ~10^{11} 7×10^7 7×10^3 $^3(BChl)BPh\ Q_A$ $4\times10^9\ (k_1)$

The work reported here focuses on two linked aspects of the electron transfer rate (k_1) from reduced bacteriopheophytin (BPh^-) to the primary ubiquinone (Q_A). First, we are endeavoring to correlate the BPh^- to Q_A reaction rate with the $-\Delta G°$ of the reaction and with temperature. In this study the $-\Delta G°$ values are changed by replacing the native UQ_{10} with other quinones with widely differing in situ electrochemistry. It is evident from this work (1) that Q_A replaced RCs provide the means to systematically change the BPh^- to Q_A electron transfer rate by a significant amount and hence control the lifetime of the $(BChl)_2^+BPh^-$. This leads us to the second aspect of study which correlates the lifetime of $(BChl)_2^+BPh^-$ with the magnetic properties in the ensuing $(BChl)_2^+BPh\ Q_A^-$ state.

Some years ago, Hoff (2) revealed by EPR a spin polarized $(BChl)_2^+$ signal mixed with another narrow spin polarized signal at g≈2 which was identified with Q_A^-; further analysis has recently been presented (3). The phenomenon was only evident in reaction centers in which magnetic effects of the high spin Fe++ were uncoupled from that of the Q_A^-. In the work reported here the Q_A^- Fe magnetic coupling was intact.

MATERIALS AND METHODS

The native Q_A, UQ-10, in RC isolated from *Rps. sphaeroides* strain R26 (4) was extracted (5) and Q_A function reconstituted with a varity of quinones (1,6). Flash-induced electron transfer reaction kinetics were analyzed (1) from signals obtained from a Varian E109 spectrometer in the

Biggens, J. (ed.), Progress in Photosynthesis Research, Vol. I. ISBN 90 247 3450 9
© *1987 Martinus Nijhoff Publishers, Dordrecht. Printed in the Netherlands.*

absorption mode without the 100kHz field modulation in order
to increase the time resolution.

Quantum yield (Φ) for $(BChl)_2^+Q_A^-$ formation was measured
at 298K by determination of the fraction of $(BChl)_2^+$ formed
as a function of the intensity of a $10\mu s$ activating flash of
light. Thus $\Phi=k_1(k_1+k_2)^{-1}$ where k_1 represents the electron
transfer rate from BPh^- to Q_A and k_2 is the sum of all
competing rate constants for the decay of the $(BChl)_2^+BPh^-$
state. k_2 is taken to be $7 \times 10^7 s^{-1}$, and is assumed to be
independent of the chemical nature of Q_A and temperature.
Thus measurement of Φ provides a value of k_1 (see ref.1 for
further discussion). The room temperature Φ values correlate
well with the fractional yield of $(BChl)_2^+Q_A^-$ obtained at
lower temperatures. This is as expected if k_1 is not
significantly temperature dependent. The lifetime of
$(BChl)_2^+BPh^-$ is $(k_1+k_2)^{-1}$.

Substituted naphtho-and anthraquinones were used in this
work; their _in situ_ $E_{1/2}$ values (Q^-/Q) were measured as
decribed in (1) and were chosen to diminish the $-\Delta G^\circ$ value
of the BPh^- to Q_A electron transfer reaction by up to 0.3 eV.

RESULTS AND DISCUSSION

The Φ value for $(BChl)_2^+Q_A^-$ formation decreases as
the $E_{1/2}$ of the Q_A is lowered. Hence, with diminishing
reaction $-\Delta G^\circ$, k_1 and the lifetime of $(BChl)_2^+BPh^-$ increase.
Concomitantly, several flash-induced EPR signal transients
are seen which could be kinetically resolved and measured
from 200 down to 7K. An example of these EPR spectra taken
at 7K with 2,3-dimethyl-9,10-anthraquinone as Q_A (which
yields a Φ of 0.8) is shown in Figure 1.

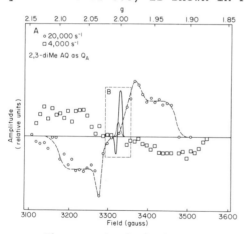 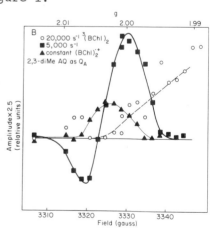

The spectral features are:
1. Centered at g=2.0026 is the conventional $(BChl)_2^+$
radical signal which decays at 30 s^{-1} as an electron returns
directly from the Q_A^-; ▲.
2. The spin polarized triplet state of the $(BChl)_2$,
$^3(BChl)_2$, decaying at 20,000 s^{-1}. This signal results from
the principle contribution to k_2 at low temperatures and is
expected to account for most of the 20% RC population that
did not form the stable $(BChl)_2^+ Q_A^-$ (7); ○.

3. A very broad signal decaying at 4000 s^{-1} is visible but not identified; it is present in all samples where the triplet is detected. This rate coincides with the rate of decay of the optically detected triplet at this temperature; □.

4. There is a signal that has a low field emissive and a high field absorptive component. It has central g value at about 2.003 and at 7K decays at ~5000s^{-1}; These features are characteristic of a chemically-induced spin-polarized radical EPR signal (see 2,3); ■.

The characteristics of the flash-activated EPR radicals are summarized in the table.

			Quinone Replacement			Temperature (K)	
			none	2,3 MAQ	UQ$_7$	200	7
1	▲	(BChl)$_2{}^+$	–	+	+	+	+
2	○	^{3}(BChl)$_2$	++	+	–	–	+
3	□	Broad signal	++	+	–	ND	+
4	■	Spin polarized radical	–	+	–	+	+

The ^{3}(BChl)$_2$ signal and the unidentified broad signal are seen in Q-depleted RCs whereas the spin-polarized radical is not; thus the spin-polarized signal is not a part of the charge recombination processes of (BChl)$_2{}^+$BPh$^-$. However, the spin polarized signal is not detectable in RCs where Φ approaches unity and the (BChl)$_2{}^+$BPh$^-$ lifetime is short (i.e.<10^{-9}s); hence this polarized signal appears to be seen if the (BChl)$_2{}^+$BPh$^-$ lifetime is longer, perhaps providing time for spin polarization. This is examined in Figure 2 which relates the yield of the spin polarized signal (normalized by accounting for Φ) as a function of the lifetime of (BChl)$_2{}^+$BPh$^-$.

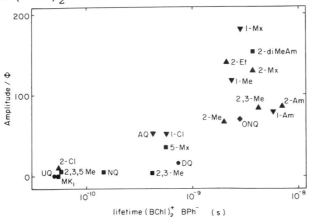

lifetime (BChl)$_2{}^+$ BPh$^-$ (s)

Although there significant scatter there is a clear trend which shows that the amplitude increases with the lifetime of (BChl)$_2{}^+$BPh$^-$. Thus, the signal can be seen in (BChl)$_2{}^+$Q$_A{}^-$ RCs only if the lifetime of the (BChl)$_2{}^+$BPh$^-$ is sufficiently long to permit polarization to occur, and if the lifetime of its spin relaxed state is longer than the relaxation time (T_1) of the polarized state. Identification of the spin-polarized signal as $(BChl)_2^+$ or Q_A^- in the long-lived $(BChl)_2^+Q_A^-$ is consistent with these criteria. However, we can rule out Q_A^- as contributing to the signal because in the Fe-containing RCs used, the Q_A signal is centered in the g=1.8 region. Thus we identify the signal as a spin polarized form of $(BChl)_2^+$.

Two points can be made in comparing this work with that of Hoff and co-workers (2,3)

1. In removing the Fe from the reaction centers, Hoff produced an RC preparation that has a $(BChl)_2^+BPh^-$ lifetime that by chance was appropriate to reveal the spin polarized $(BChl)_2^+Q_A^-$. Consistent with this is the recent finding (8) that Fe depleted RCs display a Φ of 0.47.

2. The spin polarized signal described by Hoff et.al. is a complex mixture of two emissive and absorptive signals (3) comprising spin-polarized $(BChl)_2^+$ and Q_A^- (3) The latter is seen in the g=2 region due to the absence of Fe; for the same reason its spin relaxation time is slow and is easily detected. In contrast, we anticipate that if we examine the relaxation time of the polarized state of the $Q_A^--Fe^{+2}$ signal at g=1.8 it will be much shorter due to the coupled, rapid spin relaxation of the Fe.

A final question can be asked regarding why removal of Fe from the RC causes a slowing of k_1, (8). Is this simply due to the _in situ_ $E_{1/2}$ of the native UQ_{10} which, in the absence of an interaction with Fe, is lower so as to diminish the reaction $-\Delta G°$ for the BPh^- to Q_A electron transfer? Evidence suggests not: RCs containing low potential Q_As which yield diminished $-\Delta G°$ (and Φ and k_1) are able to access an alternative, thermally activated route for charge recombination and display a rapid return of an electron from Q_A^- back to $(BChl)_2^+$ (1). For example, at 300K with 2,3 dimethyl-9,10-anthraquinone as Q_A, the charge recombination is at $19000s^{-1}$. This does not occur in the case of Fe-depleted RCs in which the recombination rate is similar to that of the native Fe containing RCs. The source(s) of the diminished Φ and k_1 must reside elsewhere (see also ref. 8).

REFERENCES
1. Gunner, M.R., Robertson, D.E. and Dutton, P.L.(1986) J. Phys. Chem. 90, 3783-3795
2. Hoff A.J. (1984) Q. Rev. Biophysics 17, 153-282
3. Hore, P.J., Watson, E.T., Boiden Peterson J. and Hoff, A.J. (1986) Biochim. Biophys. Acta 849, 70-75
4. Clayton, R.K. and Wang (1971) Meth. Enz. 23, 696-704
5. Okamura, M.Y., Isaacson and Feher, G. (1975) PNAS USA 72, 3491-3495
6. Gunner, M.R., Braun, B.S. Bruce, J.M. and Dutton, P.L. (1985) In Antennas and Reaction Centers of Photosynthetic Bacteria (Michel-Beyerle. M.E. Ed.) Springer-Verlag, Berlin pp. 298-305
7. Wraight, C.A., Leigh, J.S., Dutton, P.L. and Clayton, R.K. (1984) Biochim. Biophys. Acta 333, 401-408
8. Kirmaier, C. Holten, D. Debus, R.J. Feher, G. and Okamura, M.Y. (1986) PNAS USA (In press)
Supported by NSF and DOE

Electric Field Dependence of Electron Transfer in
Photosynthetic Reaction Centers From *Rhodopseudomonas sphaeroides*

G.A. ALEGRIA, P.L. Dutton, University of Pennsylvania, Z.
Popovic, G. Kovacs, P. Vincett, Xerox Research Centre of
Canada

INTRODUCTION
    Functional and structural analysis of photosynthetic
reaction centers has led to the view that light activates
electron transfer through the protein to generate an electric
potential difference across the supporting membrane.
    A summary of the energetics, kinetics and distances
involved in the charge separation in isolated reaction
centers devoid of the secondary quinone Q_B is given in
figures 1a and 1b.
    The work reported here focuses on the light activated
electrogenic reactions that occur in the formation of the
state $(BChl)_2^{+\bullet} Q_A^{-\bullet}$ and its subsequent dark recombination.
The goal is to alter the reactions by applying electric
fields across planar arrays of reaction centers to obtain
information on the energetics and factors that govern
kinetics of electron transfer. Application of electric
fields along the pathway of charge separation and
recombination will change the relative energy gaps between
each reaction step involved (the predicted energy shifts
introduced by the external fields are shown in figure 1c).
    Analysis of the field dependence of the formation of
$(BChl)_2^{+\bullet} Q_A^{-\bullet}$ and kinetics of charge recombination to the
ground state can provide valuable information to the
understanding of the factors that determine the high quantum
efficiency of photosynthesis as well as for testing
multiphonon tunnelling theories of electron transfer.
MATERIALS AND METHODS
    Electrical Cell Construction: The techniques used for
the preparation of the quartz substrates, involving
deposition of the conductive and blocking layers in the
sandwich cell (see fig. 2) were the same as described in ref.
1. The reaction centers were prepared using standard
chromatographic procedures and deposited as several
monolayers on this surface from a Langmuir-Blodgtett (LB)
film balance.
    Electrical and Optical Measurements: Photo-induced
electrical transients were measured using a modified RC
circuit in which noise associated with the power supply was
eliminated by using a $1\mu f$ capacitor and a set of relays (2).
As input amplifier, a Tektronix oscilloscope model 7633 was
used equipped with the 7A22 differential amplifier plug in
unit. Excitation light pulses were obtained from an
electronic flash (Metz model Mecablitz 45 Ct 1); the pulse
duration used was $50\mu s$. The applied sample bias was turned
on about 2 ms prior to the excitation light pulse and
maintained for the $10\mu s$ duration of the measurements. The

Biggens, J. (ed.), Progress in Photosynthesis Research, Vol. I. ISBN 90 247 3450 9
© *1987 Martinus Nijhoff Publishers, Dordrecht. Printed in the Netherlands.*

light intensity was adjusted to give 5% of the maximum observable voltage to insure a linear response. When measuring the time-resolved absorption recovery, the electric field was switched on within a few milliseconds after sample activation and the subsequent decay in reflectance change was recorded.

A probe beam of 860 nm light was used to determine the degree of sample bleaching after a saturating flash.

RESULTS

Direct determination of the field dependence of the quantum yield for the formation of the state $(BChl)_2^+Q_A^-$ ($\phi(E)$) was not possible due to the lack of perfect asymmetry in the orientation of the reaction centers in the LB film. Incorporation of a simple geometrical model for the LB film configuration and the plausible assumption that the quantum yeild saturates to unity for high fields assisting charge separation permitted the determination of $\phi(E)$ from measurements of the photovoltages as a fuction of positive and negative electric fields. Figure 3 shows the resultant $\phi(E)$ for two different cells (A and B). It is evident that for high negative fields (opposing charge separation) the quantum yield decreases in a hyperbolic manner and reaches a value of approximately 0.70.

The kinetic behavior of electron transfer from Q_A^- to reduce $(BChl)_2^+$ in the reaction centers in LB films showed a multiphasic character even in the absence of external fields. Determination of the field dpendence of the three rates required to fit the zero field decays was accomplished using a new technique involving the application of both high frequency bipolar fields and constant electric fields. Figure 4 shows the field dependence of the three rates after the deconvolution procedure. It is clear that all rates are approximately exponential functions of the field over the range studied.

DISCUSSION

From the predicted energy shifts introduced by the external field shown in figure 1c, it is clear that the largest relative energy perturbation occurs in the first step, the $(BChl)_2^*BPhQ_A$ to $(BChl)_2^+BPh^-Q_A$ transition. For this reaction, a field of less than -150mV/nm (opposing the separation) brings the energy gap from ~180mV (the value at zero field) to zero. Over the same range, the difference between $(BChl)_2^+BPh^-Q_A$ and $(BChl)_2^*BPhQ_A^-$ has gone from ~700mV at zero field to ~595mV and the gap between $(BChl)_2^+BPhQ_A^-$ and the ground state increases from ~500mV at zero field to ~750 mV.

From these considerations it appears that the most probable major source of quantum yield drop lies in the first step, the transition(s) involved in converting $(BChl)^*_2BPhQ_A$ to $(BChl)_2^*BPh^-Q_A$. Preliminary mathematical modeling of the observed $\phi(E)$ shows that significantly better curve fitting is obtained when assuming that the forward rate for electron transfer from $(BChl)^*_2$ to BPh, and not the reverse, is initially sensitive to the opposing external fields and slowed down. The magnitude of the sensitivity suggests the possiblilty of an intermediate step, perhaps involving the bacteriochlorophyl monomer, that under the influence of the external field is forced to go energetically uphill. More work is needed to clarify this important point.

The membrane potential in photosynthetic bacteria working under steady state conditions has been estimated to be ~200mV. Considering that the low dielectric distance across which this potential is acting is ~35Å, this represents a field of ~60mV/nm opposing charge separation. The results presented here show that the early photochemical steps catalyzed by the reaction center are capable of maintianing a quantum yield over 80% even when the membrane potential reaches its maximum value.

Taking the distance for electron transfer from $(BChl)_2$ to Q_A along the electric field direction to be 2.0nm (as illustrated in figure 1b), the range of fields used in our experiments ($\pm$150mV/nm) represents a change of 0.22eV to 0.82eV in the free energy gap between the $(BChl)_2^+BPhQ_A^-$ and $(BChl)_2BPhQ_A$ levels. Analysis of the three phases for charge recombination reveals a relatively small field dependency of the rates (the most sensitive, fastest and predominant component increases by only a factor of 10 as the energy is changed by 0.5eV). This behavior of the rates is not expected from simple electron transfer theories that have been applied to this reaction which would predict a maximum in rate would be encountered at a point close to the zero applied field. However, contributions to the observed electric field dependence of the recombination rates may come from other sources especially at high $-\Delta G^\circ$ where other reactions routes for recombination may be accessed at 300K (3). However, at $-\Delta G^\circ$ values <0.6eV the recombination kinetics will be direct and will not be sensitive to this possiblilty; in this range the results can be compared with parallel work done to alter $-\Delta G^\circ$ values by replacement of the Q_A in the site with quinones of different electrochemistry (3). Other contributions can be broadly considered as arising from electric field effects acting to induce polarization of the active components.

REFERENCES
1. Popovic, Z.D., Kovacs, G.J., Vincent, P.S. and Dutton, P.L., (1985) Chem. Phys. Lett. 116, 405-410
2. Popovic, Z.D. (1983), J. Chem. Phys., 78, 1552-1558
3. Gunner, M.R., Robertson, D.E. and Dutton, P.L. (1986) J. Phys. Chem. 90, 3783-3795

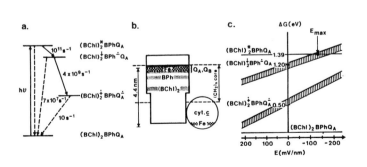

Fig. 1(a) Electron-transfer rates in RCs of R. sphaeroides. (b) Schematic representation of dielectric distances. (c) Electric field dependence of the energy

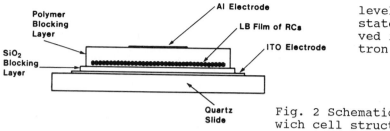

levels of
states invol-
ved in elec-
tron transfer

Fig. 2 Schematic of sand-
wich cell structure for
electrical and optical
measurements.

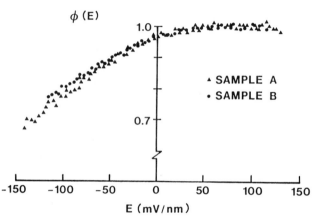

Fig.3 Dependence of the
quantum yield of charge
separation, ϕ, on the
electric field, E.

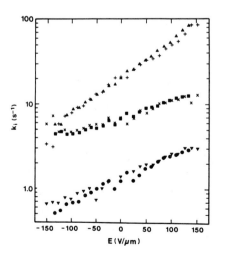

Fig.4 Electric field dependence
of the rate constants $k_i(E)$ as
determined from 2 samples:
($\times$,+, $\blacktriangledown$ sample 1), ($\blacktriangledown$,$\blacksquare$,$\bullet$ sample
2)

Hydrocarbon Tail Structure and its Effect on the Affinity and Kinetic Performance of Quinones at the Q_A Site in Reaction Centers of Rhodobacter sphaeroides R26

K. Warncke, M.R. Gunner, B.S. Braun, [+]C.-A. Yu and P.L. Dutton, Dept. of Biochemistry and Biophysics, Univ. of Penn., Phila., PA, and [+]Dept.of Biochemistry, Oklahoma State Univ., Stillwater, OK.

INTRODUCTION

The equilibrium binding and electron transfer properties of various quinones at the Q_A site in solubilized reaction centers (RC) from Rhodobacter sphaeroides R26 have been studied as a function of hydrocarbon tail structure. Dissociation constants (K_D) for quinones substituted with a variety of tail structures have been measured, revealing specificity in the interaction of the protein with the quinone hydrocarbon tail. In addition, the rate constants for Q_A reduction and oxidation have been correlated with tail structure. A clearer understanding of the binding and kinetic behavior of quinones at the Q_A site emerges.

MATERIALS AND METHODS

RCs were purified (1) and the native ubiquinone-10 extracted(2). Tailed quinones were synthesized as in (3). Tailed quinones were reconstituted in medium containing 10mM Tris-HCl (pH 8.0), less than .005% LDAO and approximately 150nM RCs. Occupation of the Q_A site was spectrophotometrically determined by the stable oxidation of $(BChl)_2$ (lifetime 1ms) measured at 602-540nm, following a 10us actinic xenon flash. K_Ds were obtained from the measurement of this activity as a function of added quinone. Binding free energies were then calculated ($\Delta G= RTlnK_D$). Electron transfer from Q_A^- to $(BChl)_2^+$ returns the system to the ground state. The first order rate constant is determined from the decay of the $(BChl)_2^+$ optical signal. Kinetics and equilibrium binding were analyzed as in (4). Conclusions from quantum yield considerations assume a BPh^- to $(BChl)_2^+$ recombination rate of $7x10^7$/s.

RESULTS AND DISCUSSION

The binding affinities of the tailed quinones were compared with their cyclohexane/water partition coefficients (P).LogP is an index of hydrophobicity. Generally, affinity appears to be an increasing function of logP, as shown in figure 1. However, marked deviations punctuate this general trend., indicating that specific interactions between the tail and protein are significant in controlling binding. For example, the homologous series of alkyl ubiquinones (UQ) exhibits an asymmetric affinity profile, whereas logP increments per added methylene are equivalent. This shows heterogeneity in the interaction between the site and the variously lengthed aliphatic chains.

A dramatic increase in relative affinity accompanies substitution on UQ_0 with 3'-methyl-2'-butene (isoprenoid) units. As shown in Figure 1, the absolute magnitude of the UQ-1 binding free energy is 1.7kcal/mole greater than the corresponding five carbon aliphatic derivative, and the affinity of UQ-2 surpasses decyl-UQ_0 by greater than 2kcal/mole. Equivalent affinities of the saturated prenyl and aliphatic derivatives indicate the important contribution of the double-bonded configuration to the observed specificity. In contrast, desaturation at carbon 4' of UQ-2 results in a marked loss of affinity. Conformance to the site apparently requires free rotation about the 4'-5' C-C single bond. Thus, the first two isoprenoid units define a binding region essential in generating enhanced affinity and specificity at a now extended Q_A site.

The conformation of the Q_A polyisoprenoid tail in the Rhodopseudomonas viridis RC structure (5,6) corroborates our conclusions based on binding free energy measurements. The hydrocarbon tail composed of 3'-methyl-2'-butene units appears particularly well suited for achieving the contorted conformation observed in the structure in the binding region sampled by this study.

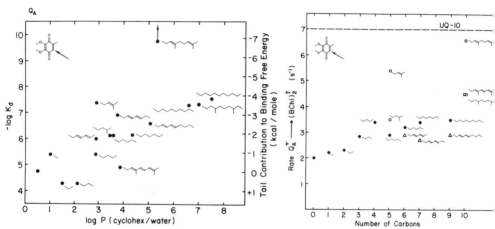

Figure 1. Dependence of affinity on logP for the substituted UQs.

Figure 2. Recombination rate as a function of the number of UQ tail carbon atoms.

The results reported here support and extend the study of the factors required for binding and function at the Q_A site. Earlier work (4) demonstrated specific interaction between various regions of the quinone head group and the site. Conclusions included: 1) additions ortho to the carbonyl add substantially to the affinity; 2) increasing the size of the ring system to naphthoquinone and anthraquinone aids binding; 3) only one carbonyl is involved in binding to the site, since removal of one carbonyl diminishes the affinity by only 3 fold, whereas removal of both weakens the affinity by more than 4 kcal/mole.

The results presented in Figure 2 demonstrate that the tail has only a minor influence on the electron transfer from Q_A^- to $(BChl)_2^+$. A small increase in rate with lengthening tail is seen and with the isoprenyl-substituted UQ-1 and UQ-2 the rate approaches that displayed by the native UQ-10. This effect may simply arise from the enhanced binding affinity of these quinones, since similar stimulated rates are seen with the larger and more tightly bound naphthoquinones and anthraquinones.

A qualitative measure of the importance of the tail in the electron transfer from BPh^- to Q_A can be obtained by determining the quantum efficiency of Q_A reduction. If the rate slowed to less than $4 \times 10^8 s^{-1}$ a decrease in the amplitude of $(BChl)_2^+$ following a subsaturating flash would be seen. None of the different tailed Q_As show less than maximal $(BChl)_2^+$ yield. Therefore, modification of the tail or its removal cannot slow the reduction of Q_A by BPh^- by more than 10 fold relative to UQ-10. Quantum yield measurements from 300 to 5K (7) and direct picosecond studies of BPh$^-$ to Q_A electron transfer at 300K (8,9) have previously shown that the tail is not a prerequisite for electron transfer reactions involving Q_A. These earlier studies, and the work presented here, show that the isoprenoid tail is not a primary conduit for electron tunnelling. This lack of major influence on Q_A electron transfer reactions contrasts with the large contributions to binding affinity and specificity at the Q_A site arising from the isoprenoid-tail structure.

Supported by NSF PCM 82-09292.

REFERENCES
1 Clayton, R.K. and Wang, R.T. (1971) Meth. Enzymol. 23, 696-704.
2 Okamura, M.Y.,Isaacson, R.A. and Feher,G. (1795) Proc. Natl. Acad. Sci. 72, 3491-3495
3 Yu,C.A., Gu,L., Lin, Y. and Yu, L.(1985) Biochemistry 24, 3897-3902
4 Gunner,M.R., Braun,B.S., Bruce,J.M., and Dutton,P.L. (1985) in Antennas and-Reaction Centers of Photosynthetic Bacteria (Michel-Beyerle, M.E. Ed.) Springer-Verlag, Berlin pp. 298-305
5 Deisenhofer,J., Epp,O., Miki,K., Huber,R., and Michel,H. (1984) J.Mol. Biol. 189 385-398
6 Deisenhofer,J., Epp,O., Mike, K., Huber, R., and Michel, H. (1985) Nature 318 618-624
7 Gunner,M.R., Robertson,D.E., and Dutton,P.L. (1986) J.Phys. Chem. 90 3783-3795
8 Laing,Y., Nagus,D.K., Hochstrasser,R.M., Gunner,M.R., and Dutton,P.L. (1981) Chem Phys. Lett. 84 236-240
9 Gunner,M.R., Laing,Y., Nagus,D.K. Hochstrasser,R.M. and Dutton, P.L. (1982) Biophys. J. 37a 226a

EXCITED STATES AND PRIMARY PHOTOCHEMICAL REACTIONS IN PHOTOSYSTEM I

A.M. NUIJS, V.A. SHUVALOV, H.W.J. SMIT, H.J. van GORKOM and L.N.M. DUYSENS
Department of Biophysics, Huygens Laboratory of the State University,
P.O. Box 9504, 2300 RA Leiden, The Netherlands

1. INTRODUCTION
     In the reaction center of photosystem I three iron-sulfur centers, de-
signated F_X, F_B and F_A, act as secondary electron acceptors [1]. The data
on the primary electron acceptor are controversial. Picosecond absorption
measurements on small PS I particles led Shuvalov et al. to the conclusion
that the primary acceptor is a chlorophyll molecule absorbing around 694 nm
[2], whereas EPR and optical measurements under continuous illumination at
progressively lower potentials provided evidence that two acceptors A_0 and
A_1 function prior to F_X [3,4]. These data suggested that A_1^- is a semiquinone and indicated that A_0 is a chlorophyll molecule absorbing around 670
nm.
     In this paper we report a picosecond absorption study on relatively in-
tact PS I from spinach (70 - 100 Chl a/reaction center). The data indicate
that the primary acceptor is a chlorophyll a species absorbing around 693
nm.

2. RESULTS AND DISCUSSION
     Fig. 1 shows the absorbance difference spectra of PS I at 40 ps (●) and
200 ps (o) after the 35 ps excitation pulse at 532 nm. The primary electron
donor, P-700, was kept in the oxidized state by ferricyanide and continuous
background illumination. Both spectra are characterized by a bleaching in
the region 660 - 740 nm, flanked by shallow increases in absorption. We
ascribe the 40 ps spectrum to the formation of singlet excited antenna

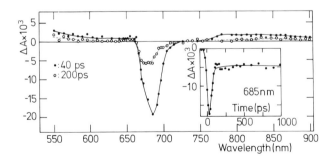

FIGURE 1. Absorbance difference spectra at 40 ps (●) and 200 ps (o) after
          the 35 ps excitation pulse at 532 nm, in the presence of 3 mM
          ferricyanide and under continuous background illumination. The
          inset shows the kinetics at 685 nm.

Biggens, J. (ed.), Progress in Photosynthesis Research, Vol. I. ISBN 90 247 3450 9
© *1987 Martinus Nijhoff Publishers, Dordrecht. Printed in the Netherlands.*

chlorophyll, Chl*a, and that at 200 ps to the formation of the triplet state, Chl¹a. The kinetics at 685 nm are given in the inset. A deconvolution procedure indicated a lifetime of Chl*a of 40 ± 5 ps

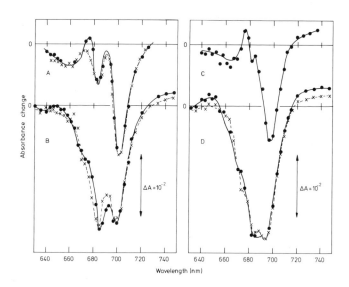

FIGURE 2. A: Comparison of the ps pulse-induced spectrum of P-700⁺ formation (●) with the spectrum under continuous illumination (x). B: The 40 ps spectrum under non-reducing conditions (●) simulated by a sum (x) of the P-700⁺ and the Chl*a spectra. C: Spectrum at 500 ps under reducing conditions. D: The 40 ps spectrum under reducing conditions (●) simulated by a sum (x) of the Chl*a spectrum and that of the radical pair.

When the reaction centers were initially open, the difference spectrum measured at 200 ps (Fig. 2A, circles) shows minima at 701, 683 and 668 nm, and is similar to the difference spectrum of P-700 oxidation induced by continuous illumination (Fig. 2A, crosses). No distinct absorbance changes due to reduction of an electron acceptor can be observed.

The difference spectrum detected at 40 ps (Fig. 2B, circles) is characterized by negative bands around 699 nm and 685 nm, which are due to the formation of P-700⁺ and Chl*a, respectively. The spectrum can be simulated (Fig. 2B, crosses) by a sum of the P-700⁺ and the Chl*a spectra. The differences between the spectra of Fig. 2B are due to the reduction of the primary acceptor, as will be made more clear below.

When the secondary acceptors are prereduced chemically, the difference spectrum observed at 500 ps after the flash (Fig. 2C) is different from that due to formation of P-700⁺ alone. Compared to the P-700⁺ spectrum, the main bleaching has shifted from 701 to 698 nm, and the steep increase around 690 nm has disappeared. We ascribe the spectrum to formation of the primary radical pair. At 40 ps after the flash this radical pair is also present, since the recorded spectrum (Fig. 2D, circles) can be well simulated by a sum (Fig. 2D, crosses) of the spectrum of Chl*a and that of the radical pair.

Subtracting the P-700⁺ spectrum from that of the radical pair yields the spectrum given in Fig. 3. These absorbance changes are characterized by a narrow bleaching around 693 nm, similar to that earlier observed by Shuvalov et al. [2] and are attributed to the reduction of the primary electron acceptor. This acceptor thus proves to be a chlorophyll a species absorbing around 693 nm, and will be designated as C-693.

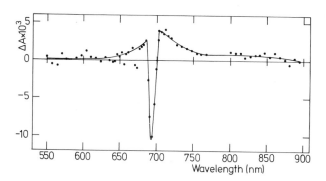

FIGURE 3. Absorbance difference spectrum calculated for the reduction of the primary acceptor.

When the secondary electron transport is inhibited, due to chemical reduction of the iron-sulfur centers, the primary radical pair decays in 20 - 50 ns (not shown) to the triplet state of P-700 (Fig. 4) with a yield of about 30 %. The lifetime of the triplet state is about 3 μs (inset).

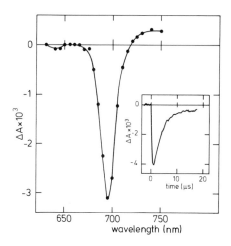

FIGURE 4. Difference spectrum and decay kinetics of the triplet state of P-700.

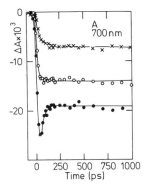 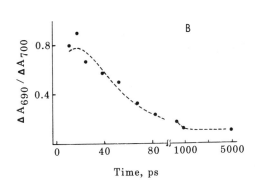

FIGURE 5. A: Kinetics at 700 nm under non-reducing conditions at different excitation intensities. B: Kinetics at 690 nm divided by those at 700 nm following excitation at 710 nm. The best fit (dashed curve) yields a time constant of 32 ps.

Fig. 5A shows the kinetics at 700 nm under conditions of normal electron transport for different excitation energy densities. At low intensity (crosses) a 50 ps risetime of P-700$^+$ formation is observed which reflects the process of trapping. At higher intensities a short-lived bleaching appears which is due to formation of Chl $\overset{*}{}$a.

In an attempt to observe the electron transfer from C-693$^-$ to a secondary acceptor, we have excited the primary donor directly at 710 nm, thus reducing antenna excitation. The kinetics at 690 nm, divided by those at 700 nm, are shown in Fig. 5B. The best simulation (dashed curve) of these kinetics was obtained using a 32 ps time constant for electron transfer from C-693$^-$ to the next acceptor.

Our data thus provide evidence that the primary acceptor in PS I is a chlorophyll molecule absorbing around 693 nm. No evidence was obtained for the participation of the chlorophyllous electron acceptor A_o [3,4]. Possibly this species is located on a sidepath. The acceptor A_1 may not cause absorbance changes in the spectral range studied and if it was in the reduced state in the conditions of Fig. 2C and D, our data do not argue against its participation in the main chain of electron transfer after C-693.

REFERENCES
1 Rutherford, A.W. and Heathcote, P. (1985) Photosynth. Res. 6, 295-316
2 Shuvalov, V.A., Klevanik, A.V., Sharkov, A.V., Kryukov, P.G. and Ke, B. (1979) FEBS Lett. 107, 313-316
3 Gast, P., Swarthoff, T., Ebskamp, F.C.R. and Hoff, A.J. (1983) Biochim. Biophys. Acta 722, 168-175
4 Mansfield, R.W. and Evans, M.C.W. (1985) FEBS Lett. 190, 237-241.

CHARACTERIZATION OF THE ELECTRON ACCEPTOR A_1 IN PHOTOSYSTEM I BY FLASH–ABSORPTION SPECTROSCOPY AT LOW TEMPERATURE : EVIDENCE THAT A_1 IS VITAMIN K_1

K. BRETTEL, P. SETIF and P. MATHIS
Service de Biophysique, CEN Saclay 91191 GIF-SUR-YVETTE CEDEX - FRANCE -

1. INTRODUCTION

According to recent studies , the iron–sulfur–type electron acceptors F_A, F_B and F_X of PS I are preceeded by two earlier acceptors, A_0 and A_1 (1,2). Spectroscopic data on A_0 are compatible with the proposal that this primary acceptor is chlorophyll–a (Chl-a) (3). A_1 is a secondary acceptor of yet unknown chemical nature, whose anion has a g–value around 2.005 (1,2) with some g–anisotropy (4). Low temperature flash absorption revealed a reaction with a half–time of 120 μs at 10 K, attributed to charge recombination between A_1^- and P-700$^+$ (5). The lack of absorption changes due to A_1 reduction in the red spectral region (5) and the high g–value of A_1^- indicate that A_1 is not a chlorophyll–like molecule. Here we present measurements of the absorption changes in the UV, blue–green and near IR due to the recombination (P-700$^+$...A_1^-)$\longrightarrow$(P-700...A_1) in digitonin–fractionated PS I particles at 10 K and demonstrate that the (P-700$^+$...A_1^-)/ (P-700...A_1) difference spectrum contains contributions in the UV due to A_1^-/A_1 which are indicative of A_1 being vitamin K_1 (phylloquinone). This is compatible with the EPR properties of A_1^- and with the recent finding that PS I contains vitamin K_1 as an integral constituent (6-8).

2. MATERIALS AND METHODS

Two different kinds of digitonin fractionated PS I particles were prepared from spinach leaves : D144 particles (9) with $\approx$150 Chl/P-700 and PS I-110 particles (10) with $\approx$110 Chl/P-700. The particles were suspended at pH 10 with 65 % glycerol and treated with either 10 mM NaBH$_4$ ("reducing conditions") or 2 mM K$_3$Fe(CN)$_6$ plus 40 μM methylviologen ("oxidizing conditions") as in (11). Measurements of flash induced absorption changes at 10 K were performed as in (11), except that a monochromator with $\Delta\lambda$ = 30 nm and a germanium photodiode were used at 800 and between 1000 and 1500 nm.

3. RESULTS AND DISCUSSION

It has been previously shown that the 120 μs relaxation phase of the absorption changes at 820 nm (center of a broad band of P-700$^+$) attributed to charge recombination between A_1^- and P-700$^+$ at low temperature, is of a larger size if F_A and F_B are prereduced than if all membrane bound acceptors are initially oxidized (5). Therefore, we treated the PS I particles with the strong reductant NaBH$_4$ at pH 10 to take advantage of the bigger signal when F_A and F_B are prereduced.

A typical signal measured at 800 nm in PS-110 particles under reducing conditions at 10 K is depicted in Fig. 1 (left, "red"). The t½$\approx$ 120 μs decay phase is followed by an $\approx$1 ms phase. This slower phase is for the most part insensitive to oxidation of P-700 prior to freezing (Fig. 1, left, "ox") and its spectrum in the 270 to 525 nm range (not shown) resembles the triplet minus singlet ground state spectrum of Chl-a. We

assume that the 1 ms phase represents mainly the decay of the triplet
state of Chl-a (^{3}Chl-a) in the antenna. After subtraction of the signal
under oxidizing conditions from the one under reducing conditions, the
remaining signal (Fig. 1, left, "red-ox") shows mainly the 120 μs phase
attributed to the recombination between A_1^- and P-700$^+$. Some remaining
absorption change decaying with t½ ≈ 1 ms may be due to the triplet state
of P-700 (^{3}P-700) in a minor fraction of reaction centers.

   In order to confirm that A_1 is not Chl-a, we extended these
measurements in the near IR up to 1500 nm since the anion of Chl-a has
some absorption between 1000 and 1700 nm (12), whereas P-700$^+$ does not
absorb in this region (13). Surprisingly, we observed absorption changes
in this spectral region which decay with complex kinetics including a
phase which can be well adapted by t½ = 120 μs. However, these signals
were independent of the redox state of P-700 (see Fig. 1, right, for an
example at 1200 nm) showing that they are not due to reaction center
photochemistry. The nature of these absorption changes in the 1000 to 1500
nm region is not yet clear. We suppose that triplet states of antenna
pigments, including Chl-a and carotenoids, are involved.
If A_1 would be Chl-a, the difference "red-ox" at 1200 nm should contain a
120 μs decay phase with approx. 10 % of the amplitude of the corresponding
signal at 800 nm. From the lack of such a phase in the "red-ox"-difference
signal at 1200 nm (Fig. 1, right, "red-ox") we conclude that A_1 is
certainly not Chl-a, in agreement with previous measurements in the red
spectral region (5).

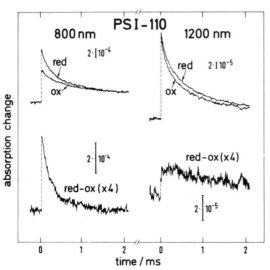

Fig. 1 : Flash-induced
absorption changes at 800
and 1200 nm in a reduced
(red) and an oxidized (ox)
sample of PS I-110 parti-
cles at 10 K and differences between them (red-ox).
OD$_{676 nm}$ (45°) = 2.4 ; the
excitation was nearly saturating the reaction center
photochemistry.

   The "red-ox" signal at 1200 nm indicates furthermore that the
recombination between A_1^- and P-700$^+$ at low temperature does not populate
a significant amount of ^{3}P-700, since this should show up in an
absorption increase arround 1200 nm due to P-700 formation with rise
kinetics corresponding to the recombination (t½ = 120 μs). Instead, all
"red-ox" signals measured between 1000 and 1500 nm showed a set-up limited
rise time below 20 μs and were rather small. These signals may reflect
some ^{3}P-700 formed in the ns-range by the recombination between A_0^- and
P-700$^+$ in centers where A_1 is not accessible (14).

Measurements as those depicted in Fig. 1 were also performed between 240 and 525 nm. In order to correct for processes which are not related to reaction center photochemistry, the signal obtained under oxidizing conditions was subtracted from the one under reducing conditions for each wavelength. The "red-ox" difference signals showed at most wavelengths a $t\frac{1}{2} \approx 150\ \mu s$ decay phase which we attribute to the recombination between A_1^- and P-700$^+$. The slightly larger half-time compared to previous studies may arise from the narrow electric bandwidth (DC - 10 kHz) that was used to improve the signal to noise ratio. The amplitudes of the 150 μs phase for D144 particles are shown in Fig. 2 (upper panel, symbols and continuous line). The $\Delta\epsilon$-scale of Fig. 2 was calculated by reference to the 150 μs phase obtained with the same pair of samples at 820 nm, assuming that P-700$^+$ alone contributes to that phase at 820 nm with $\epsilon = 6\ mM^{-1}\ cm^{-1}$ (room temperature value from (15)). Similar results were obtained with PS I-110 particles, but the signals remaining under oxidizing conditions appeared more complex so that D144-particles were prefered for a more systematic study.

In order to determine the absorption difference spectrum of A_1^-/A_1 alone, the spectrum of the 150 us phase (corresponding to the absorption difference between (P-700$^+$...A_1^-) and (P-700...A_1)) should ideally be compared with the P-700$^+$/P-700 difference spectrum at low temperature. Since the latter spectrum is not known in the UV, we have to rely on the P-700$^+$/P-700 difference spectrum at room temperature that was previously measured by Ke (15) down to 260 nm with digitonin-fractionated spinach PS-I particles (Fig. 2, dotted line). Both the spectrum of the 150 μs phase and the P-700$^+$/P-700 spectrum exhibit a major bleaching around 430 nm, which is, however, more pronounced for P-700$^+$/P-700. This difference may reflect an absorption increase due to the reduction of A_1 or a temperature dependence of $\Delta\epsilon$ for the oxidation of P-700. In any case, the reduction of A_1 appears rather related to an absorption increase than to a

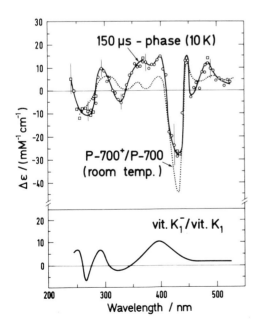

Fig. 2 : Absorption difference spectra for the 150 us phase of D144 particles at 10 K (upper panel, symbols and continuous line), for the oxidation of P-700 at room temperature (upper panel, dotted line ; redrawn from (15)) and for the reduction of vitamin K_1 to its semiquinone anion in methanol (lower panel ; as measured by Dr. E.J. Land, redrawn from (16)). The amplitudes of the 150 μs phase were extracted from "red-ox"-difference signals and calibrated as outlined in the text. Squares and circles refer to series of measurements with the quartz and plexiglass cuvettes, respectively. OD$_{676\ nm}$ (45°) = 1.4 (plexiglass cuvette) and 0.7 (quartz cuvette) ; the excitation was about two thirds saturating the reaction center photochemistry.

bleaching at 430 nm, thus confirming that the partner of P-700$^+$ in the 150 μs recombination is not the iron-sulfur center F_X (5). Below 430 nm, the deviations between the spectrum of the 150 μs phase and the P-700$^+$/P-700 spectrum correspond fairly well to the absorption changes due to the reduction of vitamin K_1 to its semiquinone anion in vitro (Fig. 2, lower panel, as measured by Dr. E.J. Land, redrawn from (16)). In the 450 to 525 nm region, electrochromic bandshifts may be superimposed. Concerning the most salient deviations between the 150 μs spectrum and the P-700$^+$/P-700 spectrum in the UV region, namely the positive ones around 380 nm and –most probably (considering the slope of the P-700$^+$/P-700 spectrum above 250 nm) – around 245 nm and the negative one around 325 nm, it should be emphasized that the correspondance to the vit.K_1^-/vit.K_1 spectrum is not much dependent on a precise calibration of the $\mathbf{\Delta\epsilon}$-scale of the 150 μs phase relative to the P-700$^+$/P-700 spectrum. From the results presented here together with previous results mentioned in the introduction, we conclude that the electron acceptor A_1 as defined by the recombination (P-700$^+$...A_1^-)$\longrightarrow$(P-700...A_1) with $t\frac{1}{2}\approx 150$ μs at low temperature is vitamin K_1.

For vitamin K_1 to be able to function as electron carrier between A_0^- and F_X under physiological conditions, its redox potential should be below – 0.7 V. Such a low potential might be achieved by the protein environment of vitamin K_1, providing possibly electron repulsive groups and aprotic conditions. However, the participation of vitamin K_1 in electron transfer under physiological conditions still needs to be established.

ACKNOWLEDGEMENTS

This study was partly supported by a long-term fellowship to K.B. from the European Molecular Biology Organization.

REFERENCES

1. Bonnerjea, J. and Evans, M.C.W. (1982) FEBS Lett. 148, 313-316
2. Gast, P., Swarthoff, T., Ebskamp, F.C.R. and Hoff, A.J. (1983) Biochim. Biophys. Acta 722, 163-175
3. Shuvalov, V.A., Nuijs, A.M., van Gorkom, H.J., Smit, H.W.J. and Duysens, L.N.M. (1986) Biochim. Biophys. Acta 850, 319-323
4. Thurnauer, M.C. and Gast, P. (1985) Photobiochem. Photobiophys. 9, 29-38
5. Sétif, P., Mathis, P. and Vänngård, T. (1984) Biochim. Biophys. Acta 767, 404-414
6. Interschick-Niebler, E. and Lichtenthaler, H.K. (1981) Z. Naturforsch, 36c, 276-283
7. Takahashi, Y., Hirota, K. and Katoh, S. (1985) Photosynth. Res. 6, 183-192
8. Schoeder, H.U. and Lockau, W. (1986) FEBS Lett. 199, 23-27
9. Anderson, J.M. and Boardman, N.K. (1966) Biochim. Biophys. Acta 112, 403-421
10. Picaud, A., Acker, S. and Duranton, J. (1982) Photosynth. Res. 3, 203-213
11. Brettel, K., Sétif, P. and Mathis, P. (1986) FEBS Lett. 203, 220-224
12. Fujita, I., Davis, M.S. and Fajer, J. (1978) J. Am. Chem. Soc. 100, 6280-6282
13. Mathis, P. and Sétif, P. (1981) Isr. J. Chem. 21, 316-320
14. Sétif, P., Bottin, H. and Mathis, P. (1985) Biochim. Biophys. Acta 808, 112-122
15. Ke, B. (1972) Arch. Biochem. Biophys. 152, 70-77
16. Romijn, J.C. and Amesz, J. (1977) Biochim. Biophys. Acta 461, 327-338

EPR EVIDENCE THAT THE PHOTOSYSTEM I ACCEPTOR A_1 IS A QUINONE MOLECULE[1]

M. C. Thurnauer, P. Gast*, J. Petersen**, and D. Stehlik**, Chemistry Division, Argonne National Laboratory, 9700 S. Cass Avenue, Argonne; Argonne, IL 60439; *Department of Chemistry, University of Chicago, Chicago, IL 60637; **Department of Physics, Free University Berlin, D-1000 Berlin 33, Federal Republic of Germany

1.   INTRODUCTION

It is believed that the electron spin polarized (ESP) EPR transients observed around g=2 in plant photosystem I (PSI) are due to the radical pair $P^+_{700} A^-_1$, the oxidized primary electron donor and reduced A_1 acceptor. In order to identify the nature of A_1 and test proposals that A_1 is a quinone molecule (1,2,3), the low temperature ESP EPR spectrum of $P^+_{700} A^-_1$ is investigated with fast time resolved EPR at K-band (~24 GHz) microwave frequency. The X-band (~10 GHz) ESP EPR signal in PSI has similar spectral characteristics as the ESP EPR signal due to the oxidized donor and reduced quinone acceptor ($P^+_{870}Q^-$) observed in bacterial reaction centers (RC's) in which the iron is uncoupled from Q (4,5). This is true also at K-band frequency. Thus, the RC's provide a relatively defined system with which we can test our methods of simulation of a quinone radical showing g-anisotropy and ESP which we then apply to the PSI ESP EPR signal. Simulations of A^-_1 based on these comparisons provide strong evidence that it is a quinone species. Furthermore, the anisotropic polarization observed for A^-_1 in ^2H PSI par-ticles should allow us to place some limits on the mechanism responsible for the ESP.

2.   MATERIALS AND METHODS

^1H and ^2H PSI particles that lack the ferrodoxin acceptors were prepared as in (2). EPR samples contained 60% ethylene glycol. Iron-depleted RC's were a kind gift from D. Budil and were prepared as in (6,7) with LDAO replaced by Brij-58. The K-band spectrometer as well as the direct detection time resolved EPR technique has been described. (8,9). Relative g-values for the ^2H PSI sample were obtained with an NMR gaussmeter, and absolute values were calibrated with the known g-values of the EPR signal from the photoexcited triplet state of acridine in fluorene. The RC samples were excited at 569 nm at a repetition rate of 20 Hz and the PSI samples at 337 nm at 30 Hz.

3.   RESULTS AND DISCUSSION

3.1   Bacterial Reaction Center Protein

K-band ESP EPR spectra obtained at a delay time, t_D =1 μs after the laser flash from iron-depleted ^1H *Rps. sphaeroides* R-26 RC's are shown in Fig. 1. Fig. 1a gives the spectrum obtained when Q is reduced prior to excitation. (The large P_{870} triplet signal obtained under these redox conditions has been subtracted numerically.) This signal corresponds to the emissive signal of Q^- which has been

[1]This work was supported by the Division of Chemical Sciences, Office of Basic Energy Science, U.S. Department of Energy under contract W-31-109-ENG-38.

Biggens, J. (ed.), Progress in Photosynthesis Research, Vol. I. ISBN 90 247 3450 9
© *1987 Martinus Nijhoff Publishers, Dordrecht. Printed in the Netherlands.*

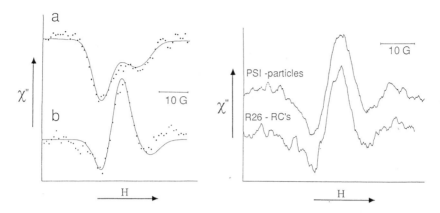

FIGURE 1 Transient K-band EPR spectra from iron-depleted ^{1}H *Rps. sphaeroides* R-26 RC's. a. Spectrum from prereduced RC sample. $t_D = 1\mu s$. Microwave power 350mW. Boxcar gatewidth 200 ns, T=85 K. b. Spectrum from unreduced sample taken under same conditions as in (a). Solid lines are the simulations described in the text.

FIGURE 2 Comparison of the ESP flash-induced EPR K-band spectrum of (top) $P^+_{700} A^-_1$ from ^{1}H PSI particles with (bottom) the ESP flash-induced EPR K-band spectrum of $P^+_{870}Q^-$. $t_D = 1\mu s$. Microwave power 350 mW. Boxcar gatewidth 200 ns,T=85 K.

observed from such samples at X-band frequency.(4,10) Figure 1b gives the spectrum obtained when Q is not reduced prior to excitation. This signal shows an ESP-pattern similar to that obtained under similar conditions at X-band frequency for the pair $P^+_{870} Q^-$.(4,5) The solid lines in Figs. 1a and 1b represent computer simulations of the observed spectra, considering only a quinone radical. For this, a powder spectrum simulation routine with g-anisotropy was employed. The g-values used for Q^- (g_{xx}=2.0067, g_{yy}=2.0056, and g_{zz}=2.0024) were taken from (10,11), and hyperfine interaction was incorporated as in (4,10). The ESP in the different principal g-value components was described by the weighting function:

$$P(\theta,\o) = P_{zz}\cos^2\theta + (P_{xx}\cos^2\o + P_{yy}\sin^2\o)\sin^2\theta$$

where $\theta,\o$ represent the polar angles of the field with respect to the principal axis frame of the g-tensor as used in the powder spectrum routine, P_{ii} (i=x,y,z) are polarization parameters in the directions defined by g_{ii}. Positive/negative values of P_{ii} represent absorptive (A)/emissive (E) character of the ESP. (A Boltzmann polarization is assumed small and not included.) The P_{nn} which gave the best fit to the EPR spectrum of the prereduced RC's (Fig. 1a) were P_{xx}= -1, P_{yy}= +0.6 and P_{zz}= -0.3, and those for the unreduced case (Fig.1b) were P_{xx}= -1, P_{yy}= +1.4, and P_{zz}= -0.3. In the latter case, the spectrum was simulated with just these parameters, although, resonances due to P^+_{870} also contribute to the high-field part of the spectrum. Yet, this portion cannot be resolved with ^{1}H samples, even at K-band microwave frequency.

3.2. PSI Reaction Center Complex

The ESP EPR spectrum $P^+_{700} A^-_1$ obtained from ^1H PSI particles at $t_D=1$ μs is shown in Fig. 2 (top). The spectrum is essentially the same in terms of overall width and ESP as that of $P^+_{870} Q^-$ (Fig. 2 (bottom)). The results of these experiments on ^2H PSI particles are shown in Fig. 3. The ESP is consistent with that of the ^1H system (compare Figs. 3a and 2 (top)), but with better resolution. The solid line of Fig. 3a represents the simulation of A^-_1 as an anisotropically polarized quinone powder spectrum. Obviously, the high field portion of the spectrum includes a now resolved P^+_{700} contribution. However, for the specific purpose of identifying the nature of A^-_1, the P^+_{700} contribution is not included here. The simulation of the spectral contribution of A^-_1 was guided by two factors: i. the g-anisotropy of A^-_1 is the same as that of Q^-, ii. for simplicity, the overall ESP pattern is that obtained from the simulations of Q^- (unreduced case- Fig. 1b). Point (i) is based on the experimental observation that the overall spectral width for the ^1H samples is the same for $P^+_{700} A^-_1$ and $P^+_{870} Q^-$ (Fig. 2). However, in order to match the experimentally obtained g-values, the simulation requires a shift of all g-values from those of Q^- to $g_{xx} = 2.0057$, $g_{yy} = 2.0046$, and $g_{zz} = 2.0014$. The parameters $P_{xx} = -1$ and $P_{yy} = +1.4$ obtained from the RC's reproduce well the two low field features in the spectrum, while a negative P_{zz} (-0.3) is one way to simulate (within the limitations of the procedure) the weak ESP in the center of the spectrum. The correspondence obtained between this simulation and the one for the bacterial system provides strong evidence that we observe in PSI a quinone-like radical as an early acceptor.

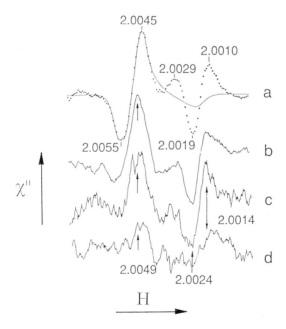

FIGURE 3. Transient K-band EPR spectra from ^2H PSI particles taken at $t_D=$ a. 0.5 μs, b. 2 μs, c. 4 μs, d. 8 μs with respect to the ~10 ns laser pulse. Microwave power 350 mW, boxcar gatewidth 100 ns, T= 85 K. g-values in (a) are within 10^{-4}

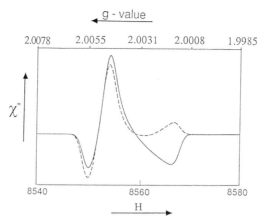

FIGURE 4. Simulations of A^-_1 component of ESP EPR spectrum from PSI particles (fig. 3a) including i.) dipolar coupling in the ESP generating radical pair with the dipolar axis parallel to g_{yy} of A^-_1 (solid line) and ii.) dipolar coupling as in (i) and consideration that $P^+_{700} A^-_1$ is the ESP generating radical pair (dashed line).

With the identity of A_1 as a quinone it is possible to consider mechanisms which may explain the anisotropic ESP which is observed for A^-_1, e.g., the oppo - site ESP when the field is oriented along the g_{xx} and g_{yy} components. Also, for a complete simulation an ESP spectrum of P^+_{700} must be included. The P^+_{700} ESP is not immediately obvious from the high field A/E/A ESP pattern which differs from the simulation in Fig.3a. The time dependence of the $P^+_{700} A^-_1$ signal from ^2H PSI particles (Fig. 3a-d) suggests that the g=2.0010 absorption is part of A^-_1. Simulations which include generation of ESP via the radical pair mechanism with inclusion of dipolar coupling in the ESP-generating pair and g-anisotropy of A^-_1 (5) can reproduce some of the features of the A^-_1 spectrum. With this approach it is found that the low field E/A can be simulated if the Z dipolar axis is not parallel to the Z-axis of the A^-_1 g-tensor, and the high field absorption can be duplicated by considering that the ESP is generated in the pair $P^+_{700}A^-_1$. Preliminary examples of such simulations are shown in Figure 4.

REFERENCES

1 Sétif, P., Mathis, P., Vänngård, T. (1984) Biochim. Biphys. Acta 767, 404
2 Thurnauer, M. C. and Gast, P. (1985) Photobiochem. Photobiophys. 9, 29
3 Mansfield, R. W. and Evans, M. C. W. (1985) FEBS Lett. 190, 237
4 Gast, P., Thesis (1983) University of Leiden, Leiden, The Netherlands
5 Hore, P. J., Watson, E. T. Pedersen, J. B. and Hoff, A. J. (1986) Biochim. Biophys. Acta 899, 70
6 Wraight, C. A. (1979) Biochim. Biophys. Acta 548, 308
7 Tiede, D. and Dutton, P. L. (1981) Biochim. Biophys. Acta 637, 278
8 Furrer, R., Fujara, F., Lange, C., Stehlik, D., Vieth, H. M. and Vollman W. (1980) Chem. Phys. Lett. 75, 332
9 Furrer, R. and Thurnauer, M. C. (1983) FEBS Lett. 153, 399
10 Gast, P., deGroot, A. and Hoff, A. J. (1983) Biochim. Biophys. Acta 723, 52
11 Feher, G., Okamura, M. Y. and McElroy, J. D. (1972) Biochim. Biophys. Acta 267, 222

This work is supported by NATO Research Grant No. 214/84.

INVESTIGATION OF THE CHEMICAL NATURE OF ELECTRON ACCEPTOR A_1 IN
PHOTOSYSTEM I OF HIGHER PLANTS

R.W. MANSFIELD, J.H.A NUGENT and M.C.W. EVANS
Department of Botany and Microbiology, University College London, Gower St.,
London, WC1E 6BT, U.K.

1.   INTRODUCTION
     The electron accepting system of Photosystem I (PSI) consists of a
chain of membrane bound components that ultimately reduce soluble ferredoxin.
It has proved possible, using progressive chemical and photochemical
reduction, to obtain characteristic EPR spectra of a number of these
components. These include the iron-sulphur centres 'X', 'A' and 'B' and two
other acceptors 'A_0' and 'A_1'. The latter two have been identified as
intermediates between the PSI donor, P700, and the iron-sulphur centres
(1,2). The primary acceptor, 'A_0', has a light-dark optical difference
spectrum that is characteristic of the formation of a chlorophyll a monomer
anion (3). Here we present an optical difference spectrum associated with
the photoreduction of the intermediate acceptor A_1. The effect of
extraction with hexane and of deuteration of PSI particles on the optical
and ESR signals accompanying reduction of A_1 will be presented.

2.   MATERIAL AND METHODS
2.1 Membrane Preparation
     PSI particles were prepared from pea seedling leaves (Feltham First)
     using Triton X-100 digestion and subsequent purification on an hydroxy-
     apatite column (3) giving a final P700:chl. ratio of 1:32.
2.2 Hexane Washing
     Freeze-dried PSI particles were gently stirred in Hexane (1.5mgs chl./
     15mls) on ice in the dark for 1hr then recovered by centrifugation at
     3000xg for 5min. This process was repeated four times and the final
     pellet was dried under a stream of nitrogen before resuspension in a
     medium consisting of 20mM glycine, pH 10 and 0.1% Triton X-100.
2.3 Deuteration
     PSI particles were diluted to 0.1mg chl./ml in 99.7% D_2O, pH 10 with
     glycine/NaOH and incubated for 5hrs on ice. The sample was then freeze-
     dried and resuspended in D_2O, pH 10, and incubated on ice for 4 days
     before freeze-drying. Subsequently the sample was resuspended in D_2O at
     pH 6.3. A control sample underwent all of these procedures in the
     presence of H_2O at the appropriate pH's.
2.4 ESR and optical studies were carried out as in (3) at 0.15mg chl./ml.
     For deuterated samples, and the relevant control, glycerol was omitted
     from the medium. Optical measurements were made on a Cary 219 spectro-
     photometer equipped with a quartz windowed cuvette in an Oxford
     Instruments cryostat.

3.   RESULTS
     Photoaccumulation of reduced intermediate PSI acceptor A_1 at 205K is
demonstrated by ESR in Fig. 1a. The superimposed traces were recorded at

Biggens, J. (ed.), Progress in Photosynthesis Research, Vol. I. ISBN 90 247 3450 9
© *1987 Martinus Nijhoff Publishers, Dordrecht. Printed in the Netherlands.*

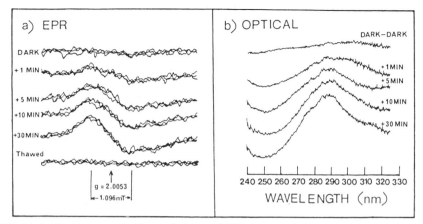

FIGURE 1.a) Photoaccumulation of A_1 at 205K in pea PSI measured by EPR.
   b) Difference spectra accompanying A_1 reduction for indicated time
      increments. Spectra are averages of 5 scans on each of 6 samples.

five minute intervals and indicate that there was no tendency for the signal
to relax at this temperature. Optical spectra were recorded under identical
conditions. In the visible region there were no changes accompanying A_1 reduction (see 3). However, in the U.V. range an absorbance increase at 290nm
and a bleaching at 250nm were detected (Fig. 1b). There was an isosbestic
point at 269nm. The amplitudes of the ESR and optical changes paralleled
each other (indicating that they were associated with the same component),
the optical absorbance increase and decrease also occurred at the same rate.
The U.V. optical difference spectrum resembles those observed upon reduction
of quinone species to the semi-quinone radical form, in particular the
neutral semi-quinone (4).

   In order to test the possibility that A_1 is a quinone PSI particles were
washed in hexane. In Fig. 2a, we show the characteristic ESR spectra associated with A_1 and A_0 reduction in freeze-dried controls.

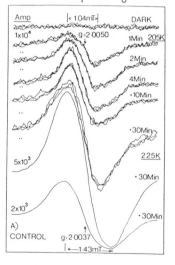

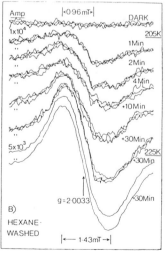

FIGURE 2.
A) ESR spectra of
A_1 (205K) and A_0
(225K) after low
temperature illumina-
tion of freeze-dried
pea PSI.
B) as above using
hexane washed PSI
Spectra recorded at
205K; power, 100 μW;
freq., 9.05 GHz;
Mod., 2G; Resp., 1s.
Chlorophyll conc. was
0.5mg/ml.
Illumination periods
were for the time
increments shown.

Following a hexane wash (Fig. 2b) the initial narrow (1.0mT) g=2.0050 signal due to A_1 was much more rapidly superseded by the broader (1.4mT) g=2.0033 signal associated with A_0 reduction. It appeared that electron flow to A_1 was inhibited and that electrons became trapped on the earlier acceptor A_0. Similarly, hexane washing severely inhibited formation of the optical difference spectrum that we have related to A_1 reduction (Fig. 3).

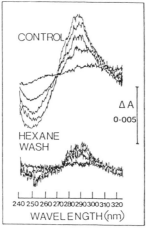

FIGURE 3.
Optical difference spectra of pea PSI particles measured at 205K following illumination for 1, 5, 10 and 30 minute increments using a 150W fibre optic light source. Samples were hexane washed as in Fig. 2. The chlorophyll concentration was 0.15mg/ml.

We also attempted to 'extract' a proton from the component A_1 and replace it with a deuteron. This should lead to a narrowing of the ESR spectrum of a quinolic component (4). Low temperature illumination of chemically reduced and deuterated PSI gave the spectra shown in Fig. 4. Illumination at 205K gave rise to a split signal with a g-value typical of A_1 (2.0053). The broad part of the signal had typical A_1 width (1.07mT) whereas the narrower part was 0.71mT wide. Upon raising the temperature to 225K and further illumination the signal broadened and shifted giving g-value and width typical of A_0 in control samples. A_0 was not affected by deuteration therefore.

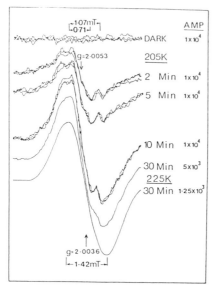

FIGURE 4.
ESR spectra of deuterated pea PSI particles recorded under the conditions described in Fig. 2. The chlorophyll concentration was 1mg/ml.
Illumination periods were for the time increments shown.

We have shown here an optical difference spectrum recorded in samples where the only change occuring was the accumulation of reduced acceptor A_1. This difference spectrum closely resembles that associated with reduction of various quinone species to semi-quinones, particularly the neutral forms. Washing samples with hexane, a treatment that should deplete lipoic components such as quinones, significantly altered the pattern of photoaccumulation of the ESR signal in the g=2.0 region due to acceptors A_1 and A_0. The signal related to A_1 appeared to have been significantly diminished in amplitude leaving an unchanged A_0. This treatment also reduced the extent of formation of the U.V. optical difference spectrum associated with A_1 reduction. Deuteration of PSI samples significantly altered the line-shape of the A_1 ESR signal in a manner consistent with that component being a quinone. The shape of the A_0 signal was unaffected by this treatment. We conclude that component A_1 in PSI is a quinone-type molecule that forms a neutral semi-quinone upon photoreduction.This conclusion is consistent with the presence of stoichiometric quantities of phylloquinone (vitamin K_1) in PSI particles from higher plants (5) and in PSI reaction centre particles from cyanobacteria (6). Further studies are required to determine whether electron acceptor A_1 is essential for forward electron flow.

REFERENCES
1 Bonnerjea, J. and Evans, M.C.W. (1982) FEBS Letts. 148, 313-316
2 Gast, P., Swarthoff, T., Ebskamp, F.C.R. and Hoff, A.J. (1986) Biochim. Biophys. Acta 722,168-175
3 Mansfield, R.W. and Evans, M.C.W. (1985) FEBS Letts. 190, 237-241
4 Kohl, D.H., Townsend, J., Commoner, B., Crespi, H.L., Dougherty, R.C. and Katz, J.J. (1965) Nature 206, 1105-1110
5 Lichtenthaler, H.K. (1969) Progress in Photosynthesis Research 1, 304-314
6 Schoeder, H.-U. and Lockau W. (1986) FEBS Letts. 199, 23-27

EVIDENCE FOR THE EXISTENCE OF ELECTRON ACCEPTORS A_0 AND A_1
IN CYANOBACTERIAL PHOTOSYSTEM 1

N.S. Smith, R.W. Mansfield, J.H.A. Nugent, M.C.W. Evans, Department of
Botany and Microbiology, University College London, London WC1E 6BT, U.K.

1. INTRODUCTION

Cyanobacteria have long been thought to have been the evolutionary
precursors to chloroplasts, originally as symbionts. They are prokaryotic
photosynthetic organisms that can carry out oxygenic photosynthesis with
two photosystems very similar to those of green plants (especially to photosystem 1), and being prokaryotic may simplify genetic analysis. Photosystem
1 (PS1) is a membrane bound complex which catalyses the light-induced
transfer of electrons from P700 to soluble ferredoxin. Two intermediary
electron acceptors (A_0 and A_1) and three iron sulphur centres (X, and B)
have been identified in the higher plant PS1 complex (1,2,3) possibly arranged
thus:

$$P700 \rightarrow A_0 \rightarrow A_1 \rightarrow Fe\text{-}S_X \rightarrow Fe\text{-}S_A \rightarrow Fe\text{-}S_B \rightarrow Fd$$

The three iron sulphur centres have also been identified in cyanobacteria
(4). Here we show the presence of A_0 and A_1 in cyanobacterial PS1 using
ESR at cryogenic temperatures to detect the components.

2. PROCEDURE

2.1 PS1 particle preparation: PS1 particles were prepared from sonicated
Synechococcus leopoliensis (Anacystis nidulans UTEX 625) cells with a 1·5%
Digitonin digestion followed by a 1·5% Triton X-100 digestion using a method
based on that in (5) without the sepharose column purification stage.
Phormidium laminosum PS1 particles were prepared from LDAO digested
membranes using a Triton X-100 digestion (J.H.A. Nugent, personal communication). Both types of PS1 particle were purified using hydroxyapatite column
chromatography. The P700:chl ratios were 1:56 and 1:90 respectively.

2.2 ESR spectroscopy: PS1 samples were prepared to a concentration of
500 µg/ml chlorophyll, and 20 mM glycine pH 10. They were then gassed
with oxygen-free nitrogen for 30 minutes followed by the addition of Sodium
dithionite to 0·1% and a further 30 minutes gassing. They were then illuminated for two minutes with a 150 W lamp filtered by a $CuSO_4$ solution
prior to freezing in the dark, with liquid nitrogen (this treatment reduces
the iron-sulphur centres, leaving A_0 and A_1 oxidised). ESR measurements
were carried out at 205 K. Illuminations to photoreduce A_1 were carried
out at 205 K, and 230 K for A_0.

3. RESULTS AND DISCUSSION

Figure 1 shows the ESR signals observed following four consecutive
illumination periods at 205 K in S. leopoliensis, followed by three consecutive
illumination periods at 230 K. Similar ESR signals were observed following
the same treatment of P. laminosum PS1.

Illumination at 205 K caused the appearance of signals 1·03 mT wide
with g = 2·005 in S. leopolensis and 1·08 mT wide with g = 2·005 in P.
laminosum. These values correspond to A_1 signals in Pea and Spinach.

Biggens, J. (ed.), Progress in Photosynthesis Research, Vol. I. ISBN 90 247 3450 9
© 1987 Martinus Nijhoff Publishers, Dordrecht. Printed in the Netherlands.

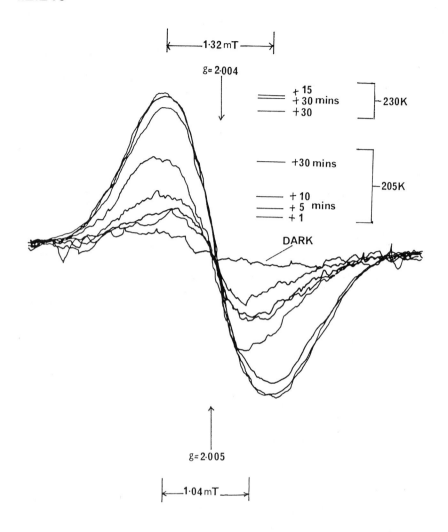

FIGURE 1. ESR spectra of A_1 and A_0 recorded after illumination (for stated time increments) of PS1 particles at 205 K (A_1) and 230 K (A_0 and A_1). The spectra were recorded with the following instrument settings: microwave power, 100 μW; frequency 9·053 GHz; modulation amplitude 2 G. Gain at: 205 K; 10^4 ; 230 K, 5 x 10^3

Subequent illumination at 230 K caused a broadening of these signals to 1·32 mT in S. leopolensis and 1·34 mT in P. laminosum, with a shift in g values to 2·004 and 2·002 respectively. These values correspond to the combined signals of A_0 and A_1 in Pea and Spinach. Table 1 shows a comparison of the signals obtained from cyanobacteria and higher plants.

TABLE 1.

Sample	Temp. (K)	Signal width (mT)	g value
Pea PS1	205	1·03	2·004
	230	1·35	2·0025
Spinach PS1	205	1·05	2·005
	230	1·34	2·003
Synechococcus PS1	205	1·03	2·005
	230	1·32	2·004
Phormidium PS1	205	1·08	2·005
	230	1·34	2·002

These results, together with the earlier identification of the three iron-sulphur proteins in cyanobacterial PS1 complexes shows a close similarity between cyanobacterial and higher plant PS1. Hence genetic studies of cyanobacterial PS1 could probably be applied to higher plant PS1 with a useful degree of accuracy, it being simpler to study cyanobacterial genetics (as they are prokaryotic) than higher plant genetics (as they are eukaryotic). This being so, genetic analysis of S. leopoliensis PS1 is now being attempted via a λgt11 expression bank.

4. ACKNOWLEDGEMENTS

This work was funded by a grant from the UK Science and Engineering Research Council.

REFERENCES

1. Bonnerjea, J. and Evans, M.C.W. (1982) FEBS Letts. 148, 313-316.
2. Gast, P., Swarthoff, T., Ebskamp, F.C.R. and Hoff A.J. (1983) Biochim. Biophys. Acta 722, 168-175.
3. Evans, M.C.W. (1982) in Iron-Sulphur Proteins (Spiro, T.G. ed.) Vol. 4, pp. 89-134, John Wiley & Sons inc.
4. Evans, E.H., Dickson, P.E., Johnson, J.D. and Evans, M.C.W. (1981) Eur. J. Biochem. 118, 81-84.
5. Evans, E.H. and Pullin, C.A. (1981) Biochem. J. 196, 489-493.

IRON X-RAY ABSORPTION SPECTRA OF ACCEPTORS IN PS I

Ann E. McDermott, Vittal K. Yachandra, R. D. Guiles,
R. David Britt, S. L. Dexheimer
Kenneth Sauer and Melvin P. Klein

Department of Chemistry and Chemical Biodynamics Laboratory
Lawrence Berkeley Laboratory
University of California, Berkeley CA 94720 USA

Introduction

The stable electron acceptors in PS I, centers A, B and X, are thought to be [2Fe-2S] or [4Fe-4S] ferredoxins [1]. Their EPR spectra resemble the spectra of the [2Fe-2S] and [4Fe-4S] ferredoxins, having anisotropic signals with $g_{ave} < 2.0$, and their EPR spectra are quite different from the spectra of the [1Fe], [3Fe-4S] or [3Fe-3S] ferredoxins. PS I contains 10-14 Fe and approximately 12 acid labile sulfide per P700 [2,3], and contains one center A, one center B and one to two X per P700 [4-7]. On the basis of these observations two models of PS I seem likely: 1) PS I contains three [4Fe-4S] ferredoxins or 2) it contains two [4Fe-4S] ferredoxins and two [2Fe-2S] ferredoxins. Evans [8,9] addressed the structure of A, B and X using Mössbauer spectroscopy and concluded that the spectra of all the reducible iron in PS I resembles spectra of [4Fe-4S] ferredoxins and not of [2Fe-2S] ferredoxins. Only approximately 65% of the iron present was reduced in that study.

In this study, we address the structures of centers A, B and X using X-ray absorption spectroscopy. Firstly we use iron K-edge spectra to assess how much of the iron is bound as ferredoxins. K-edge spectra probe bound state transitions from the 1s level and are sensitive to oxidation state and site symmetry [10]. Secondly we use Extended X-ray Absorption Fine Structure (EXAFS) to address whether the ferredoxins are [2Fe-2S] or [4Fe-4S] clusters. EXAFS provides information about types and numbers of ligands, and provides sensitive bond length information. In the case of ferredoxins it is diagnostic for [2Fe-2S] and [4Fe-4S] clusters [11].

Materials and Methods

PS I preparations from *Synechococcus* and from market spinach were prepared using Triton X-100 extraction and sucrose gradient centrifugation as detailed elsewhwere [12,13]. The preparations has 11-14 Fe and 9-13 acid labile sulfide per P700 and showed EPR signals due to A, B and X but showed no signals due to heme, soluble [2Fe-2S] spinach ferredoxin, or $g=4.3$ iron. The preparations contained peptides of molecular weight 60, 22, 19 and 18 kDa by SDS PAGE.

X-ray absorption spectra were collected as described elsewhere [12-14]. The PS I preparation was assayed for integrity both before and after X-ray exposure by EPR and optical methods. EXAFS data were analyzed by curve fitting according to the Teo-Lee method [15].

Results and Discussion

Figure 1 shows the iron K-edge spectrum of PS I from spinach and *Synechococcus* along with the spectrum of soluble spinach [2Fe-2S] ferredoxin, $(Et_4N)_2$ $Fe_4S_4(S$-benzyl$)_4$ and $(Et_4N)_2$ $Fe_2S_2(S$-o-xyl$)_2$. The iron K-edges of the PS I preparations are nearly identical to those of the soluble [2Fe-2S] ferredoxin including the size of the 1s to 3d transition, an indicator of the amount of non-centrosymmetric iron. Roe *et. al.* [10] have shown that this feature is systematically larger in tetrahedral complexes than in octahedral complexes. We

Biggens, J. (ed.), Progress in Photosynthesis Research, Vol. I. ISBN 90 247 3450 9

have found that 15% adventitiously bound iron in our soluble ferredoxin model proteins changes the edge shape dramatically, so the edge is a sensitive reporter of ferredoxin content. We estimate that over 90% of the iron in PS I is bound in ferredoxins.

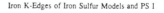

Iron K-Edges of Iron Sulfur Models and PS I

Figure 1 Iron K-edge spectra of of oxidized PS I with iron-sulfur models. From the bottom: a) PS I from spinach, b) PS I from *Synechococcus*, c) $(Et_4N)_2Fe_4S_4(S\text{-benzyl})_4$, d) spinach $[2Fe\text{-}2S]^{2+(2+,1+)}$ ferredoxin and e) $(Et_4N)_2Fe_2S_2(S_2\text{-}o\text{-xyl})_2$. The pre-edge transition has been assigned as a 1s-3d transition, and is a good indicator of non-centrosymmetric environments. The PS I from spinach and *Synechococcus* are quite similar to each other and to the [2Fe-2S] soluble spinach ferredoxin.

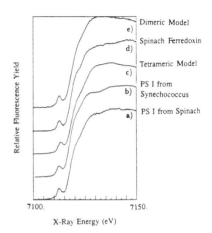

Figure 2 shows the Fourier filtered spinach PS I EXAFS spectrum plotted over spectra from soluble spinach [2Fe-2S] ferredoxin and over data from soluble *Clostridium pasteurianum* [4Fe-4S] ferredoxin. In the region of $k=7.5\text{Å}^{-1}$ the [4Fe-4S] and [2Fe-2S] ferredoxin data are out of phase owing to the much greater amount of iron backscattering in the [4Fe-4S] cluster from three equivalent Fe neighbors. The PS I spectrum, while not identical to either spectrum, more nearly resembles the [2Fe-2S] spectrum. Curve fitting analysis suggests that the PS I spectrum is better explained by a sum of [2Fe-2S] and [4Fe-4S] ferredoxins than by all [4Fe-4S] ferredoxins. Figure 3 shows the simulations of the PS I iron EXAFS spectra from spinach and *Synechococcus*. In each case the simulation was restricted to using four sulfur neighbors and either 2.3 or 3.0 iron neighbors to simulate either a system of two [2Fe-2S] clusters and two [4Fe-4S] clusters or simulate a mixture of different [4Fe-4S] clusters. All other parameters used in the simulations were chosen by comparison with our data on soluble ferredoxins or published data [11]. The central portions of these fits around $k=7.5\text{Å}^{-1}$ are most informative, because this 'beat' region is sensitive to the admixture of iron and sulfur backscattering. The lower k regions are more prone to artifacts due to contributions from other processes, and the higher k regions have poorer signal to noise ratios. For both our spinach and *Synechococcus* data the fits assuming a mixture of [2Fe-2S] and [4Fe-4S] ferredoxins had least squares errors lower by a factor of more than two. These fits agree much better with the data around $k=7.5\text{Å}^{-1}$.

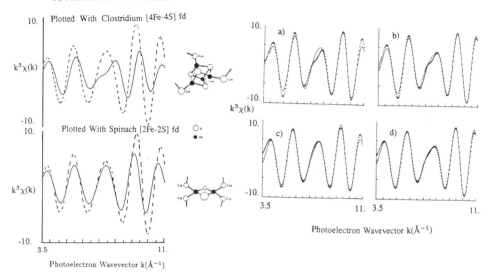

PS I Iron EXAFS Data – Dashed

Figure 2 (on the left) The Fourier filtered k^3-weighted EXAFS spectrum of PS I from spinach (dotted line) plotted with: (a) the spectrum of soluble *Clostridium pasteurianum* [4Fe-4S] ferredoxin and (b) the spectrum of soluble spinach [2Fe-2S] ferredoxin. The beat region around $k=7.5\text{Å}^{-1}$ is diagnostic of [2Fe-2S] and [4Fe-4S] ferredoxins. The beat region indicates that PS I is likely to contain some [2Fe-2S] ferredoxins.

Figure 3 (on the right) The Fourier filtered k^3-weighted EXAFS spectrum of PSI (solid lines) is plotted with calculated simulations using the Teo-Lee method [15]. The fits are as follows: a) spinach PS I simulated assuming all [4Fe-4S] ferredoxins, b) PS I from *Synechococcus* simulated assuming all [4Fe-4S] ferredoxins, c) PS I from spinach simulated assuming two [4Fe-4S] and two [2Fe-2S] ferredoxins and d) PS I from *Synechococcus* simulated assuming two [2Fe-2S] and two [4Fe-4S] ferredoxins. The quality of the fits in the beat region indicates that PS I is likely to contain [2Fe-2S] as well as [4Fe-4S] ferredoxins.

The Mössbauer results of Evans *et al* indicate that A, B and X are all [4Fe-4S] ferredoxins [8,9]. We are considering the possibility that the [2Fe-2S] ferredoxins reflected in the EXAFS spectrum may be contained in the 35% of iron which was not reduced in the Mössbauer experiment. This iron is likely to be associated with X [7], since X is more difficult to reduce quantitatively than is center A or B. Recently we have observed a near infrared absorption band near 1050 nm in PS I which is typical of [4Fe-4S] ferredoxins. This band can be titrated reversibly at -500mV. This confirms the assignment that centers A and B are the [4Fe-4S] ferredoxins in PS I, and so X is likely to be the [2Fe-2S] ferredoxin(s).

While the EPR spectrum of X does not saturate below 100mW, it has an optimum temperature of observation of approximately 8 K, below which the signal diminishes rapidly (data not shown). This indicates that the state giving rise to the EPR signal is above, but close to the ground state. This is unusual for [2Fe-2S] or [4Fe-4S] ferredoxins, and we are considering whether this be explained by an exchange coupled pair of [2Fe-2S] clusters, separated by over 3.5 Å, so that intercluster distances are not seen in the EXAFS spectrum. This exchange coupling may also explain why the power saturation and EPR linewidth of X are different from those of other ferredoxins.

Acknowledgements

We thank S. Katoh for providing the strain of *Synechococcus* and Akhihiko Yamagishi for advice concerning the preparation of PS I from *Synechococcus*. We gratefully acknowledge support from the National Science Foundation (PCM 82-16127, PCM 84-16676) and the Director, Office of Energy Research, Office of Basic Energy Sciences, Division of Biological Energy Conversion and Conservation of the Department of Energy under contract DE-AC03-76SF00098. We thank the Stanford Synchrotron Radiation Laboratory, Stanford CA for providing synchrotron radiation facilities.

References

1 Evans, M.C.W. (1982) in "Iron Sulfur Proteins", Spiro, T., ed.: pp 249-284 John Wiley and Sons, New York.
2 Golbeck, J. (1980) *Methods Enzymol.* **69**, 129-141
3 Lundell, D., Glazer, A., Melis, A. and Malkin, R. (1985) *J. Biol. Chem.* **260**, 646-654
4 Bearden, A. and Malkin, R. (1972) *Biochim. Biophys. Acta.* **283**, 456-468
5 Williams-Smith, D., Heathcote, P., Charanjit, K. and Evans, M.C.W (1978) *Biochem J.* **170**, 365-371
6 Heathcote, D., Williams-Smith, D. and Evans, M.C.W. (1978) *Biochem. J.* **170**, 373-378
7 Bonnerjea, J. and Evans, M. (1984) *Biochim. Biophys. Acta.* **767**, 153-159
8 Evans, H., Rush, J., Johnson, C. and Evans, M.C.W. (1979) *Biochem. J.* **182**, 861-865
9 Evans, H., Dickson, D., Johnson, C., Rush, J. and Evans, M.C.W. (1981) *Eur. J. Biochem.* **118**, 81-84
10 Roe, A., Schneider, R., Mayer, R., Pryz, J., Widom, J. and Que, L., Jr. (1984) *J. Am. Chem. Soc.* **106**, 1676-1681
11 Teo, B., Shulman, R., Brown, G. and Meixner, A. (1979) *J. Am. Chem. Soc.* **101**, 5624-5631
12 McDermott, A., Yachandra, V., Guiles, R., Britt, R., Dexheimer, S., Sauer, K. and Klein, M., *submitted*
13 McDermott, A. Ph. D. Dissertation, University of California, Berkeley *in preparation*.
14 Yachandra, V., Guiles, R., McDermott, A., Cole, J., Britt, R., Dexheimer, S., Sauer, K. and Klein, M. (*These proceedings*).
15 Teo, B. and Lee, P. (1979) *J. Am. Chem. Soc.* **101**, 2814-2831

PHOTOSYSTEM I CHARGE SEPARATION IN THE ABSENCE OF CENTERS A & B:
BIOCHEMICAL CHARACTERIZATION OF THE STABILIZED P700 A2(X)
REACTION CENTER

JOHN H. GOLBECK, KEVIN G. PARRETT, AND LESLIE L. ROOT
DEPARTMENT OF CHEMISTRY, ESR DOCTORAL PROGRAM[1]
PORTLAND STATE UNIVERSITY, PORTLAND, OREGON U.S.A. 97207

1. INTRODUCTION

Golbeck and Cornelius (1) recently showed that iron-sulfur Centers
A and B (P430) are functionally dissociated from the PSI reaction center
by 30s treatment with 0.5% lithium dodecyl sulfate (LDS). The resulting
1.2-ms, 700 nm absorption transient was found to correspond to the P700
A2(X) charge separation seen in the absence of reduced Centers A and B.
The continued presence of LDS, however, led to the eventual destruction
of the P700+ A2(X)- reaction; even after 20-fold dilution of the
detergent, damage to the Fe-S cluster was evident within 60 min. In
this paper, we show that the P700 A2(X) reaction is stabilized
indefinitely after 20-fold dilution of the LDS-treated reaction center
in buffer followed by ultrafiltration over an Amicon YM-100 membrane.
In addition to removing the detergent, this protocol removes totally the
19-kDa peptide and partially the 18-kDa peptide from the >110-kDa PSI
reaction center. The loss of the P700 P430 reaction correlates with
loss of the 19-kDa peptide, indicating that the Fe-S clusters
corresponding to Centers A and/or B may be closely associated with this
peptide.

2. MATERIAL AND METHODS

The PSI reaction center was isolated with Triton X-100 from spinach
chloroplasts (2). The particles were treated with 1% LDS for 60 s at
250 ug/ml chlorophyll in 0.1 M Tris buffer (pH 8.3) followed by 20x
dilution in buffer and ultrafiltration over a YM-100 membrane (Amicon).
The flash-induced absorption transient was determined at 700 nm as
described in ref. 1. The pH stability was determined in 0.1 M buffer
(HEPES, pH 7.0 to 8.0; glycylglycine, pH 8.5 & 9.0; glycine, pH 9.5 and
above). Polyacrylamide gel electrophoresis (PAGE) was performed in a
linear 10-20% gradient. The control and LDS-treated, stabilized PSI
particles (5 ug/ul protein) were incubated in a buffer system containing
0.0625 M Tris (pH 6.8), 2% SDS, 10% glycerol and 5% 2-mercaptoethanol
for 2 min at 70°C. The samples were cooled to ice temperature and
applied to the stacking gel at a protein concentration of 10 ug/well.
The electrophoresis was performed at 20 ma for 6 h at 10°C. Gels were
stained with Coomassie Brilliant Blue, overstained with silver, and
scanned with a laser densitometer.

[1] Environmental Sciences and Resources Program Publication 197.

Biggens, J. (ed.), Progress in Photosynthesis Research, Vol. I. ISBN 90 247 3450 9
© 1987 Martinus Nijhoff Publishers, Dordrecht. Printed in the Netherlands.

3. RESULTS

In-depth studies of Center A2(X) will require enhanced stability over that achieved by LDS-treatment and dilution in buffer alone (1). Fig. 1 shows the stability at pH 8.3 of the LDS-treated PSI reaction center under three experimental conditions. The ultrafiltration procedure was found to be extremely effective in removing LDS and in stabilizing the P700 A2(X) reaction; a particle treated in this manner shows no loss of signal for periods to 2 weeks when stored at 4°C.

FIGURE 1

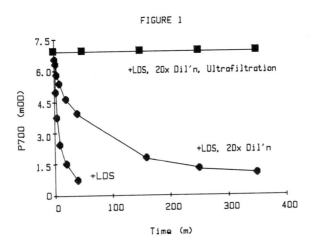

The ultrafiltered reaction center is stable over a wide range of pH values. Figure 2 shows the magnitude of the 700 nm absorption transient after 5 min of incubation at the given pH. The particle containing LDS (1) shows a steep drop in stability after 2.5 min at pH 9.0. In contrast, the LDS-treated, ultrafiltered reaction center is stable to pH 11.0 and the control particle is stable to pH 12.0 for at least 5 min. At 60 min, the LDS-treated, ultrafiltered reaction center shows no deterioration from pH values of 7.5 to 10.0.

FIGURE 2

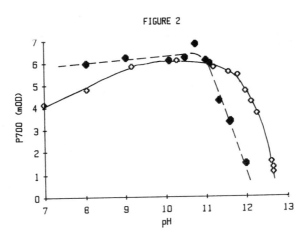

Figure 3 shows the polypeptide composition of the control PSI reaction center as determined by SDS PAGE. The spinach preparation contains 6 major peptides, with molecular weights of 64-, 19-, 18-, 17-, 15-, and 12-kDa. When the electrophoresis is performed at 10°C, we usually find a minor chlorophyll-containing peak at 110-120 kDa which may represent a dimer of the 64-kDa band. We find little evidence for the existence of an 8-kDa peptide using standard staining techniques (3).

Figure 3 also shows the polypeptide composition of the LDS-treated, ultrafiltered reaction center. The most salient feature is the near-complete loss of the 19-kDa peptide and the partial loss of the 18-kDa peptide. This result indicates that LDS removes physically one peptide (and a portion of another) from the photosystem I reaction center under conditions that lead to the loss of Centers A & B. As in the control preparation, the 64-kDa region in the LDS-treated preparation is diffuse and may represent more than one band.

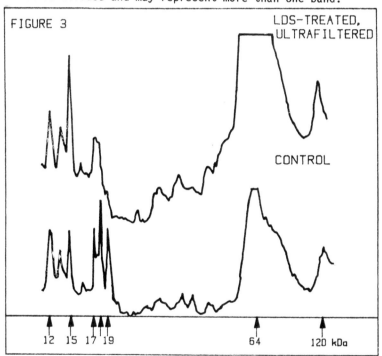

FIGURE 3 — LDS-TREATED, ULTRAFILTERED — CONTROL

12 15 17 19   64   120 kDa

The LDS-treated and stabilized reaction center is nearly devoid of functional Center A and B (4). When considered with the electrophoretic results, the data indicate that Centers A and B may be closely associated with the 19-kDa peptide in PSI. Bonnerjea et al (5) have shown that a spinach PSI particle could be substantially depleted of the 16-, 14-, 10-, and 8-kDa peptides without significant loss of Centers A and B. The 19-kDa peptide may either contain these clusters or serve to stabilize these centers on another protein.

The available biochemical and biophysical data suggest a model for the PSI reaction center (Figure 4). The 19-kDa peptide is shown to contain Centers A and B. Since Centers A and B are each 4Fe-4S clusters in a reaction center that contains 12 atoms each of iron and sulfide (6), this leaves room for only two 2Fe-2S clusters or one 4Fe-4S cluster on Center A2(X). The recent demonstration that zero-valence sulfur exists on the P700-chlorophyll a protein (1,7) indicates that the 64-kDa peptide is an Fe-S protein. The constraints on the structure of the reaction center imposed by the amino acid sequence (8) and the probability of 4 cysteine residues as ligands to the iron atoms suggests a model where A2(X) is an interpolypeptide 4Fe-4S cluster held between two 64-kDa peptides. While this simple model satisfies the apparent requirement for two high molecular weight peptides, the data do not exclude the possibility of a protein dimer composed of the 64-kDa peptide and one of the low molecular weight peptides.

FIGURE 4

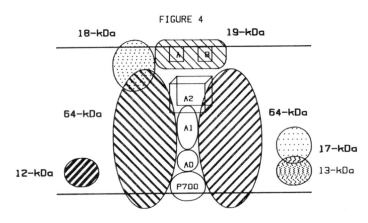

REFERENCES

1. Golbeck, J.H. and Cornelius, J.M. (1986) Biochim. Biophys. Acta 849, 16-24.
2. Golbeck, J.H. (1976) in Methods in Enzymology (San Pietro, ed), Vol 69, pp 129-141, Academic Press, New York.
3. Lagoutte, B., Setif, P. and Duranton, J. (1984) FEBS Lett. 174, 24-29.
4. Warden, J.T. and Golbeck, J.H. (1986) Biochim. Biophys. Acta 849, 16-24.
5. Bonnerjea, J., Ortiz, W. and Malkin, R. (1985) Arch. Biochem. Biophys. 240, 15-20.
6. Golbeck, J.H. and Warden, J.T. (1982) Biochim. Biophys. Acta 681, 77-84.
7. Sakurai, H. and San Pietro, A. (1985) J. Biochem. 98, 69-76.
8. Fish, L.E., Kuck, U. and Bogorad, L. (1985) J. Biol. Chem. 260, 1413-1421.

Supported by the National Science Foundation (DMB-8517391).

PICOSECOND TRANSIENT ABSORPTION SPECTROSCOPY OF PHOTOSYSTEM 1 REACTION
CENTRES FROM HIGHER PLANTS.

L.B. Giorgi, B.L. Gore, D.R. Klug, J.P. Ide, J. Barber* and G. Porter
Davy Faraday Research Laboratory, The Royal Institution, 21 Albemarle Street,
London W1X 4BS, U.K.; *Department of Pure & Applied Biology, Imperial
College, Prince Consort Road, London SW7 2BB, U.K.

1. INTRODUCTION

Picosecond time resolved absorption spectra of Photosystem 1 (PS1) reaction
centres, isolated from pea chloroplasts, have been measured at room temperature, following excitation with a 600nm optical pulse giving a time resolution of 10ps. A summary of the sequence of known electron carriers within the PS1 reaction centre is:

$$P700 \ldots\ldots A_0 \ldots\ldots A_1 \ldots\ldots A_2 \ldots\ldots F_A, F_B$$

where chemical assignments are: A_0 - a molecule of chlorophyll a; A_1 - a
quinone species (?); A_2 - a specialised iron sulphur centre; F_A, F_B - bound
iron sulphur centres, characterised by an absorption change at 430nm. Here
we have studied the excitation and decay kinetics of the antenna chlorophyll
associated with the PS1 reaction centre, and the grow in of the P700 signal.

2. EXPERIMENTAL

A conventional Coherent synchronously pumped dye laser system, CR-12 Ar-Ion
laser pumping a CR-590 folded dye laser, produces 1nJ, 7ps pulses at 600nm
and at 75MHz. A Quantel Q-switched Nd:YAG laser is used to amplify these
pulses by means of a 4 stage amplifier chain to provide 2mJ, 10ps pulses at
10 Hz. Each amplified pulse is split to provide both the excitation pulse,
up to 100μJ, and, by means of continuum generation in a 3cm cell containing
an H_2O/D_2O mixture, the probe pulse. Both pulses are combined on a beam
splitter and subsequently pass along the sample and reference arms of a dual
beam spectrograph. The pump and probe beams are colinear; they are focussed
into the sample cell, whilst they pass unfocussed through the reference cell.
The transmitted beams are then focussed into an 0.25m spectrograph and imaged
onto the two halves of a vidicon camera. Spectra are summed and processed
in a micro computer to yield absorption difference spectra. Spectra in this
investigation cover the region 625nm-765nm, with an overall time resolution
of 10ps. Each spectrum is composed of 500 data points and is the average of
2400 laser shots.

PS1 reaction centres, prepared with a chlorophyll/P700 ratio of 50 (1), were
suspended in a medium containing 50mM Tris/HCl, pH 8.0, at a chlorophyll
concentration of 25μM, giving an optical density of 1.4 in a 1cm cell at the
absorption maximum (673nm). Ferricyanide (1mM) was added to chemically oxid-
ise, and ascorbate (4mM) was added to chemically reduce, P700 in the dark.

Biggens, J. (ed.), Progress in Photosynthesis Research, Vol. I. ISBN 90 247 3450 9
© *1987 Martinus Nijhoff Publishers, Dordrecht. Printed in the Netherlands.*

3. RESULTS AND DISCUSSION

Figure 1 shows the series of transient absorption spectra obtained with PS1 reaction centres. Two obvious spectral features occur, one centred at 690nm and one at 700nm.

The feature at 690nm grows in during the excitation pulse and appears in the spectra of samples with P700 either chemically oxidised or chemically reduced. This suggests that the 690nm signal can be attributed to the excitation of antenna chlorophyll molecules to the singlet state. It decays with a lifetime of 15 ± 1ps. This lifetime value has been obtained from global analysis of the spectra from t = 6.5ps to t = 270ps, without deconvolution: this yields a good fit to the data with a single exponential decay. The decay of the 690nm signal is, within the resolution of these experiments, independent of the redox state of P700. As can be seen from the series of oxidised spectra, the 690nm feature undergoes a blue shift as it decays, finally being centred at 675nm. Previous experiments (2) have shown that the residual bleach at 675nm decays with a lifetime of between 1 and 2ns; it is believed that this signal is due to residual triplet chlorophyll molecules which are quenched by carotenoid molecules within the antenna.

The 700nm spectral feature is much narrower than the 690nm signal and is only observed in the spectra of samples with P700 chemically reduced. This suggests that it can be attributed to the photooxidation of P700 molecules. By studying the difference between the oxidised and reduced spectra, at each time delay, it is possible to follow the rise of the P700 signal as the 690nm signal decays. Despite the short antenna decay lifetime and the spectral overlap of the 690nm and P700 signals, it is clear that the P700 signal only appears once the excitation pulse is over i.e. it is delayed relative to the maximum of the 690nm signal. The rise time of the P700 signal is similar to the decay time of the 690nm signal, with the P700 signal reaching 80% of its final value 20 ± 3ps after the maximum of the excitation pulse.

Throughout these experiments it has been necessary to avoid annihilation processes within individual antenna systems caused by multiple excitation of the antenna chlorophylls. If multiple excitation of a single antenna system does occur this leads to a rapid initial decay, (this initial decay is so fast that the bleach follows the time profile of the pulse). In order to obtain some estimate of the degree of initial excitation three approaches have been taken:
1) Reduction of the pump intensity until the initial fast decay component is no longer detected. Experimentally, this fast decay component is not observed if the magnitude of the initial 690nm signal is less than 0.1 OD.
2) By interpreting the 690nm signal as a combination of ground state bleach and stimulated emission from the antenna chlorophyll, an estimate of the fraction of chlorophyll molecules excited can be made. However, since the maximum ground state chlorophyll absorption occurs at 673nm and steady state fluorescence maximum is at 675nm, we do not feel that the large red shift of the initial 690nm signal with respect to the ground state absorption can be explained solely as a combination of ground state bleach and stimulated emission from the singlet chlorophyll. Instead, it may be indicative of sub-picosecond energy transfer, within the antenna, to core chlorophyll molecules that absorb and emit significantly redder than the majority of chlorophyll molecules. As a result, we have been unable to use this as a reliable method of estimating the initial degree of antenna excitation.

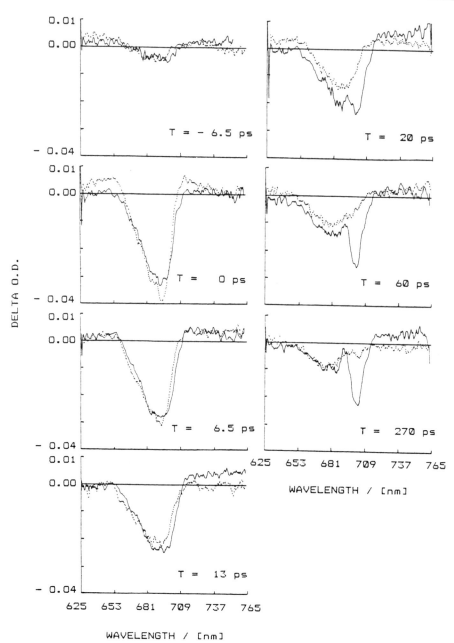

Figure 1 Transient absorption spectra of PSl reaction centres.
————, P700 chemically reduced; ....., P700 chemically oxidised.

3) By measuring the magnitude of the P700 signal at late times, e.g. 270ps, it is possible to determine the fraction of P700 molecules that have been photooxidised. At high pumping intensities, when the 690nm signal is greater than 0.1 OD, the P700 signal becomes insensitive to the pump power and has a limiting value of -0.04. In the experiment presented in figure 1 we have ensured that the P700 signal is approximately half of the limiting value and, from a simple analysis of the Poisson distribution of the number of excitations in each antenna system, it can be shown that less than 30% of excited antenna systems receive multiple excitation.

4. CONCLUSION

By combining a low antenna chlorophyll to reaction centre ratio (50:1) with a low level of pump intensity, we have reduced the occurrence of excitation annihilation within the antennae to a level where it is no longer detectable. Under these experimental conditions, the decay of the excited antenna chlorophyll signal and the concomitant grow in of the P700 signal have been resolved. The lifetime of the decay of the excited antenna chlorophyll signal is 15 ± 1ps; the P700 signal grows in at the same rate. The decay of the excited antenna chlorophyll signal appears to be independent of the redox state of P700. This is not easily explained by a simple Forster energy transfer mechanism but it is hoped that by improving the time resolution of the current apparatus the causes of this behaviour will be determined.

ACKNOWLEDGEMENTS

We thank the SERC and British Petroleum plc for financial support.

REFERENCES

1 Alberte, R.S. and Thornber, J.P. (1978) FEBS Lett. 91, 126-130
2 Giorgi, L.B., Doust, T., Gore, B.L., Klug, D.R., Porter, G. and Barber,J. (1986) Biochem. Soc. Trans. 14, 47-48

LIGHT–INDUCED FOURIER TRANSFORM INFRARED (FTIR) SPECTROSCOPIC INVESTIGATIONS OF PRIMARY REACTIONS IN PHOTOSYSTEM I AND PHOTOSYSTEM II

B.A. Tavitian[§], E. Nabedryk[§], W. Mäntele[$] and J. Breton[§].

§   Service de Biophysique, Département de Biologie, C.E.N. Saclay, 91191 Gif-sur-Yvette Cedex, France.
$ Institut für Biophysik und Strahlenbiologie der Universität Freiburg, Albertstrasse 23, D–7800 Freiburg, FRG.

1. INTRODUCTION

In photosynthesis of green plants and blue–green algae, electron transport requires the cooperation of two photosystems (PS): PS I and PS II, both located in the photosynthetic membrane. Both PS I and PS II incorporate as primary electron donor a chlorophyll (Chl) a monomer or dimer: P700 and P680 respectively, named after their visible absorption maxima. With regard to the redox components it includes and the potential it spans, the acceptor side of PS II resembles the well-characterized reaction center (RC) of purple photosynthetic bacteria.

RCs have been isolated and purified for a number of bacteria. In the case of green plants, the isolation procedures only allow membranes to be enriched in or depleted of either PS I or PS II. They lead to fractions containing as low as 40 antenna Chls per P700 or P680, while still retaining the essential function of the photosystems: efficient separation and stabilization of charges. In previous works (**1**, **2** and E. Nabedryk et al., these proceedings), we have performed light-induced FTIR difference spectroscopy on RCs of several purple photosynthetic bacteria. We have detected light-induced molecular changes associated with (i) the primary electron donor oxidation and (ii) the bacteriopheophytin (BPheo) intermediary electron acceptor I reduction in both bacteriochlorophyll (BChl) a– and BChl b–containing species. We have also obtained FTIR difference spectra on intact chromatophore membranes. Although the background absorption by components other than the RC's is much larger, the difference spectra were quite similar. This has prompted us to investigate, using the same technique, the well-characterized states that can be photoaccumulated on thylakoid subfragments enriched in either PS I or PS II, but still retaining a significant number of antenna Chls. Here, light-induced FTIR difference spectra associated (i) with the accumulation of the photooxidized primary donor in PS I particles, and (ii) with the accumulation of the photoreduced pheophytin (Pheo) intermediary electron acceptor in PS II-enriched particles, are presented.

2. MATERIALS AND METHODS

The preparation of the samples of spinach PS II-enriched particles and pea PS I particles is described in (**3**) and references therein. Suspensions of these were deposited on CaF_2 windows and air-dried overnight at room temperature in the dark. FTIR absorbance spectra were taken before, during and after continuous illumination on a Nicolet 60SX FTIR spectrophotometer. For more details, see (**3**).

Biggens, J. (ed.), Progress in Photosynthesis Research, Vol. I. ISBN 90 247 3450 9
© *1987 Martinus Nijhoff Publishers, Dordrecht. Printed in the Netherlands.*

3. RESULTS AND DISCUSSION

1.   Photooxidation of the primary electron donor in PS I.

Fig. 1 shows FTIR difference spectra of a film of PS I particles covered with a 10 mM Na ascorbate solution. Fig. iA shows the 'light-minus-dark' spectrum: it was obtained by subtracting an absorbance spectrum of the sample recorded just before illumination from one recorded during continuous illumination with light (530 nm$<\lambda<1100$ nm). One can see that even in the region of strong water and amide absorption, between 1500 and 1700 cm^{-1}, significant molecular changes can be detected. Fig. 1B shows the difference between an absorbance spectrum recorded immediately after illumination and the one recorded just before.

The most prominent features in fig. 1 are observed in the region between 1600 and 1760 cm^{-1}. The largest negative band in the spectra, at 1700 cm^{-1}, could correspond to a decrease in absorption of the 9-keto carbonyl of the pimary donor Chl $\underline{a}$ molecule(s). This decrease might be connected to the appearance of the positive band at 1689 cm^{-1}. A possible interpretation for this differential feature would be a change in binding type for the 9-keto C=O, due to a different ligation to the protein and/or to the influence of a charge in its vicinity.

The two small but very reproducible differential features at $1748/1753$ cm^{-1} and $1734/1742$ cm^{-1} in fig. 1A could be due to a shift to higher frequencies of the 7c-propionic acid and the 10-a carbomethoxy ester groups, respectively. Similar features have also been observed on FTIR difference spectra of the primary electron donor photooxidation in bacterial RCs and chromatophores (1 and E. Nabedryk et al., these

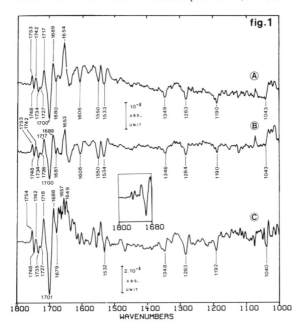

fig.1

Light-induced FTIR difference spectra of films of A,B) pea PS I particles and C) synechococcus PS I particles air-dried with 70 μM methyl viologen and hydrated with 5 mM Tris buffer. A) light-minus-dark spectrum, B) difference spectrum obtained from spectra recorded before and after illumination. C) light-minus-dark spectrum.
Inset: light-minus-dark spectrum obtained on a film of intact spinach thylakoids.
For all measurements, temperature was 22°C, resolution 4 cm^{-1}, and 512 interferogram scans were added.

proceedings) and on FTIR difference spectra of the electrochemically generated Chl $\underline{a}$ cation (data not shown) where two differential signals are present at $1736/1750$ cm^{-1} and $1687/1720$ cm^{-1}. The positive band at 1717 cm^{-1} lies between the domains of keto and ester C=O stretching vibrations and cannot be ascribed precisely to either carbonyl group. The large positive band at 1654 cm^{-1} still lies within the range where a hydrogen-bonded keto group may absorb. However, an alternative interpretation would be a change in the amide I absorption. As $\Delta A/A$ does not exceed 10^{-3} at the amide I band, and assuming 1000-2000 peptide bonds per P700, such a change would affect at most one or two peptide bonds per PS I RC.

The most characteristic FTIR differential signals observed on pea PS I particles can also be detected on PS I particles from synechococcus sp. (fig. 1C) and on intact spinach thylakoid membranes, especially in the carbonyl stretching frequency region (inset of fig. 1).

2. Photoreduction of the intermediary electron acceptor in photosystem II.

Dithionite is known to reduce the electron acceptors of PS II up to the primary quinone. Illumination of a prereduced sample at temperatures ≽-40°C with actinic light (665 nm$<\lambda<1100$ nm) leads to a trapped P680-Pheo$^-$ configuration.

Fig. 2A shows the light-minus-dark spectrum of a film of PS II particles prereduced with 150 mM Na dithionite: the 'Pheo$^-$' spectrum. Fig. 2B represents a control experiment in wich only dithionite was omitted. The absence of signal demonstrates (i) that the signals in spectrum 2A are inherent to Pheo reduction and (ii) the absence of PS I activity in these particles.

In a previous work, light-induced FTIR difference spectra of the photoreduction of I in Rps. viridis RCs and chromatophores and in quantasomes of Ch. vinosum were presented (2 and E. Nabedryk et al., these proceedings). These 'I$^-$' spectra can be compared to those obtained here, although one has to be careful when comparing BPheo $\underline{a}$ or $\underline{b}$ molecules to a Pheo $\underline{a}$. However, several similarities can be pointed out, especially in the $1600-1750$ cm^{-1} region, where stretching vibrations of the carbonyl groups are expected to contribute. The Pheo$^-$ spectrum of fig. 2A shows two negative bands, at 1740 and 1720 cm^{-1}, which can be attributed to a decrease in absorbance of the 7c-propionic and the 10a-carbomethoxy ester C=O groups of the Pheo molecule, respectively, upon its photoreduction (compare to the negative bands at 1746 and 1732 cm^{-1} in (2)). An alternative interpretation for these bands could be in terms of a deprotonation of amino acid carboxyl side chain groups. In this case, the broad positive band centered at 1589 cm^{-1} could be interpreted in terms of the appearing ionized carboxyl group.

In the remainder of the spectrum, a large number of bands appear at reproducible frequencies, but variations in their relative amplitudes were observed from sample to sample, or depending on the length of the illumination period. Since the signal-to-noise ratio, background amplitude and maximum band amplitude are quite comparable to those observed in ($1, 2$) and in fig. 1, the larger variations in the Pheo$^-$ signals could be due to differences in the state trapped under our experimental conditions, especially in the redox state of the primary and secondary quinone acceptors, or could indicate some photodegradation of the pigments. Further experiments carried out at cryogenic temperatures, as

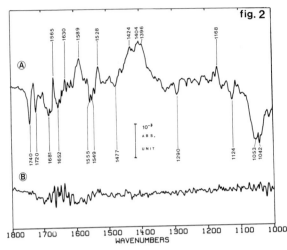

Light-minus-dark spectra obtained on films of spinach PS II particles. A) film hydrated with 150 mM Na dithionite in 150 mM Na borate buffer, pH 9.5, and B) film hydrated with pure buffer (no dithionite).
Measuring conditions as in fig. 1.

well as optical controls, will be necessary to determine precisely the contribution of these acceptors and the part of possible photodegradation in the difference spectra.

As in the case of the 'P700[+]' spectra discussed above, a contribution of the peptide backbone cannot be definitely discarded, but if there is indeed such a contribution, it can only involve one or two peptide bonds.

4. CONCLUSIONS

It appears from this work and (**1, 2**) that the lack of large protein conformational changes is a general concept in the primary reactions of both plant and bacterial photosynthesis. Moreover, this study demonstrates that small but specific molecular changes occurring at the pigments and their binding sites upon primary charge separation and stabilization are clearly identified by FTIR difference spectroscopy. In this respect, studies involving model compounds (in progress) and isotope labeling will be necessary to assign each band in the difference spectra to specific molecular bonds in the pigment molecules and the proteins. Alteration of the vibrations of these bonds, which occurs upon separation and stabilization of charges, and which is monitored by FTIR difference spectroscopy, should thus allow a better description at the molecular level of the primary processes of green plant photosynthesis.

REFERENCES

1   Mäntele, W., Nabedryk, E., Tavitian, B.A., Kreutz, W. and Breton, J. (1985) FEBS Lett. 187, 227-232.
2   Nabedryk, E., Mäntele, W., Tavitian, B.A. and Breton, J. (1986) Photochem. Photobiol. 43, 461-465.
3   Tavitian, B.A., Nabedryk, E., Mäntele, W. and Breton, J. (1986) FEBS Lett. 201, 151-157.

CHLOROPHYLL ORGANIZATION IN PHOTOSYSTEM–I REACTION–CENTER OF SPINACH CHLOROPLASTS

ISAMU IKEGAMI[a] and SHIGERU ITOH[b]
[a]Faculty of Pharmaceutical Sciences, Teikyo University, Sagamiko, Kanagawa 199–01 and [b]National Institute for Basic Biology, Okazaki 444 (Japan)

1. INTRODUCTION

P700 is enriched to antenna pigments by the ether-treatment of PS-1 particles[1]. The lowest chl-a/P700 ratio of 8(sometimes 7 or 6) was achieved with ether of higher water saturation, in which 90-95% of the antenna chlorophyll is extracted without appreciable loss of P700 activity (Fig. 1). Further enrichment of P700 was not successful even by repeating the ether extraction. This suggests that the chlorophylls in the P700-enriched particles, including P700, might be in a special site in (or near) PS-1 reaction center.

CD and absorption spectroscopy of the P700-enriched particles in the present studies revealed three dimer-like and two monomer chlorophyll a in (or near) PS-1 reaction center.

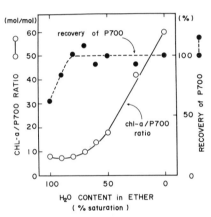

Fig.1 Effect of H_2O-content in ether on the chl-a/P700 ratio and on the recovery of P700

2. MATERIALS AND METHODS

P700-enriched particles which had a chl-a/P700 ratio of eight were prepared as described previously[1,2]. CD spectrum was determined with a JASCO-J200B CD spectrometer. Absorption and difference absorption spectra were determined with a Hitachi model 557 dual-wavelength spectrophoto-meter[3]. Curve analysis of the absorption spectrum was carried out on a Hewlett-Packard computer(HP-9845B) with the program of Mimuro et al[4].

3. RESULTS AND DISCUSSION

3.1. CD spectrum of P700-enriched particles

The solid lines in Fig. 2 show the absorption(A) and CD(B) spectra of the oxidized P700-enriched particles. Three CD bands in the long wavelength region are apparent with peaks at 688 nm(-), 678 nm(+) and 663 nm(-), where (+) and (-) indicate the signs of the CD signal(see also Fig. 3). The two major bands at 678 nm(+) and 688 nm(-) seem to be a degenerate component caused by a dimeric organization. This derivative-shaped CD crosses zero at about 684 nm, indicating that the chlorophyll responsible for this CD signal has an absorption peak at about 684 nm. The presence of chl-a-684 in P700-enriched particles has already been suggested from the excitation spectrum for variable fluorescence from PS-1[2]. The addition of reducing

Biggens, J. (ed.), Progress in Photosynthesis Research, Vol. I. ISBN 90 247 3450 9

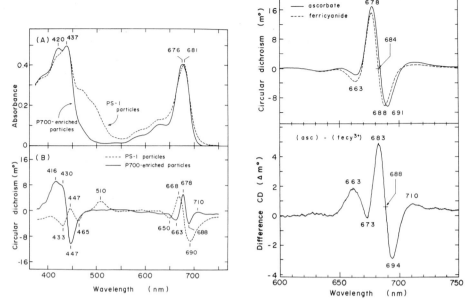

Fig.2 Absorption(A) and CD(B) spectra of the oxidized P700-enriched
particles(chl-a/P700=8) and of PS-1 particles(chl-a/P700=142)

Fig.3 CD spectra obtained in the presence of ferricyanide(1 mM) or
ascorbate(5 mM) and the reduced-minus-oxidized difference CD
spectra(lower part) of P700-enriched particles

agents such as ascorbate or dithionite to the oxidized P700-enriched
particles, induced obvious changes in the CD spectrum(Fig. 3). The
reduced-minus-oxidized difference CD spectrum showed two major bands at
694 nm(−) and 683 nm(+), in addition to a small band at 663 nm(+). The
profile of the difference CD spectrum changed progressively during a
reductive titration with a ferri-ferrocyanide couple. At the low ferro-
to ferricyanide ratio, the band at 663 nm was hardly recognizable, so that
the CD spectrum was composed of only the two peaks at 683 nm(+) and at 694
nm(−). This derivative-shaped CD spectrum seems to be owing to P700. It
has a zero-crossing point of 688 nm, which may correspond to the absorption
maximum of the reduced form of P700. On the other hand, the 663 nm CD
component was spectrally isolated by subtracting the CD signal of P700.
It has a counterpart trough at 678 nm with a zero-crossing point of about
673 nm. The difference CD spectrum, thus, can be decomposed into two
derivative-shaped species; the one with the CD peaks at 694 nm(−) and 683
nm(+) which is due to P700, and the other with minor peaks at 663 nm(+) and
678 nm(−), which may be ascribed to another dimeric chlorophyll locating
near P700. The midpoint potentials determined at 663 and 694 nm were about
400 mV and 420 mV, respectively, each with an one electron reaction.
3.2. Redox-potential-dependent change of the difference absorption
        spectrum of P700-enriched particles
        Fig. 4 shows difference absorption spectra obtained during a reductive,
or an oxidative titration with a ferri-ferrocyanide couple. The spectral
profile around 675 nm changed depending on the redox potential(Fig. 4(B)).
These results suggest that the 675 nm band is a mixture of a satellite band
of P700 and a component showing an absorption change around 675 nm. Fig. 5A

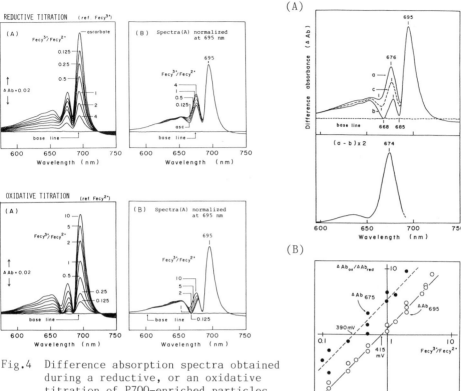

Fig.4 Difference absorption spectra obtained
during a reductive, or an oxidative
titration of P700-enriched particles

Fig.5 (A) Difference absorption spectrum of
P700 with the highest(a), the lowest(b),
or the intermediate(c) satellite band. The lower part, the difference spectrum between curves a and b. (B) Redox potential dependency
of the 695(open circles) and the 675(solid circles) nm bands.

shows the two typical difference spectra, each with the highest(curve a) and
the lowest(curve b) satellite band. The difference spectrum between the
curves a and b has a chlorophyll-like absorption with a peak around 674 nm.
The absorption change of chl-a-674 obtained after subtraction of the
contribution due to P700 gave a midpoint potential of about 390 mV with one
electron reaction (Fig. 5B), while a midpoint potential of P700 was about
25 mV higher than that of chl-a-674. These values of midpoint potentials
were consistent with those obtained by the CD signal changes of the two CD
components, CD694(-)/683(+) and CD678(-)/663(+), respectively. This confirms
that the major CD component(CD694(-)/683(+)) originates from P700 and that
the minor one(CD678(-)/663(+)) originates from chl-a-674. The difference
absorption spectrum mainly due to P700 thus obtained at the low ferro- to
ferricyanide ratio on the reductive titration(curve a in Fig. 5A) shows a
high satellite band at 676 nm with a shallow dip around 662 nm. On the
other hand, the maximum contribution of chl-a-674 reflects a very low
satellite band in the difference absorption spectrum(curve b in Fig. 5A).
These observations suggest that chl-a-674 which is probably a neutral form
in the oxidized state is reduced during the reductive titration giving the

absorption decrease at 674 nm and is reoxidized during the oxidative titration. At the end of the reductive and of the oxidative titration, P700 and chl-a-674 were both fully developped so that the resultant spectra became similar to each other(curve c in Fig. 5A).

3.3 Curve analysis of the absorption spectrum of P700-enriched particles

The best fit to the observed spectrum was obtained with three major gaussians peaking at 669 nm, 675 nm and 684 nm, in addition to three minor gaussians peaking at 650 nm, 660 nm and 698 nm. The P700-enriched particles used here contains only about 8 antenna chlorophyll $\underline{a}$ per one P700. We can classify these eight antenna chlorophylls into four species, i.e., chl-a-660, chl-a-669, chl-a-675 and chl-a-684 with a quantitative ratio of about <1:2:2:2.

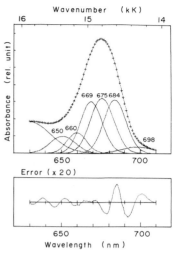

Fig.6 The absorption spectrum of the oxidized P700-enriched particles(cross marks) fitted by the sum of gaussian components(dotted line)

4. CONCLUSION

The results show the presence of three dimer-like chlorophyll $\underline{a}$(P700, chl-a-684 and chl-a-674(5)) and two monomer chlorophyll $\underline{a}$(chl-a-669)(A_0 or/and A_1) in PS-1 reaction center. Their strong resistance to the ether-extraction suggests the close location of one with each other in a special site of the reaction center.

Table Tentative classification of the eight chlorophylls in P700-enriched particle

chl-a/P700=8

chl-a-688	2 chl	CD694(-)/683(+)	dimer	P700 (E_0=420 mV)
chl-a-684	2 chl	CD688(-)/678(+)	dimer	emitter of variable fluorescence(F694)
chl-a-674	2 chl	CD678(+)/663(-)	dimer-like	e^--carrier around P700(?)(E_0=400 mV)
chl-a-669	2 chl	no CD signal	monomer	primary electron acceptor(?)
chl-b-650	<1 chl	no CD signal	monomer	e^--acceptor around A_2(X)(?)

ACKNOWLEDGEMENTS

We thank Drs. M. Mimuro and Y. Fujita (National Institute for Basic Biology(NIBB)) for generously allowing us the use of their curve analysis program and for their valuable discussions. This work was supported in part by grants from Ministry of Education, Science and Culture in Japan, and by Co-operative Research Program of NIBB in Japan.

REFERENCES
1 Ikegami, I. and Katoh, S. (1975) Biochim. Biophys. Acta 376, 588-592
2 Ikegami, I. (1976) Biochim. Biophys. Acta 449, 245-258
3 Ikegami, I. and Ke, B. (1984) Biochim. Biophys. Acta 764, 70-79
4 Mimuro, M., Murakami, A. and Fujita, Y. (1982) Arch. Biochem. Biophys. 215, 266-273

BOUND QUINONES IN THE REACTION CENTRES OF BACTERIA AND PLANTS

M.C.W. Evans
Department of Botany and Microbiology, University College
London, London WC1E 6BT, U.K.

INTRODUCTION
   Quinones have classically been defined as the first stable
electron acceptors in the reaction centres of purple bacteria
and photosystem 2 of oxygenic organisms. In photosystem 1
and green bacteria this role is taken by iron-sulphur centres,
however there have been a number of reports of the association
of a naphthoquinone, vitamin K1, with reaction centre preparations (1,2). More recently two intermediary electron carriers
have been identified in photosystem 1, A_1 and A_0 (3,4). One
of these, A_0, is probably a chlorophyll (5). There is now
strong evidence that the other, A_1, is a quinone. It does
not appear to be coupled to an iron atom as are the quinones
in the other reaction centres. If it is a carrier in the
electron transport chain it must operate at an extremely low
potential as it is not reduced until all of the iron sulphur
centres are reduced. Our recent results on the properties of
this photosystem 1 quinone are reported elsewhere in this volume (6).
   In purple bacteria and photosystem 2 there are strong analogies in the properties of the electron acceptor complex.
Both have a pheophytin as the intermediary carrier between
the reaction centre chlorophylls and the stable acceptors.
The stable acceptors are quinones, coupled to an iron atom,
which are bound to sites which stabilise the semiquinone state
following one electron reduction by the reaction centre. At
least two quinones are required for electron transfer, one Qa,
is tightly bound and cycles between the oxidised and semiquinone state. The second, Qb, has a very stable semiquinone
and acts as a gate between the one electron transfers of the
reaction centre and the two electron transfer to the quinone
pool. Qb is thought to bind transiently to the reaction
centre.
                   P ---→ Pheophytin ---→ Qa ---→ Qb

BACTERIAL CHROMATOPHORES AND PHOTOSYSTEM 2
   Experiments on the redox relationships of the components of
the reaction centre in both plants and bacteria suggest that
the acceptor system is probably more complex than this model.
In photosystem 2 many titrations of the redox dependence of
fluorescence yield have shown the presence of two components
Qh, Em = 0 mV., and Ql, Em = -250 mV. More recently we have
shown that these steps are observed in the reduction of the

Biggens, J. (ed.), Progress in Photosynthesis Research, Vol. I. ISBN 90 247 3450 9
© *1987 Martinus Nijhoff Publishers, Dordrecht. Printed in the Netherlands.*

iron-quinone complex and the forced reduction of the pheophytin by 200 K illumination, Fig. 1 (7). In purple bacteria
these titrations give very complex results (8,9). These titrations could be interpreted as either showing the presence
of an additional component or changes in the magnetic interactions of the components following double reduction of Qb.
It should be possible to distinguish between these alternatives by carrying out titrations in the presence of o-phenanthroline which is thought to displace Qb. Titration of Rhodopseudomonas viridis chromatophores in the presence of o-
phenanthroline results in the loss of the characteristic
spectrum of Qb, but two steps are seen on the titration of Qa
and the ability to photoreduce the pheophytin at 200 K, Fig.
2 (10). The interaction between chemically reduced Qb and
photoreduced Qa which results in loss of the epr signal also
disappears and the chemical and light induced signals are
additive.

   These results in both photosystem 2 and purple bacteria
support the proposal that there are two tightly bound quinones
(Qa) and a transiently bound quinone (Qb) (11). However there
is a large body of evidence showing that in isolated reaction
centres only one Qa molecule is present and that primary electron transfer to Qa occurs in crystals with a single Qa. The
redox properties of isolated reaction centres have not been
extensively investigated because lauryldiethylamine oxide
(LDAO) which is used to prepare the reaction centres is an
oxidising agent which interferes with titrations. We have now
prepared reaction centres removing the LDAO by reduction with
sodium dithionite followed by solubilisation in Triton X-100
for use in titrations. These experiments are incomplete and
the results presented here are preliminary.

ISOLATED BACTERIAL REACTION CENTRES
   It has in fact proved difficult to obtain reaction centres
which contain only one quinone and it has also become apparent
that a very wide range of quinones will form stable semiquinone-iron complexes when bound to the reaction centre. The
photochemical properties of the reaction centres are also
altered as compared to chromatophores, in that there is apparently electron donation by the high potential cytochrome c
in some centres at low temperature (6 K). Fig. 3 shows a
redox titration of the iron-quinone and the photoreduction at
200 K of the pheophytin in a reaction centre preparation. The
iron quinone shows essentially a single wave with Em = -150 mV.
The photoreduction of the pheophytin however shows two waves,
one at Em = 0 mV, and the other at Em = -150 mV. At 0 mV the
iron quinone is photoreduced by donation from the cytochromes.
The experiment suggests that more than one electron is available at 200 K to reduce the acceptor complex, however it is
not clear why there should be two steps unless both pheophytin
molecules can become reduced if Qa is chemically reduced first.
Reconstitution with low concentrations of ubiquinone 10 (10 µM)
results in reconstitution of the Qb signal and the apparent
movement of the Qa wave to Em = -250 mV. The pheophytin induction shows two waves one paralleling Qb reduction, the other

the low potential Qa wave. Addition of o-phenanthroline and
UQ10 results in a titration the same as in the unsupplemented
reaction centres, i.e. two Qa waves are not seen as they are
in the chromatophore experiments described above. If the
reaction centres are supplemented with vitamin K1 (phytyl
naphthoquinone) very little Qb like spectrum is seen, however
two steps are observed on the Qa like spectrum with Em =
-100 mV and -250 mV. The ability to photoreduce the pheophytin at 200 K parallels these two steps. These steps are still
seen when o-phenanthroline is added after the vitamin K1.

CONCLUSIONS

A simple model of the electron acceptor complex in photosystem 2 and purple bacteria suggests that electrons from the
reaction centre chlorophyll pass through a pheophytin to a
tightly bound quinone Qa and then to the transiently bound
gating quinone Qb. This model is based on kinetic measurements, but not on direct observation of redox changes in the
quinones. Crystals of reaction centres used for x-ray
crystallographic studies contain only one quinone leading to
further speculation on mechanisms involving only one quinone.
However redox studies of both chromatophores and photosystem
2 preparations suggest that there are more electron acceptors
present in the reaction centres. They suggest that there may
in fact be two "Qa" type quinones present in both types of
reaction centre. The preliminary experiments with purified
bacterial reaction centres described here show that while
preparations can be made which have a single Qa there is probably a second naphthoquinone binding site. The quinone bound
to this site is not displaced by o-phenanthroline and has
properties similar to Qa. Ubiquinone bound to the reaction
centre behaves like Qb and is displaced by o-phenanthroline.
Experiments of the type shown here do not show if the components detected are involved in the normal electron transport
chain. They show that the components exist and that kinetic
experiments must be carried out under conditions which allow
the detection of all the components which may be involved.

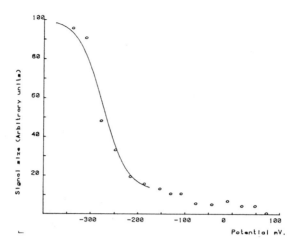

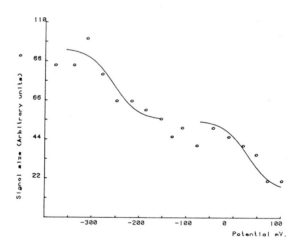

FIGURE 1. Redox titration of formate washed Photosystem 2 particles
Top. The split pheophytin radical induced by 200K illumination.
Bottom. The iron-quinone signal in samples prepared in the dark.
Epr spectra were recorded at 6K.

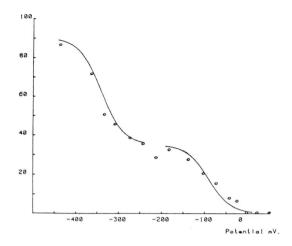

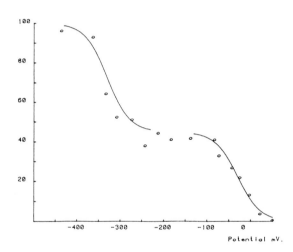

FIGURE 2.   Redox titration of <u>Rhodopseudomonas</u> <u>viridis</u> chromatophores in the
presence of 10mM o-phenanthroline.
Top. The iron-quinone signal in samples prepared in the dark.
Bottom. The split pheophytin signal induced by 200K illumination.

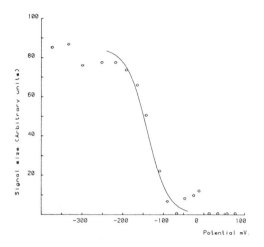

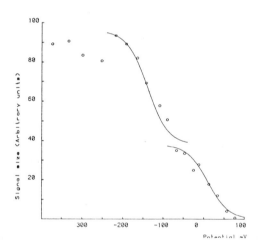

FIGURE 3.   Redox titration of purified <u>Rhodopseudomonas</u> <u>viridis</u> reaction
centres.
Top. The iron-quinone signal in samples prepared in the dark.
Bottom. The split pheophytin radical signal induced by 200K
illumination.

REFERENCES
1 Lichtenthaler, H.K. (1969) Progress in Photosynthesis Research, 1, 304-314
2 Thornber, J.P., Alberte, R.S., Hunter, F.A., Shiozawa, J.A. and Kan, K.S.
  (1977) Brookhaven Symposia in Biology, 28,132
3 Bonnerjea, J. and Evans, M.C.W. (1982) FEBS Letts. 148, 313-316
4 Gast, P., Swarthoff, T., Ebskamp, F.C.R. and Hoff, A.J. (1983) Biochim.
  Biophys. Acta 722, 168-175
5 Mansfield, R.W. and Evans, M.C.W. (1985) FEBS Letts. 190, 237-241
6 Mansfield, R.W. and Evans, M.C.W. This volume
7 Evans, M.C.W. and Ford, R.C. (1986) FEBS Letts. 195, 290-294
8 Rutherford, A.W., Heathcote, P. and Evans, M.C.W. (1979) Biochem. J. 182,
  515-523
9 Rutherford, A.W. and Evans, M.C.W. (1980) FEBS Letts. 110, 257-261
10 Evans, M.C.W. and Ford, R.C. (1986) Biochem. Soc. Trans. 14, 12-15
11 Evans, M.C.W. (1985) Physiol. Veg. 23, 563-569

HOW CLOSE IS THE ANALOGY BETWEEN THE REACTION CENTRE OF PSII AND THAT
OF PURPLE BACTERIA? 2. THE ELECTRON ACCEPTOR SIDE

A.W. RUTHERFORD, SERVICE BIOPHYSIQUE, DEPARTEMENT DE BIOLOGIE, C.E.N.
SACLAY, 91191 GIF SUR YVETTE, CEDEX, FRANCE.

1.   INTRODUCTION

In a recent paper, I discussed the EPR studies of the PSII
reaction centre triplet in comparison to that of the purple bacterial
reaction centre (1) . Here, I shall briefly compare the EPR data from
the electron acceptor complex in these two systems, pointing out the
striking similarities and one or two interesting and perhaps
functionally significant differences.

For several years it was apparent that the quinone complex of
PSII and purple bacteria were functionally similar, both having a
quinone acceptor, Q_A, which functions as a one electron acceptor
and also a secondary quinone acceptor, Q_B, which acts as a two
electron gate. In addition, evidence has gradually accumulated
indicating that pheophytin plays a role as an electron acceptor
functioning prior to Q_A in PSII (2) . Such a role for bacteriopheophytin in purple bacteria was previously well established (3) .
The most striking evidence for the closeness of the structural analogy
has come however from EPR measurements of the electron acceptors in
PSII.

2.   EPR STUDIES OF THE ACCEPTOR SIDE

2.1 The split EPR signal from the pheophytin anion

The first PSII EPR signal pointing to a close structural
similarity between PSII and bacteria was discovered by
Klimov et al (4) . Using reducing conditions and continuous
illumination at 200K, an EPR signal split by approximately
50G around g=2.002 could be generated in PSII particles. The
spectral characteristics of the signal and the conditions of
its formation were almost identical to a signal well
characterized in some species of purple bacteria (5) . The
signal in bacteria was attributed to a bacteriopheophytin
anion which magnetically interacted with the primary
semiquinone-iron complex, $Q_A^- Fe^{2+}$ (3) . The observation of
the split signal in PSII indicated not only that pheophytin
was present as an electron acceptor prior to Q_A but also
that Q_A^- was associated with a ferrous ion.

Biggens, J. (ed.), Progress in Photosynthesis Research, Vol. I. ISBN 90 247 3450 9
© *1987 Martinus Nijhoff Publishers, Dordrecht. Printed in the Netherlands.*

2.2 The semiquinone-iron complex from $Q_A^-\cdot Fe^{2+}$

The semiquinone-iron complex in bacteria gives rise to an
EPR signal at g=1.82. The signal arises from an
antiferromagnetic interaction between the semiquinone and
the high spin ferrous ion of the reaction centre (6). After
the discovery of the split Ph^- signal in PSII (4) and the
demonstration that both iron and quinone were required for
formation of this signal (7) , a $Q_A^-\cdot Fe^{2+}$ signal was looked
for in PSII. A signal close to g=1.82 was observed in PSII
preparations from Chlamydomonas (8) and from spinach (9) and
this was attributed to $Q_A^-\cdot Fe^{2+}$. The poor resolution of these
signals was due, not only to the greater lability of the
iron in PSII but also to the fact that the $Q_A^-\cdot Fe^{2+}$ gave rise
to an alternative EPR signal (10) . The new signal at g=1.9,
although unlike the "normal" g=1.82 signal found in most
species of bacteria, was very similar to the $Q_A^-\cdot Fe^{2+}$ signal
previously reported in R. rubrum (11). The two EPR signals
were found to be interconvertable and were attributed to two
specific conformations. In most preparations a mixture of
both forms is present indicating a heterogenous population
of centres (10). The interconversion of the two forms can be
achieved by changing the pH of the medium, with the g=1.9
form being predominant at high pH. If the g=1.9 form
represents an unprotonated state, it might be expected that
Q_A is more difficult to reduce in this form. This structural
heterogeneity could be related to functional heterogeneity
monitored by redox titrations (9,12) and by low temperature
charge stability (9).

2.3 The effects of inhibitory treatments on $Q_A^-\cdot Fe^{2+}$

The $Q_A^-\cdot Fe^{2+}$ signal in both PSII and purple bacteria is
affected by the binding of herbicides which block electron
transfer from Q_A^- to Q_B. In bacteria, ortho-phenanthroline
inhibits electron transfer by binding in the Q_B site (13) .
This binding is sensed by the $Q_A^-\cdot Fe^{2+}$ EPR signal which
becomes modified. In Rhb. sphaeroides, the signal at g=1.82
becomes narrower and consequently has a greater amplitude
(6). In R. rubrum, the signal at g=1.9 broadens considerably
(14). In PSII, a range of herbicides exert a range of
effects (15). DCMU increases the amplitude of the g=1.82
form by a factor of approximately two, while dinoseb, a
phenolic herbicide, markedly decreases the amplitude of the
g=1.9 form, and narrows the g=1.82 form (15). Dinoseb has no
effect on the $Q_A^-\cdot Fe^{2+}$ signal in R. rubrum despite this
species having the g=1.9-type signal from $Q_A^-\cdot Fe^{2+}$ (14).

Despite the sensitivity of the $Q_A^-\cdot Fe^{2+}$ signals to herbicide
binding, the signal (in this case the g=1.9 form
predominated) was found to be largely unaffected by a
mutation resulting in herbicide resistance in PSII (16).

Formate ions inhibits electron transfer in the region of the
quinone acceptors in PSII, probably by replacing a
bicarbonate ion from a specific site (17). When the effect
of formate was monitored by EPR, we were astounded to find
that the amplitude of the g=1.82 signal had increased by a
factor of twelve (16). Clearly the formate drastically
perturbs the semiquinone-iron interaction. Since R. rubrum
exhibits the g=1.9 form of $Q_A^-Fe^{2+}$, the effect of formate
was investigated in chromatophores of this species. No
effect was observed (14).

2.4 The orientation dependence of $Q_A^-Fe^{2+}$

The $Q_A^-Fe^{2+}$ is highly anisotropic and in bacteria it has
been shown that g=1.82 signal had a maximum amplitude when
two dimensionally ordered membranes were oriented
perpendicular to the magnetic field. The high field trough
was maximal when the oriented membranes were parallel to the
magnetic field. This was found to be the case in
Rhb. sphaeroides (18) and also in Rps. viridis   (19 unpublished), and Chloroflexus (unpublished). Both the g=1.9
and the g=1.82 forms of $Q_A^-Fe^{2+}$ in PSII show an orientation
dependence like that of the g=1.82 signal in bacteria (20).

2.5 The double reduction of Q_A

Unlike normal quinones, Q_A functions as a one electron
acceptor. Presumably, the tight binding site in the reaction
centre prevents the semiquinone from undergoing a further
reduction step. However, under extreme conditions it was
shown that $Q_A^-Fe^{2+}$ could be further reduced chemically (21)
or photochemically (22) in purple bacteria. In PSII the
absence of the splitting in the pheophytin EPR signal, when
the pheophytin anion was photoaccumulated at room
temperature, was proposed to be due to a low quantum yield,
second reduction of Q_A (4). This was confirmed by the direct
observation of the disappearance of both g=1.9 and g=1.82
signals from $Q_A^-Fe^{2+}$ under these conditions (10). As in the
bacterial system, having been double reduced, the $Q_A^-Fe^{2+}$,
could only be reformed after complete oxidation of the
quinone. This probably indicates a modification of the Q_A
site due to the double reduction, possibly leading to loss
of the quinol from the site.

2.6 Redox titrations of Q_A/Q_A^-

Redox titrations of Q_A/Q_A^- in both bacteria and PSII are
comparable. In some reports, a single pH-dependent redox
curve with values scattered widely around 0mV at pH 7.0 was
obtained but the Em at the pK was close to -160mV for most
species (23,24). Direct measurements of the EPR signal from
QA^-Fe^{2+} have given similar results (25,26). In both systems

however, two-step titration curves have also been reported
using a range of measurement techniques including EPR
(12,27). The significance of the two steps is not clear and,
although it has led to the proposal that an extra acceptor
may be present (see e.g. 28), it seems more likely that this
is due to heterogeneity where, in some centres, Q_A either is
less accessible (i.e. a titration artifact) or has a truly
different redox potential. True structural heterogeneity has
been demonstrated for $Q_A^{-} Fe^{2+}$ (10) (see section 2.2. above).

2.7 $Q_B^{-} Fe^{2+}$

When a single electron is present on Q_B, an EPR signal
similar to that from $QA^{-}Fe^{2+}$ was found in Rhb. sphaeroides
(29). Rps. viridis (30) and Chromatium vinosum (31). Redox
titrations of the $Q_B^{-} Fe^{2+}$ signal showed that $Q_B^{-}Fe^{2+}$ could
be formed stably before becoming double reduced (21).
Difficulties have been encountered in studying $QB^{-}Fe^{2+}$ in
PSII due to the lability of Q_B in the PSII preparations
normally used. Nevertheless a $Q_B^{-}Fe^{2+}$ signal was reported in
PSII under conditions in which $Q_B2^{-}Fe^{2+}$ was expected to be
formed (i.e. 77k hv to form $QA^{-}Fe^{2+}$, followed by thawing in
the dark to form $Q_B^{-} Fe^{2+}$) (31,33). In addition to a feature
at g=1.82, the signal, however, included a major feature at
a g-value (g≃1.94) higher than that seen previously for
semiquinone-iron complexes.

Having found that the $Q_A^{-}Fe^{2+}$ signal normally present in
PSII was unlike that found in most species of bacteria but
was similar to that found in R. rubrum, we looked for a
$Q_B^{-}Fe^{2+}$ signal in this species of bacteria. This signal was
found and characterised and it turn out to be remarkably
similar to the $Q_B^{-}Fe^{2+}$ signal in PSII, thus strengthening
the assignment of the signal in PSII (14). R. rubrum clearly
wins the PSII-look-alike competition in terms of its
semiquinone-iron EPR characteristics.

2.8 The redox activity of the iron

It was shown by Mössbauer spectroscopy that the redox state
of the iron did not change when untreated bacterial reaction
centres were reduced (34). Similar data was obtained in PSII
centres (35) however under oxidizing conditions it was found
that the iron could be oxidized to form Fe^{3+}. EPR signals
arising from the oxidized iron were found at g~8 and g~6
(36). The iron could be photoreduced by continuous
illumination at low temperature (36) or by a saturating
flash at room temperature (33). The Fe^{2+}/Fe^{3+} couple could
be titrated with a pH-dependent midpoint with an $Em_{pH\,7.0}$ of
400mV (36). The work of Petrouleas and Diner (36) provided
a tidy explanation for the long standing PSII anomaly, Q_{400},
the high potential acceptor. This acceptor had been

discovered years earlier by measurements of fluorescence and had been the cause of some doubts and reservations for the bacterial analogists. With this work, however, Q_{400} was painlessly integrated into the bacterial analogy model.

Iron oxidation in PSII has also been demonstrated when certain quinones are used as electron acceptors (33). The semiquinone formed after a flash is unstable and can oxidize the iron to form the stable, fully reduced quinol. This reaction, a photoreductant-induced oxidation, gives iron oxidation after odd numbered flashes and the iron is photoreduced on even numbered flashes (33).

The two EPR signals close to g=8 and g=6.0 are thought to represent iron in different environments (36), again indicating a heterogeneity of centres. This probably corresponds to the same conformational heterogeneity associated with the g=1.82 and g=1.9 forms the $Q_A^- Fe^{2+}$ signal (10).

Oxidation of the Fe^{3+} in purple bacteria has not been reported and attempts in this laboratory to observe signals at g-values similar to those seen for Fe^{3+} in PSII have so far been unsuccessful in Rhb. sphaeroides and Rps. viridis (A.W. Rutherford and I., Agalidis, unpublished) and even in R. rubrum (14).

3.   DISCUSSION

     From the EPR data, briefly reviewed above, and from other lines of evidence (37), the chemical nature of the electron acceptor side of PSII appears to be extremely similar to that of purple bacteria, being made up of a pheophytin, two quinones and an iron atom. The electron transfer reactions between these components in PSII are also comparable to those in purple bacteria (2). The magnetic interactions between these components are extremely sensitive to environmental differences and yet they are vitrually identical to those found in the bacterial reaction centre. Therefore, it is clear that the structural relationships between these components are common to both systems. For this reason also, it is difficult to imagine that extra acceptors function between pheophytin and Q_A. The similar orientation dependence of the $Q_A^- Fe^{2+}$ signal in both systems indicates that the geometry of the semiquinone-iron interaction relative to the membrane is also the same in both systems. It is likely then, that the structural model for the bacterial reaction centre recently obtained by X-ray crystallography (38) can be directly applied to the acceptor side of PSII. The application of this analogy has already provided useful insights particularly with regard to sequence analogies between the bacterial reaction centre proteins and the D1 and D2 polypeptides of PSII (38,39,40).

Despite these similarities there are a number of properties of
the PSII acceptor complex which are not in common with those of
purple bacteria: 1) formate binding (16), 2) binding of a wider range
of herbicides (39), 3) a more labile iron (41), 4) an easily
oxidizable iron atom (33,36). Some or all of these phenomena could be
explained by the iron being more exposed to the outside in PSII. In
Rps. viridis the iron is liganded to 4 histidines and to a glutamate
(13). Although the histidines appear to be conserved in the analogous
PSII proteins, D_1 and D_2, (38-40), the glutamate is absent (13).
Free or exchangeable ligands to the iron could explain the different
properties of the quinone-iron complexes in the two systems.

ACKNOWLEDGEMENTS

I would like to thank my co-authors on papers cited from this
laboratory: Drs.C. Beijer, P. Heathcote, P. Mathis, W.F.J.
Vermaas and J.-L., Zimmermann. In addition I have benefited from
lengthy discussions with Drs. B.A. Diner, G.W. Brudwig and H.
Michel. Thanks also to Agnès Rutherford for typing the
manuscript. I am supported by the CNRS.

REFERENCES

1   Rutherford, A.W. (1986) Biochem. Soc. Trans. 14, 15-17.
2   Crofts, A.R. and Wraight, C.A. (1983) Biochim. Biophys. Acta 726,
     149-183
3   Dutton, P.L., Prince, R.C. and Tiede, D.M. (1978) Photochem.
     Photobiol. 28, 939-949
4   Klimov, V.V., Dolan, E., Ke, B., (1980) FEBS Lett 112, 97-100
5   Tiede, D.M., Prince, R.C., Reed, G.H. and Dutton, P.L. (1976)
     FEBS Lett 65, 301-304
6   Butler, W.F., Calvo, R., Fredkin, D.R., Isaacson, R.A., Okamura,
     M.Y. and Feher, G. (1984), Biophys. J. 45, 947-973.
7   Klimov, V.V., Dolan, E., Shaw, E.R. and Ke, B. (1980) Proc. Natl.
     Acad. Sci. USA 77, 7227-7231.
8   Nugent, J.H.A., Diner, B.A. and Evans, M.C.W. (1981) FEBS Lett
     124, 241-244.
9   Rutherford, A.W. and Mathis, P. (1983) FEBS Lett. 154, 328-334
10   Rutherford, A.W. and Zimmermann, J.-L. (1984) Biochim. Biophys.
     Acta 767, 168-175
11   Prince, R.C. and Thornber, J.P. (1977) FEBS Lett 81, 233-237
12   Evans, M.C.W. and Ford, R.C. (1986) FEBS Lett 195, 290-294
13   Michel, H., Epp, O. and Deisenhofer, J. (1986) EMBO J. in press
14   Beijer, C. and Rutherford, A.W. (1986) Biochim. Biophys. Acta
     submitted.
15   Rutherford, A.W., Zimmermann, J.-L. and Mathis, P. (1984) FEBS
     Lett 165, 156-162
16   Vermaas, W.F.J. and Rutherford, A.W. (1984) FEBS Lett 175,
     243-248

17 Vermaas, W.F.J. and Govindjee (1981) Photochem. Photobiol. 34, 775-793

18 Tiede, D.M. and Dutton, P.L. (1981) Biochim. Biophys. Acta 637, 278-290

19 Dismukes, G.C., Frank, H.A., Friesner, R. and Sauer, K. (1984) Biochim. Biophys. Acta 764, 253-271.

20 Rutherford, A.W. (1985) Biochim. Biophys. Acta 807, 189-201.

21 Rutherford, A.W. and Evans, M.C.W., (1980) FEBS Lett 110, 257-261;

22 Okamura, M.Y., Isaacson, R.A. and Feher, G. (1979) Biochim. Biophys. Acta 546, 394-417.

23 Prince, R.C. and Dutton, P.L. (1976) Arch. Biochem. Biophys. 172, 329-334

24 Knaff, D.B. (1975) FEBS Lett 60, 331-335

25 Evans, M.C.W., Nugent, J.H.A., Tilling, L.A. and Atkinson, Y.E. (1982) FEBS Lett 145, 176-178

26 Evans, M.C.W., Lord, A.V. and Reeves, S.G. (1974) Biochem. J. 138, 177-183.

27 Rutherford, A.W., Heathcote, P. and Evans, M.C.W. (1979) Biochem. J. 182, 515-523

28 Evans, M.C.W., Atkinson, Y.E., Ford, R.C. (1985) Biochem. Biophys. Acta 806, 247-254.

29 Wraight, C.A. (1978) FEBS Lett 93, 283-288

30 Rutherford, A.W. and Evans, M.C.W.(1979) FEBS Lett 104, 227-230.

31 Heathcote, P. and Rutherford, A.W. (1987) these volumes.

32 Rutherford, A.W., Zimmermann, J.-L. and Mathis, P. (1984) in Advances in Photosynthesis Research (ed. Sybesma, C.) Vol. 1 pp. 445-448 Nijhoff/W. Junk, The Hague.

33 Zimmermann, J.-L. and Rutherford, A.W. (1986) Biochim. Biophys. Acta in press.

34 Petrouleas, V. and Diner, B.A. (1982) FEBS Let 147, 111-114

35 Boso, B., Debrunner, P., Okamura, M.Y. and Feher, G. (1981) Biochim. Biophys. Acta 638, 18 73-177

36 Petrouleas, V. and Diner, B.A. (1986) Biochim. Biophys Acta 849, 264-275.

37 Diner, B.A. (1986) in Encyclo. Plant. Physiol. Photosynthesis III (eds. Staehelin L.A. and Arntzen C.J.) pp 422-436 Springer-Verlag, Berlin.

38 Deisenhofer, J., Epp, O., Miki, K., Huber, R. and Michel, H. (1985) Nature 318, 618-624

39 Michel, H., Weyer, K.A., Gruenberg, H. Dunger, I. Oesterheldt, D. and Lottspeich, F. (1986) EMBO J. 5 1149-1158

40 Trebst, A. and Draber, W. (1986) Photosynth. Res. in press

41 Rutherford, A.W. (1983) in The oxygen evolving system of photosynthesis (Inoue, Y. et al eds) pp 63-69 Academic Press

DEPLETION AND RECONSTITUTION OF THE QUINONE AT THE Q_B SITE IN
PHOTOSYSTEM II: A THERMOLUMINESCENCE STUDY

T. Wydrzynski[§] and Y. Inoue
Solar Energy Research Group, RIKEN, Wako, Saitama, 351-01 JAPAN

INTRODUCTION: The standard protocol to remove quinones from thylakoid
membranes by extraction of lyophilized samples with dry organic solvents
leads to the loss of not only the intersystem plastoquinone (PQ) pool and
the secondary quinone acceptor (Q_B), but also the primary quinone acceptor
(Q_A) and a quinone probably functioning on the oxidizing side of Photosystem II (PSII). In this communication we report a relatively mild
heptane/isobutanol extraction (HIE) procedure of aqueous suspensions of
thylakoid membranes which removes only the functional PQ pool and the quinone at the Q_B site, without severe damage to Q_A or the oxidizing side of
PSII. Such HIE samples can be reconstituted in Q_B with simple quinone
molecules, but with modified characteristics as determined by thermoluminescence measurements.

EXPERIMENTAL: Thylakoid membrane samples were isolated from spinach by
standard procedures and suspended at 300 ug Chl/ml in a buffer medium containing 20 mM HEPES (pH 7.5), 400 mM sucrose, 10 mM NaCl, 5 mM $MgCl_2$, 30%
glycerol and 3% (v/v) isobutanol. Three parts heptane was added to one
part of the thylakoid suspension (usually 3 ml) in a 50 ml capped tube and
shaken vigorously on a rotory shaker (300 rpm) in the dark at room temperature for 45 min. After extraction, the pale yellow/green organic phase
was separated from the sample suspension. The extracted (HIE) samples
were then illuminated (unless otherwise noted) for 1 min in continuous
light and then allowed to dark adapt for 7-10 min at room temperature before being stored on ice before measurements. Freshly prepared samples
were used in all experiments. For extraction of lyophilized samples
and reconstitution with plastoquinone A (PLQ-A), the protocol of Ref. 1
was used. PLQ-A was kindly prepared by H. Koike. The protocols for the
thermoluminescence, fluorescence induction and flash O_2 measurements are
given in Ref. 2.

DEPLETION OF THE QUINONE AT THE Q_B SITE: The integrated area over the
msec range Chl a fluorescence induction curve can be used to estimate the
pool size of the intersystem electron acceptors. For a normal dark-adapted thylakoid sample the area is usually about 10-14 times greater compared
to a sample containing DCMU, which blocks electron flow after Q_A. This is
usually interpreted to indicate that in addition to Q_A and Q_B, roughly 5-7
PQ molecules function to accept electrons in the intersystem pool (eg. see
Ref. 3), although the membranes may contain considerably more PQ. The
Table in Fig. 1 shows the loss of the relative integrated area over the
fluorescence induction curve divided by the area for a sample containing
DCMU under different extraction conditions. Extraction with only heptane
(HE) reduces the area but the sample does not reach the DCMU limit. How-
ever, if 3% isobutanol is included during the extraction in the dark (HIE),

[§]CURRENT ADDRESS: Chalmers Institute of Technology, Department of
Biochemistry and Biophysics, S-412-96 Göteborg, SWEDEN

Biggens, J. (ed.), Progress in Photosynthesis Research, Vol. I. ISBN 90 247 3450 9
© *1987 Martinus Nijhoff Publishers, Dordrecht. Printed in the Netherlands.*

SAMPLE	$\dfrac{\text{FLUORESCENCE AREA}}{\text{FLUORESCENCE AREA + DCMU}}$
Control	10.8
+DCMU	1.0
HE	4.7
HIE (dark)	2.2
HIE (illuminated)	1.1

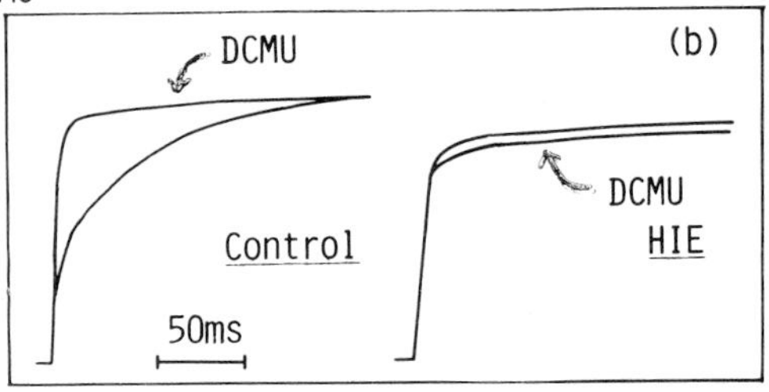

FIG. 1   Fluorescence induction measurements of control, heptane extracted (HE), and heptane/isobutanol extracted (HIE) thylakoid samples.

the area is reduced to about twice the DCMU limit. Upon illumination of
the HIE sample with 1 min continuous light and a subsequent dark-adapta-
tion for 10 min, the area reaches the DCMU limit. These results indicate
that the combined extraction/preillumination treatment leads to the loss
of the functional PQ pool. Fig. 1 shows the fluorescence induction
tracings for the control and illuminated HIE samples ± DCMU.
        Thermoluminescence (TL) techniques are used to measure the inherent
probability for charge recombination between the electrons and holes stabilized on the reducing and oxidizing sides of PSII (Ref. 4). In uninhibited thylakoid samples, a major flash-induced TL band occurs with a peak
temperature position of about 35 C (ie. the B-band), which has been analyzed in terms of $S_2Q_B^-$ charge recombination. In DCMU inhibited samples, the
TL band shifts to a lower peak temperature position at about 5-10 C (ie.
the D-band), which has been analyzed in terms of $S_2Q_A^-$ charge recombination.
In Fig. 2, Part a compares the flash-induced TL bands between the illuminated HIE sample and the control ± DCMU. The TL band of the HIE sample
is virtually identical in peak position and amplitude with the TL band of
the control + DCMU. Addition of DCMU to the HIE sample has no further
effect on the TL. These results, thus, indicate that Q_A remains intact,
but that Q_B^- no longer functions in HIE samples.

RECONSTITUTION OF THE QUINONE AT THE Q_B SITE: The remainder of Fig. 2
shows the flash-induced TL bands of extracted samples in the presence of
various quinones ± DCMU. Since high concentrations of quinones quench the
overall TL yield, in Fig. 2 quinone concentrations were used in which this
quenching effect was minimal, as evidenced by the amplitudes of the TL
bands in the presence of DCMU. Part b shows that reduced HQ has no effect
on the TL band of the HIE sample, but that oxidized BQ generates a TL band
with a peak temperature near the position of the control B-band. However,
the reconstitution is not complete at the BQ concentration used and the TL
yield of the reconstituted band is much lower than can be accounted by the
general quenching by the quinone on the TL in the presence of DCMU. Part
c shows that MMQ produces an effect similar to BQ, but that DMBQ generates
a modified band with a peak temperature position near 22 C, intermediate
between the control D- and B-bands and with little loss in the TL yield.
Part d shows that TMBQ has little effect on the TL, except perhaps for a
small shift in the peak position to a higher temperature, while DCBQ generates a broad band around 20 C with a low TL yield. Part e shows that
PhBQ restores a TL band at the same position as the control B-band, but
again with a low TL yield, and that FeCN has little effect on the TL of
the HIE sample.

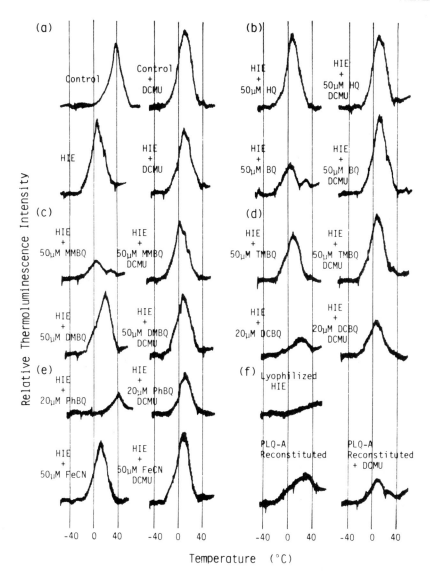

FIG. 2   Flash-induced thermoluminescence measurements of control, extracted and variously reconstituted thylakoid samples. HIE=heptane/iso-
butanol extracted samples, DCMU=3-(3,4-dichlorophenyl)-1,1-dimethylurea,
HQ=p-hydroquinone, BQ=p-benzoquinone, MMQ=monomethyl-p-benzoquinone, DMQ=
2,6-dimethyl-p-benzoquinone, TMQ=tetramethyl-p-benzoquinone, DCBQ=2,6-
dichloro-p-benzoquinone, PhBQ=phenyl-p-benzoquinone, PLQ-A=plastoquinone
A, FeCN=potassium ferricyanide

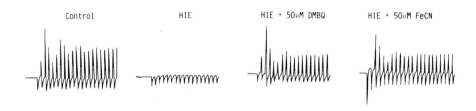

Control   HIE   HIE + 50µM DMBQ   HIE + 50µM FeCN

FIG. 3 O_2 flash yield measurements of control, HIE, and HIE samples restored with DMBQ and FeCN. Samples were allowed to equilibrate on the electrode for 45 min in the dark to compensate for interferences by the excess electron acceptors.

Since PLQ-A can not be simply added to aqueous samples, we had to resort to lyophilized samples in order to reconstitute with native PQ. In Part f, lyophilized samples extracted with heptane show complete loss of the TL in the normal range of the D- and B-bands, consistent with the loss of Q_A and the oxidizing side quinone by this method. Upon reconstitution with PLQ-A, a broad band is generated with a peak position near the normal B-band maximum. This broad band, however, is probably multicomponent and indicates a range of modified quinone binding sites upon reconstitution of extracted lyophilized samples.

Fig. 3 shows the O_2 flash yield measurements of control, HIE, and a representative reconstituted HIE sample. All O_2 activity is lost upon extraction, but is restored to about 75% of the control upon reconstitution with DMBQ. This indicates that the HIE extraction procedure does not severely damage the oxidizing side of PSII. Surprisingly, FeCN also restores O_2 activity, although it does not affect the TL of the HIE sample. Apparently, FeCN can act as an electron acceptor from Q_A when the Q_B site is depleted of the native quinone.

CONCLUSIONS: The functional PQ pool and the quinone at the Q_B site can be selectively removed from thylakoid samples by using a simple heptane/isobutanol extraction procedure on aqueous suspensions of samples, with little damage to the Q_A site or the oxidizing side of PSII. A DCMU sensitive Q_B site can be restored after extraction upon the addition of simple quinone molecules, but with modified properties as determined by thermoluminescence measurements. In a subsequent paper we will report on the modified kinetic and thermodynamic properties of the reconstituted Q_B sites.

REFERENCES:
1 Sadewasser, D.A. and Dilley, R.A. (1978) Biochim. Biophys. Acta 501, 208-216
2 Vass, I., Koike, H. and Inoue, I. (1985) Biochim. Biophys. Acta 810, 302-309
3 Chapman, D.J. and Barber, J. (1986) Biochim. Biophys. Acta 850, 170-172
4 Rutherford, A.W., Renger, G., Koike, H. and Inoue, Y. (1984) Biochim. Biophys. Acta 767, 548-556

ACKNOWLEDGEMENTS: T.W. is sincerely grateful to the Science and Technology Agency of Japan for financial support through the Versailles Summit Cooperation Program.

CHEMICALLY-INDUCED DYNAMIC ELECTRON POLARIZATION IN PHOTOSYSTEM 2 REACTION CENTERS.

Joseph T. Warden, Nathan M. Lacoff and Károly Csatorday
Department of Chemistry, Rensselaer Polytechnic Institute, Troy, New York

1. INTRODUCTION

The application of time-resolved electron spin resonance (esr) to probe electron transfers in photosynthetic reaction centers has considerable potential for detailing the identity of transient intermediates, kinetics and the nature of exchange and dipolar interactions among reaction-center components. To date the most extensive time-resolved esr investigations have focussed on the bacterial reaction center and on green plant photosystem 1 [1-6 and references within]. In this report we present a preliminary study of light-induced electron transfer in photosystem 2 (PS 2) particles utilizing laser flash-photolysis esr at room and cryogenic temperatures. Time-resolved spectra are illustrated for the primary donor, P_{680} and the spin-polarized, reaction-center triplet of PS 2. Additionally the depopulation rate constants for the PS 2 triplet state sublevels at high field have been determined and are reported for the first time. Examination of PS 2 particles, poised at potentials lower than -400 mV, has revealed the presence of transient esr signals in the g~2 region exhibiting anomalous intensities at times shorter than 500 μs after the actinic flash. Such anomalous signals are characteristic of Chemically-Induced Dynamic Electron Polarization (CIDEP) born via nascent, intermediate radical pairs. Tentative analysis of the transient esr spectrum at 100 μs post flash suggests the participation of an acceptor in PS 2 in addition to pheophytin.

2. MATERIALS AND METHODS

Time-resolved electron spin resonance spectroscopy was performed with a Varian E-9 spectrometer modified for rapid 100 kHz phase-sensitive detection. Instrument time-response (1/e) was 30 μs for kinetic experiments. Data collection and analysis, transient signal averaging and spectrometer magnetic field control were implemented by a PDP-11/23 computer. Actinic flashes (600 ns Full Width at Half Maximum [FWHM]) were provided by a Chromatix CMX-4 flashlamp-pumped dye laser. Nominal flash intensity at the sample was circa 4 mJ at 665 nm. Cryogenic temperatures were obtained with an Air Products LTD-3-100 cryostat. Sample temperatures were monitored with a calibrated carbon resistor situated directly below the 3 mm inner diameter quartz sample tube. Triplet state measurements were made by the actinic light modulation technique. A Princeton Applied Research 5240 Lock-in Analyzer in combination with a Rolfin programmable chopper was utilized for phase-sensitive detection of the photoexcited triplet. Actinic light (400-700 nm at 6000 W m^{-2}) for triplet observation was provided by a 1000 Watt Tungsten-halogen source (Oriel).

Absorption transients at 820 nm were acquired with a purpose-constructed flash-photolysis spectrometer interfaced to the PDP-11/23. Actinic illumination was provided by a Phase-R DL 1100 flashlamp-pumped dye laser

Biggens, J. (ed.), Progress in Photosynthesis Research, Vol. I. ISBN 90 247 3450 9
© *1987 Martinus Nijhoff Publishers, Dordrecht. Printed in the Netherlands.*

(200 ns FWHM, 660 nm, 40 mJ per flash). These flash-photolysis studies employed a cuvette with a 10 mm pathlength and the chlorophyll concentration was typically 20 μg mL^{-1}.

Triton photosystem 2 particles (TSF-2) were isolated by the procedure reported previously [7]. All preparations were stored frozen in 20% glycerol prior to use. Samples for electron spin resonance studies were typically at chlorophyll concentrations of 4.8 mg mL^{-1}.

2. RESULTS AND DISCUSSION

2.1. The spin-polarized reaction center triplet from photosystem 2

When photosystem 2 particles are poised at -500 mV, a transient, light-induced esr signal is observed at 8 K (Figure 1), a signal which exhibits spin-polarization and which has been attributed to the P_{680} triplet. This triplet is proposed to originate via charge recombination between P_{680}^+ and pheophytin anion [8,9], hence the observed polarization pattern of **aeeaae**. Zero-field splitting parameters for this triplet have been determined to be $|D|$=0.0287 cm^{-1} and $|E|$=0.0042 cm^{-1} after correction for magnetic field sweep nonlinearity. Although these structural parameters are similar to those obtained for the P_{700} triplet [10], we have established that these particles have no significant P_{700} contamination.

Time-resolved electron spin resonance spectroscopy has been utilized to characterize the spin dynamics of the PS 2 radical-pair triplet at cryogenic temperature. A portion of the kinetic data is presented in Figure 2 and the corresponding time-resolved spectrum for the low-field half of the triplet is displayed in Figure 3. Preliminary analysis of a series of kinetic experiments, as illustrated in Figure 2, yields the following estimates for the depopulation rates of the triplet sublevels: k_z=180 s^{-1}; k_y=1050 s^{-1}; k_x=1000 s^{-1}. These values are similar in magnitude to those measured at zero field [11] and to data reported recently for the photosystem 1 reaction-center triplet [12,13]. A further refinement of this data, utilizing the formalism of McGann and Frank [14] is in progress.

2.2 Time-resolved ESR Spectrum of P_{680}^+.

Photo-oxidation of the primary electron donor, P_{680}, is characterized by the appearance

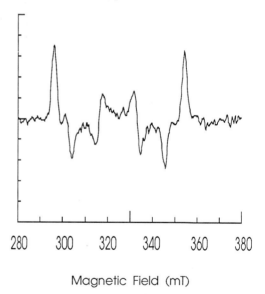

Magnetic Field (mT)

Figure 1. The spin-polarized reaction-center triplet of Photosystem 2. Triplet detection via modulated light (34 Hz, phase 299°). Temperature 8 K; microwave power 20 μW; microwave frequency 9240 MHz. Modulation amplitude 3.2 mT; gain 5000. Sweep rate 0.2 mT s^{-1}; Time constant 1 s.

of an absorption band in the region of 820 nm [15] and an esr signal in the vicinity of g~2 [16]. Photosystem 2 particles, during photolysis with a 600 ns laser flash exhibit a transient resonance centered at g=2.0025 ±.0002 with a peak-to-peak line-width of 0.80 mT (data not shown). The half-life of this Gaussian resonance is circa 4.5 ms at 11 K and 200 μs at 295 K. Treatment of the PS 2 prepara-tion with the fatty acid, lino-lenic acid, eliminates the 4.5 ms esr transient at 11 K and abolishes the optical absorption change associated with P_{680}^+ (Csatorday et al., these Proceed-ings).

When PS 2 particles are poised at potentials below -400 mV, time-resolved esr studies at 11 K reveal a complex pattern of mixed emission and absorption kinetic components in the g~2 region. The time-resolved spec-tra obtained at two sampling times are presented in Figure 4. When the spectrum is reconstruct-ed at 800 μs post flash, a symmetric resonance centered at g=2.0026±.0003 is observed with a nominal peak-to-peak linewidth of 0.76 mT. However, at earlier times (100 μs) a spectrum is obtained that is an admixture of absorption and emission compo-nents. This asymmetric resonance has significant intensity to low magnetic field and in addition a small emissive component at higher field.

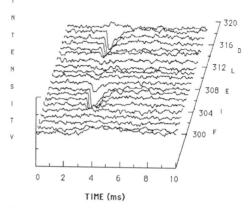

Figure 2. Stacked plot rpresentation of esr kinetics for the low-field x and y components of the PS 2 triplet. Each trace is an average of 256 flashes given at 5 Hz. Temperature 9 K.

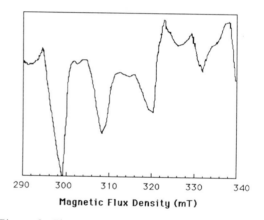

Magnetic Flux Density (mT)

Figure 3. Time-resolved spectrum for the PS 2 triplet at 9 K. This spectrum was constructed from a series of kinetic traces, utilizing a magnetic field step size of 0.5 mT and an integration window of 50 μs coincident with the actinic flash.

We interpret the spectrum obtained at 800 μs to reflect P_{680}^+ in the absence of spin polarization. In contrast, we cannot account for the spectrum at 100 μs in terms of a single spin-polarized species. Thus we regard this spectrum as reflecting contributions from both the primary donor, P_{680}^+, and an acceptor moiety (g~2.0050-2.0060), perhaps via polarization transfer from pheophytin anion. An observation of CIDEP in iron-extracted PS 2 particles, subjected to continuous illumination at cryogenic temperatures, has been reported recently by Hoff and Proskuryakov [17]; however the transient spectrum illustrated in Figure 4 is spectrally distinct from that described by the Dutch laboratory (g~2.0034-2.0038). Indeed the composite spectrum at 100 μs is similar to transient spectra obtained in photosystem 1 [3-5] and in bacterial reaction centers [1,6]. These data, combined with data presented elsewhere in these Proceedings (Csatorday et al.), suggest the existence of an additional acceptor in PS 2 ($E_m < -400$ mV), intermediate between pheophytin and Q_a [18].

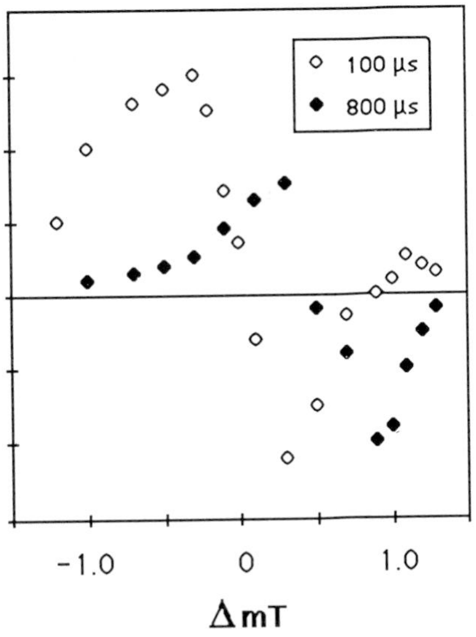

Figure 4. Time-resolved esr spectrum of PS 2 particles ($E_h \sim -500$ mV) sampled at 100 μs (o) and 800 μs ($\bullet$) post flash. Magnetic field is centered about g=2.0055.

REFERENCES

1. Hoff, A.J., Gast, P. and Romijn, J.C. (1979) FEBS Lett. 73, 185-190.
2. Norris, J.R., Bowman, M.K., Budil, D.E., Tang, J., Wraight, C. and Closs, G.L. (1982) Proc. Natl. Acad. Sci. USA 79, 5532-5536.
3. Furrer, R. and Thurnauer, M.C. (1983) FEBS Lett. 153, 399-403.
4. McCracken, J.L. and Sauer, K. (1983) Biochim. Biophys. Acta 724, 83-93.
5. Manikowski, H., McIntosh, A.R. and Bolton, J.R. (1984) Biochim. Biophys. Acta 765, 68-73.
6. Hore, P.J., Watson, E.T., Pedersen, J.B. and Hoff, A.J. (1986) Biochim. Biophys. Acta 849, 70-76.
7. Golbeck, J.H. and Warden, J.T. (1985) Biochim. Biophys. Acta 806, 116-123.
8. Thurnauer, M.C., Katz, J.J. and Norris, J.R. (1975) Proc. Natl. Acad. Sci. USA 72, 3270-3274.
9. Rutherford, A.W., Patterson, D.R. and Mullet, J.R. (1981) Biochim. Biophys. Acta 635, 205-214.
10. Nechushtai, R., Nelson, N., Gonen, O. and Levanon, H. (1985) Biochim. Biophys. Acta 807, 35-43.
11. den Blanken, H.J., Hoff, A.J., Jongenelis, A.P.J.M. and Diner, B.A. (1983) FEBS Lett. 157, 21-27.
12. Setif, P., Quaegebeur, J.P. and Mathis, P. (1982) Biochim. Biophys. Acta 681, 345-353.
13. Gast, P., Swarthoff, T., Ebskamp, F.C.R. and Hoff, A.J. (1983) Biochim. Biophys. Acta 722, 163-175.
14. McGann, W.J. and Frank, H.A. (1985) Chem. Phys. Letts. 121, 253-261.
15. Conjeaud, H. and Mathis, P. (1980) Biochim. Biophys. Acta 590, 353-359.
16. Pulles, M.P.J., van Gorkom, H.J. and Verschoor, G.A. (1976) Biochim. Biophys. Acta 440, 98-106.
17. Hoff, A.J. and Proskuryakov, I.I. (1985) Biochim. Biophys. Acta 808, 343-347.
18. Csatorday, K. and Warden, J.T. (1986) Biochim. Biophys. Acta (in press).

ACKNOWLEDGEMENTS

This work was supported in part by a grant from the National Institutes of Health (GM26133). We acknowledge also the technical assistance of Dr. S. Kumar.

THE MECHANISM OF FATTY ACID INHIBITION IN PHOTOSYSTEM 2.

Károly Csatorday, Claire Walczak and Joseph T. Warden
Department of Chemistry, Rensselaer Polytechnic Institute, Troy, New York 12180-3590

1. INTRODUCTION

The mechanism(s) associated with fatty acid inhibition of photosystem 2 (PS 2) has been the subject of a number of investigations since the pioneering study of Krogmann and Jagendorf [1]. As capsulized by Golbeck et al. [2] this topic is of considerable import since membrane damage attributed to aging or stress has been associated with the release of free fatty acids, presumably induced by endogenous lipases. A precise biochemical description for the effect of unsaturated fatty acids has not been proffered, however recent investigations by Vernotte et al. [3] and Golbeck and Warden [4] have focussed on the origin and mechanism of the fatty acid induced inhibition of PS 2 electron transport.

We have proposed recently [4,5], based on spectroscopic and kinetic analysis of electron transport components associated with the PS 2 reaction center, that linolenic acid affects multiple inhibition sites in PS 2, in particular, those sites associated with bound quinones (eg. the donor species Signal 2_s and Signal 2_f (D_1) on the oxidizing side of PS 2 and Q_a, the primary quinone acceptor, on the reducing locus of the reaction center). Furthermore, invoking the hypothesis of Klimov et al. [6] that variable fluorescence in PS 2 is a monitor of the backreaction between P_{680}^+ (the primary electron donor) and pheophytin anion (pheo$^-$, the transient, primary electron acceptor), the development of the high F_i in the presence of the inhibitor is considered to be the consequence of the absence of the acceptor, Q_a [5].

The results presented in this paper are an extension of our previous study [4] and address specifically the nature of linolenic acid inhibition at the reaction center. Utilizing time-resolved optical spectroscopy in conjuction with electron spin resonance (esr) spectroscopy at cryogenic temperatures, we have examined the effect of a variety of fatty acids on the photoreduction of the primary quinone acceptor, Q_a. Additionally we have examined the capability of fatty acid inhibited PS 2 preparations to perform charge separation, triplet state production and secondary electron transfer at low temperature. Finally we have achieved a redox titration of the fluorescence yield in control and

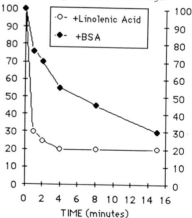

Figure 1. Time dependence for inhibition of P_{680} photo-oxidation in D-10 particles treated with 100μM linolenic acid and after subsequent addition of 0.5% bovine serum albumin.

Biggens, J. (ed.), Progress in Photosynthesis Research, Vol. I. ISBN 90 247 3450 9
© *1987 Martinus Nijhoff Publishers, Dordrecht. Printed in the Netherlands.*

inhibited samples. Our results demonstrate that linolenic acid inhibition in photosystem 2 does not suppress primary photochemistry. We propose, instead, that unsaturated fatty acids inhibit secondary electron transfers in PS 2 via competitive displacement of bound quinoidal electron transport intermediates from their protein binding sites [5].

2. RESULTS AND DISCUSSION

A unique aspect of linolenic acid inhibition is that the inhibition is completely reversible in spinach chloroplasts upon addition of 100 μM bovine serum albumin (BSA) [2]. In spinach digitonin particles (D-10) reversal of inhibition occurs only after relatively short incubation periods (30 min. or less, Figure 1); however in triton particles, the fatty acid inhibition is irreversible. If the mechanism of fatty acid inhibition involves competition between fatty acid and quinone for the quinone binding site, the degree of reversibility will be dependent on (i) the relative affinity of the quinone and fatty acid for the binding site. Note that in PS 2 particles the detergent may compete also for the binding site. The lack of reversibility in triton preparations may be due to such a synergy between linolenic acid and detergent. (ii) the relative concentrations of the quinone, fatty acid and, if present, membrane associated detergent. (iii) the duration of incubation with the inhibitor. If the quinone is displaced from its native binding site, at short incubation times reinsertion would be more probable in contrast to lengthy incubation times during which the quinone can diffuse and be diluted out in membrane lipid or become nonspecifically bound to membrane protein. This postulate is supported by the data in Figure 1.

To examine further the hypothesis that quinones are displaced competitively during fatty-acid

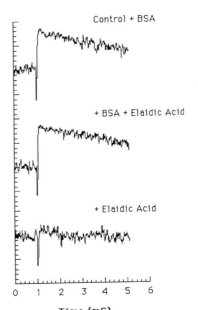

Control + BSA

+ BSA + Elaidic Acid

+ Elaidic Acid

Time (mS)

Figure 2. Reoxidation kinetics of Q_a in control and fatty-acid inhibited PS 2 particles. Vertical scale, $6 \cdot 10^{-4}$ absorbance units per major division. Chlorophyll, 100 μg mL^{-1}; elaidic acid , 100 μM; BSA, 0.5%.

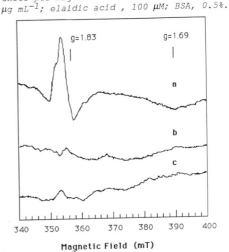

g=1.83   g=1.69

a

b

c

Magnetic Field (mT)

Figure 3. Electron spin resonance spectra of Q^-Fe. (a) control after illumination at 200 K (b) control + 7mM linolenic acid (c) same as (b) except particles were incubated with 50 mM dithionite. Chlorophyll, 6.8 mg mL^{-1}; temperature, 6.3 K.

mediated inhibition of PS 2 electron transport, we have examined the absorption change at 320 nm, a wavelength diagnostic for the PS 2 acceptor quinones (Figure 2). When PS 2 particles are monitored at 320 nm during microsecond flash photolysis, a characteristic absorption increase is observed which has been assigned to the formation of the plastoquinone anion of Q_a [7]. Incubation of the particles with 100 μM elaidic acid suppresses the photoreduction of Q_a (Figure 2); however the inclusion of 0.5 % BSA negates the action of the fatty acid inhibitor. Similar inhibition of Q_a^- formation is observed also with linolenic acid, linoleic acid, oleic acid and petroselenic acid.

The inhibition of the 320 nm absorption change of Q_a is necessary, but not sufficient in itself, to verify the hypothesis that fatty acids disrupt PS 2 charge transfer by displacing Q_a from its binding site. In order to establish whether Q_a is still resident in the acceptor complex after fatty acid inhibition, we have utilized electron spin resonance as a probe for the g=1.82 resonance (Q⁻Fe) which signifies an interaction of the semiquinone anion of Q_a and a Fe^{2+} ion in close proximity[8].

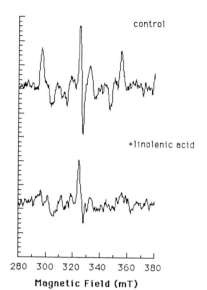

control

+linolenic acid

280 300 320 340 360 380
Magnetic Field (mT)

Figure 4. The spin-polarized reaction-center triplet in PS 2 particles maintained at 8 K. Control particles (chlorophyll, 4.8 mg mL⁻¹) were incubated with 50 mM dithionite prior to freezing. Linolenic-acid treated particles were incubated with 7mM fatty acid for 30 min and were photolyzed at 8 K in the absence of reductant. Actinic light modulation frequency, 41 Hz; phase 291 degrees.

The Q-Fe signal in Figure 3a is characterized by features at g=1.82 and 1.69 and can be generated to equal extent by photoreduction of the sample at 200 K or by chemical reduction of the sample by dithionite prior to freezing. Incubation of the PS 2 particles with linolenic acid induces an abolition of the Q-Fe resonance as assayed after illumination at 200 K (Figure 3b) or by chemical reduction (Figure 3c). These esr studies corroborate the optical studies at 320 nm and indicate that the Q_a binding site is no longer functional for photochemically or chemically-mediated reduction.

Suppression of secondary electron transfer in green plant photosystem 2 by reduction of Q_a results in the formation during illumination of a spin-polarized reaction-center triplet [9]. This triplet is proposed to originate via charge recombination between P680+ and pheophytin anion. Reduction of D-10 particles with 50 mM dithionite yields the characteristic esr spectrum (Figure 4) of the reaction center triplet with a polarization pattern **aeeaae**. This triplet is characterized by zero-field splitting parameters of $|D|=0.0287$ cm⁻¹ and $|E|=0.0041$ cm⁻¹. In accord with the recent observation of Evans et al. [10] we do not observe significant production of the triplet (<10%) at potentials above -400 mV. Incubation of the PS 2 particles with linolenic acid in the absence of a strong reductant, contrary to our expectation, does not induce a marked increase in triplet yield; although these samples were determined (in parallel experiments) to exhibit maximal fluorescence yield and to lack microsecond-detectable P680+ reduction kinetics, both criteria characteristic of linolenic acid inhibition. However, reduction of the linolenic acid-treated sample with dithionite (data not shown) results in the appearance of the spin polarized triplet at an intensity comparable to the control. The observation of the spin polarized reaction-center triplet in reduced samples demonstrates incisively that charge separation in the PS 2 reaction center is not incapacitated in the presence of linolenic acid. Indeed redox titrations of the maximum fluorescence yield (F_m) in control and fatty-acid inhibited samples demonstrate that the observed midpoint potentials for the decrease in fluorescence are identical (-585 mV) in both cases (Csatorday and Warden (1986) in the press). The absence of the triplet signal in linolenic-acid inhibited samples frozen at moderate potential (~0 mV) is unexpected, since these samples exhibit a high fluorescence yield attributed to recombination luminescence. We interpret this data as indicating the presence at cryogenic temperatures of another acceptor intermediate between pheophytin and Q_a.

In this report we have presented various data to support our assertion that fatty acids can function as quinone-binding site inhibitors in PS 2. In contrast to secondary electron transport, primary photochemical charge separation in the reaction center is not inhibited by the fatty acid. We assert that the unique, reversible inhibitory activity of the unsaturated fatty acids has its origin in a displacement of endogenous quinones via competitive binding to quinone-binding sites in PS 2.

REFERENCES

1 Krogmann, D.W. and Jagendorf, A.J. (1959) Arch. Biochem.Biophys. 80, 421-430.
2 Golbeck, J.H., Martin, I.F. and Fowler, C.F. (1980) Plant Physiol. 65, 707-713.
3 Vernotte,C., Solis, C., Moya, I., Maisson,B., Briantais, J., Arrio, B., and Johannin, G. (1983) Biochim. Biophys. Acta 725, 376-383.
4 Golbeck, J.H. and Warden, J.T. (1984) Biochim. Biophys. Acta 767, 263-271.
5 Warden, J.T. and Csatorday, K. (1986) Biochim. Biophys. Acta (in the press).
6 Klimov, V.V., Klevanik, A.V., Shuvalov, N.A. and Krasnovskii, A.A. (1977) FEBS Lett. 39, 337-340.
7 van Gorkom, H.J. (1974) Biochim. Biophys. Acta 347, 439-442.
8 Butler, W.F., Johnston, D.C., Shore, H.B., Fredkin, D.R., Okamura, M.Y. and Feher, G. (1980) Biophys. J. 32, 967-992.
9 Rutherford, A.W., Paterson,D.R. and Muller, J.E. (1981) Biochim. Biophys. Acta 635, 205-214.
10 Evans, M.C.W., Atkinson, Y.E. and Ford, R.C. (1985) Biochim Biophys. Acta 806, 247-254.

ACKNOWLEDGEMENTS

This work was supported partially by a grant from the National Institutes of Health (GM26133). Data collection, data analysis, figure and text preparation were performed with a PDP 11/23 networked to a Macintosh 512K (Apple Computer Inc.). Manuscript production on the Apple LaserWriter was accomplished using PageMaker (Aldus Corp.).

IN VIVO SPECTRAL PEAKS RELATED TO NEW CHEMICAL SPECIES OF CHLOROPHYLLS :
4-VINYL-4-DESETHYL .

MAARIB B.BAZZAZ , Dept. of Cellular and Developmental Biology,Harvard
University,16 Divinity Ave.,Cambridge,Massachusetts,02138.

1.ABSTRACT

The red shift observed for Soret band in extracted pigment is
preserved in vivo, which means that it is due to the structure 4-vinyl -
4-desethyl Chlorophylls. It may be associated to Photosystem (PS)I core
antennae as its maximum emission is 705 which is close to emission
observed in 40 Chl PSI particles; then if it is true these Chl forms
exist in (normal) plant.

2.INTRODUCTION:

It has been previously established that a mutant of maize (ON 8147)
can photosynthesize in vitro efficiently and that the mutant had
spectral and biochemical characteristics similar to those of immature
chloroplasts (1). Structural characterization of the total extracted
Chlorophyll (Chl) pigments by nuclear magnetic resonance spectroscopy
confirmed the different chemical nature of the pigment pools in the
mutant from that of the wild type maize (2). Furthermore these studies
showed that the chemical difference lies in the substitution of a vinyl
group at position 4 of the tetrapyrole macrocycle of the mutant
(Structure B,Fig.1) for the ethyl group usually present in Chl a and b.
Mass spectroscopic characterization of the mutant chlorophylls (earlier
designated as Chl a1A436, Chl b1A462) and their pheophytin derivatives
(Pheo a1A417 and Pheo b1A440) indicated a difference of 2 mass units
only from Chl a,b and their Pheophytin derivatives (3,4). The purified
pigments differ spectroscopically from Chl a,b, Pheo a and b by their
distinct red shift (8-10 nm) in the Soret region of the visible spectrum
(5).

DARK PIGMENTS (A)   LIGHT PIGMENTS (B)

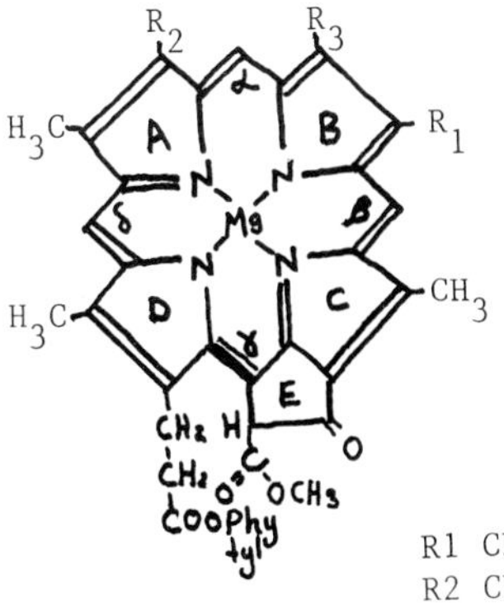

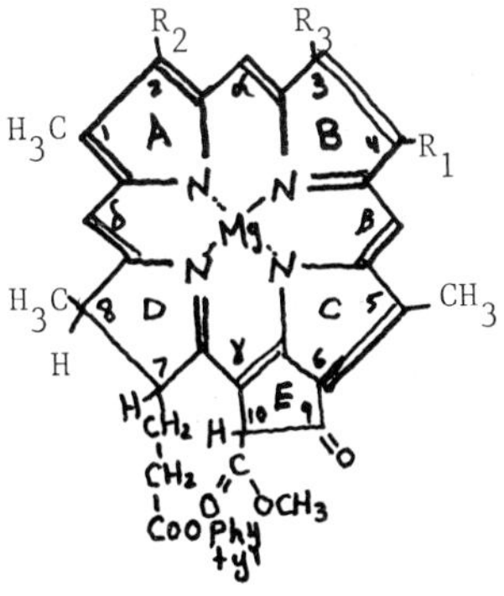

R1 CH-CH$_2$
R2 CH=CH2
R3 CHO

The pigment pool extracted and purified from mutant plants grown in dark
contain 4-vinyl Protochlorophy(llid e)(Structure A,Fig. 1). The size of
this pool is 1/3 of that of the etiolated wild type pool which is made
up mainly of Protochlophy(llide). A photoactive Protochlorophyllide
reductase is present in the mutant (M.B.Bazzaz and W.T.Griffiths,
unpublished).

Chloroplasts isolated from mutant plants grown under photoperiodic
regime of 16 hr light and 8 hr dark is capable of performing PSI and
PSII reactions; the photosynthetic unit (estimated on basis of P700 and
Cyt f) is smaller than that in chloroplasts isolated from wild type
maize. In vivo spectroscopic characterization of pigments in isolated
chloroplasts indicated an enrichment of the mutant in PS I relative to
PSII pigments as it was evident from the high a/b ratio, the higher
relative fluorescence yields at room temperature at 710 compared to 685
nm and the higher ratio of 77 K fluorescence at 735 (mainly PSI)
emission relative to 685 nm (mainly PSII)(1) Taken all these results in
consideration it became necessary to reinvestigate the fluorescence
emission and excitation spectral characteristics of isolated
chloroplasts in order to obtain new information on in vivo absorption
peaks expected to be associated with the new chemical species of the
mutant; 4-vinyl-4-desethyl chlorophylls and their Pheophytin pools. The
results presented here clearly show the presence of a distinct Soret
excitation peak at 450 nm that sensitizes a broad emission band with a
peak at 705 nm in the mutant. Three major complexes with distinct
fluorescence emission (F) and excitation (E) peaks are also detected and
enriched in the mutant chloroplasts. These are F705 E680 E665 E450;
F740 E683 E660 E450 and F710, 720 ,750 E440 & E450.

3 . PROCEDURE
3 . 1. Material and Methods
Chloroplasts were isolated from mutant and wild type maize plants
grown under similar photoperiodic regime of 16 Hrs light and 8 hrs dark.
Leaves were homogenized in 50 mM Tricine buffer pH 8.0 containing 0.2 M
sucrose,3mM MgCl2 and 3mM KCl. The homogenate was filtered through Mira
cloth and the filtrate was centrifuged at 1500 xg for 1 min and
supernatant was recentrifuged for additional 7 mins. The pellet was
suspended in homogenizing buffer. For spectrofluorometric measurements
at 77K,aliquots of chloroplast suspension containing 0.00001 ugm Chl/ml
was used .

3 . 2. Spectrofluorometry
Fluorescence emission and excitation spectra were measured by the
dual channel ratiometric acquisition method using a System 4000 scanning
spectrofluorometer (SLM instruments,Urbana,Ill) as described earlier (6)
Ratio spectra of fluorescence excitation and emission spectra recorded
on chloroplasts suspensions from the mutant/WT were obtained by
utilizing the capability of the microprocessor to divide the stored
spectra with the resultant spectrum being plotted at the recorder.
Other mathematical manipulations of the spectra were also made by
similar procedure.

4 . RESULTS:

The absorption spectrum of the major chlorophyll pigment of the mutant ON 8147 has a room temperature Soret absorption maximum in ether at 436 nm compared to 428 nm for Chl a extracted from WT plants (6) This red Soret shift in the extracted pigment was reflected also in <u>in vivo</u> absorption and fluorescence excitation spectra. To detect these differences fluorescence emission spectra excited by wavelength of 440 nm were recorded at 77 K on mutant(___,Fig.2) and WT (_ _ ,Fig.2) chloroplast suspensions. The WT spectrum showed a typical 3 banded spectrum with peaks at 685 and 733 nm and a broad shoulder around 696nm. The normalized spectra of the mutant and normal chloroplast suspensions at 733 nm show increased emission between the wavelengths 688 and 733 nm and a decrease in the short wavelength emission in the mutant relative to the WT. This is consistent with earlier detected deficiency in Pigment system (PS) II compared to PSI(1). Although the shape of the emission spectra recorded using 450 nm excitation were similar to Fig.2 (spectrum not presented), the relative fluorescence emission intensity at 733 nm was excited to a higher intensity by 450 nm compared to 440 nm light in the mutant. In the mutant a prominent emission band between 688 nm and 733 nm with a peak around 705 nm is evident from the ratio spectrum obtained by 440 nm excitation of mutant and WT samples (insert Fig.2)

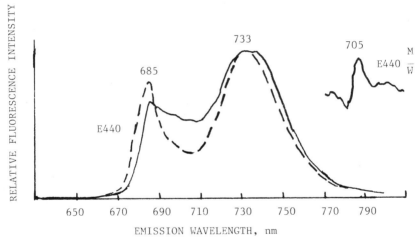

FIGURE 2:Fluorescence emission spectra,measured at 77K excited by 440 nm light,of chloroplast suspecnsions from mutant(——)and WT(-- --). (insert) ratio spectrum of mutant/WT emission excited by 440 nm light.

Other emission peaks enriched in the mutant samples were also detected by recording the difference spectrum of normalized emission spectra of the chloroplast samples recorded by 440 nm excitation (Fig,3). The spectrum shows two major broad bands with maxima around 710 nm ,720 nm and 750 nm enriched in the mutant relative to those peaks at 683 nm, 687 nm and 732 nm in the WT.

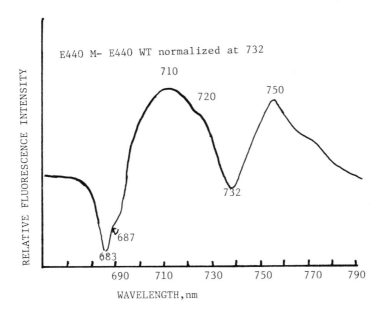

FIGURE 3: Difference fluorescence spectrum of mutant minus wild type emission spectra of isolated chloroplasts. Spectra were normalized at 732nm before subtraction. Temp. of measurement is 77K.

To find out whether the emission peak observed at 705 nm was sensitized by specific Soret excitation and red emission bands, fluorescence excitation spectra were measured between 380-700 nm for both mutant and WT samples and the ratio spectrum was obtained. These results revealed a major Soret peak at 450 nm which was accompanied by 2 red excitation bands with peaks at 680 and 665 nm. This observation indicates that the Soret Peak at 450 nm which sensitizes emission at 705 nm is not likely to be due to carotenoid origin but it is rather due to a chlorin pigment. Similar fluorescence excitation spectra were measured for fluorescence emission at 740 nm . The ratio spectrum showed a major Soret peak at 450 nm but the red excitation bands have maxima at 683 nm and 660 nm . A red excitation band with a peak at 630 seems to sensitizes F740 to higher extent than emission at705 nm in the mutant sample. The sensitization of fluorescence emission by 630 nm band is consistent with the presence of a pool of Chlorophyll-c type pigment in the mutant The peak at 497 is probably due to carotenoids. Furthermore, the lack of Chl b excitation peak around 645 nm is consistent with relative deficiency of the mutant in PSII compared to PSI and the high a/b ratio reported earlier(1).

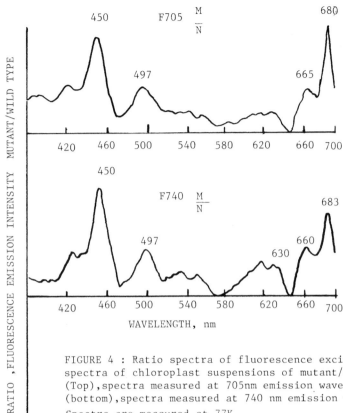

RATIO, FLUORESCENCE EMISSION INTENSITY MUTANT/WILD TYPE

WAVELENGTH, nm

FIGURE 4 : Ratio spectra of fluorescence excitation spectra of chloroplast suspensions of mutant/normal. (Top),spectra measured at 705nm emission wavelength. (bottom),spectra measured at 740 nm emission wavelength. Spectra are measured at 77K.

4 . DISCUSSION

The exclusive accumulation of 4-Vinyl-4-desethyl Chl a and b and their Porphyrin precursors in light grown leaves of the mutant ON 8147 is reflected in the presence of a broad fluorescence emission band between 688-730 nm with a prominent peak observed at 705 nm (insert, Fig.2) and relative decrease in the short wavelength emission at 680 nm.

The emission band at 705nm is sensitized mainly by a red shifted Soret peak at 450 nm. The contribution of red excitation bands at 683 and 665 detected in ratio spectrum (Fig. 4,top) is a good evidence that the Soret peak at 450 originate from Chlorophyll species rather than carotenoids. The absorption peak at 665 nm is similar to a peak detected earlier in bundle sheath chloroplasts (7). It remains to be seen if both peaks are related. It is of interest to note that the emission bands observed at 710 nm and 720 nm (Fig .3) are enriched in the mutant and similar bands were also observed in chloroplasts isolated from several partially developed higher plants(M.Bazzaz,unpublished) and PSI antennae-depleted particles. Thus,it is possible that these pigments with Soret excitation at 450nm are present and related to PSI 40Chl core-antennae. The results presented here are consistant with earlier report(8)and proposal(5).

6.REFERENCES
1.Bazzaz,M.B.,Govindjee and D.J.Paolillo,Jr. (1974) Z.Pflanzenphysiol.
72, 181-192.

2.Bazzaz,M.B.,R.G.Brereton (1982) FEBS.Letts.138,104-108.

3.Bazzaz,M.B.,Bradley and R.G.Brereton (1982)Tet.Letts.v.23.No 11.1211-
1214.

4.Brereton,R.G.,M.B.Bazzaz,S.Santikarn and D.H.Williams (1983) Tet.Letts.
v.24.No.51,5775-5778.

5.Bazzaz,M.B. (1981) Nturwissenschaften 68,94.

6.Bazzaz,M.B. (1981) Photobiochem.and Photobiophys.2,199-207.

7.Bazzaz,M.B. and Govindjee (1973) Plant Physiol.,32,257-262.

8.Bazzaz,M.B.(1980) Fed.Proc.,39,No.6.1802.

CHLOROPHYLL *a'* IN PHOTOSYNTHETIC APPARATUS: REINVESTIGATION

Tadashi WATANABE*, Masami KOBAYASHI*, Masataka NAKAZATO**, Isamu IKEGAMI†, and Tetsuo HIYAMA ††

* Institute of Industrial Science, University of Tokyo, Roppongi, Minato-ku, Tokyo 106
** Nampo Pharmaceutical Co., Ltd., Nihonbashi-honcho, Chuo-ku, Tokyo 103
† Laboratory of Chemistry, Teikyo University, Sagamiko, Kanagawa 199-01
†† Department of Biochemistry, Saitama University, Urawa 338 (Japan)

1. INTRODUCTION

In a previous study (1) we verified, by high-performance liquid chromatographic (HPLC) analyses, that chlorophyll *a'* (Chl *a'*. cf. Fig. 1) is contained in leaves of higher plants at a fairly unified concentration (Chl *a* / Chl *a'* ≃ 300). A more recent work (2) on PS1 and PS2 particles revealed that Chl *a'* is associated exclusively with PS1 at a Chl *a'* / P700 molar ratio around 2; this led us to speculate that a Chl *a'* dimer might constitute P700. In that work, however, some PS1 preparations gave Chl *a'*/ P700 values between 1 and 2. In view of this, the Chl *a'*/ P700 stoichiometry has been reinvestigated here in more detail, by carefully quantifying Chl *a'* in various plant materials (leaf tissues, blue-green algae, and PS1 particles prepared by a variety of methods) with an improved HPLC setup.

Dörnemann and Senger (3,4) recently reported isolation of a Chl *a* derivative ("Chl RC1") from several plants, and proposed that this pigment be an integral component of PS1. Examination of the nature of "Chl RC1" is another objective of the present work.

2. MATERIALS AND METHODS

Four species of blue-green algae were supplied from Dr. T. Matsunaga, Tokyo University of Agriculture and Technology. PS1 particles with Chl *a*/ P700 molar ratios ranging from 8 to 150 were prepared by I. Ikegami (5) through Digitonin fractionation of spinach chloroplasts followed by progressive ether wash. Another series of PS1 particles with Chl *a* /P700 ratios 50 - 1000 were prepared by T. Hiyama through a variety of treatment methods. The attached "PROCEDURE" gives the general method for HPLC analyses of these samples.

3. RESULTS AND DISCUSSION

3.1 Identification of Chl *a'* extracted from spinach

By means of preparative-scale HPLC, we isolated Chl *a* 8.13 mg, Chl *b* 2.77 mg, and Chl *a'* 0.035 mg from spinach leaf tissue 10 g. The CD spectrum of the Chl *a'* thus extracted

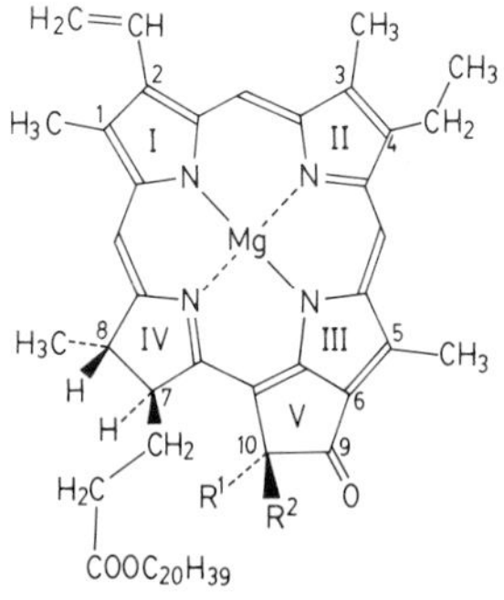

	R^1	R^2
Chl a	$COOCH_3$	H
Chl a'	H	$COOCH_3$

Fig. 1. Chl *a* and Chl *a'*

Biggens, J. (ed.), Progress in Photosynthesis Research, Vol. I. ISBN 90 247 3450 9

<u>PROCEDURE</u>

SAMPLE: Leaf Tissues, PS1 Particles, Blue-Green Algae

(1) Grind with Na_2HPO_4
(2) Add $CHCl_3$, then Sonicate for 30 s   } 10 - 15 min
(3) Filter
(4) Evaporate the Filtrate
(5) Dissolve in 0.1 mL $CHCl_3$
(6) HPLC (Analytical & Preparative)

HPLC Conditions

Column: 5 μm Silica, 200 mm x 8 mm I.D.
       Packing Pressure 250 kg/cm²
Column temperature: 0 °C
Detection wavelength: 430 nm
Elution: Isocratic, 0.8 - 1.0 mL/min
Eluent: (a) Hexane/i-PrOH (98.5/1.5)
           (Order: Chl a' - Component X - Pheo a)

       (b) Hexane/i-PrOH/MeOH (100/1/0.3)
           (Order: Pheo a - Chl a' - Component X)

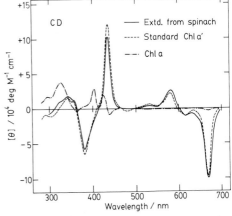

Fig. 2.   CD spectra

agreed well with that of authentic Chl a' prepared by base-catalyzed isomerization of Chl a (Fig. 2).   Further, the epimerization rate constants (6) of these two Chl a' samples were practically the same ($k + k' = 2.10$ x 10^{-5} s^{-1}) in dimethylformamide at 28 °C.

3.2   HPLC analysis of PS1 particles

Figure 3 illustrates typical HPLC traces for high-purity PS1 particles prepared by I. Ikegami. For each sample the Chl a/P700 and Chl a/Chl a' molar ratios are close to each other. This indicates that a stoichiometry Chl a'/ P700 = 1 holds at least for PS1 particles with Chl a/P700 ratios down to ca. 8.

The Chl a' - Pheo (pheophytin) a peak separation is greater than 3 min in Fig. 3, and a third component (Component X, see below) is clearly detected between these peaks. In previous measurements (1,2), this separation was less than 0.5 min and the Component X, if any, may have eluted together with Chl a'. Probably this was one of the causes for observing higher Chl a'/P700 molar ratios. Moreover, the use of prolonged and heavier treatments (Triton X-100 fractionation, PAGE, DEAE Sepharose column chromatography, etc.) for sample preparation in previous works may have caused Chl $a \rightarrow a'$ epimerization, giving apparent Chl a'/P700 ratios higher than 1.0. Figure 4 indeed shows that the stoichiometry Chl a'/P700 = 1 is established over a wide concentration range

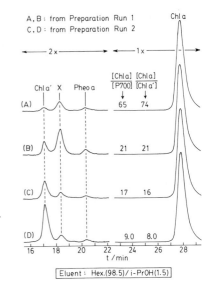

Fig. 3.   HPLC traces of four PS1 particles prepared by Digitonin fractionation followed by progressive ether wash (5)

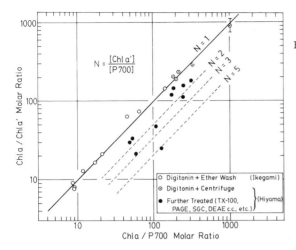

$$N = \frac{[\text{Chl }a']}{[\text{P700}]}$$

○ Digitonin + Ether Wash   (Ikegami)
⊙ Digitonin + Centrifuge   }
● Further Treated (TX-100,   } (Hiyama)
   PAGE, SGC, DEAE c.c., etc.)

Chl a / P700 Molar Ratio

Fig. 4. Summary of HPLC analytical results on PS1 particles, showing that a stoichiometry Chl a'/P700 = 1 is established for samples prepared by milder treatments
(Digitonin fractionation,
ether wash). Use of heavier
treatments in purification of
PS1 samples and/or lower-
resolution HPLC (●) tends to
enhance the apparent Chl a'/
P700 molar ratio due to Chl a
epimerization and/or Component X formation (see text).

as long as the PS1 particle preparation has been carried out under mild
conditions.

3.3 HPLC analysis on blue-green algae

Four species of blue-green algae have been analyzed for pigment composition, and the results are given in Table 1. By assuming the above-
mentioned stoichiometry Chl a'/P700 = 1, the observed Chl a/Chl a' molar ratios (120 - 160) are in accordance with Chl a/P700 molar ratios (130 - 170)
assayed by other methods (7,8). Also, the P700/P680 stoichiometry evaluated here from HPLC analyses is, except for the case of *Synechocystis*, well
within the range of reported PS1/PS2 molar ratios (1.1 - 2.1) (7,8).

3.4 Component X

Component X (see Fig. 3) is occasionally detected, *in amounts out of
proportion to that of P700*, in extracts of plant materials. A series of experiments showed that this pigment is absent in fresh leaves or freshly harvested blue-green algae, but grows gradually at the expense of Chl a in *dead*
plants and in subchloroplast particles during fractionation and purification
treatments. The Chl a → Component X conversion is especially fast in dead
blue-green algae. Thus, for *Anabaena*, being free from Component X just on
harvesting, Component X was detected in an amount comparable to that of
Chl a' after 12-h storage at 4 °C. For subchloroplast particles, even prolonged centrifuging was enough to produce Component X to a measurable level
(Fig. 5). The visible spectrum of Component X (Fig. 6) isolated from such

Table 1. CHLOROPHYLL a' AND PHEOPHYTIN a IN BLUE-GREEN ALGAE

Algae	[Chl a]/[Chl a']	[Pheo a]/[Chl a']	[P700]/[P680] *
Synechococcus sp.	136 ± 4	1.25 ± 0.05	1.54 ~ 1.67
Anabaena sp.	127 ± 7	1.22 ± 0.13	1.48 ~ 1.83
Spirulina platensis	113 ± 3	1.46 ± 0.10	1.28 ~ 1.47
Synechocystis	151 ± 9	3.20 ± 0.20	0.59 ~ 0.67

* On the assumption that [Chl a']/[P700] = 1 and [Pheo a]/[P680] = 2.

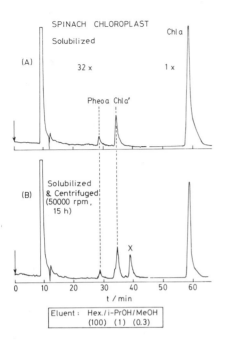

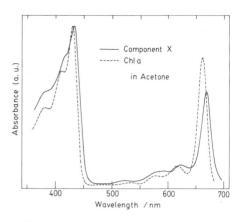

Fig. 5 (Left) Generation of Component X during centrifuging of detergent-solubilized chloroplasts. See PROCEDURE for the pigment elution order.

Fig. 6 (Right) Visible spectra for Component X and Chl a .

degraded materials is different from that of Chl a regarding peak wavelengths and blue/red absorbance ratio, but is practically the same as that of "Chl RC1" reported by Dörnemann and Senger (3,4). A low fluorescence quantum yield of Component X (1/4 to 1/5-fold lower than Chl a) also suggests that it corresponds to "Chl RC1". Since "Chl RC1" is identified as a hydroxylated and chlorinated Chl a (4), we suppose that Component X, or "Chl RC1", is a post-mortem alteration product of Chl a formed by the action of OH^- and Cl^- contained in chloroplasts. *In vitro*, even a single contact of purified Chl a with a Cl^--treated filter paper readily gives Component X.

3.5 Preliminary study on the reconstitution of P700

Photo- and redox-induced spectral changes reminiscent of P700 were observed after treating a 65-kD PS1 apoprotein with Chl a'. No such effects were noted by treatment with Chl a.

REFERENCES
1. Watanabe, T. et al. (1985) Biochim. Biophys. Acta 807, 110–117.
2. Watanabe, T. et al. (1985) FEBS Lett. 191, 252–256.
3. Dörnemann, D. and Senger, H. (1982) Photochem. Photobiol. 35, 821–826.
4. Dörnemann, D. and Senger, H. (1986) Photochem. Photobiol. 43, 573–581.
5. Ikegami, I. and Ke, B. (1984) Adv. Photosynth. Res. Vol. II, 73–76.
6. Watanabe, T. et al. (1984) Chem. Lett. 1411–1414; (1986) ibid. 253–256.
7. Mimuro, M. and Fujita, Y. (1977) Biochim. Biophys. Acta 459, 376–389.
8. Wang, R. T. et al. (1977) Photochem. Photobiol. 25, 103–108.

ACKNOWLEDGMENT
The authors are grateful to Dr. T. Matsunaga, Tokyo University of Agriculture and Technology, for supplying us with the blue-green algae.

ARE CHLORINATED CHLOROPHYLLS COMPONENTS
OF PHOTOSYSTEM I REACTION CENTERS?

J. FAJER, E. FUJITA, BROOKHAVEN NATIONAL LABORATORY, UPTON, NY 11973
H. A. FRANK, B. CHADWICK, UNIV. OF CONNECTICUT, STORRS, CT 06268
D. SIMPSON, K. M. SMITH, UNIV. OF CALIFORNIA, DAVIS, CA 95616

INTRODUCTION
    Chlorinated chlorophyll (Chl) derivatives, extracted from photosystem
I preparations of spinach, cyanobacteria and green algae, have been proposed by Dornemann and Senger (1) to be integral parts of PS I reaction
centers. 10-Hydroxy, δ-chloro-Chl, a candidate (2) for these chromophores, is also readily synthesized by reaction of Chl a derivatives with
mild oxidizing agents and chloride. The possibility that the compound
functions as either primary donor (P700) or acceptor (A_0 or A_1) in PSI is
considered by comparing its properties with those of Chl a in vitro and
those of P700, $P700^+$, A_0^- and A_1^- in vivo. The triplet zero field splittings, and redox, optical and ESR characteristics of the anion and cation
of the chlorinated compound do not differ widely from those of Chl a.
These results do not offer, therefore, clear diagnostic signatures that
would unambiguously identify the putative chlorinated species in vivo,
nor do they provide obvious advantages (or disadvantages) for a biological role for chlorinated chromophores.

RESULTS AND DISCUSSION
    The synthetic compound used as a model for RCI was 10-hydroxy,
δ-chloro, methylchlorophyllide a, prepared by successive introduction of
10-hydroxy, δ-chloro and Mg into methylpheophorbide a. For convenience,
it will be abbreviated as Cl-Chl in the text.
    Redox data for Cl-Chl and Chl a were obtained by cyclic voltammetry
in CH_2Cl_2 containing $(C_3H_7)_4NClO_4$ as electrolyte (3). Potentials were
measured versus a Ag/AgCl reference electrode and converted to the normal
hydrogen electrode. Mid-point potentials for oxidations of Chl and
Cl-Chl are 0.83 and 0.90V whereas the reductions occur at -1.08 and
-0.96V. These trends are consistent with the introduction of the electronegative chloride in Cl-Chl, which should make it harder to remove and
easier to add an electron to the macrocycle.
    For comparison, the oxidation potential of P700, the primary donor of
PSI, is estimated at 0.48V. The reduction potential for the "primary
acceptor" is presumed to be more negative than -0.7V (3).
    Clearly, without some additional in vivo interactions such as liga-
tions, special pair formation, etc., neither the chlorinated Chl nor Chl
a itself are good models for $P700/P700^+$. The reduction potentials in
vitro are compatible with estimates of the in vivo potentials of the primary acceptor (A_0).
    Triplet zero field splittings for the two compounds are identical:
D = 279 x 10^{-4} and E = 41 x 10^{-4} cm^{-1} at 90°K in 2-methyltetrahydro-
furan. For comparison, the splittings for the spin polarized triplet of
P700 range between 278-283 x 10^{-4} (D) and 39-40 x 10^4 (E) (4). Either
compound thus easily fits the in vivo results. Indeed, the close analogy

between the triplet parameters observed in PSI and those of Chl a have been used to argue for a monomeric Chl in P700 (4).

Cation and anion radicals of Cl-Chl are readily prepared electrochemically. The resulting optical spectra and the difference spectra obtained on oxidation and reduction are shown in Figures 1 and 2. Again, these spectra of Cl-Chl$^+$ and Cl-Chl$^-$ do not differ significantly from those of Chl$^+$ and Chl$^-$ (3).

The introduction of the chloro groups does not affect the g-values of the radicals, either. The values are listed in Table 1. Note that the g-value reported for A_1^- is clearly inconsistent with a chlorophyllous molecule. In fact, A_1 has recently been assigned to a quinone (5). The ESR line widths for the anions and cations of Chl, Cl-Chl, P700$^+$ and A_0^- are also presented in Table 1. There exist perhaps better agreement between the Chl$^-$ or Cl-Chl$^-$ and A_0^- than between either cation and P700$^+$. ENDOR data do not offer any better agreement between in vivo and in vitro species. The major coupling constants in frozen matrices attributed to P700$^+$ are 3.3-3.4 and 5.0-5.2 MHz versus 2.83, 3.72 and 7.56 for Chl$^+$ and 2.74 and 7.32 for Cl-Chl$^+$ (in CH_2Cl_2/CH_3OH to prevent aggregation and increase resolution). For A_0^-, the major coupling constants are 4.8 and 14 MHz versus 1.78, 5.41 and 10.77 for Chl$^-$ and 1.63, 3.38 and 10.93 for Cl-Chl$^-$ (3). Some of the disparities in coupling constants in the acceptor have been attributed to hydrogen bonding and various orientations of the 2-vinyl group whereas a variety of protein and aggregation factors have been considered for P700$^+$. (See references cited in 3)

In conclusion, the in vitro results for the chlorinated species presented above do not offer any consistent one-to-one correspondence with the properties found for either primary donors and acceptors in vivo. These results do not, therefore, offer conclusive evidence for or against chlorinated pigments, as opposed to Chl a itself.

Nonetheless, since some additional interactions such as aggregation, ligation and/or protein effects are required to explain the properties of the chromophores in vivo, whatever they are, the introductions of the 10-hydroxy group and of the electronegative and bulky chlorine do offer the potential for additional interactions with the protein and for conformational variations not available with Chl a.

REFERENCES
1 Dornemann, D. and Senger, H. (1981) FEBS Letters 126, 323-327. (1982) Photochem. Photobiol. 35, 821-826. (1986) Photochem. Photobiol. 43, 573-581.
2 Scheer, H., Gross, E., Nitsche, B., Cmiel, E., Schneider, S., Schafer, W., Schiebel, H. M. and Schulten, H. R. (1986) Photochem. Photobiol. 43, 559-571.
3 Fajer, J., Fujita, I., Davis, M. S., Forman, A., Hanson, L. K. and Smith, K. M. (1982) Adv. Chem. Ser. 201, 489-513.
4 Rutherford, A. W. and Mullet, J. E. (1981) Biochem. Biophys. Acta 635, 225-235.
5 Evans, M. C. W. These Proceedings.

ACKNOWLEDGMENTS
Work supported by the U.S. Department of Energy and the National Science Foundation.

TABLE 1. ESR parameters
           Line widths, ΔH, in Gauss; g-values

		Cation	Anion
Chl a,	T = 300°K 123°	ΔH = 9.2 G 9.5 g = 2.0026	11.0 13.0 g = 2.0028-9
Cl-Chl,	T = 300° 123°	ΔH = 7.0 8.5 g = 2.0026-7	10.2 13.7 g = 2.0029
P700	T = 77°	ΔH = 7.5 g = 2.0026	
A$_0$	" "		ΔH = 13.5 g = 2.002-2.003
A$_1$	" "		ΔH = 11 g = 2.0045

Cations in dichloromethane, anions in dimethoxyethane.

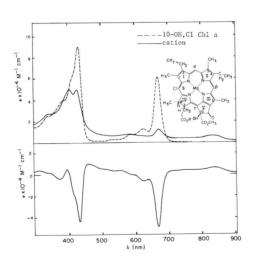

Fig. 1. Optical spectra of Cl-Chl
(- - -) and its cation radical (———)
in CH_2Cl_2 (top). Difference spectrum:
cation minus parent (bottom).

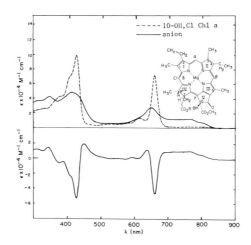

Fig 2. Optical spectra of Cl-Chl (- - -) and its anion radical (———) in dimethoxyethane (top). Difference spectrum: anion minus parent (bottom).

ENVIRONMENTAL EFFECTS ON THE PROPERTIES OF CHLOROPHYLLS IN VIVO.
THEORETICAL MODELS

L. K. HANSON, M. A. THOMPSON AND J. FAJER
BROOKHAVEN NATIONAL LABORATORY, UPTON, NY 11973

We consider here theoretical aspects of the interactions of photosynthetic chromophores with their environment. The close proximity of the donors and acceptors within the reaction centers (R.C.) could account for some of the optical changes observed on short time scales (10^{-12} to 10^{-13} secs) following the primary electron transfer (1). In bacteria this event generates a dimeric bacteriochlorophyll (BChl) cation and a bacteriopheophytin (BPheo) anion (2). According to the x-ray structure of the Rhodopseudomonas viridis R.C., these two radicals are positioned on opposite sides of a BChl b molecule within van der Waals contacts (3).

INDO calculations predict that point charges within 3-4 angstroms (A) of BChls could induce significant shifts of the red (Q_y) band (4). The magnitude and sign of the shifts depend upon the placement and sign of the point charge: positive charges placed near ring I or negative charges situated near ring III are predicted to cause large red shifts whereas reversing the charges should result in large blue shifts (Table 1).

A point charge affects the optical spectrum by modifying the magnitudes of the state dipole moments of the molecules--the larger the initial difference between the ground and excited state values, $\Delta\mu$, the larger the perturbation. For a charge at a given position, the magnitude of the induced shifts parallels the $\Delta\mu$ values and usually follows the order BChl g > BChl b > BChl a > Chl a. The Q_y state dipole moment lies along the y-axis (which runs through rings I and III), hence the sensitivity of the red band to charges placed near the y-axis.

The primary charge separation in a R.C. should therefore induce a significant electrochromic effect on the optical spectrum of the BChl that "bridges" the primary donor and the BPheo. Table 2 lists the shifts calculated using the R. viridis coordinates (3) for bridging BChl b with a positive charge placed at the donor (BChl b)$_2$ centroid and a negative charge at the center of the acceptor BPheo b. Also presented are the results of distributing the point charges over the pyrrole nitrogens of the donor and acceptor. Individually or together, these charges induce sizeable blue shifts for the bridging BChl b. Further delocalization of the charges over (BChl b)$_2$ and BPheo b, as is actually observed (5), may reduce the calculated shift. (At a given distance, the shift is proportional to the charge.) The calculations also do not reflect shielding effects by the protein and ring substituents. Experimentally, spectral changes in the bridging BChls have indeed been observed on subpicosecond or picosecond time scales (1) and they have been variously attributed to the formation of a BChl anion, to a charge transfer state or to electrochromic effects, as suggested here (see references cited in 1, 4 and 5). These effects would be susceptible to additional modulation if changes in distances or orientations between the chromophores follow electron transfer (4).

In the reaction center, the BChls are ligated to histidine residues and their various carbonyl oxygens could hydrogen (H) bond to other

Biggens, J. (ed.), Progress in Photosynthesis Research, Vol. I. ISBN 90 247 3450 9

residues (3,6). As shown in Tables 2 and 3, axial ligation or H-bonding are predicted to produce much smaller spectral shifts than point charges, in agreement with experimental studies (7).

Lastly, we consider H-bonding effects on the acceptor BPheo b. The BPheo b in the active electron transport pathway in R. viridis R.C.'s is H-bonded at the 9-keto position (3). Both extended Hückel and INDO calculations indicate that H-bonding at this position would make the BPheo easier to reduce by ~40mV. The largest change in the unpaired spin density distribution of the BPheo anion occurs at the C5 position, where an increase of 8% is predicted. These calculated effects appear rather small to be the major factors that determine the path of electron flow (L versus M) in the two arms of the reaction center. However, similar but asymmetric H-bonding of the special pair BChls (6) could result in small differences within the special pair sufficient to induce vectorial electron transport at the very onset of charge separation.

TABLE 1. Effect of point charges on the Q_y (red) band of
BChl g (INDO calcs).

The magnitudes of the calculated shifts are largest near the y-axis and fall off towards the x-axis.

Point charge	Position (charge is 3.5A above the atom)	Frequency shift of Q_y relative to no charge	
+1	N(I) (y-axis)	-1510 cm^{-1}	(red shift)
-1	N(I)	$+2291$	blue
+1	N(II) (x-axis)	-205	red
-1	N(II)	$+21$	blue
+1	N(III) (y-axis)	$+1502$	blue
-1	N(III)	-1109	red
+1	N(IV) (x-axis)	-551	red
-1	N(IV)	$+439$	blue

TABLE 2. Effect of donor and acceptor charges on the Q_y (red) band of the bridging BChl b of R. viridis reaction centers (INDO calcs).

Point charge	Position	Frequency shift of Q_y relative to no charge (all blue shifts)
+1	Center of the (BChl b)$_2$ 5.25, -9.85, 2.50 A*	940 cm^{-1}
-1	Center of the BPheo b -0.74, 5.84, 9.23 A*	554.5
+1, -1	Cation and anion located as above	1608.5
10 charges of +0.1 a.u. distributed over the 2 Mg and 8N of (BChl <u>b</u>)$_2$		858
4 charges of -0.25 a.u. distributed over the 4N of BPheo <u>b</u>		578
+, - Cation and anion distributed as above		1506

*x, y and z coordinates of the radicals (BChl)$_2$ or BPheo⁻ relative to the center of the bridging BChl.

Table 3. Effect of axial ligation, hydrogen bonding, and acetyl orientation on the Q_x and Q_y bands of BChl a (INDO calcs).

BChl a species	Frequency shifts in cm^{-1} relative to BChl <u>a</u> with C2-acetyl 25° out-of-plane	
	Q_x	Q_y
H$_2$O ligand	-676 (red shift)	9 (blue shift)
Imidazole ligand	-821 red	61 blue
H$_2$O H-bonded to C9-keto oxygen (ring V)	62 blue	96 blue
H$_2$O H-bonded to C2-acetyl oxygen (ring I)	-51 red	-101.5 red
C2 acetyl oriented 90° to macrocycle plane	245 blue	341 blue

REFERENCES
1 Breton, J., Martin, J.-L., Migus, A., Antonetti, A. and Orszag, A. (1986). Proc. Natl. Acad. Sci. USA 83, 5121-5125.
2 Fajer, J., Brune, D. C., Davis, M. S., Forman, A. and Spaulding, L. D. (1975). Proc. Natl. Acad. Sci. USA 72, 4956-4960.
3 Deisenhofer, J., Epp, O., Miki, K. Huber, R. and Michel, H. (1985). Nature 318, 618-624.
4 Fajer, J., Barkigia, K. M., Fujita, E., Goff, D. A., Hanson, L. K., Head, J. D. Horning, T., Smith, K. M. and Zerner, M. C. (1985). Antennas and Reaction Centers of Photosynthetic Bacteria (Michel-Beyerle, M. E., Ed.) pp 324-338, Springer-Verlag, Berlin.
5 Davis, M. S., Forman, A., Hanson, L. K. Thornber, J. P. and Fajer, J. (1979). J. Phys. Chem. 83, 3325-3332.
6 Lutz, M. Private communication. (See also these Proceedings.)
7 Hanson, L. K., Chang, C. K., Ward, B., Callahan, P. M., Babcock, G. T. and Head, J. D. (1984). J. Am. Chem. Soc. 106, 3950-3958.

ACKNOWLEDGMENTS
We thank J. Deisenhofer and H. Michel for the coordinates of the R. viridis chromophores. This work was supported by the Division of Chemical Sciences, U.S. Department of Energy.

EFFECTS OF STRUCTURE AND GEOMETRY OF PIGMENT-PROTEIN COMPLEXES ON EXPERIMENTAL QUANTITIES IN PRIMARY PROCESSES OF PHOTOSYNTHESIS

K. VACEK, M. AMBROZ, O. BILEK, J. HALA, V. KAPSA, P. PANCOSKA, I. PELANT, L. SKALA and L. SOUCKOVA
FACULTY OF MATHEMATICS AND PHYSICS, CHARLES UNIVERSITY, 121 16 PRAGUE 2, CZECHOSLOVAKIA

Modern methods of time-resolved and site-selection laser spectroscopy including hole-burning technique were used to study model photosynthetic systems consisting of porphyrins, aminoacids, polypeptides and quinone. Measured fluorescence lifetimes, CD spectra, frequencies of normal vibrations, site-distribution functions as well as homogenous widths of optical transitions give a detailed picture of energy structure, kinetics and interactions occuring in photosynthetic systems. Especially, we studied a set of isolated porphyrins: pheophorbide (PHEO), tetraphenylporphyrine (TPP) as references for complex model photosynthetic systems. The porphyrin – short-chain aminoacid interaction on TPP-phenylalanine and porphyrin – polypeptide interactions on PHEO – (Lys-Ala-Ala)$_{\sim}$ polypeptide were examinated. Furthemore,aggregation effects of porphyrin molecules were investigated,including fast energy transfer. The obtained results can be summarized in following:

1) Reference isolated TPP and PHEO porphyrins show major nanosecond lifetime in all used solutions and measured temperatures /1/.

2) The frequencies of normal vibrations of TPP and PHEO molecules are connected with vibrations of TPP and PHEO skeletons. These frequencies can be found in model TPP and PHEO photosynthetic systems as well /2,3/.

3) The site-distribution functions of isolated TPP and PHEO molecules embeded in low temperature matrices are gaussian-like curves with FWHM$\sim$100 – 280 cm-1 depending on solvents /3,4,5/.

4) The TPP – short-chain aminoacid (phenylalanine) interaction does not produce any significant changes in lifetimes, frequencies of normal vibrations and site-distribution functions. Much more intense is the effect of aggregation of TPP molecules in frozen n-octane. This aggregation poduces significant broadening of site-distribution functions (FWHM$\sim$100 – 240, 250 cm-1 and the red shift $\sim$110 cm-1). Similar effect was observed for n-(PHEO) – long chain polypeptide (/Lys-Ala-Ala/n, $\overline{mw}$. 10 000) model system where FWHM of particular site-distribution function was changed from 280 to 460 cm-1 and $\sim$150 cm-1 red shifted. The folding of polypeptide chain enables here the porphyrin – porphyrin interactions and fast energy transfer from monomeric PHEO molecules to aggregated PHEO followed by fast intersystem crossing. (Long wavelength fluorescence above 730 nm exhibit very intense $\sim$50 ps fast component /1/.

5) The porphyrin (TPP) – quinone interaction at 5K in model system * does not produce any significant changes in fluorescence lifetimes and frequencies of normal vibrations similarly as in /6/. Also lifetime determined from hole-burning experiments $\mathcal{T}_{HB} \geq$ 2 ps is much higher than those published in /7/ and /8/ on in vivo bacterial photosynthetic reaction centers. These results support the idea of cited authors that very fast electron transfer in fs scale is strongly coupled with protein environment.

6) The protein matrix influence on the physical properties of bonded

Biggens, J. (ed.), Progress in Photosynthesis Research, Vol. I. ISBN 90 247 3450 9
© 1987 Martinus Nijhoff Publishers, Dordrecht. Printed in the Netherlands.

porphyrins was studied /9/ on model systems with TPP fixed on twelve polypeptide matrices (-R⊕-X)n, where R⊕ = Lys, Arg, X = Ala-Ala, Ala-Gly, Leu-Ala, Ala-Leu or Ala sequences, differing in chirality of the created perturbing molecular field. The physical interpretation of the polypeptide influence on a fixed porphyrine can be as follows : The collective appearance of strong colinear components of electric and magnetic transition dipole moments in Soret region is proved by the induction of the optical activity in $\pi \to \pi^*$ transitions of the achiral porphyrine. The analysis of the corresponding selection rules for the D_{2h} chromophore showed (within the first order perturbation theory) that the necesary perturbational field should have the transformational properties as A_u -irreducible representation of D_{2h} symmetry group. It was shown that the fields of charge superhelices formed by the protonated R⊕ side chains on our sequential helical polypeptides contain the component of these properties. The calculations in /10/ localized the electrically forbidden transitions in Soret region. Thus, the spectral changes observed within the set of our model systems are the manifestation of the removing of the electrically forbidden character of these transitions. The dominant role of the molecular field helicity is demonstrated a) by the decrease of the observed changes by an order of magnitude for complex of TPP with nonhelical R⊕-Ala-Gly matrix; b) by the opposite CD band signs and shifts in spectra of R⊕AA (left-handed charge superhelix) and R⊕AAA or R⊕A (right-handed charge superhelix) complexes. The geometrical selectivity of this effect is demonstrated by the opposite CD sign patterns and Soret absorption maxima shifts for R⊕AA (-+, ϕ >0), R⊕LA (+-, ϕ <0) and R⊕AL (-+, ϕ >0) complexes, where the steric influence of leucine (L) side chain allow to manipulate with the porphyrine – helix angle ϕ (see Fig. 1).

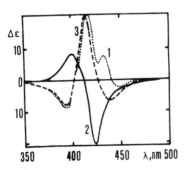

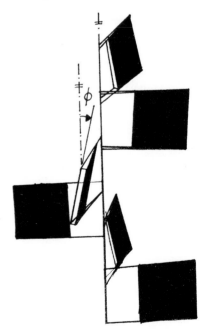

Fig. 1.
a) CD spectra of complexes
TPP + R⊕AA (1), R⊕LA (2), R⊕AL (3)

b) scheme of complex conformation:
—··—··—   – α-helix axis
black   – TPP planes
positive ϕ is shown.

Interpretation of photophysical measurements depends on a suitable treatment of the kinetics. With the aid of some simplifying assumptions, a
suitable kinetic model /10/ is solved. We investigate planar systems
consistings of C678 antennas and a single P680 reaction center in the
middle of the system. The antennas form regular square 'lattice', with
different dimensions of the reaction center, number of chlorophyl molecules
and orientation of their transition dipole moments. We take into account
levels and processes shown for the reaction center and one antenna below.

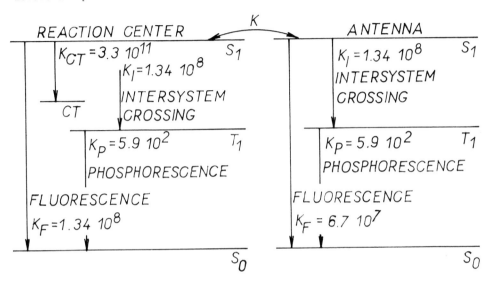

The intramolecular rate constants were obtained from the spectroscopic data
(see /11,12/); the intermolecular rate constant k is calculated from the
Forster formula /3/ for the Forster radius R_o=6.4 nm.
The most important conclusions are as follows (for details see /11/).
1) There is a characteristic time t_0 of the order of a few ps from which
all S_1 antenna levels are occupied with the same probability.
2) This allows to find for $t > t_0$ the analytic solution for the intensities
and quantum yields of the fluorescence, the phosphorescence and the
transfer to charge transfer level of the reaction center. The analytic
results are independent of the values of the intermolecular rate constants.
.3) The quantum yields and intensities of the antenna and reaction
center fluorescence and the quantum yield of the transfer to charge
transfer level have for $t > t_0$ single – exponential form. It appears that
for usual values of the rate constants the fluorescence life time is τ
=(N+1)/k_{CT} (N – number of antennas, k_{CT} – rate constant for the transfer to
charge transfer level). The measurements of τ for $t > t_0$ can be used to
calculate N or k_{CT}. Using k_{CT} =3.3 10^{11} s^{-1} and τ=492 ps (Chlorella, dark
adapted) this equation yields N=161.
4) The quantum yields and intensities of the antenna and reaction center
phosphorescence have for $t > t_0$ a two-exponential form with the corresponding life times τ and 1/k_p, k_p being the phosphorescence rate constant.
The exponential with τ can be neglected for $t > 200$ ps. Therefore, these
observables are for $t > 200$ ps single exponentials with life time 1/k_p.
5) The observables depend on 9 constants (the values of these quantities at

$t = t_0$). These values reflect the initial conditions and configuration of the system.
6) It is obvious that the most significant information on the transfer can be obtained from measurements for $t < t_0$.
7) It appears that the dimensions of the reaction center and the orientation of the transient dipole moments of the chlorophyll molecules have for $t > t_0$ very small effect on the efficiency of the transfer of the excitation to the reaction center.

* Authors would like to thank to Prof. M.R. Wasielewski for supplying the TPP-quinone sample.

REFERENCES
/1 /   J.Hala,G.F.W.Searle,T.J.Schaafsma,A.van Hoek,P.Pancoska,K.Blaha,
       K.Vacek(1986), Photochem. Photobiol., in press.
/2 /   J.Hala,I.Pelant,L.Parma,K.Vacek(1981), J.Luminisc. 26, 117-128;
         24/25,803-806.
/3 /   J.Hala,I.Pelant,M.Ambroz,P.Dousa,K.Vacek(1985),
       Photochem. Photobiol. 41,643-648.
/4 /   Hala J.,I.Pelant,M.Ambroz,P.Pancoska,K.Vacek(1986),
       Photochem.Photobiol. 41,643-648.
/5 /   Hala J.(1985) Proc.of Vth.Int. seminar on Energy Transfer,
       (Pancoska P.,Pantoflicek J. Eds.),Prague,1986.
/6 /   Wasielewski M.R.,and M.P.Niemczyk(1985) J.Am.Chem.Soc.106,
       5043-5045.
/7 /   Meech S.R.,A.J.Hoff and D.A. Wiersma(1985) Chem.Phys.Lett.
       121,287-292.
/8 /   Boxer S.G.,D.J.Lockhart,T.R.Middendorf(1986) Chem.Phys.Lett.
       123,476-482.
/9 /   Petke J.D.,Maggiora G.M.,Shipman L.L.,Christofersen R.E.
       (1978), Mol.Spect. 71,64-84.
/10/   Pancoska P.,Malon P.,Blaha K.(1986) in Proc. Vth. Int.
       Seminar on Energy Transfer,(Pancoska P.,Pantoflicek J.,Eds.)
       pp.92-100, Prague.
/11/   V. Kapsa, O. Bilek, L. Skala, Chem. Phys. Lett., to be published.
/12/   L. L. Shipman (1980), Photochem. Photobiol. 31, 157.
/13/   T. Forster (1965), in: Modern Quantum Chemistry, Istanbul Lectures,
       Part III., ed. Sinanoglu, (Academic Press, New York).

Molecular models of TPP – R⊕-Ala-Ala and R⊕-Leu-Ala polypeptide complexes
Central tube – backbone in α-helical conformations
only side chains are shown in detail
White planes – TPP

Chirality of side chain charge superhelices ⇄
Chirality of molecular fields of PP matrices

A) Elecrostatic field of PP matrix 20Å above and bellow TPP plane
at the connecting edge of TPP (12Å from the α-helix axis)
B) A_u-component of this field

Influence of angle ϕ on TPP spectral properties

Influence of reversed chirality of molecular field on TPP induced CD

Importance of PP conformation regularity for observed effects

Dependence of A_u-component of the PP molecular field on TPP-α-helix angle ϕ

INFRARED STUDY OF SOLID CHLOROPHYLL a ABSORBING NEAR 700 nm AT ROOM TEMPE-RATURE

Camille Chapados, Centre de recherche en photobiophysique, Université du Québec à Trois-Rivières, Québec, Canada, G9A 5H7

1. INTRODUCTION

Amorphous chlorophyll a (Chl a) absorbs at ~ 678 nm (Chl-678). Under the influence of diverse alcohols, the red band maximum is displaced to ~ 700 nm (Chl-701) [1]. The IR spectra of a similar sample showed three large carbonyl bands. Following solvation, the free ketone band is decreased to the profit of the coordinate ketone band. Full width at half-height (FWHH) of the bands varied from 17 to more than 38 cm^{-1} [1].

The FWHH of a carbonyl band of a sample in solution is between 10 and 15 cm^{-1} and the band shape is a Voigt profile [2]. For the carbonyl bands of Chl a, in mono- and multilayer organizations, we have obtained a value of ~ 13 cm^{-1} for the FWHH, a Voigt profile, and, ~ 14 bands [3]. Obviously, the IR bands observed for Chl-678, and Chl-701 are composed of many overlapping components. To separate these components, the Fourier self-deconvolution technique is used.

2. EXPERIMENTAL

Microcrystalline Chl a, dispersed in isooctane, is electrodeposited on Ge ATR plates and are rendered amorphous under high vacuum to give Chl-678. Vapors of 1-propanol is flushed on the sample to give Chl-701 [1].

Details of the Fourier self-deconvolution technique are given in refs. 3, and 4. The band shape used is a product function containing .75 Cauchy and .25 Gauss with a FWHH of 13 cm^{-1}; the apodization function is a triangular square, and ℓ varied from .15 to .25 cm.

3. RESULTS AND DISCUSSION
3.1 IR spectra of amorphous and 1-propanol solvated Chl a

The IR spectra of amorphous and solvated Chl a are shown in Figs. 1A and 1B, respectively. The difference between these two spectra and the integrated intensity of the difference spectra are shown in Figs. 1C, and 1D, respectively. The assignment of the IR bands is the following: the two ester carbonyl bands are situated at 1735 cm^{-1}; the free and associated ketones are situated at 1691 and 1653 cm^{-1}; the C=C and C=N bands are situated at 1608, 1550, and 1534 cm^{-1} [5]. The latter bands are not much affected by the solvation process and will not be discussed.

The most drastic modifications are occuring on the carbonyl bands. Upon solvation, the ester carbonyl band at 1735 cm^{-1} and the associate ketone carbonyl at 1653 cm^{-1} increase in intensity while the band situated at 1691 cm^{-1} decreases in intensity. Because the integrated intensity (Fig. 1D) falls to zero at ~ 1705 cm^{-1}, which is within the limit of the 1691 cm^{-1} band, the latter, in Chl-678, cannot be assigned solely to the free carbonyl group and must contains some coordinate ester groupings. The absorption around 1625 cm^{-1} is also decreased upon solvation.

3.2 Fourier self-deconvolution

The Fourier self-deconvoluted spectra are given in Figs. 2 and 3 with ℓ =.175 and .210 cm, respectively. The spectra have been normalized using

Biggens, J. (ed.), Progress in Photosynthesis Research, Vol. I. ISBN 90 247 3450 9
© *1987 Martinus Nijhoff Publishers, Dordrecht. Printed in the Netherlands.*

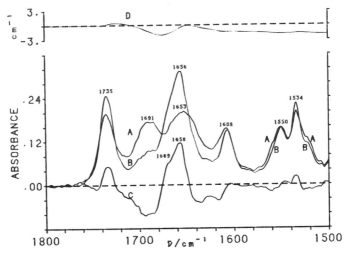

Fig.1. IR spectra of: A) Chl-678; B) Chl-701; C) Difference
between A and B; D) Integrated Intensity of C.

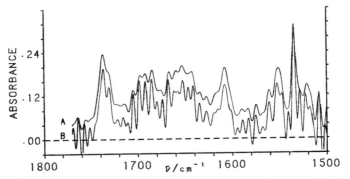

Fig.2. Fourier self-deconvoluted IR spectra of Chl-678
A) l=.175 cm; B) l=.210 cm. A displaced .04 abs.

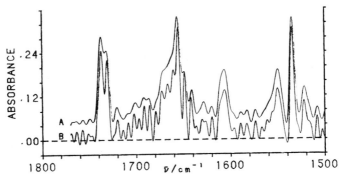

Fig.3. Fourier self-deconvoluted IR spectra of Chl-701
A) l=.175 cm; B) l=.210 cm. A displaced .04 abs.

Table 1. Self-deconvoluted position (in cm^{-1}) of the bands of Chl-678 and Chl-701. Band shape ratio = .75, FWHH = 13 cm^{-1}.

ℓ →	Ch1-678		Ch1-678		Assignment
	.175	.210	.175	.210	
1	~ 1749	1751, ~ 1745	1749	1751, 1747	} free ester CO
2	1737	1736	1737	1736	
3	~ 1731	1730	1731	1730	
4	1722	1721	1720	1719	asso. ester CO
5	1714	1715, 1710	~ 1708	1714, 1708	} asso. ester and
6	1705	1704	1701	1702	free ketone
7	1697	1698		1697	
8	1692	1692	1692	1691	} free ketone CO
9	1685	1685	1687	1685	
10	1676	1679, 1674	~ 1677	1679, 1672	
11	1666	1667	~ 1670	1666	
12	1659	1661	~ 1662	1660	asso. ketone CO
13		1657	1656	1655	first type
14	1653	1652			
15		1646	1649	1648	
16	1643	1641	1641	1642	
17	1634	1634		1630	asso. ketone CO
18	1627	1627	1628	1626	second type
29	~ 1616	~ 1619, ~ 1615	1620	1619	

the 1735 cm^{-1} band of the original spectra. The position of the bands are given in Table 1. The limit of the self-deconvolution is obtained with ℓ=.175 cm because negative lobes are starting to appear. The spectra with ℓ=.210 cm are also given to determine more precisely the position of the bands and to enhance certain spectral regions where overlapping shoulders make the determination of the number of bands difficult.

The first region, situated between 1770 and 1700 cm^{-1}, contains the free and associate ester groups. Upon solvation, the positions of the two strong bands, situated at 1736 and 1730 cm^{-1}, are not changed but their intensities are increased. This increase comes from the lowering of the intensities of the bands situated at higer and lower frequencies than these two bands. The bands situated higher than 1725 cm^{-1} are assigned to the free ester groups in different environments, those situated lower than 1725 cm^{-1} are assigned to Mg coordinated ester groups. Upon solvation, the coordinated bands are ruptured which liberates the ester and Mg groups. The bands situated at ~ 1710 and ~ 1700 cm^{-1} had previously been assigned to free ketones [6], since it is necessary to include these bands with the ester groups in order to explain the increase of the 1737 and 1731 cm^{-1} bands, the former bands must be composite bands containing associate ester groups and free ketone groups.

The second region, situated between 1700 and 1615 cm^{-1}, contains the free and associate ketone groups. Upon solvation the bands situated between 1670 and 1645 cm^{-1}, assigned to the coordinate ketones with different intermolecular bond lengths (first type), increase in intensity at the expense of: first, the free ketone bands situated between 1700 and 1670 cm^{-1}; and, secondly, the coordinate ketone bands situated between 1645 and 1615 cm^{-1} (second type). Although, the bands situated in the last region are more displaced from the free ketone position than those situated in the 1670-1645 cm^{-1} region, which would indicate a strong coordinate bond, the intensities of the bands in the 1645-1615 cm^{-1} region are decreased, whereas, the ones in the 1670-1645 cm^{-1} region, are increased. This is an indication that there are two types of coordinate ketone bonds: the second type are weaker than the first type. One way to explain the coordinate bonds of the second type is to consider that, because of the steric organization of the molecules, these are forced in unnatural positions and make bonds that are easily broken when the Chl molecules become more labile by the alcohol moiety.

To explain the experimental results, Chl-678 must have two types of coordinate carbonyls beside the free carbonyls. One type of coordinate carbonyls, which is responsible for the absorption in the 1670-1645 cm^{-1} region, makes strong coordinate bonds. The second type, which is responsible in part for the absorption in the 1725-1700 cm^{-1} region and for the absorption in the 1645-1615 cm^{-1} region, makes weak coordinate bonds. The molecular organization that best describe Chl-678 is a polymeric dimer where the intermolecular bonds within the dimers are stronger than those joining the dimer. Upon solvation, the weak bonds are ruptured to obtain the free dimer on which one molecule of alcohhol is coordinate bonded with the Mg, on one side and hydrogen bonded, with the available ketone carbonyl, on the other side to make Chl-701.

Because of the many bands in the 1670-1645 cm^{-1}, it is difficult to sort out, in the spectrum of Chl-701, a band which could be assigned exclusively to the hydrogen bonded ketone, although, the band that best describe this situation is the strong one situated at 1655 cm^{-1} (fig. 3B). In the spectrum of Chl-678 (Fig. 2B), only two weak bands (1657, 1652 cm^{-1}) are situated in this region. These bands do not coincide exactly with the 1655 cm^{-1} band whereas the position of the other bands of both spectra match.

A dimer model has been proposed for an organization of a mixture of Chl a and ethanol in toluene at 175 K that absorbs at 702 nm [7]. This model consists of two molecules of Chl a, in a C_2 symmetry coordinated to two molecules of alcohol which are hydrogen bonded to the ketone carbonyls. In our system, this model is unlikely because it would require that all the coordinate bonds are ruptured to make new ones with the alcohol. We do not observe that such is the case.

REFERENCES

1. Dodelet, J.P., Le Brech, J., Chapados, C., and Leblanc, R.M. (1980) Photochem. Photobiol. 31, 143.
2. Jones, R.N., and Sandorfy, C. (1956), Chap. IV in "Chemical applications of spectroscopy", ed. W. West, Interscience Pub., N.Y; Jones, R.N., Ramsay, D.A., Keir, D.S., and Drobiner, K. (1952) J. Am. Chem. Soc. 74, 80.; Seshadri, K.S., and Jones, R.N. (1963) Spectrochim. Acta 19, 1013.
3. Chapados, C., Béliveau, J., Trudel, M. and Lévesque, C. (1986) App. Spectros. 40, XXX.
4. Kauppinen, J.K., Moffatt, D.J., Mantsch, H.H., and Cameron, D.G. (1981) App. Spectros. 35, 271.
5. Ballschmitter, K., Katz, J.J. (1972) Biochim. Biophys. Acta 256, 307; Chapados, C., and Leblanc, R.M. (1977) Chem. Phys. Lett. 49, 180; Chapados, C., Germain, D., and Leblanc, R.M. (1980) Biophys. Chem. 12, 189.
6. Chapados, C., and Leblanc, R.M. (1983) Biophys. Chem. 17, 211.
7. Shipman, L.L., Cotton, T.M., Norris, J.R., and Katz, J.J. (1976) Proc. Natl. Acad. Sci. USA 73, 1791.
8. Supported by the Fonds FCAR du Québec and by the NSERC of Canada. The figures were done by M. Trudel.

BOROHYDRIDE REDUCTION OF BACTERIOCHLOROPHYLL a IN THE LIGHT HARVESTING
PROTEIN OF RHODOSPIRILLUM RUBRUM

Patricia M. Callahan, Therese M. Cotton and Paul A. Loach[*]
Department of Chemistry, University of Nebraska, Lincoln, NE 68588-0304
*Department of Biochemistry, Molecular Biology and Cell Biology,
Northwestern University, Evanston, IL 60201

INTRODUCTION

The absorption spectra of all bacteriochlorophyll-protein complexes
are red-shifted relative to isolated, monomeric Bchl a. The exact cause
of these spectral shifts is not known but they arise from combined
BChl-BChl and BChl-protein interactions. The magnitude of the BChl-Bchl
interactions has been studied by circular dichroism spectroscopy(1,2,3)
but additional BChl-protein interactions must be invoked to account for
the full magnitude of the absorption red-shift(4). Similar to the
retinylidene pigment of rhodopsin, hydrogen bonding, point charges in the
vicinity of the chromophore(5) and protonated Schiff base (SBH+) formation
at the ketone functionalities of BChl a(6,7) have been suggested as
possible mechanisms to perturb the BChl a spectra. Model compounds have
been developed that show that these perturbations do indeed effect chlorin
and bacteriochlorin optical spectra(8,9,10). The SBH+ models display the
largest spectral shifts(9,10) and are highly sensitive to solvent and
counterion effects thus providing a means of modulating the pigment
absorption spectrum by its immediate protein environment. To investigate
the possibility of such a protonated Schiff base linkage between the
peripheral substituent C=O groups of BChl a and an amino group of the
antenna protein from R. rubrum, we have studied the effects of $NaBH_4$ and
$[CH_3(CH_2)_3]_4NBH_4$ (tetrabutylammonium borohydride) on the native and
detergent modified forms of this protein. Borohydride reduction of both
the free C=O groups of BChl a and the $C=N^+HR$ bond of the putative iminium
linkage will result in a dramatic blue-shift in the Qy absorption maximum
of BChl a. To distinguish between these two forms we examined the organic
solvent extractability of pigment from protein because only the reduction
product of a Schiff base linkage will be covalently linked to the
polypeptide backbone.

MATERIALS AND METHODS

Bacteriochlorophyll a and chlorophyll a were isolated by the
procedures of Strain and Svec(11). Borohydride reduction of the
chlorophyll samples was carried out by addition of excess sodium
borohydride or tetrabutylammonium borohydride to a degassed solution of
the pigment. The protein samples used in this study were prepared
according to the methods described by Loach et.al.(12). A brief summary
is as follows. R. rubrum G-9 chromatophores were treated with 1% Triton
X-100, 5mM EDTA in phosphate buffer, pH 7.5 and spun at 100,000xg to
isolate photoreceptor complexes. These subchromatophore fragments are
washed with water to remove detergent and lyophilized to dryness. Further
purification to B818 antenna and reaction center complexes is possible but
was not carried out on the protein spectra recorded here. Similar
spectral shifts were observed for the partially purified antenna protein

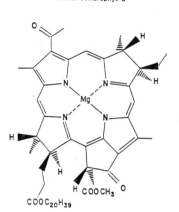

Bacteriochlorophyll a

and its isolated form. The native, B873, antenna protein was converted to an 818nm absorbing form, B818, by addition of approximately 1.0% octyl glucoside. Reversible conversion of B818 into a native 873nm absorbing species was accomplished by dilution of octyl glucoside below its critical micelle concentration. Extraction of BChl a from the native and modified antenna proteins was carried out by addition of diethyl ether to the aqueous protein solutions. The ether layer was separated and its absorption spectrum taken. The aqueous layer was also monitored spectroscopically for the presence of protein and residual pigment.

RESULTS AND DISCUSSION

The effects of the aldehyde and ketone specific reductant, $NaBH_4$, on the BChl a pigments in the antenna protein of R. rubrum are shown in Figure 1. There is no bandshift of the native B873 absorption spectrum after addition of $NaBH_4$ (not shown), but there is a slight narrowing of the Qy maximum at 873nm from 340 to 300 cm^{-1} (HWHM). The reaction of the octyl glucoside-converted 818nm form with $NaBH_4$ is more dramatic with ~100nm blue-shift in Qy from 818nm to 720nm and a distinct splitting of the Soret. There is also a large decrease in intensity of the Qy absorption maximum relative to BChl a and and increase in the Soret/Qy absorption intensity ratio. The Qx band shifts from 595nm to 565nm. These absorption shifts are consistent with reduction of one or both of the BChl a ketone functionalities at positions 2 and 9. The tetrahydroporphyrin nature of the bacteriochlorin ring seems to be still intact because of the red-shifted Qy absorption maximum of the 720nm product relative to Chl a or other chlorin species at <700nm.

The reaction of both B873 and B818 with the organic solvent soluble reductant $[CH_3(CH_2)_3]_4NBH_4$ yields the same product with a Qy absorption maximum at 720nm, similar to the spectrum shown in Figure 1. The same borohydride reduction product is formed in this case but now BChl a is accessible to exogenous reductants. The reversed 870nm antenna species formed by dilution of octyl glucoside from B818 displays the same reactivity pattern as the native antenna protein, i.e., no reaction with $NaBH_4$ and a dramatic absorption shift from 870 to 720nm after addition of $[CH_3(CH_2)_3]_4NBH_4$. The rate of the $[CH_3(CH_2)_3]_4NBH_4$ reduction is very slow, requiring approx. 8 hours for completion while the $NaBH_4$ reactions are much faster (1-2 hours).

With the exception of B873, all of the pigments are extractable with organic solvent (Et_2O and CH_2Cl_2). B818 yields 770nm absorbing BChl a upon extraction from the protein and the 720nm species absorbs at 725nm in a non-aqueous environment.

The reaction of isolated BChl a and $NaBH_4$ in detergent/buffer solvent system results in a shift of the 770nm Qy maximum to 720nm. This bacteriochlorin product can be pheophytinized by addition of HCl to yield

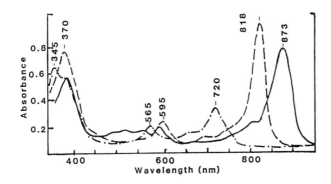

FIGURE 1   Absorption spectra of R. rubrum Triton-washed chromatophores in potassium phosphate buffer, pH 7.5 in the presence of --- 0.5% octyl glucoside, - - - 1.0% octyl glucoside and - - 1.0% octyl glucoside + NaBH$_4$.

a species with Qy at 715nm. Similar spectra of the borohydride reduction product of BChl a have been reported previously (13).

The large absorption spectral shift upon reduction of BChl a and the BH$_4$- reduction of Chl a(14), lead us to conclude that the C2 acetyl group and possibly the C9 ketyl group are reduced in the 720nm product. Ditson et. al.(13) concluded from ^{1}NMR data that only the C2 acetyl group is reduced upon borohydride reduction. NaBH$_4$ in alcohol is known to reduce aldehydes much faster than ketones(15) so the unhindered acetyl group of position 2 may react faster than the C9 ketone. Therefore, the different solvent conditions EtOH vs. H$_2$O and different reaction times (30 min. vs. several hours) may explain the different structural conclusions.

As stated in the introduction, the borohydride reduction product of an iminium linkage between a peripheral substituent ketone group of BChl a and an amino group of the protein backbone should result in a covalently linked pigment-protein complex. Because the 720nm bacteriochlorin reduction product is extractable from the protein, we conclude that borohydride does not form or lock-in a covalent linkage between BChl a and the amino acid backbone in the antenna protein and therefore an SBH+ linkage is not present in either B873 or B818. Because protonated Schiff base linkages are highly susceptible to hydrolysis, it is possible that borohydride may first modify the protein structure to expose a R$_2$C=N$^+$R$_2$ linkage to the aqueous environment, hydrolyze the linkage to R$_2$C=O and then reduce the ketone functionalities. However, because no intermediate spectral form is seen during the time required for reduction and because of the different reactivity of NaBH$_4$ and [CH$_3$(CH$_2$)$_3$]$_4$NBH$_4$ in the 873nm form we do not favor a reaction scheme of hydrolysis followed by carbonyl reduction.

The differing accessibility of exogenous reductants to B873 and B818 has implications for the organization of BChl a in the antenna protein. The detergent modified 818nm form is readily accessible to NaBH$_4$ consistent with its smaller subunit size(12). Also, the facile NaBH$_4$ reduction of the C2 acetyl group of BChl a in B818 relative to B873 is reflected in the modified BChl a C2 acetyl stretching frequency in B818 vs. B873(16). The necessity of a hydrophobic reductant in the reaction with B873 is consistent with the more tightly bound and excitonic nature of BChl a in the native antenna protein structure.

ACKNOWLEDGEMENTS
     This work was supported by: NIH 09867 (P.M.C.), NIH GM35108 (T.M.C.)
and NIH GM11741 (P.A.L.).

REFERENCES
1 Sauer,K. and Austin, L.A. (1978) Biochemistry 17,2011-2019.
2 Kramer, H.J.M., Pennoyer, J.D., Van Grondelle, R., Westerhuis, W.H.J.,
   Neiderman,R.A. and Amesz, J. (1984) Biochem. Biophys. Acta 767, 335-344.
3 Cogdell, R.J. and Scheer,H. (1985) Photochem.Photobiol. 42,669-678.
4 Pearlstein, R.M. (1982) in Photosynthesis: Energy Conversion by Plants
   and Bacteria (Govindjee, ed.) Vol. 1 pp.293-330.
5 Eccles, J. and Honig, B. (1983) Proc. Natl. Acad. Sci. USA 80,
   4959-4962.
6 Ward, B. Callahan, P.M. Young,R. Babcock, G.T. and Chang, C.K. (1983)
   J.Am Chem.Soc. 105, 634-636.
7 Hanson, L.K., Chang, C.K., Ward,B., Callahan, P.M., Babcock, G.T. and
   Head, J.D. (1984) J.Am.Chem.Soc. 106, 3950-3958.
8 Davis, R.C., Ditson, S.L., Fentiman, A.F. and Pearlstein, R.M. (1981)
   J.Am.Chem.Soc. 103,6823-6826.
9 Ward, B., Chang, C.K. and Young,R. (1984) J.Am.Chem.Soc. 106, 3943-3950.
10 Maggiora, L.L. and Maggiora G.M. (1984) Photochem. Photobiol. 39,
   847-849.
11 Strain, H.H. and Svec. W.A. (1966) in The Chlorophylls, Chapter 2,
   Academic Press, NY.
12 Loach, P.A., Parkes, P.S., Miller, J.F., Hinchigeri, S. and Callahan,
   P.M. (1985) in Molecular Biology of the Photosynthetic Apparatus (K.E.
   Steinback, et.al.,eds.)pp.197-209, Cold Spring Harbor Laboratory.
13 Ditson, S.L., Davis, R.C. and Pearlstein, R.M. (1984) Biochim. Biophys.
   Acta 766, 623-629.
14 Holt, A.S. (1959) Plant Physiol.34, 310-314.
15 Brown, Wheeler and Ichikawa (1957) Tetrahedron, 1,214-220
16 Callahan, P.M., Cotton, T.M. and Loach, P.A. (1985) in Spectroscopy of
   Biological Molecules (A.J.P. Alix et.al.,eds.) pp.354-356, John Wiley
   and Sons, NY.

FOURIER-TRANSFORM INFRARED (FTIR) SPECTROELECTROCHEMISTRY OF BACTERIOCHLOROPHYLLS

W.Mäntele*, A.Wollenweber*, E.Nabedryk°, J.Breton°, F.Rashwan", J.Heinze" ,W.Kreutz*

* Institut für Biophysik und Strahlenbiologie, Albertstr. 23, and
" Institut für Physikalische Chemie, Albertstr. 21,Universität Freiburg, D-7800 Freiburg FRG
° Service de Biophysique, Dépt. de Biologie, C.E.N. Saclay, 91191 Gif-sur-Yvette, France

1. INTRODUCTION

In the bacterial photosynthetic reaction center (RC), absorption of light leads to electron transport from a Bacteriochlorophyll (BChl) a or b dimer to a Bacteriopheophytin (Bpheo) a or b monomer and, later, to a quinone acceptor (1). Only recently, the crystal analysis of the Rp.viridis RC has allowed to visualize electron transport pathways (2). However, considerable amount of spectroscopic data has helped to obtain information on the orientation of the pigments and redox carriers, on their precise order within and on their interaction with the polypeptides (3). Many models - theoretical calculations as well as biomimetic models - have been employed to account for the chlorophyll's interactions in vivo. Among others, interaction of chlorophylls in a special pair, with one or more water molecules or with protein side chain groups has been discussed (4). Infrared spectroscopy has proved to be an extremely useful tool for the study of such interactions, since it is sensitive to all bonds of the chlorophyll molecules and not limited to the bonds involved in the π-electron system, as is resonance Raman spectroscopy. The latter technique, however, due to its selectivity, allows the pigment vibrations in the pigment-protein complexes to be studied (5).

In recent work, we have demonstrated that a similar selectivity can be obtained by using Fourier-transform infrared (FTIR) spectroscopy to obtain light-induced difference spectra in photosynthetic membranes and RC complexes (6,7,8). The sensitivity of FTIR spectroscopy is high enough to detect single bonds in a RC complex. Using different techniques, the difference spectra of the photo-oxidized primary electron donor and of the reduced intermediary electron acceptor could be obtained. All these IR difference spectra show a variety of characteristic difference bands highly reproducible in frequency and in amplitude. In order to determine the contribution of the pigment bands to the spectra (which might contain absorption changes from the polypeptides, lipids and water as well), a comparison to model compounds is necessary. Present infrared studies on model compounds, however, only involve the neutral chlorophyll species. A detailed discussion should involve cations and anions of the pigments, since abstraction or addition of an electron to the π-electron system changes bond character and thus may result in changes in intensity and position of the absorption bands.

Principally, it is possible to study the chemically oxidized or reduced pigments. However, to avoid spectral contributions of chemical oxidants, radicals generated in an electrochemical cell have to be studied. For that

Biggens, J. (ed.), Progress in Photosynthesis Research, Vol. I. ISBN 90 247 3450 9

purpose, we have developed a spectroelectrochemical cell which allows in situ electrolysis under spectroscopic control from 200 nm to 10000 nm. Radicals can be formed in this cell with high yield as well as with high reversibility upon the reversal of the current. They may be either studied by their IR (and VIS/UV) absorption spectra or by their IR difference spectra to the neutral species and may thus be directly compared to the FTIR difference spectra of the pigment-protein complexes. Potentials as well as kinetic parameters of the redox reactions, both necessary prerequisites for IR spectroelectrochemical work, were determined in another cell by cyclic voltammetry.

2. PROCEDURE

Electrochemistry: Electrolysis of chlorophylls with spectroscopic control was performed in a specially designed spectroelectrochemical cell at an optically transparent electrode. Working and counter electrode consisted of a Pt mesh, while the reference electrode was formed by a Ag/AgCl electrode. Optical path length was 0.25 mm. Cyclic (CV) and AC voltammetry of the compounds were performed in electrochemical cell designed for smallest possible sample quantities ($\leqslant 10^{-5}$ Mol). Details of both cells and the equipment will be reported elsewhere.

Pigment preparation: BChl a and BChl b were purified and assayed as described earlier (9).

Solvents and supporting electrolyte: All solvents were of analytical grade. THF was dried as described in (10). Dry D_8-THF was condensed under vacuum onto Al_2O_3 three times. $TBAPF_6$ was purified as described in (11).

Spectroscopy: IR spectra (obtained from 64 interfermeter scans) were recorded on a Nicolet 60 SX FTIR spectrophotometer equipped with a MCT-A detector.

3. RESULTS AND DISCUSSION

Fig.1a shows the FTIR difference spectrum of the cation formation of BChl a in CD_3OD at U=+0.75 V. Deuterated methanol was used to obtain an IR "window" from about 1900 cm^{-1} to about 1200 cm^{-1}. Before starting the electrolysis, a reference spectrum was taken of the neutral compound. During electrolysis, spectra were recorded every minute and referenced to the spectrum of the neutral compound (for a series of cation and anion evolution see fig. 1c/1d). Thus, in the difference spectrum, positive bands belong to the radical formed, while negative bands belong to the neutral species. As controlled by VIS absorbance spectra, the formation of the cation is almost complete after about 5 minutes. Prolonged electrolysis allows approx. 95 % of the cation to be formed, with more than 90% of it being reversible. This is also indicated in fig. 1b, where the same bands appear with opposite sign upon re-reduction of the cation. The most prominent bands in the difference spectrum are positive peaks at 1754 cm^{-1}, at 1720 cm^{-1} and at 1521 cm^{-1}, as well as a broad, intense negative band centered around 1660 cm^{-1} and a sharp negative band at 1338 cm^{-1}. In addition, a large number of small positive and negative bands appear that are higly reproducible as well as reversible upon re-reduction. If the absorbance spectrum and the electrochemically induced difference spectrum are superimposed at the same scale (not shown here), it is evident that the difference bands can reach the full size of the absorbance band, i.e. that bond character and thus the oscillator strength are drastically changed upon radical formation. This seems reasonable in the case of the bonds involved in the π-electron system, such as the bands at 1521 cm^{-1} and at 1338 cm^{-1}, which are most likely due to C=C and C=N vibrations. The band at 1754 cm^{-1}, however, which appears upon cation formation, can be assigned to

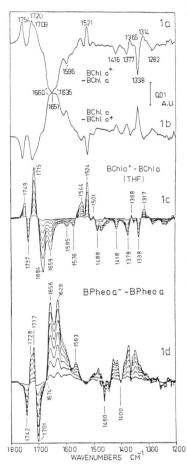

the vibration of the ester carbonyl groups, which are not involved in conjugation (5). The fact that no corresponding negative band is observed may be explained by assuming that the ester C=O group, in the neutral state, is hydrogen-bonded and thus gives rise to a broader negative band shifted to lower wave-numbers. In the difference spectrum, only the sharp positive band of the localized C=O group would show up. In the case of the band at 1720 cm^{-1}, which can be assigned to a free 9-keto carbonyl group, the corresponding nega-tive peak is centered around 1650 cm^{-1}. This extreme downshift could arise from an extensive C=O...D-O-CD$_3$ bonding to the deu-terated methanol. If the cation is formed in an aprotic solvent like THF instead (fig. 1c), this negative band is shifted to 1684 cm^{-1} and a negative band at 1737 cm^{-1} associated with the positive band at 1749 cm^{-1}, appears. All other bands are found at a similar position in THF as compared to CD$_3$OD.

The BChl b cation can also be formed electro-chemically. Due to the extra out-of-cycle double bond the π-cation formation is accom-panied by irreversible oxidation. Its IR dif-ference spectrum in the carbonyl frequency region (see insert of fig. 2a) shows small band shifts with respect to the BChl a cation. The BPheo a anion radical can be easily formed in THF at U=-1 V. The difference spectrum in fig 1d shows very strong intensity changes in the region from 1600 cm^{-1} to 1750 cm^{-1}. Two negative bands at 1742 cm^{-1} and at 1701 cm^{-1} seem to be associated with the free ester C=O and free keto C=O, respectively, which both undergo changes in their bond character. The strong positive bands at 1656 cm^{-1} and at 1629 cm might arise from acetyl C=O groups and/ or associated keto C=O groups that would point to a possible oligomerisation in the anion form. Because of the instability of BChl b pheophytinisation is not quantitative. However, HPLC-purified BPheo b can be reduced to the anion radical. Similar IR difference spectra (see insert of fig. 2d) were obtained with cha-racteristic shifts in band position as compared to BPheo a.

In fig. 2 a, series of FTIR difference spectra of the P$^+$ and of the I$^-$ state in BChl a and BChl b-containing pigment-protein complexes is shown together with the difference spectra of the corresponding radical formed electrochemically. A remarkable similarity, especially in the C=O frequency region, is observed for BChl cation formation as well as for Bpheo anion formation. A common feature in both processes of radical formation seems to be a strong involvement of the ester and keto C=O groups. Their binding to external ligands seems to be strongly influenced by abstraction or addition of an electron from the conjugated system. In the P$^+$ spectra and their BChl a and b counterparts, only one band in the ester C=O frequency region points to equivalence of the two ester groups (or just one of them being influenced by oxidation). In the I$^-$ spectra and the BPheo counterparts,

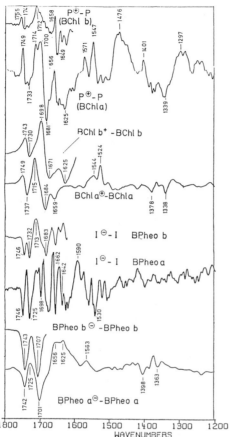

however, a single band in the ester C=O frequency region is observed for the model compounds, whereas the I^- spectrum shows two split esters, indicating a different type of ligation (4). It is interesting to note that, in the model compound IR difference spectra, bands assigned to C=O, C=N and C=C vibrations appear at comparable strength; whereas, in the difference spectra of the pigment-protein complexes, the carbonyl intensity dominates by far whilst vibrations of the tetrapyrrole moiety appear weaker and with large shifts with respect to the model compounds. However, further studies involving a number of different solvents in the IR spectroelectrochemical studies will be necessary to probe more closely the possible interactions of the peripheric groups of the chlorophyll molecules with their environment. The comparison of such IR model compound studies should allow conclusions to be drawn on the situation in vivo. Studies on the Chl a cation and the Pheo a anion radical are in progress and may help to elucidate on the nature of the primary donor in P_{700}. Finally, it appears to us that electrochemistry, supplemented by the extreme sensitivity of FTIR spectroscopy, may be of general use for the study of redox reactions, for biological, organic and inorganic compounds.

Acknowledgements: The authors would like to thank B. A. Tavitian for his help with FTIR spectroscopy, K. O. Lorenz for purifying $TBAPF_6$ and solvents and Dr. G. Berger & J. Kleo for their help in preparing pigments.

References

1. Parson,W.W.,(1982), Annu. Rev. Biophys. Bioeng. 11,57-80
2. Deisenhofer,J. et al,(1984), J. Mol. Biol. 180, 385-398
3. Breton,J. and Vermeglio,A. (1982) in: Photosynthesis: Energy conversion by Plants and Bacteria (Govindjee,ed.) vol.1, pp.331-385
4. Ballschmitter,K. and Katz J.J. (1969) J. Am. Chem. Soc. 91, 2661-2677
5. Lutz, M. (1984) in: Advances in Infrared and Raman Spectroscopy (Clark, R.J.H. and Hester, R.E., eds) vol. 11, pp. 211-300, Wiley-Heyden
6. Mäntele, W. et al (1985) FEBS Lett. 187, 227-232
7. Nabedryk, E. et al (1986) Photochem. Photobiol. 43, 461-465
8. Travitan, B.A. et al (1986) FEBS Lett. 201,151-156
9. Wollenweber,A. et al Colloque de Photosynthèse, Saclay 24-25 April 1986
10. Heinze J. Angew. Chem. int. ed (1984) 23, 831-847
11. Süttinger, R. (1979) PhD Thesis, Universität Freiburg, FRG

SOLVENT EFFECTS ON THE ELECTRON TRANSFER KINETICS OF BACTERIOCHLOROPHYLL OXIDATION

THERESE M. COTTON and RANDALL L. HEALD, Department of Chemistry, University of Nebraska-Lincoln, Lincoln, NE 68588-0304

1. INTRODUCTION

In the photosynthetic reaction center photoinduced electron transfer from the excited singlet state of the special pair BChl donor to the intermediate acceptor bacteriopheophytin (BPh) is known to occur in picoseconds with a quantum yield $\Phi \approx 1$. Because of the high quantum efficiency of the photosynthetic process, there have been numerous attempts to utilize chlorophylls as photosensitizers in photoelectrochemical cells. However, Φ has generally been quite low in these applications. Undoubtedly, the highly organized structure of the reaction center is crucial to its photochemistry. However, some improvement in the efficiency of chlorophyll photoelectrochemistry should be possible through the optimization of its electron transfer kinetics at electrode surfaces. To the best of our knowledge, there have been no detailed studies of the heterogeneous electron transfer rate constant (k_s) of the chlorophylls. Our research goal is to determine those experimental parameters (solvent, electrolyte, and electrode material) which enhance electron transfer between the chlorophylls and metal or semiconductor electrodes and minimize undesirable side reactions.

2. PROCEDURE

The BChl and BPh were prepared by sucrose chromatography and dried by established procedures. The solvents used in these experiments were HPLC grade and were dried over molecular sieves and distilled under vacuum. An IBM EC 219 Pt rotating disk electrode (RDE) was used for all the electrochemical experiments. A BAS 100 potentiostat was used to acquire the data.

3. RESULTS AND DISCUSSION

Diffusion Coefficient Figure 1 illustrates the effect of rotation rate on i_l, the limiting value of the diffusion current resulting from the one-electron oxidation of BChl: ($BChl \rightarrow BChl^{\dot{+}} + e^-$) in ethanol as measured by rotating disk voltammetry. As the potential is scanned through the region where oxidation commences, the current increases until it reaches a plateau (the gradual increase in current near + 0.700 V is due to the second electron oxidation: ($BChl^{\dot{+}} \rightarrow BChl^{2+} + e^-$).

Figure 1

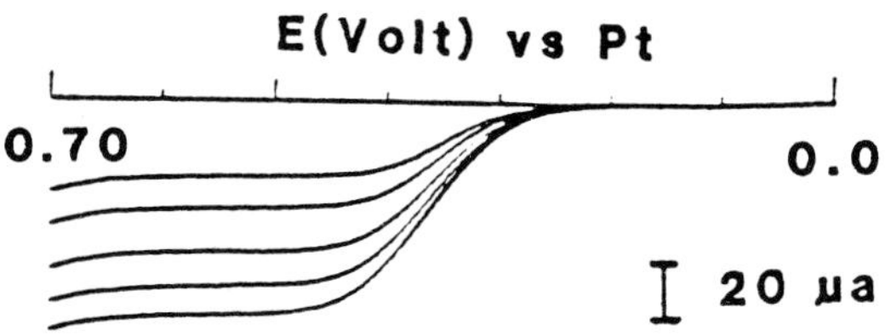
E(Volt) vs Pt
0.70
0.0
20 μa

Biggens, J. (ed.), Progress in Photosynthesis Research, Vol. I. ISBN 90 247 3450 9

The Levich equation predicts that the limiting current, i_l, should increase linearly with the square root of the rotation rate. As may be seen from Figure 2, BChl is well-behaved in this respect. Similar results were obtained for all of the solutions examined in this study.

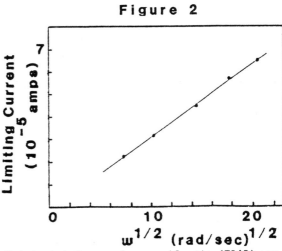

Figure 2

Limiting Current $(10^{-5}$ amps$)$

$w^{1/2}$ (rad/sec)$^{1/2}$

Table I gives the values obtained for the diffusion coefficient and heterogeneous electron transfer rate constant of BChl as a function of solvent, and counterion (0.1 M tetrabutyl ammonium perchlorate (TBAP) was used as the electrolyte unless otherwise indicated). The following conclusions can be reached from these results. First, aggregation reduces the diffusion coefficient of BChl by ca. 25 % as can be seen from a comparison of the diffusion coefficient in CH_2Cl_2 and CH_2Cl_2 with 10:1 moles of pyridine per mole of BChl. BChl is self-aggregated in dry CH_2Cl_2 and the addition pyridine results in its disaggregation and formation of five coordinate BChl. Second, the formation of six coordinate BChl (1000:1, pyridine:BChl) has only a small effect on D. Third, D decreases markedly for the longer chain alcohols. This is partially the result of their increased viscosity relative to methanol, but may also result from the enhanced solubility of the phytol tail in solvents with lower dielectric constants. Thus, the phytol tail may be in a more extended configuration, which in turn would reduce the mobility of the BChl. Fourth, the diffusion coefficient of BPh is greater than that of BChl in CH_2Cl_2. This is not unexpected because it lacks Mg and, therefore, does not self-aggregate nor undergo ligation interactions.

Heterogeneous electron transfer rate constant Figure 3 shows a typical cyclic voltammogram obtained for BChl in methanol at a scan rate of 100 mV/s. The electron transfer rate constant was evaluated from the separation in the anodic and cathodic peak potentials (ΔE_p) according to the treatment of Nicholson [1]. These values are listed in column 2 of Table I.

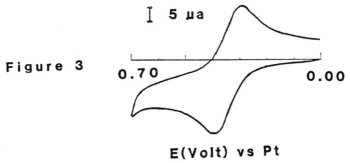

I 5 µa

Figure 3

0.70   0.00

E(Volt) vs Pt

The value of k_s is observed to vary by one order of magnitude with solvent and counterion. The fastest electron transfer rate constant for BChl was observed in methanol solutions (7.21×10^{-2} cm/s) and the slowest in the presence of TBAHFP (5.8×10^{-3} cm/s). The decrease in rate constant in the latter case results from specific adsorption of the anion at the electrode surface. The reduction in electron transfer rate constants in the presence of specifically adsorbed anions has been observed for other electroactive species. The smaller rate constant results from the increased distance between the electroactive species and the electrode surface as well as alterations in the double layer structure [6].

Aggregation and ligation interactions have only very small effect on the kinetics. From a comparison of the values observed for five and six coordinate BChl in CH_2Cl_2 with that in methanol, it can be concluded that the rate is governed primarily by the solvent polarity, with polar solvents providing the fastest rates. Also, the counterion may be important in this respect because the results obtained with $LiClO_4$ are somewhat less than with TBAP. This probably reflects a stronger tendency for ion pairing in the former which reduces anion stabilization of the BChl cation radical.

In summary, solvent polarity appears to have the greatest effect on the k_s for the one-electron oxidation of BChl. A similar conclusion was reached in electrochemical studies of the electron transfer kinetics of iron porphyrins [2,3]. The k_s values for most of the porphyrins examined to date [2-4] fall within the quasi-reversible regime and are low in comparison to homogeneous, self-exchange electron transfer (k_h) rates. Calculation of the expected value of k_h for BChl can be made using Marcus theory [5] relating the electochemical value of k_s to k_h. Although k_h has not been measured for BChl, it is expected to be much higher than calculated from our results. However, poor agreement between calculated k_h values and experimental results for various transition metal complexes has been noted by Weaver and Hupp [6] and may be due, at least in part, to the greater distance between the reactant and the electrode in the electrochemical experiment than between two reactant molecules in solution. Thus, it is expected that the electron transfer kinetics for adsorbed BChl should be markedly greater than the solution values. The marked increase in k_s for upon adsorption has already been observed iron protoporphyrin IX [7].

ACKNOWLEDGEMENTS
The financial support of the Department of Energy (Grant # ER13261) is gratefully acknowledged. We thank Rebecca K. Schmidt for her skillfull assistance in obtaining part of the data.

Table I: Solvent Effects on the Diffusion Coefficient and Heterogeneous
Electron Transfer Kinetics of BChl a

Solvent	D (cm^2/s) x 10^6	k_s (cm/s) x 10^2
CH_2Cl_2	4.71	1.19
CH_2/Cl_2 + pyridine (10:1)	5.80	1.11
CH_2Cl_2 + pyridine (1000:1)	5.57	1.00
MeOH	5.75	7.21
EtOH	3.17	2.06
2-PrOH	1.45	0.82
MeOH; $LiClO_4$	5.71	5.03
BPh a in CH_2Cl_2	6.53	1.60
CH_2Cl_2,TBAHFP	3.71	0.58
MgTPP in MeOH	1.12	2.64
MgTPP in MeOH + pyridine (1000:1)	2.30	3.02

REFERENCES

1. Nicholson, R.S. (1965) Anal. Chem. 37, 1351-1354.

2. Davis, D.G. and Bynum, L.M. (1975) Bioelectrochem. Bioenerg. 2, 184-190.

3. Feinberg, B.A., Gross, M., Kadish, K.M., Marano, R.S., Pace, S.J. and Jordan, J. (1974) Bioelectrochm. Bioenerg. 1, 73-86.

4. Kadish, K.M., Morrison, M.M., Constant, L.A., Dickens, L., and Davis, D.G. (1976) J. Am. Chem. Soc. 98, 8387-8390.

5. Marcus, R.A. (1963) J. Phys. Chem. 67, 853-857.

6. Weaver, M.J. and Hupp, J.T. (1982) in Mechanistic Aspects of Inorganic Reactions (Rorabacher, D.B. and Endicott, J.F., Eds.), ACS Symposium Series, Vol. 198, pp. 181-208.

7. Brown, A.P. and Anson, F.C. (1978) J. Electroanal. Chem. 92, 133-145 (1978).

X- AND Y-POLARIZED ABSORPTIONS OF CHLOROPHYLL A AND PHEOPHYTIN A ORIENTED IN A LAMELLAR PHASE OF GLYCERYLMONOOCTANOATE/WATER

M. Fragata[1], T. Kurucsev[2] and B. Nordén[3]. [1]Centre de recherche en photo-biophysique, Université du Québec à Trois-Rivières, Québec, Canada, G9A 5H7; [2]Department of Physical and Inorganic Chemistry, The University of Adelaide, South Australia; [3]Department of Physical Chemistry, Chalmers University of Technology, Gothenburg, Sweden.

1. INTRODUCTION

The primary step in photosynthesis is molecular electronic excitation through the capture of photons which is then followed by conduction/conversion into forms usable by living organisms. Clearly, it is of the highest importance to understand the electronic transition spectroscopy of chlorophyll and analogs. The present work was directed to this end: we determined the polarized absorption spectra of chlorophyll a and pheophytin a embedded into an oriented liquid crystal, which were then resolved into separate X- and Y-polarized spectra. Hereunder, we show that the subsequent deconvolution of the polarized spectra into vibronic progressions provides a basis for a better understanding of electronic spectra, including the fluorescence polarization.

2. MATERIALS AND METHODS

Chlorophyll a (Chl a) was extracted from chloroplasts of spinach leaves and purified (1); pheophytin a (Pheo a) was obtained from Chl a through acid treatment (2). The liquid crystal material was glycerylmono-octanoate obtained from the Chemical Centre, University of Lund.

The liquid crystalline phases incorporating the pigments were aligned macroscopically between silica plates (3) and the linear dichroism determined in a JASCO J-500 spectropolarimeter as considered and described in detail in previous publications (4, 5). The quantities determined as functions of the wavelength are the linear dichroism $LD = A_{\parallel} - A_{\perp}$, where $A_{\parallel}$ and $A_{\perp}$ are the absorbances of light parallel and perpendicular, respectively, to the plane of polarization of the incident radiation; the isotropic absorbance, A_r, and the reduced linear dichroism $LD^r = LD/A_r$.

3. RESULTS AND DISCUSSION

Figure 1 shows typical LD spectra of Chl a and Pheo a. The result agrees broadly with previous publications of similar data down to 350 nm (6) as well as with fluorescence polarization (7). Note, however, that our measurements extend to near 250 nm and it is important to remark that the systems show significant LD signals in this range which may not be neglected when considering the spectral characteristics of whole chloroplasts or thylakoid fragments (see, e.g., (8)).

In the liquid crystal the orientation of the pigments is due to the alignment of the phytyl chain in the hydrophobic lamellae. Accordingly,

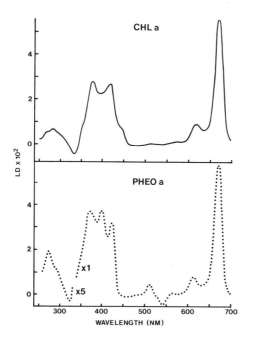

LD x 10²

CHL a

PHEO a

x1

x5

300   400   500   600   700

WAVELENGTH (NM)

Figure 1. Linear dichroism (LD) spectra of chlorophyll a (full line) and pheophytin a (dots) oriented in a lamellar phase of GMO/H_2O.

the "orientation axis" (9), X, which is taken to be in the plane of the chromophore, passes through the position of the phytyl substitution while the Y axis, orthogonal to X, is in the same plane. This coordinate system is very nearly that used by others (10, 11) for Chl a and Pheo a. In Fig. 2a we show the isotropic absorption spectrum and the reduced linear dichroism of Chl a; the latter is used to decompose the former into X- and Y-polarized spectra (12) shown in Fig. 2b and 2c, respectively. The corresponding spectra for Pheo a are shown in fig. 3a-c.

Deconvolution of the X- and Y-polarized spectra was carried out in terms of vibronic progressions used and described in detail before (13). The (0,0) bands of each progression are indicated by vertical lines with heights that are proportional to their intensities in Fig. 2b, c and 3b, c. For Chl a the X- and Y-polarized spectra could be fitted to 7 and 4 vibronic progressions, respectively, two of which coincided as indicated by their numbering in Fig. 2b and c. The fit of the isotropic spectrum to these 9 progressions is shown in Fig. 4. According to these results we may conclude first that above 480 nm the absorption spectrum of Chl a cannot be described in terms of just two progressions (Q_X and Q_Y) as has been attempted to date (10, 11). There are at least three separate electronic transitions in this region even if we allow for the possibility that the two lowest energy progressions belong to the same electronic transition. The second conclusion concerning the lower wavelength Soret region is that the absorption spectrum may be adequately described invoking fewer transitions then those predicted theoretically (11). The corresponding results for Pheo a lead to similar conclusions; the fit of the isotropic spectrum to the 9 independent progressions obtained from the deconvolution of the polarized spectra (Fig. 3b and c) is shown in Fig. 5.

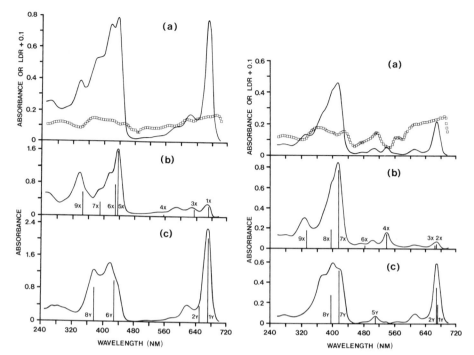

Figure 2 (left). Chlorophyll <u>a</u> spectra. a: isotropic spectrum (A_r), full line; reduced linear dichroism (LD^r), squares. b: resolved X-polarized spectrum. Vertical lines give the positions of the (0,0) bands of the various progressions with heights proportional to intensity. c: resolved Y-polarized spectrum with vertical lines defined as above.

Figure 3 (right). Pheophytin <u>a</u> spectra. a: isotropic spectrum and LD . b: resolved X-polarized spectrum. c: resolved Y-polarized spectrum. Symbols defined as in Fig. 2.

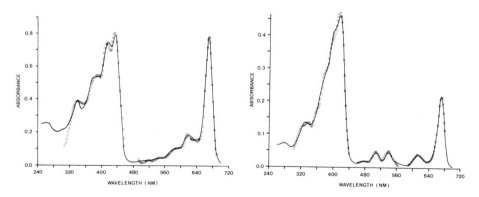

Figure 4 (left). Chlorophyll <u>a</u>, comparison of isotropic spectrum (full line) with the simulated spectrum (squares).

Figure 5 (right). Pheophytin <u>a</u>, comparison of the isotropic spectrum (full line) with the simulated spectrum (squares).

Finally, in Fig. 6 we show the energies of the (0,0) bands and the polarization directions of the progressions found for Chl a and Pheo a with a suggested correlation between relevant levels. Progressions 1 and 2 for each of the compounds are bracketed together since, as already referred to above, they may belong to the same electronic transitions.

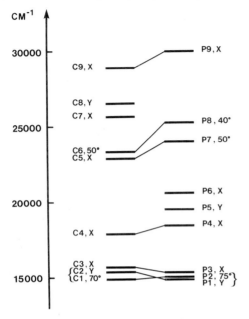

Figure 6. Energies and polarizations of the progressions used to fit the spectra of Chl a and Pheo a and their suggested correlation.

ACKNOWLEDGMENTS

This work was supported by grants from the N.S.E.R.C. Canada (A6357, E2631) and the Fonds F.C.A.R. du Québec (EQ-3186).

REFERENCES

1   Omata, T. and Murata, N. (1980) Photochem. Photobiol. 31, 183–185
2   Perkins, H.J. and Roberts, D.W.A. (1962) Biochim. Biophys. Acta 58, 486–498
3   Larsson, K. (1967) Z. Phys. Chemie New folge 56, 173–198
4   Nordén, B., Lindblom, G. and Jonas, I. (1977) J. Phys. Chem. 81, 2086–2093
5   Johansson, L. B.-A., Davidsson, A., Lindblom, G. and Nordén, B. (1978) J. Phys. Chem. 82, 2604–2609
6   Breton, J., Michel-Villaz, M. and Paillotin, G. (1971) in Proc. IId Internatl. Congress on Photosynthesis, Stresa, pp. 349–357.
7   Van Metter, R.L. (1978) Ph.D. Thesis, The University of Rochester, Rochester, N.Y
8   Breton, J., Michel-Villaz, M. and Paillotin, G. (1973) Biochim. Biophys. Acta 314, 42–56
9   Nordén, B. (1978) Appl. Spect. Reviews 14, 157–248
10 Weiss, C., Jr. (1972) J. Mol. Spect. 44, 37–80
11 Petke, J.D., Maggiora, G.M., Shipman, L. and Christoffersen, R.E. (1979) Photochem. Photobiol. 30, 203–223
12 Matsuoka, Y. and Nordén, B. (1982) Chem. Phys. Letters 85, 302–306
13 Kurucsev, T. (1978) J. Chem. Educ. 55, 128–129

THE BACTERIOCHLOROPHYLL C DIMER IN CARBON TETRACHLORIDE

J. M. OLSON, G. H. VAN BRAKEL, P. D. GEROLA, and J. P. PEDERSEN
INSTITUTE OF BIOCHEMISTRY, ODENSE UNIVERSITY, CAMPUSVEJ 55
5230   ODENSE M

1.   INTRODUCTION
     Bacteriochlorophyll (BChl) c is the principal light-harvesting pigment
of several green photosynthetic bacteria (1). It is localized in the
chlorosomes where it is associated with a specific polypeptide (2). In
vivo BChl c exists as a protein-bound aggregate of about 10-12 molecules
(3). Similar BChl c aggregates in the absence of protein have been demonstrated in CCl_4, CH_2Cl_2, and in CH_2Cl_2-hexane mixtures (3-5). All these
aggregates are characterized by a large shift of the ca. 660-nm absorption
band to between 730 and 750 nm, and the induction of large semi-conser-
vative circular dichroism (CD) signals associated with the red-shifted
absorption band (3). Recently we discovered in some model systems that
dilution of the BChl c aggregate leads to the formation of a intermediate
(λ_{max} = 706 nm) between aggregated BChl c and monomeric BChl c. This intermediate is the subject of this communication.

2.   MATERIALS AND METHODS
2.1. BChl c was extracted from wet cells of Chlorobium limicola f. thio-
     sulfatophilum 6230 with methanol/diethylether/petroleum ether (5:2:1)
     and chromatographed with acetone on polyethylene powder by HPLC (6).
     The purified and dried BChl c was washed once with methanol, dried,
     and then washed with CCl_4 and dried 3 times to remove residual water
     (7). The CCl_4 (Merck, Darmstadt) contained not more than 0.02% water.

2.2. Absorption spectra were recorded with a Perkin-Elmer 330 spectrophoto-
     meter, and CD spectra were recorded with an ISA Jobin Yvon Dichrograph
     (1975) for which the wavelength scale appears to have been 5 nm too
     high. Absorption and CD spectra were simultanenously deconvoluted by
     the computer program GAMET (8) on a Univac 1100/62 E2. Dimer proper-
     ties (exciton splitting, dipole strengths, rotational strengths) for
     fixed geometries were calculated by the computer program EXCITON (3)
     from point-dipole theory.

2.3. The absorptivity of BChl c monomer at 667 nm in CH_3OH-CCl_4 (1:100) was
     determined by preparing a solution of BChl c monomer in CH_3OH-CH_2Cl_2-
     hexane (0.4:1:200), measuring the absorbance at 662 nm (5),
     evaporating the solvent, then replacing it with an equal volume of
     CH_3OH-CCl_4, and measuring the absorbance at 667 nm.

3.   RESULTS AND DISCUSSION
3.1. When methanol is added to BChl c dissolved in CCl_4, all the chloro-
     phyll is converted to the monomeric form (λ_{max} = 667 nm, ϵ = 76 mM^{-1}
     cm^{-1}). (The monomer Qy-band can be deconvoluted into two components
     at 629 and 667 nm (see Fig. 1).) Similar results have been reported

Biggens, J. (ed.), Progress in Photosynthesis Research, Vol. I. ISBN 90 247 3450 9
© 1987 Martinus Nijhoff Publishers, Dordrecht. Printed in the Netherlands.

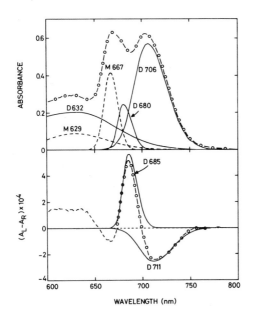

FIGURE 1. Absorption and CD spectra of BChl c in CCl₄ without CH₃OH. [BChl c] ≃ 19 µM. The CD spectrum has been corrected as explained in the text. The absorbance spectrum which represents the complete mixture of monomer, and intermediate has been deconvoluted into 2 components for the monomer (M) and 3 for the intermediate (D). The CD spectrum has been deconvoluted into 2 components for the intermediate only. The data points from 600 to 672 nm in the CD spectrum were omitted from the deconvolution procedure to simplify the analysis.

for BChl c in CH₂Cl₂ and CH₂Cl₂-hexane (5). The CD spectrum for the monomer in CCl₄ + methanol shows a single negative trough at about 677 nm with $\Delta\epsilon = -12$ M⁻¹cm⁻¹.

3.2. When BChl c is dissolved in CCl₄, CH₂Cl₂ or CH₂Cl₂-hexane mixtures in the absence of CH₃OH, the chlorophyll may exist in three forms: monomer, intermediate, and aggregate. The proportion of each form depends on the total BChl c concentration and the characteristics of the solvent. When CCl₄ is the solvent, the intermediate forms readily at BChl c concentrations between 1 and 100 µM, but the aggregate forms in appreciable amounts only at concentrations well above 100 µM. This makes CCl₄ the preferred solvent for studying the intermediate.

3.3. When a 19-µM solution of BChl c is prepared in CCl₄ (Fig. 1.), about 6 µM exists as monomer and about 13 µM exists as intermediate. In order to simplify the deconvolution of the absorbance and CD spectra of this mixture, the CD spectrum of the monomer (4.4 µM) was subtracted from the CD spectrum of the mixture. The resulting CD spectrum shown in Fig. 1 represents the CD spectrum of the mixture with 70% of the monomer contribution removed.
Simultaneous deconvolution of the absorbance and CD spectra reveals the existence of two absorption bands at 680 nm and 706 nm and two CD bands at 685 nm (+) and 711 nm (-) associated with the intermediate. These spectra strongly suggest that the intermediate is a BChl c dimer in which the exciton splitting (530 cm⁻¹) is combined with a large red shift (560 cm⁻¹) of the Qy-band.

3.4. From the relation, $D^+/D^- = (1+\cos\theta)/(1-\cos\theta)$, the angle θ between the Qy-transition dipole moments of the BChl c molecules is calculated to be 133°. We propose that the BChl c intermediate in CCl₄ is a dimer in which the BChl c rings are parallel to each other. From other work (5,10) it is clear that the 2-(1-hydroxyethyl) group of each BChl c

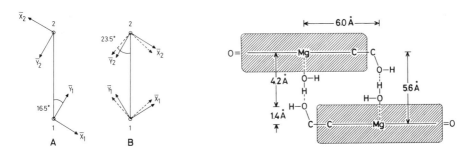

FIGURE 2. (LEFT) Orientation of BChl c molecules in various dimer configurations. (A) Piggy-back dimer in two configurations: back-to-
front (#2 BChl below the plane of the figure) and front-to-back
(#1 BChl below). (B) Face-to-face dimer in two configurations:
back-to-back (#2 BChl below) and front-to-front (#1 BChl below).
Orientation vectors ($\bar{x}$, $\bar{y}$) drawn as solid arrows indicate a
BChl c molecule in which the 2-(1-hydroxyethyl)oxygen atom is
directly over the Mg atom of the neighboring BChl c molecule.
Dashed arrows indicate a BChl c configuration consistent with
the dimer absorption and CD spectra.

FIGURE 3. (RIGHT) Face-to-face model of the BChl c dimer.

molecule is involved in binding the Mg atom of the other molecule.
Since it appears more difficult to form a "piggy-back" dimer than a
face-to-face dimer with $\Theta = 133^\circ$, the BChl c intermediate is probably
a face-to-face dimer as shown in Fig. 2 and 3. From the signs of the
CD components the z-axes (right hand rule) of the BChl c rings must
point out from the dimer. (This is the back-to-back configuration
defined in Fig. 2.)
The coordinates of the back-to-back model with $\Theta = 133^\circ$ were entered
into the computer program EXCITON with the average dipole strength of
BChl c in the dimer taken to be 5×10^{-59} C^2m^2 from the estimated
hyperchromism (90%). A distance of 5.6 Å between the chlorophyll
planes gave the best match with the experimental exciton splitting,
but the theoretical values of R_+ and R_- were about 8 x the experimental values.

3.5. In our working model for the BChl c dimer (Fig. 3) the distance of
6.0 Å between the Mg atom and the in-plane projection of the 2-(1-
hydroxyethyl) oxygen atom is calculated from the coordinates of C_2
and C_{19} in methyl pheophorbide a (9). The length of the C_{19}-O band is
assumed to be 1.43 Å (11), and the angle formed by C_2-C_{19}-O is assumed
to be 109° (tetrahedral carbon). The plane of C_2-C_{19}-O is assumed to
be nearly perpendicular to the plane of the chlorin ring. The O-Mg
coordinate bond is assumed to be about 2.0 Å (12), and the bond is
assumed to be nearly perpendicular to the chlorin plane. The large
distance between the BChl c molecules in this model prevents a direct
interaction between the hydroxyethyl oxygen atom in one molecule and
the Mg atom in the other molecule. Therefore we suggest that a water
molecule may serve as a bridge between these two atoms.

The properties of the BChl c dimer are quite different from those of
the Chl a dimer (13,14), and the model we propose for the BChl c
dimer is quite different from that proposed for the Chl a dimer (15).
The significant difference between BChl c and Chl a for dimer formation is the presence of the 2-(1-hydroxyethyl) group in BChl c in
place of the 2-vinyl group in Chl a. The dimerization constant for
BChl c is about 20 x that for Chl a and apparently reflects the
strong interaction between the 2-(1-hydroxyethyl) group of one BChl c
and the Mg atom of the other BChl c.

3.6. The significance of this work is that it is the first step in understanding higher aggregates of BChl c, since it is already clear that
the BChl c dimer is the building block of higher aggregates in CCl4,
CH2Cl2 and CH2Cl2-hexane. Of even greater importance is the possibility that BChl c dimers may exist in vivo as the building blocks
of BChl c aggregates associated with protein in chlorosomes of green
photosynthetic bacteria.

REFERENCES
1 Olson, J.M. (1980) Biochim. Biophys. Acta 594, 33–51
2 Wechsler, T., Suter, F., Fuller, R.C. and Zuber, H. (1985) FEBS Lett.
   181, 173–178
3 Olson, J., Gerola, P.D., van Brakel, G.H., Meiburg, R.F. and Vasmel, H.
   (1985) in Antennas and Reaction Centers of Photosynthetic Bacteria
   (Michele-Beyerle, M.E., ed.), 67–73, Springer-Verlag, Berlin
4 Krasnovsky, A.A. and Bystrova, M.I. (1980) BioSystems 12, 33–51
5 Smith, K.M., Kehres, L.A. and Fajer, J. (1983) J. Am. Chem. Soc. 105,
   1387–1389
6 Caple, M.B., Chow, H. and Strouse, C.E. (1978) J. Biol. Chem. 253,
   6730–6737
7 Katz, J.J., Closs, G.L., Pennington, F.C., Thomas, M.R. and Strain, H.H.
   (1963) J. Am. Chem. Soc. 85, 3801–3809
8 Olson, J.M., Ke, B. and Thompson, K.H. (1976) Biochim. Biophys. Acta
   430, 524–537
9 Philipson, K.D., Cheng Tsai, S. and Sauer, K. (1971) J. Phys. Chem.
   75, 1440–1445

10 Smith, K.M., Bobe, F.W., Goff, D.A. and Abraham, R.J. (1986) J. Am.
   Chem. Soc. 108, 1111–1120
11 Pauling (1967) The Nature of the Chemical Bond and the Structure of
   molecules and Crystals. 3rd ed. Cornell University Press, Ithaca, N.Y.
12 Strouse, C.E. (1974) Proc. Nat. Acad. Sci. USA 71, 325–328
13 Sauer, K., Lindsay Smith, J.R. and Schultz, A.J. (1966) J. Am. Chem.
   Soc. 88, 303–318
14 Dratz, E.A., Schultz, A.J. and Sauer, K. (1966) Brookhaven Symp. Biol.
   19, 303–318
15 Houssier, C. and Sauer, K. (1970) J. Am. Chem. Soc. 92, 779–791

ADDENDUM: Fluorescence emission spectra of BChl c in CCl4 show peaks at
669, ca. 690, and ca. 730–735 nm which we associate with the absorption
peaks at 667, 680, and 706 nm in Fig. 1.
These data suggest that the 680- and 706-nm absorption bands may belong
to separate species and that our dimer model may not be correct.

Superoxide Photogeneration by Chlorophyll <u>a</u> in Water/Acetone Solutions. Electron Spin Resonance Studies of Radical Intermediates in Chlorophyll <u>a</u> Photoreactions <u>In Vitro.</u>

Jun-Lin You, Karen S. Butcher, Angela Agostiano, and Francis K. Fong*, Department of Chemistry, Purdue University, West Lafayette, IN 47907

1. INTRODUCTION

Free radicals have been proposed by earlier authors as intermediates in biochemical reactions,[1] and have in fact been detected by electron spin resonance (ESR) measurements. The superoxide ion O_2^-. or its protonated form HO_2. has been postulated as an intermediate in many biological oxidation-reduction reactions.[2-4] In particular it has been established that O_2^-. is photogenerated by spinach chloroplasts in the presence of ascorbate or low potential electron acceptors such as methyl viologen.[4] In this paper we report the use of DMPO (5,5-dimethyl-1-pyrroline-1-oxide) as spin trap in the ESR observation of O_2^-. photogeneration by <u>in vitro</u> Chl <u>a</u> in water/acetone. The observed ESR signals are in qualitative agreement with those reported for the formation of O_2^-. in chloroplasts system.

2. PROCEDURE

Chlorophyll <u>a</u> was extracted from spinach and purified using the procedures described by Brace <u>et al</u>.[5] Absorption spectra were measured using a Cary 17D spectrophotometer immediately after sample preparation and at regular intervals afterwards to monitor any spectral changes. ESR measurements were carried out with a Varian E-109 Century series spectrometer equipped with Hewlett-Packard E-935 data acquisition system. DMPO purchased from Aldrich Chemical was used as spin trap agent.

3. RESULTS AND DISCUSSION

In Fig. 1A are shown the ESR spectra of photogenerated signals in a 5×10^{-5} M Chl <u>a</u> solution in 50:50% (v/v) water/acetone. In the absence of oxygen, no ESR signals were observed either in the dark (a) or in the light (b). After the sample was saturated with oxygen, the sequence of observations were made as follows: The dark-adapted (c) sample produced an ESR signal under illumination with visible light (d). After this signal had decayed completely in the dark over a period of 12 h (e), the sample was again illuminated. An ESR signal indistinguishable from that in d was observed (f). The ESR signal photogenerated in the presence of DMPO is shown in g. In addition to the Chl <u>a</u> radical cation signal DMPO spin trap signals (labelled 1,2,3 and 6) were detected. The g-value and peak width of the ESR signal in d and f are 2.0024 + 0.0002 and 1.5G, respectively, which readily identifies that signal as being attributable to the radical cation, $(Chl \underline{a}.2H_2O)_n^+$..[6] The absorption spectra of 5×10^{-4} M Chl <u>a</u> solutions corresponding to the ESR results in c-g are reproduced in Fig. 1B. The quantitative conversion of the 675-<u>nm</u> band to the remarkably broad 752-<u>nm</u> band took place over a 16-h period. The constancy in the ESR signal intensity of $(Chl \underline{a}. 2H_2O)_n^{2+}$. in d and f indicates that the amount of $(Chl \underline{a}. 2H_2O)_n$ in the broad distribution corresponding to the 755 <u>nm</u> band remains about the same in spite of the quantitative conversion.

Biggens, J. (ed.), Progress in Photosynthesis Research, Vol. I. ISBN 90 247 3450 9
© 1987 Martinus Nijhoff Publishers, Dordrecht. Printed in the Netherlands.

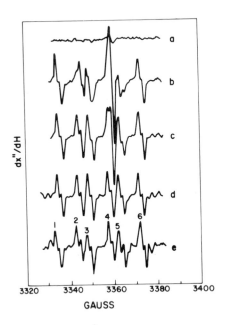

Fig. 2.   ESR spectra of a 4.2×10^{-4} $\underline{M}$ Chl $\underline{a}$ solution in water/acetone
containing 2.5×10^{-1} $\underline{M}$ DMPO.

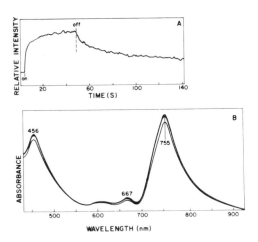

Fig. 3.   Absorption spectra and ESR kinetic response of 4.2×10^{-4} $\underline{M}$
Chl $\underline{a}$ in water/acetone.

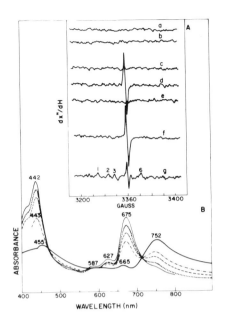

Fig. 1.   Absorption and ESR spectra of O_2-saturated 5 X 10^{-5} $\underline{M}$
           solution of Chl $\underline{a}$ in 50:50% water/acetone.

The spin trap measurements were obtained form a 4.2 X 10^{-4} $\underline{M}$ Chl $\underline{a}$
solution in an O_2-saturated 50:50% water/acetone mixture containing 2.5
x 10^{-1} $\underline{M}$ DMPO (Fig. 2). A dark-adapted sample (a) on illumination
produced a multicomponent ESR signal (b) that qualitatively resembles
the signal, g, in Fig. 1A. The signal in b persisted under continuous
sample illumination for 60 min, although detailed changes were observed
(c) indicative of a redistribution of the radical spins as the sample
approached the steady-state conditions. The Chl $\underline{a}$ radical cation siganl
decayed within 60 min after the light was turned off, leaving the DMPO
spin trap signal (d), which is indistinguishable from that (e) obtained
from a 2% H_2O_2 solution in 50:50% water/acetone after 4 min of
illumination with UV light. The ESR signals in d and e closely resemble
that reported in connection with the formation of O_2^-. in the spinach
chloroplasts under experimental conditions similar to the present
investigation. The typical kinetic response of a light-induced ESR
signal from a 5 X 10^{-4} $\underline{M}$ Chl $\underline{a}$ solution in O_2-saturated 50:50%
water/acetone is illustrated in Fig. 3A. The dark decay has a component
with a lifetime of about 3s. Fig. 3B shows the Cary 17D absorption
spectra of the Chl $\underline{a}$ sample from which the spin trap signal in Fig. 2d
was obtained.

The photogeneration by chloroplasts of an ESR signal practically identical to that shown in Fig. 2d requires the presence of oxygen, as saturating the chloroplast solu tion with pure oxygen greatly increased the signal amplitude whereas the removal of dissolved oxygen resulted in the elimination of the ESR signal.[4] The present illustration of the ability of $(Chl\ a.2H_2O)_n$ to photogenerate superoxide in a manner closely similar to that of spinach choroplasts corroborates the earlier demonstrations of the photocatalysis of water splitting[7] and carbon dioxide photoreduction[8] by $(Chl\ a.\ 2H_2O)_n$.

REFERENCES
1. Michaelis, L. (1946) in "Currents in Biochemical Research," D. Green ed., Wiley-Interscience, New York.
2. Epel, B.L., and Neumann, J. (1973) Biochim. Biophys. Acta 325, 520.
3. Allen, J. F., and Hall, D.O. (1973) Biochem. Biophys. Res. Commun. 52, 886.
4. Harbour, J.R., and Bolton, J.R. (1975) Biochem. Biophys. Res. Commun. 64, 803.
5. Brace, J.G., Fong, F.K., Karweik, D.H., Koester, V.J., Shepard, A., and Winograd, N. (1978) J. Am. Chem. Soc. 100, 5203.
6. Fong, F.K., Kusunoki, M., Galloway, L., Mathews, T.G., Lytle, F.E., Hoff, A.J., and Brinkman, F.A. (1982) J. Am. Chem. Soc. 104, 2759.
7. Fong, F.K. and Galloway, L. (1978) J. Am. Chem. Soc. 100, 3594.
8. Fruge, D.R., Fong, G.D., and Fong, F.K. (1979) J. Am. Chem. Soc. 101, 3694.

Resonant Energy Transfer Between Bulk Chlorophyll a and Chlorophyll a
Dihydrate Dimers in Water/Acetone Mixtures. A Model of Sensitized
Excitation in Plant Photosynthesis

Angela Agostiano, Karen A. Butcher, Michael S. Showell,
Jun-Lin You, Albert J. Gotch, and Francis K. Fong, Department of
Chemistry, Purdue University, West Lafayette, IN 47907

1.   INTRODUCTION

An important problem of photosynthesis research is a description
of the primary processes that follow the harvest of light by antenna
Chl a. Unfortunately, much less is known of the mechanisms for the
sensitized excitation of $\underline{P}$700 and $\underline{P}$680 than the $\underline{P}$700 and $\underline{P}$680 light
reactions themselves.[1] Consequently it seems appropriate to emphasize
the use of model systems in investigating the photophysical processes
leading to the photosynthetic light reactions. This work demonstrates
the resonant energy transfer between bulk Chl a and a Chl a
dimer/oligomer system, which provides a model for sensitized excitation
in photosynthesis.

2.   PROCEDURE

Chlorophyll a was extracted from spinach and purified according to
the method reported elsewhere.[2] A Coherent Radiation 80/2H, 80-mW He/Ne
laser, with an emission line at 632.8 nm was used as the excitation
source. Relative fluorescence quantum yields were measured[3] by
comparison of the sample fluorescence to that of a standard, a 5 X 10^{-5}
M Chl a solution in benzene.[4] The lifetime measurements were made by a
phase fluorimetry technique.

3.   RESULTS AND DISCUSSION

A freshly prepared 5 X 10^{-5} M Chl a solution in 50:50%
water/acetone has a red absorption band at 674 nm with a long-
wavelength shoulder (spectrum I). This absorption band decreased
steadily with time, as the long-wavelength shoulder converted to a
broad absorption band with a maximum at 715 nm (spectrum II). This
conversion was characterized by an isobestic point. Subtraction of
spectrum I from spectrum II yields a negative absorbance difference
($-\Delta\underline{A}$) spectrum with a maximum at 678 nm. A normalized plot of this
spectrum (Fig. 1A) is compared with that of the corrected fluorescence
spectrum of a 5 x 10-5 M Chl a solution in acetone (Fig. 1B). The
corrected fluorescence spectrum of 5 x 10^{-5} M Chl a in acetone is
compared in Fig. 2 with those in 40:60% and 50:50% water/acetone
mixtures. Fluorescence intensities, quantum yields, and lifetimes for
5 x 10^{-5} M Chl a solutions were measured. The measured quantities,
normalized with respect to those of the Chl a 676-nm fluorescence in
acetone and expressed as quantum efficiencies are plotted in Fig. 3 as
a function of the volume percent and mole fraction of water in the
solvent system.

Biggens, J. (ed.), Progress in Photosynthesis Research, Vol. I. ISBN 90 247 3450 9
© *1987 Martinus Nijhoff Publishers, Dordrecht. Printed in the Netherlands.*

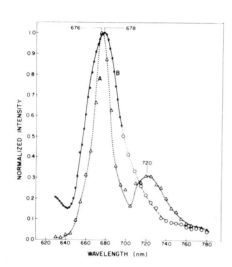

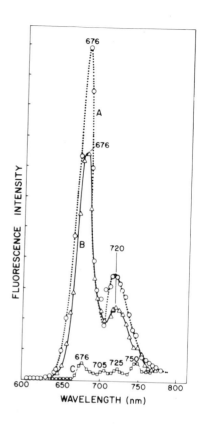

Fig. 1.   (A) normalized negative difference absorbance (−ΔA) spectrum
of $(Chl\ \underline{a}.2H_2O)_2$. The long-wavelength absorbance difference
(− − −) was corrected for the appearance of a broad long-
wavelength (755 $\underline{nm}$) absorption band; (B) normalized
fluorescence spectrum, corrected for instrumental response,
of 5×10^{-5} $\underline{M}$ Chl $\underline{a}$ in acetone.

Fig. 2.   Fluorescence spectra, corrected for instrumental response, of
5×10^{-5} $\underline{M}$ Chl $\underline{a}$ in (A) acetone, (B) 40:60% of water/acetone,
and (C) 50:50% water/acetone.

The 676-$\underline{nm}$ fluorescence band, observed from the 5×10^{-5} $\underline{M}$ Chl $\underline{a}$
solution in acetone, is assigned to that of Chl $\underline{a}$.Ac. The dependence of
the Chl $\underline{a}$.Ac fluorescence intensity, $\underline{I}$, on Chl $\underline{a}$ concentration, $\underline{C}$,
observes the numerical fit

$$\underline{I} = 2.82 \times 10^5 \underline{C} - 1.47 \times 10^9 \underline{C}^2 \tag{1}$$

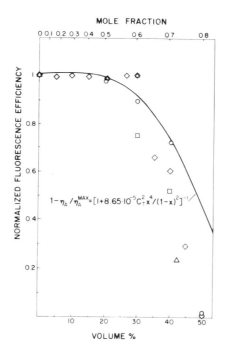

MOLE FRACTION

$1-\eta_A/\eta_A^{MAX}=[1+8.65\cdot10^{-5}C_T^2x^4/(1-x)^2]^{-1}$

NORMALIZED FLUORESCENCE EFFICIENCY

VOLUME %

Fig. 3.   Fluorescence quantum efficiencies of 5×10^{-5} M Chl a in
50:50% water/acetone: intensities in the absence (Δ) and
presence (◇) of oxygen; relative quantum efficiencies (o);
and lifetimes (□). All quantities are normalized to
corresponding values of Chl a solutions in acetone.

The resonant energy transfer problem, given in Fig. 3 as quantum
efficiencies normalized to the measured fluorescence parameters of Chl
a.Ac in acetone, is defined in a minor extension of the method given by
Fong and Diestler:[7]

$$\eta = \Phi/\Phi_m = I/I_m = \tau/\tau_m = 1/(1 + \langle w_{DA}\rangle \tau_m) \qquad (2)$$

where m denotes monomeric Chl a.Ac. In Eq.2 Φ, I, τ, and Φ_m, I_m, τ_m are
the fluorescence quantum yields, intensities and lifetimes respectively
in the presence and absence of added water. The product $\langle w_{DA}\rangle\tau_m$ is the
ensemble-averaged transition probability per donor complex Chl a.Ac, of
the transfer process to acceptor A during the lifetime τ_m of the Chl
a.Ac excited state. Assuming an electric dipole-dipole interaction[5-7]
and the equilibrium, 2Chl a.Ac (662 nm) + 4H$_2$O $\longleftrightarrow$ (Chl a.2H$_2$O)$_2$ (678
nm) + 2Ac (where 662 and 678 nm denote positions of the red absorption
maxima), for the interconversion of the acetone- and water-solvates Chl
a, we arrive at an expression for the dependence of fluorescence
quantum efficiency on the solvent composition:

$$\eta \approx [1 + \text{Const.} \times C_T^2 x^4/(1-x)^2]^{-1} \qquad (3)$$

In Eq. 3 C_T is the total concentration of water and acetone in the solvent mixture and the constant, Const. $\approx \beta K_1 C^2$, is obtained given $C_A \ll C$, where β is given in Ref. 7, K_1, C and C_A are respectively the equilibrium constant for the equilibration between the acetone- and water-solvates of Chl $\underline{a}$, the concentration of bulk Chl $\underline{a}$, and that of $(Chl\ \underline{a}.2H_2O)_2$ in the water/acetone solution. The numerical fit of Eq. 3 to the measured quantum efficiencies is given by the solid line in Fig. 3, which is in reasonable agreement with experiment for $5 \times 10^{-5}M$ Chl $\underline{a}$ solutions up to 40% water, for which the inequality, $C_A \ll C$, is experimentally verified.

Turning to Eq. 1 for the dependence of fluorescence intensity on C in the absence of added water, we write

$$I = \Phi_m \varepsilon I_0 l C (1 - n_A/n_A^{max}) \qquad (4)$$

In Eq. 4 the fluorescence intensity at infinite dilution is defined by $I_m = \Phi_m \varepsilon I_0 l C$, where Φ_m, ε, I_0 and l are respectively the Chl $\underline{a}$.Ac fluorescence quantum yield at infinite dilution, absorption extinction coefficient, incident flux and path length of the sample cell. A comparison of Eq. 4 with Eq. 1 yields $n_A/n_A^{max} = 1.9 \times 10^4 C$. In acetone without added water, in the absence of dimeric Chl $\underline{a}$ as acceptor, pairwise (two-body) resonant transfer occurs between neighboring Chl $\underline{a}$.Ac complexes. According to earlier writers,[5-7] the power dependence of n_A/n_A^{max} on C arises from the probability of finding the acceptor ions at nearest-neighbor positions. In general, for an n-body transfer process, this probability will be proportional to C^{n-1}.[7] The observed linear dependence on C, contrasted to the C^2 dependence in Eq. 3 for the monomer-dimer transfer (which is in fact a coherent "three-body" process), thus corroborates the general view.

The nearly complete overlap of the negative absorbance difference spectrum attributed to $(Chl\ \underline{a}.2H_2O)_2$ with the fluorescence spectrum of Chl $\underline{a}$.Ac (Fig. 2) supports dipole-dipole resonant energy transfer from Chl $\underline{a}$.Ac to $(Chl\ \underline{a}.2H_2O)_2$ as a probable cause for the quenching of Chl $\underline{a}$.Ac fluorescence in Fig. 3.

REFERENCES

1. Hoff, A.J. (1982) in "Light Reaction Path of Photosynthesis", edit. Fong, F.K., Springer, New York, Chap. 4.
2. Brace, J.G., Fong, F.K., Karweik, D.H., Koester, V.J., Shepard, A. and Winograd, N. (1978) J. Am. Chem. Soc. 100, 5203.
3. Parker, C.A. (1968) "Photoluminescence of Solutions", Elsevier, Amsterdam, pp.246-265.
4. Weber, G. and Teale, F.W. (1957) Trans. Faraday Soc. 53, 646.
5. Forster, Th. (1948) Ann. Physik 2, 55.
6. Dexter, D.L. (1953) J. Chem. Phys. 21, 836.
7. Fong, F.K. and Diestler, D.J. (1976) J. Chem. Phys. 56, 2875.

The Structural Organization of Photosynthetic Reaction Centers

Hartmut Michel and Johann Deisenhofer, Max-Planck-Institut fur Biochemie, Am Klopferspitz, D-8033 Martinsried, West Germany

1. Introduction

The earliest step in the conversion of light energy into the kinds of energy used by living cells is the absorption of light by photosynthetic pigments. Most frequently the light is absorbed by the pigments (chlorophylls, carotenoids, phycobilins) of the light-harvesting antenna complexes. These pigments become excited and the excitation migrates among the light-harvesting antenna pigments until it is trapped in the so-called photosynthetic reaction center. In the green photosynthetic bacteria, cyanobacteria, red algae and cryptophytes, the major light-harvesting complexes are water soluble pigment-protein complexes which are attached to the photosynthetic membranes; they are complexes of pigments and integral membrane proteins in purple photosynthetic bacteria, green algae and higher plants. In the reaction centers which are always complexes of integral membrane proteins and pigments the excited state of the "primary electron donor" is deactivated by the transfer of an electron to an electron acceptor via intermediate carriers. It is one of the functions of the surrounding proteins to keep the electron donor, intermediate carriers and acceptors in a fixed geometry in a way that electron donor and acceptor are near opposite surfaces of the photosynthetic membrane. After transfer of an electron a major part of the energy of the absorbed light is stored in the form of difference in electric potentials across the membrane ("membrane potential") and redox energy in the form of the reduced electron acceptor. The electron acceptor in the reaction centers from purple photosynthetic bacteria and photosystem II of chloroplasts and cyanobacteria is a quinone molecule, whereas it is a non-heme iron sulfur protein in the reaction center from green photosynthetic bacteria and from photosystem I of chloroplasts and cyanobacteria (for review see 1). The photooxidized primary electron donors can be re-reduced by cytochromes (mainly bacterial reaction centers), plastocyanin (photosystem I reaction centers), small molecules of sufficiently low redox potential, and - with the help of an additional water splitting complex - even by water in the case of photosystem II reaction centers.
The best known reaction centers are those from the purple photosynthetic bacteria (for review see 2, 3, 4). They are easy to isolate and relatively stable. Most of them contain

Biggens, J. (ed.), Progress in Photosynthesis Research, Vol. I. ISBN 90 247 3450 9
© 1987 Martinus Nijhoff Publishers, Dordrecht. Printed in the Netherlands.

three protein subunits which are called H (heavy), M (medium) amd L (light) subunits according to their apparent molecular weights as determined by sodium dodecylsulphate acrylamide gel electrophoresis. After the isolation of the genes encoding the reaction center proteins and sequencing the DNA the amino acid sequences of reaction center proteins from Rhodobacter (Rb.) sphaeroides (5,6), Rb. capsulatus (7) and Rhodopseudomonas (Rps.) viridis (8,9) became known. The sequences of the L, M and H subunits are published for Rb. capsulatus and Rps. viridis. In both cases the H subunit is actually the smallest one. Reaction centers from several purple photosynthetic bacteria, including Rps. viridis, contain a tightly bound cytochrome molecule, which re-reduces the photooxidized primary electron donor. The cytochrome subunit from the Rps. viridis reaction center contains four heme groups. As photosynthetic pigments four bacteriochlorophyll-bs, two bacteriopheophytin-bs, one menaquinone ("primary quinone" or "Q_A"), one non-heme-iron and one ubiquinone ("secondary quinone" or "Q_B") are found. The reaction centers from most of the other purple photosynthetic bacteria contain bacteriochlorophyll a instead of bacteriochlorphyll b, bacteriopheophytin a instead of bacteriopheophytin b, and the menaquinone is replaced by another ubiquinone.

Most importantly the reaction centers from Rps. viridis (10) and Rb. sphaeroides (11,12) could be crystallized. The crystallized reaction centers are photochemically active (13, 11, 14). In the case of Rps. viridis an electron density map of the reaction center could be calculated at 3 A resolution and the arrangement of the chromophores was determined (15). Subsequently the structure of the protein subunits (16) and details of the pigment-chromophore interactions (17) were presented. Therefore the emphasis of this article will lie on these quite recent developments.

Since the results obtained for the reaction center of the purple bacteria is also relevant for the structure of photosystem II reaction centers, we will also discuss the likely similarities between the structure of RCs from purple bacteria and photosystem II, and evolutionary relationships.

2. Results and Discussion

2.1 Pigment Arrangement

There is good spectroscopic evidence (18,19) that the arrangement of the photosynthetic pigments is the same in all species of the purple photosynthetic bacteria so far examined. Figure 1 shows the arrangement of these pigments in the reaction center from Rps. viridis as determined by X-ray crystallography (15). The four heme groups (at the top of figure 1) are related by a twofold local rotation axis which runs nearly perpendicular to the picture plain. The function of these hemes is to re-reduce the photooxidized primary electron donor which is a "dimer" of two non-covalently linked bacteriochlorophyll molecules

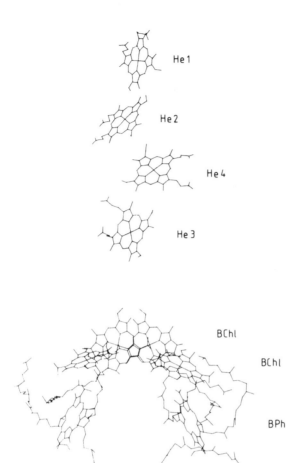

He 1

He 2

He 4

He 3

BChl

BChl

BPh

Fe

Q_a

Figure 1: Arrangement of the pigments in the photosynthetic reaction center from Rps. viridis according to ref.
15, showing four bacteriochlorophylls (BC), two
bacteriopheophytins (BP), one non-heme iron (Fe), one
quinone (MQ), and four heme groups. The twofold rotation
axis relating the photosynthetic pigments runs vertically
in the plane of the picture. The BCs, BPs, Fe, and MQ are
surrounded by the subunits L and M (not shown for
clarity).

("special pair", just below the hemes in figure 1). The
existence of such a dimer had been postulated on the basis
of EPR experiments (20), its detailed structure is now
firmly established by the X-ray structure analysis. The ring
systems of the two bacteriochlorophylls constituting the
dimer are nearly parallel and they overlap with their pyrrol
rings I. The plane to plane distance of these two
bacteriochlorophylls is about 3.1 A. The ring systems of all
the chlorine pigments are also related by a local twofold
rotation axis. This diad (vertically in figure 1) runs
between the two monomers constituting the special pair on
the periplasmic side of the membrane and through the
non-heme iron on the cytoplasmic side of the membrane. Since
the two "accessory" bacteriochlorophylls and the two
bacteriopheophytins are also related by the local diad, two
structurally equivalent branches are formed which could be
used for electron transfer across the membrane. However, the
symmetry is broken by the presence of only one quinone. This
quinone must be the primary quinone, the menaquinone, since
the ubiquinone is lost during isolation and crystallization
of the reaction centers (21). In addition, the two
bacteriopheophytins are spectroscopically inequivalent
absorbing light of different wavelengths. It is known that
only the one absorbing light at longer wavelengths is
involved in electron transfer. Comparison of absorbance
spectra of crystals, taken with plane-polarized light (13)
shows that the bacteriopheophytin absorbing at the longer
wavelengths is the one closer to the quinone. These
experiments establish clearly that only one way (the right
hand one in figure 1) is used for light driven electron
transfer across the photosynthetic membrane.
The binding site of the secondary quinone could be
determined by soaking quinones and competetive inhibitors
(o-phenanthroline , terbutryn) into the crystals and
subsequent difference Fourier analysis. These compounds bind
into an empty pocket of the protein which is symmetry
related (by the local diad) to the binding site of the
primary quinone. Thus the electron has to be transferred
from Q_A to Q_B parallel to the surface of the membrane. There
is no evidence for participation of the non-heme-iron in
this electron transfer, since it can be removed without
drastic changes in the kinetics of this electron transfer
(22).

2.2 Primary Structures

So far the sequences of the H-subunits are published for Rb.
capsulatus (7) and Rps. viridis(8). The sequence homology is
below 40 %. The sequences of the L and M subunits are
published for Rb. sphaeroides (5,6), Rb. capsulatus (7) and
Rps. viridis (8,9). The sequence homology between the M
subunits of Rps. viridis and Rb. capsulatus, as well as Rps.
viridis and Rb. sphaeroides are close to 50 %, whereas they
are close to 60 % between the L subunits. The L and M
subunits from all three species possess sequence homologies

in the order of 25-30 % indicating already a similar protein folding of both subunits and that they are derived from a common ancestor. Conserved between all L and M subunits are mainly glycines and prolines at the ends of helices and turns of the peptide chains, and the ligands to the pigments. Despite the low sequence homology the structure of all these reaction centers must be very similar.

2.3 General Architecture of the Reaction Center and the Structure of the Protein Subunits

Due to the X-ray structure analysis of the photosynthetic reaction center from Rps. viridis a detailed picture of the reaction center has emerged. The photosynthetic pigments are associated with the L and M subunits, as had already been shown by biochemical and spectroscopical experiments (2). They form the central part of the reaction center. The L and M subunits possess five long membrane spanning helices which are related by the same twofold axis as the pigments. The structural differences between the L and M subunits are mainly at the amino-terminus on the cytoplasmic side (M has a longer amino-terminus), in the connection of the first and second transmembrane helix (M shows an insertion of 7 amino acids, which give rise to a short helix parallel to the membrane), in the connection of the fourth and fifth membrane spanning helix (M possesses an additional loop providing glu M232 as a ligand to the non-heme-ferrous iron atom) and at the carboxy-terminus (the M subunit from Rps. viridis contains additional 17 amino acids, which are in contact with cytochrome subunit). The five transmembrane helices of the subunits L and M possess a remarkably open structure forming approximately half-cylinders. On both sides of the helical regions of the L and M subunits the polypeptide segments connecting the transmembrane helices and the terminal segments form flat surfaces perpendicular to the local diad. The cytochrome is bound to the surface close to the special pair on the periplasmic side. The H subunit possesses one membrane spanning helix close to its amino-terminus. Its carboxy-terminal domain is bound to the flat surface of the L-M complex on the cytoplasmic side. It is the only part of the reaction center with a significant amount of beta-sheet as secondary structure.

2.4 Pigment-Protein Interactions

The photosynthetic pigments are bound into primarily hydrophobic pockets of the L and M subunits (17). The L and M subunits provide histidine residues as ligands to the magnesium atoms of the special pair bacteriochlorophylls (L173, M200 in Rps. viridis) and the accessory bacterio-chlorophylls (L153, M180 in Rps. viridis). The Mg ions are five coordinated in agreement with recent resonance Raman data (23). The protein, besides being a scaffold for the pigments, must specifically interact with the pigments to suppress one of the two possible electron transport pathways. The choice of the pathway could be influenced already at the special pair, either by a deviation from the C2-symmetry (which would be beyond the accuracy of the present X-ray structure), or by electrostatic effects due to polar amino acids in the immediate environment of the special pair. Charged amino acids are not found in the vicinity of the special pair. However, in Rps. viridis we find hydroxyl groups of three amino acid side chains (tyr M195, thr L248, tyr M208) close to the special pair. They are located towards the branch which is used for electron transfer. Two of them form hydrogen bonds with carbonyl-oxygen atoms of the bacteriochlorophylls. On the side of the inactive branch we find only one polar side chain, his L168, in a position symmetry-related to tyr M195; his L168 also forms a hydrogen bond with the special pair. This asymmetric distibution of polar groups may cause an asymmetric electron distribution in the special pair. Sequence comparisons show that the hydrogen-bonding between special pair bacterio-chlorophylls and protein must be different between Rps. viridis, Rb. sphaeroides and Rb. capsulatus. The accessory bacteriochlorophylls are not hydrogen-bonded to the protein. That bacteriopheophytin which is an intermediate electron acceptor forms a hydrogen-bond with a, most likely protonated, glutamyl residue (L104). This glutamic acid seems to be of crucial importance for the light driven electron transfer. It is conserved between all three bacterial species and the D1 protein from photosystem II reaction centers (see below).

Another interesting amino acid is trp M250 which forms part of the binding site of the primary quinone. It also touches that bacteriopheophytin which is involved in light-driven electron transfer thereby "bridging" the pheophytin and quinone. In the symmetry related position phe L216 is found whose side chain is too small to bridge the bacteriopheophy-tin and the secondary quinone. The ferrous non-heme-iron atom is found half-way in between the primary and secondary quinone binding sites. It is bound to four histidine residues (L190, L230, M217, M264) and one glutamic acid (M232 in Rps. viridis). The primary quinone is hydrogen bonded to the histidine M217 and the peptide nitrogen of alanine M258.

2.5 Conclusions on the Structure of Photosystem II Reaction Centers

The X-ray structure analysis of the Rps. viridis reaction center shows clearly that the L and M subunits cooperate in a nearly symmetric manner to establish the primary electron donor (the special pair) and the electron accepting Q_A-Fe-Q_B complex. This feature, together with the sequence homologies of L and M subunits with the D1 and D2 proteins from photosystem II reaction centers and the herbicide binding to the D1 and L subunits has lead to the proposal that the core of photosystem II reaction centers is made up of the D1 and

D2 proteins in a similar manner as L and M form the core of the reaction center from the purple bacteria (24). Figure 2 shows a possible alignment of the L and M subunits from Rps. viridis and the D1 and D2 proteins from chloroplasts. Significant sequence homology starts with a gly,gly pair at the beginning of the second transmembrane helix. Conserved are mainly amino acids of structural importance like glycines and prolines at ends of helices and other turns of the peptide chains. Amino acids involved in binding the photosynthetic pigments like the histidine ligands to the special pair bacteriochlorphyll are frequently conserved as well as the histidine ligands to the ferrous non-heme-iron atom. Amino acids conserved specifically between L and D1 are glutamic acid L104 which participates in binding the bacteriopheophytin that is involved in the light driven electron transfer, and phe L216 which forms part of the Q_B binding site. Trp M250, which is in the symmetry related position to phe L216 is conserved between M and D2.

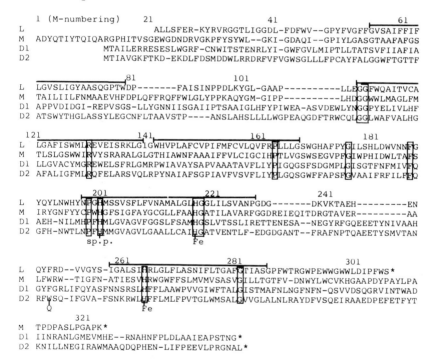

Figure 2: Sequence alignement of the L and M subunits from Rps. viridis and the D1 and D2 proteins from spinach chloroplasts (from ref. 9, slightly modified). Residues binding to the special pair (sp.p.) and to the non-heme iron (Fe), and residues forming the major part of the quinone binding sites (Q) are indicated. The transmembrane helices in Rps. viridis are indicated by bars above the L subunit sequence. Numbers refer to the M subunit from Rps. viridis.

The conservation of all these important amino acids strongly favours our view that the reaction center core of photosystem II consists of the Dl and D2 subunits. The most important difference is the absence of the histidine ligands to the accessory bacteriochlorophylls in Dl and D2. This finding means that accessory bacteriochlorophylls are either absent, or bound to the protein in a different way. The sequence homology between Dl and D2 is higher than between L and M. In particular D2 compared to Dl does not show an insertion of seven amino acids between the fourth and the fifth membrane spanning helices which provided an additional loop with glu M232 as a ligand to the ferrous non-heme-iron atom. We consider hydrogen-carbonate as a likely candidate for being a ligand to the ferrous non-heme iron atom in photosystem II reaction centers instead of the glutamic acid in the reaction centers from the purple bacteria. This proposal may help to understand the pronounced effect of hydrogencarbonate on the electron accepting side of photosystem II reaction centers (for review see 25).

The proteins CP47 and CP43 which are known to carry a considerable number of chlorophyll molecules function in our view as (energy transferring) antenna complexes surrounding Dl and D2 in the membrane.

2.6 Evolutionary Aspects

The basic symmetric arrangement of the pigments and the L and M subunits, and the need for both protein subunits to generate the special pair and the electron accepting quinone iron complex suggests that the reaction center from the purple bacteria evolved from a symmetric protein dimer containing two bacteriochlorophyll molecules forming a special pair, and possessing two equivalent pathways to transfer electrons to the cytoplasmic side of the membrane. Gene duplication and subsequent mutations should account for the origin of the asymmetric dimer. During evolution the asymmetric dimer, having switched off one of the electron transfer pathways, proved to be advantageous. A specialization of the two quinones into a more firmly bound Q_A and a loosely bound Q_B became possible. Having Q_A as an additional electron storing device may have been the most important advantage when replacing a symmetric reaction center by an asymmetric one during evolution.

The formation of a protein dimer may be the simplest way to give rise to chlorophyll dimer (special pair) as an efficient phototrap. Since photosystem I reaction centers most likely contain a dimer as primary electron donor, too, and apparantly contain two very similar proteins of 60 kD molecular weight (for reviews see 19), the motif of an asymmetric protein and pigment dimer may be repeated in photosystem I reaction centers.

3. References

1 Olson, J.M. and Thornber, P. (1979) In Capaldi, R.A. (ed), Membrane Proteins in Energy Transduction, pp. 279-340, Dekker, New York

2 Feher, G. and Okamura, M.Y. (1978) In Clayton, R.K. and Sistrom, W.R. (eds), The Photosynthetic Bacteria, pp. 349-386, Plenum Press, New York.

3 Hoff, A.J. (1982) In Fong, F.K. (ed.) Molecular Biology, Biochemistry, and Biophysics, Vol. 35, pp. 80-151, 322-326, Springer, Berlin.

4 Parson, W.W (1982). Ann. Rev. Biophys. Bioeng., Vol.11, pp. 57-80

5 Williams, J.C., Steiner, L.A., Ogden, R.C., Simon, M.I. and Feher, G. (1983) Proc. Natl. Acad. Sci. USA 80, 6505-6509

6 Williams, J.C., Steiner, L.A., Feher, G. and Simon M.I. (1984) Proc. Natl. Acad. Sci. USA 81, 7303-7307

7 Youvan, D.C., Bylina, E.J., Alberti, M., Begusch, H. and Hearst, J.E. (1984) Cell 37, 949-957

8 Michel, H., Weyer, K.A., Gruenberg, H. and Lottspeich, F. (1985) EMBO J. 4, 1667-1672

9 Michel, H., Weyer, K.A., Gruenberg, H., Dunger, I., Oesterhelt, D. and Lottspeich, F. (1986) EMBO J. 5, 1149-1158

10 Michel, H. (1982) J. Mol. Biol. 158, 567-572

11 Allen, J. and Feher, G. (1984) Proc. Natl. Acad. Sci. USA 81, 4795-4799

12 Chang, C.H., Schiffer, M., Tiede, D., Smith, U., Norris, J. (1985) J. Mol. Biol. 186, 201-203

13 Zinth, W., Kaiser, W. and Michel, H. (1983) Biochim. Biophys. Acta 723, 128-131

14 Gast,P. and Norris, J.R. (1984) FEBS-Lett. 177, 277-280

15 Deisenhofer, J., Epp, O., Miki, K., Huber, R. & Michel, H. (1984) J. Mol. Biol. 180, 385-398

16 Deisenhofer, J., Epp, O., Miki, K., Huber, R. & Michel, H. (1985) Nature 318, 618-624

17 Michel, H., Epp, O. and Deisenhofer, J. (1986) EMBO J. 5, in press

18 Breton, J. (1985) in Michel-Beyerle (ed.), Antennas and reaction centers of Photosynthetic Bacteria, Springer, Berlin, pp. 109-121

19 Staehelin, L.A. and Arntzen, C.J. (eds) (1986) Encyclopedia of Plant Physiol., New Series, Vol. 19, Springer, Berlin

20 Norris, J.R., Uphaus, R.A., Crespi, H.L. and Katz, J.J. (1971) Proc. Natl. Acad. Sci. USA 68, 625-628

21 Gast, P., Michalski, T.J., Hunt, J.E., Norris, J.R. (1985) FEBS Lett. 179, 325-328

22 Debus, R.J., Okamura, M.Y. and Feher, G. (1985) Biophys. J. 47, 3a

23 Robert, B. and Lutz, M. (1986) Biochemistry 25, 2304-2309

24 Michel, H. and Deisenhofer, J. (1986) in Staehelin, A.C. and Arntzen, C.J. (eds.), Encyclopedia of Plant Physiology, New Series, Vol. 19, pp. 371-381 Springer, Berlin

25 Vermaas, W.F.J. and Govindjee (1982) in Photosynthesis, Development, Carbon Metabolism and Plant Productivity (Govindjee, ed), Vol. 2, pp. 541-558, Academic Press, New York

RELATING STRUCTURE TO FUNCTION IN BACTERIAL PHOTOREACTION CENTERS

J. R. NORRIS, D. E. BUDIL, D. M. TIEDE, J. TANG, S. V. KOLACZKOWSKI, C. H. CHANG* AND M. SCHIFFER*
CHEMISTRY DIVISION AND *BIOLOGY DIVISION, ARGONNE NATIONAL LABORATORY, ARGONNE, IL 60439

1. INTRODUCTION

Because of the crystallization of the reaction center from *R. viridis* (1) and the subsequent determination of reaction center structure (2), a new era of photosynthetic research is possible. Now the possibility of connecting structure to function and mechanism is realistic. Two major questions arise in the primary reactions of bacterial photosynthesis. 1. Why is only one of two very similar electron transport pathways used in bacterial photosynthesis? 2. What is the role, if any, of the accessary bacteriochlorophyll bridging the gap between the primary donor and bacteriopheophytin acceptor?

With regard to the first issue, we know that the approximate C2 symmetry of the special pair is broken both in *R. viridis* and in *R. sphaeroides* on the basis of our EPR and x-ray diffraction studies of single crystals of both reaction centers. In addition, the cation of the primary donor appears to be more unsymmetrical in viridis than in sphaeroides. However, for a somewhat more detailed comparison of sphaeroides and viridis single crystal structures at the present level of refinements (1-4) see Chang et al. of these proceedings.

This paper primarily addresses the second issue. Here we present experimental data based on magnetic interactions in the radical pair formed between primary donor and primary acceptor. Our data is consistent with the bridging bacteriochlorophyll providing a site of superexchange.

The *R. viridis* crystal structure shows a bacteriochlorophyll molecule situated between primary reactants, the special pair donor bacteriochlorophyll, P, and the acceptor bacteriopheophytin, I (2). Because of this intermediate bacteriochlorophyll, B, many have postulated that B participates in charge separation. But what is the nature of its participation? Recent picosecond spectroscopy shows that $(P^+ I^-)$ is formed within ~3 ps, with no evidence for any kinetic intermediate at a time resolution of 150 fs (5-6). This admits two interpretations: 1. $(P^+ I^-)$ is formed directly from the excited singlet state of electron donor, with no intermediate; 2. an intermediate state exists between excited singlet state of the electron donor special pair and the radical pair $(P^+ B I^-)$, but lives for a much shorter time than 150 fs.

The distinction between these two possibilities involves timescales; for example, the same state $(P^+ B^- I)$ could appear as (1) a discrete intermediate with kinetic routes to both the donor and radical pair states, or, (2) as a virtual state quantum mechanically coupled coherently to both the donor state and the radical pair. The first of these is slow "hopping"; it is detectable in the magnetic properties because the expected hopping rates will cause loss of phase coherence in spin states of the radical pair. In the fast hopping limit or the coherent limit (case 2), the electronic-nuclear motion will not be discernible on the magnetic resonance timescale, and both the donor and radical pair states will appear to include "supermolecular" electronic configurations involving the B molecule. Our approach is, therefore, to use magnetic properties associated with the primary photochemistry to characterize the participation of B, or more precisely, the precursor state to the radical pair to understand more fully the primary charge separation.

Biggens, J. (ed.), Progress in Photosynthesis Research, Vol. I. ISBN 90 247 3450 9

The primary method explored in this paper is the magnetic field dependence of the radical pair state. When a magnetic field, either a static field or a resonant microwave field, is applied to the radical pair both its lifetime and the subsequent production of triplet state donor is altered. By measuring the field at which maximum triplet is formed one can determine **directly** the size of the singlet-triplet energy gap, 2J, in the radical pair. This measurement is more direct than the determination of 2J from RYDMR lineshapes. The singlet-triplet energy gap, 2J, has been previously connected with the electron transfer reactions giving rise to charge separation and the radical pair and thus is relevant to characterizing the mechanism of charge separation (7).

2. RESULTS

The radical pair lifetime of Q-depleted RCs at low fields was measured at various temperatures. As shown in Figure 1, the field maximum which occurs at 2J is invariant with temperature. Thus, the 2J of 14 ± 1 gauss is temperature independent.

3. DISCUSSION

The observed temperature independence of 2J in the range 293 K to 100 K contradicts the recent results of Moehl et al. (8), who infer indirectly from low-frequency RYDMR lineshape spectroscopy of the radical pair that 2J increases in the temperature range 293-210 K. Such an increase might be expected if contraction of the reaction center brings the interacting molecules closer together at lower temperatures; however, we find no evidence for this type of effect.

A temperature independent 2J also has strong implications for the mechanism of spin exchange in $(P^+ I^-)$, and thus, for the mechanism of primary charge transfer. A long-standing problem in the theoretical description of 2J is that its measured magnitude is much too small to account for the extremely fast rate of electron transfer forming $(P^+ I^-)$. Haberkorn et al. (7) have addressed this discrepancy by postulating that the primary radical pair is actually a combination of two configurations, one with a large exchange (the "close" site) and one with negligible electronic interactions (the "distant" site). A hopping of the radical pair state between these configurations could provide a small exchange while maintaining large interactions and fast electron transfer between adjacent chromophores in the RC. Haberkorn et al. derived an expression relating the effective exchange in the radical pair, $2J^{eff}$ to the exchange in the close site, $2J_C$:

$$2J^{eff} = \frac{(2J_C)\hbar^2 k_c k_d}{(2J_C)^2 + \hbar^2 (k_c + k_2)^2}$$

where k_c is the rate constant for hopping from close to distant sites, k_d the rate constant for the reverse process, and k_2 the rate constant for spin state phase coherence decay in the close site. Although the hopping model does not formally depend upon the identity of the close site, it does require some sort of two step charge separation, with the predominant $(P^+ B I^-)$ state identified as the distant site.

Since at least one of the rate constants k_c or k_d should be temperature dependent, the temperature independence of 2J is a strong indication that the observed spin exchange does not arise from a hopping mechanism. Viewed another way, if |2J| is to remain within 1 G of 14 G over the temperature range 293-140 K, the activation energy for k_d cannot be larger than 2 meV. But this small energy difference would predict nearly equal contributions of $(P^+ B^- I)$ and $(P^+ B I^-)$ to the room temperature radical pair optical difference spectrum, which is not observed.

This analysis by itself does not rule out a two step charge separation, however. It is possible that the amount of spin exchange mixed in by a hopping mechanism is too small to be detected by our method; thus, only an upper limit on the rate constants k_c and k_d is available from our results. Nevertheless, a hopping model cannot reconcile our observations with the picosecond kinetic studies.

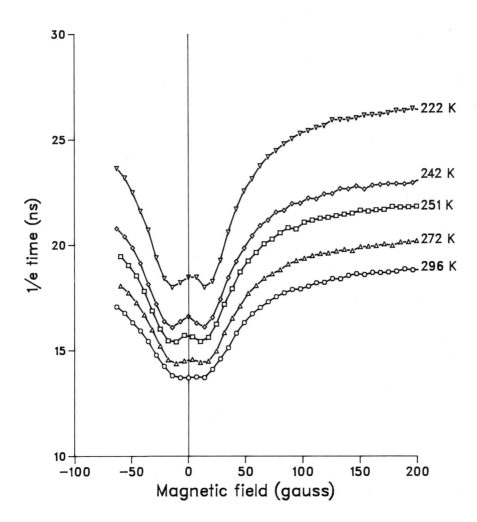

FIGURE 1. Radical pair lifetime as a function of applied field and temperature, demonstrating the appearance of the 2J (singlet-triplet energy gap) resonance structure with decreasing temperature.

Rigorous lower limits are available for some of the kinetic parameters in the above equation. Obviously, k_c must be at least as fast as the observed rate constant of 3.6×10^{11} s^{-1} for (P$^+$ I$^-$) formation. By microscopic reversibility, charge recombination in this model also takes place in two steps, so that k_d must be at least as fast as the observed triplet radical pair decay rate, or about 6×10^8 s^{-1}. Haberkorn et al. (7) estimate a $|2J_c|$ of at least 6.5×10^{-5} eV is needed to account for psec electron transfer, and a lower limit of 3.6×10^{11} s^{-1} on k_2 is estimated from the average of the singlet and triplet decay rates for the close radical pair (9).

The $2J^{eff}$ calculated with these parameters is -5.4×10^{-9} eV or about -0.45 G, much smaller than the experimental value. The estimate of J^{eff} is sensitive to the values assumed for k_c and the (P* B I)-(P$^+$ B$^-$ I) energy gap, upon which k_d and J_{PB} depend strongly. The latter parameter is not well established; estimates from *in vitro* potentiometry (10) suggest a value as large as 0.31 eV. It is noteworthy that the hopping model can accommodate a positive $2J^{eff}$ if (P$^+$ B$^-$ I) is placed above (P* B I) in energy; in any case, it is clear that the two site hopping model cannot reconcile small spin exchange with fast charge separation using magnetic and kinetic parameters consistent with electron transfer theory and experiment.

If no kinetic intermediate exists between (P* B I) and (P$^+$ B I$^-$), the problem of how to reconcile the fast rate of (P$^+$ I$^-$) formation with the small 2J in this radical pair remains. An alternative explanation is that this electron transfer takes place via a so-called superexchange interaction through B. The presence of intervening molecules has been demonstrated to enhance the rate of electron transfer over long distances (11-14). However, the contribution of electron transfer terms to spin exchange is only on the order of the fourth power of S, where S is the overlap integral between adjacent molecules (15,16). Thus, a superexchange mechanism for electron transfer could reconcile fast rates with a small 2J.

The possibility of a superexchange matrix element for primary charge separation has been suggested several times (17-21). However, complete details of the mechanism specific to photosynthesis have not been discussed. The brief considerations of the next section will show that the RC and its components differ in several significant respects from the systems for which magnetic superexchange and bridge-assisted electron transfer have been described.

Figure 2 gives a schematic representation of the electronic configurations of the primary radical pair (configuration a) and some states likely to interact with it. Two obvious possibilities are the states proposed by Haberkorn et al. (7), namely one with an electron transferred from the LUMO of I to the LUMO of B (configuration b) and one with a transfer from the HOMO of B to the HOMO of P (configuration c). It is important to re-emphasize that these configurations are included *coherently* as virtual electron transfers, rather than as a kinetic hopping of the electron or hole to B. Woodbury et al. (21) have recently suggested these configurations as virtual intermediates in primary radical pair formation; it is also noteworthy that the triplet of configuration b corresponds to the proposed intermediate state of Chidsey et al. (22). Rough estimates of the energies of b and c relative to a are available from the *in vitro* electrochemical titration of Fajer et al. (10); these place b and c 0.31 eV and 0.19 eV above the radical pair, respectively.

Another state close enough in energy to contribute substantially to both magnetic exchange and electron transfer is the configuration d, obtained by making *both* of the virtual transfers described above. The result is the first excited state of B, with P and I in their respective ground states. Although it is probably the furthest removed from (P$^+$ B I$^-$) of the states shown in figure 2, it is a more likely candidate to mediate the formation of (P$^+$ I$^-$). Whereas (P B$^+$ I$^-$) and (P$^+$ B$^-$ I) are connected to the donor state (configuration e) by normal coulombic charge transfer interactions, (P B* I) undergoes additional "excitonic" coupling between the moments of the charge distributions in P* and B*. An estimate of the magnitude of this interaction from the crystal structure of the *R. viridis* RC is about 130 cm^{-1} (23) or over five times the average value of the nearest

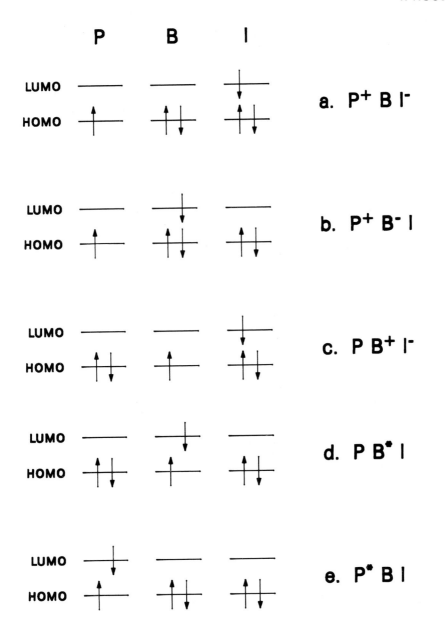

FIGURE 2. Electronic configurations likely to contribute to magnetic superexchange and superexchange electron transfer. All configurations are shown in their singlet states; the diagram is intended to represent occupancy of the various MO's involved rather than their relative energies.

neighbor electron transfer matrix element. This important feature distinguishes the RC from other systems in which superexchange-type electron transfer has been described, and it could represent a considerable enhancement of the electron transfer rate.

An alternative mechanism involving B^* has been suggested by Parson and coworkers (24) who show that about 5% of the $(P\ B^*\ I)$ state is mixed in with the excited donor P^*, and speculate that $(P^+\ B\ I^-)$ could be formed directly from $(P\ B^*\ I)$. It should be noted that this mechanism is distinct from the virtual coupling scheme just discussed since the $(P\ B^*\ I)$ state is apparently a detectable physical intermediate. Furthermore, despite the fact that the transition $(P\ B^*\ I) \rightarrow (P^+\ B\ I^-)$ requires a concerted transfer of two electrons, it is probably not a superexchange reaction as defined above. Nearest neighbor interactions are likely to be large enough for the first order matrix element between these states to be the dominant term.

Magnetic resonance cannot distinguish between the two ultrafast reaction mechanisms just discussed; however, it may be able to implicate the participation of B^* in $(P^+\ I^-)$ interactions. The configuration $(P\ B^*\ I)$ is likely to make a strong contribution to magnetic exchange in the radical pair. Although it is further away in energy than $(P\ B^+\ I^-)$ or $(P^+\ B^-\ I)$, it mixes in the large intramolecular exchange integral J_B in B^*, which is at least three orders of magnitude larger than the J's for $(P^+\ B^-)$ and $(B^+\ I^-)$. The state $(P\ B^*\ I)$ is about 0.2 eV above $(P^*\ B\ I)$ judging from the optical absorption spectra of P and B, which places it 0.35 to 0.48 eV above the radical pair state (25,26). Using this estimate and the calculated value $J_B = +0.27$ eV (27), we predict that the contribution to radical pair exchange will be positive. A reliable determination of the sign of J in the radical pair could therefore provide another piece of evidence for participation of B^* in primary electron transfer. Configurations not involving B might account for a positive J, but their inclusion would not explain the fast forward electron transfer rate. Thus, the magnetic resonance data are very relevant to the mechanism of primary charge separation.

This work was supported by the Division of Chemical Sciences, Office of Basic Energy Science and the Office of Health and Environmental Research of the U.S. Department of Energy under contract W-31-109-ENG-38.

4. REFERENCES

1. Michel, H. (1982) J. Mol. Biol. 158, 567-572.
2. Deisenhofer, J., Epp, O, Miki, K., Huber, R. and Michel, H. (1984) J. Mol. Biol. 180, 385-398.
3. Michel, H., Epp, O. and Deisenhofer, J. (1986) EMBO-J., in press.
4. Chang, C.-H, Tiede, D., Tang, J., Smith, U., Norris, J. and Schiffer, M. (1986) FEBS Letters, in press.
5. Woodbury, N. T. W., Becker, M., Middendorf, D., and Parson, W. W. (1985) Biochemistry 24, 7516-7521.
6. Martin, J.-L., Breton, J., Hoff, A. J., Migus, A., and Antonetti, A. (1986) Proc. Natl. Acad. Sci. USA, in press.
7. Haberkorn, R., Michel-Beyerle, M. E. and Marcus, R. A. (1979) Proc. Natl. Acad. Sci. USA 76, 4185-4188.
8. Moehl, K. W., Lous, E. J. and Hoff, A. J. (1985) Chem. Phys. Lett. 121, 22-27.
9. Ogrodnik, A., Kruger, H. W., Orthuber, H., Haberkorn, R. and Michel-Beyerle, M. E. (1982) Biophys. J. 39, 91-99.
10. Fajer, J., Brune, D. C., Davis, M. S., Forman, A. and Spaulding, L. D. (1975) Proc. Natl. Acad. Sci. USA 72, 4959-4960.
11. Miller, J. R. and Beitz, J. V. (1981) J. Chem. Phys. 74, 6746-6756.
12. Kuznetsov, A. M. and Ulstrup, J. (1981) Biochim. Biophys. Acta 636, 50-57.

13. Kuznetsov, A. M. and Ulstrup, J. (1981) J. Chem. Phys. 75, 2047-2055.
14. Dogonadze, R. R., Ulstrup, J. and Kharkats, Yu. I. (1973) J. Theor. Biol. 40, 279-283.
15. Keffer, F. and Oguchi, T. (1959) Phys. Rev. 115, 1428-1434.
16. Yamashita, J. and Condo, J. (1958) Phys. Rev. 109, 730-741.
17. Jortner, J. (1976) J. Chem. Phys. 64, 4860-4867.
18. Jortner, J. (1980) J. Am. Chem. Soc. 102, 6676-6686.
19. DeVault, D. (1980) Quart. Rev. Biophys. 13, 387-564.
20. Marcus, R. A. and Sutin, N. (1985), Biochim. Biophys. Acta 811, 265-322.
21. Woodbury, N. T. W., Becker, M., Middendorf, D. and Parson, W. W. (1985) Biochemistry 24, 7516-7521.
22. Chidsey, C. E. D., Kirmaier, C., Holten, D. and Boxer, S. G. (1984) Biochim. Biophys. Acta 766, 424-437.
23. Knapp, E. W. and Fischer, S. F. (1985) in Antennas and Reaction Centers of Photosynthetic Bacteria (Michel-Beyerle, M. E., ed.) Springer-Verlag, New York.
24. Parson, W. W., Scherz, A. and Warshel, A. (1985) in Antennas and Reaction Centers of Photosynthetic Bacteria (Michel-Beyerle, ed.) Springer-Verlag, New York.
25. Woodbury, N. T. W. and Parson, W. W. (1984) Biochim. Biophys. Acta 767, 345-361.
26. Chidsey, C. E. D., Takiff, L., Goldstein, R. A. and Boxer, S. G. (1985) Proc. Natl. Acad. Sci. USA 82, 6850-6854.
27. Weiss, C., Jr. (1972) J. Mol. Spectr. 44, 37-80.

CRYSTALLOGRAPHIC STUDIES OF THE PHOTOSYNTHETIC REACTION CENTER
FROM R. SPHAEROIDES

C.-H. CHANG, D. TIEDE*, J. TANG*, J. NORRIS* AND M. SCHIFFER
DIVISION OF BIOLOGICAL AND MEDICAL RESEARCH, AND *THE CHEMISTRY
DIVISION, ARGONNE NATIONAL LABORATORY, ARGONNE, ILLINOIS, 60439

1. INTRODUCTION
    Reaction centers are intrinsic membrane proteins that carry out the first light-induced, chemical charge separations in photosynthesis. In general, the reaction centers of the purple photosynthetic bacteria are composed of three membrane-bound protein subunits, designated L, M and H, with an aggregate molecular weight of 10^5 daltons (1). The complex contains four bacteriochlorophylls, two bacteriopheophytins, two quinones, and one non-heme iron (1). The subunits L and M show significant amino acid sequence homologies with subunits obtained from the photosystem II of green plants (2). In addition to the L, M and H subunits, reaction centers from Rhodopseudomonas viridis, and several other species of photosynthetic bacteria have an additional cytochrome subunit (3). The structure of the reaction center from R. viridis has recently been determined to 3 Å resolution (4,5). However, the structure of the reaction center from R. sphaeroides is of special interest since it has been extensively examined through both spectroscopic and biochemical methods. The reaction center from R. sphaeroides is one of only very few membrane proteins for which single crystals have been obtained that diffract to high resolution.

2. EXPERIMENTAL
2.1. Data Collection
    The crystals of the R. sphaeroides reaction center, which were grown from polyethylene glycol (6,7), are orthorhombic ($P2_12_12_1$) and have unit cell dimensions a = 142.2 Å, b = 139.6 Å and c = 78.7 Å. Diffraction data to 3.7 Å resolution were collected at room temperature by oscillation photography as described previously (8).

2.2. Application of Molecular Replacement
    The structures of the four protein subunits and the chromophores of the photosynthetic reaction center of R. viridis have been previously described (5). Since the protein subunits of R. viridis and R. sphaeroides have homologous amino acid sequences (9-11), they are expected to have similar three-dimensional structures. We therefore applied the molecular replacement method (12) to determine the R. sphaeroides structure by using the known R. viridis structure; the atomic coordinates of R. viridis were kindly provided by J. Deisenhofer and H. Michel. For the search structure, a model for the R. sphaeroides reaction center was constructed from R. viridis by excluding the cytochrome subunit and by removing the amino acid side chains of residues that differed between the two complexes. The resulting model had 60% of the side chains of the R. viridis L subunit, 44% of subunit M (some of the residues in R. viridis were not known at this time), and no side chains of the H subunit (the amino acid sequence is not available in R. sphaeroides).
    Orientation of the complex in the unit cell was determined with Crowther's fast rotation function (13) with 15 to 6 Å data. The translation search to determine the location of the R. sphaeroides complex within its unit cell was carried out on a CRAY-XMP computer with data from 8 to 6 Å resolution. The translation vector was determined by maximizing the correlation factor γ as a function of three-dimensional translation

Biggens, J. (ed.), Progress in Photosynthesis Research, Vol. I. ISBN 90 247 3450 9
© *1987 Martinus Nijhoff Publishers, Dordrecht. Printed in the Netherlands.*

(Tang and Chang, unpublished work). The highest correlation factor corresponded to a minimum R value of 0.44, $R = \Sigma \mid F_o - F_c \mid / \Sigma F_o$; at other translation values the average R value was 0.52. The position and orientation of the complex were further refined by rigid-body refinement. The R-factor for reflections between 8 to 6 Å decreased to 0.39.

3. RESULTS

3.1. Difference Fourier Map: Location of Side Chains

To assess the results of the molecular replacement solution, a map with coefficients $(2F_o - F_c)$ was calculated using structure factors between 8 and 3.7 Å resolution. The chromophores were not included in the calculation of these structure factors. The correctness of the solution was suggested by the appearance of electron density in the map for those side chains and pigment molecules that were not included in the calculation of the map. A Fourier series calculated with coefficient $(2F_o - F_c)$ should have electron density for residues that were not included in the calculation of F_c. As an example, the difference electron density map correctly predicts the positions of the side-chains for tryptophan residues 125 and 128 in the R. sphaeroides subunit M. This procedure was used to incorporate the side chains for 180 residues of R. sphaeroides protein into the model, yielding an improvement in the map. The R-factor decreased for reflections between 8 and 3.7 Å from 0.42 to 0.39.

3.2. Arrangement of the Molecules in the Crystal

The packing of the R. sphaeroides reaction center complex in the unit cell is of special importance, first as a proof that the complex was correctly positioned, and second, because this is only the second crystal of a membrane protein for which the structure has been determined. The packing of the molecules in the unit cell is very reasonable. Only the carboxy terminal two residues of the M subunit interpenetrated the neighboring molecule and, therefore, required adjustment. The M subunit in R. viridis has an additional 18 amino acids in its carboxy end that are pointing away from the M subunit; these amino acids interact with the cytochrome (5). The L and M surfaces of the complex that interacts with the cytochrome in the R. viridis form lattice contacts in R. sphaeroides. As was observed in the R. viridis structure, the interprotein lattice contacts are between polar residues that are located in part of the complex that would be located on the outside of the membrane. There is no contact between the transmembrane helixes of neighboring complexes. Most of the contact between molecules is along the a axis and the least contact is along the b axis of the unit cell. This positioning may explain the observation that several heavy atoms introduced into the crystal change the b cell dimension but do not affect the other cell dimensions.

3.3. Differences in Amino-Acid Residues in the Vicinity of the Chromophores between R. sphaeroides and R. viridis.

The quaternary structure of the protein subunits and that of the chromophores is very similar in R. viridis and R. sphaeroides reaction centers. Recently the amino-acid residues that are near the chromophores in R. viridis were described (14). Although we do not yet know the exact orientation of the amino-acid side chains in R. sphaeroides, the differences in these residues are of interest because they may be correlated with the different optical and resonance properties of the reaction centers from the two species.

A large number of aromatic residues surround the special pair in R. viridis. The surrounding of the special pair in R. sphaeroides is less aromatic in character. Of the five phenylalanine residues only one (L181) is conserved, L160 is replaced by threonine; L241 by valine, M154 and M194 by leucine. Of the three tryptophan residues, M183 is conserved, and L167 and M199 are replaced by phenylalanines. Tyrosines L162 and M208 are conserved; M195 is replaced by phenylalanine. It has been pointed out (14) that two of the residues that hydrogen bond to the special pair tyrosine M195 and threonine L248 in R. viridis, are replaced by phenylalanine and methionine in R. sphaeroides and, therefore, can no longer form a hydrogen bond. Both the change in the aromatic character of the surrounding of the special pair and the change in hydrogen bonding may affect the spectral properties of the special pair in the two species.

Further, two of the residues that were identified as part of the Q_A binding site of R. viridis, alanine M216 and valine M263, are replaced in R. sphaeroides by bulkier hydrophobic residues, methionine and isoleucine, respectively, probably to accommodate the different type of quinone bound by the two reaction centers.

3.4. The Arrangement of the Quinones in the R. sphaeroides Reaction Center

In using the molecular replacement approach, initially we did not include the quinones or pigments in the molecular model. Nevertheless, in the resulting electron density map, electron density for both quinones Q_A and Q_B appears along with the pigments (bacteriochlorophyll and bacteriopheophytin). Figure 1 shows the electron density for the quinone rings at both Q_A and Q_B sites; density was also found for parts of their phytyl tails. The density found for the quinones obeys the approximate local two fold. In the R. viridis crystal there is no quinone in the Q_B position (4,5,15); the inhibitor orthophenathroline can be diffused into the crystal and was found in a position similar to the Q_B site (5).

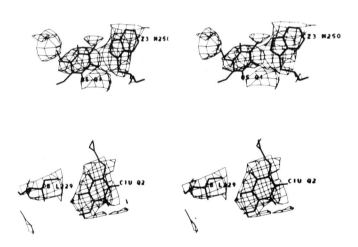

FIGURE 1.Stereo figures of the quinone rings and a neighboring residue in the Q_A (top) and the Q_B (bottom) sites. The backbones of the structures are superimposed on the electron density from the initial ($2F_o$-F_c) map that was calculated without including any of the chromophores. The orientations of the quinones are approximate; their precise orientation will have to be determined after further refinement of the structure.

The kinetics of charge recombination on the bacteriochlorophyll dimer, B_2, from the light-induced $B_2^+Q^-$ state show that the secondary quinone is fully functional in the crystals. The light-induced g=2 B_2^+ EPR signal decays with halftimes of 1 sec both in single crystals and in the reaction center solution before crystallization. This rate is characteristic of charge recombination from the secondary quinone $B_2^+Q_B^- \rightarrow B_2Q_B$ (1,16). Addition of orthophenanthroline, which blocks electron transfer from Q_A to Q_B, to the reaction center solution shortens the charge recombination halftime to about 80 msec, consistent with a recombination from the $B_2^+Q_A^-$ state.

4. DISCUSSION

The success of the molecular replacement technique in correctly revealing (i) the amino acid side-chains that are unique to R. sphaeroides and (ii) electron densities for the reaction center chromophores proves that the solution is a correct one and that the tertiary and quaternary structures of the L, M, and H subunits are essentially conserved. This finding suggests that reaction center structures from many different organisms, including the green plant photosystem II, may be highly homologous. The current structure of the R. sphaeroides reaction center at 3.7 Å resolution shows that the general features of the chromophore organization are also conserved. However, a unique feature of our crystals is that they retain the secondary quinone. Further refinements will provide details of the molecular structures, which are responsible for the different functional activities of these two quinones, and details on the small variations in the organization of the other chromophores.

We thank Drs. H. Drucker and K. L. Kliewer for program development support. This work was supported by the U.S. Department of Energy under contract W-31-109-Eng-38 and by Public Health Service grant GM36598.

REFERENCES

1. Okamura, M. Y., Feher, G. and Nelson, N. (1982) in: Photosynthesis: Energy Conversion by Plants and Bacteria, Vol. 1, pp. 195-272, Academic Press.
2. Michel, H. and Deisenhofer, J. (1986) in: Photosynthesis III. Photosynthetic Membranes and Light Harvesting Systems (L. A. Staehelin and C. J. Arntzen, eds.) Encyclopedia of Plant Physiology, New Series, Vol. 19, pp. 371-381, Springer-Verlag, Berlin.
3. Dutton, P. L. and Prince, R. C. (1978) in: The Photosynthetic Bacteria (Clayton, R. K. and Sistrom, W. R., eds.) pp. 525-570, Plenum Press.
4. Deisenhofer, J., Epp, O., Miki, K., Huber, R. and Michel, H. (1984) J. Mol. Biol. 180, 385-398.
5. Deisenhofer, J., Epp, O., Miki, K., Huber, R. and Michel, H. (1985) Nature 318, 618-624.
6. Gast, P. and Norris, J. R. (1984) FEBS Lett. 177, 277-280.
7. Chang, C.-H., Schiffer, M., Tiede, D., Smith, U. and Norris, J. (1985) J. Mol. Biol. 186, 201-203.
8. Chang, C.-H., Tiede, D., Tang, J., Smith, U., Norris, J. R. and Schiffer, M. (1986) FEBS Lett. (in press).
9. Michel, H., Weyer, K. A., Gruenberg, H., Dunger, I., Oesterhelt, D. and Lottspeich, F. (1986) EMBO J. 5, 1149-1158.
10. Williams, J. C., Steiner, L. A., Ogden, R. C., Simon, M. I. and Feher, G. (1983) Proc. Natl. Acad. Sci. USA 80, 6505-6509.
11. Williams, J. C., Steiner, L. A., Feher, G. and Simon, M. I. (1984) Proc. Natl. Acad. Sci. USA 81, 7303-7307.
12. Rossmann, M. G., ed. (1972) The Molecular Replacement Method, Gordon and Breach, Science Publishers.
13. Crowther, R. A. (1972) in: The Molecular Replacement Method (Rossmann, M. G., ed.) pp. 173-178, Gordon and Breach, Science Publishers.
14. Michel, H., Epp, O. and Deisenhofer, J. (1986) EMBO J. (in press).
15. Gast, P., Michalski, T., Hunt, J. and Norris, J. (1985) FEBS Lett. 179, 325-328.
16. Stein, R. R., Castellvi, A. L. Bogacz, J. P. and Wraight, C. A. (1984) J. Cell. Biochem. 24, 243-259.

STRUCTURE ANALYSIS OF THE REACTION CENTER FROM *Rhodopseudomonas sphaeroides*: ELECTRON DENSITY MAP AT 3.5Å RESOLUTION

J. P. ALLEN[a], G. FEHER[a], T. O. YEATES[b], AND D. C. REES[b]

[a]UNIVERSITY OF CALIFORNIA, SAN DIEGO, LA JOLLA, CALIFORNIA 92093 and [b]UNIVERSITY OF CALIFORNIA, LOS ANGELES, LOS ANGELES, CALIFORNIA 90024

ABSTRACT: *Crystals of reaction centers (RCs) from R. sphaeroides have been studied by x-ray diffraction. The molecular replacement and solvent flattening techniques have been used to obtain an electron density map at 3.5Å resolution. Comparison of the structures of the RCs from R. sphaeroides and R. viridis showed the following conserved features: the 5 membrane-spanning helices in each of the L and M subunits, and the single membrane-spanning helix in the H subunit, the two-fold symmetry axis and the approximate positions of the cofactors.*

1. INTRODUCTION

Extensive work has been done in the past to characterize RCs from *R. sphaeroides*. However, until recently there was very little information available concerning their three dimensional structure. The recent success in obtaining high quality crystals of RCs from *R. viridis* (1) and *R. sphaeroides* (2) made x-ray diffraction studies possible. The structure of the RC from *R. viridis* has been reported to a resolution of 3Å (3,4). In this work we present data of the structure of the RC from *R. sphaeroides* at a resolution of 3.5Å.

2. MATERIALS AND METHODS

2.1. Crystallization

Several different crystal forms have been grown using the vapour diffusion technique (5). The form used in the present study has the space group $P2_12_12_1$ (6). The crystals were grown using the amphiphile heptane triol and the detergent lauryl dimethyl amine oxide as previously described (7).

2.2. Data Collection

X-ray diffraction data were collected on a multi-wire area detector of the type developed by Xuong and collaborators (8), using an Elliott GX21 rotating copper anode as an x-ray source. Data collection on two native crystals produced 23,551 reflections which were scaled to give 12,237 unique intensity data with a scaling R-factor of 6.8%. Integration of the raw data was performed with the software of L. Weissman (unpublished). Further data processing utilized programs from ROCKS (9).

3. RESULTS AND DISCUSSION

3.1. Molecular Replacement Method

The diffraction data were analyzed using the molecular replacement technique as previously described (7,10). The RC model was constructed from a partially refined structure of the RC from *R. viridis* (3,4). The model contained all of the cofactors and the three subunits, L, M, and H; the fourth, nonconserved cytochrome subunit was removed. The conserved residues of L and M were retained while all other residues, including all of H, were replaced by alanine. The highest peak in the rotation function positions the two-fold symmetry axis relating the L and M subunits near the diagonal of the crystallographic yz plane. For the correctly oriented model, the largest peak in each Harker section of the translation function corresponded to a single unique solution.

Biggens, J. (ed.), Progress in Photosynthesis Research, Vol. I. ISBN 90 247 3450 9
© *1987 Martinus Nijhoff Publishers, Dordrecht. Printed in the Netherlands.*

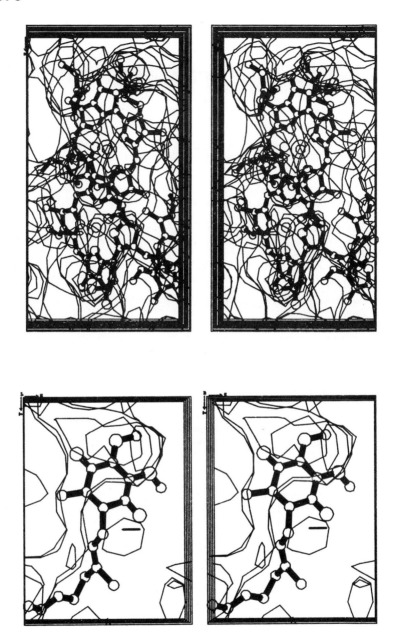

FIGURE 1. *Stereoplots of selected regions of the electron density map of the RC from R. sphaeroides at 3.5Å resolution. The top shows the electron donor and the bottom shows the secondary electron acceptor, Q_B. Each figure consists of contour plots of the electron density with the positions of the cofactors superimposed.*

To simultaneously optimize the molecular replacement parameters and the unit cell lengths, an algorithm based upon the refinement program of Jack and Levitt (11) was used. Variations in unit cell lengths were accommodated by treating the *R. viridis* model as an elastic body. The optimization procedure resulted in a rotation of the original model by 1.2 degrees and a change in 3% of one cell dimension. The R-factor between the observed data and structure factor calculated on the basis of this model was 0.42.

3.2. Solvent Flattening

Since the volume of the RC crystals consists of approximately 70% solvent and detergent that is disordered, the electron density distribution surrounding the protein is expected to be constant. A molecular boundary was defined from the results of the molecular replacement. The solvent region was set to a constant value in an iterative refinement scheme analogous to Wang's method (12). After fourteen cycles the phases converged to values that differed from the starting phases by an average of 35 degrees.

3.3 Structure Determination from the Electron Density Map

The phases obtained by the solvent flattening and the observed structure factor amplitudes were used to calculate the electron density map. Comparison of the structure of the RC from *R. sphaeroides* with that from *R. viridis* showed many conserved features, such as the 5 membrane-spanning helices in each of the L and M subunits, and the single membrane spanning helix in the H subunit, and an approximate two-fold symmetry axis between the L and M subunits. The positions and the orientations of the cofactors appear to be conserved, although slight differences between the two species cannot be ruled out at present.

Two specific regions of the electron density map are illustrated in Figure 1. They correspond to the regions around the primary donor (bacteriochlorophyll dimer) and the secondary acceptor (ubiquinone-10). The cofactor positions were determined from the RC from *R. viridis* and have not been adjusted for optimal fit to the electron density.

4. SUMMARY AND CONCLUSIONS

We have obtained a preliminary electron density map of the RC from *R. sphaeroides* at 3.5Å resolution. We are finishing data collection on both a heavy atom derivative and higher resolution native data. The exact positions of the cofactors and the polypeptide chain will be described after the analysis of these data is completed.

REFERENCES

1. Michel, H. (1982) *J. Mol. Biol.* **158**,567-572.

2. Allen, J. and Feher, G. (1984) *Biophys. J.* **45**, 256a (abst.).

3. Deisenhofer, J., Epp, O., Miki, K., Huber, R., and Michel, H. (1984) *J. Mol. Biol.* **180**, 385-398.

4. Deisenhofer, J., Epp, O., Miki, K., Huber, R., and Michel, H. (1985) *Nature* **318**, 618-624.

5. Allen, J. P. and Feher, G. (1984) *Proc. Natl. Acad. Sci. USA* **81**, 4795-4799.

6. Feher, G. and Allen, J. P. (1985) in *Molecular Biology of the Photosynthetic Apparatus* (Cold Spring Harbor Laboratory, NY) pp. 163-172.

7. Allen, J. P., Feher, G., Yeates, T. O., Rees, D. C., Deisenhofer, J., Michel, H., and Huber, R. (1986) *Proc. Natl. Acad. Sci. USA*, in press.

8. Cork, C., Hamlin, R., Vernon, W., Xuong, N. H., and Perez-Mendez, V. (1975) *Acta. Cryst.* **A31**, 702-703.

9. Reeke, G.N. (1984) *J. Appl. Cryst.* **17**,125-130.

10. Allen, J. P., Feher, G., Yeates, T. O., Rees, D. C., Eisenberg, D. S., Deisenhofer, J., Michel, H., and Huber, R. (1986) *Biophys. J.* **49**, 583a (abstr.).

11. Jack, A. and Levitt, M. (1978) *Acta. Cryst.* **A34**, 931-935.

12. Wang, B. C. (1985) *Methods in Enzymology* **115**, 90-112.

ACKNOWLEDGEMENTS

We thank E. Abresch for the preparation of the reaction centers and H. Komiya for assistance with the data analysis. This work was supported by grants from the National Institutes of Health AM36053, GM13191, and GM31875, and the Chicago Community Trust/Searle Scholars Program.

ADDENDUM : The positions of the cofactors obtained from the electron density map at a resolution of 3.3 Å are shown in Fig. 2. The RC model was refined using the restrained least squares program of Hendrickson and Konnert (Hendrickson (1985) Methods Enzymol. **115**, 252-270). At the present stage of refinement, the R factor between observed and calculated structure factors is 0.306 for data between 8 Å and 3.3 Å resolution, with the deviation of bond distances from standard values being 0.023 Å. Although the cofactors follow, in general, the two-fold symmetry axis, Q_A is located ∼ 2 Å further from the iron atom than Q_B.

The positions of several critical amino acid residues have been found to be conserved between the RC structures of *R. sphaeroides* and *R. viridis* (Michel, Epp, & Deisenhofer (1986) EMBO J., in press). Glu L104 appears to be hydrogen bonded to ring V of BPh_A and constitutes one of the differences between BPh_A and BPh_B. The plane of the ring of Trp M252 lies approximately parallel to the plane of the ring of Q_A. Phe L216 is near Q_B; the planes of the two rings are approximately perpendicular. Glu L212 is close to Q_B, but not within hydrogen bonding distance; it may, however, indirectly play a role in the protonation of Q_B.

Ile L229 and Ser L223 are close to Q_B with the structure indicating that Ser L223 is hydrogen bonded to Q_B. Mutations to either residue have been found to alter the binding of Q_B and to impart herbicide resistance to the bacteria (Paddock, et.al., these proceedings). The reduced Q_B binding follows from the structure (ie. Ser L223 → Pro causes the loss of an hydrogen bond and Ile L229 → Met produces a steric hinderance). Tyr L222 is located further from Q_B and its mutation also confers herbicide resistance, although the effect of the mutation on the binding of Q_B is weaker.

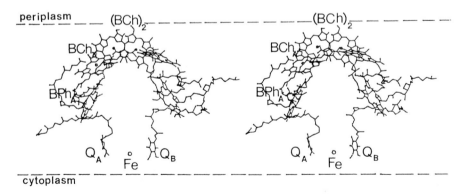

FIGURE 2.   *Stereoplot of the cofactors of the RC from R. sphaeroides at a resolution of 3.3 Å. The two-fold symmetry axis is aligned vertically in the plane of the paper and the electron transfer proceeds preferentially along the A-branch (called the L-branch in reference 4). The inferred position of the membrane is indicated by dashed lines.*

EVIDENCE OF THE PRIMARY CHARGE SEPARATION IN THE D_1D_2 COMPLEX OF PHOTOSYSTEM II FROM SPINACH: EPR OF THE TRIPLET STATE

M. Y. OKAMURA[a], K. SATOH[b], R. A. ISAACSON[a] AND G. FEHER[a]

[a]DEPARTMENT OF PHYSICS, UNIVERSITY OF CALIFORNIA, SAN DIEGO, LA JOLLA, CALIFORNIA 92093 AND [b]DEPARTMENT OF BIOLOGY, FACULTY OF SCIENCE, OKAYAMA UNIVERSITY, OKAYAMA 700, JAPAN

1. INTRODUCTION

The amino acid sequence homologies of the L and M-subunits of reaction centers from photosynthetic bacteria (1-4) and the D_1D_2 subunits from photosystem II containing organisms (5) led to the hypothesis that the site of the charge separation in photosystem II is located on the D_1D_2 subunits (5,6). Recently Satoh and coworkers have isolated a reaction center complex that contained **only** the subunits D_1D_2 and cytochrome b-559 (7-8). We report here the detection and characterization of the light induced triplet state by EPR. The results provide strong evidence for the presence of charge separation in the D_1D_2 complex.

2. THE TRIPLET STATE

2.1. General considerations: Triplet states from reaction centers from photosynthetic bacteria were first observed by Dutton *et al.* (9) and from photosystem II reaction centers by Rutherford *et al.* (10). From the shape of the signals one infers a spin polarization of the triplet state that is characteristic of a radical pair intermediate (11) i.e. the triplet is formed via the reaction

$$P_{680}^S I \xrightarrow{h\nu} P_{680}^+ I^- \rightarrow P_{680}^t I \tag{1}$$

where P_{680}^S and P_{680}^t are the singlet and triplet of the donor (Chla) respectively and I is the intermediate (pheophytin) acceptor. The spin polarization differs from that observed in triplets generated from excited states via intersystem crossing. Consequently, the polarization can be used as an assay for the presence of a charge separated precursor.

2.2 Experimental procedures and results The D_1D_2 complex was prepared from spinach as described (7,8). Samples with an absorbance of $A_{1\,cm}^{670} = 10$ were introduced into an EPR cell with dimensions 20 mm × 10 mm × 1 mm and frozen in the dark at 77°K. EPR spectra were taken at 2.1°K (for details see Figure legend). The spectrum (Fig. 1) exhibits the polarization AEE AAE (where A and E refer to absorption and emission, respectively) that is indicative of a radical pair precursor. The values of the zero-field splitting parameters D and E were obtained from the spectrum:

$$D = 287 \times 10^{-4} cm^{-1}$$
$$E = 43 \times 10^{-4} cm^{-1} \tag{2}$$

The values of these parameters as well as the amplitudes of the signals were compared with those obtained on a sample of PSII reaction centers prepared with digitonin ($\sim$ 50 Chla/RC) (12). The PSII sample was reduced with dithionite, and frozen in the dark. The sample contained approximately the same concentration of RCs as the previous (D_1D_2) sample. The incident light intensity as well as the conditions under which the EPR spectra were taken were the same as those for the D_1D_2 complex. The EPR amplitudes of the triplet signals were proportional to the light intensity and were the same in both samples. This shows that the triplet signal in the D_1D_2 preparation was not due to a residual impurity. The zero-field splittings parameters in the PSII reaction centers were determined to be $D = 286 \times 10^{-4} cm^{-1}$ and $E = 42 \times 10^{-4} cm^{-1}$. These values are within

Biggens, J. (ed.), Progress in Photosynthesis Research, Vol. I. ISBN 90 247 3450 9
© *1987 Martinus Nijhoff Publishers, Dordrecht. Printed in the Netherlands.*

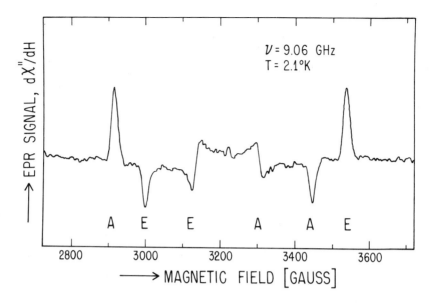

FIGURE 1. *EPR spectrum of the light induced triplet in D_1D_2 complex from spinach. A and E refer to absorption and emission. Light source: Leitz 500 W projector with 650 nm filter (BW 50 nm) microwave power 10 μ W, field modulation 100 kHz, 5 Gauss p.t.p.*

experimental error the same as those obtained in the D_1D_2 complex (eq. 2) and are in agreement with those obtained previously (10) ($D = 290 \times 10^{-4}$cm^{-1}, $E = 40 \times 10^{-4}$cm^{-1}).

3. SUMMARY AND CONCLUSIONS

We have observed and characterized the EPR spectrum of the light induced triplet state in the D_1D_2 complex. The results provide strong evidence that the D_1 and D_2 subunits are the sites of the primary charge separation between the primary donor P_{680} and the intermediate acceptor I. The D_1D_2 complex did not contain quinone; it was presumably lost during the purification. Experiments are in progress to reconstitute D_1D_2 with exogenous quinones.

REFERENCES

1. Williams, J. C., L. A. Steiner, R. C. Ogden, M. I. Simon, and G. Feher, (1983) *Natl. Acad. Sci. U.S.A.* **80**, 6505-6509.

2. Williams, J. C., L. A. Steiner, G. Feher, and M. I. Simon, (1984) *Proc. Natl. Acad. Sci. U.S.A.* **81**, 7303-7307.

3. Youvan, D. C., E. J. Bylina, M. Alberti, H. Begush, and J. E. Hearst (1984) *Cell* **37**, 949-957.

4. Michel, H., K. A. Weyer, H. Gruenberg, I. Dunger, D. Oesterhelt, and F. Lottspeich (1986) *EMBO J.* **5**, 1149-1158.

5. For a review see Trebst, A. (1986) *Z. Naturforsch*, **41c**, 240-245.

6. Michel, H. and J. Deisenhofer (1986) Encyclopedia of Plant Physiology: Photosynthesis III. A. C. Staehelin and C. J. Arntzen, eds. Springer, Berlin <u>19,</u> 371-381.

7. Satoh, K. and O. Nanba, these proceedings.

8. Nanba, O. and K. Satoh, *Proc. Natl. Acad. Sci. U.S.A.* (1986) in press.

9. Dutton, P. L., J. S. Leigh and M. Seibert (1971) Biochem. *Biophys. Res. Commun.* **40**, 406-413.

10. Rutherford, A. W., D. R. Paterson and J. E. Mullet (1981) *Biochim. et Biophys. Acta* **635**, 205-214.

11. Thurnauer, M. C., J. J. Katz and J. R. Norris (1975) *Proc. Natl. Acad. Sci. U.S.A.* **72**, 3270-3274.

12. Yamada, Y., N. Itoh, and K. Satoh (1985) *Plant and Cell Physiol.* **26**, 1263-1271.

ACKNOWLEDGEMENT

This work was supported by grants from the National Science Foundation DMB 85-18922 and INT 85-10509.

CRYSTALLIZATION AND SPECTROSCOPIC INVESTIGATIONS
OF THE PIGMENT-PROTEIN COMPLEXES OF RHODOPSEUDOMONAS PALUSTRIS

T.Wacker,K.Steck,A.Becker,G.Drews*,N.Gad'on*,W.Kreutz,W.Mäntele,W.Welte

Institut für Biophysik und Strahlenbiologie,Albertstr.23 and
*Institut für Biologie II,Mikrobiologie,Schänzlestr.1,7800 Freiburg,FRG.

1.INTRODUCTION

The photosynthetic membrane of purple nonsulfur bacteria contains three pigment-protein complexes. Two antenna systems absorb light energy ,which is transferred to the reaction-centre (RC) where charge separation takes place. The light harvesting complex B875 is in close contact and in a fixed stoichiometric ratio to the RC, a second one the B800-850 light harvesting complex, is synthesized in variable amounts, depending on light intensity during growth.
By use of suitable detergents and chromatographic procedures,the RC,the RC-B875 and the B800-850 complex have been purified. Crystallization of these pigment-protein complexes was recently reported (1-7). A crystallographic analysis of the RC of R. viridis provided a detailed structure (8).
We report here the crystallization of the RC-B875 and the B800-850 complex of R. palustris, and the investigation of the crystals with spectroscopy and X-ray diffraction.

2.MATERIALS AND METHODS

The RC-B875 and B800-850 complexes were isolated in a similar way as described in (7,9). Crystals were obtained by vapor diffusion technique using polyethylene glycol (PEG). The purified protein was dialysed against 25 mM Tris-Cl pH 8 containing 0.04% LDAO (buffer A), and concentrated by ultrafiltration to 10 mg/ml. This corresponds to a optical absorbance A (1cm,875nm) of 130 for the RC-B875 complex and A (1cm,800nm) of 170 for the B800-850 complex.
The protein was mixed with the same volume of
1) 17% w/v PEG 2000, 20 mM Magnesiumchloride in buffer A, for the RC-B875 complex.
2) 35% w/v PEG 1000, 20 mM Magnesiumchloride in buffer A, for the B800-850 complex.
The mixture was equilibrated respectively with a reservoir of 1) or 2) by vapor phase equilibration.
Large crystals for X-ray diffraction of the B800-850 complex were obtained by mixing the concentrated protein with the same volume of 56% PEG 1000 w/v in buffer A, containing 5% heptanetriol and 40 mM sodium phosphate. In this case the reservoir concentration was 34% PEG 1000 w/v in buffer A.
Absorbance and linear dichroism spectra of the crystals were taken on a single beam spectrometer as described previously (7).

Biggens, J. (ed.), Progress in Photosynthesis Research, Vol. I. ISBN 90 247 3450 9
© *1987 Martinus Nijhoff Publishers, Dordrecht. Printed in the Netherlands.*

3.RESULTS AND DISCUSSION

The B800-850 complex forms square shaped crystals suitable for spectroscopic work. Fig.1 shows a series of spectra obtained by rotating and tilting the polarization vector of the measuring beam. The results are presented as the differences of spectra obtained with polarizer horizontally and vertically to the laboraty reference direction. A preferential orientation of the B800-850 Qy transition moments as well as the Qx transition moments, perpendicular to the Qy direction , are found within the crystal plane.

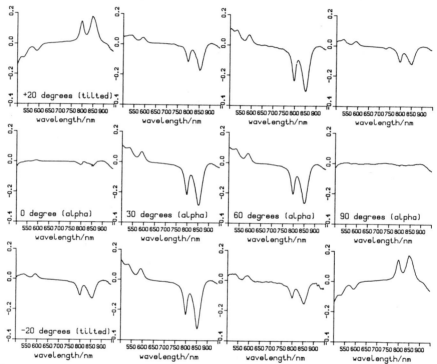

Fi.1   Rotation angle of polarization (B800-850 complex)

X-ray diffraction of large crystals showed a resolution to about 0.8 nm and revealed a orthorhombic space group $C222_1$ with unit cell dimensions of a = 12.4 nm , b = 15.5 nm , c= 11.3 nm.
The orthorhombic symmetry operations have to be applied to every transition moment in the asymmetric unit . The direction of the resulting maxima and minima of the measured linear dichroism must coincidence with the crystallographic axes (10). This has to take into account by calculating the distribution of the transition moments within the asymmetric unit.

The RC-B875 complex forms thin , rectangular crystals ,with a long (X-axis) and a short edge (Y-axis) , suitable for spectroscopy. Fig.2 shows a series of spectra obtained by rotating the polarization vector of the incident light beam in the plane of the crystal. In this plane the Qx-transition moments are aligned parallel to the X-axis , and the Qy-transitions of the B875 chlorophylls parallel to the Y-axis of the crystal.

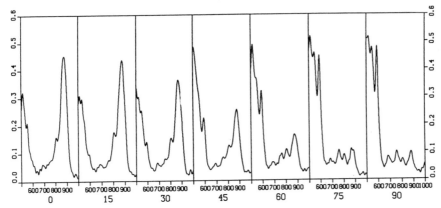

Fig.2   Rotation angle of polarization (RC-B875 complex)

Tilting the crystal around the Y-axis with polarization vector parallel to the X-axes leads to a decrease of the Qx-transition band and to the rise of the B875 Qy-transition band (not shown).
Upon tilting the crystal around the X-axis with polarization vector parallel to the Y-axis (Fig.3), the absorbance of the Qx-transition band remains zero , while the Qy absorption decreases.

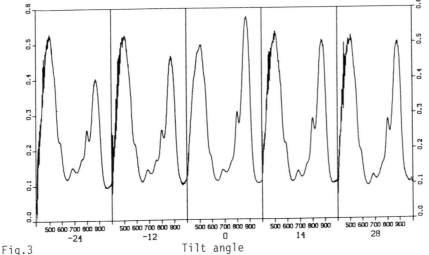

Fig.3   Tilt angle

This demonstrates the almost perfect alignement of the Qx-transition moments in the crystal and therefore in the RC-B875 complex, too. The Qy transition moments are distributed in the plane perpendicular to this direction having a preverential orientation along the Y-axis (decrease of Qy in Fig. 3 by tilting).

I.4.**386**

The porphyrin planes of the B875 antenna chlorophylls are therefore aligned with their Qx transitions parallel to each other, while the Qy-transitions are distributed in the Y-Z plane but having a maximum parallel to the Y-axis.

ACKNOWLEDGEMENTS

We are grateful to Professor J.Breton for his help with the linear dichroism measurements.

REFERENCES

(1) Michel.H.(1982) J. Mol.Biol. 158, 567-572.
(2) Allen,J.and Feher,G. (1984) PNAS 81 4795-4799.
(3) Gast,P. and Norris,J.R. (1984) FEBS LETT. 277-280.
(4) Welte,W.,Wacker,T.,Leis,M.,Kreutz,W.,Shiozawa,J.,Gad'on,N.and Drews,G. (1985) FEBS LETT. 182,260-264.
(5) Allen,J.P.,Theiler,R.and Feher (1985) Biophys.J.,47,M-AM-A8
(6) Cogdell,R,(1985) Springer Series in Chemical Physics 42,Springer Verlag Berlin, 85-87.
(7) Wacker,T., Gad'on,N., Becker,A. ,Mäntele,W. ,Kreutz,W., Drews,G., and Welte,W., (1986) FEBS LETT. 197 ,267-273.
(8) Deisenhofer,J.,Epp,O.,Miki,K.,Huber,H. and Michel,H. (1985) J.Mol.Biol. 180, 385-398.
(9) Firsow,N.N.,and Drews,G. (1977) Arch. Microbiol. 115,299-306.
(10) Wahlstrom,E.E.,Optical Crystallography,J.Wiley and Sons,New York

SPECTROSCOPY, STRUCTURE AND DYNAMICS IN THE REACTION CENTER OF RHODOPSEUDOMONAS VIRIDIS

J. BRETON, J. DEPREZ, B. TAVITIAN and E. NABEDRYK

Service de Biophysique, CEN Saclay 91191 Gif-sur-Yvette cedex - FRANCE -

1. Introduction

Most of the present knowledge on the organization of the photosynthetic pigments in vivo has been derived from two different approaches. On the one hand, biochemical techniques have been used to isolate, purify and characterize the various building blocks constituting the photosynthetic membrane. On the other hand, a variety of spectroscopic techniques (both static and dynamic) have been used to investigate the physical properties of the components (pigments, charge carriers, proteins) in the intact membrane and in the isolated building blocks as well as in model systems which were thought to mimic the in vivo environment. These techniques have revealed a very high level of organization of the membrane components in vivo. Although a wealth of spectroscopic data has been collected over the years, the problem of inferring from these measurements alone the most probable configuration of the pigments and proteins in vivo is extremely complex due to the large intrinsic heterogeneity of the system and to the number of unknown or uncontrolable parameters which are involved.

X-ray crystallography is the only general technique which allows a description of complex biological macromolecules with enough precision so that realistic atomic models can be built. Up to recent years no membrane protein had yielded three-dimensional crystals and the hope of achieving a high resolution structural study of the building-blocks of the photosynthetic membrane was rather limited. This situation is however changing rapidly as crystals of reaction centers (RCs) and antenna complexes from photosynthetic bacteria have now been obtained. A major breakthrough in photosynthesis research has been achieved when X-ray crystallography at 3 Å resolution has yielded a detailed description of the molecular architecture of the RC from R. viridis (1, 2). The main interest is twofold : firstly it gives the rough picture which was so cruelly lacking in the previous spectroscopic investigations ; secondly, by providing a unique opportunity to test the results of the detailed spectroscopic investigations done on the same system, it constitutes an invaluable reference for analyzing the structural properties of the other systems. Further improvement of the interpretation of the spectral data can be expected when refinements of the initial model are produced and also when other building blocks of the photosynthetic membrane will be amenable to high resolution X-ray crystallography. Although it might seem of little use to pursue the understanding of the spectroscopy before these detailed molecular models have been obtained from X-ray data, we would like to emphasize that it would be quite surprising if these models would directly allow an explanation of the primary processes of photosynthesis in a way similar to the one that has allowed a detailed description of the genetic code from the model of the DNA molecule. The primary processes of photosynthesis (exciton migration in the antenna, trapping by the RC,

This paper is dedicated to the memory of Dr. A. GAGLIANO

separation and stabilization of electric charges) are essentially governed by the energy levels of the pigments involved in these processes and these energy levels will not be obtained directly from the X-ray data. However, such data, by giving us the precise location and environment of the pigments, constitute a considerable help in understanding their spectral properties and thus in rationalizing the energetics and the kinetics of the primary processes of photosynthesis. It is with this perspective in mind that we will discuss some results recently obtained in our group on the structural and functional aspects of the organization of the pigments and proteins within the RC of R. viridis.

2. Organization of the protein secondary structures

In 1982, investigating membrane reconstituted RCs of R. sphaeroides by UVCD and polarized IR spectroscopies, a quantitative analysis of the protein orientation in the RC was described (3). From our IR dichroism data, we concluded that the average orientation of the α-helical segments (which constitute more than 50 % of the secondary structures in the RC) is 30° with respect to the membrane normal. Such a transmembrane organization of α-helices was later corroborated by the sequence analysis of RC polypeptides which predicted hydrophobic domains that could fit to transmembrane α-helices (4, 5) and by the recent X-ray analysis of crystallized RC of R. viridis. In the case of R. viridis, a 47 % α-helical content has been measured (6). A comparable value (42 %) can be deduced from the X-ray data (J. Deisenhofer and H. Michel, personal communication). In contrast, the UVCD spectrum of the isolated H subunit from R. sphaeroides is different, exhibiting an α-helical content of 31 % (E.N., J.B., R. Debus, M. Okamura and G. Feher, unpublished).

The X-ray data indicate that the ten α-helices observed in the LM subunit are tilted on the average at a small angle with respect to the C 2 symmetry axis with eight of them tilted at less than 25° and the remaining two at about 38° from this axis (2). These data are quite in agreement with IR dichroism results of RC and LM films of R. sphaeroides leading to an average orientation for all the α-helices at less than 30-35° from the membrane normal (3, 7). In the RC of R. viridis, an experimental average tilt angle of 38° has been estimated from IR dichroism (6). The cytochrome part of the RC protein accounting for about one third of the molecular weight of the total complex and the X-ray structure showing that the α-helical segments of this subunit have almost no preferred orientation with respect to the LM C 2 symmetry axis (1, 2), the exact tilt angle for the only α-helices of the LMH complex of R. viridis should be lower than 38° (6), and closer to the angle observed by X-ray or by IR dichroism of RC and LM films of R. sphaeroides (3, 7). Assuming a random orientation of the non transmembrane α-helices, we can apply a correction to our IR dichroism data in order to calculate the average orientation of the only transmembrane α-helices. Using (i) the 55 % α-helical content deduced from UVCD spectra of the LM subunit from R. sphaeroides (7), (ii) the assumption of an average of 25 residues per transmembrane α-helix and (iii) the ten transmembrane α-helices predicted by hydropathy plots, it can be calculated that among the 324 amino acids belonging to α-helical structures in the LM subunit of R. sphaeroides, 75 may be located in non-transmembrane α-helices leading to a 42 % content of purely transmembrane α-helices and an average tilt angle for these α-helices of 20-25° with respect to the membrane.

Preliminary X-ray data of the RC of R. sphaeroides indicate that the overall structure of its LM subunit is quite superimposable to that of the RC of R. viridis (8), implying a similar orientation of α-helices in both

RCs. Thus, our conclusion that the transmembrane α-helices in LM of R. sphaeroides are tilted on the average at 20-25° from the normal to the membrane is directly transposable to the LM subunit of R. viridis, emphasizing the similar folding and orientation of the protein in BChl a- and BChl b-containing RCs. It can also be generally concluded that there is a good agreement regarding the orientation of the α-helices as determined by X-ray or by polarized light spectroscopy. This gives confidence for interpreting the spectroscopic data obtained on other Chl-protein complexes for which no structural model is available such as the LHC and PSI of green plants (9, 10).

An important question to address is the position assumed by the C 2 symmetry axis with respect to the plane of the membrane. Deisenhofer et al. (1) have proposed that the C 2 symmetry axis evident from their X-ray crystallography data is perpendicular to the membrane. This proposal was based upon (i) the transmembrane character of the charge separation and (ii) the orientation of the chromophores as described in (11). In the following discussion we will further stress that the orientation of the pigments as seen by LD and X-ray spectroscopy is indeed strongly in favour of this positioning of the symmetry axis. However, the charge separation could in principle have also a significant component along the membrane plane while excitonic interactions among six strongly coupled pigment molecules, as recently proposed (12, 13), would lead to transition moments bearing only distant relationships with the transitions of the monomers. In such a case, the orientation of the long wavelength transition of the special pair could be significantly different from that of the vector sum of the Q_y transition moments of the two monomers, which was proposed in (11). In these conditions, the proposal of Deisenhofer et al., (1) would no longer hold. Thus, it is of interest that IR dichroism can add another observation in favor of this assignment of the orientation of the C 2 symmetry axis based upon the comparison of the average tilt angle of the α-helices with respect to either the membrane normal (determined by IR dichroism) or the C 2 axis (estimated on the model derived from X-ray diffraction). As described above, the transmembrane α-helical segments in the LM complex of R. sphaeroides R-26 are tilted on the average at 20-25° from the membrane normal. A similar estimation can also be made for R. viridis RCs. Such an average tilt angle is qualitatively in good agreement with the data from X-ray crystallography. Thus it can be concluded that in vivo the C 2 symmetry axis of the RC runs parallel to the membrane normal.

3. Organization of the pigments

Planar aggregates of isolated RCs can be oriented mechanically in polyacrylamide gels. The identical orientation for various transition moments with opposite dichroism which is observed when comparing the LD spectra of RCs from R. viridis either in the chromatophore membrane or in aggregates, clearly demonstrates that in the latter case the transmembrane axis (C 2 axis) of the RC is perpendicular to the planar aggregates (14).

LD and photoselection experiments (15 and references therein ; 16) performed on RCs from R. viridis have led to specific conclusions regarding the orientation of the pigments which can now be compared directly to the ones described in the model.The LD, absorption (A), and LD/A spectra of RCs from R. viridis and oriented in squeezed polyacrylamide gels have been obtained at 10 K, with the primary donor P either in the reduced or chemically oxidized state, in order to assign the various absorption bands to each of the twelve electronic transitions (Q_x or Q_y) of the six Chl molecules (14). In R. viridis RCs, our LD data favor a model of P in which excitonic coupling between the Q_y transitions of the two monomers leads to a main transition at 990 nm (Q_{Y_-}) and a smaller one at 850 nm (Q_{Y_+}). In the Q_X region the main component absorbs at approx. 620 nm (Q_{X_+}) and the minor one is tentatively located around 660 nm. The LD/A spectrum demonstrates that the four transitions of the two monomeric accessory BChls (B_L, B_M) are all oriented at approx. 70° from the C 2 axis. The two BPh molecules (H_L, H_M) have their Q_Y transitions oriented at approx. 40° from the C 2 axis. Their Q_X transitions appear tilted at 70-75° from this axis. These data derived from LD spectroscopy are in good qualitative agreement with the results of X-ray crystallography provided one assumes that excitonic coupling is present only in P and that the optical transitions of the B and H pigment molecules are not significantly perturbed by interactions among the pigments (14). Under these conditions, the orientations (determined by LD) for the optical transitions of the monomeric B and H molecules as well as those of the two exciton coupled BChls in P compare well with the values calculated using the atomic coordinates (J. Deisenhofer and H. Michel, personal communication) to determine the orientations of the X and Y axes of the pigments (J.B., unpublished).

A striking similarity can be observed when comparing the orientation of the pigments (as measured by the LD/A spectra) in the RCs of R. viridis and R. sphaeroides 241 (14), although the position and the overlap of the various absorption bands are very different in the two types of RCs. The Q_Y and Q_X transitions of the two H molecules exhibit almost identical orientations in the two species. While the B molecules present similar LD/A values for their Q_Y transitions in the two types of RCs, their Q_X transitions are much less resolved from the Q_{X_+} transition of P in R. sphaeroides than in R. viridis. However, the LD spectrum of oxidized RCs from R. sphaeroides clearly demonstrates that in this case also the two Q_X transitions exhibit a positive LD (J.B., unpublished). Finally, the LD spectra demonstrate that the carotenoid axis is oriented rather perpendicular to the C 2 symmetry axis in both R. viridis and R. sphaeroides, as it has also been observed in the case of RCs isolated from R. sphaeroides G1C oriented in electric field (17). This carotenoid has not been located in the X-ray study probably because, like for the secondary quinone Q_B, the conditions which are required for the crystallization of the RCs lead to a partial loss or disordering of these molecules. However, utilizing several spectroscopic observations on the carotenoid we can speculate on the location of this molecule in the RC complex. On the basis of the remarkable similarity of the geometries of BChl a and BChl b - containing RCs, we will assume that the orientation of the carotenoid relative to the long wavelength transition of the special pair (P_Y) is the same for both types of RCs. In the case of R. sphaeroides 241 this angle has been estimated to be approx. 75° (18). Accordingly, the carotenoid molecule has its transition moment quite perpendicular to both the C 2 symmetry axis and the 990 nm transition. Furthermore, it must be located quite close to the special pair in order to meet the requirements for the efficient transfer of triplet character from ^{3}P to the carotenoid. Finally, in order to take into account the extractability of the carotenoid, we will tentatively locate this carotenoid within the inner pocket between the L and M subunits. In order to maintain the remarkable C 2 symmetry of the system it is further proposed to locate this carotenoid on the symmetry axis (14).

Besides LD and CD data, the results of photodichroism studies on BChl a- and BChl b - containing RCs also point towards identical values for the relative angles between pairs of transition moments (15). For example, a 60° angle has been measured between the Q_X transition of H_L and the Q_{Y_-} transition of P in both R. sphaeroides (18) and R. viridis (16). The results from these various studies, all point towards the view that the six Chls and the carotenoid adopt very similar geometrical orientations in the two types of RCs.

The model of RC derived from X-ray crystallography shows the six Chl pigments to be buried in the hydrophobic interior of the LM subunit. More specifically, the pigments appear anchored to the protein scaffold formed by the ten transmembrane α-helices, each chromophore exhibiting specific binding sites to the protein host. In bacterial RCs, both the BChls constituting the special pair and the BPhs are oriented with their plane tilted at a large angle with respect to the membrane. In the same way, most of the antenna Chls from plants and bacteria orient with their plane strongly tilted out of the membrane (15). From spectroscopic data obtained on a variety of intrinsic antenna and RC complexes, it has been suggested (19) that α-helices tilted on the average at a large angle away from the membrane plane play a very important role in the organization of photosynthetic pigments *in vivo* and that the orientation of the Chl macrocycles perpendicular to the membrane plane may actually be determined by the preferential transmembrane orientation of the α-helices. In the RC of R. viridis, such a respective organization of α-helices and chromophores is indeed observed for the majority of the pigments. The two known exceptions to this rather general property of Chl orientation *in vivo* are the two accessory BChls in the bacterial RC (2, 14) and the B800 molecule in the B800-850 antenna complex (15). In the case of the RC, the Mg ligand for the B molecules is a His residue located in an α-helical loop connecting two transmembrane α-helices (2). In the B800-850 antenna complex, the putative His binding site for the Mg of the B800 is located at the N-terminal polar side of the polypeptide (and not within the central α-helical segment). Thus, in both cases where the plane of the Chl macrocycle is rather parallel to the membrane plane, the His binding site is found in a polypeptide segment which is different from a transmembrane α-helix and which runs rather parallel to the membrane plane. Thus, the parallel arrangement of the polypeptide segments and of the Chl macrocycles seems to be a general rule for the intrinsic Chl-protein complexes. We therefore suggest as a working hypothesis that the architecture of the RC LM core of R. viridis, with the Chls essentially buried into a hydrophobic protein pocket and anchored to the scaffold of mostly transmembranes α-helices, constitutes a good model for the organization of Chls and polypeptide chains in other membrane-embedded antenna and RC complexes (20).

4. Femtosecond spectroscopy of excitation energy transfer and initial charge separation in reaction centers

The structural organization of the pigments in the RC suggests that the charge separation initially occurs between P and B and is followed by migration of the electron to H and then to Q_A. Indeed using pulse-probe experiments on R. viridis RCs with 150-fsec pulses at 620 nm, a wavelength where all four bacteriochlorophyll molecules absorb, Zinth et al. (21) have observed a transient absorption change, which they have attributed to the $P^+B_L^-$ state. Such a state has been proposed previously on the basis of picosecond spectroscopy on R. sphaeroides RCs (22). However, doubts have also been raised regarding the existence of the $P^+B_L^-$ state (23).

In a recent femtosecond spectroscopy study of the initial charge separation in R. sphaeroides RC with direct excitation of P at 850 nm (24), performed in collaboration with the group at ENSTA (J.-L. Martin, A. Migus, A. Antonetti and A. Orszag), we have observed the generation in less than 100 fsec of an excited state of P which decays directly to $P^+H_L^-$ in 2.8 ± 0.2 psec. We found no experimental evidence for a transient state $P^+B_L^-$.

Although there are considerable analogies between the RCs of R. sphaeroides and R. viridis, one cannot exclude that subtle modifications in the arrangement of the chromophores lead to larger changes in the most primary energy transfer and electron transfer reactions. It is thus important to assess whether the differences in the initial electron transfer mechanisms and kinetics reported for R. viridis (21) and for R. sphaeroides (24,25) can be rationalized in terms of differences in the experimental conditions used in these femtosecond spectroscopy studies or if these discrepancies come from a genuine difference between the two types of reaction centers.

In a recent study (26), RCs of R. viridis have been excited at a variety of wavelengths (between 800 and 930 nm) and the photoinduced absorbance changes have been monitored in the 545 to 1310-nm spectral range with 100-fsec time resolution in order to investigate both the excitation energy transfer from the H and B molecules to P and the initial steps and kinetics of the electron transfer. The identical kinetics of the bleaching of the band of P at 960 nm observed upon excitation either directly in P at 930 nm or at several wavelengths within the B or H absorption bands (854, 837, 827, 803 and 797 nm) demonstrate that the transfer of excitation energy from the B and H molecules to the special pair occurs in less than 100 fsec. This ultrafast process of energy transfer, which to our knowledge constitutes the fastest direct measurement reported so far in a biological system, implies a very close proximity of the chromophores, as is indeed observed in the molecular model of the RC of R. viridis (1). This ultrafast transfer leads to an "instantaneous" absorbance increase observed at a variety of wavelengths which is assigned to an excited state of P, called P* for simplification (24,25). This state, characterized by a broad absorption spectrum, is capable of stimulated emission and decays in 2.8 ± 0.2 psec with formation of the radical pair $P^+H_L^-$. Upon direct excitation of P at 930 nm no fast transient bleaching is observed around 830 nm, thus excluding any significant contribution of the state $P^+B_L^-$ as a spectrally or kinetically resolvable intermediate. This conclusion contradicts the reaction scheme recently proposed by Zinth et al. (21).

Upon excitation in the H or B absorption bands the initial charge separation also proceeds from the state P* and its kinetics is unaffected when the excitation energy is varied by more than one order of magnitude. However, under these conditions, a large transient bleaching recovering in 400 ± 100 fsec is detected in the spectral range 830-850 nm and could be mistakenly interpreted as a transient state such as $P^+B_L^-$. The magnitude and spectral characteristics of this transient bleaching as well as the observation that it is essentially unaffected when the redox state of P, Q_A and H_L is altered, lead us to propose that it is due to the combination of a fast (~ 50 fsec) component of energy transfer $B^*P \longrightarrow BP^*$ which affects most of the B molecules and of a slower (~ 500 fsec) relaxation to the ground state which affects a much smaller fraction of B*, both components being probably enhanced by stimulated emission. An analogous fast transient bleaching has also been observed upon excitation and detection around 800 nm in the RC from R. sphaeroides (27). More generally, the excitation energy transfer among H, B and P as well as the characteristics of the initial charge separation (kinetics, nature of the ionized species involved) appear identical in the RCs of R. sphaeroides (24,27) and of R. viridis (26). These conclusions thus further strengthen the proposal (14) that the geometrical organization of the chromophores is essentially the same in the RCs of these two organisms.

5. Light-induced FTIR difference spectroscopy of the primary reactions

This experimental approach is aimed at gaining informations on (i) the molecular interactions between the pigments involved in the primary charge separations and their anchoring sites in the protein cage and (ii) the changes which affect these molecular interactions during the charge separation and the subsequent steps of stabilization. In collaboration with the group of W. Mäntele (University of Freiburg, FRG) we have recently investigated the light-induced changes in the IR vibrational frequencies (2000-1000 cm^{-1}) which accompany the reactions $PH_L Q_A \rightarrow P^+ H_L Q_A^-$ and $PH_L Q_A^- \rightarrow PH_L^- Q_A^-$ in R. viridis RCs and chromatophores as well as in a variety of $BChl_L$ a-containing purple bacteria (28, 29 ; E.N. et al., these proceedings). The molecular changes associated with these two photoreactions are of the order of 10^{-3} absorbance units but can be measured reproducibly at an accuracy which corresponds to the detection of single molecular bond absorption per RC. At this level of sensitivity any large conformational change of the protein backbone concommitant with these photoreactions would give rise to differential bands in the Amide I and Amide II regions. Our results clearly demonstrate that such large changes do not occur upon charge stabilization. At the present stage, possible assignments can only be given for a few carbonyl bands in the P^+ and H_L^- spectra (28,29) and it is clear that isotope-labelling will be necessary to strengthen these assignments. The recent development of FTIR spectroelectrochemistry of the radical cations and anions of the isolated pigments (Mäntele et al., and B.T. et al., these proceedings), which shows that some of the carbonyls bands in the P^+ and H_L^- spectra can be assigned to these ionized species, will be of considerable help in describing the bonding patterns of the primary donor and acceptor in their neutral and ionized states.

6. Organization of the antenna around the reaction center and the problem of energy trapping.

In the planar hexagonal lattice constituting the photosynthetic membrane of R. viridis, each RC appears surrounded by a ring of 6 antenna units (most probably constituted by $\alpha_2 \beta_2$ polypeptides binding 4 BChl molecules). The RC and its associated antenna can be isolated (quantasomes) so that one can use photoselection techniques to investigate the relative orientation between the long wavelength Q_y transition of the antenna BChls and of P-960. In vivo these transitions are all oriented approximately parallel to the plane of the membrane (15) but energy transfer amongst rotationally averaged, adjacent quantasomes could lead to extensive depolarization. The results, which allow to exclude models in which the Q_y transition moments of the antenna BChl and of P-960 are preferentially either parallel or perpendicular, are consistent with the view that the Q_y transitions of the antenna are circularly degenerate about P-960 (30).

Until this year (1986), the determination of the trapping time (antenna* + P $\rightarrow$ antenna + P*) in intact photosynthetic membranes has been inferred from fluorescence lifetime measurements which usually cannot be unambiguously interpreted. In principle more straightforward techniques, such as measuring the appearance of P^+ absorption (31), could be used. In collaboration with H.-W. Trissl (Univ. of Osnabrück, FRG), we have developped a fast electric measurement (present resolution 40 psec) of the rise of the flash-induced photovoltage generated by the transmembrane separation of electric charges. Exciting whole cells of R. viridis with 30 psec pulses at 530 or 1060 nm, we have observed a biphasic rise of the photovoltage (32 ; J.D. et al, these proceedings). The slower phase (125 $\pm$ 50 psec), which is eliminated upon reduction of Q_A, is assigned to $H_L^- Q_A \rightarrow H_L Q_A^-$. The faster phase ($\sim$ 40 psec) is assigned to the migration of excitation energy in the antenna which, upon trapping, feeds the 2.8 psec phase of initial charge separation between P and H_L (26). The identical amplitude of these two phases, which reflect the location of H_L approximately half-way in between P and Q_A (1), indicates an homogeneous dielectric constant between the different electron carriers. Thus for the study of primary photosynthetic events, fast photoelectric measurements turn out to be an alternative to measurements of absorption changes and fluorescence lifetimes.

References : 1) Deisenhofer, J., Epp, O., Miki, K., Huber, R. and Michel, H. (1984) J. Mol. Biol. 180, 385-398. 2) Deisenhofer, J., Epp, O., Miki, K., Huber, R. and Michel, H., (1985) Nature, 318, 618-624. 3) Nabedryk, E., Tiede, D.M., Dutton, P.L. and Breton, J. (1982) Biochim. Biophys. Acta 682, 273-280. 4) Williams, J.C., Steiner, L.A., Feher, G. and Simon, M.I. (1984) Proc. Natl. Acad. Sci. USA 81, 7303-7307. 5) Michel, H., Weyer, K.A., Gruenberg, H., Dunger, I., Oesterhelt, D. and Lottspeich, F. (1986) The EMBO Jour. 5, 1149-1158. 6) Nabedryk, E., Berger, G., Andrianambinintsoa, S. and Breton, J. (1985) Biochim. Biophys. Acta 809, 271-276. 7) Nabedryk, E., Tiede, D., Dutton, P.L. and Breton, J. (1984) Adv. Photosynt. Res. Vol. II. pp. 177-180. 8) Allen, J.P., Feher, G., Yeates, T.O., Rees, D.C., Eisenberg, D.S., Deisenhofer, J., Michel, H. and Huber, R. (1986) Biophys. J. 49, 583a 9) Nabedryk, E., Andrianambinintsoa, S. and Breton, J. (1984) Biochim. Biophys. Acta 765, 380-387. 10) Nabedryk, E., Biaudet, P., Darr, S., Arntzen, C.J. and Breton, J. (1984) Biochim. Biophys. Acta 767, 640-647. 11). Paillotin, G., Vermeglio, A. and Breton, J. (1979) Biochim. Biophys. Acta 545, 249-264. 12) Parson, W.W., Scherz, A. and Warshel, A. (1985) in Antennas and Reaction Centers of Photosynthetic Bacteria, (Michel-Beyerle, M.E., ed.) Springer-Verlag, pp 122-130. 13) Knapp, E.W., Fischer, S.F., Zinth, W., Sander, M., Kaiser, W., Deisenhofer, J. and Michel, H. (1985) Proc. Natl. Acad. Sci. USA 82, 8463-8467. 14) Breton, J. (1985) Biochim. Biophys. Acta, 810, 235-245. 15) Breton, J. and Vermeglio, A. (1982) in Photosynthesis : Energy Conversion by Plants and Bacteria (Govindjee, ed.), Vol. 1 pp. 153-194, Acad. Press, New York. 16) Tiede, D.M., Choquet, Y. and Breton, J. (1985) Biophys. J. 47, 443-447. 17) Gagliano, A., Breton, J. and Geacintov, N.E. (1986) Photobiochem. Photobiophys. 10, 213-221. 18) Vermeglio, A., Breton, J., Paillotin, G. and Cogdell, R. (1978) Biochim. Biophys. Acta 501, 514-530. 19) Breton, J. and Nabedryk, E. (1984) FEBS Lett. 176, 355-359. 20) Breton, J. and Nabedryk, E. (1986) in : Topics in Photosynthesis (Barber, J., ed.) Elsevier, Amsterdam, in press. 21) Zinth, W., Nuss, M.C., Franz, M.A., Kaiser, W. and Michel, H. (1985) in Antennas and Reaction Centers of Photosynthetic Bacteria, ed. Michel-Beyerle, M.E. (Springer, Berlin), pp. 289-291. 22) Shuvalov V.A. and Klevanik, V.A. (1983) FEBS Lett. 160, 51-55. 23) Kirmaier, C., Holten, D. and Parson, W.W. (1985) FEBS Lett. 185, 76-82. 24) Martin, J.-L., Breton, J., Hoff, A.J., Migus, A. and Antonetti, A. (1986) Proc. Natl. Acad. Sci. USA 83, 957-961. 25) Woodbury, N.W., Becker, M., Middendorf, D. and Parson, W.W. (1985) Biochemistry 24, 7516-7521. 26) Breton, J., Martin, J.-L., Migus, A., Antonetti, A. and Orszag, A. (1986) Proc. Natl. Acad. Sci. USA 83, pp. 5121-5125. 27) Breton, J., Martin, J.-L., Migus, A., Antonetti, A. and Orszag, A. (1986), in press. 28) Mäntele, W., Nabedryk, E., Tavitian, B.A., Kreutz, W. and Breton, J. (1985) FEBS Lett. 187, 227-232. 29) Nabedryk, E., Mäntele, W., Tavitian, B.A. and Breton, J. (1986) Photochem. Photobiol. 43, 461-465. 30) Breton, J., Farkas, D.L. and Parson, W.W. (1985) Biochim. Biophys. Acta 808, 421-427. 31) Nuijs, A.M., Van Grondelle, R., Joppe, H.L.P., Van Bochove, A.C. and Duysens, L.N.M. (1986) Biochim. Biophys. Acta 850, 286-293. 32) Deprez, J., Trissl, H.-W. and Breton, J. (1986) Proc. Natl. Acad. Sci. USA 83, 1699-1703.

INTERSPECIFIC STRUCTURAL VARIATIONS OF THE PRIMARY DONOR IN BACTERIAL REACTION CENTERS

Qing ZHOU, Bruno ROBERT and Marc LUTZ
Département de Biologie, CEN Saclay, 91191 Gif sur Yvette Cedex, France

1. INTRODUCTION

In reaction centers of purple and green photosynthetic bacteria, the photoinduced primary charge separation appears to generally involve two bacteriochlorophylls (BChl) molecules, which assume strong mutual electronic coupling. Understanding the mechanism of the primary charge separation requires knowledge about the actual structure of this molecular assembly, and in particular about the ground-state interactions which stabilize it in its host protein.

Resonance Raman (RR) spectroscopy of the BChl molecules can provide such information (1). We developed difference methods permitting selective observations of RR spectra of the primary donor in the neutral, ground-state (2). Three distinct methods, applied to reaction centers from Rhodobacter sphaeroides, wild type, yielded almost identical RR spectra (3). This permitted to exclude any artefactual origin for these spectra.

These difference spectra (fig 1c), which expectedly arise from two distinct BChl molecules, indicate that the magnesium atoms of both these molecules each bind a single axial ligand. These ligands most likely are side chains of histidine 173 (L subunit) and 200 (M subunit) of the protein (3, 4).

Difference RR spectra of the primary donor also show that the conjugated acetyl carbonyl group of one BChl is free from intermolecular bonding, and that the same group in the second BChl is strongly bound. The binding sites of this latter group is most probably not the magnesium of the other BChl molecule, but rather a proteic site. The keto carbonyls of both BChls are each interacting with chemically distinct, likely proteic sites.

The topological model of the primary donor of Rhb sphaeroides which can be proposed from these observations (fig 2a) hence does not involve any direct bonding, in the ground-state, between the two BChls, at variance with the so-called 'special pair' model (5). The precise relative positioning of the two BChls shown, in particular, by X-ray crystallography (6), essentially results from bonding of each of the BChls with the protein, as well as from π-π interaction between their conjugated macrocycles.

2. RESULTS AND DISCUSSION

In order to test the variability of this structure, we studied the primary donor of Rhodospirillum rubrum, using the same methods. Figure 1a reproduces a difference, low temperature (20 K), RR spectrum obtained by subtracting a RR spectrum of untreated Rsp rubrum RCs recorded under high illumination conditions (363.8 nm) from a spectrum recorded under low illumination. This spectrum is extremely close to the difference spectrum of Fig 1b, which was obtained in the same conditions, but from dithionite-reduced centers. This similarity, as well as the fact that these spectra clearly arise from two BChl molecules, allows to unambiguously assign the observed features to the ground, neutral state of the primary donor (P) alone (2, 3).

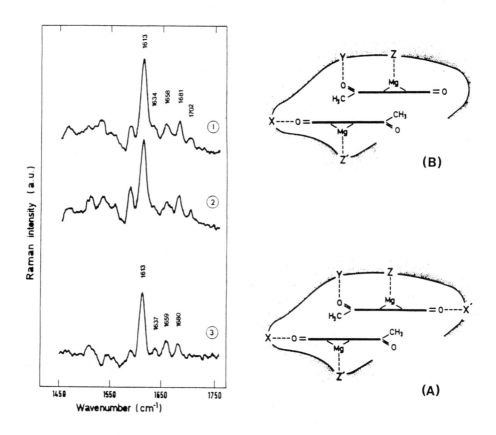

Fig.1: Differences between RR spectra recorded under low and high illumination conditions, respectively.
      1) Spectrum of untreated RCs from Rsp. rubrum S1.
      2) Spectrum of dithionite-treated RCs from Rsp. rubrum S1.
      3) Spectrum of untreated RCs from Rhb. sphaeroides 2.4.1.
(1450-1750 cm-1 region; T=30K; excitation wavelength: 364 nm).
Fig.2: Molecular models for the ground state binding interactions assumed by the Mg atom and the keto and acetyl C=O groups of the primary donor Bchls of Rhb. sphaeroides wild type (A) and of Rsp. rubrum wild type (B).

The 1613 cm^{-1} frequency of the stretching mode of the methine bridges indicates that the magnesiums of both P_A and P_B molecules constituting P each bind a single external ligand, likely a histidine side-chain as for Rps viridis (4) and Rhb sphaeroides (3). The 1681 and 1702 cm^{-1} bands unambiguously arise from the keto carbonyls of P_A and P_B. These frequencies indicate that these groups assume a weak intermolecular interaction and a free-from-bonding state, respectively.

Difference RR spectra obtained for Rsp rubrum are extremely close to those obtained for Rhb sphaeroides (Fig 1c). This indicates the same basic structures for both primary donors (Fig 2). However, the presence of the 1702 cm^{-1} band in the spectra of Rsp rubrum only demonstrates that the keto carbonyl of either P_A and P_B is free from interaction in this species, while it is strongly bound (vibrating at 1662 cm^{-1}) in Rhb sphaeroides. This shows that, in the latter species, this bond is non essential in determining the basic structure of the primary donor.

Further resonance Raman investigations are currently conducted on the primary donors of Rps viridis (8) and Rhb capsulatus (9). Difference RR spectra of Rhb capsulatus indicate a primary donor structure which is probably very close to that of the Rhb sphaeroides primary donor, with both keto groups intermolecularly bonded, one acetyl free and the second bonded. The primary donor of Rps viridis should have its two acetyl groups intermolecularly bonded (stretching modes at 1628 and 1634 cm^{-1}), one keto group bonded (1667 cm^{-1}) and the other weakly perturbed, e.g. by a van der Waals interaction (1687 cm^{-1}) (10) As those of the other three species studied, the magnesium atoms of both P_A and P_B molecules of Rps viridis each bind a single external ligand, which must be the side chain of His L173 and M200 alone, respectively (3, 6).

3. CONCLUSIONS

The present results demonstrate that, in RCs of the four bacterial species studied, the environmental interactions assumed by each of the two primary donor molecules are not identical, thus breaking the overall C_2 symmetry which most probably relates P_A and P_B in all of them. This asymmetry concerns substituents which are largely conjugated with the π electron systems of P_A and P_B. Hence, it is likely to influence intermediate fuctional states of P. It may first result in a sizeable asymmetry of the time-averaged unpaired charge densities which occur on P in the P^+ and P^R states, irrespective of a possible hopping of these unpaired charges between P_A and P_B on optical time scales (11, 12). Secondly the environmental asymmetry may break the C_2 symmetry of P in the lowest singlet state as well, hence permitting charge separation within the primary donor itself, as has been recently postulated from hole burning experiments (13).

It is moreover clear from the present resonance Raman results that the environmental asymmetry around P is species dependent : for example, with one free keto carbonyl, and the other one bonded, the primary donor of Rsp rubrum assumes a still higher degree of asymmetry than that of Rhb sphaeroides.

On the other hand, the primary donor of Rps viridis may assume a higher environmental symmetry (considering the states of the conjugated carbonyl alone) than Rhb sphaeroides. Indeed, the latter retains one acetyl C=O free from bonding, while both acetyls of the former are involved in intermolecular bonds, the energies of which are similar, as estimated from the similar downshifts (40 and 45 cm^{-1}) that they induce on the $vC=O$ frequencies.

Hence, the amounts of asymmetry of the P^+, P^R and perhaps P^{+-} states should be sizeably species-dependent.

ACKNOWLEDGMENTS

We thank Dr W. Welte for his kind gift of reaction centers of Rhb capsulatus.

REFERENCES

1) LUTZ, M. (1984) Adv. Infrared Raman Spectroc. 11, 211-300
2) LUTZ, M. and ROBERT, B. (1985) in Antenna Complexes and Reaction Centers of Photosynthetic Bacteria (Michel Beyerle, M.E. ed) pp 138-146, Springer Verlag, Berlin
3) ROBERT, B. and LUTZ, M. (1986) Biochemistry 25, 2303-2309
4) DEISENHOFER, J., EPP, O., MIKI, K., HUBER, R. and MICHEL, H. (1985) Nature 318, 618-624
5) KATZ, J.J., OETTMEIER, W., and NORRIS, J.R. (1976) Phil. Trans. R. Soc. London B 273, 227-253
6) DEISENHOFER, J., EPP, O., MIKI, K. HUBER, R. and MICHEL, H. (1984) J. Mol. Biol. 180, 385-398
7) ZHOU, Q. (1985) Diplôme d'Etudes approfondies, Université Pierre et Marie Curie, Paris
8) ROBERT, B., STEINER, R., ZHOU, Q., SCHEER, H. and LUTZ, M. (1986) these Proceedings
9) ZHOU, Q. and WELTE, W (1986) unpublished results
10) LUTZ, M., HOFF, A.J. and BREHAMET, L. (1982) Biochim. Biophys. Acta 679, 331-341
11) LUTZ, M. and KLEO, J. (1979) Biochim. Biophys. Acta 681, 365-374
12 DEN BLANKEN, H.J., HOFF, A.J. and WIERSMA, D.A. (1985) Chem. Phys. Lett. 121, 287-292

LINEAR-DICHROIC ABSORBANCE DETECTED MAGNETIC RESONANCE (LD-ADMR) SPECTRO-
SCOPY OF THE PHOTOSYNTHETIC REACTION CENTER OF RHODOPSEUDOMONAS VIRIDIS.
SPECTRAL ANALYSIS BY EXCITON THEORY

E.J. LOUS and A.J. HOFF
Department of Biophysics, Huygens Laboratory of the State University,
P.O. Box 9504, 2300 RA Leiden, The Netherlands

1. INTRODUCTION

     The photosynthetic reaction center (RC) of Rhodopseudomonas (Rps.)
viridis is a pigment-protein complex in which six pigments are uniquely
arranged [1]. Four bacteriochlorophylls b (BC) and two bacteriopheophytins
b (BP) are divided over two C-2 symmetric branches L and M, coming together
at the primary donor P, a dimer formed by BC_{MP} and BC_{LP}. At both sides P is
accessed by BC_{MA} and BC_{LA} followed by BP_M and BP_L, respectively. The L
branch is terminated by a quinone molecule.

     In the explanation of the optical spectra of the RC, transition dipole-
dipole interactions are essential as is shown in [2-4]. In addition we applied exciton theory to interpret and simulate the difference spectrum of
the absorption of the RC with P in the triplet (P^T) state, minus that with
P in the singlet state, shortly (T - S) spectrum, as well as the linear-dichroic (LD) (T - S) spectra. Both types of spectra are accurately measured
and mutually calibrated at 1.2 K by the LD-ADMR technique [5,6] which uses
resonant microwaves to change the P^T concentration in the sample. It proves
to be a good technique for determining the angle between the microwave
($\vec{\mu}^{mw}$) and the optical transition moment ($\vec{\mu}^o$), even in randomly oriented
samples.

2. MATERIALS AND METHODS
2.1. The LD-ADMR method

     At 1.2 K the non-degenerate triplet sublevels T_x, T_y and T_z of P^T are
no longer coupled by spin-lattice relaxation processes. Furthermore,
the equilibrium population of T_z is greater and its decay slower than
of T_x and T_y (Fig. 1). Application of resonant microwaves couples T_z
to T_x or T_y (the $|D| + |E|$ and $|D| - |E|$ transitions, respectively;

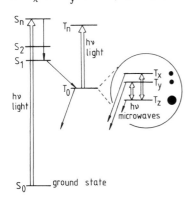

FIGURE 1. Energy level scheme of P.

Biggens, J. (ed.), Progress in Photosynthesis Research, Vol. I. ISBN 90 247 3450 9
© *1987 Martinus Nijhoff Publishers, Dordrecht. Printed in the Netherlands.*

D and E are the zero field splitting parameters) which decreases the P^T concentration in the sample, while the total P concentration is constant. By switching the microwaves off and on with ~ 315 Hz and using lock-in detection the absorbance differences ΔA are measured with high accuracy. If the microwave field $\vec{B}_1$ is polarized perpendicular on the optical probe beam (serving also as excitation beam), it selects an anisotropic RC ensemble in the sample from which the (T - S) changes are measured. This anisotropy is analyzed on LD with respect to $\vec{B}_1$, using the combination of a photoelastic modulator (PEM) performing at 100 kHz and a polarizer. After selecting the wavelength by a monochromator the transmitted light intensity is detected by a Si-photodiode (Fig. 2).

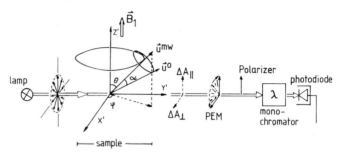

FIGURE 2. Schematic drawing of the macroscopic and microscopic picture of the LD-ADMR experiment.

The diode voltage is simultaneously lock-in detected by two separate channels, one at 315 Hz that gives the (T - S) signal and one at 100 kHz followed by 315 Hz demodulation that produces the LD-(T - S) signal. Both signals are divided by the DC voltage of the total transmitted light to produce the (T - S) and LD-(T - S) spectra. Both types of spectra are calibrated with respect to each other. The anisotropies caused by the light source and the optical components are checked to be negligible. The LD-ADMR apparatus is further described in [5,6].

2.2. Analysis of the (T - S) and LD-(T - S) spectra
Provided the photoexcitation of the P^T state is isotropic and the applied microwave power is such that the (T - S) and LD-(T - S) signals are far below their saturation point, the observed band intensity for a certain dipole strength $|\vec{\mu}^o|^2$, where $\vec{\mu}^o$ makes an angle α with $\vec{\mu}^{mw}$, is [6,7]:

$$(T - S) = |\vec{\mu}^o|^2 \cdot \frac{1}{5} \cdot (3 + \cos^2\alpha) \qquad (1)$$

$$LD-(T - S) = |\vec{\mu}^o| \cdot \frac{1}{5} \cdot (3 \cos^2\alpha - 1) \qquad (2)$$

Independently of $|\vec{\mu}^o|^2$ we can define the ratio R_i for each triplet axis i:

$$R_i = \frac{LD-(T - S)}{(T - S)} = \frac{(3 \cos^2\alpha_i - 1)}{(3 + \cos^2\alpha_i)} \qquad i = \{x,y,z\} \qquad -\frac{1}{3} \leq R_i \leq \frac{1}{2} \qquad (3)$$

R_i (Fig. 3) is for a free, non-overlapping band directly experimental-

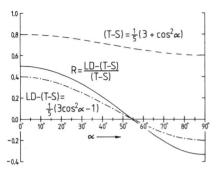

FIGURE 3. Dependence of the (T - S) and LD-(T - S) spectra on the angle α between $\vec{\mu}^o$ and $\vec{\mu}^{mw}$ for a single optical transition.

ly to determine. LD-ADMR is thus very sensitive to the orientation of $\vec{\mu}^o$ with respect to the two $\vec{\mu}^{mw}$ directions.

2.3. Sample preparation

RC of <u>Rps</u>. <u>viridis</u> were isolated and prepared as described in [8].

3. RESULTS AND DISCUSSION

The measured (T - S) and LD-(T - S) spectra for both microwave transitions $|D| - |E|$ and $|D| + |E|$ are shown in Fig. 4. The (T - S) spectra exhibit a bleaching of P at 1007 nm upon P^T formation, 2 negative peaks at 832 and 851 nm and 3 positive peaks at 823, 839 and 871 nm. Nearly no absorbance changes of the BPs, absorbing around 790 and 810 nm, are observed due to their small exciton interaction with P.

With aid of the crystal data (kindly provided by Dr. J. Deisenhofer) and by using the point-dipole approximation and Gaussian band shapes we

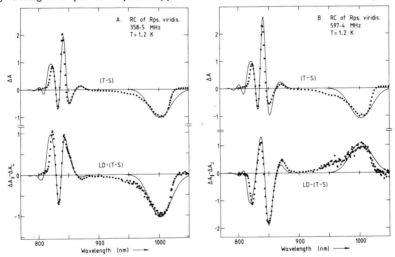

FIGURE 4. The measured (ooo) and simulated (——) (T - S) and LD-(T - S) spectra at 1.2 K of RC of <u>Rps</u>. <u>viridis</u>. The scale of the LD-(T - S) spectra is enlarged with respect to the (T - S) spectra and normalized at the top of the experimental 1007 nm band.

have simulated the (T - S) and LD-(T - S) spectra to understand the effect of the exciton coupling between the RC pigments on these optical difference spectra. Therefore, besides the simulation of the 1.2 K singlet absorption spectrum of the RC, which agrees well with that of [3], the 1.2 K absorption spectrum of the RC with P in the triplet state have to be simulated. This can be done by localizing the triplet state on BC_{LP} so that 3 BC pigments are left to produce the strong positive changes around 830 nm. The positive peak at 871 nm cannot be explained by a singlet-singlet absorption of a BC because of the strong exciton coupling of the Q_y transition between BC_{MP} and BC_{LA} [3], which always leads to a large transfer of intensity of the high energy to the low energy transition. The latter is at 871 nm in the triplet spectrum and will so get too much oscillator strength. We thus suppose that the 817 nm peak is a triplet-triplet absorption. In Fig. 4 the simulated and experimental (T - S) spectra are scaled to equal surfaces of their 1007 nm band.

The LD-(T - S) spectral simulations are found with respect to the directions of the two mutually perpendicular microwave transition moments, using formula (2). The scaling of the calculated LD-(T - S) spectrum is fixed with respect to the calculated (T - S) spectrum and lies within the error of the experimental scaling factor, determined by interpolation to zero microwave power. The success of simulating the scaled experimental LD-(T - S) spectra illustrates that, by combining LD-ADMR spectroscopy and exciton theory (with the crystal data), it is possible to give a consistent unambiguous spatial picture of the exciton coupled μ^0's (including their composition out of the uncoupled $\vec{\mu}^0$'s) and the $\vec{\mu}^{mW}$'s in the RC.

These results and more details about the simulations will be published elsewhere.

The atomic coordinates of the reaction center pigment molecules were kindly provided by Dr. J. Deisenhofer. This investigation was supported by the Netherlands Foundation for Chemical Research (SON), financed by the Netherlands Organization for the Advancement of Pure Research (ZWO).

REFERENCES
1 Deisenhofer, J., Epp. O., Miki, K., Huber,R. and Michel, H. (1985) J. Mol. Biol. 180, 385-395
2 Zinth, W., Knapp, E.W., Fischer, S.F., Kaiser, W., Deisenhofer, J. and Michel, H. (1985) Chem. Phys. Lett. 119, 1-4
3 Knapp, E.W., Fischer, S.F., Zinth, W., Sander, W., Kaiser, W., Deisenhofer, J. and Michel, H. (1985) Proc. Natl. Acad. Sci. USA 82, 8463-8467
4 Vasmel, H., Amesz, J. and Hoff, A.J. (1986) Biochim. Biophys. Acta, in press
5 Den Blanken, H.J., Meiburg, R.F. and Hoff, A.J. (1984) Chem. Phys. Lett. 105, 336-342
6 Hoff, A.J. (1985) in Antennas and Reaction Centers of Photosynthetic Bacteria (Michel-Beyerle, M.E., ed.), pp. 150-163, Springer-Verlag, Berlin
7 Meiburg, R.F. (1985) Orientation of Components and Vectorial Properties of Photosynthetic Reaction Centers, Doctoral Thesis, University of Leiden
8 Den Blanken, H.J. and Hoff, A.J. (1982) Biochim. Biophys. Acta 681, 365-374

OPTICAL PROPERTIES OF THE REACTION CENTER OF CHLOROFLEXUS AURANTIACUS AT
LOW TEMPERATURE. ANALYSIS BY EXCITON THEORY

H. VASMEL[*], R.F. MEIBURG, J. AMESZ and A.J. HOFF
Department of Biophysics, Huygens Laboratory of the State University,
P.O. Box 9504, 2300 RA Leiden, The Netherlands

[*]Present address: Koninklijke/Shell-Laboratorium Amsterdam, P.O. Box 3003,
1003 AA Amsterdam, The Netherlands

1. INTRODUCTION

The photosynthetic reaction center of purple non-sulfur bacteria is a
pigment-protein complex that contains three polypeptides, labeled L, M and
H (for light, medium and heavy, respectively), four BChl and two BPh molecules, bound to the M and L subunits, and two quinone molecules (ubi- or
menaquinone) [1]. The reaction center of the filamentous green bacterium
Chloroflexus aurantiacus has the same composition as those of purple bacteria, except that it contains three BChl and three BPh molecules; moreover, the H subunit seems to be lacking [2].

We have earlier investigated the optical properties of the reaction
center of C. aurantiacus at low temperature [3]; in this paper these optical properties are analyzed in the near-infrared region with the aid of
exciton theory. The coordinates obtained from the X-ray analysis of the reaction center of Rhodopseudomonas viridis [4] were used for the geometry of
the reaction center of C. aurantiacus, with the replacement of one of the
'accessory' BChl molecules by BPh. Throughout this paper BChl a and BPh a
are labeled BC and BP, respectively. The designations M and L refer to pigments bound to the M and L subunit, respectively; P stands for a primary
donor (P-865) pigment, A for a so-called accessory pigment. Thus the primary donor e.g. consists of BCMP and BCLP.

2. RESULTS AND DISCUSSION

According to exciton theory the exciton transition dipole moments are
linear combinations of the unperturbed original optical transition dipole
moments of an aggregate. The coefficients of this combination depend on the
mutual interaction between the chromophores, which we have approximated by
a point dipole-point dipole coupling potential [5]. The correctness of our
computational method was checked by comparing our results with those obtained with reaction centers of Rps. viridis [6]. These principles are illustrated in Figs. 1A and 1B, where the exciton transition energy of the
pigments of the reaction center of C. aurantiacus is plotted as a function
of the magnitude of the unperturbed transition dipole moments. The choice
of the monomeric (unperturbed) transition energies merits some comment. The
transition wavelength of the BPh a molecules (752 nm) is very close to
their in vitro absorption maximum. A 600 cm^{-1} red shift, relative to the in
vitro situation, had to be taken into account for the single accessory BChl
a molecule (BCMA or BCLA, at 809 nm) for which histidine-magnesium interactions may be responsible. The origin of the larger (1040 cm^{-1}) red shift of
the BChl a molecules that constitute P-865 (BCMP and BCLP) is probably due
to extensive p_σ overlap of their pyrrole rings, that are separated by about
3 A. Fig. 1B refers to a situation where the third BPh is bound to the M
subunit. As the magnitude of the dipole moments in Fig. 1B gradually in-

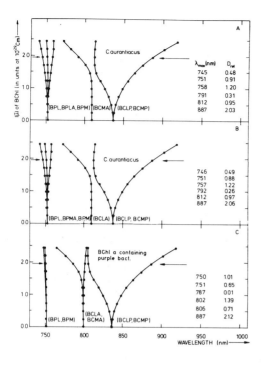

FIGURE 1. Calculated absorption maxima of reaction centers of photosynthetic bacteria, using exciton theory. The origin of the exciton bands is clarified by gradually increasing the magnitude of the dipole moment $|\vec{\mu}|$ of all pigments in the RC by the same factor. Best fits with experimental spectra are indicated by the arrows, corresponding to $|\vec{\mu}| = 1.95 \times 10^{-29}$ Cm for BChl a (1.63×10^{-29} Cm for BPh a). For these values of $|\vec{\mu}|$ the position of the absorption maxima (λ_{max}) and the value of the exciton dipole strength, relative to the monomeric dipole strength (D_{rel}) are listed on the right hand side of the figure. Energies and origins of the monomeric transitions are indicated at the bottom of the graphs. Note that the two sets of data for C. aurantiacus (A and B) originate from two different assumptions: in (A) BCLA is replaced by a BPh a molecule (BPLA), in (B) BCMA is replaced by BPMA.

creases, the shifts of the exciton bands increase, due to the increasing dipole-dipole interaction of the monomeric transitions. Clearly, the two strongest interacting transition dipoles are those of P-865, while the other four bands display relatively minor shifts. The behavior of the high energy exciton band of P-865 (starting off at 837 nm) has some unusual features: as $|\vec{\mu}|$ increases its energy approaches that of the accessory BChl transition, but does not cross it, as exciton theory does not allow degeneracy of coupled energy levels. What actually does occur is the mixing in of the antisymmetric BCMP and BCLP wavefunctions of the high energy exciton component with the wavefunction of BCLA and vice versa, until at the highest value of $|\vec{\mu}|$ the accessory BChl transition has essentially the character of the high energy exciton band of P-865 (with antisymmetric contributions of BCMP and BCLP), while the band that was originally the high energy exciton band of P-865 obtains predominantly an accessory BChl-like character. These effects are illustrated in Table 1A, which lists the contributions of the monomeric wavefunctions to the exciton states.

Similar results are obtained for C. aurantiacus reaction centers with the third BPh molecule located in the L chain (Fig. 1A). In Fig. 1C and Table 1B results are shown for reaction centers that contain 4 BChl a and 2 BPh a molecules, as is e.g. the case for Rhodobacter sphaeroides.

Fig. 2 shows a Gaussian dressing of the calculated stick spectrum (Fig. 1B) of reaction centers of C. aurantiacus, compared with the experimentally obtained absorption spectrum at 77 K (dashed line) [3] for both oxidized (for which the transitions of BCMP and BCLP were assumed to vanish) and reduced reaction centers. A good agreement exists between calculated and experimental spectra, which also extends to linear and circular dichroism and fluorescence polarization spectra [H. Vasmel, J. Amesz and A.J. Hoff, to be published].

Summarizing, our results lead to the following conclusions for the optical properties of the reaction center of C. aurantiacus: the allowed, low energy exciton component of the primary donor P-865 is located at 887 nm and carries the dipole strength of approximately two BChl a monomers; the high energy exciton transition, around 790 nm, is mixed with wavefunctions of other pigments (Table 1A), which explains its relatively small angle (~ 40°) with respect to the 887 nm transition, as observed by linear dichroism and fluorescence polarization [3]. The optical transition of the accessory BChl a molecule near 812 nm has some contribution of the BChls that constitute P-865. This accounts for the reorientation of this transition upon oxidation of P-865, as observed experimentally [3] and partly also for its blue-shift. Two of the BPh molecules are located on the same polypeptide subunit, which on basis of the CD fit (not shown) is probably the M subunit. These two molecules show a clear splitting of absorption bands (11 nm) due to exciton coupling (Figs. 1B and 2A); the single BPh on the opposite branch shows hardly any exciton shift.

Similar calculations for reaction centers containing four BChl a and two BPh a molecules result in a very low dipole strength for the high energy exciton component of the primary donor (Fig. 1C) due to antisymmetric mixing with both accessory BChl a wavefunctions (Table 1B). These reaction centers are predicted to show little splitting of the BPh a absorption band, which is confirmed by experiment.

TABLE 1. Contributions of the monomer transitions to the exciton dipole strength in percent. Bold numbers are main contributions

Exciton state (nm)	Unperturbed transition wavelength (nm)						$D_{rel}^{(2)}$
$A^{(1)}$	752 BPM	752 BPMA	837 BCMP	837 BCLP	809 BCLA	752 BPL	
746	**47**	**53**	0	0	0	0	0.49
751	0	0	0	0	2	**98**	0.88
757	**54**	**45**	0	1	0	0	1.22
792	0	2	**47**	**41**	9	0	0.26
812	0	0	2	8	**87**	2	0.97
887	0	0	**50**	**49**	0	0	2.06
$B^{(1)}$	752 BPM	800 BCMA	837 BCMP	837 BCLP	800 BCLA	752 BPL	
750	6	0	0	0	3	**91**	1.01
751	**92**	2	0	0	0	6	0.65
787	0	17	**34**	**34**	14	0	0.01
802	1	**45**	1	0	**51**	2	1.39
806	1	**35**	15	17	**31**	1	0.71
887	0	0	**50**	**49**	1	0	2.12

Notes: (1) A: C. aurantiacus RC, corresponding to Fig. 1B.
B: Purple bacterial RC, corresponding to Fig. 1C.
(2) D_{rel} is the exciton state dipole strength relative to the dipole strength of the monomeric transition.

Our results indicate that the arrangement of the chromophores in reaction centers of C. aurantiacus is very similar to that in purple bacteria. The functional L subunits of the reaction centers of purple and filamentous green bacteria bind pigments of the same type in a probably very similar arrangement.

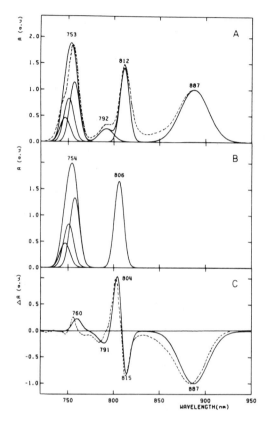

FIGURE 2. Calculated absorption spectra (solid lines) of reaction centers of C. aurantiacus, with a BPh a molecule located in the M chain (BPMA). Reaction center structure as for Rps. viridis.
(A) Reaction centers with reduced P-865. Dashed line: experimental spectrum at 77 K, normalized at 887 nm.
(B) Reaction centers with oxidized P-865.
(C) Oxidized-minus-reduced difference spectrum. Dashed line: experimental spectrum.
For the oxidized reaction centers an additional electrochromic blue shift of the 812 nm band was incorporated.
Two different batches were used for the experimental spectra of (A) and (C).

The investigation was supported by the Netherlands Foundation for Chemical Research (SON), financed by the Netherlands Organization for the Advancement of Science (ZWO).

REFERENCES
1 Okamura, M.Y., Feher, G. and Nelson, M. (1982) in 'Photosynthesis. Vol. 1, Energy Conversion by Plants and Bacteria' (Govindjee, ed.) pp. 195-274, Academic Press, New York
2 Pierson, B.K. and Thornber, J.P. (1983) Proc. Natl. Acad.Sci. U.S.A. 80, 80-84
3 Vasmel, H., Meiburg, R.F., Kramer, H.J.M., de Vos, L.J. and Amesz, J. (1983) Biochim. Biophys. Acta 724, 333-339
4 Deisenhofer, J., Epp, O., Miki, K., Huber, R. and Michel, H. (1984) J. Mol. Biol. 180, 385-398.
5 Pearlstein, R.M. (1982) in 'Photosynthesis. Vol. 1, Energy Conversion by Plants and Bacteria' (Govindjee, ed.) pp. 293-330, Academic Press, New York
6 Knapp, E.W., Fischer, S.F., Zinth,W., Sander, M., Kaiser, W., Deisenhofer, J. and Michel, H. (1985) Proc. Natl. Acad. Sci. USA 82, 8463-8467.

THE PHOTOCHEMICAL REACTION CENTER OF CHLOROFLEXUS AURANTIACUS: ISOLATION AND PROTEIN CHEMISTRY OF THE PURIFIED COMPLEX

Judith A. Shiozawa, Friedrich Lottspeich, and Reiner Feick
Max Planck Institut für Biochemie, Martinsried, Fed. Rep. Germany

INTRODUCTION

The primary photochemical events of Chloroflexus aurantiacus have been studied recently. Although there are some differences, overall, the primary processes resemble that of the better characterized members of the Rhodospirillaceae.

The Chloroflexus reaction center (RC) most likely contains three bacteriochlorophyll a, three bacteriophaeophytin a and one or two menaquinone molecules. As in purple bacteria, an electron passes from the donor, P860 (a bacteriochlorophyll dimer), via bacteriophaeophytin a to the primary electron acceptor, menaquinone (for a review, see Ref. 1).

Although the reaction center has been characterized on a spectral basis, the protein composition of this pigment-protein complex is still the subject of some controversy. In this work, a purification procedure for the isolation of highly purified Chloroflexus RC and the initial protein chemical analysis of this preparation are presented.

RESULTS

A detailed description of the RC isolation will be presented elsewhere. In brief, crude membranes were solubilized with LDAO according to Pierson and Thornber (2) and applied onto a DEAE cellulose column. After extensive washing with a buffer containing 1% LDAO, the RC was eluted with 60 mM NaCl. This step was followed by molecular sieve chromatography on Fractogel HW 50. Final purification was accomplished by FPLC on a Mono Q column in the presence of nonanoyl-N-methylglucamide.

Pure RC exhibited a characteristic absorption ratio 280/813 nm of 1.3 to 1.4. Only two polypeptides with very similar M_r (24,000 and 24,500 D) were present after heating the RC in SDS-containing buffer (3) for 30 min at 60°C (Fig. 1, lanes 3 and 4). The presence of a faint smear at the 52,000 D position resulted from incomplete denaturation of the sample. Under less stringent conditions, a functional pigment-protein complex of M_r 52,000 D could be observed (Fig. 1, lane 9). Re-electrophoresis of the excised and denatured band produced the two aforementioned polypeptides (Fig. 1, lane 10). The 52,000 D molecular weight of the intact RC complex was further confirmed by Fractogel chromatography. Hence, the Chloroflexus RC is approximately half the size of purple bacteria reaction centers. Two-dimensional PAGE (4) revealed that the upper and lower subunits have pI at approximately 6.5 and 6.7, respectively.

Reaction center preparations were subjected to N-terminal amino acid sequencing. The sequence of the first 21 amino acid residues was determined (see Fig. 2), this oligopeptide showed no similarities to the primary structures of the H, M, and L subunits of both Rps. viridis and Rb. capsu-

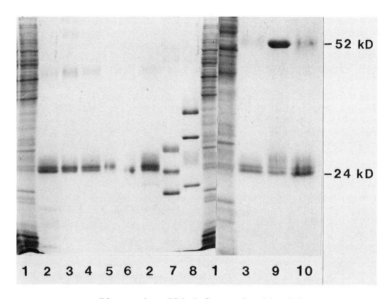

Figure 1: SDS Polyacrylamide Gel

1) Crude membranes, starting material for RC isolation
2) RC-enriched fraction
3) Purified RC
4) Purified RC, another preparation
5) Isolated 24,500 D-polypeptide
   (Upper)

6) Isolated 24,000 D-polypeptide
   (Lower)
7) RC of Rb. sphaeroides R-26
8) RC of Rps. viridis
9) Chloroflexus RC, not heated
10) Re-electrophoresis of the excised
    and heat denatured 52,000 D-
    pigment-protein band

latus reaction centers. Since just one major sequence was obtained, partial
peptide mapping was employed to assess the homology between the two polypeptides. The individual subunits (Fig. 1, lanes 5 and 6) were isolated by
SDS-PAGE of denatured RC. After staining the gel, the appropriate bands
were cut out and subjected to chemical or enzymatic cleavage according to
references 5-7. With the exception of one fragment (approximate M_r 19,000)
generated by S. aureus V8 digestion, the cleavage patterns of the two polypeptides were identical by all methods employed (see Figs. 3 and 4). This
result is indicative of a high degree of sequence and structural homology
between the two polypeptides.

Figure 2: N-terminal amino acid sequence of the Chloroflexus reaction
           center complex

                                        10
             N-Ser-Arg-Ala-Lys-Ala-Lys-Asp-Pro-Arg-Phe
                                        20
             Pro-Asp-Phe-Ser-Phe-Thr-Val-Val-Glu-Gly-Ala

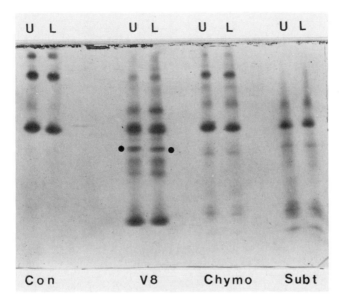

Figure 3: Cleveland Partial Peptide Mapping
The following proteases were used: S. aureus V8, chymotrypsin a, and sub-
tilisin. The peptides were detected by Western blotting analysis. Immuno-
chemicals used were rabbit anti-Chloroflexus RC and horseradish peroxidase-
conjugated goat anti-rabbit immunoglobulin. The dots indicate the position
of the two nonidentical protein fragments.

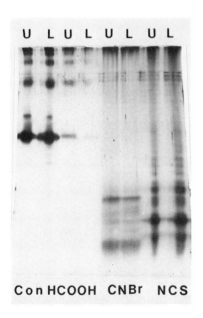

Figure 4: Chemical cleavage of the RC
subunits using formic acid, cyanogen
bromide and N-chlorosuccinimide.
Protein was detected by silver staining.

I.4.**410**

SUMMARY

a. Highly purified RC of Chloroflexus could be prepared by a combination of anion-exchange and molecular sieve chromatographies; the final purification step was anion-exchange by FPLC. Pure RC had an absorption ratio at 280/813 nm of 1.3 to 1.4. The RC was comprised of only two polypeptides with apparent M_r of 24,000 and 24,500 D. Their isoelectric points were approximately 6.7 and 6.5, respectively.

b. Molecular sieve chromatography and SDS-PAGE revealed that the intact reaction center complex had a M_r of about 52,000 D. It was noteworthy that this was less than half the M_r of purple bacteria reaction centers. The 52,000-complex was probably comprised of one copy each of the 24,000 and 24,500 D polypeptides and one set of pigments (three molecules each bacteriochlorophyll a and bacteriophaeophytin a and one or two molecules of menaquinone).

c. One major N-terminal amino acid sequence was obtained from the isolated complex. No similarities to the primary structure of purple bacteria RC was observed.

d. The individual subunits yielded identical peptide patterns regardless of how the fragments were generated. Only a very slight difference in the mobility of one fragment obtained by S. aureus V8 digestion was detected. Thus the two subunits seemed to have a high degree of structural similarity. Whether the small difference in mobility reflected a variance in the amino acid sequence or was due to a chemical modification is under current investigation.

ACKNOWLEDGEMENTS

This research was supported by a grant from the Sonderforschungsbereich (SFB 143). We wish to thank Drs. T. Redlinger and R. C. Fuller for providing the rabbit antiserum against Chloroflexus reaction center.
Special thanks to Dr. D. Oesterhelt for his generous support.

REFERENCES

1 Blankenship, R. E. and Fuller, R. C. (1986) in The Encyclopedia of Plant Physiology (Staehelin, L. A. and Arntzen, C. J., eds.), Vol. 19, pp. 390-399, Springer Verlag, Berlin.
2 Pierson, B. C. and Thornber, J. P. (1983) Proc. Nat. Acad. Sci. 80, 80-84.
3 Leammli, U. K. (1970) Nature 227, 680-685.
4 O'Farrell, P. H. (1975) J. Biol. Chem. 250, 4007-4021.
5 Cleveland, D. W., Fischer, S. G., Kirschner, M. W., and Laemmli, U. K. (1977) J. Biol. Chem. 252, 1102-1106.
6 Lische, M. A. and Ochs, D. (1982) Anal. Biochem. 127, 453-457.
7 Zingde, S. M., Shirsat, N. V., and Gothoskar, B. P. (1986) Anal. Biochem. 155, 10-13.

STRUCTURES OF ANTENNA COMPLEXES AND REACTION CENTERS FROM
BACTERIOCHLOROPHYLL B-CONTAINING BACTERIA :
RESONANCE RAMAN STUDIES

Bruno ROBERT, Robert STEINER*, Qing ZHOU, Hugo SCHEER* and Marc
LUTZ
Département de Biologie, CEN Saclay 91191 Gif sur Yvette Cedex
France
and * : Botanisches Institut der Universität München, D 8000
München 19, GFR

Resonance Raman (RR) spectroscopy yields detailed
information about the structure and ground-state environmental
interactions assumed by bacteriochlorophyll a (BChl a) and
bacteriophaeophytin a (Bpheo a) within bacterial pigment-protein
complexes (1-3). Recent successes in crystallyzing reaction centers
(RC) from Rhodopseudomonas viridis renewed interest in BChl b- and
Bpheo b-containing complexes (4). We here report the first RR
spectra of isolated BChl b and Bpheo b, as well as of BChl b-
containing antenna and reaction centers. Difficulties due to the
high photooxydability of those pigments have been overcome by
working at 20 K in anoxic conditions , and by selectively avoiding
resonance of decay products (363.8 nm excitation).

RR SPECTRA OF ISOLATED BCHL B AND BPHEO B

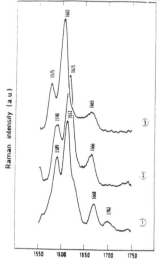

Raman intensity (a u)

Wavenumber (cm⁻¹)

Fig 1 : 1550-1750 cm^{-1} regions
of RR spectra of
1) Bpheo b in methanol
2) BChl b in hexane
3) BChl b in methanol
(T = 20 K ; exc. wv : 364 nm)

Comparison of BChl a and BChl b
RR spectra shows that the
presence of the ethylidene
grouping conjugated to cycle II
induces some large (> 6 cm^{-1})
frequency shifts, principally
of bands at 697, 952, 1161,
1218, and 1444 cm^{-1} (BChl a).
Except the 1161 cm^{-1} one,
these modes are weakly sensitive
to the ^{14}N / ^{15}N substitution
and thus should involve motions
of the macrocycle periphery (3).
BChl b yields two strong bands at
650 and 1351 cm^{-1} (Bpheo b :
1347 cm^{-1}), which are missing
in BChl a spectra. These modes do
not appear to predominantly arise
from the ethylidene grouping.
Differences observable between
RR spectra of BChl b and of Bpheo b
are very similar to those observed
for the a derivatives (3). In
particular, characteristic bands of
phaeophytins at 270, 777, 1106,
1131 and 1590 cm^{-1} are present in

Biggens, J. (ed.), Progress in Photosynthesis Research, Vol. I. ISBN 90 247 3450 9
© *1987 Martinus Nijhoff Publishers, Dordrecht. Printed in the Netherlands.*

Bpheo b RR spectra.
    Free carbonyl stretching modes at 1678 and 1700 cm^{-1} (fig
1.3) indicate that the 9-keto carbonyl stretching frequency is not
affected by the presence of the additional C=C bond conjugated with
cycle II, whereas the stretching mode of the 2-acetyl C=O is
upshifted by about 15 cm^{-1} .
    Fig 1.2 and 1.3 compare the higher frequency regions of RR
spectra of BChl b in a polar solvent (central Mg 6-coordinated) and
self-aggregated in a non polar solvent (central Mg 5-coordinated).
This clearly shows that the methine bridge stretching mode of BChl
b is sensitive to the coordination state of the central Mg of the
molecule, being located, as in BChl a, around 1614 cm^{-1}, when 5-
coordinated, and around 1600 cm^{-1} when 6-coordinated.

INTERACTION STATES OF BCHL B IN ANTENNA COMPLEXES

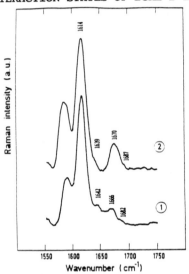

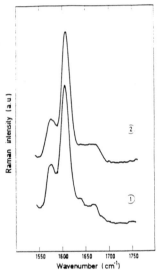

Fig 2 : RR spectra of :
1 : B 800-1020 complexes
from Ectothiorhodospira
halochloris
2 : Chromatophores from
Rhodopseudomonas viridis

Fig 3 :
1 : cf fig. 2.1
2 : RR spectra of the
same complex after
HCl treatment

    Fig 2.2 displays the higher frequency region of RR spectra
of whole chromatophores of Rps viridis : because of the low (0.08)
RC : antenna ratio, these spectra essentially arise from B 1015
antenna complexes alone. Fig 2.1 shows the same frequency range for
purified B 800-1020 antenna complexes from Ectothiorhodospira
halochloris. In both of these spectra, the frequency of the methine
bridge stretching mode, being located around 1613 cm^{-1}, clearly
demonstrates that, as BChl a, BChl b is preferentially 5-
coordinated when bound to protein.
    In the carbonyl stretching frequency region, RR spectra of
B 1015 complexes from Rps viridis are quite different from those of
the supposedly (5) homologous BChl a-containing B 880 complexes

from Rhodospirillales (1). Indeed, if the number of conjugated C=O vibrators observable in the spectra (three) is consistent with the stoichiometry proposed for these complexes (2 BChl b / complex), the frequencies of these vibrators (1639, 1670 and 1687 cm^{-1}) are different from those observed in B 880 complexes (1). Indeed, the present spectra show that one acetyl carbonyl of one of the two BChl b is intermolecularly bound, vibrating at 1639 cm^{-1}, whereas that of the other is free, vibrating at 1670 cm^{-1}. Moreover, the latter Raman band is very likely degenerate, and most probably involves one keto carbonyl, then intermolecularly bound. The second keto group vibrates at 1687 cm^{-1} and is only weakly interacting. One thus has to conclude that ground-state molecular interactions assumed by BChl b in B 1015 complexes differ from those assumed by BChl a in B 880 complexes : in the latter, which form a homogeneous class from a structural point of view (1), both of the acetyl carbonyls of the BChl a molecules are intermolecularly bound, and vibrate around 1643 cm^{-1}.

RR spectra of B 800-1020 of halochloris show that at least four unequivalent BChl b are present in these complexes (table 1). This result agrees with the stoichiometry deduced from biochemical data (5 BChl b / complex)(6). Acid treatment of this complex induces a shift of the 1020 nm transition to 960 nm (6). This treatment affects the stretching frequencies of no more than two acetyl and two keto carbonyl groups (Fig 3 and Table 1). This confirms the hypothesis according to which (7) two out of the five BChl b molecules participate to the 1020 nm absorption band.

B 880 (Rsp rubrum)	B 1015	B 800-1020	viridis RC	sphaeroides RC
		1630	1628	1628
	1639		1634	1633
1644		1643 ↕		
		1651	1654	
		1657		
		1664		1660
1667	1671	1668 ↕	1664 (?)	
1674		1677	1671	
	1681	1686 ↕	1684	1678
			1709 (?)	1684
				1705

Table 1 : compared frequencies of BChl b- and BChl a-containing complexes. (↕ : intensity variations induced by HCl treatment)

BCHL B-CONTAINING REACTION CENTERS

We obtained RR spectra of reaction centers from Rps viridis (fig 4). In the lower and medium frequencies regions of these spectra, the main bands predominantly arise from Bpheo b and BChl b appears to poorly contribute. However, in the higher frequency region, the low 1590 : 1615 cm^{-1} intensity ratio indicates a strong participation of BChl b. Moreover, the carbonyl stretching frequency region does not contain all of the 4 frequencies that are observed when resonance is with the 535-545 nm transition , hence selectively enhancing Bpheo contributions(8). Such a frequency-dependent balance of BChl / Bpheo contributions is also observed in RR spectra of RCs from Rhb sphaeroides R 26 excited at 364 nm (Robert, B. unpublished results).

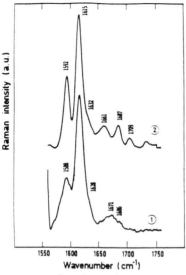

Fig 4 : RR spectra of reaction
centers (1550-1750 cm^{-1} region)
from : 1, Rps viridis
2, Rhb sphaeroides
T = 20 K
Excitation wavelength : 364 nm

The frequencies observed in the carbonyl stretching region of the spectra only partially match with those observed for Rhb sphaeroides (table 1). Some of these differences may not indicate differences in interaction states of conjugated carbonyls, but may arise from the above-mentioned difference in stretching frequencies of free acetyl C=O in BChl a and BChl b. For example, the 1670 and 1660 cm^{-1} features observed for Rps viridis and Rhb sphaeroides RCs, respectively, may both arise from interaction-free acetyl groups. On the other hand, the 1628 cm^{-1} frequency, which, in RR spectra of Rhb sphaeroides RCs arises from the 535nm-absorbing Bpheo (8), most probably arises , in RR spectra of Rps viridis RCs, from the primary donor (6, 9). A more detailed account of RR spectroscopy of the primary donor of Rps viridis is given in these Proceedings (9).

REFERENCES
1) ROBERT, B. and LUTZ, M. (1985) Biochim. Biophys. Acta 807, 10-23
2) ROBERT, B. and LUTZ, M. (1986) Biochemistry 25, 2303-2309
3) LUTZ, M. and ROBERT, B. in : Biological Applications of Raman Spectrosc. (SPIRO, T.G. ed) John Wiley, New York (in press)
4) DEISENHOFER, J., EPP, O., MIKI, K. HUBER, R. and MICHEL, H. (1984) J. Mol. Biol. 180, 385-398
5) THORNBER, J.P., COGDELL, R.J., SEFTOR, R.E.B., PIERSON, B.K. and TOBIN, E.M.(1983) in Adv. in Photos. Res. (SYBESMA, C. ed) Vol. 2, pp 25-32
6) STEINER, R. and SCHEER, H. (1985) Biochim. Biophys. Acta 807, 278-284
7) STEINER, R. (1984) Thesis, Ludwig Maximilian Universität, München
8) LUTZ, M. and ROBERT, B. (1985) in Antenna Complexes and Reaction Centers of Photosynthetic Bacteria (Michel Beyerle, M.E. ed) pp 138-146, Springer Verlag, Berlin
9) ZHOU, Q., ROBERT, B. and LUTZ, M. these Proceedings

STRONG ORIENTATIONAL ORDERING OF THE NEAR-INFRARED TRANSITION MOMENT
VECTORS OF LIGHT-HARVESTING ANTENNA BACTERIOVIRIDIN IN CHROMATOPHORES
OF THE GREEN PHOTOSYNTHETIC BACTERIUM Chlorobium limicola, STRAIN C

Z.G. Fetisova, S.G. Kharchenko and I.A. Abdourakchmanov*

A.N. Belozersky Laboratory of Molecular Biology and Bioorganic Chemistry,
Moscow State University,Bldg."A", Moscow 119899 and *Institute of Soil Sciences and Photosynthesis, USSR Academy of Sciences, Pushchino 142292,USSR

1. INTRODUCTION

It was shown theoretically that the photosynthetic unit (PSU) structure
should be strongly optimized in vivo to operate with a 90% quantum yield of
primary charge separation in the reaction center (RC) [1]. The basic principles of the structural organization of an optimal model PSUs have been
considered by us in [2,3]. The mutual orientation of the transition moment
vectors of PSU molecules is a major factor making the optimization of energy transfer from the antenna to the RC possible. It was shown by us that in
optimal model PSUs these vectors are parallel to each other and to either
the long or the short axis of an "elementary" PSU (see below). It is possible that in some natural PSUs the co-operative effect of several optimizing factors ensures the high efficiency without requiring a strong orientational ordering of these vectors. We believe that if an "ideal" orientational ordering of transition dipoles in vivo occurs, then it would be advisable to investigate the ordering in large and efficient PSUs, for which
the requirements for their structure optimization are more rigorous than
those for small PSUs [1]. This is why the green sulfur bacteria were chosen
as the test object as their light-harvesting antenna is an order of magnitude larger than that of purple bacteria, and several-fold greater than that
of higher plants. PSUs of green sulfur bacteria contain about 1000 bacterioviridin (BVr) molecules and about 80 bacteriochlorophyll a (BChl a) molecules (per RC P840) [4,5], with their main near-infrared absorption peaks at
730-750 nm and 810 nm, respectively. At the same time, as shown by us previously for C. limicola, its PSU is highly efficient: the energy transfer
from light-harvesting BVr superantenna to BChl a antenna takes place within
20-50 ps with an efficiency >95% [6], and that from BChl a to RCs within
20-60 ps with an efficiency >92% [5,7]. Our aim was to investigate dipole
orientations in C.limicola BVr superantenna in the so-called chlorosomes,
the rod-shaped structures containing all the cell BVr (about 10 000 BVr molecules per a chlorosome [4]).

2. MATERIALS AND METHODS

Chlorobium limicola cells were grown anaerobically in 1.5 l bottles at 28°C
under illumination intensity about 500 lx. The cells were harvested 36 h
after inoculation, washed with 0.05 M sodium phosphate buffer (pH 8.0) and
sonicated during 10 min. All the stages were performed at 4°C. Unbroken
cells and large debris were sedimented by centrifugation at 12 000 xg for
10 min. The chromatophores were isolated from the supernatant by differential centrifugation at 50 000 xg for 35 min and purified by centrifugation on
discontinuous sucrose gradients (10-50% (w/v) sucrose over a 60% (w/v) pad)
in L2-75 Beckman SW-27 rotor at 27 000 rpm for 18 h. Corresponding fractions were collected, resuspended in the same buffer and analyzed for purity
by spectrophotometry and electron microscopy. The buffer suspension of

chromatophores wasmixed with the components of the polyacrylamide gel (acrylamide, 12% (w/v), N,N'-methylenebisacrylamide; glycerol, 67.5% (v/v)). The samples were polymerized by the addition of 0.03% (v/v) N,N,N',N'-tetramethylethylenediamine and 0.05% (w/v) ammonium persulfate.

The direction of the transition moments of pigments was studied by linear dichroism. Orientation of chromatophores was achieved by uniaxial stretching a polyacrylamide gel in which they were packed [8]. The absorbances of the measuring light polarized parallel ($A_{/\!/}$) and perpendicular ($A_\perp$) to the direction of sample stretching (i.e. to the orientation axis) were measured at room temperature in the region 600-900 nm with a Specord M-40 spectrophotometer. Data were analyzed proceeding from statistics on distribution of rod-shaped particles in the stretched sample. If the sample deformation is symmetric with respect to the z-axis, then $1'_x = 1_x/\!\!\sqrt{N}$; $1'_y = 1_y/\!\!\sqrt{N}$; $1'_z = 1_z \cdot N$, where 1_x, 1_y, 1_z, $1'_x$, $1'_y$, $1'_z$ are the sample dimensions before and after deformation, respectively, and N is the degree of sample deformation. The dependence of the degree of dichroism, $P = (A_{/\!/} - A_\perp) / (A_{/\!/} + A_\perp)$, on the angle, α, between the transition moment vector and the long axis of the rod-shaped particle in a stretched polymer, for a given degree of sample deformation, N, is described by the following set of equations:

$$P(\alpha,N) = [(3\cos^2\alpha - 1)(3T(N)-1]/[3-\cos^2\alpha + T(N)(3\cos^2\alpha - 1)] \tag{1}$$

$$T(N) = N^3/(N^3-1)[1 - \arctan\sqrt{N^3-1} /\sqrt{N^3-1}] \tag{2}$$

3. RESULTS AND DISCUSSION

Chromatophore is the photoactive chlorosome-membrane complex whose absorption spectrum does not differ from that of whole cell. The shape of the individual chromatophore was examined in a Hitachi H-12 electron microscope and found to be rod-like. The maximal dimension of chromatophores of all known green bacteria is determined by the dimension of the long axes of their chlorosomes [4,9]. Therefore, the linear dichroism spectra yield information about the orientation of the oscillators relative to the long axis of the chlorosome of each individual chromatophore. Figs 1a and 1b show typical absorption spectra $A_{/\!/}$ and $A_\perp$ and calculated spectra of the degree of dichroism P. In the sole BVr absorption region in the near-infrared (710-770 nm) the P values are constant with an accuracy ±0.01. At wavelengths λ <710 nm and λ > 770 nm BVr is not the only absorbing oscillator [4-7]. In these regions the degree of dichroism varied significantly; this indicates the presence of absorption bands with different orientations of the transition moments. Therefore, the exact values of the α angles between the transition dipoles and the long axis of chlorosome may be calculated only for the main fraction of BVr absorbing at 710-770 nm. To examine the adequacy of the theory employed we measured the P values for 3 different values of N: 1.16; 1.60 and 1.98. Fig. 2 shows the parametric family of theoretical $P(N,\alpha)$ curves calculated with eqns 1 and 2 and the experimental P(N) values obtained from $A_\perp$ and $A_{/\!/}$ spectra for 3 values of N. It is clearly seen that the model of rod-like particles well described the orientation of chromatophores of C.limicola in the gel. The mean value of α was calculated to be

$$\alpha = 0^\circ,$$

the mean square deviation being 7°. Thus, in each individual chromatophore of C.limicola, the Q_y transition moment vectors of light-harvesting superantenna bacterioviridin are essentially parallel to each other and practically ideally oriented along the chlorosome long axis.

Our spectra reveal qualitative agreement with those obtained by Betti et al.[11] who studied linear dichroism of isolated chlorosomes of gliding fi-

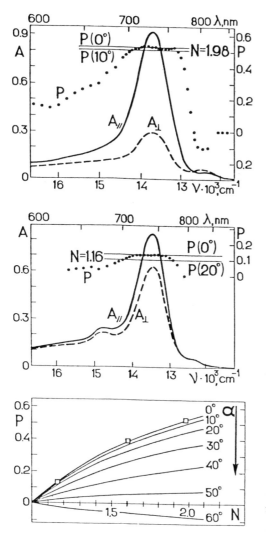

Fig. 1a. Absorption ($A_\parallel$ and $A_\perp$) and degree of dichroism (P) spectra of oriented C.limicola chromatophores. Spectra measured for N = 1.98. The horizontal lines, P(0°) and P(10°), are the theoretical P values for α = 0° and α = 10°, respectively; N = 1.98.

$$P = (A_\parallel - A_\perp)/(A_\parallel + A_\perp).$$

Fig. 1b. The same spectra as in Fig.1a. Spectra measured for N = 1.16. The horizontal lines, P(0°) and P(20°), are the theoretical P values. The long-wavelength maximum of BVr in living cells of C.limicola can vary between 730 and 740 nm normally (similar phenomenon was shown for BChl e in brown bacteria: A. Gloe et al., Arch. Microbiol. 102 (1975) 103). The sample of Fig.1b was prepared from the cells grown in the medium with a light pH change.

Fig. 2. Theoretical dependences of the degree of dichroism P on the degree of sample deformation N for 7 α values, shown near the respective curves, P(N). (□), region of P(N) values with experimental P(N) values taking into account the accuracy of P (1–2%) and N (∼1%) measurement. Experimental P(N) values measured at the BVr absorption maximum for 3 N values: 1.16; 1.60; 1.98. The experimental P(N) value for unstretched sample, P(1) = 0.

lamentous bacterium Chloroflexus aurantiacus and calculated, by equ 2, the angle α = 40±2°. Unfortunately, the α determination in [11] is based on the assumption about the ideal orientation of chlorosomes in stretched polymer film (i.e. N → ∞ in eqn 2). We believe that this assumption is not fulfilled in a real experiment as the film deformation is not symmetric with respect to the stretching axis (it is known that under the stretching load the film width decreases less than 20%, while its thickness decreases 4-fold). This assumption results in significant overestimation of N and, hence, α values (see Fig. 2). Thus, it is possible that the orientation of the BVr Q_y-transition dipoles in all green bacteria is the same as obtained by us in C.limicola [10]. This orientation was predicted by us on the basis of a model computer calculations of the optimal antenna structure. In prin-

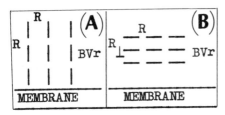

Fig. 3. Models with different
strong orientational ordering of
BVr transition dipoles: (A) normal to membrane plane; (B) parallel to it.

ciple, such orientation permits the simultaneously utilization of two optimizing
factors in C.limicola chlorosome structure – the ordering of mutual orientations
of dipoles (OMOD) and the specific any-
sotropy of intermolecular distances
(SAID) [2]. In accordance with the conclusions of our works [2] it may be expected that all the BVr transition dipoles
should be parallel to each other and directed either (i) perpendicular to the
membrane plane (if the BVr antenna structure is optimized by OMOD only; Fig.3,
model A) or (ii) parallel to it (if the

BVr antenna structure is optimized by both optimizing factors, OMOD and
SAID; Fig. 3, model B). Indeed, our calculations show that for simplest
cubic antenna lattice ($R_\perp = R$), the excitation energy transfer time from
BVr antenna to BChl a one in model A (t_A) is smaller than that in model B
(t_B) about 4-fold ($t_A < 0.25\ t_B$). If R = const and $R_\perp$ decreases, then model
B becomes more effective than model A at $R/R_\perp > 1.26$; for example, t_B =
0.5 t_A at $R/R_\perp$ = 1.4 (i.e. t_B decreases $\sim$ 8-fold at this $R_\perp$ decrease). Note,
that for the case when all the BVr transition dipoles lie in the planes
(which are parallel to the membrane plane) but are randomly oriented in
these planes, the orientational factor for energy transfer along $R_\perp$ is equal to $\overline{k^2}$ =1/2 and that along the membrane, $\overline{k^2}$ = 5/4 [2,3]. It is clear
that such a model is less effective than model B. Thus, for such a large
antenna as in green sulfur bacteria just the same orientational ordering
of transition dipoles as in model B should be expected. This theoretical
prediction was supported by experimental results [10]. Thus, this strict
orientation is one of the optimizing factors ensuring such a small time
of heterogeneous energy transfer (20-50 ps) that was measured in [6]. Note,
that the data of the authors' earlier work [6] are supported by our preliminary experimental results obtained with a picosecond laser spectrochronograph: in intact C.limicola cells (at physiological conditions), the excitation energy transfer from BVr superantenna to BChl a antenna (upon excitation at $\sim$ 720 nm) occurs within $\sim$ 30 ps. These experimental facts, both
the strong orientational ordering of BVr transition dipoles and the small
time of BVr $\longrightarrow$ BChl a energy transfer, confirm the conclusion of our earlier theoretical work [1]: the antenna structure should ensure a directed
(not random!) excitation energy transfer from antenna to RCs.

REFERENCES
1. Fetisova, Z.G. and Fok, M.V. (1984) Molek.Biol. (Russ.) 18, 1651-1656.
2. See the cycle of 6 papers by Z.G.Fetisova et al., Molek.Biol.(Russ.),
   18, 1651; 1657 (1984); 19, 974; 983; 1476; 1489 (1985).
3. Fetisova, Z.G., Borisov, A.Yu., Fok, M.V. (1985) J.Theor.Biol. 112,41-75
4. Olson, J.M. (1980) Biochim.Biophys.Acta 594, 33-51.
5. Barsky, E.L., Borisov, A.Yu., Fetisova, Z.G. and Samuilov, V.D. (1974)
   FEBS Lett. 42, 275-278.
6. Fetisova, Z.G. and Borisov, A.Yu. (1980) FEBS Lett. 114, 323-326.
7. Borisov, A.Yu., Fetisova, Z.G. and Godik, V.I. (1977) Biochim.Biophys.
   Acta 461, 500-509.
8. Abdourakhmanov, I.A., Ganago, A.O., Erokhin, Yu.E., Solov'ev, A.A. and
   Chugunov, V.A. (1979) Biochim.Biophys.Acta 546, 183-186.
9. Staehelin, L.A., Golecki, J.R. and Drews, G. (1980) Biochim.Biophys.
   Acta 589, 30-45.
10. Fetisova,Z.G., Kharchenko,S.G., Abdourakhmanov,I.A.(1986) FEBS Lett.
    199,234-236.
11. Betti, J.A., Blankenship, R.E., Natarajan, L.V., Dickinson, L.C. and
    Fuller, R.C. (1982) Biochim. Biophys. Acta 680, 194-201.

LIGHT ABSORPTION AND FLUORESCENCE OF BCHL C IN CHLOROSOMES FROM
CHLOROFLEXUS AURANTIACUS AND IN AN IN VITRO MODEL

DANIEL C. BRUNE AND ROBERT E. BLANKENSHIP, DEPARTMENT OF CHEMISTRY,
ARIZONA STATE UNIVERSITY, TEMPE, AZ 85287

1. INTRODUCTION
    Chloroflexus aurantiacus is a thermophilic green gliding bacterium
that combines a purple bacterial type reaction center with a BChl c -
containing antenna typical of the green sulfur bacteria (1,2). All of the
BChl c in C. aurantiacus occurs in chlorosomes, subcellular structures
about 100 nm x 30 nm x 10 nm that are appressed against the inner surface
of the plasma membrane (3). Isolated chlorosomes exhibit absorption
maxima at 460 nm and 740 nm due to BChl c. Chlorosomes also contain
carotenoid and a small amount of BChl a (1 per 25-30 BChl a) (4-6) with an
absorption maximum at 792 nm. BChl a_{792} is thought to be located in the
baseplate through which the chlorosome is attached to the membrane. It
mediates energy transfer from BChl c to a membrane-located BChl a antenna
protein (absorption maxima at 808 and 866 nm), which in turn transfers
energy to the reaction center.
    Each type of antenna BChl fluoresces at a particular wavelength:
BChl c at 750 nm, BChl a_{792} at 802 nm, and BChl $a_{808-866}$ at 891 nm
(5,6,7). One objective of our work was to investigate energy transfer
from BChl c and from BChl a_{792} by measuring fluorescence lifetimes at 750
nm and at 802 nm after exciting BChl c.
    Previous work showed that BChl c extracted from green sulfur bacteria
forms aggregates in nonpolar solvents with absorption spectra very similar
to that of BChl c in vivo (8-10). BChl c was isolated from C. aurantiacus
and its suitability for forming a similar aggregate was investigated. The
absorption spectrum as well as the lifetime and spectrum of fluorescence
emission of aggregated BChl c were determined and compared with those of
BChl c in chlorosomes.

2. PROCEDURE
    C. aurantiacus was grown as described previously (6). Chlorosomes
and membranes with attached chlorosomes were obtained as described by
Feick and Fuller (4). BChl c was extracted with methanol and precipitated
from a methanol-dioxane-water mixture (10:1:2.5) using a modification of
the procedure of Watanabe et al. (11). The precipitated BChl was purified
by reverse phase HPLC on a C-18 column (Whatman ODS-3) using 92.5%
methanol - 7.5% H_2O as the eluting solvent. Aggregated BChl c was
prepared by diluting a concentrated methylene chloride solution of the
pigment with hexane as described by Smith et al. (9).
    Fluorescence spectra were measured using a home-built fluorimeter
with an RCA C31034 photomultiplier or Si photodiode detector.
Fluorescence lifetime measurements at 298°K were performed at the Center
for Fast Kinetics Research. A Spectra Physics dye laser system produced
727 nm, 12 ps exciting pulses with an energy of 2.5 nJ and a frequency of
800 KHz. Fluorescence was measured using a photon-counting detector
system and lifetimes were obtained by deconvolution analysis.
Fluorescence lifetimes at 77°K were determined by T. Steiner and M. L. W.
Thewalt at Simon Fraser University using a similar instrument (12).

Biggens, J. (ed.), Progress in Photosynthesis Research, Vol. I. ISBN 90 247 3450 9
© *1987 Martinus Nijhoff Publishers, Dordrecht. Printed in the Netherlands.*

3. RESULTS AND DISCUSSION

Purified BChl $\underline{c}$ from $\underline{C}$. aurantiacus was found to form an aggregate in hexane with absorption maxima at 740 and 460 nm (Fig. 1). These absorption maxima differ slightly from those reported by Smith et al. (748 and 452 nm) (9). A second difference between our results and those of Smith et al. was that we invariably observed an increase in intensity as well as a bathochromic shift of the red absorption maximum of BChl $\underline{c}$ during aggregation. A possible reason for these differences is that the BChl $\underline{c}$ from Prosthecochloris aestuarii used by Smith et al. is a mixture of compounds which differ in structure from BChl $\underline{c}$ from $\underline{C}$. aurantiacus (13,14). An attempt to determine the extinction coefficient of aggregated BChl $\underline{c}$ at 740 nm gave a value of 91 $mM^{-1}cm^{-1}$, assuming the value to be 74 (5) for absorbance at 664 nm of monomeric BChl $\underline{c}$ in hexane + methanol. This value is probably not exact because aggregated BChl $\underline{c}$ tends to precipitate and to become adsorbed on the sides of the cuvette, resulting in an inhomogeneous solution. The mM extinction coefficient at 742 nm of BChl $\underline{c}$ in chlorosomes is 102±2 (5).

Although the positions of the absorption maxima of aggregated BChl $\underline{c}$ are identical to those of isolated chlorosomes, carotenoids also contribute to light absorption by chlorosomes in the blue part of the spectrum. Carotenoids were extracted from dried films of chlorosomes on the inner faces of a cuvette by filling the cuvette 3 times with fresh hexane. The absorption spectrum of the chlorosome film after extraction was practically identical to that of aggregated BChl $\underline{c}$ over the entire range from 310 nm to 900 nm (Fig. 1). The only significant differences between the spectra are due to BChl $\underline{a}$ in chlorosomes, which has absorption maxima at 792 nm, 600 nm, and 370 nm. There is also a slight difference between the red absorption maximum of aggregated BChl $\underline{c}$ (742 nm) and that of the chlorosome film (740 nm), but absorption maxima at both wavelengths (and in between) have been observed in different chlorosome and aggregate preparations. This remarkable agreement between the absorption spectra of BChl $\underline{c}$ in chlorosomes and in the aggregate strongly implies that the interaction of BChl $\underline{c}$ molecules with each other in chlorosomes is very similar to their interaction in the in vitro aggregate. Although Wechsler et al. (15) have proposed that BChl $\underline{c}$ in chlorosomes occurs as a specific chlorophyll-protein complex, there is no experimental evidence for such a complex. We consider it an open question whether the protein in chlorosomes binds BChl $\underline{c}$ to form a specific complex or merely provides a suitable structural environment for the formation of a BChl $\underline{c}$ oligomer like that proposed in Refs. 8 and 9.

Smith et al. (9) reported that aggregated BChl $\underline{c}$ was converted to a monomeric form by adding small amounts of methanol. We obtained similar results both with aggregated BChl $\underline{c}$ and with a hexane-extracted chlorosome film. It was also possible to convert BChl $\underline{c}$ in chlorosomes in aqueous buffer to a monomeric form by adding small amounts of a long chain alcohol such as hexanol or octanol.

Both whole cells and isolated chlorosomes exhibit fluorescence emission maxima at 750 nm due to BChl $\underline{c}$ (5,6). The fluorescence emission spectrum of aggregated BChl $\underline{c}$ also has a maximum at 750 nm. The lifetime of fluorescence of aggregated BChl $\underline{c}$ (measured at 760 nm) was found to be 50-100 ps in most determinations (Table 1). For comparison, the lifetime of 760 nm fluorescence from chlorosomes or membranes with attached chlorosomes determined with the same instrument was 30-70 ps. These lifetimes are at the instrumental limit of resolution and should be

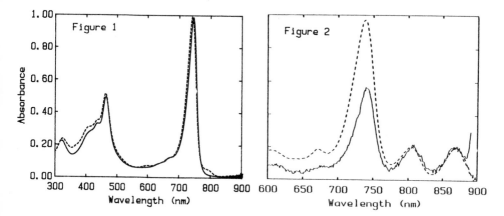

Figure 1.   Absorption spectra of aggregated BChl c (solid line) and of a
            hexane-extracted chlorosome film (dashed line).
Figure 2.   Absorption (dashed line) and fluorescence excitation (solid
            line) spectra of C. aurantiacus membranes.

Sample	Wavelength	Fluorescence Lifetime at	
		298°K	77°K
BChl c	>650 nm	6.4 ns	--
BChl c Agg	760 nm	50-140 ps	--
Chlorosomes	760 nm	33-70 ps	<30 ps
	800-810 nm, T_1	140-190 ps	200 ps
	T_2	500-1000 ps	489 ps
	R 2/1	0.03-0.1	0.13
Membranes	760 nm	35 ps	<30 ps
	800-810 nm, T_1	65-100 ps	245-255 ps
	T_2	550 ps	900-1000 ps
	R 2/1	0.02-0.05	0.001-0.02

Table 1.   Fluorescence lifetimes for BChl c, chlorosomes, and membranes
           from C. aurantiacus. T_1 and T_2 are the lifetimes of the first
           and second components of fluorescence emission and R 2/1 is the
           ratio of the amplitude of the second component to that of the
           first when more than one component was observed. The excitation
           wavelength for monomeric BChl c was 600 nm.

considered upper limits rather than actual values. The extremely short
fluorescence lifetimes of BChl c in chlorosomes and membranes are probably
due to very fast energy transfer to BChl a_{792}. The reason for the short
lifetime of fluorescence from aggregated BChl c in vitro is not known. It
may be that extensive delocalization of excitation energy in the aggregate
results in a high probability of trapping by minor impurities or defects.
    The lifetime of fluorescence at 800-810 nm from BChl a_{792} is also
shown in Table 1. Separating chlorosomes from the plasma membrane was

expected to increase the lifetime of 800-810 nm fluorescence by blocking energy transfer from BChl a_{792}. However, only a slight effect was noted. Thus the fluorescence lifetimes found at 298°K for chlorosomes (T_1 = 140-190 ps) were only about twice as long as those observed for membranes (T_1 = 65-100 ps). The difference at 77°K was even smaller. The surprisingly small effect of detaching chlorosomes from photosynthetic membranes on the lifetime of 800-810 nm fluorescence is not due to inefficient energy transfer from chlorosomes to the pigments of the attached membrane. An excitation spectrum for fluorescence of BChl $a_{808-866}$ in membranes with attached chlorosomes is compared to the absorption spectrum of the same sample in (Fig. 2). Light absorbed by BChl c in chlorosomes is at least 60% as efficient at exciting fluorescence of BChl $a_{808-866}$ as is light absorbed by BChl $a_{808-866}$ itself.

Acknowledgements: This work was supported by a grant from the Biological Energy Storage Division of the U.S. Department of Energy to R.E.B. We would like to thank T. Moore and G. King for assistance with fluorescence measurements. T. Steiner and M. L. W. Thewalt provided fluorescence lifetime measurements at 77°K, and S. Atherton assisted with the experiments and data analysis performed at the Center for Fast Kinetics Research at the University of Austin, Texas. The CFKR is supported jointly by the Biomedical Research Technology Program of the Division of Research Resources of NIH (RR 00886) and by the University of Texas at Austin.

REFERENCES
1 Blankenship, R.E. (1984) Photochem. Photobiol. 40, 801-806
2 Olson, J.M. (1980) Biochim. Biophys. Acta 594, 33-51
3 Staehelin, L.A., Golecki, J.R., Fuller, R.C. and Drews, G. (1978) Arch. Microbiol. 119, 269-277
4 Feick, R. and R.C. Fuller (1984) Biochem. 23, 3693-3700
5 van Dorssen, R.J., Vasmel, H. and Amesz, J. (1986) Photosynth. Res. 9, 33-45
6 Betti, J.A., Blankenship, R.E., Natarajan, L.V., Dickinson, L.C. and Fuller, R.C. (1982) Biochim. Biophys. Acta 680, 194-201
7 Vasmel, H., van Dorssen, R.J., de Vos, G.J. and Amesz, J. (1986) Photosynth. Res. 7, 281-294
8 Bystrova, M.I., Mal'gosheva, I.N. and Krasnovskii, A.A. (1979) Mol. Biol. 13, 440-451
9 Smith, K.M., Kehres, L.A. and Fajer, J. (1983) J. Am. Chem. Soc. 105, 1387-1389
10 Olson, J.M., Gerola, P.D., van Brakel, G.H., Meiburg, R.F. and Vasmel, H. (1985) in Antennas and Reaction Centers of Photosynthetic Bacteria (Michel-Beyerle, M.E., ed.), pp. 67-73, Springer-Verlag, Berlin
11 Watanabe, T., Hongu, A., Honda, K., Nakazato, M., Konno, M. and Saito, S. (1984) Anal. Chem. 56, 251-256
12 Bruce, D., Biggins, J., Steiner, T. and Thewalt, M. (1983) Biochim. Biophys. Acta 806, 237-246
13 Smith, K. M., Craig, G. W., Kehres, L. A. and Pfennig, N. (1983) J. Chromatog. 281, 209-223
14 Risch, N., Brockman, H. and Gloe, A. (1979) Liebigs Ann. Chem. 1979, 408-418
15 Wechsler, T., Suter, F., Fuller, R.C. and Zuber, H. (1985) FEBS Lett. 181, 281-294

SERRS AS A PROBE FOR PIGMENTS LOCATED NEAR THE SURFACES OF BACTERIAL
PHOTOSYNTHETIC MEMBRANES

RAFAEL PICOREL, RANDALL E. HOLT*, THERESE M. COTTON*, AND MICHAEL
SEIBERT**. SOLAR ENERGY RESEARCH INSTITUTE, GOLDEN, CO 80401, USA,
*UNIVERSITY OF NEBRASKA, LINCOLN, NE, 68588-0304, USA, AND **UNIVERSITY OF
DENVER, DENVER, CO 80208, USA

1.   INTRODUCTION
     Structural and functional aspects of biological macromolecules have
been investigated over the past five years using surface-enhanced
resonance Raman scattering (SERRS) spectroscopy, but the technique has
been applied only recently to biological membrane problems (1,2). The
unique feature of SERRS spectroscopy is the fact that resonance Raman
scattering at the interface between a potentiostated Ag electrode that has
been anodized and a sample adsorbed onto the surface of the electrode is
greatly enhanced over resonance Raman scattering of the sample in
suspension. This study reports on the use of SERRS to probe for the
presence of the carotenoid, Spirilloxanthin (Spir.), on the exposed
membrane surface of chromatophores (cytoplasmic side out) and spheroplast-
derived vesicles (periplasmic side out) isolated from the photosynthetic
bacterium, Rhodospirillum rubrum. The results provide evidence for the
location of Spir. on the cytoplasmic side of the membrane. However, the
potential importance of the work lies in the ability of this rather new
spectroscopic technique to provide structural information about intact
membrane systems on a non-distructive basis.

2.   PROCEDURE
2.1 Materials and methods
     2.1.1. Experimental Samples: Rs. rubrum S1 (wild type) and G9
            (carotenoidless mutant) were grown aerobically in the dark at
            32°C. Cells were harvested at the beginning of the stationary
            phase of growth. Chromatophores were isolated by differential
            centrifugation from French press cell homogenates.
            Spheroplast-derived vesicles were isolated by osmotic lysis of
            spheroplasts prepared by EDTA-lysozyme treatment (6 mM EDTA, 2
            mg/ml lysozyme, 10% [w/v] sucrose, 100 mM Tris-HCl, pH 8.0,
            for 1 hr at 35°C). During the osmotic lysis, 0.02% Brij-58
            was added to the suspension. Sidedness of chromatophores and
            vesicles was determined by NADH-dehydrogenase activity using
            ferricyanide as electron acceptor (3).
     2.1.2. SERRS Spectroscopy: The SERRS spectra were recorded using 10
            mW of 488.0 nm laser excitation. Raman scatter was analyzed
            using a Spex Triplemate 1877 spectrometer/spectrograph.
            Spectra were recorded with an optical multichannel analyzer
            (OMA II and 1420 photodiode array detector, Princeton Applied
            Research Corporation). The signal was integrated for 1 s and
            16 scans were averaged. Indene was used for frequency
            calibration of all spectra. Samples were added to N_2 degassed
            electrolyte solution (20 mM HEPES, pH 7.5, 100 mM Na_2SO_4, and

Biggens, J. (ed.), Progress in Photosynthesis Research, Vol. I. ISBN 90 247 3450 9
© *1987 Martinus Nijhoff Publishers, Dordrecht. Printed in the Netherlands.*

300 mM sucrose) in the SERRS cell with the electrode
maintained at -200 mV vs SSCE. After purging for 10 minutes
with N_2, resonance Raman (RR) spectra were recorded from the
smooth silver surface. The electrode was then anodized in the
presence of the sample by stepping the potential to +440 mV,
allowing ca. 25 mC/cm^2 of charge to pass, and stepping back to
-600 mV to reduce completely the Ag$^+$ formed at the positive
potential. Spectra were then recorded at an applied potential
of -200 mV. RR spectra of more concentrated samples were
recorded in 3 mm i.d. glass tubes.

3.   RESULTS AND DISCUSSION
     Figure 1 shows absorption spectra of chromatophores from Rs. rubrum
S1 and G9. The arrow indicates the laser wavelength (488 nm) chosen to
induce resonance Raman spectra. The resonance Raman spectrum of chromatophores from S1 exhibits peaks at 1003, 1150, and 1508 cm^{-1} while no peaks
are observed in chromatophores from G9 (Fig. 2). These data demonstrate
that 488 nm light induces only Spir. Raman spectra in these samples. Fig.
3 shows SERRS spectra of S1 and G9 chromatophores; again Spir. peaks at
1001, 1154, and 1508 cm^{-1} are seen with S1 while the peaks are absent in
G9. Similar results were obtained when laser excitation at 457.9 nm was
used (data not shown). Note the enormous difference in bacteriochlorophyll (BChl) concentration used in Fig. 2 and Fig. 3. In an attempt

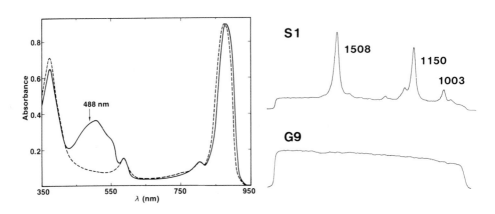

FIGURE 1. Absorption spectra of Rs.
rubrum chromatophores isolated
from S1 (——) and G9 (---), in
10 mM Tris-HCl, pH 8.0.

FIGURE 2. Resonance Raman spectra
of Rs. rubrum chromatophores
isolated from S1 and G9. λ_{EX} = 488
nm, BChl = 0.9 mg/ml, and peaks are
in cm^{-1}.

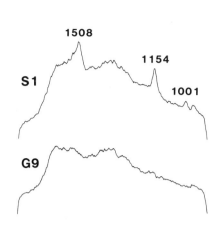

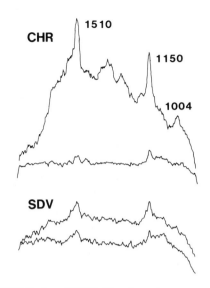

FIGURE 3. SERRS spectra of Rs. rubrum chromatophores isolated from S1 and G9. λ_{EX} = 488 nm, BChl = 0.33 µg/ml, and peaks are in cm^{-1}.

FIGURE 4. SERRS (top) and resonance Raman (bottom) spectra, respectively, of S1 chromatophores (CHR) and spheroplast-derived vesicles (SDV) at equivalent BChl concentration.

TABLE 1. A comparison of the percentage of cytoplasmic side out surface in chromatophores and spheroplast-derived vesicles as determined by the NADH-dehydrogenase enzyme assay (3) with the area under the 1150 and 1510 cm^{-1} SERRS peaks in Rs. rubrum S1 samples.

Sample	Cytoplasmic-side-out surface by enzymatic assay (%)	SERRS peak areas (%)	
		1510 cm^{-1} peak	1150 cm^{-1} peak
Chromatophores	100	100	100
Spheroplast-derived vesicles	≤36	16	19

to determine on which side of the membrane Spir. is located, SERRS spectra of chromatophores and spheroplast-derived vesicles are compared in

Fig. 4. As can be seen very little SERRS attributable to Spir. can be seen in spheroplast-derived vesicles while large signals are seen with chromatophores. Table 1 compares the percentage of cytoplasmic side out surface in our chromatophores and spheroplast-derived vesicle preparations as estimated by the enzymatic assay with the integrated areas under the 1510 and 1150 cm^{-1} SERRS peaks (corrected for resonance Raman scattering) in the respective preparations. The percentage of cytoplasmic side out membranes in the spheroplast-derived vesicles (36%) is an upper estimate due to limitations with the enzymatic assay.

SERRS signals attributable to Spir., at least 99% of the carotenoid found in Rs. rubrum, have been observed in chromatophores, but little signal is apparent in spheroplast-derived vesicles. These spectroscopic data demonstrate that Spir. is preferentially located on the cytoplasmic side of the photosynthetic membrane in this bacterium. This conclusion supports previous predictions from antenna complex sequencing (4), antibody (5), and pronase treatment (6) studies. Furthermore, the results are indicative of the extreme distance sensitivity of this particular SERRS technique (1) since the percentage of cytoplasmic side out surface in the two preparations determined by the two techniques (Table 1) are comparable. Finally, this work portends a future role for the SERRS technique in probing the surface topography of biological membranes.

ACKNOWLEDGMENTS

TMC and MS acknowledge the support of the NSF Chemistry of Life Processes Program (Grant #CHE-8509594).

REFERENCES

1 Seibert, M. and Cotton, T.M. (1985) FEBS Letts. 182, 34-38.
2 Nabiev, I.R. (1985) in Spectroscopy of Biomolecules (Alix, A.M.P., Bernard, L., and Manfait, M., eds.), pp. 348-352, John Wiley & Sons, N.Y.
3 Futai, M. (1974) J. Membrane Biol. 15, 15-28.
4 Brunisholz, R.A., Viemken, V., Suter, F., Bachofen, R., and Zuber, H. (1984) Hoppe-Seyler's Z. Physiol. Chem. 365, 689-701.
5 Brunisholz, R.A., Zuber, H., Valentine, J., Lindsay, J.G., Woolley, K.J. and Cogdell, R.J. (1986) Biochim. Biophys. Acta 849, 295-303.
6 Webster, G.D., Cogdell, R.J., and Lindsay, J.G. (1980) FEBS Lett. 111, 391-394.

Request # 1 of   1.   ILL record updated to IN PROCESS.
Screen 1 of 2
CAN YOU SUPPLY ?   YES   NO   FUTURE DATE:
ILL: 3825847   Borrower: DRB   ReqDate: 891117 Status: COND/PEND 891120
OCLC: 15077139   NeedBefore: 891213 RecDate: RenewalReq:
Lender: WEL, *DLM, DUP, FMN, NJR   DueDate: $30   NewDueDate:

1 CALLNO:
2 AUTHOR: International Congress on Photosynthesis (7th : 1986 : Providence
R.I.)
3 TITLE: Progress in photosynthesis research : proceedings of the VIIth
International Congress on Photosynthesis, Providence, Rhode Island, USA, Augu
10-15, 1986
4 EDITION:
5 IMPRINT: Dordrecht ; Boston : M. Nijhoff Publishers ; Hingham, MA, USA :
6 ARTICLE:
7 VOL: ↓42↑
8 VERIFIED: OCLC   DATE: PAGES:
9 PATRON: sirum,k.;grad,biochem
10 SHIP TO: Dana biomedical Library/DHMC/Hanover/NH 03756
11 BILL TO: Same
12 SHIP VIA: Library Rate   MAXCOST: 12.00   COPYRT COMPLIANCE:
13 BORROWING NOTES: THANKS! Pls. send what you own. Thanks

OPTICAL EXCITED TRIPLET STATES IN ANTENNA COMPLEXES OF THE
PHOTOSYNTHETIC BACTERIUM RHODOPSEUDOMONAS CAPSULATA A1a[+]
DETECTED BY MAGNETIC RESONANCE IN ZERO-FIELD

A. ANGERHOFER, J.U. VON SCHÜTZ AND H.C. WOLF
PHYSIKALISCHES INSTITUT, TEIL 3; PFAFFENWALDRING 57;
D-7000 STUTTGART 80; FEDERAL REPUBLIC OF GERMANY

1. INTRODUCTION
Zero-field optically detected magnetic resonance (ODMR) has
become a widely used tool for the study of excited triplet
states of biological molecules such as proteins [1], nucleic
acids [2] and porphyrins in photosynthesis [3]. In the last
years a number of publications dealt with the triplet state
of the primary donor in reaction centers of purple photosynthetic bacteria [4,5,6,7,8]. However the triplet state of the
antenna complexes was found only in a reaction center deficient mutant (mutant A1a[+] pho[-]) of Rhodopseudomonas capsulata
[9]. The problem was the impossibility of the preparation of
stable carotinoid free B 870-complexes of purple photosynthetic bacteria. ODMR-measurements of chromatophores yielded magnetic resonance signals of the antenna triplet states which
were obscured by the reaction center triplet signals. By oxidation of the reaction centers within the chromatophores or
slight dilution with low concentrations of Triton the energy
transfer between reaction center and antenna was disrupted.
Thus the antenna triplet states of Rhodopseudomonas sphaeroides R26, R26.1 and Rhodopseudomonas capsulata could be measured [10].
Recently a reaction center free preparation of the B870 complex from Rhodopseudomonas capsulata A1a[+] could be obtained
and was measured for comparison with the results of the chro-
matophores. Special attention was put on the |D-E|-region of
the ODMR-spectrum for the distinction and explanation of different triplet states in the complex.

2. PROCEDURE
2.1. Materials and Methods
2.1.1. Preparation
 The B870-complex was prepared from chromatophores by
 Li-Dodecylsulfate gel chromatography at low temperatures
 (4°C) and immediately frozen at T = 77 K.
2.1.2. Apparatus
 The zero-field resonance was recorded with the technique of conventional slow passage ODMR [11]. For the
 experimental set-up see Ref. [10]. Absorption- and
 Emission-spectra were recorded with a home-built
 spectrometer.

3. RESULTS AND DISCUSSION

3.1. Absorption

The absorption of the sample between 500 and 1000 nm at
T = 1,2 K (see Fig. 1) shows the antenna Bacteriochlorophyll (BChl) bands at 590 and 894 nm and additionally
two bands at 767 and 688 nm which are ascribed to free
BChl and oxidized BChl respectively. It was not possible
to remove these pigments from the preparation by further
steps of purification. Probably the very unstable antenna
complex decays rapidly into these components at room temperature.

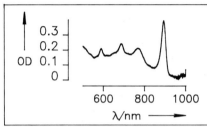

Fig. 1
Absorption of the B870 antenna
preparation from Rhodopseudomonas capsulata A1a⁺ at T =
1,2 K at 2 mm of optical path
length.

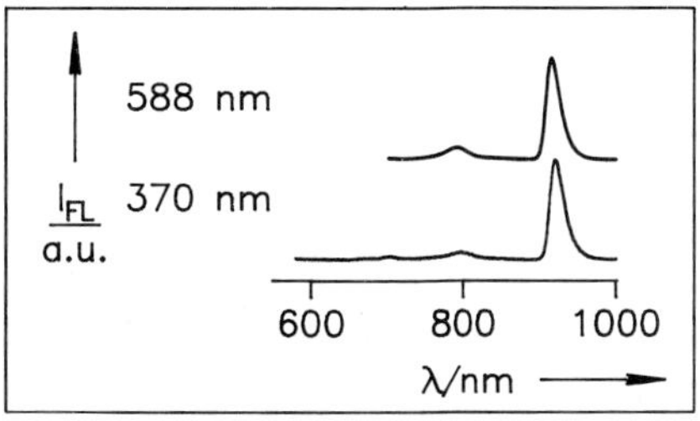

Fig. 2
Fluorescence emission spectra of the B 870 complex at
T = 1,2 K. The excitation
wavelengths are denoted left
of the spectra.

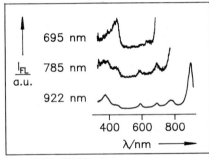

Fig. 3
Fluorescence excitation
spectra of the B870 complex
at T = 1,2 K. The detection
wavelengths are denoted left
of the spectra. The intensities are not comparable.

3.2. Fluorescence

The fluorescence emission spectra (see Fig. 2), excited
in the Soret-band at 370 nm or in the Q_x-band at 588 nm
shows the main emission maxima in the near infrared
region at 922 nm, which is emission out of the antenna
complex. Two smaller emission bands at 798 and 702 nm
are ascribed to the emission of free BChl and oxidized
BChl. This can be proofed by the fluorescence excitation
spectra in Fig. 3. Detection at 695 nm yields a Soret-
band at 447 nm and much weaker bands at 600 and 630 nm
which is characteristic of 2-Devinyl-2-Desacetyl-Chla,
an oxidation product of BChla [12]. Detection at 785 nm
yields excitation bands at 365, 393, 444, 587 and 691 nm.
They can be explained by a superposition of the excitation bands of free BChl a (365, 393 and 587 nm) with
those of oxidized BChl a. The excitation spectrum of the

antenna emission shows peaks at 375, 452, 590, 692, 779
and 897 nm which we ascribe to a superposition of bands
of antenna-BChl, free BChl and oxidized BChl. This demonstrates that the pigments, which emit at shorter wavelengths, transfer their energy to the antenna complex.

3.3. Fluorescence detected Magnetic Resonance

Fig. 4 shows the F-ODMR spectra in dependence of the wavelength of detection, whereas in Fig. 5 the signals are
shown in dependence of the wavelength of excitation.
Besides the spectrum of free BChl with a positive signal
at 488 MHz and a negative signal at 958 MHz, all the
spectra show mainly the same resonance bands, slightly
differing in their spectral localization. A negative band
in the region between 380 and 390 MHz is seen in the
spectra which are recorded at long detection wavelengths
and at short excitation wavelengths. The main band in the
ιD-Eι-region lies between 429 and 434 MHz, whereas a high
frequency shoulder between 479 and 488 MHz changes its
intensity from spectrum to spectrum. The ιD+Eι-signal
shows a systematic shift of its center from 821 MHz to
836 MHz on variation of the detection wavelength from
938 nm to 907 nm. Variation of the excitation wavelength
does not cause such a systematic shift.

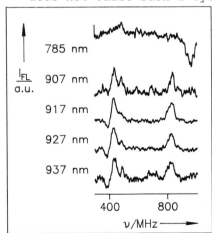

Fig. 4
F-ODMR spectra of the B870
complex at T = 1,2 K, taken
at different detection wavelengths as denoted left of
the spectra. The intensities
of the different spectra are
not comparable.

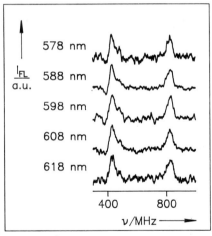

Fig. 5
F-ODMR spectra of the B870
complex at T = 1,2 K taken
with different excitation
wavelengths as denoted left
of the spectra. The intensities of the different
spectra are not comparable.

3.4. Zero field splitting (zfs) values

With exception of the negative signal at 390 MHz, the
data corresponde very well with those measured in Ref.[10].
By assigning the $|D-E|$-signal at 430 MHz to the $|D+E|$-
signal at 830 MHz one gets a range of zfs-values of the
antenna triplet state of $209 < |D| < 212$ and $65 < |E| < 67$ ($|D|$
and $|E|$ measured in 10^{-4} cm^{-1}). Fig. 4 shows, that the $|D+E|$-

signal of free BChl a does not appear in the ODMR-spectra of the antenna fluorescence despite the energy transfer between both components, which was demonstrated by the fluorescence excitation spectra.

Therefore the shoulder at 480 MHz can not be assigned to the $|D-E|$-signal of free BChl a. If it belongs to the antenna triplet and all up to now investigated complexes show this signal [10], the corresponding $|D+E|$-signal must fall together with the $|D+E|$-signal of the other triplet state, discussed above. Then the zfs-values compute to $218 < |D| < 220$ and $56 < |E| < 59$ ($|D|$ and $|E|$ measured in 10^{-4} cm^{-1}). This $|D|$-value is slightly higher than that of antenna complexes from other bacterial strains, whereas the $|E|$-value falls within this range [13].

For the explanation of the reduction of $|D|$ of the antenna pigments as compared with free BChl a, Beck already suggested the effect of selective spin-orbit-coupling of the τ_x^- and τ_y^-substate of the triplet with the singlet state, provided, that the redshift of the antenna fluorescence is at least partially, due to a crystal field effect at each antenna BChl pigment [9].

4. ACKNOWLEDGEMENT

We are very grateful to Prof. G. Drews for supplying us with the B870 antenna preparation of Rhodopseudomonas capsulata A1a+.

REFERENCES

[1] Kwiram, A.L. (1982) in Triplet State ODMR Spectroscopy (Clarke, R.H.,ed.) pp.427-478, John Wiley & Sons, New York
[2] Maki, A.H. (1982) in Triplet State ODMR Spectroscopy (Clarke, R.H.,ed.) pp.479-557, John Wiley & Sons, New York
[3] Schaafsma, T.J. (1982) in Triplet State ODMR Spectroscopy (Clarke, R.H.,ed.) pp.291-366, John Wiley & Sons, New York
[4] Clarke, R.H.; Connors, R.E.; Frank, H.A. and Hoch, J.C. (1977) Chem. Phys. Lett. 45, 523-528
[5] Hoff, A.J. (1976) Biochim. Biophys. Acta 440, 765-771
[6] den Blanken, H.J.; van der Zwet, G.P. and Hoff, A.J. (1982) Chem. Phys. Lett. 85, 335-338
[7] Beck, J.; von Schütz, J.U. and Wolf, H.C. (1983) Chem. Phys. Lett 94, 141-146
[8] Angerhofer, A.; von Schütz, J.U. and Wolf, H.C. (1984) Z. Naturforsch. 39c, 1085-1090
[9] Beck, J.; von Schütz, J.U. and Wolf, H.C. (1983) Chem. Phys. Lett. 94, 147-151
[10] Angerhofer, A.; von Schütz, J.U. and Wolf, H.C. (1985) Z. Naturforsch. 40c, 379-387
[11] Schmidt, J. and van der Waals, J.H. (1968) Chem. Phys. Lett. 2, 640-642.
[12] Smith, J.R.L. and Calvin,M. (1966) J. Am. Chem. Soc. 88, 4500-4506
[13] Angerhofer, A.; von Schütz, J.U. and Wolf, H.C. (1985) in Proceedings of the 5th International Seminar on Energy Transfer in condensed Matter (Pancoska, P. and Pantoflicek, J.,ed.) pp. 59-66, Society of Czechoslovak Mathematicians and Physicists, Prague

SINGLET ENERGY TRANSFER IN PHOTOSYNTHETIC BACTERIA:
ABSORPTION AND FLUORESCENCE EXCITATION OF B800-850 COMPLEXES

BARRY W. CHADWICK, HARRY A. FRANK, CHAOYING ZHANG and SHAHRIAR S. TAREMI,
DEPARTMENT OF CHEMISTRY, U-60, 215 GLENBROOK RD., UNIVERSITY OF
CONNECTICUT, STORRS, CT 06268; RICHARD J. COGDELL, DEPARTMENT OF BOTANY,
UNIVERSITY OF GLASGOW G12 8QQ U.K.

1. INTRODUCTION

Clayton and Clayton reported that the dissociation of 800nm absorbing
monomeric bacteriochlorophyll (BChl) molecules from the B800-850 complex of
Rb. sphaeroides 2.4.1 could be achieved by isolating the complex in lithium
dodecyl sulfate (LDS) instead of lauryl dimethylamine oxide (LDAO) [1].
Kramer et al. extended this work and probed spectroscopic differences
between B800-850 complexes isolated in LDAO and LDS, focusing primarily on
the LDS-induced spectral changes which occur near 800nm and 850nm (BChl
Qy), 590nm (BChl Qx) and in the carotenoid region [2]. The present study
explores the dynamics of the changes which occur in the absorption spectrum
of B800-850 complex from Rps. acidophila 7750 upon addition of LDS.

2. MATERIALS AND METHODS

Rps. acidophila 7750 cells were grown as previously described [3]. The
B800-850 complexes were isolated and purified in LDAO using DEAE Sephacel
[4] and Sephadex G200. All samples were adjusted to an 850nm optical
density of $1.0cm^{-1}$.

3. RESULTS

Figure 1 shows the effect on the absorption spectrum of adding 0.1% LDS
to the B800-850 complex isolated from Rps. acidophila 7750. The BChl Qy
absorption band at 857nm has been red-shifted by approximately 2nm,
attenuated by approximately 10% and broadened by approximately 0.7nm
(fwhm). The BChl Qy band near 800nm has been reduced to less than 15% of
its original intensity and two small broad peaks have grown in at
approximately 680nm and 760nm. The BChl Qx absorption band near 590nm has
decreased in intensity by approximately 40% and has been red-shifted by
approximately 4nm. The carotenoid absorption bands have been blue-shifted
by approximately 5nm. The blue-shift in the carotenoid absorption spectrum
upon LDS treatment is also accompanied by a 10-15% decrease in signal
intensity of the long-wavelength vibronic band and a comparable increase in
intensity of the short-wavelength band. The BChl Soret absorption bands at
373nm lose approximately 13% of their intensity with no shift or broadening
in their absorbance. The protein absorption peak near 281nm increases in
intensity by approximately 9% upon LDS addition.

The dynamics of the LDS-induced changes are shown in Fig. 2. The loss
in intensity of the 800nm BChl Qy, the BChl Qx and the Soret bands, and the
increase in the protein absorbance were found to occur at the same rate

Biggens, J. (ed.), Progress in Photosynthesis Research, Vol. I. ISBN 90 247 3450 9
© 1987 Martinus Nijhoff Publishers, Dordrecht. Printed in the Netherlands.

($t_{1/2}$ = 40 ± 10min.). The BChl absorption band at 857nm exhibited an immediate 1nm red-shift followed by an additional slower 1nm red-shift and 10% attenuation.

Figure 3 shows the detailed carotenoid absorption changes. There is an initial (time=295s) rapid increase in intensity of all three carotenoid absorption bands followed by a slower, more gradual decrease in intensity of the two long-wavelength vibronic bands. The short-wavelength vibronic band increases in intensity throughout the time course of the experiment. This increase and the blue-shift of the entire carotenoid absorption spectrum occur concomitantly with the decrease in BChl absorption intensity at 800nm.

Figure 4 shows the change in the protein absorption region (250-305nm). After addition of LDS an absorption band develops between 265 and 295nm having a maximum at 281nm.

Fluorescence excitation spectra of the lowest energy BChl emission (monitored at 860nm) for the Rps. acidophila 7750 B800-850 complex were used in conjunction with the absorption spectra (Fig. 1) to determine the effect of LDS addition on the efficiency of energy transfer from carotenoid to BChl. The efficiency of singlet-singlet energy transfer from carotenoid to BChl was found to decrease from 51±3% to 33±3% after addition of LDS. A parallel study was performed on Rb. sphaeroides 2.4.1 revealing a reduction in the efficiency of energy transfer from 95±5% to 72±3% after addition of LDS, in agreement with previous studies by Kramer [2].

4. DISCUSSION

The spectral observations may be explained by changes in pigment-pigment and pigment-protein interactions.

4.1. Bacteriochlorophyll Qy region: The red-shift, attenuation and broadening of the absorption band near 850nm is probably due to a slight structural change of the BChl dimer as previously suggested [1]. Accompanying this is an apparent loss of excitonic interaction between the monomeric BChl which absorbs at 800nm and the BChl dimer which absorbs near 850nm [2].

4.2. Bacteriochlorophyll Qx region: Previous linear and circular dichroism analysis of the BChl Qx (587nm) absorption spectral region of B800-850 complex from Rb. sphaeroides 2.4.1 revealed that this band is comprised of two components: A 584nm contribution attributed to the Qx transition of monomeric BChl whose Qy is at 800nm, and a larger 592nm contribution attributed to the Qx of the dimeric BChl whose Qy is at 848nm [5]. The red-shift and 40% attenuation of the BChl Qx absorption band upon LDS treatment can be interpreted as a loss of the monomeric BChl contribution to this band [2].

4.3. Carotenoid region: The initial rapid increase in intensity of the carotenoid absorption bands of B800-850 complex upon exposure to LDS is most likely the result of a protein configurational change. Slight changes in the distance between the carotenoid and the charged amino acid residues which are responsible for the red-shift of its absorption spectrum compared to its solution spectrum could account for the rapid increase in intensity and the blue-shift in carotenoid absorbance maxima [6]. The slower changes in absorbance (i.e. the gradual blue-shift in absorbance maxima and changes in vibronic structure) are probably due to a change in the carotenoids environment and/or structure.

4.4. Bacteriochlorophyll Soret and protein region: As previously mentioned the BChl Soret also shows a loss in intensity when exposed to LDS. Thus no evidence of intensity borrowing from the BChl Soret transitions is apparent here [7]. However, an increase in intensity in the protein spectral region (250-305nm) occurs concomitantly with the attenuation of the monomeric BChl absorption bands at 800nm, 590nm and 373nm. An interaction between the π electron systems of the amino acid residues which absorb in this spectral region and monomeric BChl may account for the hyperchromism in the absorption spectrum of this molecule. The fact that these changes in protein absorbance occur between 250nm and 305nm (Fig. 4) suggests that the amino acids responsible for the interaction are either tryptophan, tyrosine or phenylalanine residues. Recent sequence work by Theiler et al. [8] show tryptophans occupying the 7 and 40 positions, tyrosines at the 44 and 45 positions and phenylalanine at the 19 position of the α subunit of the B800-850 complex from Rb. sphaeroides R26.1 and 2.4.1.

Theiler et al. [8] have suggested that the monomeric BChl whose Qy is at 800nm is absent from the antenna complex isolated from Rb. sphaeroides R26.1 because a change in the primary structure of the α-apoprotein may have induced a conformational change sufficiently large to prevent binding of the monomeric BChl molecule. It is possible that addition of LDS to B800-850 complex from purple photosynthetic bacteria induces a change in the conformation of the protein which subsequently dislocates the 800nm absorbing BChl.

It has previously been suggested that the LDS-induced reduction in singlet-singlet energy transfer (from carotenoid to BChl) efficiency is a result of a dislocation of the monomeric BChl and its associated carotenoid pool from the B800-850 complex [2]. Recent picosecond absorption experiments performed on B800-850 complex from Rps. acidophila 7750 have shown that singlet excitation is transferred directly from the carotenoid to the BChl dimer in 5.8 ± 0.9ps [9]. The transfer dynamics are monophasic and do not show an involvement of the monomeric 800nm absorbing BChl in the energy transfer process. This suggests that the reduction in carotenoid to BChl singlet-singlet energy transfer efficiency observed here for the LDS treated complex is not caused by the removal of the 800nm absorbing monomeric BChl molecule and its associated carotenoid molecules. Protein structural changes induced by the presence of LDS may alter the distance and/or orientation between the carotenoid and dimeric BChl resulting in reduction in energy transfer efficiency.

REFERENCES

1. Clayton, R. K. and B. J. Clayton (1981) Proc. Natl. Acad. Sci. USA 78, 5583-5587
2. Kramer, H. J. M., Van Grondelle, R., Hunter C. N., Westerhuis, W. H J. and J. Amesz (1984) Biochim. Biophys. Acta 765, 156-165
3. Angerhofer, A., Cogdell, R. J. and M. F. Hipkins (1986) Biochim. Biophys. Acta 848, 333-341
4. Cogdell, R. J., Durant, I., Valentine, J., Lindsay, J. G. and Schmidt, K. (1983) Biochim. Biophys. Acta 722, 427-455
5. Bolt, J. and K. Sauer (1979) Biochim. Biophys. Acta 546, 54-63
6. Kakitani, T., Honig, B. and A. R. Crofts (1982) Biophys. J. 39, 57-63

7. Scherz, A. and W. W. Parson (1984) Biochim. Biophys. Acta 766, 666-678
8. Theiler, R., Suter, F., Zuber, H. and R. J. Cogdell (1984) FEBS Lett. 175, 231-237
9. Wasielewski, M. (1986) Ultrafast Phenomena V (G. R. Fleming and A. Siegmann eds.) Springer-Verlag, Berlin (in press)

ACKNOWLEDGEMENTS

This work is supported by grants from the National Science Foundation (PCM-8408201) and the University of Connecticut Research Foundation.

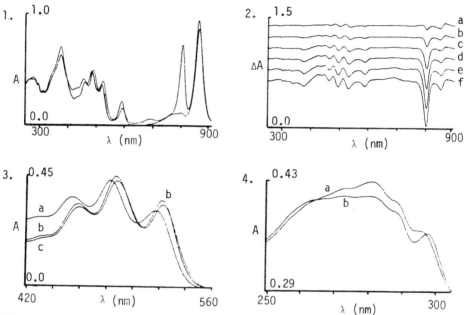

FIGURE CAPTIONS

Figure 1. The effect of LDS on the absorption spectrum of B800-850 complex from Rps. acidophila 7750. The spectra have not been normalized and were taken from samples having identical protein concentrations.

Figure 2. The dynamics of the LDS-induced changes in the absorption spectrum. The spectra represent the difference in absorbance of the sample after addition of LDS minus the absorbance of the sample prior to LDS addition. Times elapsed from the addition of 0.1% LDS are: (a) 105s; (b) 563s; (c) 1517s; (d) 2417s; (e) 3727s; (f) 8327s. The subsequent spectra have been offset for clarity.

Figure 3. The effect of LDS on the carotenoid absorption. Times elapsed after addition of LDS are: (a) 8517s; (b) 295s; (c) prior to LDS addition.

Figure 4. The effect of LDS on the protein absorption. (a) 8627s after addition; (b) prior to LDS addition.

PROPERTIES OF THE CORE COMPLEX OF PHOTOSYSTEM II

J.J. PLIJTER, R.J. van DORSSEN, J.P. DEKKER, F.T.M. ZONNEVELD, H.J. van GORKOM and J. AMESZ

Department of Biophysics, Huygens Laboratory of the State University, P.O. Box 9504, 2300 RA Leiden, The Netherlands

1. INTRODUCTION
    A considerable fraction of the total pigment and protein content of photosystem II grana preparations is made up by the light-harvesting Chl a/b protein complex (LHCP). The remaining part, which we shall call the core complex consists of at least five intrinsic proteins. Some of these proteins contain Chl a, which acts as the light harvesting pigment for the reaction center. In the present communication we describe the photochemical and spectroscopic properties of a purified oxygen evolving the core complex in comparison with those of a PS II grana preparation.

2. MATERIAL AND METHODS
    Photosystem II complexes were prepared from spinach chloroplasts according to Berthold et al. [1] except that the second Triton X-100 step was omitted. This so-called BBY complex was used as the starting material for the isolation of an oxygen evolving PS II core complex by solubilization with 1-O-n-octyl-β-D-glucopyranoside, by an adaption of the method of Ikeuchi et al. [2]. A considerable increase in purity was obtained by improved sucrose gradient centrifugation with a linear gradient (27 - 35 % w/v) on top of a 55 % layer. Oxygen evolution was measured with a Clark type electrode in a buffer containing 50 mM MES, 15 mM NaCl and 5 mM $MgCl_2$ (pH = 6.0). The concentration of the secondary acceptor Q_A was measured by the light-induced absorbance change at 325 nm in the presence of 2.5 mM ferricyanide and 10 μM DCMU. The apparatus used to measure absorption and fluorescence emission and excitation spectra is described in Ref. [3].

3. RESULTS AND DISCUSSION
    Upon sucrose gradient centrifugation of the detergent-treated BBY complex seven colored bands were formed. Most of the pigmented material settled on top. This band was found to consist mainly of the light harvesting Chl a/b protein complex (LHCP) on basis of its absorption spectrum and its low Chl a/b ratio, while six green bands were formed below (Table 1). The band on top of the 55 % sucrose layer (band 7) consisted of the PS II core complex. It had a Chl/reaction center ratio of 45 as determined from Q_A reduction, and was practically devoid of Chl b. Polyacrylamide gel electrophoresis showed that the core complex contained the 47, 43, 34, 33, 32 and 10 kDa intrinsic and the 33 kDa extrinsic protein in approximately equal molar ratios, while a significant fraction of the extrinsic 24 kDa protein was also retained. Measurement of oxygen evolution with the lipophilic and hydrophilic electron acceptors DCBQ and ferricyanide, respectively, revealed that with the core complex the highest rate was obtained with ferricyanide, while the opposite was observed in the BBY complex. This indicates that the low rate of oxygen evolution exhibited by the core complex is due to a rate limiting step on the reducing side rather than on the oxidizing

Biggens, J. (ed.), Progress in Photosynthesis Research, Vol. I. ISBN 90 247 3450 9
© *1987 Martinus Nijhoff Publishers, Dordrecht. Printed in the Netherlands.*

TABLE 1. Properties of the BBY complex and of the fractions in the sucrose gradient. Oxygen evolution was measured with 300 µM DCBQ or 2.5 mM ferricyanide (FeCy) as electron acceptors.

Fraction	Chl/Q_A	Chl $\underline{a}/\underline{b}$	O_2 evolution[a] DCBQ	FeCy	Yield of Chl recovery (%)
BBY	260	2.15	290	45	100
Band 1	>5000	1.45		n.d.	75
Band 2	2500	6.5		n.d.	6
Band 3	270	7.5		n.d.	3
Band 4	170	13.5		n.d.	2
Band 5	100	>15		n.d.	1
Band 6	65	>20	n.d.	80	2
Band 7	45	>20	80	150	4

(a) Expressed in µmol mg^{-1} Chl h^{-1}; n.d., not determined

side of PS II. Bands 6 to 2 (in this order) showed decreasing reaction center activities and Chl $\underline{a}/\underline{b}$ ratios and may have consisted of associations of the core complex and LHCP.

The low-temperature absorption spectra (Fig. 1) show the strong reduction of the Chl $\underline{b}$ content in the core complex by the absence of the bands near 650 and 472 nm and a much higher proportion of the band absorbing at 683 nm as indicated by the shoulder in the spectrum. In the core complex β-carotene bands are visible at 505 and 467 nm which are obscured in the BBY complex by the carotenoids of LHCP. The band near 540 nm which has been attributed to pheophytin $\underline{a}$ is composed of two components, with maxima at 537 and 543 nm (Fig. 1B, inset), indicating that the two Phe $\underline{a}$ molecules of the reaction center are spectrally different in the Q_x region. Deconvolution with Gaussian components gave a reasonable fit for the Q_y region and revealed the presence of several absorption bands which are listed in Table 2. The positions of these bands agree quite well with those for PS II in intact chloroplasts [4] indicating that the structure of the complexes was left virtually intact by the isolation procedure, as is also confirmed by

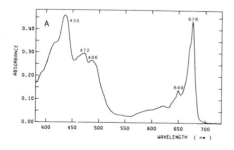
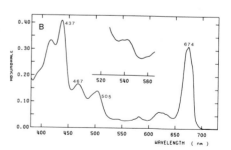

FIGURE 1. Absorbance spectra of the BBY complex (A) and of the core complex (B) at 4 K. The inset of Fig. 1B shows the region between 540 and 560 nm on a three-fold expanded scale.

TABLE 2. Gaussian deconvolution of the absorption spectra of the BBY and core complexes in the Q_y region at 4 K

band center (nm)		halfwidth (nm)		amplitude		rel. intensity	
BBY	CORE	BBY	CORE	BBY	CORE	BBY	CORE
649	649	11	10	0.13	0.04	18.9	–
659	660	9	10	0.11	0.08	12.9	12.6
669	668	9	10	0.22	0.21	26.8	33.1
676	675	8	9	0.36	0.23	37.9	33.6
684	682	7	8	0.04	0.17	3.4	19.5
–	693	–	8	–	0.01	–	1.2

The relative intensities of the bands were obtained by plotting A/ν versus ν (where ν is the frequency) and integrating this expression over the band. They are related to the total dipole strengths of each spectral component. In the case of the core complex the 649 component is excluded because of its doubtful physical significance.

their photochemistry, oxygen evolution and polypeptide composition. Absorption bands that could be attributed to PS I were virtually absent in both preparations. The intensity of the long-wave absorbing band at 683 nm in the core complex is enhanced about five-fold with respect to the BBY complex. The same increase was observed in reaction center content (Table 1). This indicates that all of the material absorbing near 683 nm is associated with the reaction center. From the Chl/Q_A ratio of 45 it follows that 8 – 10 pigments absorbing at 683 nm are present per reaction center, if all Q_y transitions have the same dipole strength. One of these pigments must be P-680 [5] another one is presumably the photoactive Phe a [6]. The core complex was also enriched in a minor component absorbing near 693 nm, which appeared to be present at a concentration of about 1 molecule per reaction center.

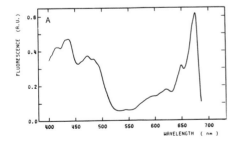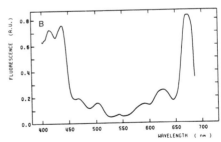

FIGURE 2. Fluorescence excitation spectra of the BBY complex (A) and the core complex (B) at 4 K. Detection wavelength: 696 nm. The absorbances of the samples were the same as in Fig. 1.

Fig. 2 shows the fluorescence excitation spectra of the 695 nm emmission of the BBY complex and the core complex. Comparision with the absorption spectra showed an efficiency of energy transfer from short wave to long wave absorbing Chl exceeding 90 %, again confirming the structural integrity of both preparations. The carotenoids of the core complex showed an average efficiency of energy transfer of 35 % to Chl a, whereas in the BBY complex the efficiency of energy transfer from the carotenoids to Chl appeared to be much higher, indicating that this energy transfer occurs mainly in LHCP.

The investigation was supported by the Netherlands Foundation for Chemical Research (SON), financed by the Netherlands Organization for the Advancement of Pure Research (ZWO).

REFERENCES
1 Berthold, D.A., Babcock, G.T. and Yocum,C.F. (1981) FEBS Lett. 134, 231-234
2 Ikeuchi, M., Yuasa, M. and Inoue, Y. (1985) FEBS Lett. 185, 316-322
3 Kramer, H.J.M. and Amesz, J. (1982) Biochim. Biophys. Acta 682, 201-207
4 Kramer, H.J.M., Amesz, J. and Rijgersberg, C.P. (1981) Biochim. Biophys. Acta 637, 272-277
5 Den Blanken, H.J., Hoff, A.J., Jongenelis, A.P.J.M. and Diner, B. (1983) FEBS Lett. 157, 21-27
6 Nuijs, A.M., van Gorkom, H.J., Plijter, J.J. and Duysens, L.N.M. (1986) Biochim. Biophys. Acta 848, 167-175

PIGMENT ARRANGEMENT IN PHOTOSYSTEM II

R.J. van DORSSEN, J.J. PLIJTER, A. den OUDEN, J. AMESZ and H.J. van GORKOM

Department of Biophysics, Huygens Laboratory of the State University,
P.O. Box 9504, 2300 RA Leiden, The Netherlands

1. INTRODUCTION
     Photosystem II of green plant photosynthesis is located in the grana of
the chloroplasts. Its antenna consists for the major part of the light-harvesting Chl a/b protein complex (LHPC). The remaining part, which we shall
call the core complex, contains mainly Chl a, which transfers its excitation energy to the reaction center. This communication describes the spectroscopic properties of a purified oxygen-evolving core complex in comparison with those of a PS II grana preparation.

2. MATERIALS AND METHODS
     Starting from a PS II grana preparation (which will be called BBY com-
plex) as described by Berthold et al. [1] an oxygen-evolving core complex
was obtained by an adaptation of the method described by Ikeuchi et al. [2]
using ß-octylglucoside as a detergent. The preparation contained 45 Chl a
[Plijter et al., these proceedings] per reaction center (RC).

3. RESULTS
     The low temperature absorption spectra (Fig. 1) showed that the core
complex has an approximately five-fold enhanced absorption band in the
long-wave region (683 nm) as compared to the BBY complex. From a Gaussian
deconvolution it can be calculated that the three major absorption bands,
due to Chl a 669, Chl a 676 and at 683 nm correspond to 15, 15 and 9 pigments per RC, respectively, if equal dipole strengths for those transitions
are assumed.

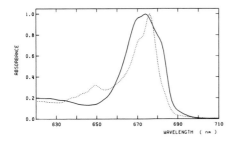

FIGURE 1. Absorbance spectra of the
Q_y region of the BBY complex (- - -)
and of the core complex (——) at 4
K.

     Upon excitation at 440 nm the room temperature fluorescence spectra of
both preparations (Fig. 2) showed a band located at 685 nm (F-685). Upon
cooling a second emission developed at 695 nm (F-695) and at 4 K practically all emission occurred from F-695. Both these bands also appear in the
emission spectra of intact chloroplasts.The characteristic PS I band at 735
nm was lacking in the low-temperature spectra of both preparations and the

Biggens, J. (ed.), Progress in Photosynthesis Research, Vol. I. ISBN 90 247 3450 9
© *1987 Martinus Nijhoff Publishers, Dordrecht. Printed in the Netherlands.*

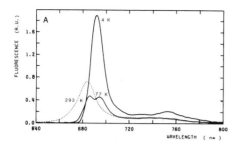 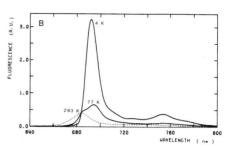

FIGURE 2. Emission spectra of the BBY complex (A) and of the core complex (B) excited at 440 nm and measured at various temperatures.

remaining minor bands probably reflect vibrational sub-bands of F-685 and F-695.

Fig. 3 shows the temperature dependence of the relative intensities of F-685 and F-695 obtained by deconvolution of the emission spectra. Semilog-arithmic plots of the ratios of the intensities versus the reciprocal of the temperature yielded straight lines for both preparations down to 70 K, indicating a thermal equilibrium between these two emissions. The slope of both lines yielded an energy difference $\Delta E = 155 \pm 5$ cm^{-1}, corresponding to a difference in the absorption maxima of the two states of approximately 7 nm. If equal intrinsic fluorescence yields are assumed for F-685 and F-695 the intercepts with the y-axis (31 for the BBY complex and 12 for the core complex) give the relative numbers of emitting dipoles for the two states.

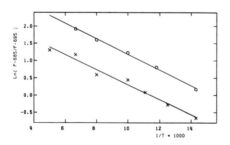

FIGURE 3. Temperature dependence of the ratio of the F-685 and F-695 emission yields. Excitation at 440 nm. (xxx) Core complex, and (ooo) BBY complex.

The linear dichroism spectra obtained by uniaxially pressing a poly-acrylamide gel between two parallel prisms [3] from an initial thickness of 5 mm to a final one of 2 mm are shown in Fig. 4. As the Q_y transitions of Chl a are oriented mainly parallel to the plane of the thylakoid membrane the positive LD of the BBY complex indicates that upon sqeezing the partic-les are aligned with their membranes parallel to the face of the prisms. The dichroic ratio, defined as $(A_{//} - A_{\perp})/(A_{//} + A_{\perp})$, of the BBY complex rises steeply across the Q_y band region and reaches a maximum at about 690 nm. This indicates that the Q_y transitions of the long-wave chloro-phylls are oriented more parallel to the membrane than those absorbing at shorter wavelengths. The large dichroic ratio near 690 nm for the BBY com-plex indicates a high degree of orientation of the membranes. A linear relationship between the dichroic ratio and the degree of pressing was observed to values of up to 0.5 at this wavelength. This suggests that the dichroic ratio at perfect orientation is at least 0.6 as compared to the maximum value of 0.90 for our geometry, corresponding to an angle of at least 70° with the normal to the plane of the membrane. The linear dichroism spectrum of the core complex (Fig. 4b) shows roughly the same features as the spectrum of the BBY complex, but the lower dichroic ratio suggests that much less orientation was obtained with this preparation. The polarized emission spectra of the biaxially oriented BBY complex indicated that the pigments emitting F-685 are oriented preferentially parallel to the plane of the membrane, whereas F-695 is emitted by pigments making a substantial angle with this plane.

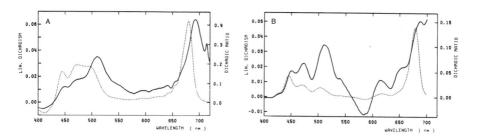

FIGURE 4. Linear dichroism spectra (---) of the BBY complex (A) and of the core complex (B) obtained by uniaxial pressing at 293 K. The solid lines indicate the dichroic ratio. Absorbances in the Q_y maxima were 0.24 and 0.37, respectively.

The LD and fluorescence measurements indicate that F-685 must be emitted by Chl a 676. The orientation of F-695 shows that it cannot be attributed to the Chl a transition at 683 nm, which is almost parallel to the membrane. This tends to support the hypothesis [4] that F-695 may be fluorescence from the photoactive Phe a that functions as an electron acceptor in the reaction center of PS II. The Q_y transition dipole of this Phe a makes a large angle with the membrane and is located at 680 - 685 nm at room temperature [5]. From the fact that 15 Chl a 676 molecules are present per RC in the core complex and from the value of 12 for the y-axis intercept of the temperature dependence it follows that the species emitting at 695 nm may occur in a concentration of only one molecule per RC, which would support the conclusion mentioned above.

Our results do not support the hypothesis recently made by Tapie et al. [6] that the weak 693 nm transition (visible as a tail on the long-wave side of the spectra of Fig. 1) would be responsible for the 695 nm emission. First of all, one would then have to assume that the absorption band of the photoactive Phe a shifts by about 10 nm upon cooling and that the Stokes shift is unusually small. Moreover, the temperature equilibrium of F-685 and F-695 indicates an energy difference corresponding to 7 nm for the two absorption dipoles.

The investigation was supported by the Netherlands Foundations for

Chemical Research (SON) and for Biophysics, financed by the Netherlands Organization for the Advancement of Pure Research (ZWO).

REFERENCES
1 Berthold, D.A., Babcock, G.T. and Yocum, C.F. (1981) FEBS Lett. 134, 231-234
2 Ikeuchi, M., Yuasa, M. and Inoue, Y. (1985) FEBS Lett. 185, 316-322
3 Vasmel, H., van Dorssen, R.J., de Vos, G.J. and Amesz, J. (1986) Photosynth. Res. 7, 281-294
4 Breton, J. (1982) FEBS Lett. 147, 16-20
5 Ganago, I.B., Klimov, V.V., Ganago, A.O., Shuvalov, V.A. and Erokhin, Y.E. (1982) FEBS Lett. 140, 127-130.
6 Tapie, P., Choquet, Y., Wollman, F., Diner, B. and Breton, J. (1986) Biochim. Biophys. Acta 850, 156-161

THREE-DIMENSIONAL CRYSTALS OF THE LIGHT-HARVESTING CHLOROPHYLL a/b
PROTEIN COMPLEX FROM PEA THYLAKOIDS

W.Kuehlbrandt, AFRC Photosynthesis Group, Department of Pure and Applied
Biology, and Biophysics Section, Department of Physics, Imperial College
of Science and Technology, London SW7 2BB, England

High-resolution structure analysis of the photosynthetic apparatus requires
well-ordered crystals of the chlorophyll protein complexes involved in
photosynthesis. Three-dimensional crystals of the light-harvesting
chlorophyll a/b protein complex (LHC II) from pea thylakoids have been
grown. The complex was crystallized from detergent solution by vapor
diffusion at low ionic strength without addition of small amphiphilic
molecules. Two forms of three-dimensional crystals have been obtained
in this way.

Octahedral crystals grew to a size suitable for X-ray crystallography
(body diagonals up to 0.7 mm). The crystals were dark green to black in
colour, depending on thickness, and did not display any detectable
birefringence. Absorption spectra of thin slabs showed the characteristic
absorption bands of LHC II. One-dimensional SDS polyacrylamide gels
of single crystals indicated the same polypeptide composition as that
normally found in isolated, purified pea LHC II, a main band at 24,000
apparent molecular weight and two satellite bands at 23,000 and 23,500
apparent molecular weight, respectively. The crystals thus contained
at least three different LHC II polypeptides, suggesting that the various
polypeptides of pea LHC II were structurally similar enough to co-crystal-
lize interchangeably.

X-ray diffraction patterns of octahedral crystals indicated a very large
cubic unit cell (a = 390 Å). Assuming the protein density of octahedral
crystals is similar to that of two-dimensional crystals of LHC II (1),
a unit cell of this size may contain up to 400 LHC II monomers. Reflections on X-ray diffraction patterns extended to about 20 Å resolution.
Crystals of this type were therefore too disordered for high-resolution
data collection.

A second type of three-dimensional crystals of LHC II, thin hexagonal
plates, formed under similar conditions. Unlike the octahedral crystals,
the hexagonal plates tended to be unstable. When seen face-on, they did
not show any optical activity, but when seen edge-on, they displayed bright
green birefringence. Crystals of this type were too small for X-ray
crystallography, measuring up to 50 μm across and a few μm in thickness.
However, examining thin specimens in the electron microscope yielded
high-resolution electron diffraction patterns with sharp reflections at
3.7 Å resolution, demonstrating that the hexagonal plates were highly
ordered. The symmetry of the diffraction pattern (6mm) and the dimension
of the hexagonal unit cell in projection, calculated from the spacing

Biggens, J. (ed.), Progress in Photosynthesis Research, Vol. I. ISBN 90 247 3450 9
ⓒ *1987 Martinus Nijhoff Publishers, Dordrecht. Printed in the Netherlands.*

of the reciprocal lattice (a = 127 Å) suggested that the hexagonal plates
were, in fact, stacks of two-dimensional crystals which have a similar
unit cell dimension in projection (a = 125 Å) and the same symmetry of the
diffraction pattern (1). Hexagonal plates thus seem well-suited for high-
resolution structure analysis by image processing (2,3) and electron
diffraction.

References:

(1) Kuehlbrandt,W. (1984). Nature 307, 478-480
(2) Henderson,R. and Unwin, P.N.T. (1975). Nature 257, 28-32
(3) Henderson,R., Baldwin,J., Downing,K., Lepault,J. and Zemlin,F. (1986).
    Ultramicroscopy, in press.

INTERPRETATION OF TRANSIENT LINEAR DICHROISM SPECTRA OF LHC PARTICLES

ROBERT S. KNOX AND SU LIN, DEPARTMENT OF PHYSICS AND ASTRONOMY,
UNIVERSITY OF ROCHESTER, ROCHESTER, NY 14627, USA

1. INTRODUCTION

   Gillbro et al. [1] have measured the anisotropy associated with the
electronically excited states of chlorophyll b (Chlb) in the light-
harvesting chlorophyll protein LHC-II. No decay in anisotropy was
observed over a time of 5-25 ps, while the Chlb excited-state population
itself decayed rapidly (in 6±4 ps). From these facts, the authors
concluded that no transfer of excitation energy occurs among molecules
of Chlb if they are differently oriented.

   The interpretation given by Gillbro et al. tends to imply that strong
coupling of Chlb's would be inconsistent with their data. The model
developed during the past few years by Van Metter, Shepanski, Gülen, and
one of the present authors [2-4] relies upon such strong coupling to
explain the circular dichroism spectrum of this antenna protein. In
particular, whereas the Chlb transition dipole moments in our trimeric
model are never taken to be parallel, the authors of [1] believe that
only the case of parallel transition moments would be consistent with
their data if there were intermolecular excitation coupling.

   This note points out that the absorption anisotropy data is consistent
with, and therefore supports, the trimeric model. Other data obtained by
Gillbro et al. are also consistent with this model.

2. REVIEW OF THE TRIMER MODEL

   On the basis of absorption, circular dichroism, and fluorescence
polarization data, a model of the relative chromophore orientations in
the Chla/b protein has been developed [2-4] in which the Chlb molecules
are strongly coupled into states α, β, γ, of which two at 1.902 eV (652
nm) are degenerate (β,γ) and split from the third at 1.864 eV (665 nm).
In addition, the three Chla molecules assumed to exist in the complex are
weakly coupled with one being shifted to 1.832 eV (677 nm) from the other
two at 1.851 eV (670 nm). These numbers refer to the SDS-solubilized [5]
low-molecular weight derivative of LHC, but the spectral features are
reasonably similar to those of LHC and the two spectra can be related [6].

3. CHLOROPHYLL-b ANISOTROPY

   Gillbro et al. find a virtually time-independent absorption anisotropy
during the recovery period of Chlb bleaching. This might occur in at
least three ways. First, if the three Chlb molecules do not interchange
excitation energy, those which are photoselected remain excited and there
is no depletion of absorption in a different polarization. Second, if the
transition dipoles of three nearest-neighboring Chlb's are parallel, then
even if there is local interchange of excitation energy there will be no
depletion of absorption in a different polarization because no other

Biggens, J. (ed.), Progress in Photosynthesis Research, Vol. I. ISBN 90 247 3450 9
© *1987 Martinus Nijhoff Publishers, Dordrecht. Printed in the Netherlands*

exists. Third, if the Chl$\underline{b}$ molecules are so strongly coupled as to act virtually as a single molecule (Chl$\underline{b}$)$_3$, this entire molecule is photo-selected and remains so. If excitation transfer takes place among α, β, and γ it does not matter because this molecule is no longer participating in upward transitions until energy is transferred out by radiation or radiationless processes (to Chl$\underline{a}$ or other sinks). When that occurs, the initial upward transition can again occur, but with the same polarization.

Gillbro et al. considered only the first two of these possibilities, concluding that their results were contrary to available CD information. When the third possibility is brought out, the contradiction disappears and the trimer model is supported by the anisotropy data.

The aforementioned degeneracy of the states β, γ would appear to imply that the maximum Chl$\underline{b}$ absorption anisotropy should have the canonical value 0.10 (p=1/7 in Perrin's formalism [7]). The measured value is 0.42±0.05. However, this value refers to LHC and not the smaller CP-II particle. It has been noted [6,8] that in LHC, Chl$\underline{b}$ has a complex structure in the region of 650 nm, part of which must be due to splitting of the β, γ pair. Each would act as a nondegenerate oscillator with r=0.40.

Although we are confident that the foregoing analysis resolves the apparent contradiction, we are not entirely happy with either analysis, since excited state absorption is ignored. In the case of Chl$\underline{a}$ this would not be a good assumption, and it should be examined carefully in future research.

4. ABSORPTION RECOVERY NEAR 665 nm

In the data under discussion there are two interesting characteristics of the recovery of absorption at 665 nm. One is the existence of a 20-ps component and the other is the need for assuming fast equilibration between postulated Chl$\underline{a}$ species at 670 and 677 nm. Since the authors of [1] do not discuss the details of the existing model, we point out how it is also consistent with their data.

The Chl$\underline{b}$ trimer component at 665 nm is one possible cause of the 20-ps component. However, the absorption by this state is quite weak and it is rather more likely that transfer from all of the (Chl$\underline{b}$)$_3$ states occurs on the 2-ps to 10-ps scale envisioned by Gillbro and coauthors. We therefore prefer their interpretation, wherein the 670-nm Chl$\underline{a}$ species is initially preferentially bleached and then achieves thermal equilibrium with the 677-nm species. What they do not go on to point out is that the energy difference between these species is of the appropriate order of magnitude for such an equilibration. It is 0.019 eV or 0.77 k_BT at room temperature. Therefore the back and forward rate constants have a ratio of $e^{-0.77}$=0.47. Complete equilibration is therefore eminently reasonable on the model, not contrary to it, as has been stated. A low-temperature experiment could easily check this hypothesis. Figure 1 indicates one possible energy-level relationship for the $\underline{a}/\underline{b}$ protein.

To establish more firmly the equilibration hypothesis we have evaluated the populations of upper (N_1) and lower (N_2) level Chl$\underline{a}$'s as a function of time, using simple first-order kinetics,

$$dN_1(t)/dt = I(t)\sigma_1 N_{10} - (K_{21}+T_1^{-1})N_1(t) + K_{12}N_2(t)$$
$$dN_2(t)/dt = I(t)\sigma_2 N_{20} - (K_{12}+T_2^{-1})N_2(t) + K_{21}N_1(t),$$

an excitation pulse $I(t)$ of 12-ps FWHM, and rate parameters $K_{21}=(20\text{ ps})^{-1}$, $K_{12}=(43\text{ ps})^{-1}$, $T_1=1500$ ps, and $T_2=1500$ ps. Only the ratio $\sigma_1 N_{10}/\sigma_2 N_{20}$ (the products of cross-section at 665 nm and ground-state chromophore density) is significant to the linear kinetics. The ratio σ_1/σ_2 was taken to be 17/6, estimated by making simple shifts of a Chla solution absorption curve to account for excitation of two species centered at 670 and 677 nm. The ratio N_{10}/N_{20} was taken as 2. Predicted absorption recoveries at 665 nm, assuming probing at the same wavelength as that of excitation, depend on weighting of the N_1 and N_2 curves, as might be expected from different polarization geometries. The two-component nature is quite evident in our predicted time dependence and the combination $2N_1(t)+N_2(t)$ compares favorably with the data of ref. [1], Figure 3. Full details will be submitted for publication elsewhere.

5. CONCLUDING REMARKS

Some further aspects of the experimental results discussed here remain to be analyzed in terms of any model of Chl a/b. The large value of anisotropy in the Chlb region (0.42) is unusual, despite our rationalization of it presented above. Also, while the effects occurring upon denaturation of Chlb are consistent with our own denaturation results [3], those involving Chla are not.

Research was supported in part by USDA grant 82-CRCR-1-1128.

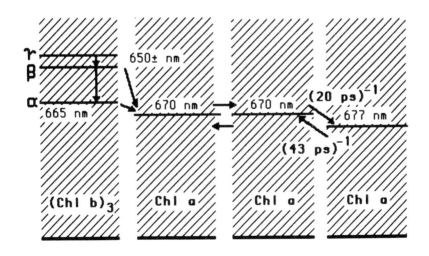

Fig. 1. Energy levels of the distinct energy-transferring species appearing in the model of the chlorophyll-a/b protein discussed in the text [2-4]. Only the chlorophyll levels corresponding to or deriving from the usual Qy transition are shown. The linear arrangement is only a convenient depiction. There may be some transfer from the Chlb group to any of the Chla's.

RESONANCE RAMAN SPECTROSCOPY OF CHLOROPHYLLS AND THE LIGHT-HARVESTING
CHLOROPHYLL a/b PROTEIN

H.N. FONDA AND G.T. BABCOCK, DEPARTMENT OF CHEMISTRY, MICHIGAN STATE
UNIVERSITY, EAST LANSING, MI 48824-1322

1. INTRODUCTION

    The technique of resonance Raman (RR) spectroscopy holds great
potential for the investigation of chlorophylls in photosynthetic systems
in vivo. However, successful application requires a thorough
understanding of the RR scattering properties of chlorophylls in solution.
Of key importance is an accurate assignment of the vibrational modes of
the chlorophyll molecule. Vibrational assignments for chlorophyll by
comparison with the well-studied porphyrins are complicated by a) the
lower symmetry of the chlorin macrocycle and b) the presence of ring V in
chlorophyll.
    Naturally-occurring chlorophylls contain magnesium as the central
metal atom. A number of vibrational modes in metalloporphyrins are metal-
sensitive. The vibrational frequency of these modes can be correlated
with the core-size of the metalloporphyrin [1]. In the work described, we
have prepared a series of metal-substituted chlorophylls and examined
their RR spectra. Correlations have been made for the core-size sensitive
modes. Furthermore, this allows diagnosis of the coordination number of
the magnesium atom in chlorophyll in various solvents and in the light-
harvesting chlorophyll a/b protein (LHC).

2. MATERIALS AND METHODS

    Chlorophyll a was isolated from spinach by column chromatography with
DEAE-Sephanose CL-6B and Sepharose CL-6B [2]. Pheophytin a was obtained
by acidification of chlorophyll a with 1N HCl. The zinc, copper, cobalt
and nickel-substituted chlorophylls were prepared by addition of a
five-fold excess of the metal acetate to pheophytin a in
chloroform/acetic acid. LHC was isolated from pea thylakoids by the
method of Burke et al. [3].
    RR spectra were recorded by using the 406.7 nm line of a Kr^+ laser in
a backscattering geometry. Chlorophyll a and LHC samples were run at
-140°C. Solutions of the metal-substituted chlorophylls in ether were
obtained at room temperature.

3. RESULTS AND DISCUSSION

    Figure 1 shows the electronic absorption spectrum of copper-
substituted chlorophyll a. Both the Qy and Soret bands shift to higher
energy in the order Co > Ni > Cu > Zn > Mg. These data are summarized in
Table 1.

Biggens, J. (ed.), Progress in Photosynthesis Research, Vol. I. ISBN 90 247 3450 9
© *1987 Martinus Nijhoff Publishers, Dordrecht. Printed in the Netherlands.*

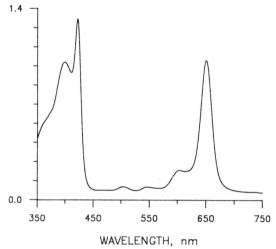

TABLE 1. Absorption Maxima
of metal-substituted
chlorophylls in ether

Metal	Absorption Soret	Maximum Qy
Mg	430	662
Zn	423	654
Cu	422	650
Ni	418	647
Co	417	641

WAVELENGTH, nm

FIGURE 1. Electronic absorption spectrum
of copper-substituted chlorophyll in ether.

Increasing the size of the central metal ion causes an expansion in
the core-size of the metalloporphyrin. This in turn weakens the C_α-C_m
bonds resulting in a lower frequency for those modes with a contribution
from C_α-C_m stretching. The order of core-size as determined by X-ray
crystallography for model porphyrin systems is Mg > Zn > Cu > Co > Ni.
Figure 2 shows the RR spectrum of copper-substituted chlorophyll a. Five
core-size sensitive bands have been identified at 1641, 1581, 1554, 1510
and 1451 cm^{-1}. Data for the other metal-substituted chlorophylls are
summarized in Table 2.

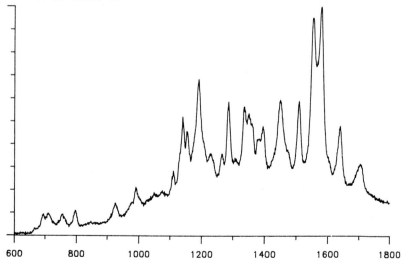

FIGURE 2. RR spectrum of copper-substituted chlorophyll at room
temperature; λ_{ex} = 406.7 nm.

TABLE 2. Core-size sensitive frequencies for metal-substituted chlorophylls.

Metal	Core-size (Å) (ref. 1)	core-size sensitive modes (cm^{-1})				
		a	b	c	d	e
Mg	2.055	1611	1551	1530	1490	1434
Zn	2.047	1615	1561	1539	1499	1439
Cu	2.000	1641	1581	1554	1510	1451
Co	1.976	1638	1593	1566	1512	1455
Ni	1.958	1657	1596	1570	1521	1463

The correlation between core-size and frequency for modes a to e is shown in Figure 3. The modes labelled a,b,c and d may correlate with the ν_{10}, ν_2, ν_{11} and ν_3 modes of Ni(OEP) [4], respectively.

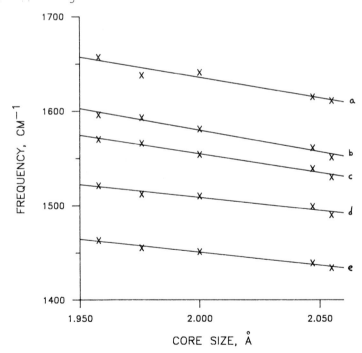

FIGURE 3. Correlation between core-size and frequency of core-size sensitive modes for metal-substituted chlorophylls.

Assignment of the core-size sensitive modes is supported by work in this lab and by Fujiwara and Tasumi [5] on the effect of solvent coordination to chlorophyll in vitro. With five-coordination, the magnesium atom is out-of-plane. On going to six-coordination, the magnesium atom would presumably move into the plane of the ring, resulting in an expansion of the core. The core-size sensitive modes in chlorophyll should therefore occur at lower frequencies. Fujiwara and Tasumi showed that in n-hexane, CCl_4, CS_2, diethyl ether, acetone, ethyl acetate and ethanol solutions (five coordination) three bands are observed at 1612-1606, 1554-1551 and 1529-1527 cm^{-1}. In tetrahydroforan, dioxane, pyridine and methanol solution (six coordination) these bands shift to 1599-1596, 1548-1545 and 1521-1518 cm^{-1} respectively. Additionally, we have observed that the 1490 cm^{-1} band shifts to 1485-1483 cm^{-1} on going to six-coordination.

RR spectra of LHC show chlorophyll a modes at 1610, 1552, 1492 and 1435 cm^{-1} (the ~ 1530 cm^{-1} mode is obscured by a carotenoid band). The chlorophyll a molecules in LHC are therefore five-coordinate.

ACKNOWLEDGEMENTS

This research was supported by grants from N.I.H. (GM 25480) and the McKnight Foundation.

REFERENCES

1 Spaulding, L.D., Chang, C.C., Yu, N.-T., Felton, R.H. (1975) J. Am. Chem. Soc. 97, 2517-2525.
2 Omata, T., Murata, N. (1983) Plant Cell Physiol. 24, 1093-1110.
3 Burke, J.J., Ditto, C.L., Arntzen, C.J. (1978) Arch. Biochem. Biophys. 187, 252-263.
4 Abe, M., Kitagawa, T., Kyogoku, Y. (1978) J. Chem. Phys. 69, 4526-4534.
5 Fujiwara, M., and Tasumi, M. (1986) J. Phys. Chem. 90, 250-255.

OXYGEN-EVOLVING COMPLEX OF PHOTOSYSTEM II IN HIGHER PLANTS

NORIO MURATA AND MITSUE MIYAO
NATIONAL INSTITUTE FOR BASIC BIOLOGY, MYODAIJI, OKAZAKI 444, JAPAN

1. INTRODUCTION

The development of techniques to prepare oxygen-evolving Photosystem II (PSII) particles (membrane fragments) by treating thylakoids with Triton X-100 (1,2) and inside-out PSII vesicles by aqueous phase partition (3) has made it possible to study the detailed biochemistry of the oxygen-evolving PSII complex (4). Also, advances in molecular biology have provided new techniques to explore the protein structures in the complex (5). Polypeptide analysis by sodium dodecyl sulphate gel electrophoresis indicates that at least 15 protein components exist in the oxygen-evolving PSII complex. The chemical characteristics and functional roles of some of these components have been clarified. Manganese atoms and calcium and chloride ions are essential inorganic components for oxygen evolution. The components of the oxygen-evolving PSII complex are listed in Table 1, while the model for the complex is presented in Fig. 1 and will be described in Section 5.

2. PROTEIN COMPONENTS OF OXYGEN-EVOLVING PSII COMPLEX

Several groups have attempted to disintegrate the oxygen-evolving PSII complex into smaller active complexes. By treating of thylakoids with digitonin followed by ion exchange chromatography, Satoh (1983) isolated the PSII core complex composed of five proteins of 47 kDa, 43 kDa, 34 kDa, 32 kDa and 9 kDa, which was active in the photochemical reaction, but not in oxygen evolution (6). Later, Tang and Satoh (1985) using the same preparation procedures but at a different pH, prepared the oxygen-evolving PSII core complex composed of the five PSII core proteins and in addition the extrinsic 33-kDa protein (7). Similar complexes were prepared from spinach PSII particles (8) and cyanobacterial thylakoids (9) by treatment with octylglucoside followed by centrifugation on a sucrose gradient; these preparations contained 3-4 Mn atoms per oxygen-evolving unit. The finding that this type of PSII core complex can oxidize water to produce molecular oxygen under certain conditions of ions and solutes suggests that these proteins constitute the minimum machinery for oxygen evolution. The other protein components in the oxygen-evolving PSII complex (Table 1) are assumed to function as light-harvesting, regulatory and ion concentrating components. However, Ljungberg et al. (10) reported that the oxygen-evolving PSII core complex additionally contains polypeptides of molecular masses of 7, 6.5, 5.5 and 5 kDa, which have not been characterized.

Amino acid sequences of the intrinsic PSII core proteins have been determined from the nucleotide sequences of their genes in the chloroplast genome, and their secondary structures have been postulated from the

Biggens, J. (ed.), Progress in Photosynthesis Research, Vol. I. ISBN 90 247 3450 9
© *1987 Martinus Nijhoff Publishers, Dordrecht. Printed in the Netherlands.*

hydropathy plot (11). The total amino acid sequence of the extrinsic 33-kDa protein has been determined by partial digestion and Edman degradation (12). Sequences of the extrinsic 18-kDa and 23-kDa proteins and some other not so well characterized proteins have been partially determined at the N-terminal side (Table 1; 13).

TABLE 1. Components in the oxygen-evolving PSII complex of spinach.

Component	Characteristics	Stoi-chio-metry	Amino acid sequence determination	Gene location***
PSII Core Complex Proteins				
47 kDa*	Chl-binding	1 (14)	Total; CtDNA**(15)	C (16)
43 kDa*	Chl-binding	1 (14)	Total; CtDNA (17)	C (16)
34 kDa*	D2 protein	1 (14)	Total; CtDNA (17)	C (16)
32 kDa*	Q_B (D1) protein	1 (14)	Total; CtDNA (18)	C (18)
9 kDa*]	Cyt b-559	2 (19)	Total; CtDNA (20,21)	C (20)
4 kDa*		2	Total; CtDNA (20)	C (20)
PSII Extrinsic Protein				
33 kDa*		1 (19)	Total (12)	N (22)
23 kDa		1 (19)	Partial (13)	N (22)
18 kDa		1 (19)	Partial (23)	N (22)
Light-Harvesting Pigment Proteins				
25-28 kDa	Chl a and b	?	Total; cDNA** (24)	N (25)
Other Protein Components				
24 kDa	Hydrophobic	?	---	?
22 kDa	Hydrophobic	?	Partial (13)	N (13)
10 kDa(I)	Hydrophobic	?	Partial (13)	N (13)
10 kDa(II)*	Phosphoprotein	?	Partial (26)	?
5 kDa *	Hydrophilic	?	Partial (13)	N (13)
Pigments and Other Components				
Chl a*+b		200 (19)		
Pheophytin a*		2 (27)		
Plastoquinone*		3 (19)		
Lipids		300 (28)		
Mn *		4 (19)		

*These components constitute the oxygen-evolving PSII core complex (7,8). **The amino acid sequences of the proteins were predicted from the nucleotide sequences of their genes in the chloroplast DNA or cDNA. ***C and N stand for chloroplast and nuclear genomes, respectively.

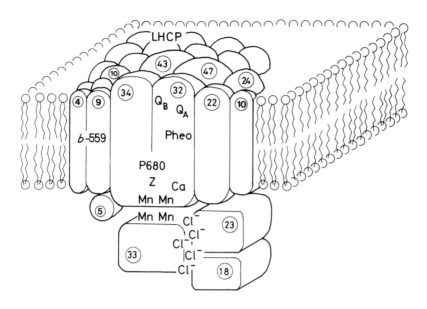

FIGURE 1. A model for the oxygen-evolving PSII complex. Q_A and Q_B are the primary and secondary quinone acceptors, respectively. ^AZ is the primary electron donor to P680. LHCP is the light-harvesting chl protein.

3. FUNCTIONS OF THE PROTEIN COMPONENTS IN THE OXYGEN-EVOLVING PSII COMPLEX

There has been some dispute as to the identity of the reaction center proteins in the PSII core complex. In early studies involving fragmentation of the PSII core complex with detergents, only the 47-kDa and 43-kDa proteins were found to carry chlorophyll (Chl) a. With this limited information, it was first proposed that the 47-kDa protein was the site of the reaction center (29-32). In a chlorophyll-bound form this protein fluorescenced at 695 nm at 77 K, which seemed to be characteristic of the reaction center.

Another argument concerning the identity of the reaction center protein(s) is based on the homology of amino acid sequences and nucleotide sequences of their genes between the 32-kDa and 34-kDa proteins of PSII core complex and the L and M subunits of the bacterial reaction center (see 33). Similarity between them is also suggested by the herbicide-binding characteristics of the 32-kDa protein and the L subunit (34,35). Recently, Satoh has isolated a complex containing the 34-kDa, 32-kDa and 9-kDa proteins, and two pheophytin a and five Chl a molecules, in which pheophytin a was reduced by light in the presence of dithionite (36). These findings strongly suggest that the 32-kDa and 34-kDa proteins constitute the PSII reaction center, as do the L and M subunits of the bacterial reaction center, whereas the 47-kDa and 43-kDa proteins are the antenna pigment proteins. The 9-kDa and 4-kDa proteins in the PSII core complex are apoproteins of cytochrome b-559 (20,21). No function of this cytochrome has been well demonstrated in either photochemical reaction or oxygen evolution.

The three extrinsic proteins of 33 kDa, 23 kDa and 18 kDa contribute to oxygen evolution (2,37). They have been purified and their physico-chemical characteristics well characterized (38-40). A technique for selective removal and rebinding of these proteins has made it possible to discriminate the functional roles of individual proteins (4). The 33-kDa protein is necessary to preserve Mn in the functional site; at low Cl^- concentrations below 100 mM, two of the four Mn atoms are lost without this protein (41-43). It is also necessary for the oxygen-evolution acti-vity at Cl^- concentrations lower than 100 mM (43). Its function in preserv-ing the Mn and oxygen-evolution activity can be partially substituted for by 150 mM Cl^-. The kinetic analysis suggests that a dark step in the oxygen-evolving reaction is slowed down by the removal of the 33-kDa pro-tein; the step of oxygen release, i.e., the S_3 to S_0 transition is retarded from 5 ms to 15 ms (44). The change in the time of the S_3 to S_0 transition was confirmed by the kinetics of UV absorption change due to oxidation-reduction reactions of Mn (H.J. van Gorkom, unpublished).

Two functions of the 23-kDa protein have been proposed. One is the trapping of Ca^{2+}, an essential component in oxygen evolution, at the func-tional site (45). In the PSII complex depleted of this protein, Ca^{2+} is released from the functional site under illumination (46,47). Another function of this protein is to reduce the requirement for Cl^-, also an essential factor in oxygen evolution, from 30 mM to 10 mM (48,49). Whether this effect in reducing the Cl^- requirement is caused by the protein itself or by the binding of Ca^{2+} can be resolved by step-wise addition

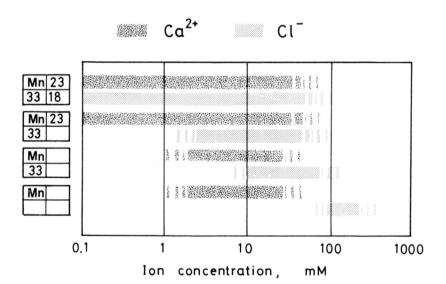

FIGURE 2. Changes in optimal concentrations of Ca^{2+} and Cl^- in the oxygen-evolving PSII complex upon removal of the extrinsic proteins.

of Ca^{2+} and the 23-kDa protein. When the PSII particles depleted of Ca^{2+} and the 23-kDa protein were supplemented with Ca^{2+} alone, the oxygen-evolution activity was enhanced with no change in the Cl^--requirement, while further supplementation of the 23-kDa protein reduced the optimal Cl^- concentration from 30 mM to 10 mM. Therefore, the effects of the 23-kDa protein to tightly trap Ca^{2+} at the functional site and to reduce the Cl^- requirement are two independent functions of the 23-kDa protein.

The 18-kDa protein sustains oxygen evolution at low Cl^- concentrations such as below 1 mM (49,50). Since the 18-kDa protein binds to the PSII complex via the 23-kDa protein (51), the effect of the 18-kDa protein is observed only when the 23-kDa protein is present (50). Since the Cl^- effect is observed in the absence of the 18-kDa protein as well as the other two extrinsic proteins, the functional binding site of Cl^- is supposed to be one of the intrinsic proteins. How the 18-kDa protein changes the binding affinity of Cl^- to the functional site is still in question.

In recent years, there have been a great number of studies published concerning the role of the three extrinsic proteins, sometimes contradictory. Most confusion has arisen from incorrect selection of concentrations of Ca^{2+} and Cl^- in the reaction medium. Fig. 2 summarizes the concentration ranges of Ca^{2+} and Cl^- suitable for biochemical and physicochemical measurements related to oxygen evolution using PSII complexes with different compositions of the extrinsic proteins.

TABLE 2. Proposed functions of the proteins

Protein	Function	Ref.
PSII Core Complex		
47 kDa]	Core antenna	29-31
43 kDa		29-31
34 kDa]	Reaction center	36
32 kDa		
9 kDa]	Cyt b-559	20
4 kDa		
Extrinsic Proteins		
33 kDa	Mn stabilizer	41-43
	Decrease in Cl^- requirement	49
	Acceleration of $S_3 \longrightarrow S_0$	44,58
24 kDa	Ca^{2+} trap	45,47
	Decrease in Cl^- requirement	48,49
18 kDa	Decrease in Cl^- requirement	49,50
Hydrophobic Proteins		
24 kDa	Minor Chl-binding protein	52
22 kDa	Binding of 23-kDa protein	53

Functions of other protein components are not clear. Dunahay et al. (52) reported that there are at least two minor chlorophyll-binding proteins with molecular masses in the region of 20-24 kDa. Ljungberg et al (53) proposed that the hydrophobic 22-kDa and 10-kDa proteins constitute the binding site of the extrinsic 23-kDa protein. Proteins of 10 kDa (phosphoprotein) and hydrophilic 5 kDa have also been isolated (10,26), but their functions are not known.

The mode of action of Cl^- and Ca^{2+} in oxygen evolution has been investigated by kinetic measurements. Chloride ions are required for the transition from S_2 to S_3 (54,55). Calcium ions are required for the reduction of Z^+ by the S states (46,56). From a study of delayed fuorescence, Boussac et al. (57) proposed that Ca^{2+} depletion inhibits the transition from S_3 to S_0, that is, the Z^+ reduction by S_3.

Proposed functions of the protein components and ions are listed in Table 2 and Fig. 3.

4. MANGANESE CLUSTER IN THE OXYGEN-EVOLVING PSII COMPLEX

The oxygen-evolving PSII complex contains four Mn atoms which are supposed to form a cluster which acts as a catalytic center for oxidation of water. The characteristics of the Mn cluster have been studied by physical techniques such as EPR, NMR, EXAFS and optical measurements. However, the Mn cluster has not been biochemically characterized, e.g., its binding site is not known.

The finding that only the S_2 state gives rise to a multiline EPR signal centered at g=2 (58-60) suggests that the physical properties of this state are distinct from those of the other S states. A similar multiline signal is observed in synthesized compounds, and Dismukes and his colleagues suggest that the S_2 state corresponds to either a Mn

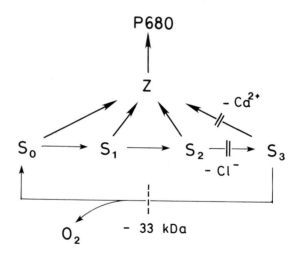

FIGURE 3. Electron transport between the electron carriers and the S-state transitions in the oxygen-evolving PSII complex and the site of inhibition by elimination of ions and the 33-kDa protein.

dimer in the form of Mn(III)Mn(IV) or a Mn tetramer in the form of
3Mn(III)Mn(IV) (58,61). Another EPR signal at g=4.1 related to the S_2
state was discovered by two groups (62,63). This signal was first regarded
as originating from an intermediate electron carrier between the S state
and the reaction center at a transient state between S_1 and S_2 (62,63).
However, Zimmermann and Rutherford (64) have recently suggested that the
multiline and g=4.1 signals both arise from the S_2 state but at different
magnetic states of Mn atoms.

The absorbance change in the infrared region (65) and the rate of
spin-lattice relaxation of proton NMR (66) have also suggested that the
transition from S_1 to S_2 results from an oxidation of Mn(III) to Mn(IV).

In a study of flash-induced absorbance change in the ultraviolet
region, Dekker et al. (67) depicted the absorbance change due to the oxidation and reduction of Mn atoms. In their model they proposed that each
of the three transitions from S_0 to S_3 corresponds to an oxidation of
Mn(III) to Mn(IV) and the transition from S_3 to S_0 to reduction of 3Mn(IV)
to 3Mn(III); an oxidation sequence of [1,1,1,-3] upon each S-state transition from S_0 to S_3 and back to S_0. However, Lavergne (68) claimed that
the experimental data of Dekker et al. can be better explained by an oxidation sequence of [0,1,0,-1].

The ligand environment of Mn has been studied by EXAFS (extended X-ray
absorption fine structure) spectrometry (69). The results suggest that
at the S_1 state at least two Mn atoms exist as a dimer separated by 2.7
Å and that N or O atoms form a ligand bond with the Mn. No evidence for
the binding of Cl^- to Mn was found by this technique.

5. ORGANIZATION OF OXYGEN-EVOLVING PSII COMPLEX

Fig. 1 shows a model for the oxygen-evolving PSII complex embedded
in the thylakoid membrane. The amino acid sequences and the hydropathy
plots of the proteins indicate that all proteins in the PSII core complex
(47-kDa, 43-kDa, 34-kDa, 32-kDa, 9-kDa and 4-kDa proteins) span the membrane (15,17,26). Although the amino acid sequences of the hydrophobic
proteins of 24 kDa, 22 kDa, 10 kDa (I), 10 kDa (II) have been only partially determined, they are assumed to span the membrane. The amino acid
sequence of the hydrophilic 33-kDa protein (12) indicates that this protein
does not contain a trans-membrane region. The other hydrophilic proteins
of 23 kDa and 18 kDa are assumed not to span the membrane, since they
are released from the complex by a high-salt medium. Therefore, these
three hydrophilic proteins are regarded as extrinsic proteins. The partial
amino acid sequence of the 5-kDa protein (13) suggests that this protein
is also an extrinsic protein. It is most likely that there are some other
polypeptides in the PSII complex (10). However, they are not included
in the model in Fig. 1, since they have not yet been isolated or characterized.

The arrangement of the proteins in the oxygen-evolving PSII complex
is not well understood. The core antenna pigment proteins of 47 kDa and
43 kDa are thought to be located close to the reaction center composed
of the 34-kDa and 32-kDa proteins, since they transfer the electronic
excitation energy from the light-harvesting chlorophyll a/b protein to
the reaction center. The site of binding of the Mn cluster, which constitutes the most essential part of the oxygen-evolving machinery, is not
known at all. However, it is reasonable to assume that the Mn cluster
is located close to the photochemical reaction center. The 33-kDa protein
is thought to be located close to the Mn cluster, since this protein has

a strong influence on the state of the Mn (41–44). The extrinsic 23-kDa protein is also considered to be located close to the Mn cluster, since without this protein the Mn cluster becomes accessible to reducing reagents (70). The hydrophobic 22-kDa and/or 10-kDa protein constitutes an anchor for the extrinsic 23-kDa protein (53). The extrinsic 18-kDa protein is located on the 23-kDa protein, since the 18-kDa protein can specifically and functionally bind to the complex only when the 23-kDa protein is present (51). The localization of Ca^{2+} and Cl^- is not yet known. It is clear, however, that Ca^{2+} does not bind functionally to any of the extrinsic proteins (41), nor to the hydrophobic 10-kDa and 22-kDa proteins (53), since the effect of Ca^{2+} is observed in PSII preparations depleted of these proteins. Probably, Ca^{2+} binds to one of the instrinsic proteins in the PSII core complex.

REFERENCES

1. Berthold, D.A., Babcock, G.T. and Yocum, C.F. (1981) FEBS Lett. 134, 231–234
2. Kuwabara, T. and Murata, N. (1982) Plant Cell Physiol. 23, 533–539
3. Andersson, B. and Åkerlund H.-E. (1978) Biochim. Biophys. Acta 503, 462–472
4. Murata, N. and Miyao, M. (1985) Trends Biochem. Sci. 10, 122–124
5. Cramer, W.A., Widger, W.R., Herrmann, R.G. and Trebst, A. (1985) Trends Biochem. Sci. 10, 125–129
6. Satoh, Ki. (1983) in the Oxygen Evolving System of Photosynthesis (Inoue, Y., Crofts, A.R., Govindjee, Murata, N., Renger, G. and Satoh, K., eds.), pp. 27–38, Academic Press, Tokyo
7. Tang, X.-S. and Satoh, Ki. (1985) FEBS Lett. 179, 60–64
8. Ikeuchi, M., Yuasa, M. and Inoue, Y. (1985) FEBS Lett. 185, 316–322
9. Satoh, Ka., Ohno, T. and Katoh, S. (1985) FEBS Lett. 180, 326–330
10. Ljungberg, U., Henrysson, T., Rochester, C.P., Åkerlund, H.-E. and Andersson, B. (1986) Biochim. Biophys. Acta 849, 112–120
11. Kyte, J. and Doolittle, R.F. (1982) J. Mol. Biol. 157, 105–132
12. Oh-oka, H., Tanaka, S., Wada, K., Kuwabara, T. and Murata, N. (1986) FEBS Lett. 197, 63–66
13. Murata, N., Kajiura, H., Fujimura, Y., Miyao, M., Murata, T., Watanabe, A. and Shinozaki, K. (1986) in these proceedings
14. Satoh, Ki. (1985) Photochem. Photobiol. 42, 845–853
15. Morris, J. and Herrmann, R.G. (1984) Nucl. Acids Res. 12, 2837–2850
16. Westhoff, P., Alt, J. and Herrmann, R.G. (1983) EMBO J. 2, 2229–2237
17. Alt, J., Morris, J., Westhoff, P. and Herrmann, R.G (1984) Curr. Genet. 8, 597–606
18. Zurawski, G., Bohnert, H.J., Whitfeld, P.R. and Bottomley, W. (1982) Proc. Natl. Acad. Sci. USA 79, 7699–7703
19. Murata, N., Miyao, M., Omata, T., Matsunami, H. and Kuwabara, T. (1984) Biochim. Biophys. Acta 765, 363–369
20. Herrmann, R.G., Alt, J., Schiller, B., Widger, W.R. and Cramer, W.A. (1984) FEBS Lett. 176, 239–244
21. Widger, W.R., Cramer, W.A., Hermodson, M., Meyer, D. and Gullifor, M. (1984) J. Biol. Chem. 259, 3870–3876
22. Westhoff, P., Jansson, C., Klein-Hitpaβ, L., Berzborn, R., Larsson, C. and Bartlett, S.G. (1985) Plant Mol. Biol. 4, 137–146
23. Kuwabara, T., Murata, T., Miyao, M. and Murata, N. (1986) Biochim.

Biophys. Acta 850, 146-155
24. Coruzzi, G., Broglie, R., Cashmore, A. and Chua, N.-H. (1983) J. Biol. Chem. 258, 1399-1402
25. Kung, S.D., Thornber, J.P. and Wildman, S.G. (1972) FEBS Lett. 24, 185-188
26. Farchaus, J. and Dilley, R.A. (1986) Arch. Biochem. Biophys. 244, 94-101
27. Murata, N., Araki. S., Fujita, Y., Suzuki, K., Kuwabara, T. and Mathis, P. (1986) Photosyn. Res. 9, 63-70
28. Farineau, N., Guillot-Salomon, T., Tuquet, C. and Farineau, J. (1984) Photochem. Photobiol. 40, 387-390
29. Camm, E.L. and Green, B.R. (1983) J. Cell. Biochem.. 23, 171-179
30. Camm, E.L. and Green, B.R. (1983) Biochim. Biophys. Acta 724, 291-293
31. Nakatani, H.Y., Ke, B., Dolan, E. and Arntzen, C.J. (1984) Biochim. Biophys. Acta 765, 347-352
32. Yamagishi, A. and Katoh, S. (1984) Biochim. Biophys. Acta 765, 118-124
33. Kyle, D.J. (1985) Photochem. Photobiol. 41, 107-116
34. de Vitry, C. and Diner, B. (1984) FEBS Lett. 167, 327-331
35. Brown, A., Gilbert, C., Guy, R. and Arntzen, C.J. (1984) Proc. Natl. Acad. Sci. USA 81, 6310-6314
36. Satoh, Ki. (1986) in these proceedings
37. Yamamoto, Y., Doi, M., Tamura, N. and Nishimura, M. (1981) FEBS Lett. 133, 265-268
38. Kuwabara, T. and Murata, N. (1979) Biochim. Biophys. Acta 581, 228-236
39. Kuwabara, T. and Murata, N. (1982) Biochim. Biophys. Acta 680, 210-215
40. Jansson, C. (1984) in Advances in Photosynthesis Research (Sybesma, C., ed.) Vol. 1, pp. 375-378, Martinus Nijhoff/Dr. W. Junk Publishers
41. Miyao, M. and Murata, N. (1984) FEBS Lett. 170, 350-354
42. Ono, T. and Inoue, Y. (1984) FEBS Lett. 168, 281-286
43. Kuwabara, T., Miyao, M., Murata, T. and Murata, N. (1985) Biochim. Biophys. Acta 806, 283-289
44. Miyao, M., Murata, N., Maison-Peteri, B., Boussac, A., Etienne, A.-L. and Lavorel, J. (1986) in these proceedings
45. Ghanotakis, D.F., Topper, J.N., Babcock, G.T. and Yocum, C.F. (1984) FEBS Lett. 170, 169-173
46. Dekker, J.P., Ghanotakis, D.F., Plijter, J.J., van Gorkom, H.J. and Babcock, G.T. (1984) Biochim. Biophys. Acta 767, 515-523
47. Miyao, M. and Murata, N. (1986) Photosyn. Res. in press
48. Andersson, B., Critchley, C., Ryrie, I.J., Jansson, C., Larsson, C. and Anderson, J.M. (1984) FEBS Lett. 168, 113-117
49. Miyao, M. and Murata, N. (1985) FEBS Lett. 180, 303-308
50. Akabori, K., Imaoka, A. and Toyoshima, Y. (1984) FEBS Lett. 173, 36-40
51. Miyao, M. and Murata, N. (1983) Biochim. Biophys. Acta 725, 87-83
52. Dunahay, T.G. and Staehelin, L.A. (1986) Plant Physiol. 80, 429-434
53. Ljungberg, U., Åkerlund, H.-E. and Andersson, B. (1986) Eur. J. Biochem., in press
54. Itoh, S., Yerkes, C.T., Koike, H., Robinson, H.H. and Crofts, A.R. (1984) Biochim. Biophys. Acta 766, 612-622
55. Theg, S.M., Jursinic, P.A. and Homann, P.H. (1984) Biochim. Biophys. Acta 766, 636-646
56. Ghanotakis, D.F., Babcock, G.T. and Yocum, C.F. (1984) FEBS Lett. 167, 127-130
57. Boussac, A., Maison-Peteri, B., Vernotte, C. and Etienne, A.-L. (1985) Biochim. Biophys. Acta 808, 225-230
58. Dismukes, G.C. and Siderer, Y. (1980) FEBS Lett. 121, 78-80

59. Hansson, Ö and Andréasson, L.-E. (1982) Biochim. Biophys. Acta 679, 261-268
60. Brudvig, G.W., Casey, J.L. and Sauer, K. (1983) Biochim. Biophys. Acta 723, 366-371
61. Dismukes, G.C., Ferris, K. and Watnick, P. (1982) Photobiochem. Photobiophys. 3, 243-256
62. Casey, J.L. and Sauer, K. (1984) Biochim. Biophys. Acta 767, 21-28
63. Zimmermann, J.-L. and Rutherford, A.W. (1984) Biochim. Biophys. Acta 767, 160-167
64. Zimmermann, J.-L. and Rntherford, A.W. (1986) Biochemistry, in press
65. Dismukes, G.C. and Mathis, P. (1984) FEBS Lett. 178, 51-54
66. Srinivasan, A.N. and Sharp, R.R. (1986) Biochim. Biophys. Acta 850, 211-217
67. Dekker, J.P., van Gorkom, H.J., Wensink, J. and Ouwehand, L. (1984) Biochim. Biophys. Acta 767, 1-9
68. Lavergne, J. (1986) Photobiochem. Photobiophys. 43, 311-317
69. Yachandra, V.K., Guiles, R.D., McDermott, A., Britt, R.D., Dexheimer, S.L., Sauer, K. and Klein, M.P. (1986) Biochim. Biophys. Acta 850, 324-332
70. Ghanotakis, D.F., Topper, J.N. and Yocum, C.F. (1984) Biochim. Biophys. Acta 767, 524-531

KINETICS AND STRUCTURE ON THE HIGH POTENTIAL SIDE OF PHOTOSYSTEM II.

G.T. BABCOCK[a], T.K. CHANDRASHEKAR[a], D.F. GHANOTAKIS,[a,b]
C.W. HOGANSON,[a] P.J. O'MALLEY,[a] I.D. RODRIGUEZ[a] AND C.F. YOCUM[b]
[a]Department of Chemistry, Michigan State University, East Lansing,
Michigan 48824 [b]Division of Biological Sciences, University of Michigan,
Ann Arbor, Michigan 48109

1. Introduction

    Recent research clearly indicates that photosystem II (PSII) and the
oxygen evolving complex (OEC) occur as a single multisubunit complex, the
PSII/OEC, incorporated into the photosynthetic membrane (see refs. 1-3 for
reviews). Compared to other integral membrane proteins, the polypeptide
structure of the PSII/OEC is moderately complex. Considerable progress
has been made in identifying and characterizing the principal electron
transport cofactors which catalyze primary charge separation, intermediate
electron transfer and charge storage and water-splitting chemistry.
Several aspects of structure and kinetics in the PSI/OEC are considered
here including the stabilization of stored oxidizing equivalents in the
OEC and the structure, kinetics and local protein environment of the
intermediate redox species, Z which catalyzes electron transfer between
P680 and the Mn cluster(s) of the OEC.

2. Kinetic stabilization of the Stored Oxidizing Equivalents in the OEC.

    A remarkable aspect of the state S_2 and S_3 in the OEC is that they are
reasonably stable (tens of seconds) despite their necessarily high redox
potentials. This implies that their stabilization is achieved by kinetic
means. Recent work implicates the peripheral polypeptides with molecular
weights of 17, 23 and 33 kDa in this process. Thus, in addition to their
well-known roles in facilitating manganese, calcium and chloride binding
to the intrinsic polypeptides of the PSII/OEC, the peripheral polypeptides
appear to form a protective cap over the OEC which limits access of
endogenous and most exogenous reductants.
    Figure 1 shows EPR kinetic data on the Z^+ species which implicate the
33 kDa polypeptide in this capping mechanism. In part 1 of the Figure Z^+
decay traces in tris-washed PSII particles, which had the 17, 23 and 33
kDa polypeptides, are shown as a function of the concentration of the
endogenous reductant, benzidine (BZ). As the concentration of BZ
increases, the decay halftime decreases. This behavior is typical of a
classic second order reaction in which the donor BZ has direct access to
the Z^+ species [4]. From the data, a second order rate constant of 1.1 x
10^6 $\underline{M}^{-1}$ s^{-1} may be calculated [5]. With only the 17 and 23 kDa
polypeptides removed, part 2 of Fig. 1, much different behavior is
observed. The decay of Z^+ is short, but added BZ has no effect on the
decay half time; only the signal amplitude progressively declines as the
BZ concentration increases. This indicates that the access of the
reductant to the Z^+ species is sharply attenuated, presumably by the 33
kDa polypeptide which remains bound after salt-washing. If one assumes a
model in which BZ is able to reduce higher S state in the dark time
between flashes, a rate constant of ~3 x 10^3 $\underline{M}^{-1}$ s^{-1} for BZ accessibility

Biggens, J. (ed.), Progress in Photosynthesis Research, Vol. I. ISBN 90 247 3450 9
© *1987 Martinus Nijhoff Publishers, Dordrecht. Printed in the Netherlands.*

to the water side of PSII is calculated [5]. This represents an increase by a factor of 3% in the kinetic stabilization of high potential oxidizing equivalents in the PSII/OEC by the 33 kDa polypeptide. Tamura et al have reported similar stabilization by the peripheral polypeptides to the reductant TMPD. [6]. From their data, a shielding factor of 100 may be estimated.

In addition to the 33 kDa polypeptide, the 17 and 23 kDa species are also inplicated in the capping function. Ghanotakis et al reported that manganese in the OEC is susceptible to reduction and extraction when salt washed PSII's are incubated with exogenous donors [7]. The reductant-induced extraction is not inhibited by $CaCl_2$ addition, even though this salt restores O_2 evolving activity. Similarly, Schröder and Åkerlund have found that the 17 and 23 kDa polypeptides limit access of H_2O_2 to the OEC in everted thylakoid membrane vesicles [8].

3. Time-Resolved EPR spectrum of $Z^{+}_{\cdot}$ in O_2-evolving PSII particles.

The $Z^{+}_{\cdot}$ species has been characterized spectrally most extensively in inhibited preparations, where it has a lineshape identical to the so-called Signal II spectrum [3]. Only in the original publication on its

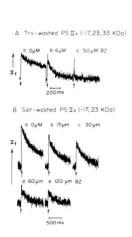

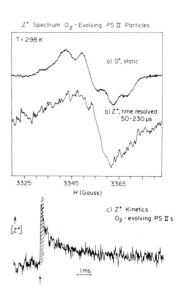

FIGURE 1: Z^{+} kinetic transients in tris-washed (upper) and salt-washed (lower) PSII particles at the indicated benzidine (BZ) concentrations.

FIGURE 2: a) Static Signal IIs (D^{+}) EPR spectrum; b) time-resolved Z^{+} EPR spectrum in O_2-evolving PSII fragments and c) Z^{+} kinetic trace in O_2-evolving PSII fragments.

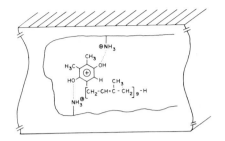

FIGURE 3: Postulated model for the structure and environment for the D^+/Z^+ species.

observation in untreated preparation is a kinetic spectrum of this species under O_2-evolving conditions presented [9]. However, this work was done in unfractionated thylakoids and severe overlap with $P700^+$ obscures a considerable part of the spectrum. Because $P700^+$ decay transients occur over the same time range as $Z^{+\cdot}$, deconvolving the contributions of the two is particularly difficult. With the development of O_2-evolving PSII membrane fragments, the $P700^+$ spectral contribution is eliminated and the way is clear to obtaining an unobscured, time-resolved $Z^{+\cdot}$ spectrum. This is of particular importance owing to the recent observation that the shape of the optical difference spectrum, $Z^{+\cdot}-Z$, is dependent upon the integrity of the OEC. [10].

Figure 2 shows the approach and presents a low resolution $Z^{+\cdot}$ EPR spectrum in O_2 evolving PSIIs (Fig. 2b). A higher resolution spectrum of the Signal IIs species, $D^{+\cdot}$, is also shown for comparison (Fig. 2a). The kinetic trace for $Z^{+\cdot}$ generation and decay in Fig. 2c is multiphasic with an overall halftime of ~800 μs. By using boxcar techniques, the signal intensity in the time window 50-230 μs following the light pulse was integrated as shown by the boxed area in Figure 2c. The time-resolved spectrum in Figure 2b was acquired by monitoring the integrated area as a function of magnetic field [11]. A comparison of the time resolved $Z^{+\cdot}$ with the static Signal IIs spectra shows reasonable agreement, particularly in overall line width and g-value, considering the low resolution conditions of the kinetic spectrum. Analogous experiments at higher resolution are underway.

The preliminary result in Fig. 2 that the EPR spectrum of $Z^{+\cdot}$ in O_2 evolving PSII preparations is similar to the $Z^{+\cdot}$ spectrum in inhibited preparations does not necessarily contradict the optical data noted above which indicate preparation dependent $Z^{+\cdot}-Z$ optical difference spectra [10]. The EPR experiment monitors only a single state, $Z^{+\cdot}$, whereas the optical experiment is dependent on the properties of at least four states: the ground and excited states of both Z and $Z^{+\cdot}$. A shift in the excited state energy of Z upon inhibition of O_2 evolution, for example, would influence the optical difference spectrum but have no noticeable effect on the $Z^{+\cdot}$ EPR spectrum. Given the well-known occurrence of electrochromic shifts in membrane bound chromophores such a situation is likely and may rationalize the apparent difference between optical and EPR results on $Z^{+\cdot}$.

4. ENDOR spectroscopic characterization of the Signal II species.

An analysis of the EPR and ENDOR properties of the two PSII radicals, $D^{+\cdot}$ and $Z^{+\cdot}$, along with considerations of the high redox potential requirement for $Z^{+\cdot}$ in its reaction with the Mn cluster(s) in the OEC, led to the suggestion that both radicals were plastosemiquinone cations in a protein environment designed to prevent deprotonation of the radical intermediates [12], Figure 3. We have investigated this hypothesis by carrying out a more detailed ENDOR analysis of the $D^{+\cdot}$ radical and a study of its H_2O/D_2O exchange properties.

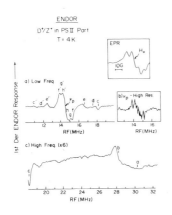

ENDOR
D'/Z' in PS II Part
T = 4K

a) Low Freq

b)νp - High Res

c) High Freq (x6)

FIGURE 4: ENDOR spectrum
of the D⁺ species at 4 K.
Insets show EPR and high
resolution ENDOR spectra.

Figure 4 shows the ENDOR spectrum of D⁺ at 4.2 K in PSII particles the resonances in the 28-30 MHz range have been reported previously and attributed to the methyl group at the 2 position on the PQH₂⁺ ring [13]. Controversy exists over this assignment [14], but we consider it to be the most likely. Further details on this point will be presented elsewhere (in preparation). In addition to the large – CH₃ coupling, hyperfine interaction with other 'H on the PQH₂⁺ ring are expected and a number of resonances are observed in the 10-20 MHz range. (a-e in Fig. 4) Resonances a-c occur in the matrix region around ν_p, which is seen to be highly structured in the high resolution inset in Fig. 4, and arise primarily from dipolar interactions with the protons in the surrounding protein environment. Resonances d and e have complex lineshapes and we have used temperature variation and H/D exchange to study these more thoroughly.

Figure 5 shows the D⁺ ENDOR spectrum at 115K before (Fig. 5a) and after (Fig. 5b) H₂O/D₂O exchange. The exchange process is slow, as predicted [12,13], and is described in more detail by Chandrashekar et al [15] and Rodriguez et al [16].

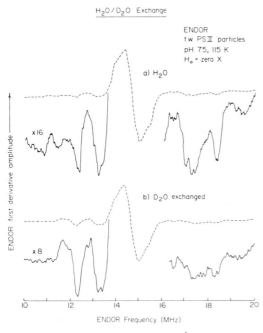

H₂O / D₂O Exchange

ENDOR
t.w. PS II particles
pH 7.5, 115 K
H₀ = zero X

a) H₂O

x16

b) D₂O exchanged

x8

ENDOR Frequency (MHz)

FIGURE 5: ENDOR spectra of D⁺ before (a) and after (b) H₂O/D₂O exchange. T=115 K.

In Fig. 5a, the d and e resonances of Fig. 4 are seen to consist of several lines; moreover, a new feature is observed as a trough near 19.6 MHz (coupling = 9.4 MHz). Upon H₂O/D₂O the resonance at 11.2 MHz (coupling = 7.1 MHz) is absent and the derivative shape feature at ~ 13 MHz decreases in intensity relative to the axially symmetric set of resonances at 11.7 and 12.2 MHz ($a_{||}$ = 5.9 MHz, $a_\perp$ = 5.0 MHz) to reveal a rhombic set of three resonances (a_1 = 2.5 MHz, a_2 = 3.2 MHz, a_3 = 3.9 MHz). We interpret the spectrum in the 11-13.5 mHz region to indicate that resonances due to hyperfine coupling with (a) an exchangeable, hydrogen-bonded proton, (b) a methyl group and (c) as alpha proton occur. The hydrogen-bonded proton and the methyl group are both expected to have axially symmetric hyperfine interactions, as observed, whereas the alpha-proton interaction is expected to be rhombic [17, 18]. For the spin density distribution postulated for the D^+/Z^+ radicals, weak coupling to the alpha proton at the ring 6 position and to the methyl at the ring 3 position is expected [12, 13], consistent with the small hyperfine couplings in Fig. 5. Our proposed assignments for the resonances in this region, for the resonance at 19.6 MHz (coupling = 9.4 MHz) and for the higher frequency resonances are given in Table 1. Model compound data which support these assignments, particularly that of the alpha proton and of the -OH resonance, are given by Chandrashekar et al in these proceedings [19].

TABLE 1

Tentative Assignments of D^+ ENDOR lines within the PQH_2^+ Model[a]

1,4-OH	H-bonded	2-CH$_3$	3-CH$_3$	6-H
9.4[b]	3.5 $(a_\perp)$[c] 7.1 $(a_\parallel)$[c]	27.2 $(a_\perp)$ 31.4 $(a_\parallel)$	5.0 $(a_\perp)$ 5.9 $(a_\parallel)$	2.5 (a_1) 3.2 (a_2) 3.9 (a_3)

a) couplings in MHz
b) may be exchangeable
c) exchangeable

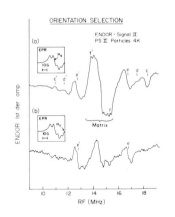

ORIENTATION SELECTION

ENDOR - Signal II
PS II Particles 4K

(a)
EPR

(b)
EPR

Matrix

ENDOR 1st der amp

RF (MHz)

FIGURE 6: Orientation of selected ENDOR spectra of D^+. The fields at which the two spectra were recorded are shown in the insets.

The observation of a hydrogen-bonded proton in the Z^+ ENDOR spectrum is predicted by the model in Fig. 3. Moreover, we expect the proton involved to lie in the plane of the quinone ring. Owing to the g-anisotropy of quionone radicals, an orientation selection experiment is possible to test this prediction [18]. The g tensor of a quionone radical is oriented with respect to the plane of the ring such that the g_x and g_y components are in-plane and g_z is perpendicular to the ring plane. Moreover, the g_z value is near the free electron value (2.0023) whereas g_x and g_y are larger. This leads to a situation in which radicals frozen such that their ring planes are perpendicular to the applied field contribute most strongly to the EPR spectrum as the high field side. By carrying out an ENDOR experiment at these magnetic field values, perpendicularly oriented molecules are selected. If the H-bond is in plane we expect to see only its perpendicular component. Figure 6 shows the experiment. The spectrum in Fig. 6a was recorded at the zero crossing and both perpendicular (here labelled aa') and parallel (cc') components of the H-bond are observed. Figure 6b is the spectrum recorded on the high field side and only the aa' perpendicular component is observed. The hydrogen bond we observe lies in the plane of the quinone ring.

The data presented above indicate that the ENDOR spectrum of the Signal II species can be interpreted well within the $PQH_2 \cdot$ model. Resonances due to both ring methyl groups, to the alpha proton and to the -OH proton can be observed and assigned and an in plane, exchangeable hydrogen bond is apparent. Despite this good agreement, several caveats must be made. (1) The ENDOR work has been done on the $D \cdot$ species, not the $Z \cdot$ species. Given the similar EPR spectra of these two radicals, however, we expect that the extrapolation to $Z \cdot$ is reasonable. (2) Hoff and coworkers [14] have noted difficulties in simulating the Signal II line shape within the above interpretation, although they agree with the $PQH_2^+ \cdot$ origin of the radical. The basis for this is being explored. (3) de Vitry et al [20] find insufficient amounts of PQ-9 in PSII preps to allow a ready identification of both D and Z with readily extractable PQH_2-9. This observation as well is under study.

Acknowledgement

This research was supported by NIH GM 37300, by the U.S. Department of Agriculture Competitive Research Grants Office and by the McKnight Foundation.

References

1. Ghanotakis, D.F. and Yocum, C.F. (1985) Photosynthesis Research 7, 97-114.
2. Govindjee, Kambara, T. and Coleman, W. (1985) Photochem. Photobiol. 42, 187-210.
3. Babcock, G.T. (1986) in New Comprehensive Biochemistry: Photosynthesis (Amesz, J., ed.) Elsevier-North Holland, submitted.
4. Yerkes, C.T. and Babcock, G.T. (1980) Biochim. Biophys. Acta 590, 360-373.
5. Ghanotakis, D.F., Babcock, G.T. and Yocum, C.F. (1984) Biochim. Biophys. Acta 765, 388-398.
6. Tamura, N., Radmer, R., Lantz, S., Cammarata, K. and Cheniae, G. (1986) Biochim. Biophys. Acta 850, 369-379.
7. Ghanotakis, D.F., Topper, J.N. and Yocum, C.F. (1984) Biochim. Biophys. Acta 767, 524-531.
8. Schroder, W.P. and Akerlund, H.-E. (1986) Biochim. Biophys. Acta 848, 359-363.
9. Blankeship, R.E., Babcock, G.T. Warden, J.T. and Sauer, K. (1975) FEBS Lett. 51, 287-293.
10. Weiss, W. and Renger, G. (1986) Biochim. Biophys. Acta 850, 173-183.
11. Hoganson, C.W., Demetriou, Y. and Babcock, G.T. (1986) These Proceedings.
12. Ghanotakis, D.F., O'Malley, P.J., Babcock, G.T. and Yocum, C.F. (1983) Academic Press, Tokyo, pp. 91-101
13. O'Malley, P.J., Babcock, G.T. and Prince, R.C. (1984) Biochim. Biophys. Acta 766, 283-288.
14. Brok, M., Ebskamp, F.C.R. and Hoff, A.J. Biochim. Biophys. Acta (1985) 809, 421-428.

15. Chandrashekar, T.K., Rodriguez, I.D., O'Malley, P.J. and Babcock, G.T. (1986) Photosynthesis Research in press.
16. Rodriguez, I.D., Chandrashekar, T.K. and Babcock, G.T. (1986) These Proceedings.
17. O'Malley, P.J. and Babcock, G.T. (1984) J. Chem. Phys. 80, 3912-3913.
18. O'Malley, P.J. and Babcock, G.T. (1986) J. Am. Chem. Soc. 108, 3995-4001.
19. Chandrashekar, T.K., O'Malley, P.J. and Babcock, G.T. (1986) These Proceedings.
20. de Vitry, C., Carles, C. and Diner, B.A. (1986) FEBS Lett. 196, 203-206.

ENDOR CHARACTERIZATION OF H_2O/D_2O EXCHANGE IN THE D^+Z^+ RADICAL IN PHOTOSYNTHESIS.

I. D. RODRIGUEZ, T. K. CHANDRASHEKAR, and G. T. BABCOCK, Department of Chemistry, Michigan State University, East Lansing, MI 48824

1. INTRODUCTION

EPR as well as ENDOR studies of model quinones in conjunction with in vivo experiments on chloroplasts and PSII particles support the idea that Signal II is a plastoquinone cation radical (1), consistent with optical data which lead to the same conclusion (2). The components of Signal II related to the primary donor to P680 can be divided as follows. Signal II_s is a stable radical and is designated as D^+. A less stable radical called Signal II_{vf} in oxygen evolving systems and Signal II_f in inhibited systems is designated as Z^+. Even though Z^+ and D^+ are functionally different, their EPR spectra as well as orientation in the membrane are identical. In the studies reported here, we used ENDOR spectroscopy to explore the structure and environment of the D^+/Z^+ radical. This technique is excellent for these studies because it can detect hydrogen bonded protons directly and is able to provide insight into the solvent accessibility of a radical site. This aspect of ENDOR is useful in light of the hydrogen-bonded structure postulated for the D^+/Z^+ radical.

2. MATERIALS AND METHODS

D_2O/H_2O exchange was done with tris washed PSII particle, (t.w. PSII) as follows:
a) Incubation of t.w. PSII particles in a buffer containing D_2O at pD = 7.5 (Hepes 50 mM; NaCl 10 mM) for up to 12 hours in the dark at 4°C.
b) Lyophilization of t.w. PSII particles in a buffer containing D_2O at pD = 7.5 and resuspension in D_2O.
c) Incubation of t.w. PSII particles in a buffer containing D_2O at pD = 6.0 (MES, 50 mM; NaCl 10 mM) for three days. The buffer was changed twice during the course of the exchange. Brief periods of room light (2-3 min) were given at 6 hours intervals during the incubation. We have found that freeze-thaw cycles accelerate the exchange process; this was done three times for the samples reported here.
d) Combination of a and b
The Signal II EPR lineshape did not change following these procedures in the D_2O exchange sample or in the H_2O control. Additional free radicals are generated in substantial amounts after long periods of dark, cold incubation (24 hrs. at pD = 7.5 and about 9 days at pD = 6.0). EPR and ENDOR spectra were recorded on samples that did not show any additional free radical. Spectra were recorded as described previously (3,4-7) with samples containing 3-6 mg Chl/ml.

3. RESULTS and DISCUSSION

The partially resolved lineshape of the Z^+/D^+ EPR spectrum has been assigned to the methyl group in position 2(6). PQH_2 hydroxyl protons and hydrogen bonded protons in the binding site should exchange with solvent water, if the site is accessible, whereas the $-CH_3$ protons, α-protons and $-CH_2$-protons from the isoprenoid chain on the ring are not expected to exchange under mild conditions. Fig. 1 shows the ENDOR spectrum of t.w. PSII particles incubated for 6 hours at pD = 7.5, freeze dried and resuspended in D_2O. The resonance at 7.1 MHz is absent in the D_2O exchangedsample compared to that of H_2O control. Another resonance at 3.5 MHz is also absent in the D_2O exchanged sample and an underlying rhombic

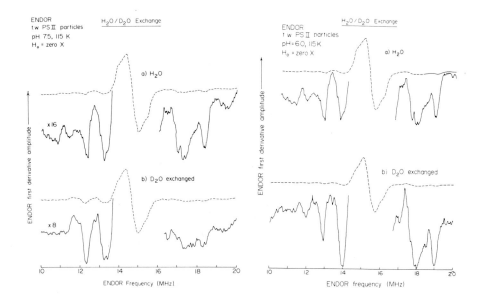

Figure 1. (Left) ENDOR spectrum of Signal II at pH/pD = 7.5. Exchange was carried out as described in methods.

Figure 2. (Right) ENDOR spectrum of Signal II at pH/pD = 6.0. Exchange was carried out as described in methods.

set of resonances becomes apparent. With the help of model quinones (5), we have assigned the 3.5 and 7.1 MHz resonances to hydrogen bonded protons. The axial nature of the hyperfine tensor and the fact that is is essentially traceless support this conclusion. Another resonance at 9.5 MHz is diminished in intensity in the D_2O exchanged sample. This has been tentatively assigned to the protons of the hydroxyl groups (see Chandrashekar et al, These Proceedings). If the sample is incubated at pD6 = 6.0 (Fig. 2), the resonance at 7.1 MHz decreases in intensity compared to that of the H_2O control, but is not completely absent as it is in the case of pD = 7.5. At lower pD's the exchange seems to be slower. It is apparent that the pD has an effect on the exchange and therefore on the configuration or protein structure of the system. At higher pD's (pH's) the system seems to be more open and susceptible to exchange than at lower pD's (pH's).

In the region around the free proton frequency (~ 14.7 MHz) a complex structure is observed (Fig. 3). In the protein binding site the radical is in a well-defined, highly structured environment. Such a situation will produce specific dipole-dipole interactions between the radical and the amino acid protons in the binding site and will lead to the resolved matrix spectrum (7). It appears that changes in line shape occur more quickly in the matrix region once the D_2O incubation has begun. Under conditions in which the D_2O exchange is not complete, as judged by the resonances in the 10.5 - 13 MHz region discussed above, several resonances are absent in the matrix region, indicating rapid exchange. A pH effect is also observed. Samples at different pH's (e.g. pH = 6 compared to pH = 7.5) show different

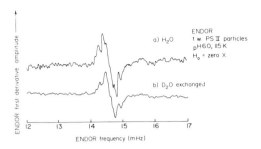

Figure 3. ENDOR of t.w. PSII particles (matrix region) at pH/pD = 6.0.

intensities and ENDOR transitions in this region. Because the matrix region contains resonances from protons more remote from the radical, it isnot too surprising that it responds more rapidly to the conditions in bulk water.

If the samples are incubated for long periods of time (up to two weeks at pD = 6, and up to 24 hrs. at pD = 7.5), another free radical is observed (Fig. 4). This is not a result of deuterium substitution or exchange, because it can be also seen in the H_2O control sample. If a sample containing this new radical is illuminated an EPR spectrum with g = 2.0035 and line width (ΔH) of 11 G is observed. If the sample is incubated in the dark for a period of about 3 hrs. the contribution of Signal II is absent and an EPR spectrum with g = 2.0026 and ΔH = 10G is obtained. The characteristics of this free radical correlate with those of a chlorophyll cation radical. The formation of the radical is a result, most likely, of denaturation of the protein structure. This radical can be also formed by using high concentrations of K_2IrCl_6 (~15 mM) and by illuminating t.w. PSII particles at -154°C in the presence of 5 mM K_2IrCl_6 or $K_3Fe(CN)_6$.

In conclusion, the results presented here indicate the following for the Z^+/D^+ free radical: a) exchange is slow (indicating a well shielded binding site for Z^+/D^+ species). b) the exchange is pH/pD dependent; and c) there is a hydrogen bond present in the system, in agreement with the proposed model for the Z^+/D^+ species.

REFERENCES

1) O'Malley, P.T.; Babcock, G. T. (1984) Biochem. Biophys. Acta, 765, 370-379.
2) Dekker, J. P., Van Gorkon, H. J.; Brok, M. and Ouwehand, L. (1984) Biochem. Biophys. Acta, 764, 301-309.
3) O'Malley, P. T., Babcock, G. T. and Prince, R. C. (1984) Biochem. Biophys. Acta, 766, 283-288.
4) O'Malley, P. T., Babcock, G. T. (1984) J. Chem. Phys. 80, 3912-3913.
5) O'Malley, P. T., Chandrashekar, T. K. and Babcock, G. T. (1985) in: "Antennas and reaction centers of photosynthetic Bacteria", (Michele-Beyerle, M. E. ed.) Vol. 42, pp. 359-344, Sprunger-Verlag, Berlin.

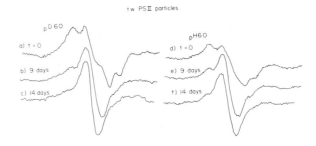

tw PSII particles

Figure 4. Signal II EPR spectra of samples incubated at pD/pH = 6.0 for the time indicated in the figure.

6) O'Malley, P. T. and Babcock, G. T. (1985) J. Am. Chem. Soc., Submitted.
7) Chandrashekar, T. K., O'Malley, P. T., Rodriguez, I. D. and Babcock, G. T. (1986) Photosynthesis Research, in press.

ACKNOWLEDGMENTS

We thank Professor C. F. Yocum for helpful discussions. This research was supported by NIH GM 37300, the Competitive Research Grant Office of the U.S. Dept. of Agriculture and the McKnight Foundation.

ENDOR CHARACTERIZATION OF THE $Z^{\cdot+}/D^{\cdot+}$ SPECIES IN PHOTOSYSTEM II AND RELEVANT MODEL COMPOUNDS.

T.K. CHANDRASKEKAR, P.J. O'MALLEY, I.D. ROGRIGUEZ, AND G.T. BABCOCK, DEPARTMENT OF CHEMISTRY, MICHIGAN STATE UNIVERSITY, EAST LANSING, MICHIGAN, 48824

1. INTRODUCTION

The free radical species which are involved on the oxidizing side of PSII, $Z^{\cdot+}$ and $D^{\cdot+}$, give rise to Signal II EPR spectra (see Ref. 1 for review). Both optical [2] and EPR [3] data suggest that these species are plastoquinone cation radicals in a specialized protein environment, although concerns over this assignment in terms of amounts of plastoquinone-9 in PSII preparations has been raised [4]. More detailed study of the $Z^{\cdot+}/D^{\cdot+}$ species is necessary to address this uncertainty as well as to characterize more extensively the structure of the radicals and their interactions with the local protein environment. ENDOR spectroscopy is well-suited to this task owing to its high information content. To date, most of the Signal II ENDOR work has focused on the large couplings which give rise to the partially-resolved EPR spectrum [5]. In the experiments reported here we have used H_2O/D_2O exchange to resolve congested areas in the 10-20 MHz region of the ENDOR spectrum in order to assign several of the observed resonances.

2. MATERIALS AND METHODS

PSII preparations were carried out according to well established procedures and D_2O exchanged as described by Rodriguez et al [6]. Model semiquinones were prepared by methods given in [5,7] and ENDOR spectroscopy was carried out on a Bruker ER200/ENDOR system [3].

3. RESULTS AND DISCUSSION

The 10-20 MHz ENDOR spectrum of $D^{\cdot+}$ at 115 K is shown in Figure 1a. In the 10.5-13.5 MHz a number of resonances are apparent. This region is simplified by H_2O/D_2O exchange (Fig. 1b) which eliminates a line at ~ 11 MHz and decreases the intensity of the derivative shape feature at 13 MHz considerably. We attribute the H_2O/D_2O exchange sensitive resonances to a hydrogen-bonded proton dipolar coupled to unpaired spin density in the radical. Fig. 2 shows a model system which exhibits a similar dipolar-coupled hydrogen bonded proton. The upper spectrum is of the 2-methyl-benzoquinone anion radical in isopropanol. At least four resonances are observed in the region between 10.5 and 14 MHz. The lower spectrum is of the same radical but now in isopropanol in which the -OH solvent proton is replaced by deuterium (isopropanol-d_1). The resonances at ~ 11.8 and 13.6 MHz are absent in the lower spectrum which indicates that they arise from the parallel and perpendicular components of the solvent hydrogen-bonded proton hyperfine tensor (see also [7]). Once the hydrogen-bonded resonances are removed from the spectrum in Fig. 2b, the characteristic axial hyperfine components of the 2 -CH_3 group are easily assigned at ~ 11 and 12.3 MHz.

Biggens, J. (ed.), Progress in Photosynthesis Research, Vol. I. ISBN 90 247 3450 9
© 1987 Martinus Nijhoff Publishers, Dordrecht. Printed in the Netherlands.

Returning to Fig. 1, we note two well-resolved features at ~ 12 and ~ 13 MHz once the H-bonded resonances are removed (Fig. 1b). The resonance at ~ 12 MHz has the axially symmetric line shape of a methyl group [8] and we attribute this to the methyl at the $PQH_2^+\cdot$ 3 position. The ring carbon at this position is expected to have low unpaired electron spin density and hence a small hyperfine coupling. Analysis of the spectrum indicates that a = 5.9 MHz and a = 5.0 MHz. From this we calculate a_{iso} = 5.3 MHz = 1.9 G, a typical small coupling value.

For the rhombic set of resonances in the 13 MHz region in Figure 1b, we have used the 1,2,4,5 tetrahydroxybenzene cation radical (Figure 3) as a model. The substitution pattern is expected to favor the antisymmetric benzenoid orbital as is postulated in $D^+\cdot/Z^+\cdot$ [13] and thus the ring protons in the model may be expected to have an ENDOR spectrum similar to the α-proton at the 6 position in $PQH_2^+\cdot$. In H_2SO_4, there are resonances near 13 MHz (and 16.5 MHz) as well as a trough at ~ 20 MHz (Fig. 3a). The same radical in D_2SO_4 (Fig. 3b) retains the 13 MHz features but the trough near 20 MHz disappears indicating that it arises from the hydroxyl protons. Interestingly, there is a weakly H_2O/D_2O exchange sensitive

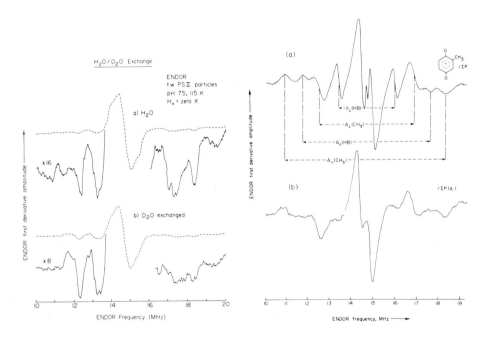

FIGURE 1: ENDOR spectra of the $D^+\cdot$ free radical at 115 K before (a) and after (b) H_2O/D_2O exchange at pD = 7.5. The spectra were recorded at the zero crossing of the EPR spectrum.

FIGURE 2: ENDOR spectra of the 2-methylbenzoquinone anion radical in isopropanol (a) and in isopropanol-d_1 (b). Temperature = 123 K.

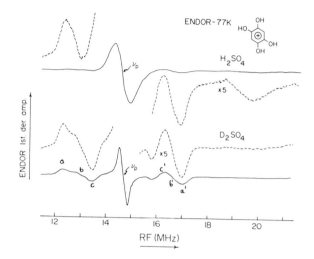

FIGURE 3: ENDOR spectra of 1,2,4,5 tetrahydroxybenzene cation radical recorded at 123 K in (a) H_2SO_4 and (b) D_2SO_4.

resonance in the $\overset{+}{D\cdot}$ spectrum at ~ 20 MHz in Fig. 1 (compare spectrum a with b in the 20 MHz region) which suggests that this may arise from the $PQH_2\cdot$ hydroxyl groups. The ring α-protons of 1,2,4,5 hydroxybenzene cation radical clearly contribute to the rhombic pattern near 13 MHz (a_1 = 2.5 MHz, a_2 = 3.2 MHz, a_3 = 4.7 MHz; a_{iso} = 3.5 MHz) in Figure 3b and is very similar to the rhombic set of resonances near 13 MHz in the $D\cdot$ spectrum (a_1 = 2.5 MHz, a_2 = 3.2 MHz, a_3 = 3.9 MHz; a_{iso} = 3.2 MHz). we assign these features in the in situ radical to the α proton at the $PQH_2\cdot$ ring 6-position. We conclude from the brief analysis above, that H-bonded, $-CH_3$, $-H$ and $-OH$ protons may be detected in the $\overset{+}{D\cdot}$ ENDOR spectrum and that their couplings may be interpreted within the $PQH_2\overset{+}{\cdot}$ model.

ACKNOWLEDGEMENTS

This research was supported by NIH GM 37300, the Competitive Research Grants Office of the U.S. Department of Agriculture and the McKnight Foundation.

REFERENCES

1. Babcock, G.T. (1986) in New Comprehensive Biochemistry: Photosynthesis (Amesz, J., ed.) Elsevier-North Holland, submitted.
2. Dekker, J.P., Van Gorkom, H.J., Brok, M., and Onwehand, L. (1984) Biochim. Biophys. Acta 764, 301-309.
3. O'Malley, P.J. and Babcock, G.T. (1984) Biochim. Biophys. Acta 765, 370-379.
4. de Vintry, C., Carles, C. and Diner, B.A. (1986) FEBS Lett. 196, 203-206.
5. O'Malley, P.J., Babcock, G.T. and Prince, R.C. (1984) Biochim. Biophys. Acta 766, 283-288.

6. Rodriguez, I.D., Chandrashekar, T.K. and Babcock, G.T. (1986) These Proceedings.

7. O'Malley, P.J. and Babcock, G.T. (1986) J. Am. Chem. Soc. 108, 3995-4001.

8. O'Malley, P.J. and Babcock, G.T. (1984) J. Chem. Phys. 80, 3912-3913.

TIME-RESOLVED ESR SPECTRUM OF Z^+ IN OXYGEN-EVOLVING PHOTOSYSTEM II
MEMBRANES.

C.W. HOGANSON, Y. DEMETRIOU AND G.T. BABCOCK, DEPARTMENT OF CHEMISTRY,
MICHIGAN STATE UNIVERSITY, EAST LANSING, MICHIGAN 48824-1322

1. INTRODUCTION

Oxygen evolution occurs at a catalytic site containing manganese. The
power to oxidize water is created in the excited state of P680. To
transport oxidizing equivalents from the site of their generation to the
catalytic site, an electron donor, Z, probably a plastohydroquinone,
exists. It is most easily studied in its oxidized state by electron spin
resonance. Its ESR signal is induced by light and is transient. Previous
work leaves unanswered several questions about the role and nature of Z.
The ESR lineshape is not well established, since the only previous
determination used thylakoid membranes necessitating subtracting the
kinetics of $P700^+$, which has decay components not too different from the
decay of Z^+ [1]. Knowledge of the dependence of Z^+ kinetics on flash
number is limited by the 100 μs time resolution of the one reported
experiment [2]. From that result it is not clear if Z functions on all
four flashes or on only two of four. Indeed, some authors have
hypothesized that a second electron carrier might operate in series of in
parallel with Z (for a review, ref. [3]).

To address some of these problems we have begun to use ESR with 50
microsecond time resolution. Because our measurements require substantial
signal averaging, necessitating many turnovers of the system and addition
of an exogenous electron acceptor, we have looked for effects related to
the exogenous electron acceptor. We have examined two electron acceptors:
ferricyanide and 2,5--dichlorobenzoquinone (DCBQ). We also present a
preliminary, low-resolution spectrum of Z^+ in oxygen evolving photosystem
II particles.

2. MATERIALS AND METHODS

Photosystem II membranes were obtained from spinach [4]. Electron
spin resonance was measured with a Bruker ER 200D spectrometer. The ESR
signal was either plotted directly, stored in a microcomputer for
kinetics, or sampled by a Stanford Research Systems model SR250 gated
integrator for the time-resolved spectrum. Samples were contained in a
Scanlon flat cell and, for long experiments, were pumped through the cell
by a Gilson Minipuls 2 peristaltic pump. 17 μs Xenon flashes excited the
sample. The gated integrator was controlled to sample the ESR signal for
a 160 μs interval beginning 50 μs after the flash. The baseline was
sampled 0.1 S before each flash to subtract the signal from stable
radicals.

3. RESULTS

Kinetics of the ESR transient signal were measured at the low field
peak of the stable radical D^+ (signal IIs, fig. 1b, g=2.011). When DCBQ
was the added acceptor, the signal decay was not single exponential

(fig. 2a); 1st and 2nd half times ranged from 400 μs to 1ms. The
amplitude of the signal corresponded to 25% to 40% of the D^+ signal.

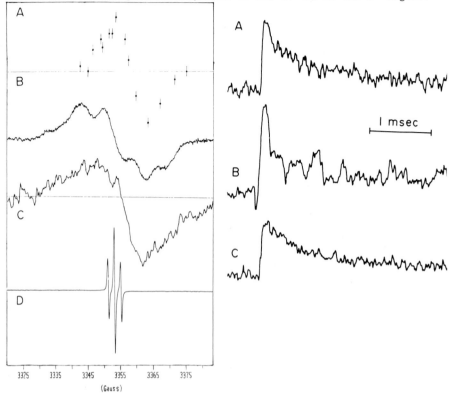

FIGURE 1.   ESR spectra of PSII membranes.
a)   Initial amplitudes of 100 μs decay phase with 100 mM $K_3Fe(CN)_6$.
b) Dark spectrum showing D^+ (signal IIs).   c) Time-resolved spectrum of Z^+
in oxygen-evolving PSII membranes, acquired using a gated integrator
sampling between 50 μs and 230 μs following a flash.   d) Dark spectrum
with 1 mM DCBQ.   a,b and c were recorded with 4 Gauss field modulation; c
was recorded with 0.6 Gauss modulation.

FIGURE 2.   ESR kinetic traces of PSII membranes recorded a) at IIs low
field peak, 1 mM DCBQ, b) at IIs low field peak, 10 mM $K_3Fe(CN)_6$, c) at
DCBQ peak, 1 mM DCBQ.

These observations are in accord with optical absorbance data [5] which
suggest that in the S_0-->S_1, and S_1-->S_2 transitions, Z is reduced in 30
μs and 110 μs, respectively, and in the S_2-->S_3 times).   The 85 μs time
constant we used would prevent observation on the two faster components.
    When ferricyanide was used as acceptor at concentrations from 0.5 mM
to 2 mM, similar decay times and amplitudes were observed.   When the
ferricyanide concentration exceeded 3 mM, however, a fast decay, 70 μs to
200μs, replaced the slow decay and its amplitude was 50% of the D^+ signal

(fig. 2b). With 10 m$\underline{M}$ ferricyanide, the initial amplitudes of kinetic traces at different field settings yield a spectrum for the 100 µs radical that is 8-10 Gauss wide and centered at g=2.002 (fig. 1a). The dependence of the amplitude on microwave power shows that the ESR transition does not saturate below 200mW. These properties are similar to those of a transient radical observed in PSII particles inhibited by NH_2OH, which was attributed to $P680^+$ [6]. Therefore we believe the present radical also to be $P680^+$, and the observed decay to represent the return of an electron from Q_A to $P680^+$. It seems ferricyanide prevents Z from reducing $P680^+$, but not by oxidizing Z to Z^+.

2,5-Dichlorobenzoquinone semiquinone is quite stable and its ESR spectrum is observed in the dark when DCBQ is added to PSII membranes (fig. 1d). Flash illumination produces a transient increase in the DCBQ signal (fig. 2c). The rise represents the reduction of DCBQ, and decay of the disproportionation of the semiquinone. That a transient signal is observed suggests that DCBQ semiquinone does not bind tightly to the Q_B^-, binding site, otherwise its ESR spectrum would be broadened by Fe^{2+}, as is the case for the native plastoquinone. The unusual stability of the DCBQ semiquinone and this lack of binding to the Q_B site may explain why DCBQ did not promote the phenomenon recently termed photoreductant-induced oxidation [7]. In contrast, benzosemiquinone and dimethyl benzosemiquinone, produced on a flash of light, so oxidize the acceptor side Fe^{2+}. Another possible explanation is that DCBQ might oxidize the acceptor side Fe^{2+} in the dark.

We present also a spectrum of Z^+ (fig. 1c) obtained from oxygen-evolving PSII membranes (500µ mole O_2/mg Chl hr). The spectrum was obtained by using a gated integrator that sampled the transient for a 180 µs period beginning 50 µs after a flash. 18,000 flashes were required. $K_3Fe(CN)_6$ was the 1 m$\underline{M}$. The zero crossing is at g=2.003 and the line width is 15Gauss.

The time-resolved Z^+ spectrum in Figure X is preliminary and may deviate from the true spectrum for several reasons:

1) A longer lived decay component (detected by simultaneously recording kinetic traces) contributes about 25% to the spectrum. This could be Z^+ in inhibited reaction centers or residual P700 (<5%) in the preparation.

2) The time constant-sweep rate product is 0.65 Gauss. This may distort the spectrum slightly, especially at the zero crossing.

3) The microwave power used was 40mW. Previous work [8] indicates this power does not saturate the Z^+ ESR signal, but since this was determined only at one g value, power broadening of the lineshape cannot be excluded.

4) The spectrum is noisy. More averaging will be necessary. We are vigorously pursuing a high resolution spectrum of the Z^+ radical.

4. CONCLUSIONS

1. Ferricyanide, at concentrations above 3 m$\underline{M}$, prevents reduction of P680$^+$ by Z.
2. 2,5-Dichlorobenzoquinone differs from other quinones at accepting electrons from photosystem II by having a more stable semiquinone. DCBQ functions kinetically as a one-electron acceptor that does not bind tightly to the Q_B binding site.
3. The preliminary time-resolved ESR spectrum of Z$^+$ uninhibited PSII membranes is broad, as it is in inhibited material, and suggests that the unpaired electron density distribution in the Z$^+$ radical is not strongly perturbed upon inhibiting O_2 evolution.

ACKNOWLEDGEMENTS

For helpful discussion, we thank D. Ghanotakis and C.F. Yocum. This work was supported by USDA Competitive Research Grants Office, NIH Grant GM 37300 and the McKnight Foundation.

REFERENCES

1. Blankenship, R.E., Babcock, G.T., Warden, J.T. and Sauer, K. (1975) FEBS Letts. 51, 287-293.
2. Babcock, G.T., Blankenship, R.E. and Sauer, K. (1976) FEBS Letts. 61, 286-289.
3. Bouges-Boucquet, B. (1980) Biochim. Biophys. Acta 594, 85-103.
4. Ghanotakis, D.F., Babcock, G.T. and Yocum, C.F. (1984) Biochim. Biophys. Acta 765, 388-398.
5. Dekker, J.P., Plijter, J.J., Ouwehand, L. and VanGorkom, H.J. (1984) Biochim. Biophys. Acta 767 176-179.
6. Ghanotakis, D.F. and Babcock, G.T. (1983) FEBS Letts. 153, 231-234.
7. Zimmerman, J.-L. and Rutherford, A.W. (1976) Biochim. Biophys. Acta, in press.
8. Warden. J.T., Blankenship, R.E. and Sauer, K. (1976) Biochim. Biophys. Acta 423, 426-478.

SPATIAL RELATIONSHIP BETWEEN THE INTRAMEMBRANE COMPONENTS (D^+, Z^+) WHICH GIVE RISE TO SIGNAL II AND THE MEMBRANE PERIPHERAL PROTEINS WORKING IN PHOTOSYSTEM II OXYGEN EVOLUTION STUDIED BY THE EFFECT OF SPIN-RELAXING REAGENT DYSPROSIUM

SHIGERU ITOH, [+]YASUHIRO ISOGAI,[++]XIAO-SONG TANG AND [++]KIMIYUKI SATOH. National Institute for Basic Biology, Okazaki 444, [+]Department of Biology, Faculty of Science, Kyushu Univ., Fukuoka 812 and [++]Department of Biology, Faculty of Science, Okayama Univ., Okayama (Japan).

1. INTRODUCTION

Localization of spin-active components inside membrane can be studied by analyzing the interaction of spins [1]. Blum et. al [2,3] developed a method to use a paramagnetic strong spin-relaxer Dysprosium (Dy), which relieves the power saturation of nearby spins depending on $r^{-3\sim-6}$ (where r is a distance from the Dy binding sites). This method was adapted to Signal II i. e., to the components D^+ and Z^+ [4,5], to get the information for their intramembrane localization and the spatial relationship between D/Z, Mn and the three (33,24 and 18 kDa) polypeptides which cover inner surface of PSII reaction center [6] and work in oxygen evolution [6,7]. Distance between D^+(Z^+) and the membrane surface was estimated in
(a) sealed thylakoids vesicles
(b) PSII membrane fragments (which have all the three polypetptides)
(c) NaCl-treated PSII fragments (lack 24 and 18 kDa polypeptides)[8]
(d) CaCl$_2$-treated PSII fragments (lack all the three polypeptides)[9]
(e) Tris-treated PSII fragments (lack Mn and the three polypeptides)[10]
(f) oxygen evolving PSII core complex (lacks LHCP, 18 and 24 kDa poly- peptides but has most of 33 kDa polypeptides and Mn)[11].

Charge distribution on the membrane surface was also estimated from the comparison of the effects of Dy^{3+} and Dy-EDTA^{-1} ions.

2. MATERIALS AND METHODS

PS II particles were prepared from spinach chloroplasts according to Kuwabara and Murata [8] and suspended in 0.3 M sucrose , 25 mM Mes-NaOH (pH 6.5), 10 mM NaCl and 30 % (w/v) glycerol (medium A) for use or to store in liquid nitrogen. NaCl- and CaCl$_2$ treatments were done by washing the PSII particles (at 0.2 mgChl/ml) in 0.3 M sucrose, 25 mM Mes-NaOH (pH 6.5) and 1 M NaCl or CaCl$_2$, respectively. Tris-treatment was done by washing the PSII particles in 0.8 M Tris-Cl (pH 8.0). Oxygen evolving PSII core complex was prepared according to Tang and Satoh [11]. These preparations were incubated with Dy(NO$_3$)$_3$ or Dy-EDTA in medium A for 30 min at 0 °C, frozen in the dark and illuminated at 200K for 2 min by white light from a projector. This illumination had only a small effect on the signal shape, height (less than 5 %) and the power saturation profile compared to those induced by Dy addition itself. Incubation with La^{3+} or Dy^{3+}, under the present conditions released almost no peripheral proteins and Mn (checked by SDS-PAGE and EPR) (it may occur at the higher ionic strength at room tempera-ture [12]). EPR spectra were measured with a Bruker ER-200 X-band spectro-meter with an Oxford Instrument ESR-900 cryostat as reported[13].

3. RESULTS AND DISCUSSION

Spatial relationship between D^+ and peripheral polypeptides.

Fig. 1 shows the first derivative EPR spectrum of Signal II(mainly IIs=D^+) of various PSII preparations at 20 K. Without Dy^{3+} (La^{3+}, added as control, had no effect) each PSII preparation (normal, NaCl- and CaCl$_2$-treated) showed almost the same intensity both at low and high (see Fig. 2)

Biggens, J. (ed.), Progress in Photosynthesis Research, Vol. I. ISBN 90 247 3450 9
© *1987 Martinus Nijhoff Publishers, Dordrecht. Printed in the Netherlands.*

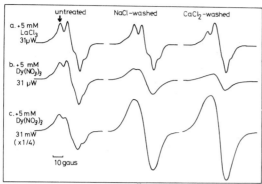

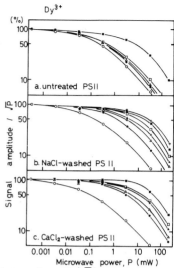

Fig. 1. Effects of Dy^{3+} on Signal II_s EPR spectra in normal, NaCl- and $CaCl_2$-treated PSII preparations. Modulation amplitude 4 gauss. 20 K. 4 mg Chl/ml sample was suspended in medium A.

Fig. 2. Dependence of the normalized signal intensity $(I/\sqrt{P})$ on microwave power (P). Signal heights at the magnetic field shown by an arrow in Fig. 1 were divided by $\sqrt{P}$. $-o-$, 5 mM $laCl_3$ added. $-\bullet-$, $-\triangle-$, $-\blacktriangle-$, $-\square-$, $-\blacksquare-$, and $-+-$ correspond to 0.5, 1, 2. 4.8. 9.1 and 16.7 mM $Dy(NO_3)_3$.

microwave powers. Due to the spin relaxing effect of Dy^{3+}, peak broadening and relieve of power saturation are expected to be stronger in a case with closer approach of Dy^{3+} to the Signal II EPR center[1-3]. In the normal PSII preparation Dy^{3+} had only a small effect. Peak broadening and the signal increase at the high power by Dy^{3+} became more prominent in the NaCl- and $CaCl_2$-washed PSII preparations.

Power saturation profiles in the presence of varied concentrations of Dy^{3+} are shown in Fig. 2. The signal intensities divided by square roots of microwave power $(I/\sqrt{P})$ were plotted against power according to Blum et al. [2]. This plot gives a horizontal straight line only in a case of no power saturation. Without Dy^{3+}, each PSII preparation showed almost the same curve indicating no change of D^+ itself. In each preparation the power range at which the saturation begins, shifted to the higher side with the higher concentration of Dy^{3+}. The sensitivity to Dy^{3+}, however, became higher as the depletion of peripheral proteins. These results suggest that the depletion of the polypeptides results in shorter distances between D^+ and Dy^{3+}, i.e., these proteins are disturbing access of Dy^{3+} to D^+.

For the measure of Dy^{3+} effect, P values at which $I/\sqrt{P}$ becomes 50% $(P_{0.5})$ are calculated from Fig. 2 and plotted against concentration of Dy (Fig. 3). The slope of the plot is then quantitatively related to the accessibility of Dy^{3+} to D^+. NaCl- and $CaCl_2$-treated PSII preparations gave higher slopes indicating shorter distances between Dy^{3+} and D^+. Similar plot was also done with a negative Dy-EDTA$^-$ (Fig. 3 left). In this case, however, NaCl-treatment did not enhance the accessibility. This strongly suggests that the surface newly exposed is rich in negative charges but lacks positive charges which attract Dy-EDTA$^-$. Removal of 33 kDa protein seems to expose new domain which is accessible for both positive and negative (although favorable for positive) ions.

The slope values obtained as in Fig. 3 are summarized in Table I. The slope value is expected to be proportional to $r^{-3\sim-6}$ [1-3]. Blum et al. [2] proposed an experimental equation assuming r^{-6} dependence for small

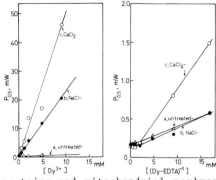

Fig. 3. Dependence of $P_{0.5}$ value on Dy^{3+} and Dy-EDTA$^-$ concentrations in various PSII preparations. $P_{0.5}$ values are calculated from the P value at which $I/\sqrt{P}$ becomes 50 % as shown in Fig. 2. Tris (D^++Z^+) represents Tris-washed PSII sample frozen under illumination to induce Signal II$_f$ (Z^+, about 1/3) in addition to Signal II$_s$ (D^+).

proteins and mitochondrial membrane [2,3]. The distances between D^+ and Dy calculated by using this equation , however, gave a too large membrane thickness (90 ångstrom). Therefore, another way of getting a correction factor was introduced. We assumed 60 ångstrom for the membrane thickness at first and then, adjusted the correction factor to meet this by assuming the r value obtained with sealed thylakoids to be a distance from the outer surface and that with the surface-protein-depleted (CaCl$_2$-treated) PSII particles from the inner surface to D^+. By assuming r^{-6} dependency, the distances are calculated (Table I). D^+ is estimated to be about 15 angstrom from the inner surface to which 24 (also 18) kDa and 33 kDa proteins are attached and interrupting access of Dy ions. Binding sites of these proteins, especially of 24 kDa one, are assumed to be negative by comparing the r value obtained with Dy^{3+} and Dy-EDTA$^-$.

Localization of Mn and Z^+

Results obtained in Tris-washed PSII preparation, which lacks Mn as well as three peripheral proteins [10], are also shown in Table I. The slope value was higher than that obtained with the CaCl$_2$-PSII preparation suggesting the exposure of new binding sites (probably Mn binding sites) for Dy^{3+} in the vicinity of D^+. Preliminary study suggested that 2 of the total 4 Mn/RC are closer to D^+. The results with the Tris-PSII preparation frozen under illumination are also shown (Tris- D^++Z^+). Signal II in this case was about 1.5 times larger than the dark frozen sample, i. e., a mixture of about 1/3 Signal II$_f$ (Z^+) and 2/3 Signal II$_s$ (D^+). Dy^{3+} showed a little stronger effects in this case suggesting that Z^+ is at a slightly shorter (similar) distance than D^+ from the surface. Table I also includes the distance calculated with the oxygen evolving reaction center core

Table I. Estimation of distance between $D^+(Z^+)$ and membrane surface in various membrane preparations.

	$\Delta P_{0.5}$ (mW/mM)		r (Å)	
	Dy^{3+}	Dy-EDTA$^-$	Dy^{3+}	Dy-EDTA$^-$
Thylakoid membrane	0.0061		45	
PSII membranes				
untreated	0.048	0.035	32	34
NaCl-treated	2.0	0.036	17	34
CaCl$_2$-treated	4.7	0.097	15	29
Tris-washed(D^+)	8.4		14	
(D^++Z^+)	9.9		13	
PSII RC complex	71		9.5	

$$r = 19.3 \times 10^{-8} \times \Delta P_{0.5}^{-1/6} \quad \text{(cm)}$$

complex [11] which preserves more than 75 % of the 33 kDa protein and Mn but lacks LHCP (an intermediate condition between NaCl- and CaCl$_2$-treated PSII preparations as for the surface proteins). It seems that the depletion of LHCP also exposes new binding sites closer to D$^+$ for Dy^{3+} than the sites on the inner surface.

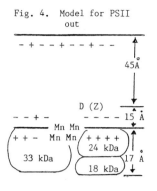

Fig. 4. Model for PSII

4. SUMMARY

Intramembrane localization of the PSII secondary donor (D and Z) and the binding sites of Mn, 33, 24 and 18 kDa membrane peripheral proteins on the membrane surface were studied by analyzing the effects of spin relaxing reagent Dy on EPR Signal II$_s$ and II$_f$ in various membrane preparations.
(1) D as well as Z locates inside membrane at 15 ångstrom (assuming membrane thickness to be 60 ångstrom) from the inner surface.
(2) These three polypeptides are covering inner membrane surface close to D/Z and constitute a 17 angstrom barrier for Dy ions.
(3) Depletion of 24 and 18 kDa proteins decreases the barrier to be only 2 ångstrom and exposes a new surface rich in negative charges. Therefore, these proteins are situated close to D/Z and their binding sites (and probably Ca^{2+} binding site) on the membrane surface are strongly negative and lack positive surface charges.
(4) 33 kDa protein also seems to bind to a negative surface situated a little more distant from D/Z and interrupt the access of Dy to D/Z.
(5) Mn-binding site seems to be on the inner surface close to D/Z.
(6) Depletion of LHCP also increase the accessibility of D/Z to Dy^{3+}.

ACKNOWLEDGMETNS

The authours thank Dr. Y. Fujita for his discussion, Miss M. Iwaki for her technical assitance, and financial aids by NIBB Cooperative Research Program and from Japanese Ministry of Education, Science and Culture.

REFERENCES
1 Hyde, J. S. and Rao, K. V. S. (1978) J. Mag. Resonance 29, 509–516
2 Blum, H., Cusanovich, M. A., Sweeney, W. V. and Ohnishi, T. (1981) J. Biol. Chem. 256, 2199–2206
3 Blum, H. and Ohnishi, T. (1980) Biochim. Biophys. Acta 626, 31–40
4 Babcock, G. T., Blankenship, R.E. and Sauer, K. (1976) FEBS Lett. 61, 286–289
5 Berthold, D. A., Babcock, G. T. and Yocum, C. F. (1981) FEBS Lett. 134, 231–234
6 Murata, N., Miyao, M. and Kuwabara, T. (1983) In The Oxygen-evolving System of Photosynthesis (Inoue, Y. et al. eds), pp. 213–222. Academic Press, Tokyo
7 Akabori, K., Imaoka, A. and Toyoshima, Y. (1984) FEBS Lett. 173, 36–40
8 Kuwabara, T. and Murata, N. (1982) Plant Cell Physiol. 23, 533–539
9 Ono, T. and Inoue, Y. (1984) FEBS Lett. 168, 281–286
10 Yamamoto, Y., Doi, M., Tamura, N. and Nishimura, M. (1981) FEBS Lett. 133, 265–268
11 Tang, X.-O., and Satoh, K. (1985) FEBS Lett. 179, 60–64
12 Ghanotakis, D. F., Babcock, G. T. and Yocum, C. F. (1985) Biochim. Biophys. Acta 809, 173–180
13 Itoh, S., Tang, X.-O and Satoh, K. (1986) FEBS Lett. in press

THE EFFECTS OF CHEMICAL OXIDANTS ON THE ELECTRON TRANSPORT
COMPONENTS OF PHOTOSYSTEM II AND THE WATER-OXIDIZING COMPLEX

J. Tso, D. Hunziker and G.C. Dismukes. Department of Chemistry, Princeton University,
Princeton, NJ 08544 (U.S.A.).

INTRODUCTION

Attempts at the direct chemical oxidation of the water-oxidizing complex in spinach
Photosystem II (PS-II) particles has revealed several classes of oxidants. These classes differ in
the degree to which they release manganese and the 17, 23 and 33 kDa extrinsic polypeptides, as
well as in their reactivity with the PS-II electron donors such as Signal II_s (D^+) and cytochrome
b-559 (cyt. b-559). In particular, this study has focused on a previously unexplored oxidant,
$Ce(NH_4)_2(NO_3)_6$. The aim of utilizing powerful oxidants such as $Ce(NH_4)_2(NO_3)_6$ was to
determine the redox properties of the electron donor components in PS-II. The effects of
selective removal of the three extrinsic polypeptides (17, 23 and 33 kDa) and Mn on the EPR
lineshape of Signal II_s has been examined, along with the identity of the species responsible for
the altered lineshape upon treatment with $Ce(NH_4)_2(NO_3)_6$. The redox midpoint potential of the
D/D^+ species was also measured using $Ce(NH_4)_2(NO_3)_6$.

MATERIALS AND METHODS

Spinach PS-II membranes were prepared from fresh market spinach according to the
procedure of Berthold, et al. (4), modified to include only one Triton X-100 treatment. The
particles were stored in a buffer solution containing 200 mM sucrose, 20 mM MES, 20 mM
NaCl pH 6.5 and 30% glycerol at -80 C until further use.

The electrochemical titrations were carried out in a thermally regulated cell at 4 C using a
saturated calomel electrode (SCE) as the reference and a platinum mesh as the working electrode.
In order to cover a wide potential range, a solution containing $K_3Fe(CN)_6$, K_2IrCl_6 and 10
mM $Ce(NH_4)_2(NO_3)_6$ was added to a suspension of PS-II membranes in 50 mM MES, 25 mM
NaCl and 10 mM $CaCl_2$ pH 6.5 to attain the highest achievable potential of 580 mV vs. SCE.
Reduction of the potential was achieved by addition of small amounts of 100 mM sodium
ascorbate and aliquots were removed from the cell at stable potential readings and placed in EPR
tubes. The samples were then initially frozen at 210 K and subsequently at 77 K until EPR
measurements were performed. The reverse titration was performed on 1 M $CaCl_2$-washed
PS-II membranes resuspended in a buffer containing 50 mM MES, 25 mM NaCl and 10 mM
$CaCl_2$ pH 6.5. Initial addition of 5mM ascorbate was needed to bring the potential down to 90
mV vs. SCE. Subsequent addition of the oxidants followed the order of $K_3Fe(CN)_6$, K_2IrCl_6
then $Ce(NH_4)_2(NO_3)_6$, thereby covering their respective, effective potential ranges.

The effects of other exogenous oxidants on PS-II membranes were investigated and prepared
as follows. Fresh stock solutions of oxidants were added to a suspension of PS-II particles at a
[Chl] of 10 mg/ml to give a final oxidant concentration of 10 mM. The oxidation reaction was
carried out in an EPR tube for 1 min. in the dark followed by rapid freezing at 210 K and then at
77 K.

Biggens, J. (ed.), Progress in Photosynthesis Research, Vol. I. ISBN 90 247 3450 9
© *1987 Martinus Nijhoff Publishers, Dordrecht. Printed in the Netherlands.*

The Ca^{2+}-induced suppression of inhibition by Ce^{4+} was investigated by addition to a suspension of PS-II membranes and 2 mM DCBQ as an electron acceptor, a fixed concentration of $CaCl_2$ followed by the addition of various concentrations of $Ce(NH_4)_2(NO_3)_6$. The mixture was incubated in the dark for 3 min. before illumination. Oxygen evolution was measured with Clark-type oxygen electrode.

RESULTS AND DISCUSSION

The effects of a variety of hydrophilic and hydrophobic oxidants on dark-adapted PS-II membranes are summarized in Table 1. Several classes of behavior are suggested based upon their effects on EPR Signal II_s, cyt. b-559 and the release of protein and Mn.

TABLE 1 - Characteristics of EPR Signals for PS-II components after treatment with the oxidizing reagents.

CLASS	REAGENT	SIGNAL II_S	CYT. b-559	Mn
1	NH_4ReO_4 $K_3Fe(CN)_6$ H_2O_2	No detectable effects on the donor components of PS-II.		bound
2	K_2IrCl_6 $KMnO_4$ KIO_3 $Ru(bpy)_3Cl_3$	Partially reduces the intensity of Signal II.	No change in oxidation state of cyt. b-559	bound
3	$Ce(IV)(NH_4)_2(NO_3)_6$* $Cp_2Fe(FeCl_4)$	Signal II lineshape replaced by a symmetric EPR signal. No change in lineshape of Signal II but the intensity is reduced.	Conversion of LS -> HS heme at $g_\perp =6.0$	released
4	DCDCBQ	Signal II is masked by a free radical signal with $\partial H=6.5$ G.	Conversion of LS -> HS heme at $g_\perp =6.0$	bound

* Releases both the three extrinsic polypeptides (17, 23 and 33 kDa) along with EPR detectable Mn^{2+}.
All oxidant concentration=10 mM. [Chl]=10 mg/ml.

The group of compounds comprising Class 1 exhibited no detectable effects on EPR Signal II_s lineshape or intensity and did not release Mn from the Oxygen-Evolving Complex (OEC). The oxidation state of cyt. b-559, which exists predominantly in the oxidized low-spin heme form as seen via EPR, was not altered (data not shown). Illumination of these membranes at 200 K produced the S_2 multiline EPR signal. The Class 2 oxidants caused partial loss of Signal II ranging from 30 to 50%. There was no evidence for Mn release nor changes in cyt. b-559. The lipophilic ferricinium cation and the hydrophilic Ce^{4+} cation in the Class 3 category caused both release of Mn^{2+} and conversion of cyt. b-559 from a low-spin heme to a high-spin heme. Additionally, Ce^{4+} reacts with PSII to form a symmetric free radical signal in place of the asymmetric Signal II lineshape seen in the dark (Fig. 1). The reagent belonging to the Class 4 oxidants, 2,3-dichloro-5,6-dicyano-1,4-benzoquinone (DCDCBQ), also converts cyt. b-559 from the low spin to the high spin $g_\perp=6.0$ form, but Signal II is masked by a free radical signal that is 10^3 greater in intensity and having a linewidth $\partial H=6.5$ G. DCDCBQ inhibits at 100-fold lower concentration.

The symmetric signal which replaces Signal II_s, as seen in Fig. 1, appears to be the same species that was observed by Boussac and Etienne upon treatment of Tris-washed PS-II membranes with K_2IrCl_6 (1). Unlike K_2IrCl_6, the ceric ion releases the extrinsic proteins, thus removing the need for pretreatment with Tris. The g-value of this symmetric signal, as determined with the use of DPPH (1,1-diphenyl-2-pikryl-hydrazyl) as a g-factor, gave g=2.0031 and that of the native Signal II_s gave g=2.0033 by the same method. The structureless signal has a linewidth ∂H=9G.

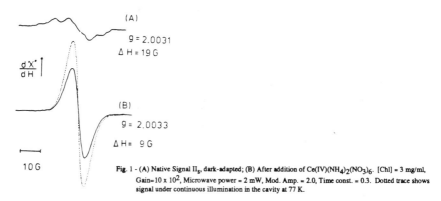

Fig. 1 - (A) Native Signal II_s, dark-adapted; (B) After addition of Ce(IV)(NH$_4$)$_2$(NO$_3$)$_6$. [Chl] = 3 mg/ml, Gain= 10 x 10^2, Microwave power = 2 mW, Mod. Amp. = 2.0, Time const. = 0.3. Dotted trace shows signal under continuous illumination in the cavity at 77 K.

By comparison with the plastoquinone-9 (PQ-9) cation radical species that was generated in a 50:50 mixture of CF_3COOH/H_2SO_4, as outlined by Sullivan and Bolton (5), gave a symmetric signal with linewidth ∂H=11 G and g=2.0039. These observations point to an identity of the signal produced upon addition of Ce(NH$_4$)$_2$(NO$_3$)$_6$ as a modified form of EPR Signal II_s. However, the possibility that this new signal belongs to either a chlorophyll radical or a tightly bound carotenoid radical cannot currently be ruled out. Further experiments are underway to understand how the modification of the molecular environment surrounding the D^+ species might give rise to the altered lineshape of Signal II_s.

Treatments which selectively remove two or more of the three extrinsic polypeptides and Mn were found to have no effect on the lineshape of Signal II_s (data not shown). These treatments include washing the PS-II membranes at pH 6.5 with 1 M NaCl (removes 17 & 23 kDa), washing in 1 M CaCl$_2$ followed by resuspension in 200 mM Cl$^-$ (removes 17, 23 & 33 kDa), washing in 1 M CaCl$_2$ followed by resuspension in 20 mM Cl$^-$ (removes all three proteins & Mn) and washing in 0.8 M Tris at pH 9.5 (removes all three proteins & Mn). These observations suggest that the immediate environment surrounding Signal II does not directly involve the three extrinsic polypeptides nor Mn, as has been previously observed by Berthold, et al. (4).

Treatment of the PS-II membranes with Ce(NH$_4$)$_2$(NO$_3$)$_6$ results in the removal of the 17, 23 and 33 kDa polypeptides along with Mn as detected by the EPR hexaquomanganese(II) signal. Following this, Signal II_s became accessible to hydrophilic oxidants such as $K_3Fe(CN)_6$, K_2IrCl_6 and Ce(NH$_4$)$_2$(NO$_3$)$_6$ and the altered signal titrated with a midpoint potential of 490 mV vs. SCE (Fig. 2b). Illumination at 77 K of the sample at the highest, stable potential produced an increase in the intensity of the symmetric signal with a linewidth identical to that of the chemically oxidized signal. This observation may correspond to an additional Signal II species that is light-induced and having a midpotential higher than was accessible under the present conditions, possibly the P_{680} donor species Z.

The behavior of the Ce^{+4} ion on oxygen activity in the presence of Ca^{2+} is shown in Fig. 3. The inverse plot of oxygen activity vs. inhibitor concentration demonstrates an inhibition of O_2 evolution by Ce^{4+}, or to its reduction product Ce^{3+}, at sites which are competed for by Ca^{2+}.

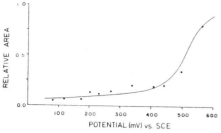

Fig. 2a - Electrochemical titration of 1 M CaCl₂ treated PS-II membranes with oxidants. Area was computed in comparison to the signal obtained at 575 mV. Titration buffer contained 50 mM MES, 25 mM NaCl, 10 mM CaCl₂ pH 6.5. [Chl]=2.5 mg/ml.

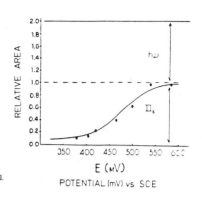

Fig. 2b - Electrochemical titration of $Ce(NH_4)_2(NO_3)_2$ with sodium ascorbate. Area was computed in the same manner as with Fig. 2a. [Chl]=10.0 mg/ml. Upper arrow shows an additional component of Signal II that is light-induced.

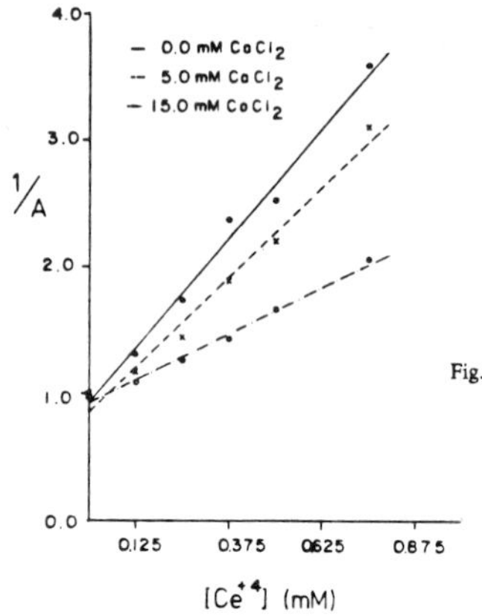

Fig. 3 - Plot of Activity⁻¹ vs. [Ce⁴⁺]. Oxygen-evolution data showing competitive behavior between Ce^{4+} and Ca^{2+} for the same sites on the Oxygen-Evolving Complex (OEC). PS-II membranes were initially resuspended in buffer solution containing 200 mM sucrose, 20 mM MES and 20 mM NaCl pH 6.5, and had activities of 300 μmol O₂/mg Chl-hr.

The effects of $Ce(NH_4)_2(NO_3)_6$ on the water-oxidizing complex combine the effects seen for K_2IrCl_6 on the lineshape of EPR Signal II_s (1) and the competition for calcium sites on the oxidizing side of PS-II as seen with lanthanum salts (2). The species giving rise to EPR Signal II_s has been attributed to a plastoquinone cation radical and its lineshape is thought to arise from the stabilization of an asymmetric orbital in the benzoquinone ring through H-bonding with the surrounding amino acid residues (3). Treatment of intact PS-II membranes and Tris-treated membranes with $Ce(NH_4)_2(NO_3)_6$ and K_2IrCl_6, respectively, disrupt this environment and produce a symmetric EPR signal in place of the asymmetric Signal II_s seen in dark-adapted membranes. The changes seen in the lineshape cannot strictly be attributed to the release of the three extrinsic polypeptides nor Mn. Further experiments are needed to determine the origin of this signal.

REFERENCES

1. Boussac, A. and Etienne A.L. (1984) Biochim. et Biophys. Acta, 766, 576-581.
2. Ghanotakis, D., Babcock, G. and Yocum, C. (1985) Biochim. et Biophys. Acta, 809, 173-180.
3. O'Malley, P. and Babcock, G. (1984) Biochim. et Biophys. Acta, 765, 370-379.
4. Berthold, D., Babcock, G. and Yocum, C. (1981) FEBS Lett, 134, 231-234.
5. Sullivan, P. and Bolton, J. (1968) JACS, 90, 5366-5370.

ON THE MECHANISM OF PHOTOSYNTHETIC WATER OXIDATION

Gary W. Brudvig and Julio C. de Paula
Department of Chemistry, Yale University, New Haven, CT   06511

With the increasing knowledge of the structure and function of Mn and other components involved in photosynthetic water oxidation, the possibility now exists to define the molecular mechanism of this process. Recently, a number of mechanisms have been proposed (representative models are given in 1-3, for reviews see 4-6). In considering these various mechanisms, one must keep in mind the energetic constraints. The energetics of the individual steps in each proposed mechanism must be considered with respect to the redox potential available from photosystem II (PSII). Many mechanisms can be ruled out based on this comparison and, moreover, one can obtain further insight into possible mechanisms by a detailed evaluation of the energetic constraints taken in light of the available chemical and biophysical data on PSII. In this contribution, we evaluate the energetics for a variety of possible mechanisms of water oxidation. Our calculations are specific for a Mn-oxo active site, based on the evidence which indicates that catalysis of water oxidation occurs at a Mn-oxo complex (5). Several important factors are noted which are necessary for favorable energetics of water oxidation. Based on these results, we have extended our proposed mechanism for water oxidation (1).

Krishtalik (7) has recently outlined the energetics for multielectron reactions with specific reference to photosynthetic water oxidation. In Krishtalik's work, the concentration dependence of the free energy change of each elementary reaction was factored out to arrive at a configurational component of the free energy change, ΔG_c. In short, ΔG_c represents the part of the free energy that does not depend on the entropy of mixing. From an evaluation of ΔG_c, it is possible to evaluate the probability of given single- or multi-step reactions in PSII. If ΔG_c is significantly positive, then one can conclude that the elementary act in question is energetically unfavorable.

ΔG_c can be calculated according to Eq. 1, where n is the number of

$$\Delta G_c = n(E_c - E_r)e \qquad (1)$$

electrons involved in the reaction, e is the electron charge, E_c is the configurational electrode potential for the reaction, and E_r is the reduction potential of the oxidant. In PSII, the primary oxidant is P680$^+$, with a reduction potential estimated to be 1.17V (8). This value was used by Krishtalik (7) as the reduction potential of the oxidant in water oxidation (E_r in Eq. 1) and based on this value the energetics of various elementary acts were evaluated. The reduction potential of the oxidant in water oxidation, however, is significantly lower than 1.17V;

Biggens, J. (ed.), Progress in Photosynthesis Research, Vol. I. ISBN 90 247 3450 9
© *1987 Martinus Nijhoff Publishers, Dordrecht. Printed in the Netherlands.*

the Mn complex in PSII is the oxidant and a portion of the available
redox potential in PSII is expended in order to stabilize the
intermediate oxidation states of the O_2-evolving complex. The reduction
potentials for each of the intermediate S state transitions have been
estimated by Bouges-Bocquet (9). The values are listed in Table 1,
although the reduction potential for $S_0 \rightarrow S_1$ may not be as low as
estimated. The reduction potential for $S_0 \rightarrow S_1$ must, however, be
significantly lower than for $S_1 \rightarrow S_2$. The average potential is only
about 0.75V per electron and, thus, the energetic constraints on
possible mechanisms of water oxidation are, in fact, significantly more
severe than would be if P680$^+$ was the direct oxidant of water.

Table 1. Reduction Potentials for S State Transitions (taken from 9).

Transition	Reduction Potential (V)
$S_0 \rightarrow S_1$	≤ 0.24 (?)
$S_1 \rightarrow S_2$	0.89
$S_2 \rightarrow S_3$	0.93
$S_3 \rightarrow S_4$	0.87

E_c is calculated for each elementary reaction as shown in Eq. 2

$$E_c = E_0 - (RT/nF)[\; \Sigma \ln X_i^0 - \Sigma \ln X_f^0 \;] \tag{2}$$

where n is the number of electrons, X_i^0 and X_f^0 are the mole fractions of
the initial and final reagents in their respective standard states, E_0
is the standard potential for the reaction, and $R \ln X_i$ is the transpo-
sitional entropy of component i for an ideal solution. Note that only
the entropy of mixing is factored out of E_0 in the calculation of E_c.
In order to take account of the enthalpy of mixing, it is necessary to
include binding energies for each component in both the initial and
final states. It is the estimation of these binding energies that leads
to the greatest uncertainty in a calculation of E_c. In general, such
binding energies are not known for the reactants and products involved
in photosynthetic water oxidation because they depend on the unknown
structure and composition of the active site of water oxidation.
However, it is widely believed that the active site for water oxidation
consists of an oxo-bridged Mn complex (5). Hence, binding energies of
the various reactants and products can be estimated from data that are
available on inorganic metal-oxo complexes. The values we have used in
our calculations of E_c's are given in Table 2. These values have been
used consistently in all of our calculations. Even though the
magnitudes of the binding energies may be off slightly, it is still
possible to compare alternate mechanisms for water oxidation in order to

ascertain which schemes are most favorable.

It is also necessary to evaluate pK_a's for ligands to Mn and for any protein groups which may play a role in binding the protons that are released upon oxidizing water. The energetics of water oxidation can be significantly altered if a basic group bound to or near the Mn complex undergoes a significant change in pK_a during a given step in the water oxidation process. However, the pK_a's of such groups are not known. Therefore, we have chosen one of two pK_a values for basic groups that are involved in the water oxidation process: if the basic group is to be deprotonated in the reaction, a pK_a of 4 is assigned; if the basic group is to be protonated in the reaction, a pK_a of 8 is assigned. These values, although arbitrary, are chosen to be consistent with the optimal internal pH of 5-6 for water oxidation by PSII (4). The pK_a values for specific ligands to Mn that were used in the calculations are given in Table 3.

Based on the data listed in Tables 2 and 3, we have evaluated the configurational reduction potentials, E_c, for various water oxidation reactions (Table 4). Only $2e^-$ and $4e^-$ reactions are included because it has already been demonstrated by Krishtalik (7) that $1e^-$ reactions are inherently unfavorable; the redox potential needed for a $1e^-$ oxidation of water to produce the hydroxyl radical oxidation level is not available from PSII. Hence, mechanisms involving sequential $1e^-$ oxidations of water can be immediately ruled out based on a consideration of the energetics. It is known that $2H^+$ are released along with O_2 from PSII in the $S_4 \rightarrow S_0$ transition (10). Therefore, we have only considered reactions in which $2H^+$ are released in the step associated with O_2 release, although these protons are not necessarily directly related to dissociation of protons from oxidized water molecules.

Table 2. Binding Energies used to Calculate E_c (taken from 7).

	Species	$\Delta G_{cB}(eV)$
A. $(Mn)_n(L) \rightarrow (Mn)_n + L$:	H_2O	0.3
	OH^-	0.75
	O^{2-}	1.3
	H_2O_2	0.6
	O_2^{2-}	1.0
	O_2	0.4
B. $\quad BH^+ \rightarrow B + H^+$:	$H^+ \ (B = H_2O)$	0.2
	$H^+ \ (B = X^-)$	0.2

Table 3.   Assumed pK_a Values for Ligands of Mn.

S State	Species (see Figure 1)	pK_a
S_0	μ_3-hydroxo (two of the four O ligands)	4
S_0	μ_3-hydroxo (remaining two of the four O ligands)	8
S_4	μ_2-hydroxo (two of the six O ligands)	4
S_4	μ_2-hydroxo (remaining four of the six O ligands)	8
S_4'	μ_2-hydroxo (two of the four O ligands)	8
S_4'	μ_3-hydroxo (remaining two of the four O ligands)	8
S_4'	μ_2-peroxo (in reactions involving H_2O_2)	8
S_4'	μ_2-peroxo (in reactions involving O_2^{2-})	4

We will begin by discussing the energetics of $4e^-$ reactions (Table 4, part I).   As was noted by Krishtalik (7), it is more favorable to oxidize deprotonated forms of water.   In the absence of other factors, it is most difficult to oxidize H_2O, then OH^-, and then O^{2-} (compare reactions Ia, Ib, and Ic).   It is also favorable to couple a change in pK_a of a hydroxo group (or other deprotonatable group) associated with the Mn complex to the release of O_2 such that the hydroxo group is deprotonated as O_2 is released (compare reactions Ic and Ie).   We denote hydroxo groups, whose deprotonation is coupled to the redox reaction, with asterisks in Tables 4 and 5.   It becomes even more favorable if the protons released along with O_2 are initially bound to a nearby basic group on the protein (designated as X^-) rather than releasing the protons directly into the bulk aqueous phase (compare reactions Ib and Id or reactions Ie and If).   Note, however, that the coupling of a change in pK_a of a basic group associated with the Mn complex to O_2 release must be consistent with the observed release of two protons in the $S_4 \rightarrow S_0$ transition (10).   Consequently, only the energetics for oxidation of O^{2-} can be made more favorable by coupling a change in pK_a to O_2 release because only in the oxidation of O^{2-} are no protons involved.

Overall, the most favorable $4e^-$ oxidation of water involves oxidizing two O^{2-} ligands of a Mn complex and concomitantly releasing two protons from basic groups associated with the Mn complex.   In order for protons to be released from basic groups associated with the Mn complex, it is necessary that these groups undergo a significant decrease in pK_a in the $S_4 \rightarrow S_0$ transition.   We have equated such basic groups with hydroxo ligands of the Mn complex, which may, in fact, be expected to undergo a change in pK_a in the $S_4 \rightarrow S_0$ transition due to a structural rearrangement of the Mn complex upon release of O_2 (see below).

Table 4. Energetics for Oxidation of Water to O_2.

I. 4e⁻ Reactions		$E_c(V)$
a.	$2H_2O \rightarrow O_2 + 4H^+ + 4e^-$	1.55
b.	$2OH^- \rightarrow O_2 + 2H^+ + 4e^-$	1.06
c.	$2O^{2-} \rightarrow O_2 + 4e^-$ $2HX \rightarrow 2H^+ + 2X^-$	0.92
d.	$2OH^- + 2X^- \rightarrow O_2 + 2HX + 4e^-$ $2HX \rightarrow 2H^+ + 2X^-$	0.77
e.	$2O^{2-} + 2^*OH^- \rightarrow O_2 + 2^*O^{2-} + 2H^+ + 4e^-$	0.71
f.	$2O^{2-} + 2^*OH^- + 2X^- \rightarrow O_2 + 2^*O^{2-} + 2HX + 4e^-$ $2HX \rightarrow 2H^+ + 2X^-$	0.53

II. 2e⁻ Reactions

A. Oxidation of Water to Peroxide.

a.	$2OH^- \rightarrow H_2O_2 + 2e^-$	1.35
b.	$2OH^- \rightarrow O_2^{2-} + 2H^+ + 2e^-$	0.93
c.	$2O^{2-} \rightarrow O_2^{2-} + 2e^-$ $2OH^- + O_2^{2-} \rightarrow H_2O_2 + 2O^{2-}$	1.12
d.	$2O^{2-} \rightarrow O_2^{2-} + 2e^-$	0.88
e.	$2O^{2-} + 2^*OH^- \rightarrow H_2O_2 + 2^*O^{2-} + 2e^-$	0.65
f.	$2O^{2-} + 2^*OH^- \rightarrow O_2^{2-} + 2^*O^{2-} + 2H^+ + 2e^-$	0.47

B. Oxidation of Peroxide to O_2.

a.	$O_2^{2-} \rightarrow O_2 + 2e^-$	0.95
b.	$H_2O_2 \rightarrow O_2 + 2H^+ + 2e^-$	0.77
c.	$O_2^{2-} + 2^*OH^- \rightarrow O_2 + 2^*O^{2-} + 2H^+ + 2e^-$	0.54
d.	$H_2O_2 + 2X^- \rightarrow O_2 + 2HX + 2e^-$ $2HX \rightarrow 2H^+ + 2X^-$	0.40
e.	$O_2^{2-} + 2^*OH^- + 2X^- \rightarrow O_2 + 2^*O^{2-} + 2HX + 2e^-$ $2HX \rightarrow 2H^+ + 2X^-$	0.17

Table 5. Energetically Favorable Water Oxidation Reactions.

Reaction	ΔG_c (eV)[a]
1. $\qquad 2O^{2-} + 2^*OH^- \rightarrow O_2 + 2^*O^{2-} + 2H^+ + 4e^-$	-0.16
2. $\quad 2O^{2-} + 2^*OH^- + 2X^- \rightarrow O_2 + 2^*O^{2-} + 2HX + 4e^-$ $\qquad\qquad\qquad 2HX \rightarrow 2H^+ + 2X^-$	-0.88
3. $\qquad\qquad\quad 2O^{2-} \rightarrow O_2^{2-} + 2e^-$ $\quad O_2^{2-} + 2^*OH^- \rightarrow O_2 + 2^*O^{2-} + 2H^+ + 2e^-$	-0.04 -0.12
4. $\qquad\qquad\quad 2O^{2-} \rightarrow O_2^{2-} + 2e^-$ $\quad O_2^{2-} + 2^*OH^- + 2X^- \rightarrow O_2 + 2^*O^{2-} + 2HX + 2e^-$ $\qquad\qquad\qquad 2HX \rightarrow 2H^+ + 2X^-$	-0.04 -0.86
5. $\quad 2O^{2-} + 2^*OH^- \rightarrow H_2O_2 + 2^*O^{2-} + 2e^-$ $\quad H_2O_2 + 2X^- \rightarrow O_2 + 2HX + 2e^-$ $\qquad\qquad\qquad 2HX \rightarrow 2H^+ + 2X^-$	-0.50 -0.40

[a]E_r, the reduction potential available from the Mn complex, is taken to be 0.75V per electron for a 4e$^-$ reaction, 0.90V per electron for the first 2e$^-$ reaction, and 0.60V per electron for the second 2e$^-$ reaction.

    With the estimate that the reduction potential available from the Mn complex is 0.75V per electron for a 4e$^-$ reaction, only two of the 4e$^-$ reactions listed in Table 4 are energetically favorable (reactions 1 and 2 in Table 5). Both of these reactions involve oxidation of two O^{2-} ligands of the Mn complex coupled with dissociation of two protons from hydroxo ligands which change their pK_a in the $S_4 \rightarrow S_0$ transition. The energetics are somewhat more favorable if the protons released along with O_2 are initially bound to other basic groups of the protein (the pK_a of these groups is chosen to be 6.0 in our calculations) before being released into the bulk aqueous phase. Such a binding of protons is, however, not necessarily required to arrive at an energetically favorable reaction; both reactions 1 and 2 (Table 5) are predicted to be energetically favorable. Nonetheless, there is evidence that protons released from PSII are, in fact, initially released into an internal proton pool before being released into the bulk aqueous phase (11). Hence, the initial binding of protons by basic groups on the protein may be utilized by PSII to make the oxidation of water energetically more favorable.

    Many of the same considerations discussed above with regard to 4e$^-$ water oxidation reactions also pertain to the 2e$^-$ reactions. However, for a Mn complex, it is expected that the redox potential available for the first 2e$^-$ reaction will be significantly greater than that for the second 2e$^-$ reaction. We estimate 0.90V and 0.60V per electron, respectively. Even with 0.90V per electron available from the Mn complex, it is not energetically favorable to form peroxide from H_2O. Furthermore, the formation of peroxide from OH$^-$ must be coupled to the

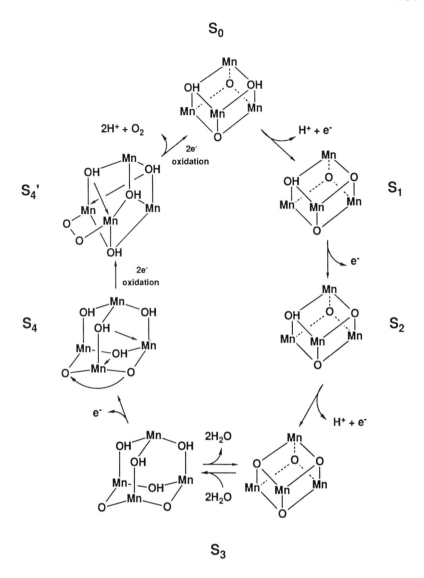

Figure 1.   A proposed scheme for water oxidation.   The mechanism depicted is an elaboration on the mechanism proposed by Brudvig and Crabtree (1).   The protonation states of the O ligands to Mn are consistent with the pK$_a$ values for ligands of Mn given in Table 3 which were chosen to give the most favorable energetics for water oxidation. The energetics for this scheme are given by reaction 3 or 4 in Table 5, depending on whether the protons are released directly into the aqueous phase (reaction 3, Table 5) or are initially bound by a basic group on the protein, X$^-$ (reaction 4, Table 5).

release of two protons in order to arrive at an energetically favorable reaction (compare reaction IIAa and IIAb, Table 4). However, the formation of O_2 from peroxide is also energetically favorable only if the reaction is coupled to the release of two protons (compare reactions IIBa and IIBc, Table 4). Because only two protons are released in the $S_4 \rightarrow S_0$ transition, it is not possible to combine two energetically favorable 2e⁻reactions that convert two OH⁻ into O_2. Therefore, the 2e⁻ oxidation of both H_2O and OH⁻ can be ruled out.

Overall, three combinations of 2e⁻ reactions are found to give favorable energetics (Table 5). Of these, reactions 3 and 4 are equivalent, differing only in whether the protons are released directly into the bulk phase (reaction 3) or are initially bound by basic groups on the protein (reaction 4). Reaction 5 is also closely similar to reactions 3 and 4, differing only in the pK_a of the peroxo intermediate. It can be seen that the energetics for water oxidation via two sequential 2e⁻ reactions are only favorable when the oxidation reaction is coupled to a change in pK_a of a hydroxo group (or other basic group) such that a deprotonation occurs. This requirement is common to all of the energetically favorable 2e⁻ and 4e⁻ water oxidation reactions.

In order for a significant drop in the pK_a of a basic group associated with the Mn complex to occur, the Mn complex must undergo a structural change in the $S_4 \rightarrow S_0$ transition. Such a drop in pK_a of hydroxo ligands of Mn is predicted in the model recently proposed by Brudvig and Crabtree (1). The change in pK_a is due to the structural rearrangement from a Mn_4O_6 adamantane-like structure to a Mn_4O_4 cubane-like structure that is proposed to occur in this step. A μ_3-hydroxo bridge in a cubane-like structure (S_0 state) is expected to have a significantly lower pK_a than a μ_2-hydroxo bridge in an adamantane-like structure (S_4 state). The scheme shown in Figure 1 provides a molecular description for the oxidation of water by PSII which incorporates the energetic considerations outlined in this paper. We have chosen to depict a 2e⁻ scheme, although, based on the energetics, the 2e⁻ scheme shown cannot be discriminated from the equivalent 4e⁻ scheme (reaction 1 in Table 5).

This work was supported by NIH (GM32715). GWB is the recipient of a Searle Scholar Award, a Camille and Henry Dreyfus Teacher/Scholar Award, and an Alfred P. Sloan Fellowship.

REFERENCES
1. Brudvig, G.W. and Crabtree, R.H. (1986) PNAS **83**, 4586.
2. Kambara, T. and Govindjee (1985) PNAS **82**, 6119.
3. Webber, A.N., Spencer, L., Sawyer, D.T., and Heath, R.L. (1985) FEBS Lett. **189**, 258.
4. Babcock, G.T. in "New Comprehensive Biochemistry - Photosynthesis" (J. Amesz, ed.) Elsevier, Amsterdam, in press.
5. Brudvig, G.W. (1986) J. Bioenerg. Biomemb., in press.
6. Renger, G. and Govindjee (1985) Photosyn. Res. **6**, 33.
7. Krishtalik, L.I. (1986) BBA **849**, 162.
8. Klimov, V.V., Allakhverdiev, S.I., Demeter, S., and Krasnovsky, A.A. (1980) Dokl. Akad. Nauk. SSSR **249**, 227.
9. Bouges-Bocquet, B. (1980) BBA **594**, 85.
10. Förster, V., Hong, Y.-Q., and Junge, W. (1981) BBA **638**, 141.
11. Laszlo, J.A., Baker, G.M., and Dilley, R.A. (1984) J. Bioenerg. Biomemb. **16**, 37.

COORDINATION OF AMMONIA, BUT NOT LARGER AMINES, TO THE MANGANESE SITE OF THE O_2-EVOLVING CENTER IN THE S_2 STATE

Warren F. Beck and Gary W. Brudvig,
Department of Chemistry, Yale University, New Haven, CT 06511 U.S.A.

1. Introduction

The Mn site of the O_2-evolving center (OEC) of photosystem II (PSII) consists of a tetramer of exchange-coupled Mn ions. Electron paramagnetic resonance (EPR) studies have suggested that the Mn ions in the OEC are arranged in a "cubane"-like configuration in the S_2 state. The proposal that the Mn site functions as the substrate-binding site of the OEC has been based primarily upon the inhibition of O_2-evolution activity in the presence of primary amines, which might be expected to coordinate as Lewis bases to Mn ions (2). To investigate this possibility, we have monitored ligand-exchange reactions at the Mn site in both the S_1 and S_2 states in the presence of several primary amines, employing the S_2 state multiline EPR signal as a structural probe (3).

2. Procedure

PSII membranes were isolated from market spinach as previously described (4). The methods employed for amine treatments and for sample illuminations have been reported previously (3). EPR spectra were obtained at 8.0K using the apparatus described previously (3). The spectra shown in this paper were obtained through computer subtraction of the dark background spectrum from the post-illumination or post-treatment spectrum obtained under the same measurement conditions.

3. Results

Figures 1a-e show S_2 state multiline EPR signals produced using illumination at 0°C of untreated and amine-treated PSII membranes at pH 7.5. The hyperfine line spacing and lineshape of the signals exhibited by PSII membranes treated with 2-amino-2-ethyl-1,3-propanediol (Figure 1b), Tris (Figure 1c), or $CH_3NH_2 \cdot HCl$ (Figure 1d) are the same as those observed in the spectrum from untreated PSII membranes (Figure 1a). However, the signal produced in PSII membranes treated with 100 mM NH_4Cl (Figure 1e) has a markedly different lineshape and hyperfine line spacing. Compared to the 87.5 G average hyperfine line spacing observed in the signal from untreated membranes, the S_2 state multiline EPR signal present in NH_4Cl-treated PSII membranes has an average hyperfine line spacing of 67.5 G. As argued previously (3), the observation of the novel S_2 state multiline EPR spectrum exhibited by NH_4Cl-treated PSII membranes is direct spectroscopic evidence for coordination of one or more NH_3 molecules directly to the Mn site in the S_2 state. That the S_2 state multiline EPR signals obtained in PSII membrane samples treated with larger amines, even CH_3NH_2, are indistinguishable from those

Biggens, J. (ed.), Progress in Photosynthesis Research, Vol. I. ISBN 90 247 3450 9
© 1987 Martinus Nijhoff Publishers, Dordrecht. Printed in the Netherlands.

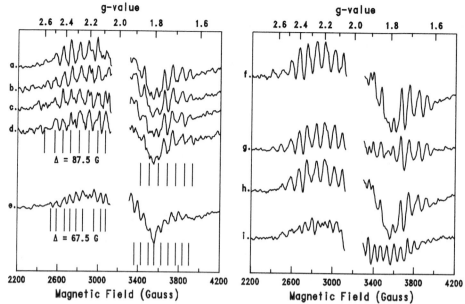

Figure 1. S_2 state multiline EPR signals produced in PSII membranes at pH 7.5. The g=2.0 region, which is obscured by interference from EPR Signal II_s, is not shown. The samples in a) through e) contained 100 μM DCMU and were illuminated for 30 sec at 0°C; in f) through i) the samples contained 250 μM DCBQ. EPR spectrometer conditions: microwave frequency 8.9 GHz, microwave power 200 μW, field modulation amplitude 20 G; temperature 8.0K; a) untreated PSII membranes (total $[Cl^-]= 0.5$ mM); b) PSII membranes treated with 100 mM 2-amino-2-ethyl-1,3-propanediol (total $[Cl^-]= 0.5$ mM); c) PSII membranes treated with 100 mM Tris (total $[Cl^-]= 0.5$ mM); d) PSII membranes treated with 100 mM $CH_3NH_2 \cdot HCl$ (total $[Cl^-]= 115$ mM); e) PSII membranes treated with 100 mM NH_4Cl (total $[Cl^-]= 115$ mM); f) untreated PSII membranes illuminated at 200K for 2 min; g) the sample in f), after warming to 0°C for 1 min in darkness; h) PSII membranes treated with 100 mM NH_4Cl, illuminated at 200K for 2 min; i) the sample in h), after warming to 0°C for 1 min in darkness.

obtained in untreated samples indicates that the larger amines do not readily coordinate to the Mn site in either the S_1 or S_2 states.

When illumination at 0°C is employed to produce the S_2 state, the electron transfer and ligand binding steps that produce the NH_3-inhibited S_2 state occur simultaneously (3). In order to observe these steps separately, the PSII membrane samples are first illuminated at 210K to produce the S_2 state and then the samples are warmed to 0°C in darkness, allowing the ligand binding chemistry to proceed. Untreated PSII membranes illuminated at 210K produce S_2 state multiline EPR signals (Figure 1f) that do not change in either lineshape or hyperfine line spacing upon warming to 0°C in darkness (Figure 1g); the presence of 2,5-dichloro-p-benzoquinone (DCBQ) in the samples causes re-oxidation of the reduced primary quinone electron acceptor Q_A^- upon warming, as evidenced by the collapse of the g=1.9 Fe(II)Q_A^- EPR signal. In comparison, NH_4Cl-treated PSII membranes produce S_2 state multiline EPR signals after illumination at 210K (Figure 1h) that are indistinguishable from those obtained in untreated PSII membranes. After warming to 0°C in darkness, the S_2 state multiline EPR signal observed in NH_4Cl-treated PSII membranes is transformed to the type having an altered lineshape and hyperfine line spacing (Figure 1i). The results shown in Figures 1f-i demonstrate that NH_3 binds to the Mn site after a structural change has occurred during the S_1 to S_2 state transition. Moreover, the fact that the S_2 state multiline EPR spectrum obtained in untreated PSII membranes does not change upon warming from 210K to 0°C shows that the coordination sphere and configuration of Mn in the OEC is the same in the S_1 and S_2 states in untreated PSII membranes.

4. Discussion

The action of NH_3 as an inhibitor of photosynthetic O_2 evolution apparently occurs through the ability of NH_3 to coordinate in lieu of H_2O to the Mn site. From our EPR results, it is clear that NH_3 binds to the Mn site in the S_2 state and not in the S_1 state. These results are in concord with the findings of Velthuys (5), who concluded using results of luminescence experiments that an NH_3-binding event occurs in both the S_2 and S_3 states. Since both EPR (6,7) and X-ray absorption edge (8) results indicate that the Mn site functions to accumulate oxidizing equivalents, it is probable that the Mn site becomes increasingly electron deficient as it is advanced through the S states. It is likely, then, that NH_3 binds to Mn in a nucleophilic addition reaction that is triggered by the oxidation of the Mn site in the S_1 to S_2 state transition. However, the EPR data on untreated samples indicate that H_2O does not bind to Mn in the S_1 to S_2 state transition. This result can be easily accounted for by the greater basicity of NH_3 relative to that of H_2O. We propose that one molecule of NH_3 coordinates to the Mn site in the S_2 state and that a second NH_3 molecule binds in the S_3 state. The inhibition of H_2O oxidation by NH_3 can thus be readily attributed to the greater basicity of NH_3.

Brudvig and Crabtree (9) recently proposed a model for the structure of the Mn site which attempts to account for the available EPR (1) and X-ray absorption (8) data. Figure 2 shows a possible mechanism for the binding of a primary amine NRH_2 to the Mn site based on the Brudvig and Crabtree hypothesis and the above discussion.

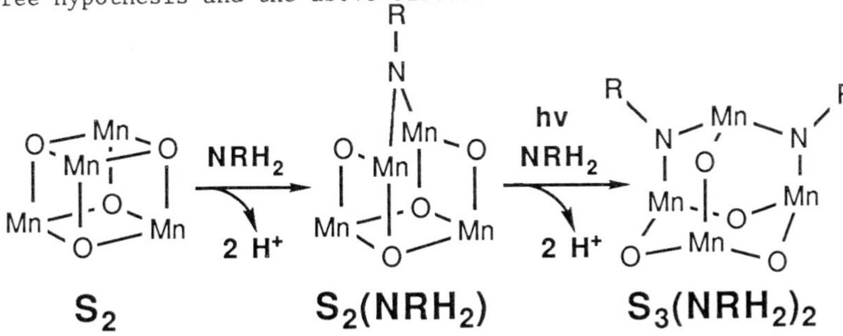

Figure 2. Proposed mechanism for the nucleophilic addition of a primary amine NRH_2 to the Mn site in the S_2 and S_3 states, based on the proposed mechanism for photosynthetic O_2 evolution of Brudvig and Crabtree (9).

NRH_2 is proposed to coordinate to the "cubane" form of the Mn site in the S_2 state through a nucleophilic addition reaction. The resultant structure for the $S_2(NRH_2)$ form of the Mn site incorporates a μ-alkylimido bridging ligand. Through the same mechanism, a second NRH_2 binds to the Mn site, forming a $Mn_4O_4(NR)_2$ "adamantane"-like structure. The antiferromagnetic and ferromagnetic superexchange couplings between the Mn ions in the $S_2(NRH_2)$ complex are expected to be intermediate in strength between that of the untreated S_2 state "cubane" structure (strongly coupled)(1) and that of the $S_3(NRH_2)_2$ "adamantane" structure (weakly coupled)(10), which is consistent with the currently known magnetic properties of the $S_2(NH_3)$ form of the Mn site (1). As explained previously, the weaker antiferromagnetic and ferromagnetic superexchange couplings present in the NH_3-inhibited S_2 state cause the observed reduction in the hyperfine line spacing in the S_2 state multiline EPR signal (1,3).

An additional attractive feature of the model shown in Figure 2 concerns the geometry of the μ-alkylimido bridging ligand in the $S_2(NRH_2)$ form of the Mn site. The alkyl group of the bound NRH_2 projects outward from the Mn tetramer, which suggests that it would be severely sterically hindered by the surrounding protein matrix. The nucleophilic addition of amines larger than NH_3 to the Mn site, then, is expected from this model to be prohibited by steric constraints, which is consistent with our EPR results (3) and with the O_2-evolution inhibition studies of Sandusky and Yocum (2). That amines larger than NH_3 are incapable of binding to the Mn site in the S_2 state shows that the Mn site is extremely sterically selective for small Lewis bases, making it even more attractive to assign the substrate-binding site of the OEC to the Mn site, as was proposed by Sandusky and Yocum (2).

5. References

1 de Paula, J.C., Beck, W.F., and Brudvig, G.W. (1986) *J. Am. Chem. Soc.* *108*, 4002-4009.
2 Sandusky, P.O., and Yocum, C.F. (1984) *Biochem. Biophys. Acta 766*, 603-611; Sandusky, P.O., and Yocum, C.F. (1986) *Biochem. Biophys. Acta 849*, 85-93.
3 Beck, W.F., de Paula, J.C., and Brudvig, G.W. (1986) *J. Am. Chem. Soc.* *108*, 4018-4022; Beck, W.F., and Brudvig, G.W. (1986) *Biochemistry 25*, in press.
4 Berthold, D.A., Babcock, G.T., and Yocum, C.F. (1981) *FEBS Lett. 134*, 231-234; Beck, W.F., de Paula, J.C., and Brudvig, G.W. (1985) *Biochemistry 24*, 3035-3043.
5 Velthuys, B.R. (1975) *Biochem. Biophys. Acta 396*, 392-401.
6 Dismukes, G.C., and Siderer, Y. (1981) *Proc Natl. Acad. Sci. USA 78*, 274-278.
7 Zimmermann, J.L., and Rutherford, A.W. (1984) *Biochem. Biophys. Acta 767*, 160-167.
8 Goodin, D.B., Yachandra, V.K., Britt, R.D., Sauer, K., and Klein, M.P. (1984) *Biochem. Biophys. Acta 767*, 209-216.
9 Brudvig, G.W., and Crabtree, R.H. (1986) *Proc. Natl. Acad. Sci. USA 83*, 4586-4588.
10 Wieghardt, K., Bossek, U., and Gebert, W. (1983) *Angew. Chem. Int. Ed. Engl. 22*, 328-329.

EPR STUDIES OF THE OXYGEN-EVOLVING SYSTEM. THE INTERACTION WITH AMINES

LARS-ERIK ANDRÉASSON AND ÖRJAN HANSSON

DEPARTMENT OF BIOCHEMISTRY AND BIOPHYSICS, CHALMERS UNIVERSITY OF
TECHNOLOGY AND UNIVERSITY OF GÖTEBORG, S-412 96 GÖTEBORG, SWEDEN

INTRODUCTION

Amines react with the photosynthetic oxygen-evolving system in
different ways depending on the chemical nature of the amine and the
experimental conditions. Hydroxylamine and hydrazine in millimolar
concentrations inhibit oxygen evolution irreversibly by releasing
manganese (1) and extrinsic proteins (2,3) from the photosynthetic
membrane. At lower concentrations these amines are oxidized in the place
of water and shift the familiar flash-dependent oxygen-yield pattern by
two units (4,5) presumably by binding to and reducing the S_1 state of the
oxygen-evolving complex. Because of their structural similarities with
water it was suggested that these amines act as water analogs with one
amine replacing two molecules of water at the water-splitting site (6).
    Recently, however, it has been shown that the reactions of
hydroxylamine and hydrazine are more complicated than originally thought.
Studies of the proton release associated with the oxidation of
hydroxylamine (7) and the oxygen flash yield in the presence of hydrazine
(8) indicate that more than one amine molecule bind in the S_1 state.
    The interaction of ammonia with the oxygen-evolving system is funda-
mentally different from that of hydroxylamine and hydrazine. From
luminescence studies Velthuys (9) found that ammonia inhibits the oxygen
release step by binding to the S_2 and S_3 states and proposed direct
binding to the manganese. Ammonia is not expected to act as a reductant in
this process. There are indications that more than one binding site for
ammonia exist. At a second site essential chloride competes with
ammonia (10).
    The discovery of EPR signals from the S_2 state (11-15) has opened up
new possibilities for studying directly the interaction of amine compounds
with the oxygen-evolving system and with manganese in particular. Recent
examples of such studies have shown that water (16) and probably also
ammonia (17) are directly coordinated to manganese in the S_2 state whereas
no firm evidence for the direct interaction of chloride with manganese has
been found so far (18).
    In the present study the reactions of ammonia, hydroxylamine and
hydrazine with the oxygen-evolving system are further analyzed to gain
additional information about their modes of interaction with the manganese
and their possible interference with the binding of water and chloride.

MATERIALS AND METHODS

Oxygen-evolving photosystem II-enriched membranes (600 μmol O_2/mg Chl·
h) were prepared from spinach as described in (19) and suspended in Mes
(20 mM Mes, pH 6.3) or Hepes buffer (20 mM Hepes, pH 7.5) containing 400
mM sucrose, 15 mM NaCl and 5 mM $MgCl_2$. Oxygen evolution and manganese were
measured as in (19). PSII membranes in $H_2{}^{17}O$-enriched media were prepared

Biggens, J. (ed.), Progress in Photosynthesis Research, Vol. I. ISBN 90 247 3450 9

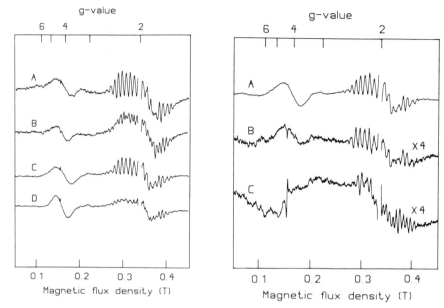

Figure 1. Effect of NH$_3$ on the EPR signals from the S$_2$ state. The S$_2$ state was generated by illumination of PSII membranes (10 mg Chl/ml). Conditions: A, dark-adapted PSII membranes in Hepes buffer, pH 7.5, with 80 µM DCMU (added dissolved in dimethylsulfoxide) illuminated at 273 K for 30 s before freezing. B, as (A) but with 50 mM (NH$_4$)$_2$SO$_4$ present. C, PSII particles in Mes buffer, pH 6.3, illuminated at 200 K. D, as (C) but with 300 mM (NH$_4$)$_2$SO$_4$ added. The spectra are the average of 4 scans. The spectra of the samples taken before illumination have been subtracted. Microwave frequency, 9.46 GHz; power, 20 mW; modulation amplitude, 2.5 mT; temperature, 11 K.

Figure 2. Binding of NH$_3$ to the S$_2$ state of the oxygen-evolving system. A, EPR spectrum of the S$_2$ state produced by illumination of dark-adapted PSII membranes (10 mg Chl/ml in Hepes, pH 7.5) at 200 K. B, thawed at 273 K for 120 s and refrozen. C, thawed at 273 K for 90 s after which 50 mM (NH$_4$)$_2$SO$_4$ was added and allowed to react for 30 s before freezing. All handling of the samples after the initial illumination was carried out in the dark. Conditions for EPR as in Fig. 1.

as in (16). The S$_2$ state was generated by illumination of dark-adapted (1 h) PSII membranes at 200 K (13) for 5 min or at 273 K for 30 s in the presence of 80 µM DCMU followed by freezing at 200 K. EPR measurements were made as described in (16).

RESULTS

Effect of NH$_3$

It has recently been shown that ammonia induces a perturbation of the S$_2$ state (17). This is seen as a change in the appearance of the multiline

EPR signal characterized by a reduced spacing between the hyperfine lines and a redistribution of intensity among these (Fig. 1 A,B). Here we can show that the EPR signal from the S_2 state around $\underline{g}=4$ is also affected by the presence of ammonia. The center of the signal shifts from $\underline{g}=4.1$ to $\underline{g}=4.2$ and the width of the line decreases from 40 mT to 30 mT (Fig. 1 A,B).

Beck et al. (17) showed that the modification of the multiline signal induced by ammonia could be seen after generation of the S_2 state at 273 K, a temperature where ligand substitution is expected to be possible. When ammonia was added to PSII membranes in the S_1 state and the S_2 state was produced by illumination of solid samples at 200 K, the normal multiline signal was detected. This was interpreted as binding of ammonia to the S_2 state only. We have obtained additional evidence for binding to the S_2 state by generating this state in the absence of ammonia by low-temperature illumination and observation of the effect on the multiline EPR signal after addition of ammonia to the thawed PSII membranes (Fig. 2). The addition of ammonia leads to a rapid formation of a modified EPR signal (within 30 seconds). The lower amplitude of the signals in the ammonia-treated sample (Fig. 2 C) and in the control (Fig. 2 B) compared to the original signal after illumination at 200 K (Fig. 2 A) is caused by the decay of the S_2 state by charge recombination before the samples are refrozen.

When the PSII membranes were illuminated at 200 K in the presence of ammonia (Fig. 1 D) a multiline signal with largely normal properties was produced in terms of the spacing between the hyperfine lines as observed before (17). However, the yield of the multiline signal was considerably lower than in the absence of ammonia (Fig. 1 C). In addition, the effects on the EPR signal around $\underline{g}=4$ were similar to those observed after illumination at 273 K, i.e. a decrease in width and a shift in $\underline{g}$-value.

Yocum et al. (10) have found two ammonia-binding sites which affect the oxygen-evolving capacity of PSII. One of these is also capable of binding chloride. If this site was the same as that where ammonia binds and perturbs the multiline EPR signal. the presence of chloride would be expected to prevent the ammonia-induced effect. However. even at concentrations of chloride as high as 200 mM the modified multiline signal is seen (Fig. 3) which argues against chloride binding at this site.

The effect of ^{17}O-labelled water on the NH_3-modified S_2 state

We have earlier shown that one or several oxygen atoms derived from water is coordinated to manganese in the S_2 state of the oxygen-evolving system (16) by the broadening of the multiline signal observed in the presence of ^{17}O-labelled water. The broadening is the result of interaction between the magnetic moment of the oxygen nucleus ($\underline{I}=5/2$) and the unpaired electron spin localized on the manganese. When the multiline EPR signal is generated by illumination of DCMU-treated PSII membranes at 273 K in the presence of ammonia in a medium enriched in ^{17}O-labelled water, a similar broadening of the individual lines of the ammonia-perturbed multiline signal is observed (Fig. 4 B,B'). This shows that at least one water-derived oxygen ligand can bind to the manganese in the presence of bound ammonia.

The interaction with NH_2OH and N_2H_4

Oxygen-evolving PSII membranes were treated with different

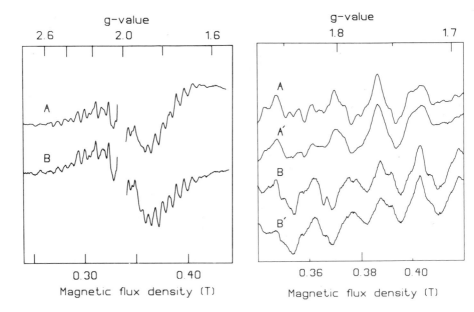

Figure 3. The effect of chloride on the ammonia-modified multiline EPR signal from the S_2 state. The S_2 state was produced as in Fig. 1B. A, control particles. B, in the presence of 200 mM NaCl. Spectrometer conditions as in Fig. 1.

Figure 4. Effect of $H_2{}^{17}O$ on the ammonia-modified multiline EPR signal from the S_2 state. The figure shows a section of the signal on the high-field side. The S_2 state was produced by illumination of dark--adapted PSII particles (14 mg Chl/ml in Mes. pH 6.3 with 80 μM DCMU dissolved in ethanol) in the absence (A,A′) or presence (B,B′) of 300 mM $(NH_4)_2SO_4$. In A′ and B′ the medium contained 42% $H_2{}^{17}O$. Dark spectra have not been subtracted. Average of 32 scans. Modulation amplitude 0.8 mT, other spectrometer conditions as in Fig. 1.

concentrations of hydroxylamine (50 μM to 5 mM) at 273 K for one hour in darkness. Then the hydroxylamine was removed by washing and the oxygen-evolving activity. the manganese content and the amplitude of the multiline and g=4.1 signals were measured. The results show that the activity and EPR signal amplitudes and the total manganese content follow each other closely (Fig. 5). Similar results were obtained when hydrazine was used instead of hydroxylamine.
    In order to investigate the behaviour of the oxygen-evolving system in the presence of low. non-destructive concentrations of hydroxylamine (<100 μM), PSII membranes were incubated in the dark with the amine for 30 seconds on ice and frozen to stop further reaction. EPR spectra of the multiline signal taken before and after illumination at 200 K did not differ from those of samples where hydroxylamine was omitted, i.e. no additional signals were seen in the sample before illumination and the amplitude of the light-induced signal in the hydroxylamine-treated sample was the same as that in the control. Thus. since the normal S_2 state was formed by an illumination procedure which allows the transfer of only one

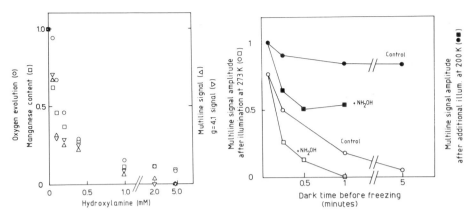

Figure 5. The effect of NH$_2$OH on the oxygen-evolving system. PSII membranes (2 mg Chl/ml in Mes, pH 6.3) were treated with NH$_2$OH for 1 h at 273 K as described in the text and the properties measured: Circles, oxygen evolution (control, 580 µmol O$_2$/mg Chl·h); squares, manganese content (control, 3.9/250 Chl); triangles, multiline signal amplitude; inverted triangles, amplitude of the g=4.1 signal.

Figure 6. The reaction of NH$_2$OH with the S$_2$ state. PSII membranes (9 mg Chl/ml in Mes, pH 6.3 with 80 µM DCMU) were illuminated at 273 K to generate the S$_2$ state and allowed to react with NH$_2$OH as described in the text. The amplitude of the multiline EPR signal was taken as a measure of the concentration of the S$_2$ state. Circles, amplitude in the absence of NH$_2$OH (control); squares, amplitude in the presence of 100 µM NH$_2$OH. Filled symbols represent amplitudes after the final illumination at 200 K.

electron from the water-splitting site, no reaction of the initial S$_1$ state had taken place during the 30 s incubation. When higher concentrations of hydroxylamine (>100 µM) or hydrazine (>300 µM) or longer incubation times were used, a lower yield of the S$_2$ EPR signal was seen. This could be accounted for by a loss of manganese as described above.

The reaction of hydroxylamine with the S$_2$ state was examined as follows. Dark-adapted PSII membranes were illuminated at 273 K in the presence of DCMU to stop the reaction at the S$_2$ state. Hydroxylamine (80 µM) was then added to the PSII membranes and allowed to react with S$_2$ in the dark. The reaction was quenched by freezing after various times. Figure 6 summarizes the effect of hydroxylamine on the rate of decay of the multiline signal of the S$_2$ state. It is clear that the rate of decay of the signal was significantly faster in the hydroxylamine-treated samples than in the control, an indication that the S$_2$ state is reactive towards the amine.

The samples were given an additional illumination at 200 K to analyze the state of the PSII membranes after the decay of the S$_2$ state. In the control samples the multiline signal was regenerated to a level exceeding the signal level in the sample with the shortest reaction time (about 5 seconds). The excess signal represents a recombination of S$_2$Q$_A^-$ during the transfer of the samples from 273 K where they were illuminated. The decrease in the ability to regenerate the signal seen at longer times is possibly due to a slow leakage of electrons from the acceptor side. In the hydroxylamine-treated samples. however. the signal level after the additional 200 K illumination was significantly lower, a further indica- tion that a reaction with hydroxylamine had taken place. Experiments with hydrazine as the reactive amine gave similar results, i.e. a faster decay of the S_2 state than in the control. The g=4.1 signal is more difficult to measure accurately. but its amplitude seemed to follow that of the multiline signal quite closely under the various conditions.

DISCUSSION

Hanssum and Renger (8) have recently investigated the reactions of the different S-states with hydroxylamine and hydrazine by analyzing the oxygen yield pattern in the presence of these amines. EPR provides an independent and more direct method for studying these reactions through the observation of the ability to form the paramagnetic S_2 state under various conditions.

The observation that the formation of the multiline EPR signal is not affected after incubation with low concentration of hydroxylamine or hydrazine at 273 K is fully consistent with the results by Hanssum and Renger who found a halftime of about 1 minute for the reaction with S_1 at room temperature (8). The reaction is expected to be even slower at 273 K. Longer reaction times or higher concentrations of the amines invariably cause a loss of manganese which in turn leads to a decrease in oxygen evolution activity and a parallel impairment of the ability to form the multiline and g=4.1 EPR signals (Fig. 5). As probably more than one amine molecule are required for reaction with the S_1 state (7), the high concentrations needed for EPR lead to loss of manganese. This makes it difficult to study a reduction of the S_1 state and the reaction which is assumed to be responsible for the shift in the oxygen-evolution pattern. However, the results in Fig. 5 show that formation of these signals are equally dependent on an intact manganese-binding site and a further indication that also the g=4.1 signal originates in this site.

The results of Hanssum and Renger also indicated that the S_2 state reacts much faster than the S_1 state with hydroxylamine. The direct measurement of the decay of the S_2 state by observation of the disappearance of the multiline signal in the presence of hydroxylamine supports this notion. The halftime of the S_2 state decreased from about 30 to about 15 seconds with hydroxylamine present.

Our results also give an indication of the mechanism of the reaction between hydroxylamine and the S_2 state. In the control samples (Fig. 6) the illumination at 200 K following the decay of the multiline signal led to high recovery of the amplitude of the signal. This shows that the decay in this case must have been the result of charge recombination between the acceptor and donor sides in PSII, regenerating the S_1 state. In the hydroxylamine-treated samples, however. the low-temperature illumination to a large extent failed to regenerate the multiline signal. The recovery which did occur is easily explained by assuming that two processes compete when hydroxylamine is added to PSII membranes in the S_2 state. In some of the PSII centra normal charge recombination takes place to regenerate the S_1 state. The low-temperature illumination of this population is responsi- ble for the partial reappearance of the multiline signal. In a second population a reaction between the S_2 state and hydroxylamine appears to have taken place which prevents formation of the S_2 state in the sub- sequent low-temperature illumination. This finding eliminates a mechanism in which hydroxylamine only induces an accelerated recombination of charges, i.e. acts as an ADRY agent (20). Instead a reduction of the manganese site is likely to have occurred. Our present results do not allow us to say whether the reduction by hydroxylamine proceeded to the S_1 or S_0 state of the donor side. The reason for this ambiguity is that, since DCMU was present, the acceptor side would still be reduced so that further electron transfer from the donor side would be impossible in both cases.

The rate of decay of the S_2 state was accelerated also with hydrazine. Conceivably, a similar type of interaction occurs as with hydroxylamine. Hanssum and Renger (8) failed to detect any interaction of the S2 state with hydrazine during a flash sequence. This difference in results can probably be traced to the different concentrations of photosynthetic material used in the oxygen yield and the EPR measurements. It is also possible that the PSII membranes are slightly modified in their behaviour towards amines compared to that of chloroplasts.

From the structural requirements for inhibition of oxygen evolution found with water analogues (6) one would expect two ammonia molecules to bind, thus replacing two water molecules. Our results show that the ammonia-modified multiline EPR signal is broadened in the presence of ^{17}O-labelled water. The broadening must be the result of a direct coordination of oxygen from water to the manganese responsible for the multiline EPR signal and a strong indication that water and ammonia are simultaneously bound to the manganese in the S_2 state. The broadening of the ammonia-modified multiline EPR signal induced by the labelled water does not permit an unambiguous determination of the number of bound oxygens. However, if the above picture of the oxygen-evolving site is correct, our observations indicate that only one of the two water molecules has been replaced by ammonia. This agrees with the mechanism of inhibition of oxygen evolution by ammonia suggested by Velthuys (9).

High concentrations of chloride do not affect the formation of the ammonia-modified multiline signal. Thus, this binding site for ammonia does not appear to be accessible to chloride. Recent EPR studies also seem to exclude the binding of chloride to the manganese responsible for the multiline EPR signal (18).

Some earlier experiments have suggested that binding of ammonia takes place only after the S_2 state has formed (9). The modification of the g=4.1 EPR signal and the change in yield in the otherwise normal multiline signal when the S_2 state is generated in frozen samples indicate that some sort of interaction with ammonia occurs already in the S_1 state. We have proposed that the g=4.1 and the multiline signals arise from separate PSII electron donors which are in redox equilibrium in the S_2 state (21). One possibility would be that ammonia binds already in the dark to the g=4.1 component, resulting in a modification of its spectral properties and affecting the ability to form the multiline signal from the other component. If ammonia competes with chloride at this site (10), the use of a high concentration of chloride in (9) may explain why no binding of ammonia to S_1 was observed previously.

ACKNOWLEDGMENTS

This work was supported by the Swedish Natural Science Research Council. We are indepted to Professor Tore Vänngård for his encouragement and helpful critizism.

REFERENCES

1 Cheniae. G.M. and Martin, I.F. (1971) Plant Physiol. 47, 568-575
2 Ghanotakis. D.F., Topper, J.N. and Yocum, C.Y. (1984) Biochim. Biophys. Acta 767, 524-531
3 Tamura. N. and Cheniae, G. (1985) Biochim. Biophys. Acta 809, 245-259
4 Bouges, B. (1971) Biochim. Biophys. Acta 234, 103-112
5 Kok. B. and Velthuys. B.R. (1977) in Research in Photobiology (Castellani, ed.), 111-119, Plenum, N.Y.
6 Radmer. R. and Ollinger. O. (1983) FEBS Lett. 152, 39-43
7 Förster, V. and Junge, W. (1985) FEBS Lett. 186, 153-157
8 Hanssum, B. and Renger. G. (1985) Biochim. Biophys. Acta 810, 225-234
9 Velthuys, B.R. (1971) Biochim. Biophys. Acta 396, 392-401
10 Sandusky. P.O. and Yocum, C.F. (1984) Biochim. Biophys. Acta 766, 603-611
11 Dismukes. G.C. and Siderer, Y. (1981) Proc. Natl. Acad. Sci. USA 78, 274-278
12 Hansson. Ö. and Andréasson, L.-E. (1982) Biochim. Biophys. Acta 679, 261-268
13 Brudvig, G.W., Casey, J.L. and Sauer, K. (1983) Biochim. Biophys. Acta 723, 366-371
14 Casey. J.L. and Sauer, K. (1984) Biochim. Biophys. Acta 767, 21-28
15 Zimmermann, J.L. and Rutherford. A.W. (1984) Biochim. Biophys. Acta 767, 160-167
16 Hansson. Ö., Andréasson, L.-E. and Vänngård, T. (1986) FEBS Lett. 195, 151-154
17 Beck. W.F., de Paula, J.C. and Brudvig, G.W. (1986) J. Am. Chem. Soc. 108, 4018-4022
18 Yachandra. V.K., Guiles. R.D., Sauer, K. and Klein. M.P. (1986) Biochim. Biophys. Acta 850, 333-342
19 Franzén, L.-G., Hansson, Ö. and Andréasson. L.-E. (1985) Biochim. Biophys. Acta 808, 171-179
20 Renger. G. (1972) Biochim. Biophys. Acta 256, 428-439
21 Hansson. Ö. Aasa, R. and Vänngård. T. (1986) Biophys. J. submitted

COOPERATIVE BINDING OF HYDROXYLAMINE AND HYDRAZINE TO THE
WATER-OXIDIZING COMPLEX

VERENA FÖRSTER AND WOLFGANG JUNGE, BIOPHYSIK, UNIVERSITÄT OSNABRÜCK,
POSTFACH 4469, D-4500 OSNABRÜCK, GERMANY (F.R.G.)

1. INTRODUCTION

The active site of the water-oxidizing complex (WOC) consists of a
binuclear manganese complex, which acts as buffer of oxidizing
equivalents. Some small molecules interact with the WOC. NH_2OH and NH_2NH_2
cause a two-digit shift in the four-step sequential progress of
one-electron oxidations which are observed when dark-adapted thylakoids
are subjected to a series of light flashes. This was first observed via
the oxygen evolution pattern (1,2). Correspondingly, we found that the
pattern of proton release was shifted from $OH^+(S_1-S_2)$: $1H^+(S_2-S_3)$:
$2H(S_3-S_0)$: $1H^+(S_0-S_1)$:... in the absence of these agents to $2H^+(S_*-S_0)$:
$1H^+(S_0-S_1)$: $OH^+(S_1-S_2)$: $1H^+(S_2-S_3)$: $2H^+(S_3-S_0)$:... in the presence of
NH_2OH or NH_2NH_2.

2. MATERIALS AND METHODS

Thylakoids were prepared from peas (pisum sativum), frozen and stored
under liquid nitrogen until use (3). pH changes in the thylakoid lumen
were monitored via absorption changes of neutral red, the external phase
buffered by BSA (4). Proton release due to plastohydroquinone oxidation by
PS I was inhibited by DNP-INT. Flashphotometric measurements were carried
out as described in detail in (3).

3. RESULTS AND DISCUSSION

Fig.1 shows proton release into the lumen of dark-adapted thylakoids upon
excitation by a series of flashes in the presence of various amines. The
time resolution was 2ms/point. Rapid acidifications ($200\mu s$-1.2ms, for time
resolution see ref.3) were followed by slower phases. The amplitude of the
rapid acidifications was plotted as function of flash number in Fig.2
(except traces d and e of Fig.1). Traces a in Fig.1 and 2 show the normal
damped oscillatory pattern of proton release which is explained by the
proton release stoichiometry $1H^+(S_0-S_1)$: $OH^+(S_1-S_2)$: $1H^+(S_2-S_3)$:
$2H^+(S_3-S_4-S_0)$ (3). Among different amines NH_2OH, NH_2NH_2 and NH_2O-SO_3H were
most effective in causing a two-digit shift. CH_3NHOH was less effective
and NH_2OCH_3 showed no effect. With NH_2OH, NH_2O-SO_3H as well as NH_2NH_2 $2H^+$
were released upon the first excitation (3.1ms half-rise time, see ref.5).
From the second flash on the pattern seemed to proceed normally with
S_0-S_1, S_1-S_2,... (6).

Fig.3 shows the increase of the rapid proton yield upon the first flash
and the decrease of the proton yield upon the third flash as function (a)
of NH_2OH and (b) of NH_2NH_2 concentration. The r e l a t i v e increase on
the first flash and the r e l a t i v e decrease on the third flash were
taken as a measure of the "normalized reactivity", i.e. the percentage of
WOCs which were modified. We calculated the Hill coefficient n at half
saturation. It was 2.43 for NH_2OH and 1.48 for NH_2NH_2. Thus, the WOC had
at least 3 cooperative binding sites for NH_2OH and at least 2 binding
sites for NH_2NH_2 (for details see refs.5,7). The concentration dependences

Biggens, J. (ed.), Progress in Photosynthesis Research, Vol. I. ISBN 90 247 3450 9
© *1987 Martinus Nijhoff Publishers, Dordrecht. Printed in the Netherlands.*

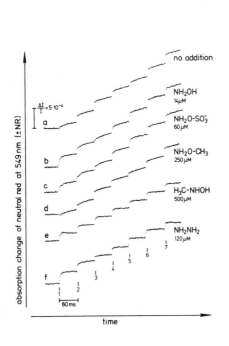

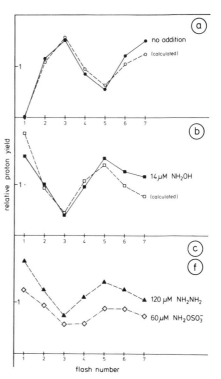

Fig.1:
Absorption changes of neutral red
at 549nm (±NR) in response to
a series of saturating flashes,
measured in the absence and
in the presence of NH_2OH, NH_2OSO_3H,
NH_2OCH_3, CH_3NHOH and NH_2NH_2 at
concentrations as indicated.

Fig.2:
Rapid internal pH changes as
function of flash number,
obtained from the respective traces
in Fig.1. Open symbols: calculated
under assumption of (a) $100\%S_1$
and (b) $100\%S_*$ in the dark,
10% misses and 10% double hits.

were fitted assuming the sequential interaction model for multisite
enzymes, as described by the following equilibria (8):

$$W \underset{K_d}{\overset{+D}{\rightleftharpoons}} WD_1 \underset{aK_d}{\overset{+D}{\rightleftharpoons}} WD_2 \underset{bK_d}{\overset{+D}{\rightleftharpoons}} WD_3 \underset{cK_d}{\overset{+D}{\rightleftharpoons}} WD_4 \rightleftharpoons \ldots$$

where W is the WOC, D the binding molecule, K_d the dissociation constant
and a,b,c the sequential interaction factors. For simplicity, we assumed
a=b=c. Occupation of 1,2 or more equivalent binding sites lead to one and
the same modification of the proton-release pattern. The experimental
curves could be fitted by various parameter sets. Table 1 lists possible
sets. Binding of 3 or 4 NH_2OH and of 2,(3) or 4 NH_2NH_2 explained the
concentration dependence equally well (solid lines in Fig.3). The
assumption of cooperativity, however, was indispensible. Via measurements
of oxygen evolution cooperative binding of NH_2NH_2 has also been observed
by Hanssum and Renger (9).
It was assumed that binding of NH_2OH/NH_2NH_2 to the WOC in the dark was
reversible. ~70% of the bound NH_2OH was washed off by three centrifugation

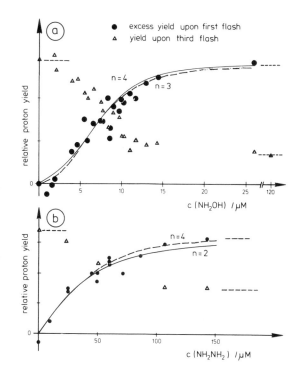

Proton yields upon the
first and third flash
as function of the
concentration of
NH_2OH (a) NH_2NH_2 (b).
The solid curves were
calculated under
assumption of
cooperative binding
of 3 and 4 molecules of
NH_2OH and of 2 and 4
molecules of NH_2NH_2
(see text).

steps. Incubation and washing were carried out in total darkness and
lasted about 90 minutes. NH_2NH_2 seemed to bind much more tightly to the
WOC.
We tested whether the anions Cl^- and OH^- (i.e. pH) had an influence on the
binding of NH_2OH (Fig.4). The "low Cl^- concentration" was between 1 and
10mM due to contamination from BSA-bound Cl^-. (Note: BSA was indispensable
for the neutral red technique. 2.6g/l Sigma A-7906 were used). In order to
adjust micromolar Cl^- concentrations BSA was dialysed before measurement
against the suspension medium. Proton-release patterns measured with the
dialysed BSA showed, dependent on the ionic strength (varied via sulfate
conc.), different deviations from the normal patterns (not documented
here) and were difficult to interpret. Thus we were confined to a somewhat
limited statement: Variation of Cl^- concentration between 10 and 100mM (at
ionic strength 100mM) and variation of pH between 6.2 and 7.7 had no
effect on the interaction of NH_2OH with the WOC.
The necessary assumption of 3 if not 4 equivalent binding sites for NH_2OH
on the WOC implied a certain symmetry. This is conceivable at the
binuclear manganese center. In Fig.5 we present a hypothesis for the
interaction of NH_2OH and NH_2NH_2 with a binuclear center of the WOC. A
twice bridged binuclear manganese complex (A) provides a symmetrical
arrangement of 4 equivalent ligand sites at the O bridges. NH_2OH and
NH_2NH_2 substitute the bridging oxygen ligands binding end-on to Mn^- (B,C).
All ligand sites are occupied by amino groups. This implied binding of
4 NH_2OH or 2 NH_2NH_2 at similar cooperativity (a=b=c=0.2, see Table 1 in
(5); NOTE printing error in (5): 2nd line in Table 1 refers to 4 NH_2NH_2
instead of 3.). That binding of NH_2OH was unaffected by changes of
chloride concentration is in line with recent work of Yachandra et

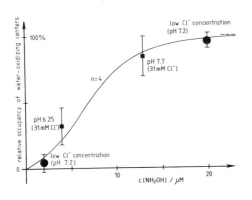

Fig.4
Percentage of modified
proton-release patterns
at different chloride
concentrations and pH
as function of NH_2OH
concentration. (solid
line taken from Fig.4a).

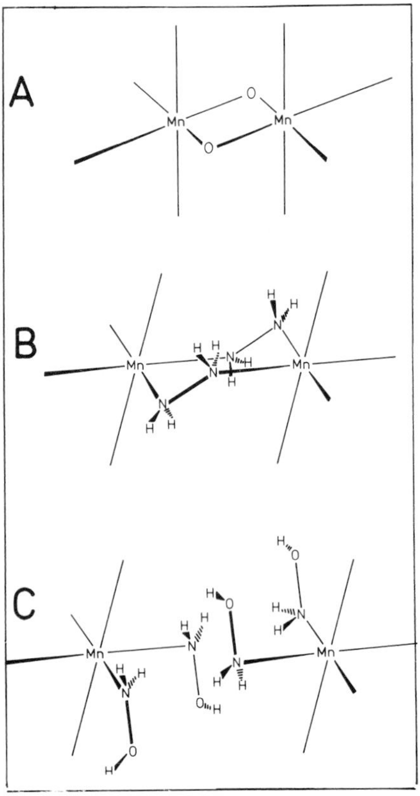

Fig.5:
Hypothetical structural
elements of the WOC.

al.(10), who find N, but not Cl in the first coordination sphere of Mn.
Insertion of the artificial ligands would cause an enlargement of the
Mn–Mn distance from 2.7Å to more than 5Å, followed by conformational
changes. These predictions will be subjected to experimental tests.

ACKNOWLEDGEMENTS
Financial aid has been provided by the Deutsche Forschungsgemeinschaft
(Sonderforschungsbereich 171-84, Projekt A2).

REFERENCES
1. Bouges,B. (1971) Biochim. Biophys. Acta 234, 103-112
2. Radmer,R. and Ollinger,O. (1982) FEBS Lett. 144, 162-166
3. Förster,V. and Junge,W. (1985) Photochem. Photobiol.41, 183-190
4. Junge,W., Ausländer,W., McGeer,A. and Runge,Th. (1979)
        Biochim. Biophys. Acta 546, 121-141
5. Förster,V. and Junge,W. (1986) Photosynth. Res.9, 197-210
6. Förster,V. and Junge,W. (1985) Photochem. Photobiol.41, 191-194
7. Förster,V. and Junge,W. (1985) FEBS Lett. 186, 53-57
8. Segel,I.H. (1975) Enzyme Kinetics, John Wiley and Sons Inc., New York
9. Hanssum,B. and Renger,G. (1985) Biochim. Biophys. Acta 810, 225-234
10.Yachandra,V.K., Guiles,R.D., McDermott,A., Britt,R.D., Dexheimer,
        S.L. Sauer,K. and Klein,M.P. (1986), Biochim. Biophys. Acta 850,
        324-332

REACTION MECHANISMS OF H_2O SUBSTRATE ANALOGUES AT THE PS II-DONOR SIDE IN THYLAKOIDS AND PS II-PARTICLES

B. Hanssum and G. Renger, Max-Volmer-Institut für Biophysikalische und Physikalische Chemie, Technische Universität Berlin, Strasse des 17. Juni 135, 1000 Berlin 12, FRG

1. INTRODUCTION

The mechanistic details of photosynthetic water oxidation are unresolved (for a recent review see Ref. 1). Among different ways to attack this problem, a well defined modification of the reaction pattern by selectively acting agents, e.g. ADRY-agents (2) or substrate analogues like hydroxylamine (3) can provide valuable information. Hydroxylamine (NH_2OH) at low concentrations (µM-range) was found to shift the caracteristic oscillation pattern of flash induced oxygen evolution in dark adapted chloroplasts by a number of two (3). A similar effect has been observed for hydrazine (NH_2-NH_2) and related compounds (4). Recently it was shown that the reaction pattern is more complex, especially in the case of NH_2OH (5). Accordingly, an unambiguous interpretation of experimental data obtained in the presence of water substrate analogues requires detailed knowledge about the underlying reaction mechanism of these substances. In this study experiments were performed in order to analyse the effect of NH_2OH and NH_2NH_2 in PS II-particles, as well as the action of some hydrazine derivatives in thylakoid preparations.

2. PROCEDURE
2.1 Material and Methods

PS II-particles with oxygen evolving capacity were prepared from market spinach as described in (6) with some modifications as outlined in (7), using a Triton X 100/chlorophyll ratio of 20:1. The oscillation patterns of oxygen yield in dark adapted thylakoids were detected by an unmodulated Joliot-type electrode (8) as described in (5). The buffer solution contained: 10 mM NaCl, 5 mM $MgCl_2$ and 50 mM Tricine/OH^- at pH 7.6. The suspension of PS II-particles was mixed with NH_2OH and NH_2NH_2 at concentrations given in the figures before being used in experiment, if not indicated otherwise.

3. RESULTS AND DISCUSSION

In order to check structural requirements for substances that act as H_2O-substrate analogues in terms of the two flash phase shift of the oxygen yield pattern, experiments were performed with a set of different NH_2NH_2-derivatives compiled in fig. 1.

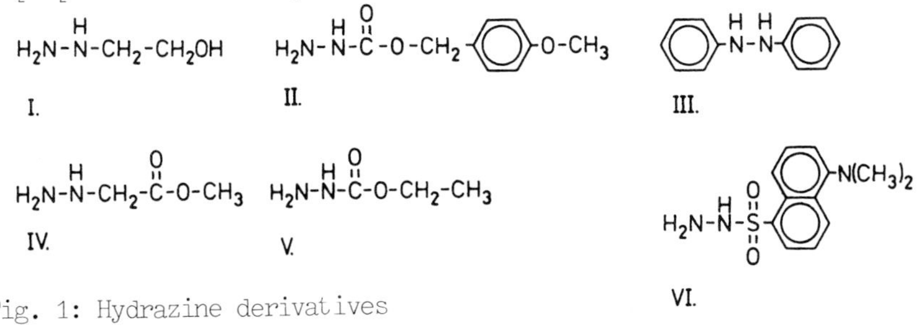

Fig. 1: Hydrazine derivatives

Biggens, J. (ed.), Progress in Photosynthesis Research, Vol. I. ISBN 90 247 3450 9
© *1987 Martinus Nijhoff Publishers, Dordrecht. Printed in the Netherlands.*

It could be shown that 2-hydroxyethylhydrazine (compound I) is still active, but higher concentrations are required to achieve the same effect as with hydrazine in normal thylakoids (900 μM versus 200 μM). Chemical substitution of one or more hydrogens in the NH_2NH_2 molecule with more bulky groups prevented the induction of the above metioned two flash phase shifts. Some compounds caused an increase in the damping of the oscillation pattern due to a markedly higher miss parameter α (data not shown). This phenomenon could be explained either by an ADRY-type effect (2) or by binding to system Y that is sterically strongly hindered but at very high concentrations (∿1 mM), some residual access to the catalytic site might still be possible (compounds II-IV). In this latter case the classical two-flash-shift might then be scattered statistically over all systems Y of the thylakoid membrane during the flash sequence. The damping effect at 1 mM concentration decrease in the order of compound II > compound III > compound IV. Sterical and/or electrostatical reasons seem to be responsible for the failure of compounds V and VI to cause either a two flash phase shift or a damping effect of the O_2-yield pattern even at very high concentrations (∿50 mM). The data obtained with the bulky derivatives of NH_2NH_2 lead to a general conclusion that either the access to system Y manganese cluster might be shielded by the extrinsic regulatory proteins of 17, 23 and 33 kDa including a possible cleft at the catalytic site (9) or on the other hand that stabilisation of t bound species at the catalytic site is determined by strong sterical restric tions.

These considerations raised questions about possible structural modification caused by Triton X 100 treatment used in the preparation of PS II-particles. Substitution of lipids by detergent molecules could lead to conformational changes close to the catalytic site which might be reflected by modified reaction pattern of water substrate analogues. As an attempt to attack this problem experiments were performed in PS II-particles which were incubated in the dark with increasing concentrations of NH_2NH_2 and NH_2OH. It has to be mentioned that in contrast to thylakoids the oscillation pattern of O_2-yield rapidly drops with flash # due to the lack of PQ-pool capacity.

It was found that 100 μM NH_2NH_2 and 20 μM NH_2OH, respectively, induced the two-flash-shift, yielding a maximum in the fifth flash (Fig. 3).

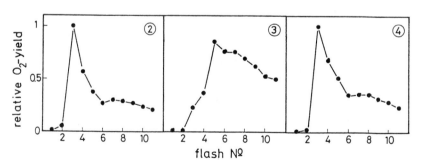

Figure 2: Oxygen yield in PS II-particles. Yield normalized to third flash. Control.

Figure 4: Flash yield after 40 preflashes in presence of 100 μM NH_2NH_2 and subsequent dark adaption.

Figure 3: O_2-pattern of PS II-particles incubated with 100 μM hydrazine.

In most of the experiments PS II-particles appeared to be more susceptible by a factor of two to NH_2NH_2/NH_2OH incubation than thylakoids. However, it has to be emphasized that different preparation conditions for isolating PS II-particles cause considerable variation in the susceptibility of the catalytic site to H_2O-substrate analogues, thermodynamic values for NH_2NH_2/NH_2OH binding are changed and these compounds really provide a sensitive tool for monitoring structural changes in the environment of the catalytic site in the water oxidizing enzyme system Y. Surprisingly, the dark equilibration kinetics of NH_2NH_2 with system Y are retarded in PS II-particles compared with thylakoids. The kinetics is reflected by the Y_3/Y_5 ratio (difficulties in the normalization procedure using PS II-particles prevented an exact quantitative calculation). The rate constant of the monoexponential decay is approx. $3.3 \cdot 10^{-3} \, s^{-1}$, i.e. the half life time is 3.5 min. If one takes into account that in PS II-particles both membrane faces are exposed to the aqueous suspension the binding kinetics appear to be limited by sterical conditions of the active site cleft (Fig. 5).

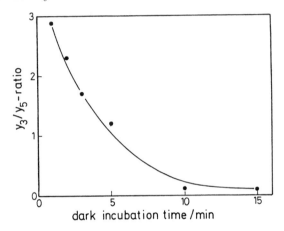

Figure 5: Y_3/Y_5-ratio as a function of dark incubation time with 100 µM NH_2NH_2 in PS II-particles, as outlined in Materials and Methods

After the washing of dark incubated samples with buffer solution, a dissociation of the NH_2NH_2-system Y-complex takes place. This process is completed in 10 min as reflected by the restoration of the O_2-yield maximum in the third flash (data not shown here).

In analogy to our previous findings with thylakoids (5) we found that PS II-particles do not rebind NH_2NH_2/NH_2OH after a set of preflashes and subsequent dark equilibration of 15 min. The detecting flash train after this dark time renders almost the same O_2-pattern as with control samples without NH_2NH_2/NH_2OH incubation (Fig. 1 and 3). The light induced oxidative turnover of water-analogues seems to change the microenvironment at the catalytic site of the water oxidizing enzyme system. This structural modification drastically reduces the susceptibility to NH_2NH_2 or NH_2OH without affecting the oxygen evolving capacity. The implications of this phenomenon for the spectroscopic properties of a structurally altered catalytic site remain to be clarified.

ACKNOWLEDGMENT
The financial support by the Deutsche Forschungsgemeinschaft is greatfully acknowleged.

REFERENCES

1 Renger, G. and Govindjee, (1985), Photosynth. Res. $\underline{6}$, 33-55

2 Renger, G., Bouges-Bocquet, B. and Delosme, R., (1973),
Biochim. Biophys. Acta $\underline{292}$, 796-807

3 Bouges, B., (1971), Biochim. Biophys. Acta $\underline{234}$, 103-112

4 Kok, B. and Velthuys, B.R., (1977), in Research in Photobiology
(Castellani, A. ed), pp 111-119, Plenum Press, New York

5 Hanssum, B. and Renger, G., (1985), Biochim. Biophys. Acta $\underline{820}$, 225-234

6 Berthold, D.A., Babcock, G.T. and Yocum, C.F., (1981), FEBS Lett. $\underline{134}$,
231-234

7 Völker, M., Ono, T., Inoue, Y. and Renger, G., (1985), Biochim. Biophys.
Acta $\underline{806}$, 25-34

8 Joliot, P., (1972), Methods Enzymol. $\underline{24}$, 123-134

9 Radmer, R., (1983), in: The Oxygen Evolving System of Photosynthesis
(Inoue, Y., Crofts, A.R., Murata, N., Govindjee, Renger, G. and
Satoh, K., eds), pp 135-144, Academic Press, Tokyo

PROTON RELEASE BY PHOTOSYNTHETIC WATER OXIDATION

Ralf DIEDRICH-GLAUBITZ, Manfred VÖLKER, Gernot RENGER and
Peter GRÄBER

Max-Volmer-Institut für Biophysikalische und Physikalische
Chemie, Technische Universität Berlin, Strasse des 17. Juni
135, 1000 Berlin 12, FRG

1. INTRODUCTION

   Photosynthetic water oxidation occurs via a four-step univalent redox-reaction sequence which gives rise to a characteristic oscillation pattern of oxygen evolution (1,2). Among
different unresolved mechanistic questions (for recent review
see ref. 3) the stoichiometry of the proton release which is
coupled with individual redox steps within the water oxidizing
system is of special interest. Accordingly, this problem has
been attacked in several laboratories (for review see ref. 4).

   In thylakoids the protons are released into the lumen and,
therefore, the pH changes due to the proton release have to be
measured with a membrane-permeable pH indicator. For such
measurements mostly neutral red has been used (4). One diffi-
culty in the interpretation of such measurements is that
neutral red indicates pH changes in the outer and inner
aqueous phase as well as in the membrane. In this work we
measured the flash-induced proton release due to water oxidation with "inside-out", ISO, thylakoids. In these ISO
vesicles the protons are directly released into the outer
aqueous phase and can be measured with bromcresol purple, a
pH indicator which is not membrane permeable. Therefore, the
interpretation of the measurements should be greatly facilitated.

2. RESULTS

   Fig. 1, top, shows the pattern of oxygen evolution
measured with dark-adapted ISO thylakoids in a sequence of 15
flashes with a dark interval of 500 ms between the flashes.
On the left side, the electric signals of a Joliot-type electrode (1) are shown; on the right, these data are normalized.
In control ISO vesicles the conventional oscillation pattern
is observed with maxima in the 3rd, 7th and 11th flash. Fig.1,
center, shows that in the presence of 20 μM NH_2OH the oscillation pattern is modified. In this case the maximum is shifted from the 3rd to the 5th flash, in agreement with previous
findings in chloroplasts (5). Fig. 1, bottom, shows the oxygen-yield pattern of ISO vesicles which were preilluminated
with three flashes (dark time between the three flashes 500 ms,
dark time between pre-flashes and the analyzing flash group
1 min). A maximal oxygen evolution is now observed in the 4th
flash and, additionally, the damping of the oscillation is
more pronounced. The data of Fig. 1 show that the functional

Biggens, J. (ed.), Progress in Photosynthesis Research, Vol. I. ISBN 90 247 3450 9
© *1987 Martinus Nijhoff Publishers, Dordrecht. Printed in the Netherlands.*

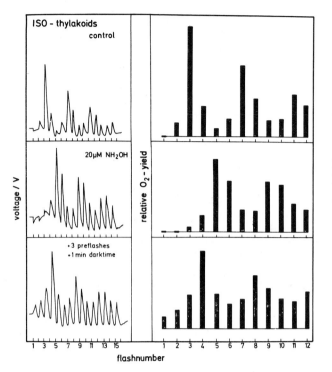

FIGURE 1.
Oxygen yield of
dark-adapted inside-
out-thylakoids in a
sequence of 15
single turnover
flashes spaced
500 ms apart. Top:
control; center:
in the presence of
20 μM NH$_2$OH;
bottom: after 3
pre-flashes and
1 min dark adapta-
tion

organization of the photosystem II and the coupled water oxidizing enzyme system in ISO thylakoids is the same as in
chloroplasts. Therefore, ISO vesicles appear to provide
proper samples for the analysis of the stoichiometry between
the redox transitions in the water oxidizing enzyme system and
proton release.

Fig. 2 shows the absorption changes of bromcrsol purple
within ISO thylakoids excited with a sequence of 8 flashes
(dark interval 500 ms). The pattern exhibits maxima of proton
release in the 3rd and 7th flash. The signal amplitude does
not relax within the measuring time; i.e., the proton release
is practically irreversible. Only after the 1st flash, a fast
rebinding is observed. This shows that the proton release observed in the 1st flash might not be related with water oxidation (see Discussion).

3.   DISCUSSION
Fig. 3 shows the relative proton yield as a function of
the flash number (solid line). Data are from Fig. 2 and
similar measurements. It has been attempted to describe these
data within the framework of the Kok model (2). First, the
Kok parameters, misses, α, double hits, β, and the S_0/S_1 ratio
have been determined from a fit of the oxygen evolution
pattern (Fig. 1, top). The best fit resulted in: $\alpha = 0.1$, $\beta =$
0.03 and $S_1 = 1.0$. Since for the proton release measuring

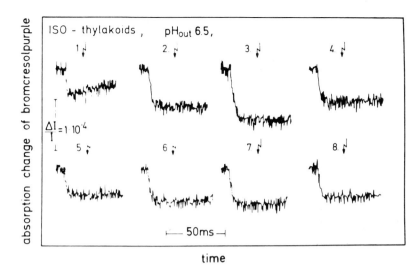

FIGURE 2. Absorption change of bromcresol purple at 570 nm of inside-out thylakoids in a sequence of 8 single turnover flashes 500 ms apart

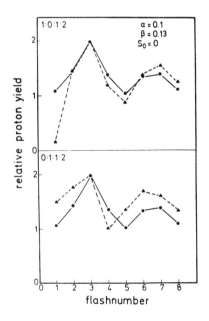

FIGURE 3. Relative proton yield in a sequence of 8 single turnover flashes. Data are from Fig. 2 (solid lines). The dotted lines connect calculated data.

light also leads to some excitation, the double hit probability must be corrected. The calculation gives β= 0.13. In Fig. 3 the experimental results are compared with the calculated proton release pattern for the stoichiometry 1:0:1:2 (top) and 0:1:1:2 (bottom). These calculated data are connected by the dashed lines. It can be seen that except for the first flash the agreement between measured and calculated data is satisfactory. This finding raises questions about the origin of the proton release in the 1st flash. In the 1st flash the transition $S_1 \rightarrow S_2$ should be the dominating reaction at the donor side. If DCMU is added the Q_A^- oxidation by Q_B is blocked and Q_A is regenerated by the back reaction $Q_A^- - S_2 \rightarrow Q_A - S_1$. This reaction has a half-life time of about 2 s (6). A repetitive flash experiment (dark time between flashes 10 s) will therefore reveal the proton release in this step. It results that under these conditions only about 20% of the signal amplitude is observed. We conclude from this result that the proton release in the first flash is mainly due to photosystem I. This may also explain the different rebinding kinetics of the first proton compared to the rebinding of the following protons. We conclude from these measurements that - in accordance with earlier results - the proton release stoichiometry is 1:0:1:2 for $S_o:S_1:S_2:S_3$.

4. ACKNOWLEDGEMENTS

This work has been supported by a grant from the Deutsche Forschungsgemeinschaft

REFERENCES
1 Joliot, P., Barbieri, G. and Chabaud, R. (1969) Photochem. Photobiol. 10, 302-325
2 Kok, B., Forbush, B. and McGloin, M.P. (1970) Photochem. Photobiol. 11, 457-475
3 Renger, G. and Govindjee (1985) Photosynth. Res. 6, 33-55
4 Förster, V. and Junge, W. (1985) Photochem. Photobiol. 41, 183-195
5 Bouges, B. (1971) Biochim. Biophys. Acta 234, 103-112
6 Weiss, W. and Renger, G. (1984) FEBS Lett. 169, 219-223

ON THE CLEAVAGE OF WATER
Pattern of Charges and Protons. States of Water and Manganese.
Routes and Rate of Intermediates.

H.T. WITT, Ö. SAYGIN, K. BRETTEL and E. SCHLODDER

Max-Volmer-Institut für Biophysikalische und Physikalische
Chemie, Technische Universität Berlin, Strasse des 17. Juni
135, 1000 Berlin 12, FRG

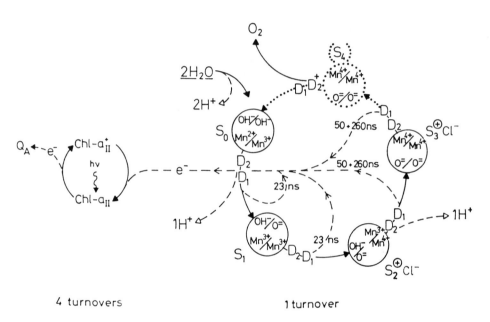

FIGURE 1.   "Clock" of water cleavage and its "driver"
Chlorophyll-a_{II} (details see text).

Fig. 1 shows the enzyme system S responsible for the cleavage
of water and its different oxidized states, induced by the
turnover of Chlorophyll-a_{II}. Fig. 2A shows measurements of
patterns of electrochromic absorption changes of chlorophyll-a
performed on a dark-adapted system in the red region coupled
with the S-state transitions. The quarternary oscillation
indicates the variation of surplus charges on S with the stoichiometry of 0:+1:0:-1 (1,2)(see Table I). The pattern provides also information about the intrinsic release of protons,
because the quarternary oscillation of charges 0:+1:0:-1 is
only understandable if the stoichiometry of the intrinsic
proton dissociation is 1:0:1:2. The absence of a proton
release during the electron extraction from S_1 to S_2 sets up

Biggens, J. (ed.), Progress in Photosynthesis Research, Vol. I. ISBN 90 247 3450 9

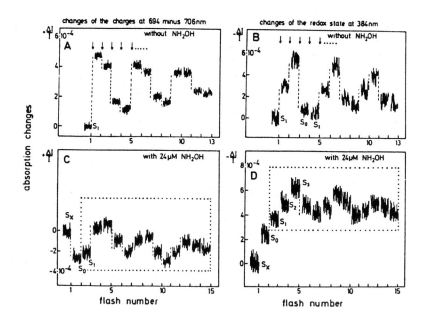

FIGURE 2. A: quarternary oscillation of the changes of
charges in dependence on the flash number (1,2)
B: quarternary oscillation of the redox changes of
the oxidizing equivalents in dependence on the
flash number (9)
C: the same as A but with addition of NH_2OH shifting the S states two steps backwards
D: the same as B but with addition of NH_2OH shifting the S states two steps backwards

the surplus charge. The release of two protons together with
the electron extraction with $S_3 \rightarrow S_0$ would cause the disappearance of this charge (see Fig. 1). (The extrinsic proton release measured in the outer water phase by pH electrodes and pH
indicators, resp., must not necessarily correspond to the intrinsic proton release, but the latter one may not be in equilibrium with the external phase.) The observed surplus charges
in states S_2 and S_3 explain why the primary electron transfers
in states S_2 and S_3 are strongly retarded (3)(see Fig. 1). The
retardation can be quantitatively explained by Coulomb attrac-
tion between the surplus charge and the electrons extracted (3).
It may be that the surplus charges in states S_2 and S_3
are in part compensated by Cl^- ions at the outer sphere of the
complex, because of the high affinity binding of Cl^- in states
S_2 and S_3 (17).

With respect to the nature of oxidizing equivalents created through the electron extractions from S, in the UV region
absorption changes were observed which have been attributed to
a valence state change of manganese (4). Dekker et al.

proposed that three successive Mn^{3+}/Mn^{4+} transitions are accumulated on each of the S_0-S_3 steps and reversed in the $S_3 \rightarrow S_0$ step; i.e., the pattern of the change of the valence states should be +1:+1:+1:-3 (5). However, according to former results from Velthuys (6) the pattern should be 0:+1:0:-1. Lavergne (7) supported the Velthuys results. Renger and Weiss reported on two patterns, 0:2:0:-2 for Mn and 1:-1:+1:-1 for an unknown species (8,24). However, the spectra published in (8,24) are superimposed by a contribution from a binary oscillation of the acceptor side (private communication by Renger). In all experiments in (5-8), through unavoidable misses and double hits, the populations of the four S states become progressively mixed with increasing number of flashes, obscuring the true pattern. Conclusions as to the true pattern are therefore only possible by disintegration of the mixed S states through theoretical calculations. The latter are based on the values of double hits, misses, the initial S_0/S_1 ratio, and the available precision. We have shown that one can fit different true patterns with practically the same misses and double hits (9). Therefore, the results in (5-8,24) which are based on such fitting procedures do not permit an unambiguous discrimination.

For clarification we have therefore additionally introduced an independent method: through a chemical modification with NH_2OH the S states are shifted backwards by two full units. In this way, heavily mixed S states are prevented (see below).

Without NH_2OH we measured oscillation patterns of the surplus charges at 690 nm and valence state changes of Mn at 384 nm as depicted in Fig. 2A and 2B (9)(binary oscillations in the UV due to Q_B were eliminated through addition of DCMU + SiMo). The transition in the 4th flash ($S_0 \rightarrow S_1$), i.e., the last step of the quarternary period, shows only a small change in Figs. 2A and 2B. This is the most critical point for two reasons: (1) this transition might be heavily mixed with simultaneous transitions from $S_3 \rightarrow S_0$ (20-45%) and even from $S_2 \rightarrow S_3$; (2) the disturbing $S_3 \rightarrow S_0$ transition has a sign opposite to all other transitions; in this transition the reduction of all oxidized equivalents takes place together with the O_2 evolution. The overlap of the $S_0 \rightarrow S_1$ transition with a large opposite transition may strongly masks the true $S_0 \rightarrow S_1$ transition. This situation can be avoided, however, when it becomes possible to observe the $S_0 \rightarrow S_1$ transition in its beginning, when practically no mixing takes place and when also the other transitions, $S_1 \rightarrow S_2$ and $S_2 \rightarrow S_3$, are located before the opposing jump, $S_3 \rightarrow S_0$, takes place. This was realized by chemical reduction of the S states with NH_2OH which shifts the S states backwards by two units to a state S_x (10).

With NH_2OH the backward shift is easily recognized in the pattern of the charges as well as in the pattern of redox changes of the oxidizing equivalents (compare the framed patterns in Figs. 2C and 2D with Figs. 2A and 2B). Therefore, with the 2nd flash, the cycle starts with S_0 followed by the $S_0 \rightarrow S_3$ transitions preceding $S_3 \rightarrow S_0$.

Regarding the chemical meaning of S_x, (a) according to (10)

NH_2OH reduces S_1 in the dark to S_0, whereby instead of water NH_2OH is bound; i.e., $S_x = S_0^!(NH_2OH)$. This explanation has been criticized (9); (b) on the other hand, it may be possible that NH_2OH is unbound but has reduced manganeses in S_1 via S_0; i.e., $S_x^2 = S_0^{\!!}(n \cdot Mn^{2+})$.

Considering now first the pattern of the change of charges: In Fig. 2C after the 2nd flash, i.e., transition from S_0 to S_1, a negligible change takes place similar to the result shown in Fig. 2A after the 4th flash. After the 1st flash (Fig. 2C) one observes a negative jump, corresponding to the formation of a negative charge. This indicates that after the removal of one electron from S_x and the formation of S_0 (water), resp., the state S_0 is left with one more negative charge than in S_x. This can be explained, if with the removal of one electron from S_x a release of two protons is coupled with the formation of S_0. The release of two H^+ has been observed too by measurement of pH changes in the presence of NH_2OH (11). Concerning the source of these 2 H^+, in case of $S_x \stackrel{?}{=} S_0^!$ (see above), one proton may be released if in the 1st flash bound NH_2OH is oxidized with the transition $S_x \rightarrow S_0$; then the second proton should be released with the binding of water in S_0. In case $S_x = S_0^{\!!}$, one of the $n \cdot Mn^{2+}$ is oxidized with the transition $S_x \rightarrow S_0$; then both protons should be released with the binding of water in S_0. This means that at least one water must be in dissociated form in state S_0. This has been first discussed by Andreasson et al. in (12) based on other arguments. The configuration of water in the further states, S_1-S_3, consequently can be obtained by considering the sequence of the intrinsic proton release. One of the two possibilities is outlined in Table I and Fig. 1.

With regard to the redox reaction in Fig. 2D after the 2nd flash a jump is observable from $S_0 \rightarrow S_1$. This is not seen in the $S_0 \rightarrow S_1$ transition after the 4th flash shown in Fig. 4B, which is heavily mixed with other transitions (see above). The three-fold absorption changes in flashes number 2, 3 and 4 indicate that with transitions $S_0 \rightarrow S_1 \rightarrow S_2 \rightarrow S_3$ a stepwise formation and stabilization of three oxidizing equivalents takes place. This points out that water may be left unoxidized during the $S_0 \rightarrow S_3$ transition. Water may be oxidized only after the 5th flash, in the $S_3 \rightarrow (S_4) \rightarrow S_0$ transition, through a concerted 4-equivalent oxidation with simultaneous O_2 release (5).

Measurements of each step of the pattern shown in Fig. 2D, i.e., in the absence and presence of NH_2OH (24 µM) as a function of wavelength between 250-400 nm have been carried out (see Saygin and Witt, these proceedings). Without NH_2OH the heavily mixed $S_0 \rightarrow S_1$ transition in the 4th flash shows practically no change in the whole spectral region from 250-400 nm (not shown here, but see Saygin and Witt, these proceedings). However, with NH_2OH addition (24 µM) the $S_0 \rightarrow S_1$ transition, created after the 2nd flash, is characterized by a spectrum as indicated in Fig. 3. The transitions $S_1 \rightarrow S_2$ and $S_2 \rightarrow S_3$ differ markedly from the spectrum of the $S_0 \rightarrow S_1$ transition. Within our measuring capacity, the $S_1 \rightarrow S_2$ and $S_2 \rightarrow S_3$ transitions show practically the same spectra. The $S_1 \rightarrow S_2$ and $S_2 \rightarrow S_3$ transi-

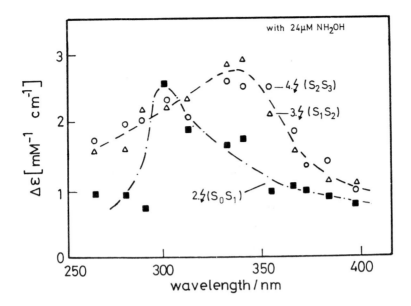

FIGURE 3. Difference spectrum of the oxidizing equivalents at different S-state transitions in the presence of 24 μM NH$_2$OH (16).

tions may be due to valence state changes of Mn^{3+} to Mn^{4+} as stated by Dekker et al. (5). The $S_o \rightarrow S_1$ transition which is shifted to a shorter wavelength may be due to a change from Mn^{2+} to Mn^{3+}. This suggestion may be supported by the fact that also in vitro the difference spectrum of Mn^{2+}/Mn^{3+} is located at a shorter wavelength than that of Mn^{3+}/Mn^{4+} (13). Furthermore, an Mn^{2+} to Mn^{3+} valence state change for $S_o \rightarrow S_1$ fits the following argumentation. With addition of the electron donor, NH$_2$OH (Fig. 2D), one may assume that all manganeses are reduced down to the state Mn^{2+}; i.e., S_x is realized as S_x(n·Mn^{2+})(see above). In the 1st flash, one electron is extracted with $S_x \rightarrow S_o$. If only two manganeses are engaged in the oxidation of water (n = 2), the S_o-state should be present as S_o(Mn^{2+}/Mn^{3+}). The other states must then be S_1(2 Mn^{3+}), S_2(Mn^{3+}/Mn^{4+}) and S_3(Mn^{4+}/Mn^{4+})(see Fig. 1 and Table I). The oxidizing equivalent created with the last transition from S_3 to the unstable (S_4) might be the electron carrier D$_2^+$ (see below) oxidized in the 4th flash and reduced immediately in unison with the oxidized manganeses present in state S_3 (see Fig. 1).

If four manganeses are engaged in the oxidation of water (n = 4) besides the two manganeses mentioned above, a third one might be oxidized in (S_4) but remains "invisible" because of the immediate re-reduction together with the 2 oxidized Mn's in S_3. The fourth Mn might function only as stabilizer in a tetrameric Mn-cluster.

The principal difference in regard to the spectra of Dekker et al. in (5) which were obtained by the fitting procedure method is given with our result, showing that the spectrum of $S_o \rightarrow S_1$ is not identical with that of $S_1 \rightarrow S_2$ and $S_2 \rightarrow S_3$. This principal difference has been demonstrated also without NH_2OH, i.e., without the backward shifting of the S state, by two pieces of evidence: (1) Using the fitting procedure and dis-integration treatment of the other authors (a method which alone is not unequivocal, see above), e.g., at 360 nm where practically no changes of other components besides manganese take place, it was found, also through this procedure, that the change with the transition $S_o \rightarrow S_1$ is different from that of $S_1 - S_3$. (2) Furthermore, analysis of the kinetics of the S-state formations shows also without NH_2OH that the $S_o \rightarrow S_1$ spectrum is different from the others (see (16) and Saygin and Witt, these proceedings).

The proposed states of manganese and of water in the different S states may be arranged in form of dimers (18)· Cuban-like structures or adamantane-like complexes (22) have been also discussed (23).

According to Fig. 1 four of the following electron trans-fer routes have to be gone through for the formation of the four S states and the evolution of one O_2, resp. (n=0,1,2,3); (m=0.1,2):

$$Chl-a_{II}^+D_1D_2S_n \underset{k_{-1}}{\overset{k_1}{\rightleftharpoons}} Chl-a_{II}D_1^+D_2S_n \underset{k_{-2}}{\overset{k_2}{\rightleftharpoons}} Chl-a_{II}D_1D_2^+S_n \underset{k_{-3}}{\overset{k_3 \quad m \cdot H^+}{\rightleftharpoons}} Chl-a_{II}D_1D_2S_{n+1}$$

D_2 is known through the ESR signal II_{vf} (14) and assumed to be a plastoquinol cationic radical. D_1 is introduced to bridge the kinetic gap between $Chl-a_{II}$ and D_2 (3). It acts as the immediate donor to $Chl-a_{II}$. The k-values depend on the S_n states. This is due to the fact that the electron transfers are retarded in states S_2 and S_3 through the Coulomb forces of the surplus charges (see above). The dependence on S_n has been taken into account by the following rate constant values which have been measured through the re-reduction kinetics of $Chl-a_{II}^+$ (3): For S_o and S_1 it is $r_+ = \ln 2/23$ ns, $r_- = \ln 2/100$ ns, $a_1 = 0.95$ and $a_2 = 0.05$; for S_2 and S_3 it is $r_- = \ln 2/50$ ns, $r_+ = \ln 2/260$ ns, $a_1 = 0.5$ and $a_2 = 0.5$. It is $k_1 = a_1 \cdot r_+ + a_2 \cdot r_-$, $k_{-1} = r_+ + r_- - k_1 - k_2$, $k_2 = r_+ \cdot r_-/k_1$ and $k_{-2} \ll k_2$. The solution of the corresponding differential equa-tions results in time courses for the redox reactions of $Chl-a_{II}^+$, D_1 and D_2 (up to the maximum D_2 oxidation) in the four electron transfer routes as illustrated in Fig. 4 (solid lines).

According to these results it follows inter alia that the as yet chemically unknown D_1 carrier is characterized by S_n-dependent rise and decay times in the 23-260 ns time range.

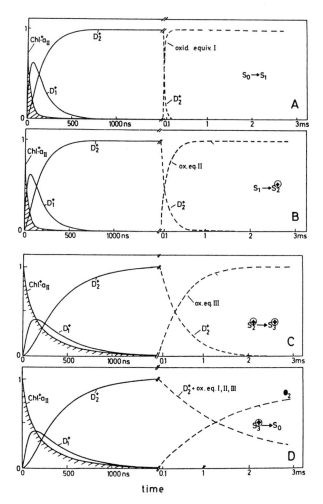

FIGURE 4. Time course of the redox reactions of the
components of the four electron transfer
routes and S states, resp. (for details
see text)

Within the S system the final steps before cleavage of
water are the re-reduction of D_2^+, the formation of the oxidizing equivalents together with the intrinsic H^+ release, and
the variation of surplus charges. D_2^+ reduction times should
correspond to the oxidation times of the oxidizing equivalents.
The latter have been measured through the time course of the Mn
absorption changes in the UV region (see dotted lines in Fig.4
and Table II).

TABLE 1.   Stoichiometry and states of five events in the turnover of the water splitting cycle. In (S_4) creation of a
further oxidizing equivalent (*) takes place with immediate
reduction, together with the three oxidizing equivalents
accumulated in S_2 (see text).   If, instead of two manganeses,
four are engaged in water oxidation, see text for the action
of the two additional Mn's.

	$S_0 \longrightarrow S_1$	$\longrightarrow S_2$	$\longrightarrow S_3$	$\xrightarrow{(S_4)} S_0$	
absorbed quanta $h\nu_{II}$	1	1	1	1	
electron extraction $Chl\text{-}a_{II}^+ \leftarrow e^- \ldots S_n$	1	1	1	1	
changes of positive surplus charges	0	1	0	-1	
intrinsic H^+ release	1	0	1	2	
possible states of water	OH^- OH^-	OH^- $O^=$	OH^- $O^=$	$O^=$ $+O_2$ $O^=$ $-2\,H_2O$	OH^- OH^-
possible states of oxidizing equivalents	Mn^{2+} Mn^{3+}	Mn^{3+} Mn^{3+}	Mn^{3+} Mn^{4+}	Mn^{4+} Mn^{4+} (*)	Mn^{2+} Mn^{3+}

TABLE 2.   Half-life times of the redox reactions of the
oxidizing equivalents coupled with the S-state transitions

$S_0 \longrightarrow S_1$	$\longrightarrow S_2$	$\longrightarrow S_3$	$\xrightarrow{(S_4)} S_0$	Lit
30 µs	110 µs	350 µs	1.2 ms	(19)
50 µs	40 µs	80–120 µs	1.2 ms	(16)
50 µs	100 µs	220 µs	1.2 ms	(8)

ACKNOWLEDGEMENTS
    This work was supported by grants from the Deutsche
Forschungsgemeinschaft (Sonderforschungsbereich 312, Teil-
projekt Al und A3).

REFERENCES
1 Saygin, Ö. and Witt, H.T. (1984) FEBS Lett. 176, 83-87
2 Saygin, Ö. and Witt, H.T. (1985) FEBS Lett. 187, 224-226
3 Brettel, K., Schlodder, E. and Witt, H.T. (1984) Biochim.
   Biophys. Acta 766, 403-415
4 Dekker, J.P., van Gorkom, H.J., Wensink, J. and Ouwehand,L.
   (1984) Biochim. Biophys. Acta 764, 301-309
5 Dekker, J.P., van Gorkom, H.J., Wensink, J. and Ouwehand,L.
   (1984) Biochim. Biophys. Acta 767, 1-9
6 Velthuys, B.R. (1981) in: Photosynthesis II (Akoyunoglou,
   ed.), pp. 75-85, Balaban Intern. Science Serv., Philadelphia, Pa.
7 Lavergne, J. (1986) Photochem. Photobiol. 43, 311-317
8 Renger, G. and Weiss, W. (1986) Biochemical Society
   Transactions, Vol. 14, pp. 17-20
9 Saygin, Ö. and Witt, H.T. (1985) Photochem. Photobiol. 10,
   71-82
10 Radmer, R. (1979) Biochim. Biophys. Acta 546, 418-425
11 Förster, V. and Junge, W. (1985) Photochem. Photobiol. 41,
   183-190
12 Andréasson, L.E., Hansson, Ö. and Vänngard, T. (1983)
   Chemica Scripta 21, 71-74
13 Sawyer, D.T., Bodini, M.E., Willis, L.A., Riechel, T.L.
   and Magers, K.D. (1977) in: Bioinorganic Chemistry
   (Raymond, K.N., ed.), Vol. 2, pp. 330-349, American
   Chemical Society, Washington, D.C.
14 Babcock, G.T., Blankenship, R.E. and Sauer, K. (1976)
   FEBS Lett. 61, 286-289
15 Dekker, J.P., Plitjer, J.J., Ouwehand, L. and van Gorkom,
   H.J. (1984) Biochim. Biophys. Acta 767, 176-179
16 Saygin, Ö. and Witt, H.T. (in preparation)
17 Preston, C. and Pace, R.J. (1985) Biochim. Biophys. Acta
   810, 388-391
18 Boucher, L.J. and Coe, C.G. (1975) Inorg. Chem. 14,
   1289-1294
19 Dekker, J.P., Plijter, J.J., Ouwehand, L., van Gorkom, H.J.
   (1984) Biochim. Biophys. Acta 767, 176-179
20 Klimov, V.V., Allakhverdiev, S.I., Shuvalov, V.A. and
   Kraskovsky, A.A. (1982) FEBS Lett. 148, 307-312
21 Yamamoto, Y., Nishimura, M. (1983) Biochim. Biophys. Acta
   724, 294-297
22 Wieghardt, K., Bossek, U. and Gebert, W. (1983) Angew.
   Chem. Int. Ed. Engl. 22, 328-329
23 Brudvig, G.W. and Crabtree, R.H. (1986) Proc. Natl. Acad.
   Sci. USA 83, 4586-4588
24 Renger, G. and Weiss, W. (1986) Biochim. Biophys. Acta 850,
   184-196

ABSORPTION CHANGES WITH PERIODICITY FOUR, ASSOCIATED WITH PHOTOSYNTHETIC OXYGEN EVOLUTION

Jan P. DEKKER[1], Johan J. PLIJTER[2] and Hans J. van GORKOM[2]

1) Max-Volmer-Institut, Technische Universität Berlin, Strasse des 17. Juni 135, D-1000 Berlin 12, Germany.
2) Dept. of Biophysics, Huygens Laboratory of the State University, P.O. Box 9504, 2300 RA Leiden, The Netherlands.

1. THE PERIOD 4 OSCILLATION IN ULTRAVIOLET ABSORBANCE

The well-known period 4 oscillation of oxygen release upon illumination of dark-adapted photosystem II (PS II) with single-turnover light flashes has indicated that every PS II reaction center stores four positive equivalents before oxygen is released (1). The chemical nature of the "charge-storing" groups functioning in oxygen evolution still is a matter of debate, although, from many points of view, it is clear that manganese plays an essential role. The S_2-state is accompanied by a multiline EPR signal, observable at very low temperature, which is ascribed to a mixed-valence Mn-dimer or -tetramer. An EPR signal near g=4.1, which probably also is specific for the S_2-state, may also be caused by manganese. The oxidation of manganese on the S_1-S_2 transition was inferred from K-edge X-ray absorption spectra, and from absorbance difference spectroscopy, both in the ultraviolet and in the infrared. Most likely, manganese is oxidized from valency +3 to valency +4 on this transition (see for a review, Ref. 2).

The ultraviolet absorbance changes probably yield relatively unambiguous information on changes of manganese redox state. The absorbance difference spectrum of the compound oxidized on the first flash in dark-adapted BBY PS II preparations was reported to consist of a broad, asymmetric absorbance increase peaking near 300 nm with $\Delta\varepsilon=6000\ M^{-1}cm^{-1}$ (3). Qualitatively, its sequence suggested an oxidation on the S_1-S_2 transition and a reduction during oxygen release (4).

Several authors have now reported on the sequence of UV absorbance changes. Dekker et al. (5) measured changes in dark-adapted BBY PS II preparations supplied with the electron acceptor 2,5-dichloro-p-benzoquinone (DCBQ). The pattern of changes measured at 541-551 nm (the bandshift "C-550" caused by a blue shift of a pheophytin a molecule upon reduction of Q_A) did not reveal any oscillation (not shown), suggesting that the amount of reduced Q_A after each flash was the same. Only after the first flash a higher value was observed. The S-states parameters were determined from the oscillation pattern of the millisecond transient accompanying oxygen release. By using this information, the absorbance difference spectra of the successive S-states transitions were calculated from the oscillating absorbance changes 30 ms after the flashes 2-8. The spectra were interpreted to consist of an acceptor side contribution oscillating somewhat irregularly with periodicity two, and of a donor side spectrum with sequence +1, +1, +1, -3 during the S-states cycle. The spectrum of the former oscillation resembled that of electron transfer from Q_B^- to DCBQ, that of the latter resembled the broad, asymmetric increase around 300 nm, attributed previously (3) to the oxidation of Mn(III) to Mn(IV).

Three independent experiments have confirmed the interpretation mentioned above. Fig. 1 (circles) shows the spectrum of the millisecond transient of the S_3-S_0 transition (5) together with a fit (line) consisting of three times the spectrum of the S_1-S_2 transition (3) and the spectrum of Z-oxidation in Tris-washed PS II (3). The striking similarity indicates that the spectral contribution attributed to the acceptor side does not occur during this kinetic phase. Fig. 2 shows the oscillating, 310 nm absorbance changes 30 ms after the flashes in 2 M NaCl washed PS II preparations (6). In these preparations, the two-electron gating mechanism at the acceptor side does not function anymore (6), and the absorbance changes due to the successive S-states transitions (inset) directly show the +1, +1, +1, -3 sequence. Recently, the difference spectrum has been presented of the ultimate electron donor oxidized after the first flash in BBY preparations at pH 8.3 (7), i.e. largely the S_0-S_1 transition. The shape of this spectrum closely resembles that in Fig. 1, suggesting that, due to inactivation, Z is the terminal donor in a minor part of the centers, and that the spectra of the S_0-S_1 and S_1-S_2 transitions are very similar, even at the very different pH values.

Renger and Weiss (8) presented absorbance difference spectra measured in trypsinized BBY PS II preparations supplied with ferricyanide. As noted by the authors, the measurements can in principle be explained by a +1, +1, +1, -3 sequence of a broad absorbance increase around 300 nm together with a period 2 oscillation from the acceptor side. In this case, the observed shift of one unit of the period 2 oscillation is explained by the dark oxidation of Q-400, the quinone bound iron, by ferri-cyanide, and by its subsequent reduction after the first flash.

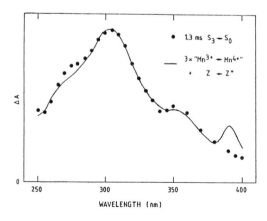

Fig. 1. Circles: spectrum of the 1.3 ms phase of the S_3-S_0 transition, measured as the difference between the third and the fifth flash (from Ref. 5). Line: three times the spectrum of the donor oxidized on the first flash, plus the spectrum of the oxidation of Z (both from Ref. 3). The spectra are normalized at 305 nm.

However, Renger and Weiss preferred to attribute the period 2 oscillation directly to the S-states mechanism. This is rather unlikely, because the spectrum of the millisecond transient accompanying oxygen release (when measured as the difference between the third and the fifth flash - see Fig. 1) does not contain the spectral contribution of the period two oscillation, and because the one-unit shift due to the addition of ferricyanide can not be explained.

Lavergne (9) measured absorbance changes at 295 nm, and analyzed the response of the pattern of the period 4 oscillation upon changing the S-states parameters. Apparently, the 0, +1, 0, -1 sequence gives a better explanation for the experimental results, but this is based primarily on the properties of the changes due to the first flash, which, to our opinion, must be excluded for the analysis of the period 4 oscillation (5). When neglecting the changes due to the first flash, Lavergne's data do not discriminate between both models.

Saygin and Witt (10) reported absorbance changes at 384 nm in PS II particles from a thermophilic cyanobacterium. Due to the use of silicomolybdate/DCMU, acceptor side oscillations can practically be excluded. The results, including those obtained with low concentrations of hydroxylamine, indicate similar absorbance changes due to the S_1-S_2 and the S_2-S_3 transitions, but a somewhat different contribution upon the S_0-S_1 transition (Saygin, Ö. and Witt, H.T., this volume).

In conclusion, the data show that all four S-states transitions cause absorbance changes in the ultraviolet, and strongly suggest that they are due to a single spectral component oscillating with a sequence +1, +1, +1, -3 during the S_0-S_1-S_2-S_3-S_0 cycle.

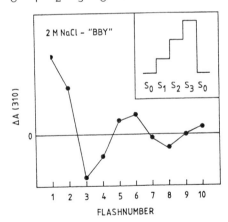 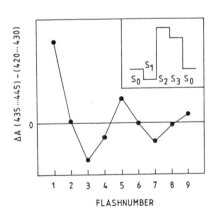

Fig. 2. (left) Sequence of oscillating absorption changes 30 ms after the flashes in dark-adapted 2 M NaCl washed PS II preparations at 310 nm (6). Inset: absorption changes due to the successive S-states transitions, calculated from flashes 2-8 as described in the text. Fig. 3 (right). Sequence of the Chl a bandshift 30 ms after the flashes in dark-adapted PS II preparations (data from Ref. 5). Inset: as for Fig. 2.

2. THE PERIOD 4 OSCILLATION OF A CHLOROPHYLL a BANDSHIFT

So, the "charge-storing" groups functioning in photosynthetic oxygen evolution are very likely three Mn(III) ions that are oxidized to Mn(IV). However, a net charge-accumulation does possibly not occur, in view of the release of a proton on the S_0-S_1 and S_2-S_3 transitions and the reported binding of a chloride ion on the S_1-S_2 transition (11).

More information about the charge-distribution within the oxygen-evolving complex may be obtained by analysis of the pattern of the electrochromic shift of Chl a that accompanies the S-states transitions (3, 12). Fig. 3 shows the pattern of this shift in the BBY PS II preparations with DCBQ, taken from the data in Ref. 5. The contamination of the period two oscillation of the acceptor side is largely removed by taking the difference between the regions around 440 and 425 nm, since the absorption changes of $Q_B{}^-$ are practically the same around both wavelengths. After applying the same mathematical analysis as performed with the UV changes (see above), a pattern of the shift due to the successive S-states transitions as in the inset of Fig. 3 is found. The results suggest that the shift due to the S_1-S_2 transition is not reversed completely on the S_3-S_0 transition, but also, to a minor extent, on the S_0-S_1 and S_2-S_3 transitions. Possibly, the bandshift observed by Saygin and Witt (12) shows a similar oscillation pattern. Thus, the charge distribution within the oxygen-evolving complex may be as follows: $S_2 > S_3 \geqslant S_0 > S_1$
This situation may help to explain the stability of the individual S-states.

REFERENCES
1 Joliot, P. and Kok, B. (1975) in Bioenergetics of Photosynthesis (Govindjee, ed.), pp. 387-412, Academic Press, New York
2 Dismukes, G.C. (1986) Photochem. Photobiol. 43, 99-115
3 Dekker, J.P., Van Gorkom, H.J., Brok, M. and Ouwehand, L. (1984) Biochim. Biophys. Acta 764, 301-309
4 Velthuys, B.R. (1981) in Photosynthesis II (Akoyunoglou, G., ed.), pp. 75-85, Balaban International Science Services, Philadelphia, PA
5 Dekker, J.P., Van Gorkom, H.J., Wensink, J. and Ouwehand, L. (1984) Biochim. Biophys. Acta 767, 1-9
6 Dekker, J.P., Ghanotakis, D.F., Plijter, J.J., Van Gorkom, H.J. and Babcock, G.T. (1984) Biochim. Biophys. Acta 767, 515-523
7 Plijter, J.J., De Groot, A., Van Dijk, M.A. and Van Gorkom, H.J. (1986) FEBS Lett. 195, 313-318
8 Renger, G. and Weiss, W. (1986) Biochim. Biophys. Acta 850, 184-196
9 Lavergne, J. (1986) Photochem. Photobiol. 43, 311-318
10 Saygin, Ö. and Witt, H.T. (1985) Photobiochem. Photobiophys. 10, 71-82
11 Preston, C. and Pace, R.J. (1985) Biochim. Biophys. Acta 810, 388-391
12 Saygin, Ö. and Witt, H,T. (1985) FEBS Lett. 187, 224-226

STATE OF MANGANESE DURING WATER SPLITTING

Ö. SAYGIN and H.T. WITT

Max-Volmer-Institut für Biophysikalische und Physikalische Chemie, Technische Universität Berlin, Strasse des 17. Juni 135, 1000 Berlin 12, FRG

During water splitting in photosynthesis four stable states, S_0-S_3, are reached successively in one turnover of the water splitting enzyme complex, S. A transition from S_i-S_{i+1} is also coupled with an accumulation of a redox equivalent and the reaction with water only occurs at the last step, S_3-S_0. This was stated first by Dekker et al. (1) and confirmed later by us at one certain wavelength (384 nm)(2). After subtracting contributions due to binary oscillations from the acceptor side, Dekker further proposed a model of +1:+1:+1:-3 for the spectrum of absorption changes in the UV, which means that the same elementary process occurs three times during the accumulation of these redox equivalents. By comparing this spectrum with that of an in vitro Mn-complex, he assigned three Mn(III)-Mn(IV) oxidations to the observed absorption changes. Using silicomolybdate as an external acceptor, we were able in (2-4) to eliminate binary oscillations from the acceptor side. Here we like to report as an extension of our previous work (2) the spectra of the absorption changes accompanying the S_0-S_3 transitions, where no correction due to binary oscillation is necessary. Fig. 1a shows a typical pattern of absorption changes measured at 367 nm under our conditions at PS II particles from Synechococcus sp. Usually, other workers in this field (1,5) do not consider the changes after the first flash since these absorption changes deviate considerably from the expected values due to unknown effects after dark adaptation. In our measurements (Fig. 1) the effect after the first flash is not significantly different from the corresponding effect after the 5th and 9th

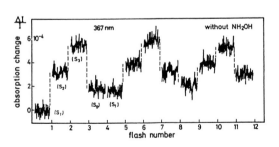

FIGURE 1. Pattern of the stable absorption changes in the presence of silicomolybdate ($2 \cdot 10^{-5}$ M) at pH 7.0

flash. Therefore, the uncorrected absorption changes of all four flashes in the first period are shown in Fig. 2 as a function of wavelength. Because no correction due to misses and double hits has been made, it is open whether the spectra after the 1st and 2nd flash are similar (but see later). While

Biggens, J. (ed.), Progress in Photosynthesis Research, Vol. I. ISBN 90 247 3450 9
© *1987 Martinus Nijhoff Publishers, Dordrecht. Printed in the Netherlands.*

the principal magnitude of the absorption changes coupled with S_1-S_2 and S_2-S_3 transitions can be observed already by direct inspection after the first two flashes, the assignment of an absorption change to the S_0-S_1 transition in the 4th flash can only be evaluated by calculation. This transition is heavily masked through mixing with the changes coupled with the opposite transition S_3-S_0 (see Witt et al., these proceedings). This requires, however, information about the misses, double hits and initial S_0/S_1 ratio. Fig.3a shows a time-resolved measurement at 362 nm where besides the stable absorption changes (Fig. 3c) also the ms-amplitude for the S_3-S_0 transition (Fig. 3b) can be seen. An analysis of the oscillation of the ms-amplitude (Fig. 3b) which corresponds to O_2-evolution yields for misses 7.7%, for double hits 5.0% and for the initial S_1 = 100%. Using the parameters we fitted stable absorption changes to each S_i-S_{i+1} transition in Fig. 3c. The result was for S_1-S_2 $\Delta \varepsilon$ = 2500, for S_2-S_3 $\Delta \varepsilon$ = 2600 and for S_0-S_1 $\Delta \varepsilon$ =550 (M cm)$^{-1}$. This means that at least at this wavelength (362 nm) S_0-S_1 transitions must differ from the other two transitions (in contrast to the Dekker model). To gather further independent information about the spectra coupled with these transitions, we measured absorption changes in the presence of low concentrations of NH_2OH (24 μM) as a function of wavelength. This compound shifts the pattern by two flashes backwards. One common point of all models which try to explain the mechanism of this shift is that the 2nd,3rd and 4th flash is assigned to the transitions S_0-S_1, S_1-S_2 and S_2-S_3, resp. (6-8). This means that now the S_0-S_1 transition is in front of the train of flashes and not mixed with the opposite S_3-S_0 transition as after the 4th flash in Figs. 1 and 2; i.e., under these conditions the absorption changes after the 2nd flash present nearly the true spectrum of S_0-S_1 transitions. By measuring the pattern at different wavelengths in the presence of NH_2OH, we observed principally two different types of patterns in the whole UV region (266-400 nm). These two typical patterns are presented at 355 nm and 315 nm in Figs. 4a and 4b. While at 315 nm the jumps after the 2nd, 3rd and 4th flashes are almost the same, the jump after the 2nd flash at 355 nm is much smaller than those after the 3rd and 4th flash. The complete spectrum of the absorption changes under these conditions in the 2nd, 3rd and 4th flash are

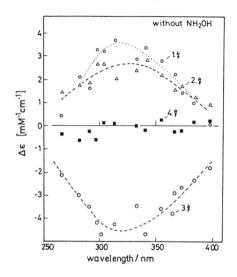

FIGURE 2. Spectrum of the first four flashes with silicomolybdate as acceptor

depicted in Fig. 5. It is obvious that, as in the fitting experiments without NH_2OH, the spectrum of the S_o-S_1 transition differs from that of S_1-S_2 and S_2-S_3 transitions, while S_1-S_2 and S_2-S_3 show practically the same spectrum. From this information we suggest that the S_o-S_1 transition is coupled with

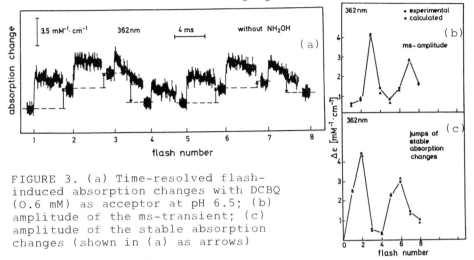

FIGURE 3. (a) Time-resolved flash-induced absorption changes with DCBQ (0.6 mM) as acceptor at pH 6.5; (b) amplitude of the ms-transient; (c) amplitude of the stable absorption changes (shown in (a) as arrows)

another manganese redox-state change, compared to the transitions S_1-S_2 and S_2-S_3. It is very likely that the S_o-S_1 transition is an Mn(II)-Mn(III) oxidation, while S_1-S_2 and S_2-S_3 transitions are due to Mn(III)-Mn(IV) oxidations (see Table I). Such a model would give straightforward and uncomplicated explanations for many of the observations in the field of water oxidation:
1. The spectrum of Mn^{4+} is in vitro always shifted to longer wavelengths compared to Mn^{3+}. This is in accordance with our absorbance difference spectrum of Fig. 5.
2. The well-known multi-line ESR signal in S_2 is explained to be due to a Mn(III)-Mn(IV) dimer. Such a constellation is in Table I only in S_2 present and would also explain the disappearance of the signal in other states.

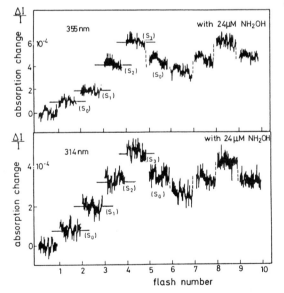

FIGURE 4. Pattern of stable absorption changes in the presence of NH_2OH and SiMo

TABLE 1. Manganese States

S_0	S_1	S_2	S_3
Mn^{2+}	Mn^{3+}	Mn^{4+}	Mn^{4+}
Mn^{3+}	Mn^{3+}	Mn^{3+}	Mn^{4+}

3. After dark adaptation S_1 is the only stable state of S. According to Table I both manganese are present in S_1 as Mn(III). As in some in vitro manganese complexes (9) Mn(II) can be unstable in the presence of oxygen and Mn(IV) should be also unstable in the neighborhood of intrinsic donors. S_1 is the only state without Mn(II) or Mn(IV).

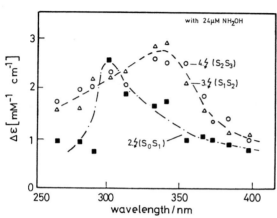

FIGURE 5. Spectrum of the stable absorption changes after the 2nd, 3rd and 4th flashes in the presence of NH_2OH and silicomolybdate

4. The reason that some reductants like NH_2OH, NH_2NH_2 or H_2O_2 shift the oscillation patterns mainly by two units, could be that they reduce all higher oxidation states of manganese in S down to Mn(II), which is also the only end product of the same reaction in vitro.

5. In the presence of NH_2OH (low conc.) according to Renger et al.(10) the desactivation of S_2 and S_3 is much faster than of S_1. This demonstrates again the different reactivity of S_1 from S_2 and S_3. This can be understood, e.g., by a fast Mn(IV) reduction in S_2 and S_3 and a slow Mn(III) reduction in S. Attempts to explain all these points by a model +1:+1:+1:-3; i.e., the same process (Mn^{3+}/Mn^{4+} oxidation) in all three transitions leads to complicated assumptions.

REFERENCES
1 Dekker, J.P. et al. (1984) Biochim. Biophys. Acta 767, 1-9
2 Saygin, Ö. and Witt, H.T. (1985) Photochem. Photobiol. 10,71-82
3 Saygin, Ö. and Witt, H.T. (1984) FEBS Lett. 176, 83-87
4 Saygin, Ö. and Witt, H.T. (1985) FEBS Lett. 187, 224-226
5 Renger, G. and Weiss, W. (1986) Biochem. Soc. Trans. 14, 17-20
6 Velthuys, B. and Kok, B. (1978) Biochim. Biophys. Acta 502,211
7 Radmer, R. (1982) FEBS Lett. 144, 162-166
8 Förster, V. (1984) PhD thesis, Osnabrück
9 Boucher, L.J. and Coe, C.G. (1975) Inorg. Chem. 14, 1289-1295
10 Hanssum, B. and Renger, G. (1985) Biochim. Biophys. Acta 810, 225-234

NEW RESULTS ABOUT THE MOLECULAR MECHANISM OF PHOTOSYNTHETIC WATER OXIDATION

G. Renger, B. Hanssum and W. Weiss, Max-Volmer-Institut für Biophysikalische und Physikalische Chemie, Technische Universität Berlin, Strasse
des 17. Juni 135, 1000 Berlin 12, FRG

1. INTRODUCTION

Photosynthetic water oxidation to dioxygen takes place in a mangano protein complex via a four step univalent redox reaction sequence (for review
see Ref. 1). Among different unresolved questions the identification of
the structure of the catalytic site and the nature of its intermediary redox states, referred to as S_i-states (2), are of paramount importance for
understanding the mechanism of this crucial process of solar energy ex-
ploitation. Spectroscopic methods provide the most proper tool to attack
this problem. In this short communication there are presented laser flash
induced UV-absorption changes in dark adapted PS II-particles and the analysis of the experimental data within the framework of Kok's model (2).
Special attention has been paid to the interference of possible contributions due to binary oscillations at the acceptor side.

2. PROCEDURE

2.1 Materials and Methods

PS II particles with oxygen evolving capacity were prepared from market
spinach as described in (3) with the modifications outlined in (4), using
a Triton X-100/chlorophyll ratio of 20:1. Limited proteolysis was performed
at room temperature with 10 μg trypsin/ml in the measuring suspension containing PS II particles (10 μg chlorophyll/ml), 10 mM NaCl, 2 mM $MgCl_2$,
20 mM MES/NaOH, pH = 6.0, addition of exogenous electron acceptor and other
substances as indicated in the figure legends or text. All samples were
completely dark adapted for 5 min prior to start of the measurements. The
absorption changes in the UV were detected by the use of a single beam
flash photometer with a pulsed measuring light beam as described in (5).
After each flash train the sample was renewed automatically. Photosystem II
was excited with flashes at 1 Hz from a Q-switched frequency doubled
Nd/YAG laser (FWHM = 7 ns).

3. RESULTS AND DISCUSSION

A recent study (6) revealed that the calculated difference extinction
coefficient $\Delta\varepsilon_{i+1,i}(\lambda)$ for the redox transitions $S_i \rightarrow S_{i+1}$ exhibit the
most remarkable differences around 325 nm. In this wavelength region,
however, any interference with binary oscillations due to Q_B^- stabilization
at the acceptor side are expected to be maximal. Therefore, a thorough analysis is required for this wavelength region. Fig. 1 shows typical traces
of flash induced absorption changes in dark adapted PS II particles. It is
obvious that the pattern of the absorption changes strongly depends on
the type of the exogenous electron acceptor. In the presence of phenyl-p-
benzoquinone (Ph-p-BQ) the extent of the absorption change due to the 1st

Biggens, J. (ed.), Progress in Photosynthesis Research, Vol. I. ISBN 90 247 3450 9
© *1987 Martinus Nijhoff Publishers, Dordrecht. Printed in the Netherlands.*

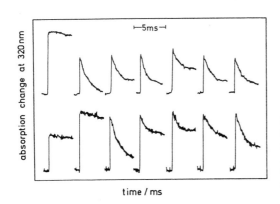

time / ms

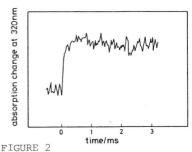

FIGURE 2

Absorption change at 320 nm
induced by the 1st flash of a
laser flash train in dark adapted PS II particles in the presence of 1 mM $K_3[Fe(CN)_6]$.

FIGURE 1   Absorption changes at 320 nm induced by a train of laser
flashes in dark adapted PS II particles upper trace:
100 μM Ph-p-BQ; lower trace: 500 μM $K_3[Fe(CN)_6]$, 10 mM $CaCl_2$,
other experimental conditions as described in Materials and
Methods.

flash is exceptionally large whereas in the presence of $K_3[Fe(CN)_6]$ it is
smaller than that of the following flashes. The relative amplitude of the
absorption changes induced by the 1st flash was found to be dependent on
the type of the exogenous quinone acceptor (data not shown). The origin
of the effect remains to be clarified. On the other hand, the $K_3[Fe(CN)_6]$
induced decrease appears likely to be due to the oxidation of Q-400,
which was recently identified as the high spin Fe^{2+} located between Q_A
and Q_B (7). This interpretation is supported by two findings: a) the relative extent of the absorption changes induced by the first flash decreases
at increasing $K_3[Fe(CN)_6]$-concentration; b) the effect disappears if PS II
particles were excited with a train of flashes (FWHM 10 μs) which permit a
sufficiently high double hit probability (Renger, G. and Kayed, A., unpublished results). If $K_3[Fe(CN)_6]$ would oxidize only Q-400 without affecting the binary oscillation due to semiquinone stabilization at the Q_B-
binding site a phase shift by one flash of this pattern is expected to
arise with serious implications for the calculation of the difference
extinction coefficients $\Delta\varepsilon_{i+1,i}(\lambda)$ as outlined in (6). In this case the
relative extent of fast decay in the 2nd flash should decrease with increasing $K_3[Fe(CN)_6]$ concentration while that of the 3rd flash is expected
to increase. An analysis of the experimental data revealed that both quantities decrease (data not shown). This phenomenon can be explained by the
assumption that $K_3[Fe(CN)_6]$ oxidizes not only Q-400 but also the semiquinone state at the PS II acceptor side. Based on this interpretation
effects due to the binary oscillation at the acceptor side are inferred to
disappear at sufficiently high $K_3[Fe(CN)_6]$-concentrations. The numerical
calculation of $\Delta\varepsilon_{i+1,i}(\lambda)$ on the basis of the Kok-scheme (2) was performed
as described in (6). All experiments performed in the presence of
$K_3[Fe(CN)_6]$ in control and in trypsinized samples led qualitatively to the
same result, i.e. $\Delta\varepsilon_{i+1,1}(\lambda)$ differs at 320 nm for the different redox
transitions in the water oxidizing enzyme system. However, the differences
for the transitions $S_0 \to S_1$ and $S_2 \to S_3$ are quantitatively more pronounced

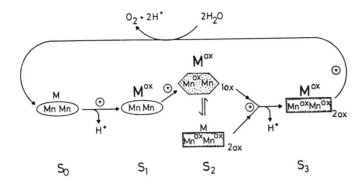

FIGURE ? Reaction scheme of photosynthetic water oxidation (for details see text).

than previously reported (6). The absolute values of $\Delta\varepsilon_{i+1,i}(\lambda)$ were found to be strongly dependent on the values of α, β and $|S_0|_0 / |S_1|_0$. ratio used for the numerical analysis. However, the basic pattern did not change qualitatively even at rather large variations of these values (data not shown). Further information should be obtainable by resolving the rise kinetics of the absorption changes. Unfortunately, at 320 nm also the turn-over of redox component Z which functionally connects system Y with the reaction center gives rise to transient absorption changes (8) thereby overlapping with the $S_i \to S_{i+1}$ redox transitions. Fig. 2 shows the absorption change at 320 nm caused by the 1st flash at an expanded time scale. The rise kinetics are biphasic with an unresolved fast rise which reflects the formation of Q_A^-. The subsequent slower rise with 50 - 100 μs half life time could be either due to S_1 (or S_0) oxidation or due to another process. For kinetical reasons (6) an assignment of the 50-100 μs rise to an S_1 and/or S_0 oxidation would be reasonable. However, we did not observe a corresponding slower rise of the 320 nm absorption change due to the 2nd flash which is expected to exhibit a half time of 200-250 μs for the $S_2 \to S_3$ transition (6). This finding could be interpreted alternatively either by the assumption that the difference spectra $\Delta\varepsilon_{i+1,i}(\lambda)$ are different for i = 1 and 2, respectively, or that the 50-100 μs rise kinetics observed in the 1st flash reflects a process which is not related to the redox transitions of the water oxidizing enzyme system Y. Further experiments are required to clarify this point.

Regardless of the quantitative details in numerical fitting of the experimental data we never obtained the same $\Delta\varepsilon_{i+1,i}(\lambda)$ values for all redox transitions i = 0, 1 and 2 under different experimental conditions. Based on our analysis we conclude that the different univalent oxidation steps $S_i \to S_{i+1}$ (i = 0,1,2) are not caused by the same underlying chemical reaction. Previous considerations led to the conclusion that there does exist a redox component M which attains its oxidized form in the dark adapted system Y (9). Recently M was proposed to undergo twice a univalent oxidation/reduction turnover during the complete 4-step oxidation sequence leading from water to dioxygen (6). In the light of the present data this model has to be modified by the assumption of a redox equilibrium between component M and the binuclear manganese cluster which is assumed to form the core of the catalytic center in the water oxidizing enzyme system Y. Therefore, the scheme depicted in Fig. 3 is postulated to describe the mechanism of photosynthetic water oxidation. In this scheme (Mn Mn) represents the catalytic binuclear manganese cluster; the unknown coordination sphere is described by different symbols which indicate possible structural changes coupled with the different redox transitions. The presented scheme is a generalized form which implies our previous considerations as the upper and lower limits of the proposed redox equilibrium in state S_2. In the case of this equilibrium being tilted towards M^{ox} the original model in (9) arises, whereas the opposite shift to M and a double oxidized binuclear manganese cluster corresponds to our recent model (6). The question about the existence of this equilibrium is of mechanistic relevance because possible shifts under different experimental conditions would give rise to different experimental findings. Detailed studies are required to clarify this crucial point.

The possible existence of another important equilibrium has to be considered too. Latest findings indicate that water participates in the first coordination sphere of the catalytic manganese (10). Accordingly, a redox equilibrium between water ligands and manganese would be consistent with formation of a peroxide like bond at S_2 and/or S_3 (6,9). An analogous equilibrium has been discovered recently in cytochrome oxidase (11). The mechanistic implications of the intermediary peroxide formation are discussed elsewhere (6).

ACKNOWLEDGEMENTS

The authors gratefully acknowledge the financial support of this work by the Deutsche Forschungsgemeinschaft (Sfb 312).

REFERENCES

1 Renger, G. and Govindjee (1985) Photosynth. Res. 6, 33-55
2 Kok, B., Forbush, B. and Mc Gloin M.P. (1970) Photochem. Photobiol. 11, 457-475
3 Berthold, D.A., Babcock, G.T. and Yocum, C.F. (1981) FEBS Lett. 134, 231-234
4 Völker, M., Ono, T., Inoue, Y. and Renger, G. (1985) Biochim. Biophys. Acta 806, 25-34
5 Renger, G. and Weiss, W. (1983) Biochim. Biophys. Acta 722, 1-11
6 Renger, G. and Weiss, W. (1986) Biochim. Biophys. Acta 850, 184-196
7 Petrouleas, V. and Diner, B.A. (1986) Biochim. Biophys. Acta 849, 193-202
8 Renger, G. and Weiss, W. (1986) Biochim. Biophys. Acta 850, 173-183
9 Renger, G. (1977) FEBS-Letters 81, 223-228
10 Hansson, G. Andreasson, L.E. and Vänngard, T. (1986) FEBS Letters 195, 151-154
11 Blair, D.F., Witt, S.N. and Chan, S.I. (1985) J. Am.Chem. Soc. 107, 1389-1399

THE MODIFICATION OF THE DONOR SIDE REACTION PATTERN IN PS II MEMBRANE
FRAGMENTS BY TRYPSIN AND CaCl$_2$

M. Völker, H.J. Eckert and G. Renger, Max-Volmer-Institut für Biophysikalische und Physikalische Chemie, Technische Universität Berlin, Strasse
des 17. Juni 135, 1000 Berlin 12, FRG

1. INTRODUCTION

Photosynthetic water oxidation takes place in system II, initiated by
photooxidation of a special chlorophyll-a, P680. Oxygen formation is
assumed to occur in the water-oxidizing enzyme system Y at a Mn-containing
catalytic site. The functional connection between system Y and P680 implies at least one further redox component Z. In samples competent in
oxygen evolution, P680$^+$ reduction predominantly occurs in the ns range;
whereas, after elimination of the oxygen-evolving capacity P680$^+$ is reduced in the μs range (for review see ref. 1).

In order to study possible correlations between P680$^+$-reduction
kinetics and the structural and functional integrity of PS II, experiments
were performed in trypsinized PS II membrane fragments with special
emphasis on the influence of Ca^{2+} ions.

2. PROCEDURE

2.1. Materials and Methods

PS II particles with oxygen-evolving capacity were prepared from
market spinach as described in (2) with the modifications described in (3),
with a Triton X-100/chlorophyll ratio of 20:1. The trypsin treatment was
carried out at room temperature at a trypsin/Chl-ratio of 2:1. The
measurements of the 830 nm absorption changes were performed with a
single beam flash photometer similar to the one described in (4). Photosynthesis was excited by repetitive (2 Hz) flashes (FWHM = 7 ns) from a
Q-switched frequency doubled Nd/YAG-laser. All experiments were carried
out with a chlorophyll concentration of 50 μg/ml in a medium containing
10 mM NaCl and 20 mM MES/NaOH (pH 6.0) with 1 mM K$_3$(Fe(CN)$_6$) as artificial
electron acceptor. Other additions as indicated in the figure legends.

3. RESULTS AND DISCUSSION

Typical traces of absorption changes at 830 nm induced by repetitive
flashes in spinach PS II particles are depicted in Fig. 1. Under these
conditions P680$^+$ is reduced via multiphasic kinetics (5,6). In control
samples ca. 65-70% of the relaxation kinetics decay in the ns-time scale.
This percentage increases up to 80% in the presence of 10 mM CaCl$_2$.
Complete elimination of oxygen evolution by incubation with NH$_2$OH (3 mM,
6 min) elicits the well-known pH-dependent reduction kinetics in the μs-
time scale.

Proteolytic degradation of PS II particles with trypsin at pH 6.0
leads to conversion of   ns- into μs-kinetics. The total amplitude of
the absorption changes remains constant, thus reflecting a still intact
reaction center of PS II.

In these mildly trypsinized PS II particles CaCl$_2$ causes a markedly
larger restoration of the ns kinetics than in control samples. It was
found that this restoration effect is mainly due to Ca^{2+} and that Cl$^-$
plays only a marginal role. It is rather Ca^{2+} specific, because only

Biggens, J. (ed.), Progress in Photosynthesis Research, Vol. I. ISBN 90 247 3450 9
© *1987 Martinus Nijhoff Publishers, Dordrecht. Printed in the Netherlands.*

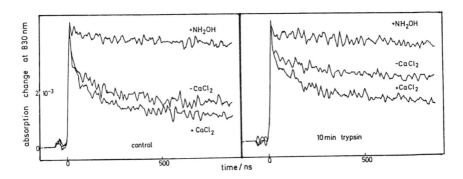

FIGURE 1 Absorption changes at 830 nm induced by repetitive laser flashes in normal and trypsinized PS II membrane fragments at pH = 6.0 in the absence and after addition of 10 mM CaCl$_2$. Experimental conditions as described in Materials and Methods

Sr^{2+} was found to induce a comparable restoration effect; whereas, Mg^{2+}, Mn^{2+} and monovalent cations are almost inefficient. This result raises the question about the nature of the Ca^{2+}-binding site. In an attempt to characterize its properties the dependence on CaCl$_2$ concentration of the restoration effect has been analyzed.

The extent of the ns kinetics can be simply determined as the difference of the amplitudes at 1 µs after the excitation flash, $\Delta A_{830}^{1\ \mu s}$, in the absence and presence of 3 mM NH$_2$OH, because $\Delta A_{830}^{1\ \mu s}$ (NH$_2$OH) was shown to be invariant to different salt additions and trypsin treatment. Taking into accout that CaCl$_2$ exhibits the maximum restoration effect at a concentration of C = 10 mM, the restoration factor ΔR_{CaCl_2} is expressed as:

$$\Delta R_{CaCl_2} = \left[\frac{\Delta A_{830}^{1\ \mu s}(\text{control}) - \Delta A_{830}^{1\ \mu s}(C)}{\Delta A_{830}^{1\ \mu s}(\text{control}) - \Delta A_{830}^{1\ \mu s}(10\ \text{mM CaCl}_2)} \right]_{\text{trypsin}}$$

The results are depicted in Fig. 2. Fig. 2 shows that the restoration effect does not exhibit a rectangular hyperbolic concentration dependence and, therefore, suggests the idea of a cooperative Ca^{2+} effect. An analogous but quantitatively slightly different sigmoidicity has been also observed for Sr^{2+} (data not shown). In order to estimate the number of binding sites per PS II, the experimental data were analyzed in terms of the classical Hill-plot. From the double logarithmic plot presented as insert in Fig. 2 a Hill coefficient of n = 2 was obtained. This result indicates that at least two Ca^{2+} cations interact to give the restoration effect of the ns kinetics of P680$^+$ reduction in mildly trypsinized PS II particles. Based on the minimum number of two binding sites, the degree of cooperativity can be calculated by using the generalized Adair equation. The analysis was performed in the same way as described for hydrazine binding in the water oxidizing enzyme system Y (7). The fitting program leads to the curve given in Fig. 2, which implies binding constants of $k_1 = 7.5 \cdot 10^{-4} \mu M^{-1}$ and $k_2 = 1.7 \cdot 10^{-2} \mu M^{-1}$. From the ratio of the binding constants the free interaction energy of the Ca^{2+}

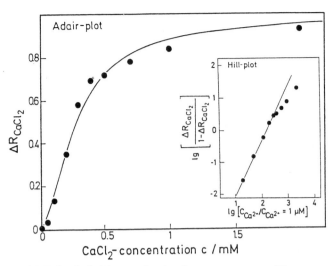

FIGURE 2   Relative extent of ns recovery, ΔR_{CaCl_2}, as a function of CaCl$_2$ concentration in PS II particles trypsinized at pH = 6.0 for 10 min. Experimental details as described in Materials and Methods. The experimental data are symbolized by dots. The curve represents the best fit Adair function with binding constants given in the text. Insert: Hill-plot of the experimental data.

binding was estimated to ΔG_{int} = -8 kJ/mol.

The present data do not permit identification of the protein which contains the Ca^{2+}-binding sites. From earlier experiments it seems unlikely that either the 17 kDa or the 24 kDa extrinsic protein provides the Ca^{2+} target, because these polypeptides are easily attacked by trypsin (3). Our results could be related to recent findings of Ca^{2+} requirement for oxygen evolution in PS II particles that were deprived of their 17 kDa and 24 kDa proteins (8). It was inferred that there does exist an intrinsic membrane protein which contains the Ca^{2+}-binding site(s) of functional relevance to water oxidation. Likewise, an analysis of the oscillation pattern of thermoluminescence (B-band) and oxygen yield in PS II particles depleted of their extrinsic regulatory subunits of the water oxidizing enzyme system led to the conclusion that Ca^{2+} is required for oxygen evolution through an all-or-none type activation. It was further found that PS II exhibits a heterogeneity towards the Ca^{2+} effect (9).

The above-mentioned findings (8,9) might suggest that the Ca^{2+} effect on restoration of the ns kinetics of $P680^+$ reduction in mildly trypsinized PS II particles is intimately correlated with the Ca^{2+} requirement for oxygen evolution. In order to test this idea, experiments were performed with different substances that are reported to affect the PS II donor side. It was found that lauroylcholine chloride (LCC) treatment or incubation with the Ca^{2+} antagonist trifluoperazine (TFP) markedly reduces the average yield per flash, while the extent of the ns kinetics of $P680^+$ reduction is much less affected (10). This result indicates that the correlation between the functional integrity of the water oxidizing enzyme system and $P680^+$-reduction kinetics is complex. Therefore, it remains to be clarified whether or not the observed Ca^{2+} effects at the

FIGURE 3

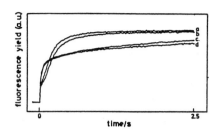

Fluorescence yield as a function of actinic illumination time in PS II membrane fragments: (a) control, (b) trypsinized for 10 min at pH = 7.5 in the presence of 10 mM $CaCl_2$, (c) trypsinized for 10 min at pH = 7.5 in the absence of $CaCl_2$, (d) as (c) but with addition of 10 mM $CaCl_2$ after trypsin treatment

PS II donor side are really caused by the same Ca^{2+}-binding protein. Another interesting mechanistic aspect is the possibility of Ca^{2+} interaction with polar head groups of lipids. This could give rise to structural modifications (probably through lipid/protein interactions) which are responsible for the observed Ca^{2+} effects.

In addition to the Ca^{2+}-specific restoration described in Figs. 1 & 2, we observed also a protection by $CaCl_2$ to tryptic attack of PS II particles. Fig. 3 shows fluorescence induction curves of differently treated samples. Trypsination at pH = 7.5 (this is a harsh attack compared with pH = 6.0, see ref. 3) causes a marked change of fluorescence induction (curve c). If, however, the sample contained 10 mM $CaCl_2$ before trypsin addition, then only marginal changes are observed (compare curves a and b). On the other hand, after starting the proteolysis $CaCl_2$ was without protective or recovery effect.

The fast rise of the fluorescence induction caused by trypsin (curves c and d) is indicative for blockage of Q_A^- reoxidation by Q_B. It was found that the protective effect is not Ca^{2+}-specific (data not shown). The origin of this protection induced by bivalent cations (monovalent cations are less efficient) at the PS II acceptor side remains to be clarified.

ACKNOWLEDGEMENTS
The authors gratefully acknowledge the financial support of this work by the Deutsche Forschungsgemeinschaft (Sfb 312).

REFERENCES
1 Renger, G. and Govindjee (1985) Photosynth. Res. 6, 33–55
2 Berthold, D.A., Babcock, G.T. and Yocum, C.F. (1981) FEBS Lett. 134, 231–234
3 Völker, M., Ono, T., Inoue, Y. and Renger, G. (1985) Biochim. Biophys. Acta 806, 25–34
4 van Best, J.A. and Mathis, P. (1978) Biochim. Biophys. Acta 408, 154–163
5 Brettel, K. and Witt, H.T. (1983) Photobiochem. Photobiophys. 6, 253–260
6 Eckert, H.J., Renger, G. and Witt, H.T. (1984) FEBS Lett. 167, 316–320
7 Hanssum, B. and Renger, G. (1985) Biochim. Biophys. Acta 810, 225–234
8 Miyao, M. and Murata, N. (1986) Photosynthesis Res. (in press)
9 Ono, T. and Inoue, Y. (1986) Biochim. Biophys. Acta 850, 380–389
10 Eckert, H.J., Wydrzynski, T., Völker, M. and Renger, G. (1986) Photochem. Photobiol. 43, Suppl., 1025

STUDIES ON WATER OXIDATION BY MASS SPECTROMETRY IN THE FILAMENTOUS
CYANOBACTERIUM *OSCILLATORIA CHALYBEA*

Klaus P. Bader, Pierre Thibault[*] and Georg H. Schmid
Universität Bielefeld, Lehrstuhl Zellphysiologie, D-4800 Bielefeld 1, FRG
[*]Centre d'Etudes Nucléaires de Cadarache, DB/SRA, F-13115 Saint-Paul-Lez-
Durance, France

INTRODUCTION

In recent publications we have analyzed the oxygen evolution pattern of
the filamentous cyanobacterium *Oscillatoria chalybea* (1,2). The peculiarity
of such a pattern, produced as the consequence of a train of short saturating light flashes, is the fact that a positive amperometric signal is
observed under the first flash despite extensive dark adaptation (1).
Further studies clearly showed that the signal was at least partially due
to oxygen since in a polarogram the signal of the first flash showed the
same dependence on the polarization voltage as that of the second and
the third one (1). This was further confirmed by Bader who showed by mass
spectrometry that under the first flash a substantial signal of mass 32
was observed (3). In conclusion the signal seemed to be due to metastable
S_3 which is not easily acceptable in the coherent Kok model. For further
verification we have carried out an analysis of the properties of the S_3^-
state of *Oscillatoria chalybea* by means of mass spectroscopy.

RESULTS AND DISCUSSION

In figure 1 the basic problem outlined in
the introduction is seen in the control
sequence which shows a substantial positive amperometric signal under the first
flash. This signal is certainly not caused by an artefactual peroxide-like substance or hydrogen peroxide produced with
our plant material on the electrode surface, since catalase does not influence
the signal amplitude. The only indication
for an influence of hydrogen peroxide on
the flash sequence might be seen in the
observation that the amount of misses
occasionally seems to be reduced in the
presence of our catalase preparation.

If a flash sequence is analysed by mass
spectroscopy, photosynthetic oxygen evolution is only observed, if the sample is
not in an anaerobic condition. Absolutely
no oxygen evolution is observed in samples
which are flushed with nitrogen (Fig. 2a)
or which have become anaerobic for other
reasons. The preparation, however, can be
fully reactivated by gassing the sample

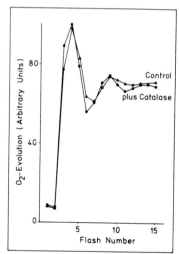

Fig. 1: Influence of catalase
on the O_2-evolving pattern in
Oscillatoria chalybea. (▲) control sequence with no additions.
(●) Sequence registered in the
presence of 1000 enzyme units
of catalase.

Biggens, J. (ed.), Progress in Photosynthesis Research, Vol. I. ISBN 90 247 3450 9
© 1987 Martinus Nijhoff Publishers, Dordrecht. Printed in the Netherlands.

with air or oxygen (Fig. 2b) As clearly seen from figure 2, the signal due to the first flash is indeed oxygen, supporting our earlier hypothesis, namely that the first amperometric signal shown in Fig. 1 should be due to metastable S_3.

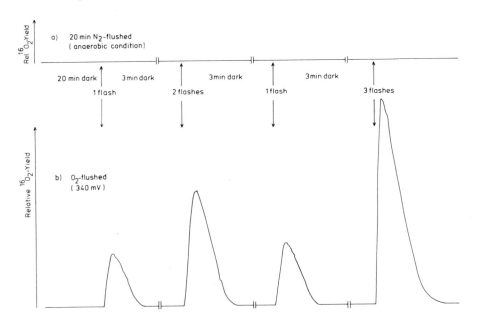

Fig. 2: O_2-evolution as the consequence of short light flashes measured as mass 32 by mass spectrometry in preparations of *Oscillatoria chaly-bea*. a) Anaerobic conditions, b) the preparation of a) reactivated by short flushing with O_2.

The experiment also shows that photosynthetic oxygen evolution requires activation or binding of molecular oxygen onto the water-splitting enzyme, before oxygen evolution really begins. In the following experiment we gene-rated S_3 by giving two flashes onto our dark adapted preparation (according to our analysis (1) the S-state population of dark adapted *Oscillatoria* contains approx. 50% S_1). Immediately upon these two flashes we injected $H_2^{18}O$ which required roughly 20 s and looked at the O_2-level in a subsequent single flash. From the life time of the S_3-state (4) it follows that at the moment of the $H_2^{18}O$ addition and also at the time of measurement a substan-tial amount of S_3 was still present in the assay. Two types of experiments were carried out. In the first experiment the third flash was given after 15 min and in the second experiment the third flash was given 3 min after the two preilluminating flashes. Fig. 3 shows the time course and the sig-nal amplitudes of the experiment in which mass 36 was registered. Two pre-illuminating flashes were given on an *Oscillatoria* preparation which was suspended in 2 ml assay buffer containing only ordinary water i.e. $H_2^{16}O$. Immediately upon these two flashes 0.45 ml $H_2^{18}O$ (98.37 atom % ^{18}O) were added. The third flash which leads to oxygen evolution was given 15 min later. It is clearly seen that the oxygen evolved at the third flash is considerably labelled. It should be remembered that the peculiarity of *Os-cillatoria* is the fact that a dark adapted preparation has a S-state popu-

lation containing 10-12% metastable S_3 (1). This type of experiment in which
15 min were allowed before giving the oxygen-evolving third flash had the
advantage that the perturbation of the assay system due to pollution of
$^{16}O_2$ dissolved in the $H_2^{18}O$ added and due to the interference with the gas
exchange equilibrium in the resting 2 ml assay had completely calmed down
at the time of the measurement.

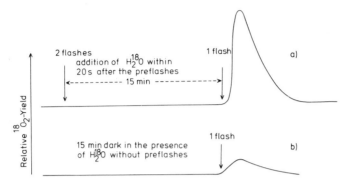

Fig. 3: Flash yield in a preparation of *Oscillatoria chalybea*. a) Generation
of S_3 by two flashes in a medium containing only $H_2^{16}O$ with addition of $H_2^{18}O$ immediately after the preflashes and analysis of the
flash yield in a subsequent third flash by measuring oxygen evolution as mass 36. b) Same preparation as a) but addition of $H_2^{18}O$ to
a dark adapted sample and dark incubation with $H_2^{18}O$ for 15 min as
in a).

In figure 4 we have generated, as in the experiment before, S_3 by preilluminating the *Oscillatoria* preparation with 2 flashes. As in the experiment
of figure 3, S_3 was generated in an assay medium containing only $H_2^{16}O$.
Immediately upon these two preflashes we added $H_2^{18}O$ and measured oxygen
evolution 3 min later. The addition itself required as in the experiment
of figure 3 approximately 15 s. The oxygen evolved under the third flash
is clearly labelled.

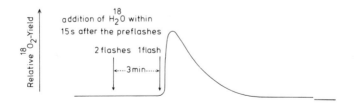

Fig. 4: Flash yield in a preparation of *Oscillatoria chalybea*. Generation
of S_3 by two flashes in a medium containing only $H_2^{16}O$ with addition of $H_2^{18}O$ and analysis of the flash yield within 3 min after
the two preflashes as $^{18}O_2$ (mass 36).

These experiments solve several questions: In the first place figs. 2a and
3a show that the amperometric signal observed earlier under the first flash
in our electrode device is oxygen and thus due to metastable S_3. Secondly,
Fig. 4 shows that the condition S_3 of the oxygen-evolving enzyme does not

contain bound non-exchangeable water or partially oxidized water (the "O_2-precursor" of the literature) which would have obstructed the labelling observed. This is in agreement with the observation by Radmer and Ollinger (5). However, figure 3 and also our earlier experiment (1) show that meta-stable S_3 can be distinguished from freshly generated S_3 (Fig. 4), because, if the $H_2^{18}O$ addition is made to dark adapted *Oscillatoria* preparations in the dark, the first flash contains no or very littly $^{18}O_2$-labels (see (1) and Fig. 3b). This together with our experiments of figs. 3a and 4 could mean that at least the metastable S_3 condition of *Oscillatoria chalybea* binds to water that is not exchanged.

REFERENCES
1. K.P. Bader, P. Thibault and G.H. Schmid (1983) Zeitschr. für Naturforsch. 38c, 778-792
2. K.P. Bader and G.H. Schmid (1980) Arch. Microbiol. 127, 33-38
3. K.P. Bader (1984) Mitteilungsband Botanikertagung, p. 29, Wien
4. K.P. Bader (1983) Proceedings of the VIth Intern. Congress on Photosynthesis (C. Sybesma, ed.) I.3.287-290, Brussels
5. R. Radmer and O. Ollinger (1986) FEBS Lett. 195, 285-289

FLASH-INDUCED ENHANCEMENTS IN THE [1]H-RELAXATION RATE OF PHOTOSYSTEM II PARTICLES

A.N. SRINIVASAN AND R.R. SHARP,* DEPARTMENT OF CHEMISTRY, THE UNIVERSITY OF MICHIGAN, ANN ARBOR, MI 48109

Flash-induced enhancements in the NMR spin-lattice relaxation rate R_1 of solvent water protons have been detected in suspensions of Photosystem II particles prepared as described in (1). The relaxation enhancements produced by trains of single turnover flashes are small (<1% of the background) and have been detected using signal averaging techniques (2,3). In all experiments described here (Figs 1-8), one preflash followed by a 6-6.5 minute dark delay was used to preset the S-state system to the S_1 state prior to the application of the flash train.

The enhancements correlate well with the known properties of the S-states of water oxidation as summarized in the experiments of Figs 1-3: [A] One and two actinic flashes cause an abrupt enhancement in the R_1 of about 0.008 s^{-1}. The one-flash (Fig 1A) and two-flash (Fig 1B) transients then

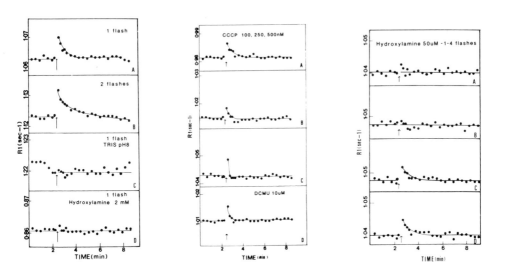

FIGURES 1-3. Effect of various treatments on the flash-induced R_1 enhancement.

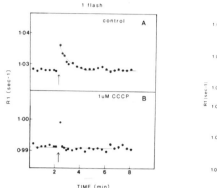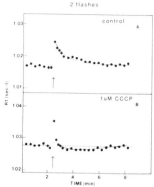

FIGURES 4&5. R_1 enhancements induced by 1 and 2 saturating
               flashes $\mp$1 µM CCCP.

decay to the baseline with halftimes of about 25 s and 40 s,
respectively, values which are in good agreement with halftime
values reported from thermoluminescence and oxygen evolution
studies. [B]   Treatment with either alkaline Tris buffer or
with 2 mM NH_2OH, which are known extractants of manganese
from the oxygen evolving complex, suppress the flash-induced
R_1 transient (Fig 1C and 1D). [C] ADRY reagents which
accelerate the decay of the S_2 and S_3 states also accelerate
the decay of the one-flash transient. Titration with the ADRY
reagent CCCP (Fig 2A-C) shows that the decay of the transient
is almost completely suppressed by 500 nM CCCP. [D]   The
transient is sensitive to but not eliminated by DCMU (Fig 2D).
DCMU blocks electron transport between Q_a and Q_b and
accelerates the charge recombination of S_2 with the acceptor
side of PS II. This effect is reflected in an accelerated
decay of the transient (halftime <10 s) with respect to the
control. [E]   Low concentrations (ca. 50 µM) of NH_2OH are
known to retard the S-state cycle of oxygen evolution by two
flashes. A corresponding effect occurs in the R_1 experiment
where the enhancement associated with the $S_1 \to S_2$ transition
is delayed from the first to the third flash. Taken together,
these results indicate that the observed R_1 transients monitor
redox chemistry associated with S-state transitions.

     The behavior of the R_1 enhancement produced by trains of
one to five flashes is shown in Figs 4-8. The one flash
relaxation transient is a positive R_1 enhancement of 0.008
sec^{-1} which decays with a half-time of about 25 s (Fig 4A);
the two flash response is a positive transient of nearly
identical amplitude to the one flash response but with a
slower decay (ca. 40 s) (Fig 5A). Both the one flash and the

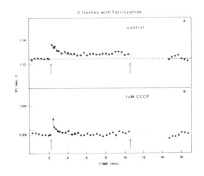

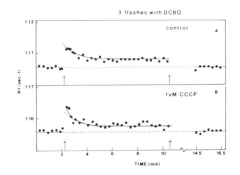

FIGURES 6&7. R_1 enhancement induced by 3 saturating flashes
              $\mp$ 1 μM CCCP.   Electron acceptor = 1.5 mM
              ferricyanide (Fig 6) and 250 μM DCBQ (Fig 7).

two flash transients are sensitive to
the presence of CCCP (Figs 4B and 5B).
The appearance of a strongly relaxing
paramagnetic center after one flash
is consistent with the expected
properties of a Mn(III) → Mn(IV)
oxidation (2,3).[1]   In contrast, the
absence of a further R_1 enhancement
following the second flash shows that
a second strongly relaxing center, in
addition to the one associated with
the S_2 state, is not formed on the
S_2 → S_3 transition.   The NMR experiment gives no indication that
manganese redox chemistry occurs on
the S_2 → S_3 transition.

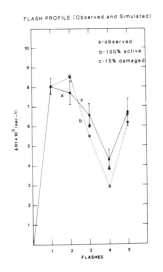

FIGURE 8.   Experimental
and simulated flash
profiles of the ampli-
tude of the R_1
enhancement.

    The three flash R_1 response,
which reflects a maximum number of
centers in the S_0 state, is a positive
transient of 0.0066 s^{-1}, which decays
with a halftime of about 50 s to a
stable R_1 value of 0.002 s^{-1} above
the preflash baseline (Fig 6A).   This
overall response is similar both when
ferricyanide and DCBQ are used as the
electron acceptor for PS II (Figs 6A
and 7A).   The kinetically stable component of the three flash
change reflects increased relaxation in the S_0 state and
indicates that the formation of the S_0 state from S_1 involves
the production of a strongly relaxing center.   The sign of the
R_1 response is consistent with the expected properties of a
Mn(III) → Mn(II) reduction for the S_1 to S_0 transition (2,3).

    The positive R_1 enhancements induced by the three flash
train persist in the presence of CCCP (Figs 6B and 7B) at
concentration levels where the one and two flash decays are

effectively suppressed (compare with Figs 4B and 5B). In the presence of ferricyanide and CCCP, the transient decays back to the original preflash baseline (Fig. 6B). In contrast, with DCBQ as the acceptor (Fig 7B), the R_1 decays to the stable, enhanced value characteristic of the strongly relaxing center associated with S_0. This indicates that in the presence of the ADRY reagent CCCP, the S_0 state can be oxidized (presumably to S_1) by ferricyanide but not by DCBQ.

Flash transients across a five flash cycle (Fig 8) exhibit the expected oscillatory behavior with a minimum on the fourth flash. Amplitudes of the R_1 transients across a five flash cycle have successfully been simulated (Fig 8) assuming that strongly relaxing centers are formed on the $S_1 \rightarrow S_2$ and $S_1 \rightarrow S_0$ transitions with specific relaxation enhancements of 0.008 s^{-1} and 0.005 s^{-1}, respectively. Simulations of the flash profile indicate that there is an R_1 contribution from the fraction of S-state centers which are capable of transitions upto S_3 but which cannot cycle past S_3 to S_0.

ACKNOWLEDGEMENT

Partial support for this research was obtained in the form of a grant from the United States Department of Agriculture Competitive Grants Research Program (No. 83-CRCR-1-1047).

REFERENCES

1 Ghanotakis, D.G., Babcock, G.T. and Yocum, C.F. (1984) FEBS Lett. 167, 127-130
2 Srinivasan, A.N. and Sharp, R.R. (1986) Biochimica et Biophysica Acta 850, 211-217
3 Srinivasan, A.N. and Sharp, R.R. (1986) Biochimica et Biophysica Acta, in press

FOOTNOTE
1 Efficient relaxation traps are provided by paragmagnetic species with long electronic times and large magnetic moments. Among the common oxidation states of manganese, Mn(II) and Mn(IV) (which are orbital singlets) have long paramagnetic relaxation times. These provide relaxation traps that are more efficient than Mn(III). Accordingly, the appearance of a strongly relaxing center on an oxidative S-state transition is indicative of a Mn(III) $\rightarrow$ Mn(IV) oxidation; the appearance of a strongly relaxing center on a reductive S-state transition suggests a Mn(III) $\rightarrow$ Mn(II) reduction.

The State of Manganese in the Photosynthetic Apparatus: An X-ray Absorption Spectroscopy Study

Vittal K. Yachandra, R. D. Guiles, Ann McDermott, James Cole
R. David Britt, S. L. Dexheimer
Kenneth Sauer and Melvin P. Klein

Laboratory of Chemical Biodynamics, Lawrence Berkeley Laboratory
University of California, Berkeley, CA 94720 USA

Introduction

The Mn-containing O_2-evolving complex (OEC) in PSII is the least understood of the light-driven electron transfer protein complexes in chloroplasts, even though it has been the subject of extensive studies [1,2]. Studies of O_2-evolution using a train of saturating light flashes have given rise to a model for the accumulation of oxidizing equivalents in which some intermediates, labeled S_0-S_4 operate in a cyclic fashion [3]. The multiline EPR signal produced by illumination at 190 K, at 277 K in the presence of DCMU or by room temperature flashes has been identified with the S_2 state [4,5,6], although it has been speculated that the state generated at 190 K is different from that generated at 277 K due to the ease of ligand exchange and conformational flexibility at the higher temperature. The g=4.1 EPR signal generated by illumination at 140 K or 200 K was attributed to an intermediate between S_1 and S_2 [7] or S_2 and S_3 states [8], but recently it has been proposed that it arises from another structural form of the S_2 state [9,10].

Detailed studies of Mn X-ray K-edges provide information about oxidation states and site symmetry, and EXAFS furnishes information about the types of ligands, the Mn to ligand distances and the coordination number of Mn. In this paper we present Mn K-edge and EXAFS results of samples illuminated at 140 K, 190 K or 277 K in the presence of DCMU and the effects of peptide release on the structure of Mn in the OEC. The Mn K-edge and EXAFS results for the S_0 and S_3 states and evidence for a tetranuclear Mn cluster are presented in the companion paper by R.D. Guiles et al [11]. Additionally, studies on the OEC in the thermophilic cyanobacterium *Synechococcus* are presented in the paper by A. McDermott et al [12].

Materials and Methods

Preparation of oxygen-evolving PSII sub-chloroplast membranes from spinach was accomplished by a Triton X-100 fractionation procedure [13]. The samples had rates of O_2 evolution of 300-400 μmol O_2 (mg Chl)$^{-1}$h^{-1} and contained ~4 Mn atoms and 260 Chl per reaction center, using the integrated EPR signal II_S for quantitation [14].

X-ray absorption samples were suspended in 50 mM MES buffer, pH 6.0, 10 mM NaCl, 5 mM $MgCl_2$ and ~30% glycerol. The samples of 20-30 mg Chl ml^{-1} concentration were mounted in lucite sample holders, and the procedure of dark adaptation, illumination, EPR and X-ray measurements was carried out directly in these sample holders. X-ray absorption edge spectra and EXAFS spectra were collected in the fluorescence mode at the Stanford Synchrotron Radiation Laboratory, Stanford, CA. Energy calibration was maintained by simultaneous measurement of the strong and narrow pre-edge feature of $KMnO_4$ at 6543.3 eV. The uncertainty in our edge energy measurements is 0.1 eV. During X-ray measurements, samples were suspended in a double walled kapton cryostat maintained at either 150 or 170 K by a liquid N_2 boil-off jet. Details of the X-ray experimental procedure and data analysis are described in Refs 13 and 15. EPR spectra were recorded on a Varian E-109 spectrometer. Samples were run at 8 K at a microwave frequency of 9.21 GHz and with 100 kHz field modulation.

The S_1 samples were prepared by dark adaptation of the PSII particles for ~2 h and then freezing the sample in liquid N_2. The other samples were initially dark adapted and then were

Biggens, J. (ed.), Progress in Photosynthesis Research, Vol. I. ISBN 90 247 3450 9
© *1987 Martinus Nijhoff Publishers, Dordrecht. Printed in the Netherlands.*

equilibrated at 140 K, 190 K or at 277 K in a Varian V6040 NMR temperature controller and illuminated with a 400 W tungsten lamp through a 5 cm water filter. The samples were then frozen in liquid N_2. The 277 K illuminated samples contained 100 μM DCMU. The production of the S_2 state by illumination at 190 K or at 277 K was monitored by observing the characteristic multiline EPR signal at 8 K. The 140 K illuminated samples were monitored by observing the g=4.1 EPR signal (about 35% of the maximal multiline signal was also induced upon illumination at 140 K).

Results and Discussion

We demonstrated earlier that a significant change occurs in the Mn K-edge energy of PSII particles upon advancing from the S_1 to the S_2 state by illumination at 190 K [15,16]. These observations established for the first time that Mn is directly involved in the light-driven storage of oxidizing equivalents. A 1s-3d pre-edge feature is seen at ~6543 eV in the edge spectra of the S_1 and S_2 samples indicating the non-centrosymmetric environment of Mn in the complex.

Figure 1. The X-ray absorption Mn K-edge spectra of spinach PSII samples produced by illumination at 140 K (..............), 190 K ($_____$) and 277 K ($-\cdot-\cdot-$) in the presence of DCMU. The spectrum of the S_1 state (——————) is overlayed in each case for comparison. A smoothed curve is drawn through the data points. The small pre-edge feature at ~6543 eV is due to the 1s-3d bound state transition. The K-edge inflection energy for the S_1 state is at 6551.3 eV and for the 140 K, 190 K and 277 K illum. samples is at ~6552.5 eV. The energy shifts from the dark adapted S_1 state to the illum. samples indicate an oxidation of Mn in the OEC.

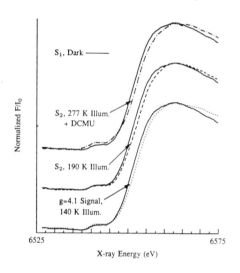

Fig 1 shows the Mn K-edge spectra of the S_1 state and S_2 states produced by illumination at 190 K, by illumination at 277 K in the presence of DCMU and by illumination at 140 K. The inflection points of the 140 K and 277 K illuminated samples are shifted to higher energy compared to the inflection point for the S_1 sample and are at an energy comparable to the 190 K illuminated S_2 sample. This shows that the S_2 state produced at 190 K is similar to that produced at 277 K in the presence of DCMU. We conclude from these results that the oxidation state of Mn in the OEC is similar in samples produced by illumination at 140 K, 190 K, or at 277 K in the presence of DCMU. The similarity of the 140 K illuminated spectrum to the spectrum of the other S_2 samples supports the model that the g=4.1 EPR signal and the multiline signal arise from different conformations of the S_2 state [9].

The Mn K-edge inflection of the S_1 sample occurs at about 6551.3 eV, which is close to the K-edge of Mn(III) complexes [17]. The K-edge inflection of the illuminated samples exihibits a shift to higher energy by about 1 eV and occurs at ~6552.5 eV, which is between that observed for Mn(III) and Mn(IV) complexes.

In Fig 2 are shown Fourier transforms of the k^1-weighted Mn EXAFS data from spinach PSII particles in the S_1 and S_2 states produced by 190 K or by 277 K illumination in the presence of DCMU. The EXAFS results show that the salient features of the Mn structure are essentially identical for samples in the S_1 and S_2 states. These features are a Mn neighbor at 2.70 Å, and N or O ligands at 1.75 Å and 2.00 Å, which are typical distances for bridging and terminal ligands respectively. Such coordination is similar to that of a di-μ-oxo bridged binuclear complex [18].

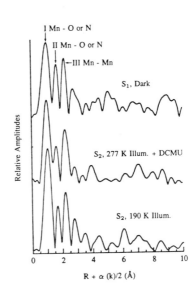

Figure 2. Fourier transforms of the k^1-weighted Mn EXAFS data from spinach S_1 and S_2 states produced by illum. at 190 K and at 277 K in the presence of DCMU. The peaks labelled I and II are characteristic of bridging and terminal N or O ligands and peak III is due to a neighboring Mn atom; the distances are typical for μ-oxo bridged binuclear Mn clusters. The peak labeled I fits best to N or O ligands at 1.75 Å, peak II fits to N or O ligands at 2.00 Å, and peak III fits to a Mn atom at 2.70 Å. We estimate the uncertainty in distances to be 0.03 Å except for the bridging ligand distances where it is 0.05 Å. The fits were performed using theoretical phase and amplitude functions by the Teo-Lee method [22].

It is possible that the coordinating ligands of the S_2 state produced at 190 K could be different from that produced at 277 K due to the ease of ligand exchange at 277 K. Chloride is essential for O_2 evolution, and it has been suggested that Cl^- is a ligand of Mn [19]. In this view a possible difference in the S_2 state produced at 190 K and at 277 K could be in Cl^- ligation to Mn. We do not find any evidence in our data for chloride coordination in the S_1 or the S_2 states produced at 190 K or 277 K. This is in accord with our EPR data of Cl^- and Br^- containing PSII particles [20]. The great similarity of the EXAFS results from samples prepared in the S_1 and S_2 states and the light-induced edge shift results support a model that the light-induced S_1 to S_2 transition involves a change in the oxidation state of Mn with no change in the coordination of Mn in the OEC.

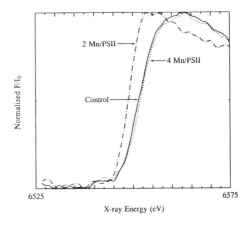

Figure 3. The Mn K-edge spectra of $CaCl_2$ washed PSII particles. Smooth curves are drawn through the data points. The control is a S_1 sample. The 4 Mn/PSII sample is depleted of the 16, 24 and 33 kDa peptides by incubation in 800mM $CaCl_2$, 50 mM MES, pH 6.0, for 2 h. The 2 Mn/PSII sample was subsequently incubated for 12 h in 50 mM MES, pH 6.0, 15 mM NaCl. The inflection energy is at 6551.9 eV for the 4 Mn/PSII sample and at 6548.9 eV for the 2 Mn/PSII sample. The invariance of the 4 Mn/PSII K-edge indicates that the 16, 24 and 33 kDa peptides are not required for maintaining the binding environment of Mn in the OEC.

CaCl$_2$ washing of PSII particles is known to release three peptides of molecular weight 16, 24 and 33 kDa, which have been shown to be involved in O$_2$ evolution [21]. The Mn K-edge spectra of samples depleted of the three peptides and containing 4 Mn/PSII and 2 Mn/PSII are shown in Fig 3. The K-edge spectrum of the 4 Mn/PSII sample is similar to that of the control S$_1$ sample, while the 2 Mn/PSII sample is distinctly different in shape and position, with an inflection point around 6548.9 eV, suggesting a +2 oxidation state for Mn. The EXAFS results (not shown) indicate that the structure of Mn in the 4 Mn/PSII sample is similar to that in intact samples, but a major change is observed in the 2 Mn/PSII sample. The Fourier transform of the EXAFS data for the 2 Mn/PSII sample lacks the Fourier peaks labeled I and III in Fig 2. This indicates that the Mn atoms are now farther than 4 Å apart, making a bridged Mn structure unlikely. We conclude that the bridged Mn complex is still intact in the 4 Mn/PSII samples in spite of the loss of the three peptides, which indicates that the 16, 24 and 33 kDa peptides are not directly involved in the binding of Mn in the OEC. However, when 2 Mn/PSII are also released, a major change occurs in the structure of the Mn complex as evidenced by the dramatic change in the edge shape, the disappearance of the pre-edge feature, the shift in position of the inflection, and the dramatic change observed in the EXAFS spectra.

Acknowledgements

We gratefully acknowledge support from the National Science Foundation (PCM 82-16127 and PCM 84-16676) and the Director, Office of Energy Research, Office of Basic Energy Sciences, Division of Biological Energy Conversion and Conservation of the Department of Energy under contract DE-AC03-76 SF00098. We thank the Stanford Synchrotron Radiation Laboratory, Stanford, CA for providing synchrotron radiation facilities.

References

1. Sauer, K. (1980) Acc. Chem. Res. 13, 249-256
2. Amesz, J. (1983) Biochem. Biophys. Acta 726, 1-12
3. Kok, B., Forbush, B. and McGloin, M. (1971) Photochem. Photobiol. 14, 307-321
4. Dismukes, G.C. and Siderer, Y (1981) Proc. Natl. Acad. Sci. USA, 78, 274-278
5. Hansson, Ö. and Andrèasson, L.-E. (1982) Biochim. Biophys. Acta 679, 261-268
6. Brudvig, G.W., Casey, J.L and Sauer, K (1983) Biochim. Biophys. Acta 723, 366-371
7. Casey, J.L. and Sauer, K. (1984) Biochim. Biophys. Acta 767, 21-28
8. Zimmermann, J.L. and Rutherford, A.W. (1984) Biochim. Biophys. Acta 767, 160-167
9. de Paula, J.C. and Brudvig, G.W. (1985) J. Am. Chem. Soc. 107, 2643-2648
10. Rutherford, A.W. (personal communication)
11. Guiles, R.D., Yachandra, V.K., McDermott, A., Britt, R.D., Dexheimer, S.L, Sauer, K. and Klein, M.P. (These proceedings)
12. McDermott, A., Yachandra, V.K., Guiles, R.D., Britt, R.D., Dexheimer, S.L, Sauer, K. and Klein, M.P. (These proceedings)
13. Yachandra, V.K., Guiles, R.D., McDermott A., Britt, R.D., Dexheimer, S.L., Sauer, K. and Klein, M.P. (1986) Biochim. Biophys. Acta 850, 324-332
14. Berthold, D.A., Babcock, G.T. and Yocum, C.F. (1981) FEBS Lett. 134, 231-234
15. Goodin, D.B., Yachandra, V.K., Britt, R.D., Sauer, K. and Klein, M.P. (1984) Biochim. Biophys. Acta 767, 209-216
16. Goodin, D.B., Yachandra, V.K., Guiles, R.D., Britt, R.D., McDermott, A., Sauer, K. and Klein, M.P. (1984) in EXAFS and Near Edge Structure III (eds., Hodgson, K.O., Hedman, B. and Penner-Hahn, J.E.), Springer-Verlag, New York, 130-135
17. Kirby, J.A., Goodin, D.B., Wydrzynski, T., Robertson, A.S. and Klein, M.P. (1981) J. Am. Chem. Soc. 103, 5537-5542
18. Cooper, S.R., Dismukes, G.C., Klein, M.P. and Calvin, M. (1978) J. Am. Chem. Soc. 100, 7248-7252
19. Sandusky, P.O. and Yocum, C.F. (1983) FEBS Lett. 162, 339-343
20. Yachandra, V.K., Guiles, R.D., Sauer, K. and Klein, M.P. (1986) Biochim. Biophys. Acta 850, 333-342
21. Ghanotakis, D.F. and Yocum, C.F. (1985) Photosynth. Res. 7, 97-114
22. Teo, B.-K. and Lee, P.A. (1979) J. Am. Chem. Soc. 101, 2815-2832

Structural Features of the Manganese Cluster
in Different States of the Oxygen Evolving Complex of Photosytem II:
An X-ray Absorption Spectroscopy Study

R. D. Guiles, Vittal K. Yachandra, Ann E. McDermott
R. David Britt, S. L. Dexheimer
Kenneth Sauer and Melvin P. Klein

Laboratory of Chemical Biodynamics, Lawrence Berkeley Laboratory
University of California, Berkeley, CA 94720 USA

Introduction

In this paper we present the results of an X-ray absorption study examining: the effect of hydroxylamine on the manganese-containing oxygen-evolving complex (Mn-OEC) in photosystem II, structural and oxidation state changes associated with advancement to the S_3 state and evidence for a tetranuclear manganese cluster within the Mn-OEC.

1. Hydroxylamine at low concentrations causes a two flash delay in the occurrence of the first maximum flash yield of oxygen [1,2]. Three mechanisms proposed to explain this effect differ in the extent to which reduction of the Mn-OEC occurs in the dark. We have used the decrease in amplitude of the light-induced multiline EPR signal as a criterion for determining the effect of hydroxylamine on S-state composition. In addition, state advancement was monitored by examining changes in the amplitude of the EPR spectrum in the g=1.9 region, which has been associated with the reduced primary acceptor FeQ_A^- [3,4]. Samples in which the mutiline amplitude decreased significantly were then examined using X-ray absorption spectroscopy.

2. There have been a number of reports of attempts to stabilize active preparations in the S_3 state [5,6]. We have poised samples in the S_3 state by a double turnover low temperature illumination procedure. Changes in the oxidation state and the ligation sphere of the manganese cluster within the Mn-OEC have been explored through an analysis of the Mn K-edge and extended X-ray absorption fine structure (EXAFS) spectra of PSII samples poised in the S_2 state and of samples partially advanced to the S_3 state.

3. X-ray absorption spectroscopy has been demonstrated to be a powerful tool for determining local site structure and variations in local site structure associated with mechanistic details of multinuclear metalloenzymes [7]. There are few examples of multinuclear metalloenzymes in which more than one metal-metal distance has been observed. Recently an EXAFS study of a copper cluster in metallothionein has revealed three different intermetal distances [8]. In this report we present evidence for the existence of two different metal-metal distances in an EXAFS study of the manganese cluster in the oxygen-evolving complex of photosystem II.

Materials and Methods

Oxygen-evolving PSII subchloroplast membranes from spinach were prepared by a Triton X-100 fractionation procedure [9]. EPR characterization and preparation of samples for X-ray absorption analysis is described in a companion paper by Yachandra et al [10]. Multiline EPR amplitudes were measured by adding peak to peak amplitudes of four of the hyperfine lines on the low field side of g=2. Because of the instability of hydroxylamine solutions, hydroxylamine treated samples were prepared by diluting volumes of freshly prepared 1mM stock solutions into PSII subchloroplast membranes suspended at 3.5 mg/mL chlorophyll. Samples containing a final concentration of 40 μM hydroxylamine were prepared. These were allowed to stand in the dark for 30 min in order to insure complete reaction with hydroxylamine. After 30 min, DCMU was added

Biggens, J. (ed.), Progress in Photosynthesis Research, Vol. I. ISBN 90 247 3450 9
© *1987 Martinus Nijhoff Publishers, Dordrecht. Printed in the Netherlands.*

to the suspension in order to limit photosystem II to one turnover. The concentration of DCMU was adjusted to 50 μM.

Photosystem II subchloroplast membrane suspensions used in attempts to generate the S_3 state by double turnover experiments contained 200 μM dichlorobenzoquinone. Photosystem II preparations were poised in the S_1 and S_2 states as described in the companion paper [10]. Samples poised in the S_2 state were advanced to the S_3 state by first warming the sample to 273 K for 30 sec to allow reoxidation of the primary acceptor [11], followed by illumination at 235 K. EPR spectra were recorded at 8 K after illumination at 195 K, following warming to 273 K and again after illumination at 235 K.

Results and Discussion

Hydroxylamine at 40 μM caused a 65% reduction of the amplitude of the multiline EPR signal generated by 190 K illumination. Figure 1 contains the EPR spectrum of a sample treated with a hydroxylamine and illuminated at 190 K (S_0^*) together with that of a sample illuminated at 190 K in the absence of hydroxylamine. The disproportionate increase of the g=1.9 signal relative to the multiline on illumination was evidence of stable charge separation and hence state advancement in the presence of hydroxylamine. Figure 2 contains the Mn K-edges of samples poised in the S_1, S_2 and S_0^* states. There is a dramatic shift to lower energy in the edge of S_0^* relative to samples in the S_1 state. The magnitude of this shift to lower energy is comparable to the light-induced shift to higher energy that we have previously reported for the S_1 to S_2 transition. The simplest interpretation of this result is that concurrent with advancement to the S_2 state, hydroxylamine causes a two electron reduction of the Mn-OEC, causing the two step delay in the state of the system. The Mn K-edge inflection of S_0^* occurs at 6550.2 eV. The Mn K-edge inflections of an S_1 state sample and a dark adapted sample containing hydroxylamine occur at 6551.4 and 6551.5 eV respectively. Thus it appears that hydroxylamine does not reduce the Mn-OEC in the dark.

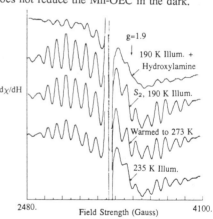

Figure 1. Multiline EPR spectra recorded at 8 K. From top to bottom the spectra are: A PSII subchloroplast preparation containing 40 μM hydroxylamine which was illuminated at 190 K; a sample illuminated at 190 K, S_2; the same sample after equilibration at 273 K for 30 sec; and again after illumination at 235 K for 10 min. Changes in the region of the second hyperfine feature on the high field side of g=2 (e.g at g=1.9) are indicative of changes in the oxidation state of the primary acceptor.

Figure 1 also contains the EPR spectrum of a PSII preparation illuminated at 190 K and subsequently warmed to 273 K for 30 sec. The decrease of the amplitude of the g=1.9 signal in this spectrum is an indication of the reoxidation of the primary acceptor. Based on fluoresence measurements in chloroplast suspensions, reoxidation of the primary acceptor should occur rapidly at 273 K [11]. The spectrum of the same sample following illumination at 235 K for 10 min is shown in fig 1. The amplitude of the multiline EPR signal decreased by 47% upon illumination at 235 K. The growth of the g=1.9 signal concurrent with the decrease in the amplitude of the multiline EPR signal is an indication of advancement to the S_3 state. Based on these criteria we have achieved a 47% S_3 state composition. The Mn K-edges of a S_1 sample and a sample advanced partially to the S_3 state are shown in figure 2. The manganese K-edge inflection of a sample advanced to the S_3 state occurs at 6552.4 which is at the same as that observed in the S_2 state. The lack of change in the Mn K-edge samples advanced to the S_3 state implies that the oxidative equivalent stored on the donor side of PSII is stabilized on some intermediate donor or that partial oxidation of bound water has occured with no net change in the oxidation state of the Mn cluster.

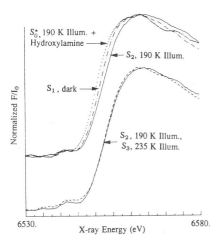

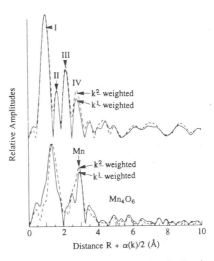

Figure 2. The X-ray absorption K-edge spectra of spinach PSII preparations. The spectra plotted above are: a sample containing 40 μM hydroxylamine which was illuminated at 190 K, S_0^* (.....); a dark adapted sample, S_1(-..-); and a sample illuminated at 190 K, S_2(—). The spectra plotted below are: the spectrum of a samples illuminated at 190 K, S_2(—), and the spectrum of a sample illuminated at 190 K, warmed to 273 K for 30 s and then illuminated at 235 K for 10 min, S_3(- - -). The K-edge inflection energies for S_0^* and S_1 are 6550.2 and 6551.4 eV respectively. The S_2 and S_3 inflection energies both occur at 6552.4 eV.

Figure 3. The upper plot contains the Fourier transforms of the k^1-weighted (—) and k^2-weighted (- - -) EXAFS of a spinach PSII subchloroplast preparation. The figure contains PSII EXAFS spectra of samples poised in both the S_1 and S_2 states. Peaks labeled I, II and III correspond respectively to a μ oxo bridge at 1.8 Å, a terminal N or O shell at 2.0 Å and the neighboring Mn atom at 2.7 - 2.8 Å. The lower plot contains Fourier transforms of the k^1 and k^2 weighted EXAFS of an adamantane-like tetranuclear Mn cluster, Mn_4O_6. Note the amplitude of the peak labeled Mn increases in amplitude with k-weighting as do the peaks labeled III and IV of the photosystem II sample EXAFS data.

We have also analysed the EXAFS spectra of an S_0^* sample and a sample advanced to the S_3 state. The Fourier transforms of the EXAFS (not shown) of these spectra contain the three dominant peaks (labelled I, II and III in figure 3) corresponding to a μ-oxo bridge at 1.8 Å, a terminal O or N ligation shell at 2.0 - 2.2 Å and a manganese atom at 2.7 - 2.8 Å. Thus it appears that the di-μ-oxo bridged Mn core structure is present in both the hydroxylamine-treated sample and the sample advanced partially to the S_3 state. This short Mn-Mn distance is indicative of a di-μ-oxo bridged structure and is inconsistent with an adamantane-like oxo-bridged cluster which has been proposed for the S_3 state [12]. Simulations of the manganese wave of the S_0^* state sample indicate a significantly longer Mn-Mn distance than the distance we have observed in the S_1 and S_2 states. It has been suggested in several studies [2,13] that ammonia and hydroxylamine bind directly to the manganese cluster as competitive inhibitors of water oxidation. The changes in the oxidation state and Mn-Mn distance of the manganese complex we have observed upon hydroxylamine treatment provide the first direct evidence of competitive binding of a water analog to manganese.

By performing an analysis on a sum of EXAFS spectra obtained during the last two years, we have obtained high quality Mn EXAFS of PSII subchloroplast membrane preparations. This analysis reveals a fourth scattering shell (labelled IV in figure 3). Simulations of a Fourier isolate of this fourth shell yield two possible interpretations. The isolated shell could be fit by a pair of carbons at 3.0 and 3.2 Å. This distance is characteristic of second shell contributions from an imidazole ring [14]. Alternatively, the shell could be fit by an additional transition metal at 3.3 Å. Based on a detailed comparison of the properties of peak IV with those of transform peaks of the EXAFS of four inorganic model compounds we believe we can assign this peak to an additional transition metal at 3.3 Å. One method which has been used to determine whether an EXAFS feature is due to an combination of low Z elements or a heavy atom is to examine the k-weighting behavior of its Fourier peak [14]. As is shown in Fig. 3, the amplitude of the third and fourth shells increase in amplitude with increasing k-weighting. This behavior is also observed for the manganese shell of the tetranuclear adamantane like cluster, Tetramanganese(IV) hexa-μ-oxotetrakis(1,4,7-triazacyclononane) perchlorate [15]. Note also that the position of the manganese peak in the tetranuclear model is similar to the position of the peak IV. The Fourier transform of the EXAFS of models containing imidazole-like rings, Dimanganese(III) μ-oxobis(acetato)bis(hydrotris(1-pyrazolylborate)) [16] and Dimanganese(III,IV)di-μ-oxotetrakis(2,2'-bipyridine) perchlorate [17], did not contain peaks which increase with k-weighting in the region near peak IV. One plausible interpretation of these results is that the fourth peak corresponds to an additional manganese at 3.3 Å, indicating a cluster of manganese atoms consisting of two binuclear centers in close proximity.

Acknowlegements We gratefully acknowlege support from the National Science Foundation (PCM 82- 16127 and PCM 84-16676) and the Director, Office of Energy Research, Office of Basic Energy Sciences, Division of Biological Energy Conversion and Conservation of the Department of Energy under contract DE-AC03-76 SF00098. We thank the Stanford Synchrotron Radiation Laboratory, Stanford, CA for providing synchrotron radiation facilities.

References
1. Bouges-Bocquet, B.(1975) Biochim. Biophys. Acta 502, 211-221
2. Radmer, R., Ollinger, O., (1982) FEBS Lett. 144, 162-166
3. Rutherford, A. W., Zimmermann, J. L. (1984) in Advances in Photosynthesis Research Vol. I (ed. Sybesma, C.) Martinus Nijhoff Publishers, The Hague, 445-448
4. de Paula, Julio, C., Innes, J. B. and Brudvig, G. W. (1985) Biochemistry 24, 8114 - 8120
5. Brudvig, G., Casey, J. and Sauer, K. (1983) Biochem. Biophys. Acta 723, 274-278
6. Goodin, D. B., Yachandra, V. K., Britt, R. D., Sauer, K. and Klein, M. P. (1984) Biochim. Biophys. Acta 767, 209-216
7. Powers, L. (1982) Biochim. Biophys. Acta 683, 1-38
8. Freedman, J. H., Powers, L. and Peisach, J. (1986) Biochem. 25, 2342-2349
9. Yachandra, V. K., Guiles, R. D., McDermott, A., Britt, R. D., Dexheimer, S.L., Sauer, K. and Klein, M. P. (1986) Biochim. Biophys. Acta 850, 324-332
10. Yachandra, V. K., Guiles, R. D., McDermott, A., Cole, J. L., Britt, R. D., Dexheimer, S. L., Sauer, K. and Klein, M. P. (in these proceedings)
11. Joliot, A. (1974) Biochim. Biophys. Acta 357, 439-448
12. Brudvig, G. W. and Crabtree, R. H. (1986) Proc. Natl. Acad. Sci. USA 83, 4586-4588
13. Warren, B. F., dePaula, J. C. and Brudvig, G. W. (1986) J. Am. Chem. Soc. 108, 4018-4022
14. Woolery, G. L., Powers, L., Winkler, M., Solomon, E. I. and Spiro, T. G. (1984) J. Am. Chem. Soc. 106, 8692
15. Wieghardt, K., Bossek, U. and Gebert, W. (1983) Angew. Chem 95, 328-329
16. Armstrong, W. H. (personal communication)
17. Cooper, S. R., Dismukes, G. C., Klein, M. P. and Calvin, M. (1978) J. Am. Chem. Soc. 100, 7248-7252

CHARACTERIZATION OF THE MN-CONTAINING O₂ EVOLVING COMPLEX FROM THE CYANOBACTERIUM *SYNECHOCOCCUS* USING EPR AND X-RAY ABSORPTION SPECTROSCOPY

Ann McDermott, Vittal K. Yachandra, R. D. Guiles
R. David Britt, S. L. Dexheimer
Kenneth Sauer and Melvin P. Klein

Laboratory of Chemical Biodynamics, Lawrence Berkeley Laboratory,
University of California, Berkeley, CA 94720 USA

Introduction

Recent EXAFS and Mn K-edge results from our laboratory elucidate the oxidation state of Mn and the structure of the Mn complex in PS II from spinach prepared in the S_1 and S_2 states [1,2]. We concluded that the Mn complex in PS II is a μ-oxo bridged cluster with a Mn oxidation state of approximately +3 in the S_1 state. We show that the Mn complex does not change in structure on advancing from S_1 to S_2, but the oxidation state of Mn does increase, confirming its direct role in the storage of oxidizing equivalents. We also concluded that there is no chloride or sulfur directly ligated to the Mn in the complex.

O₂ evolving PS II preparations have been developed recently which are highly resolved in peptide content. We have developed a preparation from a thermophilic cyanobacterium *Synechococcus* which has is lacking many of the peptides that spinach Triton X-100 grana stack preparations contain. The purification of PS II from thylakoids allows complete conservation of O₂ evolving activity. We have studied this preparation in order to compare the primary electron transfer participants in *Synechococcus* to those in spinach. We found that the structure of the Mn complex is identical in spinach and *Synechococcus*. We have found differences between the iron-quinone acceptors in spinach and *Synechococcus* and, in contrast to spinach, there is no light induced $g=4.1$ donor signal generated at low temperature in *Synechococcus*.

Materials and Methods

The preparation of PS II from *Synechococcus* is described briefly below. The X-ray Absorption spectroscopy was done at SSRL as described elsewhere in these proceedings [2].

Synechococcus was grown at 55°C using a modification of the medium described by Dyer and Gafford [3] with increased copper and iron concentrations. Whole cells were collected with a Sharples centrifuge. The cells were washed in 50 mM HEPES, pH 7.5, 10 mM NaCl and 1 mM PMSF, and centrifuged at 16,000 x g for 12 min. The cells were then incubated with 100 mg lysozyme at 37°C for 1 h in the dark in 100 m*l* of a medium containing 0.4 M mannitol, 5 mM EDTA 50 mM HEPES, pH 7.5 and 1mM PMSF. Thylakoids were isolated from whole cells in 50 mM MES, pH 6.5, 10 mM NaCl and 1 mM PMSF using a French press at 5000 psi. The DNA was then degraded with 1μg DNAase and 5 mM MgCl₂ at 0°C, and the reaction was stopped after 1 h with 10 mM EDTA. Unbroken cells were removed by centrifuging at 12,000 x g for 15 min, and thylakoids were collected by centrifuging at 300,000 x g for 1 h. The thylakoids were washed to remove phycobili proteins three times in a medium containing 0.5 M sucrose, 50 mM MES, pH 6.5, 5 mM NaCl and 0.5 mM PMSF, and collected by centrifugation at 300,000 x g for 1 h. The thylakoids were resuspended again in the same medium at 1 mg m*l*⁻¹, and PS II was solubilized from the membranes using 0.5% β-octyl glocoside. After extraction, the PS II was concentrated by a two-fold dilution in water followed by centrifugation at 300,000 x g for 3 h.

The PS II particles so obtained were further purified using Sepharose CL-6B (Pharmacia) in some cases. The column was developed in a 20 to 150 mM NaCl gradient in the presence of 0.5% lauryl maltoside. This procedure reduces the amount of phycobili protein contamination.

The PS II particles after extraction have approximately 3 to 4 Mn and 70-90 Chl per photosynthetic unit as defined by the integrated intensity of the EPR signal due to D+. These preparations exhibit rates of oxygen evolution of approximately 1000-1500 μmol O_2(mg Chl)⁻¹h⁻¹.

Biggens, J. (ed.), Progress in Photosynthesis Research, Vol. I. ISBN 90 247 3450 9
© *1987 Martinus Nijhoff Publishers, Dordrecht. Printed in the Netherlands.*

Results and Discussion
EPR Studies of Cryogenic Charge Separation in PS II from Synechococcus

We have generated a multiline EPR signal in these PS II preparations similar to the signal observed in spinach PS II and which has been assigned as a Mn species associated with the S_2 state [4]. Using the multiline signal as a reporter of the S_2 state, we have monitored the advance from S_1 to S_2 in *Synechococcus* during continuous illumination as a function of temperature of illumination. Figure 1 shows that the optimum temperature of illumination for multiline generation is 210-220K, and that the multiline signal generated is similar to that observed in spinach PS II.

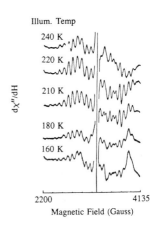

Illum. Temp

240 K

220 K

210 K

180 K

160 K

dχ''/dH

2200   4135

Magnetic Field (Gauss)

Figure 1 Light-minus-dark EPR Spectra at 8 K of PS II from *Synechococcus* after continuous illumination for 1.5 min at the indicated temperature. Spectrometer conditions were 50 mW, 32 G and 100kHz field modulation, microwave frequency 9.21 GHz, gain 5000. Spectra were collected using a Varian E-109 spectrometer with an Air-Products cryostat. The maximum amplitude of the multiline signal is generated at 210-220K.

In spinach PS II the optimum temperature for generation of the multiline signal is 190K [5]. Below this temperature an EPR signal at $g=4.1$ is generated which is believed to be due to a precursor or an alternate form of the S_2 species which gives rise to the multiline signal [6,7]. Above 190K electron transfer from Q_A to Q_B becomes facile, and the oxygen-evolving complex advances beyond S_2. Using illumination at temperatures below 180K we do not observe a signal at $g=4.1$ corresponding that in spinach, and when the sample is illuminated at 140K and warmed to 210K no multiline signal is generated. Therefore we believe that in these preparations the advance from S_1 to S_2 is blocked below 180K. In addition, the block between Q_A and Q_B seems to persist to slightly higher temperatures in this organism: up to 230K compared to approximately 200K in spinach. DCMU addition followed by continuous illumination at 215K or 290K results in a multiline signal of the same amplitude as that formed without DCMU by continuous illumination at 215K. Therefore we conclude there is an effective block for multiple transfers of electrons in *Synechococcus* below 215K.

Continuous illumination at 110K to 180K in these preparations leads to a new EPR signal at $g=1.62$ which is approximately 320 G wide. Figure 2 shows this signal, which exhibits Curie behavior between 4 and 12 K, and is not saturated below 50 mW. We suggest that this signal resembles those of quinone species interacting with a transition metal [8]. Figure 3 shows the temperature dependence of generation of this signal during continuous illumination. The signal is optimally generated at 140K, and upon warming to 215K the signal completely decays while the $g=1.89$ iron-quinone signal is formed. The generation of the $g=1.6$ signal is inhibited by a variety of chemical treatments, including the donor-side inhibitors NH_2OH and Tris, and DCMU, which replaces Q_B in its binding site. The E-1 preparation of Yamagishi and Katoh [9] does not show the $g=1.6$ signal. E-1 is known to be competent in electron transfer from Z to Q_A, but is lacking both the oxygen evolving complex and Q_B.

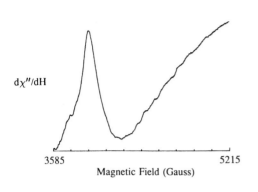

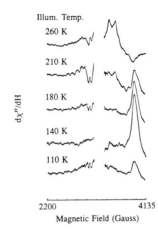

Figure 2 (on the left) Light-minus-dark EPR spectrum at 5 K of the $g=1.6$ signal generated by continuous illumination for 1 min at 140K. Spectrometer conditions as above. The signal is approx 320 G wide, exhibits Curie behaviour between 4K and 12K, and is not saturated below 50 mW.

Figure 3 (on the right) Light-minus-dark EPR spectra at 5 K following continuous illumination at the indicated temperature for 1 min. Spectrometer conditions as in figure 1. The optimum temperature for generation of the $g=1.6$ signal is 140K. No $g=1.6$ signal is generated above 210 K, and a $g=1.89$ signal is generated at that temperature.

We have considered several assignments for this signal. Based on the temperature dependence of generation and decay, and on the absence of this signal in E-1, three possible assignments are: 1) an alternate donor to P680 in parallel with the oxygen-evolving complex, 2) Q_B or 3) Q_A. Because the DCMU treatment leaves the multiline signal intact but abolishes the $g=1.6$ signal, we feel that it is more likely to be an acceptor signal than a donor signal. Its lineshape and g-value are suggestive of a quinone interacting with a metal. Recently, Takahashi and Katoh [10] have determined the number of acetone-extractable quinones present in PS II preparations from *Synechococcus*. They find only enough quinone to account for Q_A, Q_B and Z. Therefore, we are reluctant to assign the $g=1.6$ signal as an entirely new quinone species. Of the two quinone acceptors, we feel it is more likely to be Q_A because transfer of electrons to Q_B is believed to be blocked below 200K in spinach PS II, and electron transfer is also limited to one equivalent transfer in *Synechococcus* below 215K. If the signal were due to Q_B, then we would have to propose that the S_2 to S_3 advance is blocked below 200K. Whether it is due to Q_A or Q_B, it reflects a form of the quinone which is different from that observed in spinach and different from that generated by illumination at 200 K, and it relaxes to the $g=1.89$ form on warming to 215K.

Characterization of the Mn Complex by X-ray Absorption Spectroscopy

Figure 4 shows the Mn K-edge spectra of *Synechococcus* PS II prepared in the S_1 and S_2 states. The edge shape is quite similar to that of spinach PS II, including the 1s to 3d pre-edge feature, and the edge inflection point shifts from 6551.1 eV to 6552.1 eV on advancing from S_1 to S_2. This shift in edge position indicates that the oxidation state of Mn increases and that Mn stores oxidizing equivalents in the transformation from S_1 to S_2. Figure 5 shows the Fourier transformed Mn EXAFS spectra of PS II from *Synechococcus* and spinach, both prepared in the S_1 state. The neighboring atoms indicated by the Mn EXAFS spectrum, which include a transition metal neighbor at 2.7Å, oxygen or nitrogen ligands at 1.8 and 2.0 Å and the lack of any chloride or sulfur ligands, are similar in the two spectra. We conclude that the structure of the Mn complex is preserved over the estimated two billion year evolutionary gap separating the two organisms. The similarity also suggests that both preparations are free of adventitious Mn. This study strengthens our earlier findings concerning the structure and function of Mn in PS II.

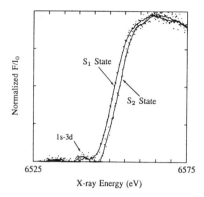 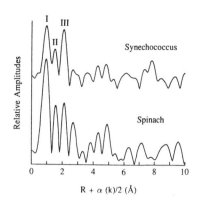

Figure 4 (on the left) The X-ray absorption Mn K-edge spectra of *Synechococcus* PS II in the S_1 and S_2 states prepared by dark adaptation and by continuous illumination at 215K after dark adaptation, respectively. A smoothed curve is drawn through the data points. The small pre-edge feature at 6543 eV is a 1s to 3d transition. The K-edge inflection energy for the S_1 state is 6551.1 eV and for the S_2 state is 6552.1 eV. This shift in edge energy indicates an oxidation of Mn.

Figure 5 (on the right) Fourier transforms of the k^1-weighted Mn EXAFS data of spinach PS II in the S_1 state and *Synechococcus* PS II in the S_1 state. The peaks labeled I and II are due to oxygen or nitrogen ligands at 1.8 and 2.0Å, and the peak labeled III is due to a transition metal at 2.7 Å, probably a Mn neighbor. (The larger first peak in the spectrum from spinach PS II is due to a background artifact.) The shorter (1.8 Å) oxygen bond and the metal bond distance are characteristic of a μ-oxo-bridged Mn cluster. The bond lengths were obtained by curve fitting analysis according to the Teo-Lee method [11].

Acknowledgements

We thank Akhihiko Yamagishi for extensive advice concerning the preparation of PS II and we thank S. Katoh for providing us with the strain of *Synechococcus*. We gratefully acknowledge support from the National Science Foundation (PCM 82-16127 and PCM 84-16676) and the Director, Office of Energy Research, Office of Basic Energy Sciences, Division of Biological Energy Conversion and Conservation of the Department of Energy under contract DE-AC03-76SF00098. We thank the Stanford Synchrotron Radiation Laboratory, Stanford, CA for providing synchrotron radiation facilities.

References
1 Yachandra, V. K., Guiles, R. D., McDermott, A., Britt, R. D., Dexheimer, S. L., Sauer, K. and Klein, M. P. (1986) *Biochim. Biophys. Acta.* **850**, 324-332
2 Yachandra, V. K., Guiles, R., McDermott, A., Cole, J., Britt, D., Dexheimer, S., Britt, R., Sauer, K. and Klein, M. (*These proceedings*)
3 Dyer, D. and Gafford, R. (1961) *Science* **134**, 616-617
4 Dismukes, G. and Siderer, Y. (1981) Proc. Natl. Acad. Sci. U.S.A. **78**, 274-278
5 Brudvig, G. W., Casey, J. L. and Sauer, K. (1983) *Biochim. Biophys. Acta.* **723**, 366-371
6 Casey, J. L. and Sauer, K. (1984) *Biochim. Biophys. Acta.* **767**, 21-28
7 Brudvig, G. L. (1986) *J. Am. Chem. Soc.* **108**, 4002-4009
8 Butler, W., Calvo, R., Fredkin, D., Isaacson, R., Okamura, M. and Feher, G. (1984) *Biophys. J.* **45**, 947-973
9 Yamagishi, A. and Katoh, S. (1983) *Arch. Biochem. Biophys.* **225**, 836-846
10 Takahashi, Y. and Satoh, S. (1986) *Biochim. Biophys. Acta.* **848**, 183-192
11 Teo, B. K., and Lee, P. A. (1979) *J. Am. Chem. Soc.* **101**, 2815-2835

The Flash Number Dependence of EPR Signal II Decay As a Probe for Charge Accumulation in Photosystem II

James Cole and Kenneth Sauer
Department of Chemistry and Chemical Biodynamics Laboratory,
Lawrence Berkeley Laboratory,
University of California, Berkeley, CA 94720 USA

1. Introduction

The O_2-evolving complex of photosystem II catalyzes the four-electron photooxidation of water to oxygen. Kok and coworkers proposed a model in which photosystem II advances during photochemical turnover through five successive oxidation states $S_i (i = 0$ to $4)$, the state S_4 spontaneously decaying to produce oxygen and S_0 [1]. The S_2 and S_3 states are not stable, and decay to S_1. Manganese is thought to be associated with the accumulation of oxidizing equivalents [2,3], and has been shown to be oxidized at the $S_1 \rightarrow S_2$ transition by a shift in the the the Mn X-ray absorption edge to higher energy[4]. Slower electron transfer from the higher S-states is thought to reflect accumulation of net positive charge at the O_2-evolving complex . A semiplastoquinone cation species, Z^+, which gives rise to EPR signal II, is directly oxidized by P_{680}^+ [5]. The rereduction kinetics of Z^+ varies with flash number, suggesting that it functions as an intermediate electron carrier between the O_2-evolving complex and P_{680} [6]. Thus, the transient response of Z^+ following a series of saturating flashes is a direct probe for accumulation of oxidizing equivalents in O_2-evolving and inhibited preparations.

Release of 16 kDa and 24 kDa extrinisic polypeptides by NaCl washing partially inhibits O_2-evolution in the absence of high concentrations of Cl^- or Ca^{2+} [7]. The inhibition of O_2-evolution caused by $CaCl_2$ washing, which releases the 16, 24 and 33 kDa peptides, is not completely reversible except upon rebinding of the 33 kDa peptide. Incubation at pH 7-8 reversibly inhibits O_2-evolution without peptide release. Utilizing the flash number dependence of Z^+ decay kinetics and X-ray edge spectroscopy we observe that accumulation of oxidizing equivalents at the O_2-evolving complex is blocked or greatly altered by these inhibitory treatments.

2. Materials and Methods

O_2-evolving PSII preparations were obtained by Triton X-100 extraction of spinach chloroplasts as previously described [8]. Salt extraction and alkaline pH incubation were performed as previously described [9,10]. EPR Samples were suspended in 0.4 M Sucrose, 15 mM NaCl and 50 mM Mes at pH 6.0 (control and salt washed samples) or 50 mM Hepes at pH 7.75 (alkaline pH samples). X-ray absorption samples were suspended in 15 mM NaCl, 5 mM $MgCl_2$, 50 mM Mes and 30% glycerol. Signal II kinetics was measured with a Varian E109 spectrometer at g=2.010 (half time of instrument rise, 0.2 ms). Saturating flashes (15 mJ/pulse) of 0.5 μs duration were provided by a Phase R DL-1400 dye laser. Immediately before EPR measurements, 1 mM ferricyanide, 1 mM ferrocyanide and 0.5 mM DCBQ were added to the sample. The sample was dark adapted in a reservoir on ice for at least 5 min prior to data collection. Fresh sample was pumped through a light-tight flow system into the EPR flat cell prior to each train of flashes. There were no systematic changes in the kinetic traces during the course of signal averaging. Transients were fit to the sum of a rising component (determined by the instrument time response) and one or two decaying components. The amplitudes of decay components with half-times shorter

Biggens, J. (ed.), Progress in Photosynthesis Research, Vol. I. ISBN 90 247 3450 9

than about 1.5 ms were not uniquely determined using this procedure; presumably, the decay components become convoluted with the instrument rise. In this case, the fast decay amplitude was estimated by taking the difference between the initial signal maximum and the amplitude of the slow component. X-ray absorption spectra were collected in the fluorescence mode at Stanford Synchrotron Radiation Laboratory, Stanford, CA.

3. Results and Discussion

The sequential patterns of signal II ms decay transients in O_2-evolving or inhibited PSII preparations following four saturating laser flashes are shown in Fig. 1. In the control (active) preparation, the overall signal amplitude and decay kinetics change as a function of flash number, with the maximal signal amplitude occuring on the third flash. The two component fits to the decay traces are summarized in Table I. Following each flash, a small 50-120 ms component is observed. As this component does not exhibit period four oscillations, it does not appear to be connected to the O_2-evolving complex . Low signal amplitude is observed following the first flash because reduction of Z^+ in S_0 and S_1 is faster than the instrument response time. The half-times for the fast decay components

Fig. 1. The transient decay kinetics of Signal II in dark-adapted O_2-evolving and inhibited PSII preparations. Four saturation flashes were given with 0.5 s between flashes. From top to bottom: control, NaCl washed, pH 7.75 incubated and $CaCl_2$ washed. Sample preparation and experimental conditions are described in Materials and Methods. Instrument conditions: microwave power, 5 mW; modulation amplitude, 5 G; modulation frequency, 100 kHz. The recorder was DC-coupled, so that baseline offsets represent signal that does not decay between flashes. Each trace represents 1400-4400 events.

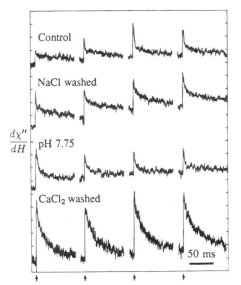

$\dfrac{d\chi''}{dH}$

Control

NaCl washed

pH 7.75

$CaCl_2$ washed

50 ms

Sample	Flash	A_f	t_f (ms)	A_s	t_s (ms)
Control	1	<0.10	—	0.15	120
	2	0.20^a	0.6	0.16	70
	3	0.40^a	1.3	0.14	50
	4	0.27^a	1.4	0.14	50
NaCl	1	0.27^a	0.9	0.30	100
	2	0.30	2.4	0.26	130
	3	0.27	4.3	0.22	80
	4	0.37	2.9	0.22	90
pH 7.75	1	0.51	3.9	0.21	230
	2	0.28	4.3	0.11	140
	3	0.34	3.0	0.08	110
	4	0.27^a	1.3	0.11	70
$CaCl_2$	1	0.54	4.2	0.49	30
	2	0.46	6.6	0.47	30
	3	0.75	10.5	0.22	80
	4	0.43	3.1	0.64	40

^{a} Amplitude was estimated as described in Materials and Methods

Table I. The transient decay kinetics of Signal II in dark-adapted O_2-evolving (control) and inhibited (NaCl washed, pH 7.75 incubated and $CaCl_2$ washed) PSII samples. Amplitudes (A) and half-times (t) were obtained by fitting the kinetic traces in Fig. 1 to the sum of a rise component and fast (f) and slow (s) decay components as described in Materials and Methods. Amplitudes are normalized such that the average total signal amplitude of the $CaCl_2$ washed sample equals 1.0. Estimated uncertainty in amplitudes and half-times, ±20%.

following the second and third flash (0.6 ms and 1.3 ms, respectively) agree well with those reported previously in chloroplasts (0.4 ms and 1 ms) [6], indicating that electron transfer from the O_2-evolving complex to Z^+ is unaltered by the detergent extraction. The signal amplitude due to the fourth flash arises from centers left in the S_2 and S_3 states, and thus is an indicator of dephasing. Assuming a dark distribution of 75% S_1, 25% S_0 and no double hits, the present data are consistent with about a 15-20% miss parameter for each flash.

Fig. 2 demonstrates that a second maximum occurs at the seventh flash, confirming the period four behavior of the Z^+ decay kinetics. The period four oscillation in the observed amplitudes and half-times of Z^+ reduction reflects successively slower electron transfer from the higher S-states due to accumulation of net positive charge at the O_2-evolving complex .

Fig. 2. Transient decay kinetics of Signal II in dark-adapted control O_2-evolving (pH 6.0) and inhibited (pH 7.75) PSII preparations. Eight saturating flashes were given at a frequency of 2 Hz. Experimental conditions were the same as in Fig. 1. Each trace represents the average of 1600 (pH 6.0) or 1300 (pH 7.75) events.

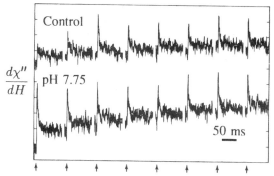

In contrast to the intact preparation, NaCl-washed samples exhibit a large signal following the first flash, and subsequent flashes induce signals of similar amplitude to the first (Fig. 1). The decay following the first flash is biphasic, with 1 ms and 100 ms components contributing about equal amplitude. On all subsequent flashes, 2-4 ms and 100 ms components are observed which are again of approximately equal amplitude. In addition, the signal does not decay to the baseline in the 0.5 s between flashes, and a large offset is accumulated after four flashes. We have previously observed similar ms kinetics in NaCl-washed samples assayed under repetitive flash conditions [8]. Incubation of samples at pH 7.75 also eliminates the periodicity of the signal II decay kinetics (Fig. 1). The largest transient is observed following the first flash. The decay fits to a sum of components with half-times of 3.8 ms and several hundred ms, with the latter contributing about 30% of the amplitude. Subsequent flashes induce similar decay transients, except that the overall amplitude is decreased to about 50-60% of that observed following the first flash, and the half-time of the slower phase decreases with each flash. In addition, a large baseline shift is induced primarily following the first flash. To determine if the large transient signal and baseline shift occuring on the first flash exhibits period-four behavior, we examined the flash transients at pH 7.75 for a train of eight flashes (Fig. 2). There is no corresponding change after the fifth flash, suggesting that this phenomenon occurs only after dark adaptation. The loss of the period four oscillations following NaCl extraction or incubation at pH 7.75 indicates that turnover of the O_2-evolving complex is blocked. Furthermore, there is no evidence for partial advancement of the lower S-states, since the first flash induces about the same transient Signal II amplitude as subsequent flashes. While for NaCl washed samples this first-flash decay is faster than on subsequent flashes, it is still much slower than expected for donation from the lower S-states. The various first flash effects may result from slow changes in the state of PSII during dark adaptation. Because we observe much slower Z^+ reduction following the first flash in the inhibited samples than in intact samples, our study suggests that the loss of the light-induced S_2 multiline EPR signal upon NaCl treatment [9,11] or alkaline pH incubation [10] occurs because advancement to the S_2 state is blocked, not because of an alteration of the Mn environment which eliminates the EPR signal but does not block advancement.

To confirm that NaCl washing blocks the formation of S_2 we have utilized X-ray absorption spectroscopy at the Mn K-edge (data not shown). As previously observed, the edge shifts about 1 eV to higher energy on illumination at 195K, which causes advance from S_1 to S_2. In contrast, the NaCl sample edge is shifted only about 0.3 eV on illumination, indicating that Mn is not oxidized and that the S_2 state is not attained. The magnitude of this shift is proportional to the amplitude of the light-induced multiline EPR signal observed in the NaCl washed samples.

Extraction of the 16, 24 and 33 kDa peptides by $CaCl_2$-washing abolishes the variation of Z^+ decay transients with flash number and also dramatically slows the ms decay (Fig. 1). Also, the average signal amplitude is increased to about twice that observed in the other preparations studied. Under repetitive flash conditions, we previously observed that $CaCl_2$ washing abolishes a $140\mu s$ Z^+ reduction phase present in the control and NaCl washed samples, resulting in an apparent two-fold increase in the amplitude of the ms phases [8]. Thus, the physiological donor to Z is not present upon removal of the 33 kDa peptide, and the reduction of Z^+ may proceed via back reaction with the reduced acceptor or donation by an exogenous species.

4. Acknowledgements

We thank Dr. Michael Boska for setting up the timing apparatus and initial data collection and John Colosi for preparation of PSII samples. This work was supported by the Director, Office of Energy Research, Office of Basic Energy Sciences, Division of Energy Conversion and Conservation of the Department of Energy, under contract DE-AC03-76SF00098, and by a grant from the National Science Foundation (PCM 82-16127). We thank SSRL for providing synchrotron radiation facilities.

References

1 Forbush, B., Kok, B. and McGloin, M.P. (1971) *Photochem. Photobiol.* **14**, 307–321
2 Sauer, K. (1980) *Acc. Chem. Res.* **13**, 249–256
3 Amesz, J. (1983) *Biochim. Biophys. Acta.* **726**, 1–12
4 Goodin, D.B., Yachandra, V.K., Britt, R.D., Sauer, K. and Klein, M.P. (1984) *Biochim. Biophys. Acta.* **767**, 209–216
5 Boska, M., Sauer, K., Buttner, W. and Babcock, G.T. (1983) *Biochim. Biophys. Acta.* **722**, 327–330
6 Babcock, G.T., Blankenship, R.E. and Sauer, K. (1976) *FEBS Lett.* **61**, 286–289
7 Ghanotakis, D.F. and Yocum, C.F. (1985) *Photosyn. Res.* **7**, 97–114
8 Boska, M., Blough, N.V. and Sauer, K. (1985) *Biochim. Biophys. Acta.* **808**, 132–139
9 Blough, N.V. and Sauer, K. (1984) *Biochim. Biophys. Acta.* **767**, 377–381
10 Cole, J., Boska, M., Blough, N.V. and Sauer, K. (1986) *Biochim. Biophys. Acta.* **848**, 41–47
11 Franzén, L.-G., Hansson, Ö. and Andréasson, L.-E. (1985) *Biochim. Biophys. Acta.* **808**, 171–179

Electron Spin Echo Studies of PSII Membranes

R. David Britt, Kenneth Sauer, and Melvin P. Klein
Laboratory of Chemical Biodynamics, Lawrence Berkeley Laboratory
University of California, Berkeley, Ca 94720 USA

Introduction

We have recently constructed a high-power pulsed EPR spectrometer for research in photosynthesis. We have applied several Electron Spin Echo (ESE) techniques to the study of PSII membrane preparations. In this paper we present some preliminary ESE results on Signal II and the multiline EPR signal.

Spectrometer Design

The design of the new spectrometer is similar to that of a conventional homodyne EPR spectrometer. The addition of a high-power travelling wave tube amplifier and fast switching circuitry enables pulsed EPR operation. We currently use 25 ns wide pulses with amplitudes as high as 1 kW. A phase-locked Gunn oscillator serves as the microwave source. A GaAs FET amplifier is used to amplify the weak spin echo signals. The pulses are controlled by a homebuilt timing module interfaced to a DEC PDP11/34 minicomputer. The instrument can also be operated as a sensitive conventional EPR spectrometer by bypassing the high-power pulse amplifier. The spectrometer has a frequency range of 8.5 to 9.5 GHz.

Most of the work reported here has been performed using a new loop-gap resonator probe which we have designed. The high filling factor and low Q of the loop-gap resonator make it ideal for pulsed EPR applications [1]. In our system a two-gap resonator is mounted near the end of a shorted section of X-band waveguide. The resonator Q is adjusted via a Gordon coupler. The probe assembly resides in a liquid He immersion dewar. Samples are contained in conventional EPR tubes and can be readily changed during cryogenic operation without removing the probe structure. With this system a 25 ns 90° pulse requires approximately 20W power. Post-pulse dead times of 130 ns are typical.

Materials and Methods

Oxygen-evolving PSII membranes were prepared from spinach using a Triton X-100 fractionation procedure [2]. The PSII preparations were mixed with equal volumes of glycerol and placed in 3.7 mm OD quartz EPR tubes. The S_1 samples were prepared by dark adaptation at 4°C for one hour followed by freezing in liquid N_2. The S_2 samples were prepared by subsequent 190K illumination of S_1 samples.

Signal II Studies

The electron spin echo spectrum of Signal II is shown in Figure 1. The figure displays the magnetic field dependance of the ESE amplitude over a 100 Gauss range about g = 2. The derivative plot matches the familiar Signal II cw lineshape. For these preparations and experimental conditions, Signal II originates from the stable radical D^+.

Table 1 provides a summary of Signal II spin-lattice relaxation time measurements. The spin-lattice relaxation time (T_1) is calculated from recovery curves obtained by measuring the ESE amplitude as a function of the repetition time t_r between sets of 2 pulse spin echo sequences. The average T_1 values obtained by fitting the recovery curves to a single exponential are similar to those previously reported by de Groot et al [3]. The two-exponential component fits provide a much better fit to the data, but we have not yet assigned roles to the two components.

Biggens, J. (ed.), Progress in Photosynthesis Research, Vol. I. ISBN 90 247 3450 9
© *1987 Martinus Nijhoff Publishers, Dordrecht. Printed in the Netherlands.*

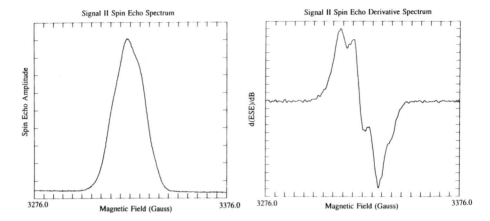

Figure 1. Signal II spin echo spectrum and its first derivative. Two pulse ESE S_2 state data taken at 9.3208 GHz; 200 ns τ; 30 ms repetition time; 4.2K.

Table 1. Summary of Signal II spin-lattice relaxation data for S_1 and S_2 states. Two pulse ESE data taken at 9.3208 GHz; 3326.0 Gauss; 200 ns τ; 4.2K.

	Relative Amplitude	T_1 (ms)
One Exponential		
S_1	1.0	10.61
S_2	1.0	8.56
Two Exponential		
S_1	0.63	25.97
	0.37	1.77
S_2	0.59	23.25
	0.41	1.75

The Nuclear Envelope Modulation (NEM) experiment is used to measure weak hyperfine couplings [4]. Figure 2 shows a 2 pulse NEM trace for Signal II. The Fourier transform of a 3 pulse NEM trace is displayed in Figure 3. The NEM experiments show a strong modulation with a period of about 675 ns. The small 70 ns modulation is due to weakly coupled protons with splittings close to the proton Zeeman frequency. Fourier analysis of the 3 pulse data reveals a 1.47 MHz frequency for the long period component. Table 2 provides a summary of experiments performed at three different sets of microwave frequency and magnetic field . The results show that the low frequency component remains constant while the Zeeman component shifts over a 0.51 MHz range. We therefore conclude that the low frequency component cannot correspond to a low frequency proton splitting where the hyperfine and Zeeman terms nearly cancel. The most likely origin of this constant component is a weakly coupled ^{14}N nucleus.

Table 2. Summary of hyperfine frequencies for Signal II as determined by Fourier analysis of 3 pulse NEM data.

Microwave Frequency (GHz)	Magnetic Field (Gauss)	Low Frequency Component (MHz)	Proton Zeeman Frequency (MHz)
9.22	3275.	1.47	13.90
9.4664	3375.	1.49	14.27
9.557	3404.	1.47	14.41

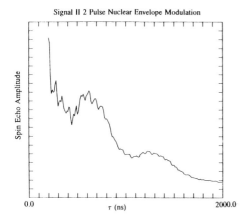

Figure 2. Signal II 2 pulse nuclear envelope modulation. S_1 state data taken at 9.557 GHz; 3403.0 Gauss; 4.2K.

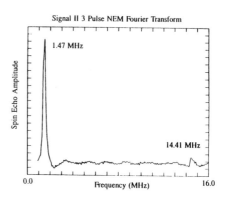

Figure 3. Fourier transform of Signal II 3 pulse nuclear envelope modulation. S_1 state data taken at 9.557 GHz; 3403.0 Gauss; 250 ns τ; 4.2K.

Multiline Signal Studies

In S_2 samples we see an electron spin echo signal corresponding to the multiline EPR spectrum associated with photosynthetic oxygen evolution [5]. Figure 4 shows the S_2 state multiline spin echo spectrum and its derivative. The derivative spectrum is very similar to the multiline EPR spectrum obtained via conventional cw spectroscopy. The narrow Signal II ESE feature has been replaced in these figures with a cubic spline interpolation. This endpoints of this region are marked by vertical bars.

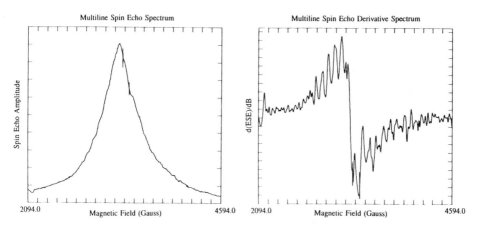

Figure 4. S_2 multiline spin echo spectrum and its first derivative. Two pulse spin echo data taken at 9.380 GHz; 150 ns τ; 0.5 ms repetition time; 4.2K.

We have performed preliminary T_1 measurements on the multiline signal. At 4.2K the spin-lattice relaxation time is approximately 1.1 ms. This T_1 value is independent of field position across the multiline spectrum.

Figure 5 shows a 2 pulse NEM scan for the multiline signal. The 2 pulse NEM experiment shows a deep modulation of the multiline spin echo envelope at the proton Zeeman frequency. No other modulation frequencies are observed.

We have compared the areas of the electron spin echo spectra for Signal II and the multiline signal. The integrals are corrected for phase memory decay by extrapolating the peak heights of the 2 pulse NEM scans back to $\tau = 0$. The resulting ratio of multiline to Signal II at 4.2K is 1.13 ± 0.2.

We also observe a large, broad signal centered near g=2 in S_1 samples. No resolved Mn hyperfine features are associated with this signal. This signal is displayed in Figure 6. The area of the S_1 signal is comparable to that of the S_2 multiline signal. Relaxation properties of the two signals are very similar.

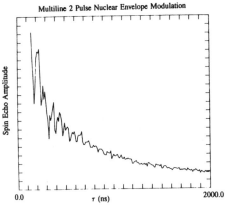

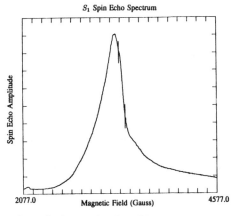

Figure 5. Multiline 2 pulse nuclear envelope modulation. S_2 state data taken at 9.380 GHz; 3200.0 Gauss; 4.2K.

Figure 6. S_1 state 2 pulse ESE spectrum. Data taken at 9.3427 GHz; 150 ns τ; 4.2K.

Acknowledgements

We thank Dr. Vittal Yachandra and James Cole for help in preparing PSII membranes. This work was supported by the Director, Office of Energy Research, Office of Basic Energy Sciences, Division of Energy Conversion and Conservation of the Department of Energy, under contract DE-AC03-76SF00098, and by a grant from the United States Department of Agriculture (85-CRCR-1-1847).

References

1 Froncisz, W. and Hyde, J.S. (1982) J. Mag. Res. **47**, 515-521.

2 Yachandra, V.K., Guiles, R.D., McDermott, A., Britt., R.D., Dexheimer, S.L., Sauer, K. and Klein, M.P. (1986) Biochim. Biophys. Acta **850**, 324-332.

3 de Groot, A., Plijter, J.J., Evelo, R., Babcock, G.T. and Hoff, A.J. (1986) Biochim. Biophys. Acta **848**, 8-15.

4 Mims, W.B. and Peisach, J. (1981) in Biological Magnetic Resonance Vol.3 (ed. Berliner,L.J. and Reuben,J.), 213-263.

5 Dismukes, G.C. and Siderer, Y. (1981) Proc. Natl. Acad. Sci. USA, **78**, 274-278.

EPR STUDIES AT 9 AND 34 GHz OF THE MULTILINE AND $g=4.1$ S_2 SIGNALS

ROLAND AASA, ÖRJAN HANSSON AND TORE VÄNNGÅRD
Department of Biochemistry and Biophysics, Chalmers Institute of
Technology and University of Göteborg, S-412 96 Göteborg (Sweden)

INTRODUCTION

An intermediate, S_2, in the oxidation of water to oxygen in higher plants and algae gives rise to two EPR signals: a multiline signal at g about 2 and a single line at $g=4.1$ (reviewed in Ref.1). Understanding the origin and nature of these signals would help greatly in the formulation of a mechanism for oxygen evolution. In this work measurements at three different X-band frequencies and at 34 GHz show that the g-tensor of the multiline species is isotropic with $g=1.982$. The intensity ratio of the $g=4.1$ signal to the multiline signal was found to be almost constant from 5 to 23 K. It is proposed that the two signals originate from separate PS II electron donors in different reaction centers and that the $g=4.1$ signal arises from monomeric Mn(IV).

MATERIALS AND METHODS

Oxygen-evolving (600 µmol O_2/mg Chl per h) PS II-enriched membranes (17 mg Chl/ml) were prepared from spinach (2) in a final buffer containing 20 mM Mes-NaOH (pH 6.3), 400 mM sucrose, 15 mM NaCl and 5 mM $MgCl_2$. EPR samples were dark-adapted for one hour. Where indicated, 4% ethanol was added before freezing the samples at 200 K. Light-induced EPR signals were generated by continuous illumination at 200 K (3).

EPR measurements at 9.46 GHz were made with a Bruker ER 200D-SRC spectrometer (3). A Varian E-231 cavity was used for studies at somewhat lower microwave frequencies. Measurements at 34 GHz were made with a Varian V-4503 spectrometer.

RESULTS

Measurements at X-band

Illumination of PS II enriched membranes at 200 K generates the multiline signal at $g=2$ (Fig.1B) which on addition of 4% ethanol is enhanced and gets a reduced intrinsic linewidth (Fig.1A). Addition of 4% ethanol inhibits the generation of the $g=4.1$ signal. These results are consistent with earlier findings (4).

The multiline signal was studied at three different X-band frequencies. The shape of the individual peaks was the same at all frequencies, but the position of the peaks changed. Ten peaks, indicated in Fig.1A, were studied in detail, and for 7 of these their peak positions are shown in Table 1. Since second-order hyperfine contributions to the frequency dependence are negligible in this frequency range (*cf*. Ref.5), the g-values of the individual peaks can be obtained from the slope in a plot of their position *versus* frequency. Within the accuracy of the determination all peaks had the same g-value with $g_{av}=1.97$.

Biggens, J. (ed.), Progress in Photosynthesis Research, Vol. I. ISBN 90 247 3450 9
© 1987 Martinus Nijhoff Publishers, Dordrecht. Printed in the Netherlands.

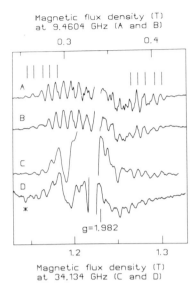

Magnetic flux density (T)
at 9.4604 GHz (A and B)

0.3   0.4

g=1.982

1.2   1.3

Magnetic flux density (T)
at 34.134 GHz (C and D)

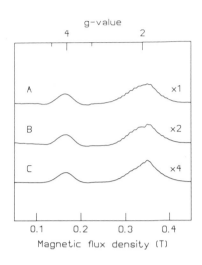

g-value

4   2

Magnetic flux density (T)
0.1   0.2   0.3   0.4

FIGURE 1. The light-induced multiline EPR S_2 signal from a PS II preparation measured at 9 and 34 GHz. Spectra from samples with (A,D) or without (B,C) 4% ethanol are shown. Spectrometer conditions for A,B (C,D in parenthesis): microwave frequency, 9.4604 (34.134) GHz; power, 20 (10) mW; modulation amplitude, 1.25 (2.1) mT; temperature, 11 (15) K. Spectra obtained before the illumination have been sub-tracted in A,B. The signal at 1.19 T in C,D is probably due to a cavity contamination. The vertical bars above A mark the peaks for which apparent g-values were determined. All spectra have been aligned on g=1.982. The feature marked with "*" was not reproducible.

FIGURE 2. Temperature dependence of the light-induced g=1.98 multi-line and g=4.1 signals from a PS II preparation. Light-minus-dark spectra at three different temperatures were corrected for a sloping baseline and then integrated once. Spectrometer conditions: microwave frequency, 9.460 GHz; power, 0.05 (A), 0.2 (B), 20 (C) mW; modulation amplitude, 2.5 mT; temperature, 5 (A), 12 (B), 23 (C) K. Relative gain after correction for differences in microwave power is indicated in the figure.

Measurements at 34 GHz

It was possible to detect the multiline signal both in samples without (Fig.1C) and with 4% ethanol (Fig. 1D). No peaks could be found outside the region shown in the figure. A good correspondence between X-band and 34 GHz spectra from samples without ethanol (taking second-order hyperfine effects into account with the formulas in Ref.5) was obtained when they were aligned on g=1.982±0.002 (Fig.1B,C). Other alignments, obtained by shifting the X-band and 34 GHz spectra an integral number of peak separations (0.01 g-value units for each shift), resulted in less good fits.

Addition of 4% ethanol gave a small shift in some of the peaks in the 34 GHz spectrum (Fig.1D).

TABLE 1. Magnetic field positions at different X-band frequencies of 7 peaks in the multiline signal (Fig.1A)

Microwave frequency (GHz)	Magnetic field (mT)						
9.0754	243.8	262.1	269.9	278.4	371.0	379.2	397.5
9.2468	250.1	268.2	276.2	284.6	377.3	385.4	404.1
9.4592	258.0	276.1	283.9	292.4	384.9	392.8	411.6
g-value	1.94	1.96	1.96	1.97	1.98	2.01	1.95

Temperature dependence of the $g=1.98$ multiline and $g=4.1$ signals

Fig.2 shows the integrated $g=1.98$ multiline and $g=4.1$ signals at three different temperatures. The area under the $g=4.1$ signal was determined by integration from 0.10 to 0.22 T, while the $g=1.98$ multiline signal was integrated from 0.24 to 0.44 T and corrected for weak underlying signals (Ö. Hansson, unpublished). The area ratio of the $g=4.1$ signal to the $g=1.98$ multiline signal was 0.37 at 23 K, 0.38 at 18 K, 0.35 at 12 K and 0.32 at 5 K.

DISCUSSION

The g-value of the multiline signal

From the measurements at three different X-band frequencies and at 34 GHz, it can be concluded that the g-tensor of the multiline signal is isotropic with $g=1.98$. Spectral similarities with low-molecular-weight complexes (6,7) point to bi- or tetra-nuclear mixed-valence manganese structures as origins for the signal. The g-value obtained here is close to the value $g=2.003$ estimated for a di-μ-oxo(phenanthroline)Mn(III,IV) dimer (6) and to the values $g_x=g_y=2.006$ and $g_z=2.00$ estimated for the Mn(II,III) dimers derived from O_2 oxidation of Mn(II) Schiff-base complexes (7).

In a recent paper de Paula *et al.* (8) obtained a good simulation of the signal only by introducing an appreciable g-anisotropy ($g_x=1.810$, $g_y=1.96$, $g_z=2.274$). Such anisotropy should have been observed in the present experiments, particularly at the high- and low-field wings of the spectrum. An alternative explanation for the shape of the spectrum might include hyperfine anisotropy and a non-Gaussian line-shape.

The ethanol-induced shift of some of the lines observed at 34 GHz is possibly due to a small g-anisotropy. However, this is probably too small to be observed at X-band and need not be included in simulations of X-band spectra. The small anisotropy observed at X-band in Ref. 9 could be due to hyperfine anisotropy.

The origin of the $g=4.1$ and $g=1.98$ multiline signals

Since the $g=1.98$ multiline signal can be observed at 34 GHz (microwave quantum, 1.1 cm^{-1}) the signal must arise from a doublet separated from other spin multiplets by *more than* a few cm^{-1}. Otherwise the signal would have been much distorted, probably beyond detection. On the other hand, if the $g=1.98$ multiline and $g=4.1$ signals arose from two spin multiplets of the same paramagnetic species, the weak temperature dependence in their area ratio shows that the energy splitting of the multiplets would be *less than* 1 cm^{-1}. Thus, the two signals must originate from different paramagnetic species. Furthermore, since both signals are produced when only one electron is transferred from the donor to the acceptor side (see also Ref.4) they cannot arise from oxidation of two different species in the same center. Most likely, they originate from different PS II centers, and the observed changes in their relative intensity on treatment with ethanol (Ref.4 and Ö. Hansson, unpublished) or with changes in temperature (10) would then be due to electron transfer between the two sites.

Recently, de Paula *et al.* (8) suggested, mainly from deviations from Curie-law behavior below 8 K, that the $g=1.98$ multiline and $g=4.1$ signals arise from an excited doublet and a ground-state quartet, respectively, from the same species. We offer no explanation for this discrepancy, but would only like to point out that our conclusions do not depend on the accurate determination of absolute concentrations and low temperatures.

As an alternative to the tetranuclear manganese cluster proposed in Ref.8 we suggest that two of the manganese ions form a spin-coupled pair that gives rise to the $g=1.98$ multiline signal and is directly involved in the oxidation of water. The other ions would be monomeric, with one of them in an axial Mn(IV) $S=3/2$ state producing the $g=4.1$ signal. This arrangement would be analogous to that in enzymes catalyzing the reverse reaction, the reduction of oxygen to water. Laccase contains two monomeric copper ions and one copper pair, and in cytochrome oxidase one heme and one copper ion are monomeric while another heme and another copper ion interact strongly. In both enzymes the interacting metal ions react directly with oxygen.

ACKNOWLEDGMENTS

We would like to thank B. Källebring for computer programming and L.-E. Andréasson and T. Wydrzynski for stimulating discussions. This work was supported by the Swedish Natural Science Research Council.

REFERENCES

1   Dismukes, G.C. (1986) Photochem. Photobiol. 43, 99-115.
2   Franzén, L.-G., Hansson, Ö., and Andréasson, L.-E. (1985) Biochim. Biophys. Acta 808, 171-179.
3   Hansson, Ö., Andréasson, L.-E., and Vänngård, T. (1986) FEBS Lett. 195, 151-154.
4   Zimmermann, J.-L., and Rutherford, A.W. Biochemistry. In press.
5   Hansson, Ö., and Andréasson, L.-E. (1982) Biochim. Biophys. Acta 679, 261-268.
6   Cooper, S.R., Dismukes, G.C., Klein, M.P., and Calvin, M. (1978) J. Am. Chem. Soc. 100, 7248-7252.
7   Mabad, B., Tuchagues, J.-P., Hwang, Y.T., and Hendrickson, D.N. (1985) J. Am. Chem. Soc. 107, 2801-2802.
8   de Paula, J.C., Beck, W.F., and Brudvig, G.W. (1986) J. Am. Chem. Soc. 108, 4002-4009.
9   Rutherford, A.W. (1985) Biochim. Biophys. Acta 807, 189-201.
10 Casey, J.L., and Sauer, K. (1984) Biochim. Biophys. Acta 767, 21-28.

STRUCTURAL AND FUNCTIONAL ASPECTS OF ELECTRON TRANSFER IN PHOTOSYSTEM 2
OF OXYGEN-EVOLVING ORGANISMS

V.V.KLIMOV, I.B.GANAGO, S.I.ALLAKHVERDIEV, M.A.SHAFIEV AND G.M.ANANYEV
INSTITUTE OF SOIL SCIENCE AND PHOTOSYNTHESIS, USSR ACADEMY OF SCIENCES,
PUSHCHINO, MOSCOW REGION, USSR, I42292

I. Polypeptide composition of Photosystem-2 reaction centers. Preparations
of reaction centers (RC) of Photosystem 2 (PS-2) isolated by different
methods, usually contain five main polypeptides with molecular weight
(m.w.) of 5I-44, 44-40, 34-32, 34-30, and Io-7 kDa [I]. There is no unani-
mous conclusion on the functional role of the polypeptides. The pigments
involved in the primary photoreactions of charge separation as well as the
primary electron acceptor Q_A are probably located at the 50-47 kDa polypeptide [2,3]. According to other authors (see ref.I) both "heavy" polypeptides (50-47 kDa and 44-40 kDa) are required for photochemical activity of
RC. On the other hand, from homology between the aminoacid sequence of the
polypeptides with m.w. 34-32 kDa in PS-2 and that of the L and M subunits
of bacterial RC it was suggested [4] that the photoactive components of RC
in PS-2 can be located on these polypeptides.
　　Earlier [5,6] we have reported on the isolation of PS-2 reaction center from spinach and wheat with m.w. nearly I50 kDa using detergents chola-
te and Deriphat-I60. The preparations contain 20-30 Chl per one RC (one
photochemically active P680, one capable of reversible photoreduction molecule of pheophytin, Pheo, and one molecule of Q_A). The ratio Chl/Pheo is
equal to I2$\pm$2, i.e. there are 2 molecules of Pheo per RC but only one of
them takes part in the primary photochemical reactions.
　　Analysis of the polypeptide composition of the preparation (carried
out by means of SDS-electrophoresis in I5% polyacrylamide gel) reveals 2

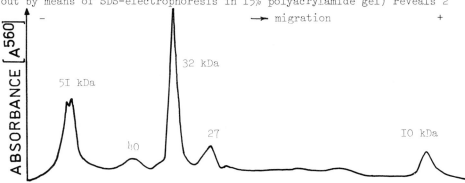

FIGURE I. A densitogramme of I5% gel after SDS-electrophoresis of the pre-
　　　　　paration of PS-2 reaction center. Before electroforesis the pre-
　　　　　paration was incubated for I5 min at 60°C in the medium contai-
　　　　　ning 3% SDS, 5% β-mercaptoethanol, 4 M urea. The gel plate was
　　　　　stained by Coomassie; density was measured at 560 nm.

Biggens, J. (ed.), Progress in Photosynthesis Research, Vol. I. ISBN 90 247 3450 9
© *1987 Martinus Nijhoff Publishers, Dordrecht. Printed in the Netherlands.*

main polypeptides with m.w.near 5I and 32 kDa (Fig.I). On the densitogramme one can also see small bands at 40,27 and IO kDa. The component IO kDa (it gives positive staining on heme iron and is evidently cytochrome b_{559} [5]) can be completely removed using a "gentle" electrophoresis in 4% PAAG without any loss of the photochemical activity [6]. The ratio between the content of the polypeptides with m.w. 5I,43 and 32 kDa is changed from one experiment to another. Judging from the intensity of staining of the proteins by Coomassie the following ratio between these polypeptides can be observed: I:I:I or I:0:2, or I:0:I. The absence of polypeptide 43 kDa in the photochemically active preparations means that it is not strictly necessary for the primary reactions of charge photoseparation in RC of PS-2. The photochemically inactive preparations obtained in a number of experiments during isolation or as a result of the long-term storing of photochemically active RC, did not contain the polypeptide(s) with m.w. 32kDa. This fact may indicate that the 32 kDa polypeptide(s) is an essential part of RC and it (they) carry the photoactive components of the RC.

Thus, the obtained results show that, the polypeptides 5I and 32 kDa are enough for functioning of the primary photoreactions in the PS-2 reaction center; the loss of the polypeptide with m.w. 32 kDa correlates with the loss of the photochemical activity of RC.

2. Photoreduction of pheophytin in Photosystem-2 of the whole cells of Chlamydomonas reinhardii. According to the present view the primary light reaction in PS-2 consists of electron transfer from excited chlorophyll P680 to an intermediary electron acceptor Pheo and further to the "stable" electron acceptor Q - a complex of plastoquinone with Fe [7]. The primary state with separated charges [$P680^{+}\cdot Pheo^{-}_{\overline{}}$] recombines during 2-4 ns when Q is reduced and the recombination is accompanied with the recombination luminescence (the so-called "variable fluorescence" or ΔF) [7]. After the prior reduction of Q in PS-2 preparation or in chloroplasts the continious illumination results in photoaccumulation of the long-lived state [$P680\ Pheo^{-}_{\overline{}}$]$Q^{-}$ due to reduction of $P680^{+}\cdot$ from the secondary electron donors,and the photoreaction is observed in the presence of dithionite as well as after creation of anaerobic conditions [7-9]. Here we report on photoreduction of Pheo in PS-2 of the whole cells of a Chl.reinhardii mutant lacking PS-I (ACC-I) [9] under anaerobic conditions.

The creation of anaerobic conditions leads to the increase of Chl fluorescence (F) of the ACC-I mutant up to F_{max} reflecting reduction of Q by the weak measuring light.   An actinic light   induces a significant decrease of F which is due to photoreduction of some substance since O_2 added during the illumination results in an increase of F. The photoinduced decrease of F is accompanied with the absorbance changes with the same kinetic [9] the spectrum of which is quite similar to the spectrum of photoreduction of Pheo in PS-2 particles in the presence of dithionite measured earlier [7].

The photoinduced decrease of F related to the photoreduction of Pheo is enhanced upon addition of I μM CCCP;it is significantly diminished upon addition of DCMU (IO μM) and strongly reactivated by the subsequent addition of ascorbate or dithionite. The effects can be explained by the influence of these agents on the S-states of PS-2 [8,9]. Addition of $NADP^{+}$(3 mM) results in a remarkable (by a factor of 20-30) increase in the rate of the dark relaxation of the photoinduced fluorescence decrease (revealing acceleration of the dark oxidation of $Pheo^{-}_{\overline{}}$) and lowers the dark level of F. Similar effects were seen upon addition of 3 mM benzyl viologen or methyl viologen. The latter two agents can also oxidize $Pheo^{-}_{\overline{}}$

in PS-2 particles [8] while NADP[+] can induce this effect (according to our results) only after its addition to PS-2 particles jontly with the system: ferredoxin/ferredoxin-NADP[+] reductase.

The results indicate an enzymatic way of the electron transfer from Pheo[-] to NADP[+] in PS-2. Taking into account the high rate of the electron transfer from Pheo[-] to Q ($\leq$200 ps [7]) one cannot expect high efficiency of NADP[+] photoreduction by PS-2 when Q is oxidized. It can probably take place (as an alternative electron path) only at a high light intensity when Q is accumulated in the reduced state.

3. Reactivation of oxygen evolution after a complete removal of Mn from PS-2 particles. Untreated pea PS-2 particles DT-20 [IO] evolve oxygen under illumination in the presence of 0.4 mM ferricyanide with the rate of I50 μmoles/mg Chl·h. After a complete extraction of Mn [IO] (accompanied with the removal of proteins of the water-oxidizing system with m.w. I7,23 and 33 kDa) the oxygen evolution is eliminated; only the consumption of oxygen is observed upon illumination (Fig.2). CaCl$_2$ (8 mM), if added alone (Fig.2) or jointly with the proteins I7,23 and 33 kDa isolated using the treatment of DT-20 particles with I M CaCl$_2$ [II] (not shown), does not restore oxygen evolution nor does it eliminate oxygen consumption. Addition of MnCl$_2$ at the concentration of I2 μM (Fig.2) or at any higher one (not shown) does not lead to the reactivation of oxygen evolution either, though an effective reactivation of the electron transfer through PS-2 has been

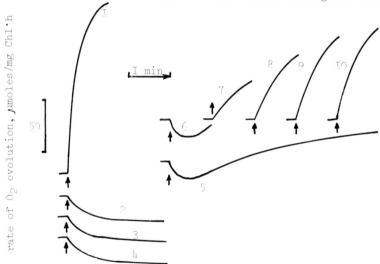

FIGURE 2. Reactivation of photoevolution of O$_2$ after a complete extraction of Mn from PS-2 particles (DT-20) by means of I M tris-HCl-buffer and 0.5 M MgCl$_2$ [IO]. I - untread DT-20 particles, 2-IO - after removal of Mn without additions (2) and after addition: I2 μM MnCl$_2$ (I2 Mn/RC of PS-2) -3, 8 mM CaCl$_2$ - 4, I2 μM MnCl$_2$ plus 8mM CaCl$_2$ (5-IO) at continious illumination for 5 min (5) or at intermittent illumination (I min of light and 2 min of dark, 5 cycles) - (6-IO). Conditions: 50 mM MES-buffer (pH 6.5)/35 mM NaCl/0.3 M sucrose / 0.4 mM ferricyanide / Chl concn. 0.I mg/ml. Intensity of light (λ > 600 nm) is 50 W/m^2.

reported earlier [IO]. Only a joint addition of $MnCl_2$ (I2 μM) and $CaCl_2$ (8 mM) results in the gradual restoration of oxygen evolution under continious illumination (Fig.2). So, a functionally active oxygen-evolving Mn-containing complex can be formed after a complete removal of Mn from the PS-2 particles in the absence of water-soluble proteins I7, 23 and 33 kDa. The most effective reactivation is observed when intermittent illumination (I min of light and 2 min of dark) is used (Fig.2). In this case the rate of oxygen evolution exceeds 50% of control. With hydrophobic electron acceptors of PS-2 (p-benzoquinone, dimethyl-p-benzoquinone, phenylen-benzoquinone) the reactivation of oxygen-evolving system is not observed. However, in the presence of the pair phenylen-p-benzoquinone (0.2 mM) and ferricyanide (0.2 mM) the reactivation takes place.

The process of reactivation of the oxygen-evolving system requires electron phototransfer through PS-2, since preliminary illumination of the PS-2 particles lacking Mn in the presence of the all above mentioned components, except ferricyanide, and subsequent addition of ferricyanide in the dark, does not restore oxygen evolution. Our results confirm a scheme put forward earlier [I2], which suggests that the process of incorporation of Mn into the oxygen-evolving complex of PS-2 may be possible as a result of the oxidation of Mn^{2+} to Mn^{3+} and Mn^{4+} by RC of PS-2.

During incubation of the reactivated particles in the dark the rate of oxygen evolution is decreased by 20-30% per hour, but it is restored upon a subsequent short-term illumination. Evidently, a slow loss of Mn from the water-oxidizing complex of PS-2 takes place in the dark.

Thus, after a complete extraction of Mn from the PS-2 particles the function of oxygen evolution can be restored stage by stage using $MnCl_2$ and $CaCl_2$. This can be used as an experimental approach for further investigation of the mechanism of photosynthetic oxydation of water.

REFERENCES
I Satoh, K. (I985) Photochem. Photobiol. 42, 845-853
2 Nakatani, H.Y., Ke, B., Dolan, E., Arntzen, C.J. (I984) Biochim. Biophys. Acta 765, 347-352
3 Yamagishi, A. and Katoh, S. (I985) Biochim. Biophys. Acta 807, 74-80
4 Cramer, W.A., Widger, W.R., Hermann, R.G., Trebst, A. (I985) Trends in Biochem. Sci. IO, I25-I29
5 Ganago, I.B., Klimov, V.V., Krasnovsky, A.A. (I985) Biofizika (USSR) 30, 8II-8I6
6 Ganago, I.B., Klimov, V.V., Maevskaya, Z.V., Krasnovsky, A.A. (I986) Doklady of USSR Academy of Sciences 286, 749-753
7 Klimov, V.V.(I984) in Advances in Photosynthesis Research (Sybesma, C., ed.).Vol.I, pp.I3I-I38, Martinus Nijhoff/Dr.W.Zunk Publishers,The Hague
8 Klimov, V.V., Shuvalov, V.A., Heber, U. (I985) Biochim. Biophys. Acta 809, 345-348
9 Klimov, V.V., Allakhverdiev, S.I., Ladygin, V.G. (I987) Photosynthesis Research, IO, N 2
IO Klimov, V.V., Allakhverdiev, S.I., Shuvalov, V.A., Krasnovsky, A.A. (I982) FEBS Lett. I48, 307-3I2
II Ono,I. and Inoue, J. (I984) FEBS Lett. I68, 28I-286
I2 Radmer, R., Cheniae, G.M. (I97I) Biochim. Biophys. Acta 253, I82-I86

THE STUDY OF EFFECTS ON STRONGLY-BOUND MANGANESE OF OXYGEN EVOLVING COMPLEX
IN WHEAT CHLOROPLASTS BY EPR

SUN QI AND LUO CHANG-MEI
LABORATORY OF LIFE SCIENCE, ZHEJIANG UNIVERSITY, HANGZHOU, PRC.
ZHANG LI-LI, FANG ZHAO-XI AND MEI ZHEN-AN
DEPARTMENT OF BIOLOGY, BEIJING UNIVERSITY, BEIJING, PRC

1. INTRODUCTION
    In 1939-1940, Emerson, R and Lewis, C. M. (1) first showed that manga-
nese deficiency reduced the quantum efficiency for photosynthesis,which
suggested that manganese was somehow involved in photosynthetic oxygen
evolution. Although during the last 15 years the role of manganese
in photosynthesis has been extensively studied, there are still many uncertainties about the mechanism of its participation (2).
    There are several pools of bound Mn in chloroplasts (3), but only
one of them is functional in oxygen evolving process. This pool is a
strongly bound pool (2/3 of the rest) which is released by NH OH, alkaline
Tris and heat treatment (4).
    In the present preliminary work we describe the effects of light, tem-
perature and some ionic treatments on the relative intensity of Mn^{2+}-EPR
signal in wheat chloroplasts with a purpose of exploring the possible role
of Mn pool in oxygen evolution in photosynthesis.

2. PROCEDURE
2.1. Materials and methods
    2.1.1. Chloroplast preparation: Chloroplast were isolated from leaves
           of wheat (Triticum aesticum L.) in a medium containing 400
           mM sucrose, 15 mM NaCl and 25 mM HEPES (pH 7.4). The free man-
           ganese was removed by washing with 0.5 mM EDTA. Chlorophyll
           concentration was determined by the method of Arnon, and it
           was controlled in 1 mg/ml.
    2.1.2. EPR measurements: EPR spectra were obtained using a Bruker
           (ER 200 E/D-SRC-series) spectrometer. All spectra except those
           described in Fig. 2 were recorded at room temperature. For
           the experiments of heat treatments a variable temperature acces-
           sory was employed. In order to observe the change of manga-
           nese signal under light exposure the sample was preilluminated
           for 30 sec with an intensity of 600μ einstein/m sec.
    2.1.3. Tris treatment: The mixture of 0.3 ml of chloroplast suspension
           and 0.3 ml of 1.6 M Tris (pH 8.0) was incubated for 25 min
           in darkened ice bath.
    2.1.4. HCl treatment: 0.2 ml of 1 N HCl was added to the chloroplast
           suspension. Then the final volume was adjusted to 0.6 ml by
           HEPES buffer. Time of treatment should be longer than two
           hours.
    2.1.5. Ca and Mg treatments: A certain quantity of $CaCl_2$ or $MgCl_2$
           was added to chloroplast suspension. Then the sample was incu-
           bated for 25 min in darkened ice bath.
    2.1.6. HCO and HCOO⁻ treatments: A certain quantity of $NaHCO_3$ or HCOONa
           was added to Tris treated sample.

Biggens, J. (ed.), Progress in Photosynthesis Research, Vol. I. ISBN 90 247 3450 9
© 1987 Martinus Nijhoff Publishers, Dordrecht. Printed in the Netherlands.

3. RESULTS AND DISCUSSION

3.1. Light effects: Fig.1 shows a typical light saturation curve of photo-synthesis. The intensity of saturated light was about 550 μeinstein/m sec. Under light Mn^{2+}-EPR signal was decreased by about 60%.

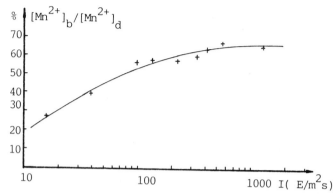

FIGURE 1. The effect of light intensity on the relative intensity of Mn Mn^{2+}-EPR signal. I=light intensity. Abscissa is in logarithmic scale, microwave power p=50 mW, 100 kHz modulated field ampli-tude 10 Gpp, time constant 20 mS, sweeping rate 10G/S, room tempe-rature, Tris treated sample. $[Mn^{2+}]/[Mn^{2+}]_b = [Mn^{2+}]$ in illu-minated/$[Mn^{2+}]_d$ in darkness.

TABLE 1. Rebound and released of Mn

Cycle	1st dark	1st light	2nd dark	2nd light
Hpp(peak to peak height, in arbitrary unit)	15.5	8.5	17.2	12.5
$[Mn]/[Mn]$		45.2		26.5

As shown in Table 1, there was a part of released Mn which would be rebound by light and released again by darkness. This cycle could be run several times. It seemed that rebound and released cycle will run as long as the photosynthetic activity of chloroplast lasts.

3.2. Temperature effects: Ratio of $[Mn^{2+}]_T/[Mn^{2+}]_A$ designates the Mn^{2+}-EPR signal relative intensity, where $[Mn^{2+}]_A$ is the total concentr-tration of released bound-Mn by acid washing, and $[Mn^{2+}]_T$ is the stable value of released bound-Mn at certain temperature(T). We observed that the amount of released bound-Mn at certain tem-perature was different as the time of heat treatment varied. In general, it increased with the increase of time of heat treatment, and finally kept at a stable value (Fig. 2). Fig. 3 shows about 20% of total bound-Mn was released by heat treatment at 50°C to 90°C, while there was only 7% at 40°C. Therefore. it can be supposed that the Mn^{2+} released at 40°C and 50°C might represent two different types of membrane-bound Mn^{2+}. They play different roles in photo-synthetic oxygen evolution (4).

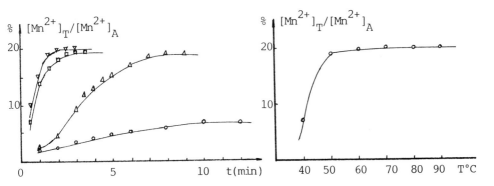

FIGURE 2. The change of the relative intensity of Mn^{2+}-EPR signal was
(left)   affected as the time of heat treatment was varied at different
           temperatures (O 40°C, Δ 50°C, □ 60°C, ▽ 70°C).
           100 kHz modulated field amplitude 8 Gpp, microwave   power
           p = 50 mW, time constant 0.5S, sweeping rate 6G/S.
FIGURE 3. The effect of temperature on the relative intensity of Mn^{2+}-
(right)   EPR signal.

3.3. Ionic effects: Effects of Ca^{2+}, Mg^{2+}, HCO_3^-, and $HCOO^-$ on light re-
     bound Mn have been investigated. Fig. 4 shows that Ca^{2+} affected
     the amount of rebound Mn in an inverse proportion to it's concentra-
     tions. In case of Mg treatment, probably, only lower concentra-
     tions had minor influences on the relative intensity of Mn^{2+}-EPR
     signals. These results indicate that Ca plays a special role in
     the rebinding of Mn and Mg is not as effect.
     During the last several years the role of calcium in the electron
     donor side have been investigated (5,6). more recently, it has been
     shown the possible relation between Ca and 18-24 kD protein or O_2-
     evolving enzyme(7). Possibly, this is the reason why only lower con-
     centration of Ca affects on light rebound Mn more efficiently.

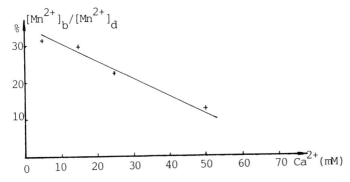

FIGURE 4. The effect of Ca concentrations on light rebound Mn.
          EDTA washed sample, other experimental conditions as shown in
          Fig. 1.

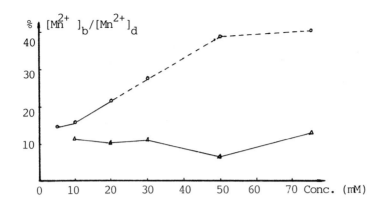

FIGURE 5. The effect of different concentrations of HCO_3^- and $HCOO^-$ on
light rebound Mn.
EDTA washed sample, time constant 200 mS, sweeping rate 5 G/S,
other experimental conditions as shown in Fig.1.
○ Na-carbonate, △ Na-formate.

Bicarbonate was shown as in Fig.5 to stimulate the rebinding of
Mn^{2+}. This stimulation was linearly proportional to the concentrations of HCO_3^- within the ranges of 5mM to 10 mM, and 10mM to 50mM.
There was a turning point at 10 mM. In consideration of the recent
advances in our knowledge of bicarbonate effects (8) it is reasonable to suppose that HCO_3^- concentration less than 10 mM affects
on the rebinding of Mn^{2+} by different mechanism from that of higher
concentrations. Formate as an antagonist of bicarbonate just has
opposite influence on the Mn^{2+} rebinding within the range of 10mM
to 50 mM.

REFERENCES
1 Emerson, R. and Lewis, C. M. (1939-1940) Carnegie Inst. Wash. Yearb.
39, 155-158
2 Govindjee, Kambara, T. and Coleman, W. (1985) Photochem. Photobiol. 42, 2
, 187-210
3 Sharp, R. R. and Yocum, C. F. (1980) Biochim. Biophys. Acta 592, 185-
195
4 Khanna, R., Rajan, S., Govindjee and Gutowsky, H. S. (1983) Bichim.
Biophys. Acta 725, 10-18
5 Yu,CM. C. and Brand J. J. (1980) Biochim. Biophys. Acta 591, 483-489
6 England, R. R. and Evans, E. H.(1983) Biochem. J. 210, 473-476
7 Ono, T. A. and Y, Inone (1983) Biochim. Biophys. Acta 723, 191-201
8 Stemler, A., Vermaas, Wim F. J. and Govindjee (1982) in Photosynthesis (Govindjee, ed.), Vol. 1, pp.513-558, Academic Press, Inc.

EVIDENCE FOR THE ROLE OF FUNCTIONAL MANGANESE IN HYDROGEN-PEROXIDE-STIMULATED OXYGEN PRODUCTION ON THE FIRST FLASH IN $CACL_2$-WASHED PHOTOSYSTEM II MEMBRANES

STEVEN P. BERG AND MICHAEL SEIBERT, SOLAR ENERGY RESEARCH INSTITUTE, GOLDEN, COLORADO 80401, USA

1. INTRODUCTION

When detergent-derived photosystem II (PSII) membranes are washed with 1 M $CaCl_2$, three extrinsic proteins associated with water oxidation are removed, leaving four functional Mn bound to the PSII complex (1). In an effort to learn more about the mechanism of water oxidation, we have studied flash-induced O_2 evolution catalyzed by this material. The membranes evolve anomalous O_2 on the first two flashes, and this O_2 emission requires, among other things, PSII electron transport capability as well as functional Mn associated with the O_2-evolving complex.

2. PROCEDURE

Control PSII membranes (K&M type) were prepared as described elsewhere (2). $CaCl_2$-washed PSII (CaPSII) membranes were made from control PSII preparations by the method of Ono and Inoue (1). Heat-treated CaPSII membranes were prepared by exposing a small aliquot of CaPSII to a 56°C water bath for 3 min and used without further treatment. Tris-treated membranes were prepared by mixing equal volumes of a CaPSII membrane suspension and a solution containing 150 mM NaCl, 5 mM $CaCl_2$, 40 mM MES, and 1.6 M Tris (pH 8.2). This mixture was incubated for 5 min, and washed once with 150 mM NaCl, 5 mM $CaCl_2$ and 40 mM MES (pH 6.5). EDTA-treated PSII membranes were prepared by adding enough 400 mM EDTA (pH 6.5) to a suspension of CaPSII to bring the final concentration of EDTA to 2 mM for at least 3 min.

Flash-induced O_2 yields (2 μs Xenon flashes spaced 1 s apart) were detected with a Joliot-type O_2 rate electrode in the absence of added acceptor. Except as noted, the flow buffer contained 150 mM NaCl, 5 mM $CaCl_2$ and 40 mM MES (pH 6.5). The PSII membranes were exposed to DCMU by adding the inhibitor to the flow buffer.

Hydrogen peroxide (3%) was diluted in 150 mM NaCl, 5 mM $CaCl_2$, and 40 mM MES (pH 6.5) buffer to 11 mM. CaPSII membranes were exposed to H_2O_2 by mixing equal volumes of dark adapted membrane suspensions and H_2O_2-containing buffer. This mixture was gently swirled (10 s) and then placed on the O_2 rate electrode for an additional 3 min of dark equilibration.

3 RESULTS AND DISCUSSION

Figure 1 shows O_2 yields when control PSII membranes are subjected to O_2 rate electrode analysis. In contrast, the flash pattern of $CaCl_2$-washed PSII membranes commences with O_2 emission on the first flash, but the relative yield is greatly suppressed. There is no clear evidence of any flash yield oscillation observable in the $CaCl_2$-washed membranes. When catalase is added to CaPSII membranes prior to placing them on the rate electrode, all evidence of the anomalous O_2 evolution associated with the first two flashes disappears. The flash pattern and flash yields of the subsequent flashes remain essentially unchanged. The catalase experiment shows that H_2O_2 is involved in O_2 evolution associated with the first two flashes. To further test this, PSII membranes were exposed to exogenous H_2O_2 (5.5 mM) prior to onset of a flash sequence. Control PSII membranes were unaffected by this treatment (data not shown), but the flash yields of the CaPSII membranes (particularly the first few flashes) were greatly stimulated relative to untreated, CaPSII membranes (Fig. 1). These experiments confirm the involvement of H_2O_2 in the anomalous O_2 evolution phenomenon. Apparently, when the three extrinsic proteins associated with water oxidation are bound to the membrane, H_2O_2 cannot interact with the O_2-evolution apparatus. This, of course, is not the case when the three extrinsic proteins are missing.

Figure 2 shows that the herbicide, DCMU (a PSII inhibitor), does not affect O_2 evolution associated with the first flash of a sequence in CaPSII membranes, but the yields of all other flashes are sensitive to the inhibitor. Since DCMU blocks electron transfer between Q_A and Q_B (3), DCMU inhibition should only be manifest after the first flash reduces Q_A. Thus, the DCMU data clearly show that electron transport which gives

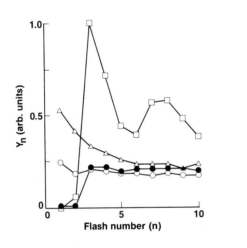 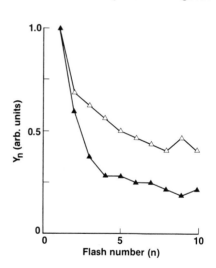

FIGURE 1. Flash yield (Y_n) sequences of control PSII (□), $CaCl_2$-washed PSII (O), $CaCl_2$-washed PSII plus catalase (●) and $CaCl_2$-washed PSII plus exogenous H_2O_2 (△). Control PSII or $CaCl_2$-washed PSII membranes (plus or minus 0.1 mg catalase/ml) equivalent to 1 mg chl/ml were mixed with an equal volume of 150 mM NaCl, 5 mM $CaCl_2$, 40 mM MES (pH 6.5) and 11 mM H_2O_2 (when present), gently stirred for about 10 s, and placed on the O_2 rate electrode. Dark incubation time on the rate electrode was 3 min.

FIGURE 2. DCMU inhibition of the flash yields associated with H_2O_2-stimulated $CaCl_2$-washed PSII membranes. $CaCl_2$-washed PSII membranes were mixed with an equal volume of 150 mM NaCl, 5 mM $CaCl_2$, 40 mM MES (pH 6.5), and 11 mM H_2O_2 (△). An identical preparation was exposed to a circulating flow buffer containing 5 µM DCMU (▲).

rise to anomalous O_2 evolution on the first two flashes is tightly coupled
to normal PSII photochemistry.

In an effort to determine whether functional Mn is directly
associated with anomalous O_2 evolution (see also Seibert, Peteri-Maison,
and Lavorel in these Proceedings), $CaCl_2$-washed membranes were subjected
to a number of inhibitory treatments, each of which has a well
characterized effect on the pool of functional Mn. As shown in Fig. 3,
heat treatment completely inhibits all flash-induced O_2 evolution in
CaPSII preparations. This treatment is known to remove almost all Mn
associated with the PSII complex (4), suggesting that at least some
functional Mn is required for the appearance of anomalous O_2 evolution on
the first two flashes. Tris treatment also removes functional Mn (4,5),
and as expected this treatment inhibits O_2 evolution (both in the presence
or the absence of exogenous H_2O_2) on the first and all subsequent flashes
(Fig. 3). EDTA treatment produced similar results, but perhaps for a
different reason as previous literature suggests. EDTA is thought to
inhibit anomalous O_2 production on the first flash by binding free Mn
associated with salt-treated inside-out vesicles (6). Furthermore, the
addition of free Mn to NH_2OH-treated PSII preparations (that retain 1
bound functional Mn) in the presence of H_2O_2 stimulates the production of
anomalous O_2 production measured by steady-state methods (7). Thus, we
suggest that the presence of non-functional (in normal O_2 evolution) or
free Mn might in addition be required to see amperometric signals on the
first two flashes in our CaPSII. Alternatively, EDTA might have a more
direct effect by interfering with functional Mn, since the chelator can
remove Mn from inside-out thylakoid vesicles (8).

In summary, the data presented here and in a companion paper
(Seibert, Peteri-Maison, and Lavorel, these Proceedings) suggest that
anomalous O_2 production on the first two flashes in CaPSII membranes

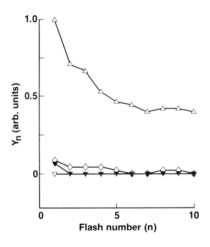

FIGURE 3. O_2 flash patterns obtained with $CaCl_2$-washed PSII membranes
(Δ), and duplicate membranes that were Tris- ($\Diamond$), heat- (∇), or EDTA-
treated ($\blacktriangledown$) in addition. After treatment, the PSII membrane suspensions
were mixed with an equal volume of 11 mM H_2O_2 in 150 mM NaCl, 5 mM $CaCl_2$
and 40 mM MES (pH 6.5).

requires the absence of the extrinsic proteins associated with normal photosynthetic O_2 evolution, the presence of at least some functional Mn associated with the O_2-evolving complex, intact coupling to PSII photochemistry, the presence of H_2O_2, and perhaps the presence of "free" Mn.

ACKNOWLEDGMENTS
    Supported by the Division of Biological Energy Research, Office of Basic Energy Sciences, U.S. Department of Energy contract FTP 006-86 (M.S.) and by the Competitive Research Grants Office, U.S. Department of Agriculture grant #85-CRCR-1-1564 (S.P.B.). Sabbatical support for S.P.B. came from Associated Western Universities and the University of Denver.

REFERENCES

1. Ono, T.-A., and Inoue, Y. (1983) FEBS Lett. 164, 225-260.
2. Dunahay, T. G., Staehelin, L. A., Seibert, M., Ogilvie, P. D., and Berg, S. P. (1984) Biochim. Biophys. Acta 764, 179-193.
3. Pfister, K., and Arntzen, C.J. Z. Naturforsch. 34c, 996-1009.
4. Miller, M. (1985) FEBS Lett. 189, 355-360.
5. Yamashita, T., and Butler, W. L. (1969) Plant Physiol. 44, 435-438.
6. Schroder, W.P., and Åkerlund, H.-E. (1986) Biochim. Biophys. Acta 848, 359-363.
7. Boussac, A., Picaud, M., and Etienne, A-L. (1986) Photochem. Photophys. 10, 201-211.
8. Mansfield, R.W., and Barber, J. (1983) Biochem. Biophys. Res. Commun. 110, 545-551.

INTERACTION BETWEEN MANGANESE AND THE 33-KILODALTON PROTEIN IN SPINACH PS II

YASUSI YAMAMOTO, DEPARTMENT OF BIOLOGY, FACULTY OF SCIENCE,
KYUSHU UNIVERSITY 33, FUKUOKA 812, JAPAN

1. INTRODUCTION

Several lines of experiment have shown that the peripheral 33-kDa protein of PS II plays a crucial role in oxygen evolution by keeping catalytic Mn atoms in situ [1-3]. Four atoms of Mn are associated with PS II, and removal of the 33-kDa protein from the oxygen-evolving PS-II particles of spinach by various treatments induced release of Mn from the membranes and concomitant inhibition of oxygen evolution. There was, however, apparent heterogeneity in the susceptibility of the Mn to the inhibitory treatments, which may be attributable to the difference in the microenvironment where each Mn atom is located. From the experiments with the PS-II particles treated with $CaCl_2$, urea and proteases [4-6], it was suggested that two atoms of Mn in PS II, being relatively labile and easily removed from the membranes were located at the interface between the 33-kDa protein and the reaction center complex of PS II. The other two atoms of Mn were tightly bound to the membranes and probably embedded in the core complex of PS-II reaction center. Recently, isolation of an Mn-carrying 33-kDa protein from spinach chloroplasts was reported [7,8]. Presence of oxidants in the isolation medium for the protein was effective in increasing the yield of the Mn-protein, and it was suggested that the Mn located at the interface between the 33-kDa protein and the PS-II reaction center complex was stabilized by the binding to the 33-kDa protein in its oxidized state [8]. In the present paper, we studied the interaction between the 33-kDa protein and Mn in spinach PS-II particles further with butanol/water phase partitioning of the PS-II particles. Our results showed that one of four Mn atoms in PS II was located at a more hydrophilic environment than others and was partitioned into the aqueous phase together with the 33-kDa protein on the butanol/water phase partitioning. This Mn atom is probably in close association with the 33-kDa protein and is required for the functional binding of the protein to the membranes. It was also shown that under the oxidizing condition, there was a significant change in environment of Mn in the oxygen-evolution enzyme complex, and a part of Mn atoms bound to the 33-kDa protein.

2. MATERIALS AND METHODS

The oxygen-evolving PS-II particles were prepared from spinach as previously described [9]. The butanol/water phase partitioning of the PS-II particles was carried out according to ref. 8 with some modifications. The PS-II particles were first treated with 1 M NaCl for 30 min at 0°C to remove the 24- and 18-kDa proteins from the membranes. For the isolation of an Mn-carrying 33-kDa protein, the NaCl-treated PS-II particles were preilluminated with white light (intensity, 1500 W/m^2) for 10 min at 0°C in the presence of 1 mM potassium ferricyanide and 0.1 mM phenyl p-quinone as the oxidants. In the case where the 33-kDa protein having no Mn was obtained, the preillumination of the PS-II particles with the oxidants was omitted. Butanol/water phase partitioning was done by adding cold n-butanol

Biggens, J. (ed.), Progress in Photosynthesis Research, Vol. I. ISBN 90 247 3450 9
© *1987 Martinus Nijhoff Publishers, Dordrecht. Printed in the Netherlands.*

to the suspension of either the dark-adapted or preilluminated PS-II particles to 50 % (v/v) and shaking the whole mixture vigorously for 30 s at $0^{\circ}C$. The aqueous phase obtained by low-speed centrifugation of the suspension contained the 33-kDa protein. The aqueous fraction was dialyzed overnight against 10 mM MES (pH 6.5) and concentrated if necessary. $CaCl_2$-treatment of the PS-II particles and rebinding of the 33-kDa protein to the $CaCl_2$-washed PS-II particles were carried out according to Ono and Inoue [10]. Mn was assayed with a flame-type atomic absorption spectrophotometer. Protein concentration was determined by the method of Lowry et al [11]. SDS-polyacrylamide gel electrophoresis (SDS-PAGE) was carried out as previously described [12].

3. RESULTS AND DISCUSSION

3.1. Phase partitioning of the 33-kDa protein and Mn in the PS-II particles

Butanol/water phase partitioning is a procedure for separation of membrane components by their hydrophilicity. By applying this procedure to the NaCl-washed and dark-adapted PS-II particles, we obtained the peripheral 33-kDa protein in the aqueous phase of the phase partitioning (Fig. 1). Based on the result that the 33-kDa protein is present in PS II with the ratio of one 33-kDa protein per reaction center [13], the amount of the 33-kDa protein obtained here was estimated to be about 80 % of the total 33-kDa protein in PS II. Analysis of Mn in the same phase partitioning showed that one of four Mn atoms associated with PS II was partitioned into the aqueous phase together with the 33-kDa protein. These results suggest that one fourth of the total Mn in PS II is located close to the 33-kDa protein and the interaction between this specific Mn atom and the PS-II reaction center complex is weak compared with that between other Mn atoms and the reaction center complex.

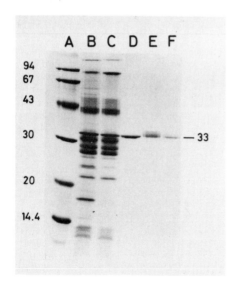

FIGURE 1. A SDS-PAGE gel showing the 33-kDa protein isolated by butanol/water phase partitioning of the NaCl-washed PS-II particles. A, marker proteins. B, PS-II particles. C, PS-II particles washed with 1 M NaCl. D, the 33-kDa protein obtained from the dark-adapted PS-II particles. E, the 33-kDa protein from the PS-II particles preilluminated with 1 mM potassium ferricyanide and 0.1 mM phenyl p-quinone. F, the 33-kDa protein obtained under the same condition as E except that 10 μM DCMU was included during the preillumination.

3.2. Binding of the 33-kDa protein to the CaCl$_2$-washed PS-II particles

The 33-kDa protein isolated by the butanol/water phase partitioning bound efficiently to the PS-II particles treated with 1 M CaCl$_2$ and reconstituted the O$_2$-evolution activity. The reconstitution required 10 mM CaCl$_2$ and 50 mM NaCl, and 60-70 % of the original activity of the PS-II particles was recovered. The binding of the 33-kDa protein to the PS-II membranes was decreased by 20-30 % when one of four atoms of Mn in PS II was released from the membranes by aging of the CaCl$_2$-treated PS-II particles for 4 h at 0°C. O$_2$-evolution activity was not recovered by the reconstitution with the aged PS-II particles. These results suggest that the loosely bound Mn in PS II, especially the Mn atom most weakly associated with PS II, is essential for the functional binding of the 33-kDa protein to the membranes and for the reconstitution of O$_2$ evolution.

3.3. Association of Mn with the 33-kDa protein

Association of a small amount of Mn with the 33-kDa protein (0.1-0.25 Mn/33-kDa protein) was observed previously when the protein was isolated from the PS-II particles with the butanol/water phase partitioning in the presence of oxidants. These results suggested that a part of Mn has tendency to bind to the 33-kDa protein in its oxidized state [8]. To ensure the oxidation of Mn in PS II, we preilluminated the PS-II particles in the presence of 1 mM potassium ferricyanide and 0.1 mM phenyl p-quinone, and then subjected the PS-II particles to the butanol/water phase partitioning to isolate the 33-kDa protein. The amount of Mn associated with the 33-kDa protein thus obtained was 0.4-0.8 atom/33-kDa protein (Table 1), which was significantly larger than that obtained from the dark-adapted PS-II particles with the oxidants (0.1 Mn/33-kDa protein). The binding of Mn to the 33-kDa protein under the oxidizing conditions was confirmed previously by EPR measurement [8]. The light-induced association of Mn with the protein was inhibited by DCMU. The presence of 10 mM CaCl$_2$ in the reaction mixture during the preillumination had no effect on the binding

TABLE 1. Effects of preillumination on the amount of Mn in the 33-kDa protein obtained by butanol/water phase partitioning of the NaCl-washed PS-II particles. Data are the average of 3 measurements. Variation of the data was within 10 %. The concentrations of potassium ferricyanide, phenyl p-quinone and DCMU used here were 1 mM, 0.1 mM and 10 μM, respectively.

Conditions	Atom Mn/33-kDa protein
Dark-adapted	0
Preilluminated	0
+ Ferricyanide, phenyl p-quinone/ Dark-adapted	0.1
+ Ferricyanide, phenyl p-quinone/ Preilluminated	0.8
+ Ferricyanide, phenyl p-quinone, DCMU/ Preilluminated	0.5

of Mn to the 33-kDa protein. The Mn-protein was, however, labile and the
amount of Mn associated with the protein was decreased gradually during
the prolonged dialysis (such as 48 h dialysis) against MES buffer (10 mM,
pH 6.5). Inclusion of 1 mM potassium ferricyanide in the dialysis medium
had no effect on the stability of the Mn-protein. When the 33-kDa protein
isolated by the butanol/water phase partitioning in the presence of oxidants
was applied to SDS-PAGE, the protein showed a diffused band in the gel
(Fig. 1), which suggested that some changes occurred in surface electric
property or conformation of the protein. On PAGE without SDS, the Mn-
carrying protein showed large mobility toward anode probably due to the
increase in negative charge on the surface of the protein. There is
a possibility that ferrocyanide produced by the preillumination of the PS-II
particles with potassium ferricyanide binds to the 33-kDa protein specifically and induces enhanced association of Mn to the 33-kDa protein. It was
shown that the addition of 1 mM potassium ferrocyanide to the suspension
of the PS-II particles during the phase partitioning induced association
of Mn to the protein to some extent (0.15 Mn/33-kDa protein). In the
presence of oxidants, the amount of Mn partitioned into the aqueous phase
on the phase partitioning of the PS-II particles increased significantly,
which suggests that there is a significant change in the environment of
Mn in PS II under these conditions. Although the artificial effects of
ferricyanide are not completely excluded here, the isolation of the 33-kDa
protein with an appreciable amount of Mn shows that the oxidizing condition
has a considerable effect on the interaction between the 33-kDa protein
and Mn.

REFERENCES

1 Cammarata, K., Tamura, N., Sayre, R. and Cheniae, G. (1984) in Advances
   in Photosynthesis Research (Sybesma, C., ed.), Vol. 1, pp. 311-320,
   Martinus Nijhoff/Dr W. Junk Publishers, The Hague
2 Govindjee, Kambara, T. and Coleman, W. (1985) Photochem. Photobiol.
   42, 187-210
3 Ghanotakis, D. and Yocum, C. F. (1985) Photosynthesis Res. 7, 97-114
4 Ono, T. and Inoue, Y. (1984) FEBS Lett. 168, 281-286
5 Miyao, M. and Murata, N. (1984) FEBS Lett. 170, 350-354
6 Isogai, Y., Yamamoto, Y. and Nishimura, M. (1985) FEBS Lett. 187, 240-244
7 Abramowicz, D. A. and Dismukes, G. C. (1984) Biochim. Biophys. Acta
   765, 318-328
8 Yamamoto, Y., Shinkai, H., Isogai, Y., Matsuura, K. and Nishimura, M.
   (1984) FEBS Lett. 175, 429-432
9 Yamamoto, Y., Hermodson, M. A. and Krogmann, D. W. (1986) FEBS Lett.
   195, 155-158
10 Ono, T. and Inoue, Y. (1984) FEBS Lett. 166, 381-384
11 Lowry, O. H., Rosebrough, N. J., Farr, A. L. and Randall, R. J. (1951)
    J. Biol. Chem. 193, 265-275
12 Yamamoto, Y., Shimada, S. and Nishimura, M. (1983) FEBS Lett. 151, 49-53
13 Yamamoto, Y., Tabata, K., Isogai, Y., Nishimura, M., Okayama, S.,
    Matsuura, K. and Itoh, S. (1984) Biochim. Biophys. Acta 767, 493-500

ACKNOWLEDGEMENTS

    This investigation was supported by a research grant from the Ministry
of Education, Science and Culture of Japan.

MANGANESE AND CALCIUM BINDING PROPERTIES OF THE EXTRINSIC 33 KDA
PROTEIN AND OF PHOTOSYSTEM II MEMBRANES[1]

D. HUNZIKER, D.A. ABRAMOWICZ, R. DAMODER AND G.C. DISMUKES,
PRINCETON UNIVERSITY, PRINCETON, NJ 08544

1. INTRODUCTION

The protein binding sites within Photosystem 2 (PS-2) membranes that are required for water
oxidation remain elusive. Evidence implicating an extrinsic 33 kDa protein and one or more of the
intrinsic membrane proteins of PS-2 has accumulated (1). This paper summarizes a study of the
correlation between the release of the 3 extrinsic proteins involved in water oxidation, the binding
of Mn, the ability to photo-oxidize Mn and oxygen evolution using a quantitative method for protein
determination (Hunziker, Abramowicz, Reddy, and Dismukes, submitted). The rebinding of Mn(II)
and Ca(II) to the isolated 33 kDa and the depleted PS-2 membrane is also reported.

2. PROCEDURE

PS-2 membranes were prepared as described previously (2). Salt-washing treatments were
based on the original procedure by Kuwabara and Murata (3). PS-2 membranes retaining specific
amounts of the extrinsic 33 kDa protein can be prepared by salt washing at various pH values with an
appropriate Good buffer (25 mM) in 1M NaCl. The membranes were then reequilibrated to pH=6.5
and assayed for oxygen evolution, Mn retention (by atomic absorption) and S2 multiline EPR signal
yield.

Proteins released by salt washing were prepared for electrophoresis by precipitation in acetone.
Supernatants were diluted to a final NaCl concentration of 0.5 M and diluted with cold acetone to a
final concentration of 80%. The solution was centrifuged at 48,000 xg for 20 min. and the pellet
was resuspended in an electrophoresis sample buffer. The amount of the 33 kDa protein remaining
on the membranes after salt washing can be determined by its release into the supernatant upon
heat shock of the depleted membranes (4). This method is more reliable than estimation of the
amount of protein in the membrane because overlapping bands cause interference, e.g. at 31-34
kDa. For these experiments the salt-washed membranes that were prepared for EPR
measurements were heat shocked at 55 C for 5 min. at 1.0 mg Chl/ml. Relative protein
concentrations were determined by absorbance of Coomassie stained slab gel electrophoresis
chromatograms by densitometry.

EPR spectra were recorded at 11 K. The intensity of the S2 multiline signal was obtained by
summing the amplitudes of the peaks after baseline subtraction. Metal binding was determined
by room temperature EPR by titration of free Mn(II) to solutions of 33 kDa protein prepared by first 1
M NaCl washing at pH=7.5 to release the extrinsic 17 and 23 kDa proteins. The washed membranes
were then 1M CaCl2 treated or 2.6 M urea treated to release the 33 kDa protein. The resultant
supernatant was then dialyzed and concentrated three times against 42 volumes of 5mM MES
pH=6.5, 5mM NaCl. The pH=7.5 membranes were also acetone precipitated, and the water
solubilized supernatant was found to be highly enriched in the 33 kDa protein. These three
protein preparations were then used in metal ion binding assays.

3. RESULTS

Figure 1, panel 1, are LDS-PAGE densitometric scans showing incremental release of the three
extrinsic proteins upon increasing the hydroxide concentration in the presence 1M NaCl [(B),
pH=4.5 to (I), pH=9.5]. Figure 1, panel 2, are scans of the resultant depleted membranes which are
then heat shocked to release the remaining 33 kDa present. By contrast, figure 1, panel 2, scan (J)
shows that (30%) 33 kDa remains bound after CaCl2 washing at pH=6.5, a treatment widely used in
the literature to prepare membranes stripped of the 33 kDa protein.

[1]Supported by the DOE/SOLERAS Grant DE-FG02-84ch10199

Biggens, J. (ed.), Progress in Photosynthesis Research, Vol. I. ISBN 90 247 3450 9
© *1987 Martinus Nijhoff Publishers, Dordrecht. Printed in the Netherlands.*

PANEL 1

PANEL 2

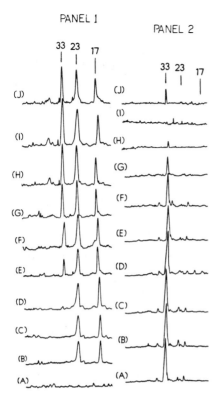

Figure 1. PAGE densitometric scans. Panel 1, proteins released from PS-2 membranes by salt washing at various pHs. A, Contol; B - I, pH 4.5,5.5,6.5,7.5,8.0,8.5,9.0,9.5. Panel 2, proteins released by heat shock of salt washed PS-2 membranes.

Figure 2. Correlation between the loss of the S2 multiline EPR signal (o), binding of the 33 kDa protein (*), oxygen evolution (x) and the amount of Mn/reaction center (■) in salt washed PS-2 membranes.

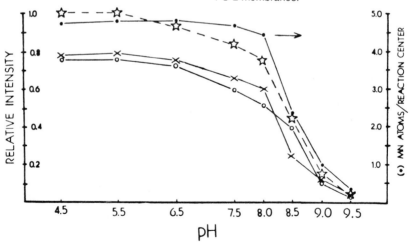

After salt treatment at differing hydroxide concentrations the reequilibrated (pH=6.5) membranes were then measured for their oxygen evolution capability and S2 multiline EPR signal yield. The data in figure 2 shows a reduction in oxygen evolution capacity (x) and multiline yield (o) as the pH is increased which parallels the loss of the 33 kDa protein (*) and retention of Mn (■) to the membrane. Mn release occurs only at higher pH, after partial loss of the 33 kDa protein. The initial 20% loss of oxygen evolution activity and multiline are most likely due to the release of the 17 and 23 kDa proteins. The Mn release data from figure 2 are reploted in figure 3 according to the Hill formulation to test for the cooperativity of Mn release. The fraction of bound Mn, Y is given by;

$$Y = \frac{([OH^-]/K)^{\alpha}}{1 + ([OH^-]/K)^{\alpha}} \qquad (1)$$

were $\alpha = 1$ refers to non-cooperative release. We find that the solid curve for $\alpha = 2$, shown in figure 3, agrees with the experimental data for the release of the first 75% (2-3 Mn atoms). This indicates that at least the first pair of Mn ions are released together, implying a closely coupled binding site, that is part of the 4 Mn ions comprising the water oxidizing comples (5).

The capacity of the 33 kDa protein to bind Mn(II) was examined and analyzed by a Scatchard plot as shown in figure 4. A dissociation constant of 2.6 mM and a stoichiometry of 0.14 Mn/33 kDa were determined for the protein prepared by CaCl2 washing (.) and acetone extraction (x), while the urea prepared protein binds considerably less Mn. This bound Mn(II) was found to be competitive with calcium as shown in figure 5 with an apparent dissociation constant of 1.0 mM. Mn(II) was also found to bind to PS-2 membranes at a pH=6.5, which had been depleted of the extrinsic proteins and Mn by Tris washing at pH=9.5. Scatchard analysis produced a binding constant of 2.0 uM and a stoichiometry of 20 Mn/PS-2 were determined (data not shown).

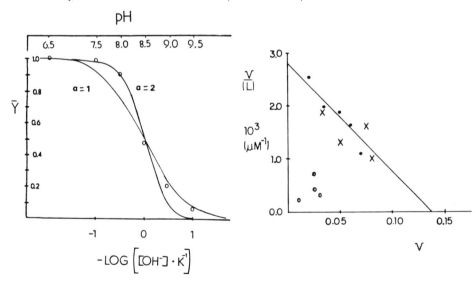

Figure 3. Comparasion of the Mn retention data taken from fig. 2 with an empirical model for cooperative binding: see eqn. (1).

Figure 4. Scatchard plot of Mn (II) binding to different preparations of the 33 kDa protein 1 M CaCl2 (.), acetone extraction (x) and 2.6 M urea (o).

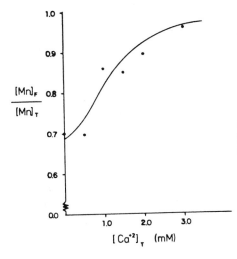

Figure 5. Ca(II)displacement of Mn(II) from the 33 kDa protein prepared by CaCl$_2$ washing of PS-2 membranes.

4. CONCLUSION

The data shows that washing PS-2 membranes in 1M CaCl2 to prepare "depleted" membranes for reconstitution studies is subject to incomplete removal of proteins. Reliable reconstitution measurements should first confirm the extent of protein depletion. The assay conditions shown here provide a reliable criterion for estimating protein content.

Using alkaline salt washing we find that the binding of the 33 kDa protein is required for photo-oxidation of the Mn complex (S2 signal) and for oxygen evolution. The binding of Mn to PS-2 membranes does not require the co-binding of the 33 kDa protein, but this protein enhances the stability of Mn binding. This may suggest the possibility that Mn binds at the interface between the 33 kDa and a PS-2 core polypeptide. Thus this protein influences both Mn binding and the efficiency of photo-oxidation.

Divalent ions like Mn(II) and Ca(II) compete for binding to the isolated 33 kDa protein at a sub-stoichiometric site having very low binding affinity. This site is probably non-specific and unrelated to the Mn pool involved in water oxidation. The tight binding of Mn(III) to the 33 kDa protein in the presence of oxidants has been previously demonstrated (6),but cannot be studied by the equilibrium techniques used here for Mn(II).

The binding of Mn(II) to PS-2 membranes depleted of the 3 extrinsic proteins is 1,000 times stronger than it is to the isolated 33 kDa protein and has about 20 sites. Thus there are many more sites than just the 4 needed for water oxidation.

REFERENCES
(1) Miyao,M. Murata, N.(1985) *Trends Biol. Sci.* March, 122-124.
(2) Ford, R. C. and Evans, M.D. (1983) *FEBS Lett.* 160,159-164.
(3) Kuwabara, T. and Murata, N. (1983) *Plant Cell Physiol.* 24, 741-747.
(4) Franzen, L. G. and Andreasson, L.E. (1984)*Biochim Biophys. Acta.* 765,166-170.
(5) Dismukes G. C. (1986) *Photochem. Photobiol.* 43, 99-115.
(6) Abramowicz, D. A. and Dismukes, G. C., (1984) *Biochim Biophys. Acta.* 765, 318-328.

THE 33 kDa EXTRINSIC POLYPEPTIDE OF PHOTOSYSTEM II IS NOT
A LIGAND TO MANGANESE IN THE O_2 EVOLVING COMPLEX.

Anne-Frances Miller, Julio C. de Paula and Gary W. Brudvig
Department of Chemistry, Yale University, New Haven, CT 06511

There are a number of claims that the 33 kDa polypeptide is the Mn protein, or forms part of the Mn site, with the Mn bound to both the core of photosystem II (PSII) and the 33 kDa polypeptide at their interface (1,2). An alternate proposal is that the 33 kDa polypeptide does not bind Mn at all but merely helps to stabilize the site under certain conditions or to maintain the O_2-Evolving Complex (OEC) in a specific configuration (3,4,5). This second proposal is supported by the observation that a number of treatments release this protein without disrupting the Mn. The low oxygen evolution activity of PSII depleted of the 17, 23 and 33 kDa polypeptides but not Mn can be reconstituted to 40% of untreated PSII activity by 200 mM NaCl and 5 mM Ca^{2+} (6). Concerning the role of the 33 kDa polypeptide in oxygen evolution activity, there are numerous reports of S state transition blockage in PSII lacking the 33 kDa polypeptide, but they do not all agree as to which transition is blocked. So far there has been relatively little direct structural or mechanistic evidence to resolve the questions of whether or not the 33 kDa polypeptide forms part of the Mn binding site, whether it is essential to the normal structure of the Mn site and whether or not the 33 kDa protein is required for normal Mn site function.

Working with buffers chosen for their ability to support maximum rates of oxygen evolution, we have used electron paramagnetic resonance (EPR) spectroscopy at low temperatures to probe the Mn site directly in the S_2 state and study the S_1 to S_2 transition in samples lacking the 33, 23 and 17 kDa polypeptides. In this communication, we report the detection of an S_2 state multiline EPR signal in PSII lacking all three extrinsic polypeptides with an amplitude as large as 70% of that of the S_2 state multiline EPR signal from untreated PSII. These results show that the 33 kDa polypeptide is not required for the S_1 to S_2 state transition, nor is the structure of the Mn site in S_2 greatly altered in its absence.

EXPERIMENTAL PROCEDURES

PSII membranes were prepared by the method in (7) and EPR spectra were collected as in (8) in the presence of 100 μM DCMU. Polypeptide composition was analyzed by polyacrylamide gel electrophoresis as described in (8) and the extent of polypeptide depletion evaluated from densitometry scan peak areas. Mn content was evaluated by EPR spectroscopy after releasing protein-bound Mn by mixing the sample with an equal volume of 1 M HCl and incubating for 20 minutes at room temperature. Oxygen evolving activity was measured as in (8). Untreated PSII typically evolved 400 to 600 μMoles O_2 /mg Chl/hr.

The three extrinsic polypeptides were removed by washing twice with

Biggens, J. (ed.), Progress in Photosynthesis Research, Vol. I. ISBN 90 247 3450 9
© *1987 Martinus Nijhoff Publishers, Dordrecht. Printed in the Netherlands.*

1.0 M $CaCl_2$ as in (5), except that for the first incubation PSII membranes at 3 mg Chl/ml were diluted to 1 mg /ml and 1 M $CaCl_2$ by adding two volumes of 1.5 M $CaCl_2$ buffer (1.5 M $CaCl_2$, 10 mM NaCl, 30% ethylene glycol, 25 mM MES-NaOH, pH = 6.5). After the second $CaCl_2$ wash the membranes were washed into high salt buffer (HSB: 200 mM NaCl, 30% ethylene glycol, 25 mM MES-NaOH, pH = 6.5 (6)) or high salt calcium buffer (HSCaB: same as HSB + 15 mM $CaCl_2$), a sufficient number of times to dilute away excess $CaCl_2$ from the 1.0 M $CaCl_2$ treatment without ever diluting them below 0.2 mg Chl/ml. In order to deplete the PSII membranes of Ca^{2+}, PSII in HSB, with $\leq$ 1 μM residual Ca^{2+} from the $CaCl_2$ treatment, were resuspended and incubated for 20 min in HSB with 5 mM added ethylene glycol bis-(β-aminoethyl ether) N,N,N',N'-tetraacetic acid (EGTA) and then washed three times with HSB or HSCaB to remove EGTA. Untreated and treated samples were stored at 77 K in buffers made up with 30% ethylene glycol at concentrations of 4-6 mg Chl/ml. All manipulations were carried out in dim green light and the samples were kept on ice at all times.

RESULTS AND DISCUSSION

Treatment with 1.0 M $CaCl_2$ releases the extrinsic 17, 23 and 33 kDa proteins, leaving essentially none of the first two and up to 10% of the 33 kDa polypeptide. We do not observe any Mn loss from PSII in the course of this treatment. Untreated PSII have on average 4.2 $\pm$0.3 Mn per 220 Chl (average of 5 measurements) and 1 M $CaCl_2$-treated PSII have on average 4.2 $\pm$0.6 Mn per 220 Chl (average of 10 measurements). Thus we conclude that our 1.0 M $CaCl_2$-treated PSII are effectively stripped of the extrinsic 17, 23 and 33 kDa polypeptides but retain all their functional Mn. Moreover, the percent oxygen evolution activity retained by our treated samples (40% in HSCaB) agrees well with that reported for PSII lacking the 17, 23 and 33 kDa polypeptides (3).

The oxygen evolving activity of polypeptide-depleted PSII has been shown to be enhanced by both high Cl^- concentration and added Ca^{2+} (3,9). Both ions have also been reported to stabilize Mn binding and this may contribute to their effect on the oxygen evolution rate of PSII lacking all three extrinsic proteins. In order to minimize Mn loss and improve the quality of our samples we decided to maintain the Ca^{2+} and Cl^- concentrations optimal for oxygen evolving activity throughout our manipulations of polypeptide depleted samples. Based on the results of titrations of O_2-evolution activity as a function of Ca^{2+} and Cl^- concentrations, we chose to adopt 200mM NaCl, 15mM $CaCl_2$ (30% ethylene glycol, 25mM MES-NaOH : HSCaB) as our standard buffer for polypeptide depleted PSII.

The fact that PSII lacking the 17, 23 and 33 kDa polypeptides evolve oxygen at a significant rate and thermoluminescence data (10) together suggest that it should be possible to generate the S_2 state and perhaps observe it by EPR at low temperatures. Nonetheless a number of attempts to generate the S_2 state multiline EPR signal in PSII samples lacking the 17, 23 and 33 kDa polypeptides have failed (4,11-13). We have found that using low Cl^- concentrations leads to loss of Mn from $CaCl_2$-treated PSII (in agreement with (9)) and also reduces the S_2 state multiline EPR signal yield. This and the Ca^{2+} dependence of the S_2 state multiline EPR signal may have been factors in the past unsuccessful attempts to detect the S_2 state multiline EPR signal in PSII samples lacking the 17, 23 and 33 kDa polypeptides.

Upon illumination at 200 K for 2 min in the presence of 100 μM DCMU we observe a large S_2 state multiline EPR signal in PSII lacking the 17, 23 and 33 kDa polypeptides. Figure 1 displays the S_2 state multiline EPR signal of CaCl$_2$-treated PSII (b) and the signal from untreated PSII for comparison (a). The intensity of the S_2 state multiline EPR signal in CaCl$_2$-treated PSII is approximately 70% that of the signal from untreated PSII and demonstrates that the Mn site can indeed undergo the S_1 to S_2 state transition in the absence of the 33 kDa polypeptide. Calcium is essential for the generation of the S_2 state multiline EPR signal as demonstrated in figure 1. Treatment with 5 mM EGTA renders PSII lacking the 17, 23 and 33 kDa polypeptides incapable of producing the S_2 state multiline EPR signal (c). When washed into (15 mM CaCl$_2$) HSCaB, EGTA-treated polypeptide-depleted membranes produce an S_2 state multiline EPR signal with an amplitude of 90% that of polypeptide-depleted PSII before

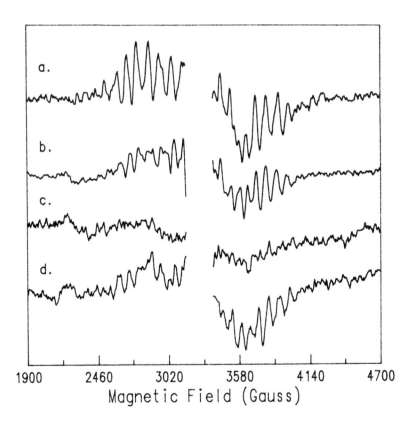

Figure 1. The S_2 state multiline EPR signal of: (a) untreated PSII membranes; (b) 1 M CaCl$_2$-treated PSII; (c) 1 M CaCl$_2$-treated PSII after treatment with EGTA; and (d) 1 M CaCl$_2$-treated PSII which was EGTA-treated and then reconstituted with 15 mM CaCl$_2$. Instrument conditions as in (8).

Ca^{2+} depletion (d). Thus the inhibitory effect of EGTA is to deplete PSII membranes of Ca^{2+}.

The striking similarity of the S_2 state multiline EPR signals from untreated PSII and $CaCl_2$-treated PSII indicates that the structure of the Mn site in the S_2 state is not significantly altered by the loss of the three extrinsic polypeptides. There does appear to be a slight difference in the ^{55}Mn nuclear hyperfine interaction in $CaCl_2$-treated PSII relative to untreated PSII. This is evident in the lower intensity of the hyperfine lines in the low field portion of the EPR signal (complicated by variable interference from $Mn(H_2O)_6^{2+}$ formed due to the $CaCl_2$ treatment). However the temperature dependence of the S_2 state multiline EPR signal's intensity still exhibits the same characteristic non-Curie behaviour with a maximum at about 7 K as in untreated PSII (data not shown). Because the temperature dependence of the signal intensity is a very sensitive probe of the geometry-dependent magnetic interactions between Mn ions, the structural change in the Mn site caused by the removal of the extrinsic polypeptides must be very minor. Thus, we conclude that in the S_2 state, the 33 kDa polypeptide does not itself bind Mn, nor is it essential for the S_1 to S_2 state transition.

This work was supported by NIH (GM32715). GWB is the recipient of a Searle Scholar Award, a Camille and Henry Dreyfus Teacher/Scholar Award, and an Alfred P. Sloan Fellowship.

REFERENCES

1. Abramowicz, D.A. and Dismukes, G.C. (1984) Biochim. Biophys. Acta 765, 318-328
2. Yamamoto, Y., Shinkai, H., Isogai, Y., Matsuura, K. and Nishimura, M. (1984) FEBS Lett. 175,429-432
3. Kuwabara, T., Miyao, M., Murata, T. and Murata, N. (1985) Biochim. Biophys. Acta 806, 283-289
4. Imaoka, A., Akabori, K., Yanagi, M., Izumi, K., Toyoshima, Y., Kawamori, A., Nakayama, H. and Sato, J. (1986) Biochim. Biophys. Acta 848, 201-211
5. Ono, T.-A. and Inoue, Y. (1984) FEBS Lett. 168, 281-286
6. Miyao, M. and Murata, N. (1984) FEBS Lett. 170, 350-354
7. Berthold, D.A., Babcock, G.T. and Yocum, C.F. (1981) FEBS Lett. 134, 231-234
8. de Paula, J.C., Li, P.M., Miller, A.-F., Wu, B.W. and Brudvig, G.W. (1986) Biochemistry, in press
9. Miyao, M. and Murata, N. (1985) FEBS Lett. 180, 303-308
10. Ono, T.-A. and Inoue, Y. (1985) Biochim. Biophys. Acta 806, 331-340
11. Blough, N.V. and Sauer, K. (1984) Biochim. Biophys. Acta 767, 377-381
12. Franzén, L.-G., Hansson, Ö. and Andréasson, L.-E. (1985) Biochim. Biophys. Acta 808, 171-179
13. Ghanotakis, D.F., Babcock, G.T. and Yocum, C.F. (1985) Biochim. Biophys. Acta 809, 173-180

EFFECT OF RELEASE OF THE 17 AND 23 kDa POLYPEPTIDES OF
PHOTOSYSTEM II ON CYTOCHROME b-559

Julio C. de Paula, Brian W. Wu, and Gary W. Brudvig
Department of Chemistry, Yale University, New Haven, CT 06511

1. INTRODUCTION

It is now believed that the 17 and 23 kDa polypeptides of Photosystem
II (PSII), which can be extracted from PSII membranes by treatment with
NaCl, do not affect the structure of the Mn-containing catalytic site of
the O_2-evolving complex (OEC) (1); they simply enhance the binding of
Ca^{2+} and Cl^- (2,3), whose depletion from PSII membranes may alter the
structure of the OEC and/or the kinetics of the S state transitions (1).
In this study, we will consider the effect of the 17 and 23 kDa
polypeptides, and Ca^{2+} on the oxidation of cytochrome b-559, a component
of PSII whose function in O_2 evolution remains an enigma.

2. EXPERIMENTAL METHODS

The following procedures used in this study were described in detail
by de Paula et al. (1): EPR measurements, O_2 evolution assay method,
isolation of PSII membranes from spinach, removal of the 17 and 23 kDa
polypeptides by treatment with 2 M NaCl, Ca^{2+} depletion by treatment with
5 mM EGTA, Ca^{2+} reconstitution by addition of 15 mM $CaCl_2$ in the presence
of 100 mM NaCl, and Cl^- depletion by suspension in Cl^--free buffer. Low-
potential cytochrome b-559 was reduced by suspending NaCl-washed PSII
membranes to 1.0 mg-chl/ml in a buffer containing 20 mM MES/NaOH, 15
mM NaCl, 5 mM $MgCl_2$, 5 mM ascorbate, 30% ethylene glycol, pH 6.0. After
incubating for 20 min in the dark at 0ºC with stirring, the suspension
was centrifuged at 30,000 x g for 20 min, and then resuspended in a
buffer similar to the one described above, with no ascorbate.
Reoxidation of cytochrome b-559 was effected by continuous illumination
at 77 K for 10 min in a transparent Dewar flask filled with liquid N_2 or
by treating samples with 2 mM K_2IrCl_6 (4).

3. RESULTS AND DISCUSSION

Cytochrome b-559 can exist in two forms which can be differentiated by
their EPR spectra. Oxidized high-potential cytochrome b-559 (E_m = 380
mV) exhibits turning points at g_z = 3.08, g_y = 2.16, and g_x = 1.36 (5),
whereas the EPR spectrum of oxidized low-potential cytochrome b-559 (E_m =
80 mV) has resonances at g_z = 2.93, g_y = 2.26, and g_x = 1.55 (6).

Illumination of PSII membranes at 77 K oxidizes one molecule of high-
potential cytochrome b-559 per PSII, as can be seen in Figures 1a and 1b
(see also Reference 4). On the other hand, it can be shown (1,7) that
removal of the 17 and 23 kDa polypeptides from PSII membranes by
treatment with 2 M NaCl causes oxidation in the dark of two molecules of
low-potential cytochrome b-559 per PSII (Figures 1c and 1d). This effect

Biggens, J. (ed.), Progress in Photosynthesis Research, Vol. I. ISBN 90 247 3450 9
© *1987 Martinus Nijhoff Publishers, Dordrecht. Printed in the Netherlands.*

was not dependent on the concentration of Ca^{2+} or Cl^- in the samples. It has also been shown (1) that the generation of the S_2 state multiline EPR signal in NaCl-treated PSII membranes was only limited by the amount of Ca^{2+} and/or Cl^- bound to the OEC and not on the oxidation state or potential of cytochrome b-559 (1).

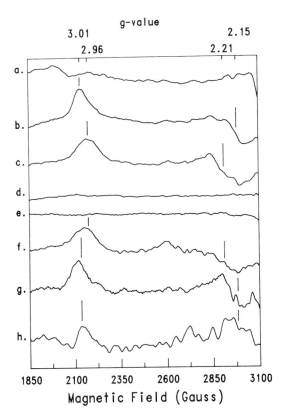

FIGURE 1. Effect of the 17 and 23 kDa polypeptides, and Ca^{2+} on the EPR spectrum of cytochrome b-559. The bars indicate the g-values of high- and low-potential cytochrome b-559. (a) dark-adapted PSII membranes; (b) 77K illuminated minus dark-adapted PSII membranes; (c) dark-adapted, NaCl-washed PSII membranes; (d) 77K illuminated minus dark-adapted, NaCl-washed PSII membranes; (e) 77K illuminated minus dark-adapted, NaCl-washed PSII membranes depleted of Ca^{2+}, and then treated with 5 mM ascorbate; (f) dark-adapted, NaCl-washed PSII membranes treated with: 5 mM EGTA, then 5 mM ascorbate, and then 2mM K_2IrCl_6; (g) 77K illuminated minus dark-adapted, NaCl-washed PSII membranes treated with 5 mM ascorbate; (h) 77K illuminated minus dark-adapted, NaCl-washed PSII membranes depleted of Ca^{2+}, then treated with 15 mM $CaCl_2$ + 100 mM NaCl, and then 5 mM ascorbate. Instrument conditions as in de Paula et al. (4).

In order to probe further the relationship between the removal of the 17 and 23 kDa polypeptides and the change in potential of cytochrome b-559, we reduced low-potential cytochrome b-559 in NaCl-treated PSII membranes with 5 mM ascorbate. We have found that the form of cytochrome b-559 present in these samples after reoxidation was dependent on the amount of bound Ca^{2+}. In samples where Ca^{2+} was removed by treatment

with 5 mM EGTA, illumination at 77 K produced no light-induced EPR
signals in the g = 3 region (Figure 1e). Oxidation with 2 mM K_2IrCl_6
produced an EPR spectrum characteristic of oxidized low-potential
cytochrome b-559 (Figure 1f). It is apparent that 77K illumination
oxidizes only the high-potential form of cytochrome b-559. These results
were not changed when the ascorbate treatment preceded the EGTA
treatment. On the other hand, illumination at 77 K of NaCl-washed PSII
which had been treated with ascorbate and retained bound Ca^{2+} (1)
produced an EPR spectrum which was indicative of oxidized high-potential
cytochrome b-559 (Figure 1g). Similarly, high-potential cytochrome b-559
was photooxidized in EGTA-treated samples where Ca^{2+} was rebound in the
presence of 100 mM NaCl (Figure 1h). Therefore, Ca^{2+} plays an essential
role in the reconstitution of high-potential cytochrome b-559 in PSII
membranes which have lost the 17 and 23 kDa polypeptides.

TABLE 1. Effect of Removal of the 17 and 23 kDa polypeptides, and Ca^{2+}
on the S_2 State Multiline and Cytochrome b-559 EPR Signals

Preparation	O_2 Evolution Activity[a] (% control)	S_2 State EPR Intensity[b] (% control)	State of cyt b-559[c] dark	dark/asc	chem ox
Control PSII	100	100	hp-red	hp-red	hp-ox
2 M NaCl-washed PSII					
a. 25 mM NaCl	34	70	lp-ox	hp-red	hp-ox
b. 25 mM NaCl, 5 mM EGTA treatment	28	40	lp-ox	lp-red	lp-ox
c. as (b), but with 15 mM $CaCl_2$ and 100 mM NaCl added	80	75	lp-ox	hp-red	hp-ox

a. The control activity for PSII membranes was 440 μmoles O_2/mg chl-hr
when assayed with 2 mM Ca^{2+}, and 370 μmoles O_2/mg chl-hr without added
Ca^{2+}; b. Continuous illumination at 200 K; c. Legend: hp = high-
potential; lp = low-potential; ox = oxidized; red = reduced; dark = dark-
adapted; dark/asc = dark-adapted, treated with 5 mM ascorbate; chem ox =
chemical oxidation by 2 mM K_2IrCl_6.

Table 1 summarizes our recent data on the reconstitution of high-
potential cytochrome b-559 and the results of de Paula et al. (1) on the
yield of the S_2 state multiline EPR signal in NaCl-washed PSII
membranes. It can be concluded that Ca^{2+} plays a dual role on the
oxidizing side of PSII: it facilitates the production of the S_2 state
multiline EPR signal, and it promotes a low- to high-potential change in
cytochrome b-559, provided that the cytochrome is first reduced. The
former effect is probably related to the stimulation of O_2 evolution
activity by Ca^{2+} in NaCl-treated PSII membranes (2). The latter effect

may be structural: Ca^{2+} may influence the structure of the cytochrome b-559 polypeptides and hence change its potential from low to high if the cytochrome is reduced first. The different electron donation properties of low- and high-potential cytochrome b-559 also suggest a structural change of cytochrome b-559 upon rebinding of Ca^{2+}. Despite the greater driving force for electron donation, low-potential cytochrome b-559 is not photooxidized at 77 K, which could be explained by an increase in the distance between P680 and cytochrome b-559. Finally, it appears that high-potential cytochrome b-559 may not be essential for optimal O_2 evolution activity, although it remains possible that cytochrome b-559 is reconstituted to its high-potential form in the O_2 evolution assay.

This work was supported by the National Institutes of Health (GM32715). GWB is the recipient of a Searle Scholarship (1983-1985), a Camille and Henry Dreyfus Teacher/Scholar Award (1985-1990), and an Alfred P. Sloan Foundation Research Fellowship (1986-1988).

REFERENCES

1 de Paula, J.C., Li, P.M., Miller, A.-F., Wu, B.W., and Brudvig, G.W. (1986) Biochemistry, in press.
2 Ghanotakis, D.F., Babcock, G.T., and Yocum, C.F. (1984) FEBS Lett. 167, 127.
3 Murata, N., Miyao, M., and Kubawara, T. (1983) in "The Oxygen Evolving Complex of Photosynthesis" (Inoue, Y. et al., Eds.) p. 213, Academic Press, Tokyo.
4 de Paula, J.C., Innes, J.B., and Brudvig, G.W. (1985) Biochemistry 24, 8114.
5 Bergström, J., and Vänngård, T. (1982) Biochim. Biophys. Acta 682, 452.
6 Babcock, G.T., Widger, W.R., Cramer, W.A., Oertling, W.A., and Metz, J.G. (1985) Biochemistry 24, 3638.
7 Ghanotakis, D.F., Yocum, C.F., and Babcock, G.T. (1986) Photosyn. Res. 9, 125.

CYTOCHROME B$_{559}$ PLAYS A STRUCTURAL ROLE IN THE OXYGEN EVOLVING COMPLEX
OF PHOTOSYSTEM II

Lynmarie K. Thompson, Julian M. Sturtevant, and Gary W. Brudvig
Department of Chemistry, Yale University, New Haven, CT 06511

INTRODUCTION
    One approach to the task of identifying the roles of the component
proteins of the multisubunit complex of photosystem II (PS II) is to study
the heat denaturation of the intact complex, using differential scanning
calorimetry (DSC). We have studied the DSC profile of PS II membranes and
have assigned several of the endothermic transitions to the denaturation
of particular polypeptides and to the loss of particular activities (1).
We have focused on a low temperature endotherm which, in a DSC study of
chloroplasts, was found to occur at the same temperature as the heat
denaturation of oxygen evolution activity (2).

EXPERIMENTAL PROCEDURES
    PS II membranes were isolated, washed, and treated as described
previously (1). Calorimetric measurements were conducted on a Russian-
built DASM-4 differential scanning calorimeter at a heating rate of
1°C/min. Activity assay and gel conditions are described elsewhere (1).
Thermal gel analysis (3) can be used to study proteins which undergo a
loss of solubility upon denaturation. Aliquots of a sample identical to
those used for DSC were heated at 1°C/min to the desired temperature,
placed on ice, and later solubilized for 20 minutes in pH 6 buffer (20 mM
MES-NaOH pH 6.0, 15 mM NaCl, 5 mM MgCl$_2$, 30% ethylene glycol) plus 5% TX-
100 at 0.5 mg Chl/ml. After centrifugation of the solubilized aliquots at
35,000xg for 15 minutes, the supernatants were prepared for gel
electrophoresis. Integration of the appropriate peaks of densitometer
scans of the gels was used to quantitate the percent of each polypeptide
which can still be solubilized after heat treatment.

RESULTS AND DISCUSSION
    The denaturation profile of PS II membranes in pH 6 buffer plus 0.01%
TX-100 is shown in Fig. 1. A theoretical fit to this curve was
calculated using the van't Hoff equation, and assuming 5 independent
denaturation steps (1). The 5 calculated component peaks and their sum
are also shown in Fig. 1. The major component of the PS II complex (about
65% by mass) is the light-harvesting Chl proteins (LHCP), and thus their
denaturation should dominate the DSC curve. Since the area of the two
largest peaks, C and D, is about 65% of the total area, we have suggested
that C and D are the denaturation of the LHCP (1). We have attempted to
assign the DSC peaks to the denaturation of particular polypeptides using
thermal gel analysis. The results for the 47 kD, 43 kD, and LHCP
polypeptides are presented in Fig. 2. The DSC peak positions under
conditions equivalent to the thermal gel analysis experiment (pH 6 buffer
plus 0.01% TX-100) are as follows: A$_2$ at 47.5°C, B at 54°C, C at 59.5°C,
and D at 66°C. The denaturation temperatures obtained from thermal gel

Biggens, J. (ed.), Progress in Photosynthesis Research, Vol. I. ISBN 90 247 3450 9
© *1987 Martinus Nijhoff Publishers, Dordrecht. Printed in the Netherlands.*

analysis (the temperature of half maximal heat-induced loss in solubility, $T_{1/2}$) suggest that the 43 kD polypeptide denatures during peak B ($T_{1/2}$=54.3°C), the 47 kD polypeptide denatures during peak C ($T_{1/2}$=57.5°C), and the LHCP denature during peak D ($T_{1/2}$=63.3°C).

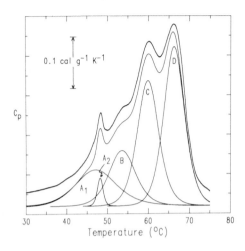

FIGURE 1. Experimental DSC scan of PS II membranes (top trace) and calculated curves.

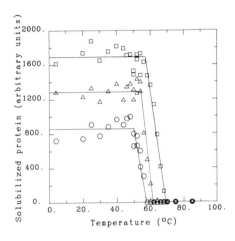

FIGURE 2. Thermal gel analysis curves: △, 47 kD; ○, 43 kD; and ☐, LHCP.

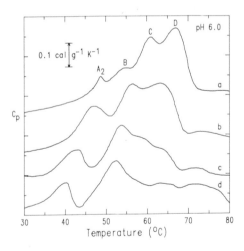

FIGURE 3. Detergent dependence of DSC curves. Samples in pH 6 buffer plus (a)0.01%, (b) 0.03%, (c) 0.05%, and (d) 0.1% TX-100.

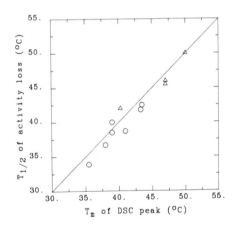

FIGURE 4. ○, correlation of DSC peak A_2 with heat inactivation of oxygen evolution. △, correlation of peak B with inactivation of DPC→DCIP activity. Note that coincidence of $T_{1/2}$ and T_m occurs along the diagonal shown.

In order to assign the roles of the components of PS II, the above polypeptide denaturation assignments must be supplemented by the assignment of the DSC peaks to heat-induced activity losses. The possibility that a DSC peak and loss of activity occur at the same temperature by coincidence can be reduced by manipulating the buffer conditions to shift the DSC peaks and determining whether the heat inactivation temperatures shift in parallel. Such shifts were induced by a range of detergent concentrations (causing no irreversible inhibition of oxygen evolution) as shown in Fig. 3. The effects of residual TX-100, which is gradually removed by the multiple wash steps of most treatments of PS II, have been ignored by various workers, and probably account for the variable heat inactivation results which have been cited recently (4). Fig. 4 shows the correlation between heat inactivation temperatures and DSC peaks under a range of conditions (pH 6 and 7.5, 0.01 - 0.1% TX-100). The heat inactivation of oxygen evolution correlates well with peak A_2; the heat inactivation of the remainder of the electron transport chain (assayed by the photoreduction of DCIP, in the presence of the artificial electron donor DPC) correlates with peak B.

The thermal gel analysis and heat inactivation results show that both the 43 kD polypeptide and electron transport denature during peak B, and the 47 kD polypeptide denatures during peak C. Since the 43 kD polypeptide can be removed without loss of activity (reviewed in 5), denaturation of another protein or domain during peak B probably causes the activity loss. The 47 kD polypeptide, which is thought by some to contain the reaction center (5), denatures at a higher temperature than the temperature of electron transport activity loss. Therefore, this polypeptide does not contain all components required for DPC→DCIP activity. Perhaps denaturation of a polypeptide containing electron acceptor components causes activity loss during peak B.

The effect of treatments which inhibit oxygen evolution on the DSC trace were studied in an attempt to identify the polypeptide(s) which denature during peak A_2. The removal of the 17 and 23 kD polypeptides with 2 M NaCl (5), which partially inhibits oxygen evolution, and the removal of the 17, 23, and 33 kD polypeptides and Mn with 0.8 M Tris-HCl pH 8.0 (5), which completely inhibits oxygen evolution, both result in the disappearance of peak A_2 from the DSC trace, and no significant changes in the other peaks. However, peak A_2 is also not resolved in the DSC scan after simply oxidizing cytochrome b_{559} with 1 mM ferricyanide (Fig. 5). Re-reduction with 1 mM ascorbate restores the peak. The removal of the extrinsic polypeptides also results in the conversion of high potential cytochrome b_{559} (which is reduced under ambient redox conditions) to its low potential form (which is oxidized under ambient redox conditions) (6). Thus the disappearance of peak A_2 is due to the oxidation of cytochrome b_{559}, and there are no further changes in the DSC trace due to the removal of the extrinsic polypeptides and Mn.

Peak A_2, which we have assigned to the functional denaturation of the oxygen evolving complex, is quite sensitive to changes in the oxidation state of cytochrome b_{559}. Such redox changes alter the heat denaturation of the oxygen evolving complex, as is also shown by the heat inactivation experiment of Fig. 6. Oxidation of cytochrome b_{559} broadens the temperature range over which heat inactivation occurs, suggesting that a parallel broadening of the very small peak A_2 is the cause of its disappearance. The heat inactivation of oxygen evolution has been shown to be accompanied by the high to low potential conversion of cytochrome

b_{559} (2). This simultaneous change, and the sensitivity of heat denaturation to the redox state of cytochrome b_{559} suggest that peak A_2 may be the denaturation of cytochrome b_{559}. This is consistent with the small size of peak A_2, which we have estimated to correspond to the denaturation of (very roughly) 20 kD of protein (1). Further evidence for this assignment may be obtained through thermal gel analysis studies.

It has often been suggested that native high potential cytochrome b_{559} is required for a functional oxygen evolving complex (7). It is clear from this work that the oxidation of cytochrome b_{559} causes some macroscopic structural change, observable by DSC, which alters the heat denaturation of oxygen evolution. This is evidence that cytochrome b_{559} plays a significant structural role in the oxygen evolving complex.

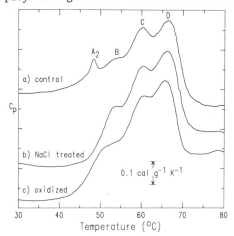

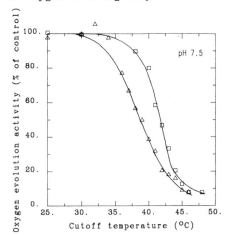

FIGURE 5. Effect of the oxidation state of cytochrome b_{559}, and of 2 M NaCl treatment on the DSC behavior.

FIGURE 6. Effect of the oxidation state of cytochrome b_{559} on heat inactivation: □, reduced; △, oxidized.

This work was supported by NIH (GM32715). GWB is the recipient of a Searle Scholar Award, a Camille and Henry Dreyfus Teacher/Scholar Award, and an Alfred P. Sloan Fellowship.

REFERENCES
1 Thompson, L. K., Sturtevant, J. M., & Brudvig, G. W. (1986) Biochemistry 25, in press.
2 Cramer, W. A., Whitmarsh, J., & Low, P.S. (1981) Biochemistry 20, 157.
3 Rigell, C. W., de Saussure, C., & Freire, E. (1985) Biochemistry 24, 5638.
4 Nash, D., Miyao, M., & Murata, N. (1985) Biochim. Biophys. Acta 807, 127.
5 Babcock, G. T. (1986) in New Comprehensive Biochemistry-- Photosynthesis (Amesz, J., Ed) Elsevier, Amsterdam, in press.
6 Larsson, C., Jansson, C., Ljungberg, U., Åkerlund, H.-E., & Andersson, B. (1984) in Advances in Photosynth. Res. I (Sybesma, C., Ed.) p.363, Martinus Nijhoff/Dr. W. Junk Publishers, Dordrecht, The Netherlands.
7 Cramer, W. A. & Whitmarsh, J. (1977) Ann. Rev. Plant Physiol. 28, 133.

EFFECT OF THE 33-kDa PROTEIN ON THE S-STATE TRANSITION IN THE OXYGEN-EVOLVING COMPLEX

M. MIYAO[a], N. MURATA[a], B. MAISON-PETERI[b], A. BOUSSAC[b], A.-L. ETIENNE[b] and J. LAVOREL[c]
[a]National Institute for Basic Biology, Myodaiji, Okazaki 444, Japan,
[b]Equipe de Recherche 307, C.N.R.S., 91190 Gif-sur-Yvette, France, and
[c]A.R.B.S., C.E.N. Cadarache, 13108 St-Paul-lez-Durance, France

1. INTRODUCTION

The oxygen-evolving complex of higher plants contains one molecule each of three extrinsic protein of 33 kDa, 23 kDa and 18 kDa and four Mn atoms (1). The 33-kDa protein is necessary to preserve the binding of two of the four Mn atoms to the complex (2-4). In addition, this protein seems to maintain the conformation of the Mn-cluster required for oxygen evolution: the oxygen-evolving complex depleted of the 33-kDa protein is inactive in oxygen evolution in 10 mM Cl^- even if all the Mn atoms remain bound to the complex (2,3). Chloride ion at concentrations higher than 100 mM can partially be substituted for the 33-kDa protein in the preservation of the Mn binding and the oxygen-evolution activity (2,3).

In this study, we used Photosystem II (PS II) particles from spinach thylakoids (5) and investigated the effect of the 33-kDa protein on the S-state transitions by comparing PS II particles depleted of the three extrinsic proteins with those reconstituted with the 33-kDa protein.

2. MATERIALS AND METHODS

PS II particles were prepared from spinach thylakoids with Triton X-100 (5) and stored at $-196^{\circ}C$ until use (6). The particles were treated with 1.0 M NaCl or with 2.6 M urea and 200 mM NaCl as described previously (2,6). The 33-kDa protein was extracted from untreated particles with 1.0 M Tris-HCl (pH 9.3 at $4^{\circ}C$), purified by column chromatography (7) and dialyzed against 10 mM Mes-NaOH (pH 6.5).

3. RESULTS AND DISCUSSION

Treatment of PS II particles with 1.0 M NaCl specifically removed the 24-kDa and 18-kDa proteins (6) and treatment with 2.6 M urea and 200 mM NaCl removed all the three extrinsic proteins (2). The NaCl-treated particles containing the 33-kDa protein were fully active in oxygen evolution in the presence of 5 mM Ca^{2+} and 30 mM Cl^-, whereas the (urea + NaCl)-treated particles depleted of the 33-kDa protein required 5 mM Ca^{2+} and 200 mM Cl^- for the maximum oxygen-evolution activity (2). In order to investigate the effect of the 33-kDa protein on the oxygen evolution apart from the change in Cl^- requirement by the elimination of the 33-kDa protein, all measurements were done in the presence of 200 mM Cl^-. Five or 10 mM Ca^{2+} was added to the reaction medium in order to maximize the activity in the absence of the 23-kDa protein (1).

Plots of oxygen-evolution activity versus activity divided by light intensity (Fig. 1) indicate that the apparent Michaelis constant (slope) and the maximum rate (ordinate intercept) were reduced by removal of the 33-kDa protein and recovered by its rebinding, while the relative quantum yield (abscissa intercept) remained constant throughout. These findings suggest that almost all the oxygen-evolving complexes in the (urea + NaCl)-treated particles are active, and that a dark, but not a light,

Biggens, J. (ed.), Progress in Photosynthesis Research, Vol. I. ISBN 90 247 3450 9
© *1987 Martinus Nijhoff Publishers, Dordrecht. Printed in the Netherlands.*

process is slowed down by the elimination of the 33-kDa protein.

Patterns of oxygen yield by a flash train are presented in Fig. 2. Catalase eliminated anomalous production of oxygen which was caused by hydrogen peroxide and weekly bound Mn present in the preparation (8). In the presence of catalase the three types of particles showed a maximum on the third flash and a damped oscillation with a period of four. This indicates that all the S-state transitions proceeded properly under flash illumination with a dark interval of 1 s between flashes.

The turnover of S_1 and that of S_2 were not affected by the removal and rebinding of the 33-kDa protein (Table 1). However, it should be noted that the turnover time was estimated as the time needed to relax the re-action center II so that the next flash could induce the subsequent charge separation in the reaction center II (10). Hence, the turnover time thus estimated may reflect the re-oxidation of Q_A^- rather than the S-state transition itself, if the S-state transition is faster than the Q_A oxida-tion. Therefore, it is still obscure whether these S-state transition are affected by the 33-kDa protein. Table 1 also shows the deactivation times of the S states in darkness. The elimination of the 33-kDa protein remarkably stabilized the S_2 state, while the S_3 state was affected only slightly. The rebinding of the 33-kDa protein accelerated the deactivation of the S_2 and S_3 states. Since the S-state deactivation is considered to arise from the back reaction in PS II (11). The change in the deactiva-tion rate upon removal and rebinding the 33-kDa protein may have resulted from the modification of the oxidation-reduction potential of the S states by the 33-kDa protein, which is supposed to interact with the Mn-cluster.

The transition from S_3 to S_0 was studied by measuring fast kinetics of oxygen release after the third flash (Fig. 3). It is obvious that the oxygen release was retarded by the elimination of the 33-kDa protein and accelerated by its rebinding. It is concluded from the time to reach the maximum level (Fig. 3) that the removal of the 33-kDa protein retarded

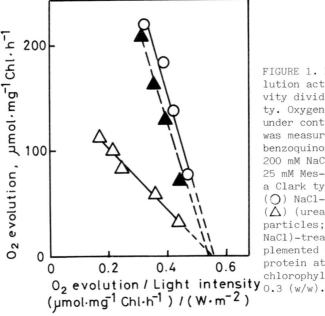

FIGURE 1. Plot of oxygen-evo-lution activity versus acti-vity divided by light intensi-ty. Oxygen-evolution activity under continuous illumination was measured with pheny-p-benzoquinone in 5 mM $CaCl_2$/ 200 mM NaCl/300 mM sucrose/ 25 mM Mes-NaOH (pH 6.5) using a Clark type electorde (5). (○) NaCl-treated particles; (△) (urea + NaCl)-treated particles; (▲) (urea + NaCl)-treated particles sup-plemented with the 33-kDa protein at a protein-to-chlorophyll (Chl) ratio of 0.3 (w/w).

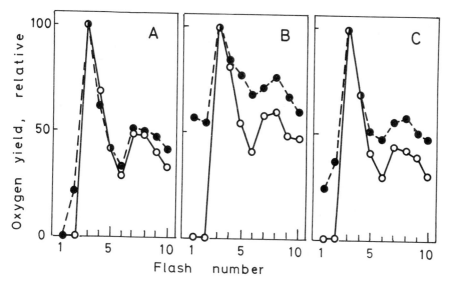

FIGURE 2. Pattern of flash-induced oxygen yield. Oxygen yield was measured in 10 mM $CaCl_2$/180 mM KCl/25 mM Mes-NaOH (pH 6.5) in the absence of artificial electron acceptors using a Joliot-type oxygen rate electrode (9). The time between flashes was 1.0 s. The data are normalized to the yield on the third flash. (●) No addition; (○) catalase was added to the particle suspension at a catalase-to-Chl ratio of 0.25 (w/w). (A) NaCl-treated particles. (B) (Urea + NaCl)-treated particles. (C) (Urea + NaCl)-treated particles supplemented with the 33-kDa protein at a protein-to-Chl ratio of 0.3 (w/w).

TABLE 1. The half-time for turnover and deactivation of the S states. The turnover time of the S state to its successive state following a charge separation in PS II and the deactivation time of each S state in darkness were estimated from the change in the oxygen yield upon varying the interval between two successive flashes in the flash train (10,11). The measurement was done in the presence of catalase as in Fig. 2.

Type of particles	Turnover ($t_{1/2}$, ms)		Deactivation ($t_{1/2}$, s)	
	$S_1 \rightarrow S_2$	$S_2 \rightarrow S_3$	S_2	S_3
NaCl-treated	6	1.5	34	28
(Urea + NaCl)-treated	5	2.2	160	40
(Urea + NaCl)-treated + 33-kDa protein	5	1.5	12	13

the oxygen release from the oxygen-evolving complex approximately 2.5 times. Since either turnover of S_1 to S_2 or of S_2 to S_3 was not detectably affected by the removal and rebinding of the protein (Table 1), it is concluded that the suppression of the oxygen evolution under continuous illumination by depleting the 33-kDa protein is caused by the slowdown of the transition from S_3 to S_0.

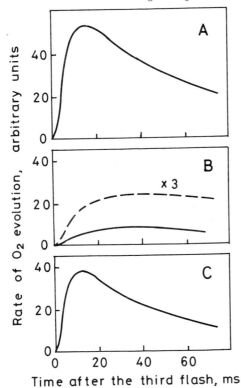

FIGURE 3. Fast kinetics of oxygen release after the third flash. Particles which had been incubated with catalase in 10 mM $CaCl_2$/180 mM KCl/25 mM Mes-NaOH (pH 6.5) were sedimented onto the platinum electrode as a thin layer by cnetrifugation. The signal after the third flash in a flash train spaced by 1.0 s was recorded. (A) NaCl-treated particles. (B) (Urea + NaCl)-treated particles. (C) (Urea + NaCl)-treated particles supplemented with the 33-kDa protein at a protein-to-Chl ratio of 0.3 (w/w).

REFERENCES

1 Murata, N. and Miyao, M. (1985) Trends Biochem. Sci. 10, 122-124
2 Miyao M. and Murata, N. (1984) FEBS Lett. 170, 350-354
3 Kuwabara, T., Miyao, M., Murata, T. and Murata, N. (1985) Biochim Biophys. Acta 806, 283-289
4 Ono, T. and Inoue, Y. (1984) FEBS Lett. 168, 281-286
5 Kuwabara, T. and Murata, N. (1982) Plant Cell Physiol. 23, 533-539
6 Miyao, M. and Murata, N. (1983) Biochim. Biophys. Acta 725, 87-93
7 Kuwabara, T., Murata, T., Miyao, M. and Murata, N. (1986) Biochim. Biophys. Acta 850, 146-155
8 Schröder, W.P. and Åkerlund, H.-E. (1986) Biochim. Biophys. Acta 848, 359-363
9 Boussac, A., Maison-Peteri, B., Vernotte, C. and Etienne, A.-L. (1985) Biochim. Biophys. Acta 808, 225-230
10 Bouges-Bocquet, B. (1973) Biochim. Biophys. Acta 292, 772-785
11 Joliot, P., Joliot, A., Bouges, B. and Barbieri, G. (1971) Photochem. Photobiol. 14, 287-305

PSII C̲a̲ ABUNDANCE AND INTERACTION OF THE 17,24 kD PROTEINS WITH THE
Cl⁻/Ca²⁺ ESSENTIAL FOR OXYGEN EVOLUTION

KIRK CAMMARATA AND GEORGE CHENIAE, University of Kentucky, U.S.A.

INTRODUCTION

The 17,24 kD PSII extrinsic polypeptides are not directly required for
ligation of PSII Mn (1-3) or for cycling of the S-states of O_2 evolution
(4-6). On the other hand, modulation of the concentrations of Cl^- and
Ca^{2+} required for the S-state transitions (1-3,7) or PSII trap/S-state
coupling (4,8) apparently are functions of these proteins. However,
controversy exists concerning the extent of inhibition of VO_2 upon 17,24
kD depletion from PSII membranes (1), as well as the mechanisms and
extents to which the extrinsic proteins influence the binding of chloride
and calcium to the PSII/S-state complex (1,7). We have measured the
abundances of specifically ligated PS̲I̲I̲ Ca (~3 Ca/reaction center) and
compared the enhancement of VO_2 by Ca^{2+} in 17,24 kD-less PSII membranes
(~1 Ca/reaction center) prepared from wheat vs spinach. Wheat PSII
membranes showing a large enhancement of VO_2 by Cl^- were used to study the
interaction of the 17,24 kD proteins with the critical protonatable
group(s) required for Cl^- activation of O_2 evolution (see model of P.
Homann in ref. 9).

METHODS

PSII membranes (TMF-2) and PSII membranes fully depleted of the 17,24 kD
proteins (NaCl-TMF-2) were prepared (5) from either wheat or spinach using
Triton X-100/Chl ratios (w/w) of 20:1 and 15:1, respectively. Depletion
of Cl^- from NaCl-TMF-2 and reconstitution with the 17,24 kD were performed
essentially as described in (10). For Ca analyses, PSII membranes were
incubated 15 min in Buffer C (0.3M sorbitol/15mM NaCl/40mM Mes-NaOH, pH
6.5/1mM EGTA/20µM A23187) then washed in Buffer C (EGTA/A23187 omitted).
All glassware was acid washed. To further deplete Ca, NaCl-TMF-2 was
extracted 1 hr with Buffer C + 2M NaCl, then washed 3 times in Buffer C
(A23187 omitted). VO_2 was assayed with 300µM PBQ and 30mM NaCl (5) or
15mM $CaCl_2$. SDS-PAGE/densitometer analyses were described in (5).

RESULTS AND DISCUSSION

In direct comparisons, spinach PSII preparations always showed lower VO_2
than wheat PSII membranes, regardless of assay conditions. For the 17,24
kD-less spinach (NaCl-TMF-2), we routinely observed: 1) ~70% inhibition
of VO_2 (relative to unextracted TMF-2) when assayed in the presence of
saturating amounts of Cl^-; and 2) ~2-fold greater enhancement of VO_2 by
Ca^{2+} than by Cl^- during assay. When the same extraction and assay
procedures were applied to wheat NaCl-TMF-2, we observed, in contrast to
spinach NaCl-TMF-2, only ~50% inhibition of VO_2 and ~2-fold greater

Biggens, J. (ed.), Progress in Photosynthesis Research, Vol. I. ISBN 90 247 3450 9
© *1987 Martinus Nijhoff Publishers, Dordrecht. Printed in the Netherlands.*

enhancement of VO_2 by Cl^- than by Ca^{2+}. Since the 17,24 kD-less PSII membranes from wheat vs spinach show similar Cl^-/Ca^{2+} concentration dependencies for VO_2 (despite the different magnitudes of these effects), Cl^-/Ca^{2+} probably affect mechanisms common to both wheat and spinach PSII. The contrasting behaviors of spinach vs wheat NaCl-TMF-2 may reflect a greater Ca^{2+} dependent rate limitation of VO_2 in spinach vs wheat, thereby diminishing the magnitude of Cl^- stimulation of VO_2 in spinach PSII membranes.

PREPARATION[1]		RATE OF OXYGEN EVOLUTION (μMOLES O_2/MG CHL/HR)			
SOURCE	TYPE	PLUS 30mM NaCl	PLUS 15mM CaCl$_2$	INCREASE BY Ca^{2+} OVER 30mM NaCl	
A. WHEAT	1. TMF-2	746	747	+1	2.9Ca/RC
	2. NaCl-TMF-2	299	415	+116	3.3Ca/RC
	3. 2M NaCl/EGTA/				
	A23187 NaCl-TMF-2	88	422	+334	1.1Ca/RC
B. SPINACH	1. TMF-2	620	605	-15	1.8Ca/RC
	2. NaCl-TMF-2	163	384	+221	1.1Ca/RC
	3. 2M NaCl/EGTA/				
	A23187 NaCl-TMF-2	41	248	+248	0.9Ca/RC

1. ALL PREPARATIONS WERE WASHED IN BUFFER C CONTAINING 1mM EGTA AND 20μM A23187 FOLLOWING DETERMINATION OF V_{O_2} (PLUS NaCl). WITHOUT THIS WASH OF WHEAT OR SPINACH TMF-2, 12-15 Ca/200 CHL WERE OBSERVED; HOWEVER, NO DECREASE OF V_{O_2} OCCURRED BY DECREASE OF THIS Ca ABUNDANCE TO THE LEVELS SHOWN FOR TMF-2.

Table I shows the PSII Ca abundances following washing in Buffer C to remove non-specifically bound Ca. Such washing did not diminish VO_2 of either wheat or spinach TMF-2. Note the low abundance of ~2.9 and 1.8 Ca/reaction center for wheat and spinach TMF-2, respectively. Whereas the EGTA/A23187 wash of either wheat or spinach NaCl-TMF-2 did not diminish their VO_2 determined in the presence of Ca^{2+}, VO_2 determined without Ca^{2+} was diminished for wheat NaCl-TMF-2 but not for spinach NaCl-TMF-2. This decreased rate, however, was fully restored by Ca^{2+} addition to assays; the magnitude of this Ca^{2+} effect was ~2-fold greater than shown in Table I for unwashed wheat NaCl-TMF-2. The EGTA/A23187 washing of 17,24 kD depleted membranes diminished the abundance of specifically ligated Ca in both spinach and wheat to only ~1 Ca/reaction center. We conclude for wheat preparations that only ~2 Ca/PSII reaction center are susceptible to release following 17,24 kD protein solubilization and that these 2 Ca/PSII are required for either coupling of the PSII trap to the S-state complex (4,8) and/or for the $S_3 \longrightarrow (S_4) \longrightarrow S_0+O_2$ transition (11). We suspect the diminished Ca abundance of spinach TMF-2 is related to the consistently lower VO_2 observed for spinach vs wheat preparations.

Exhaustive dark extraction of wheat NaCl-TMF-2 with EGTA/A23187 in 2M NaCl diminished VO_2 in the absence of Ca^{2+} addition and invoked a large (~4.8-fold) enhancement of VO_2 by Ca^{2+}. This Ca^{2+} enhanced rate of O_2 evolution was equal to that of the parent NaCl-TMF-2. With spinach NaCl-TMF-2, the same extraction also decreased VO_2 in the absence of Ca^{2+} addition, but the absolute magnitude of Ca^{2+} stimulation was unchanged and the enhanced rate was ~36% less than the value observed prior to 2M NaCl/EGTA/A23187 extraction. Apparently, 2M NaCl/EGTA/A23187 extraction caused some secondary irreversible effects in spinach, but not wheat, NaCl-TMF-2.

Wheat NaCl-TMF-2 extracted with 2M NaCl/EGTA/A23187 was preincubated with μmolar concentrations of Ca^{2+}, assayed (+1mM EGTA), then reassayed with 15mM $CaCl_2$. We observed that both high (K_m ~65μM and low (K_m ~1.5mM) affinity Ca binding occurs in 17,24 kD-less wheat PSII depleted to ~1 Ca/reaction center. These values are similar to those obtained in (11) for spinach PSII. However, preincubation of spinach NaCl-TMF-2 with 1mM

$CaCl_2$ had little effect on VO_2. Thus, our analyses suggest that a significant fraction of the high affinity Ca sites of spinach $NaCl-TMF-2$ were modified irreversibly and that the different magnitudes of Cl^-/Ca^{2+} effects on VO_2 in wheat vs spinach membranes apparently relate to different stabilities of the high affinity Ca site.

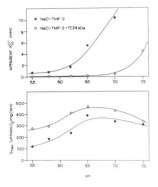

Using wheat NaCl-TMF-2 (free of any severe limitation by lack of high affinity Ca) we observed the concentration dependency of VO_2 on NaCl over the pH range of 5.5-7.5. Figure 1 shows: 1) values for V_{max} and the apparent K_m for Cl^- (estimated from the saturation profiles) increased ~3 and ~40-fold, respectively, over the indicated pH range; and 2) reconstitution of NaCl-TMF-2 with the 17,24 kD decreased the apparent K_m for Cl^- by 10 to 15-fold with a maximum of only 2.3-fold increase in V_{max}. Further evaluation of similar data from 3 experiments (in the context of the model of a protonation event(s) required for Cl^- activation of O_2 evolution) revealed that the $K_{Cl}-$ was decreased from 740μM to 52μM upon reconstitution of NaCl-TMF-2 with the 17,24 kD. However, the pKa of the critical protonatable group(s) (6.00 for NaCl-TMF-2) was not appreciably altered in the presence of the 17,24 kD proteins (6.05). This pKa difference was never more than 0.36 units in any one experiment, which compares favorably with results in (9).

The 14-fold decrease in $K_{Cl}-$ without significant change in the pKa of the protonatable group(s) indicates that the 17,24 kDa proteins function at a site removed from the required protonatable group(s). This may suggest a barrier function for these proteins, however, such a postulate seems inconsistent with observations (12,13, confirmed by us for wheat PSII) suggesting that Cl^- is electrostatically bound to its action site and remains freely accessible to the outer medium.

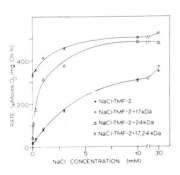

Figure 2 compares the Cl^- concentration dependencies of VO_2 for NaCl-TMF-2 reconstituted with the 17 and/or 24 kD proteins. Reconstitution with the 24 kD protein alone (but not with the 17 kD alone), increased VO_2 and decreased the K_m for Cl^- about 5-fold. However, reconstitution with both the 17 and 24 kD further decreased the K_m for Cl^- but increased VO_2 only at low Cl^- concentrations. Thus, at low Cl^- concentration, the 17 kD protein functions synergistically with the 24 kD to retain functional Cl^-.

In Figure 3, wheat NaCl-TMF-2 was extracted
20 min at pH 6.2 with the indicated
concentration of NH_2OH, reconstituted with
saturating amounts of the 17,24 kD proteins,
then washed to remove non-specifically bound
proteins. Under these conditions, VO_2 is
proportional to the PSII Mn abundance.
Depletion of PSII Mn results in a similar
decrease in the specific re-binding of the
17,24 kD proteins. This protein re-binding
was functional since the efficiency of Cl^-
retention during assay was unchanged
irrespective of the extent of inactivation of
the O_2 evolving centers (inset). Thus, PSII Mn influences the structural
parameters involved in 17,24 kD re-binding near the water oxidizing site.

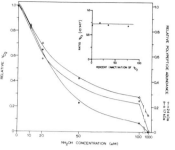

We conclude that the differences observed in direct comparisons of wheat
vs. spinach PSII membranes, and also the variable extents of inhibition and
Ca^{2+} effects on VO_2 reported in the literature (5), are the result of
different stabilities of the high affinity Ca sites. Although the 17,24
kD proteins promote the specific binding of ~2 Ca/PSII (see also 14), Ca^{2+}
binding with a K_m ~65μM is observed in their absence. The 17,24 kD
proteins also function synergistically to retain functional Cl^-, but their
site of action is apparently removed from the critical protonatable
group(s) required for Cl^- activation of O_2 evolution yet within the
proximity of structural influences by PSII Mn.

REFERENCES

1. Govindjee, Kambara, T. and Coleman, W. (1985) Photochem. Photobiol.
    42, 187-210.
2. Dismukes, G.C. (1986) Photochem. Photobiol. 43, 99-115.
3. Critchley, C. (1985) Biochim. Biophys. Acta 811, 33-46.
4. Dekker, J.P., Ghanotakis, D.F., Plijter, J.J., Van Gorkom, H.J. and
    Babcock, G.T. (1984) Biochim. Biophys. Acta 767, 515-523.
5. Radmer, R., Cammarata, K., Tamura, N., Ollinger, O. and Cheniae, G.
    (1986) Biochim. Biophys. Acta 850, 21-32.
6. Ono, T-A., and Inoue, Y. (1986) Biochim. Biophys Acta 850, 380-389.
7. Homann, P.H. and Inoue, Y. (1986) in Ion Interactions In Energy
    Transfer Biomembranes (Papageorgiou, G.C., Barber, J. and Papa, S.,
    eds.) pp 279-290, Plenum Publishing Corp., New York.
8. Ghanotakis, D.F., Babcock, G.T. and Yocum, C.F. (1984) FEBS Lett.
    167, 127-130.
9. Homann, P. (1985) Biochim. Biophys. Acta 809, 311-319.
10. Tamura, N., Radmer, R., Lantz, S., Cammarata, K. and Cheniae, G.
    (1986) Biochim. Biophys. Acta 850, 369-379.
11. Boussac, A., Maison-Peteri, B., Vernotte, C. and Etienne, A-L. (1985)
    Biochim. Biophys. Acta 808, 225-230.
12. Itoh, S. and Iwaki, M. (1986) FEBS Lett. 195, 140-144.
13. Itoh, S. and Uwano, S. (1986) Plant Cell Physiol. 27, 25-36.
14. Ghanotakis, D.F., Topper, J.N., Babcock, G.T. and Yocum, C.F. (1984)
    FEBS Lett. 170, 169-173.

PHOTOACTIVATION OF THE WATER OXIDIZING COMPLEX BY PHOTOSYSTEM 2 MEMBRANES.

N. TAMURA AND G. CHENIAE

INTRODUCTION

The conversion of the apo-S-state complex to a water oxidizing tetra-Mn
S-state complex occurs via a PS2 multiquantum photoactivation process
(1-3) in which both Mn^{2+} and Ca^{2+} are required (4-6). This photoligation
of Mn^{2+} proceeds independently of the 17/23/33 kD PS2 extrinsic proteins,
but the expression of catalytic activity of the Mn-complex requires
subsequent reconstitution of membranes with the 33 kD protein and Ca^{2+}
addition to assays (5). The 17/23 kD proteins assemble with membranes
only during photoactivation (7,8) but at rates ~3-fold faster than O_2
evolution activity (7). In efforts to obtain insights into the Mn
photoligation process, we studied the photoactivation process using PS2
membranes depleted of functional Mn, and either the 17/23 kD (NH_2OH-TMF-2)
or the 17/23/33 kD proteins (Tris-TMF-2).

METHODS

TMF-2 (PS2 unit of ~200 Chls; ~4Mn/200 Chls; e^- acceptor pool of ~2.5
equivs) was prepared from wheat (9) then subjected either to 5mM NH_2OH
extraction [500 µg Chl/ml in Buffer A (0.4 M sucrose/50 mM MES-NaOH, pH
6.5/15 mM NaCl) for 60 min] or to Tris extraction (500 µg Chl/ml in 0.8 M
Tris-Cl$^-$, pH 8.2 for 40 min). Following centrifugation/washing, the
membranes routinely were preincubated (15 min) at 4°C in Buffer A (750 µg
Chl/ml) containing 375 µg 33 kD protein/ml. Photoactivation of the
preincubated membranes routinely was done at 23°C in Buffer A containing 1
mM $MnCl_2$/50 mM $CaCl_2$/250 µg Chl/ml using continuous (24 µE m^{-2} s^{-1}) or
flash (~2 µs) illumination. VO_2 was assayed polarographically using the
assay buffer described in (9) except as noted with addition of 15 mM
$CaCl_2$. Procedures have been described (9) for measurement of DCIP
photoreduction [by Mn^{2+} (50 µM), TPB (50 µM) and DPC (1 mM)] and
functional Mn abundance (1 mM EDTA washed membranes).

RESULTS

Table I records the effects of additions of $MnCl_2$/$CaCl_2$/33 kD protein on
photoactivation (30 min) of VO_2 and photoligation of Mn by NH_2OH-TMF-2.
In darkness at any of the conditions shown, neither VO_2 nor functional Mn
increased beyond values obtained with non-incubated NH_2OH-TMF-2.
Similarly, incubations in weak light but in absence of any additions
yielded no increase of VO_2 or functional Mn. In contrast, light
incubation in the presence of $MnCl_2$ alone or with $CaCl_2$ [or additionally
the 33 kD protein (added during preincubation)] yielded significant
increases of VO_2 and functional Mn relative to dark incubated controls.

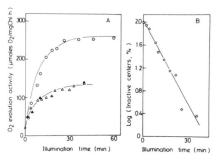

The data show photoactivation is strictly dependent on Mn^{2+} but only partially dependent on Ca^{2+} addition, if Ca^{2+} is included in the assay of VO_2. With addition of Mn^{2+} but no Ca^{2+}, a fraction of the Mn abundance values shown proved non-extractable by EDTA but exchangeable with Ca^{2+} during dark incubations. With this correction, good correlation between the increase of VO_2 and the photoligated Mn is observed. Since NH_2OH extraction of the incubated samples abolished any VO_2 and diminished Mn abundance of all samples to ~0.5 Mn/200 Chl, we conclude photoligation of Mn^{2+} and photoactivation of the apo-S-state complex was obtained. Similar magnitudes of increase of VO_2 were observed during photoactivation by Tris-TMF-2 in $MnCl_2/CaCl_2$; however, in this case, as with $DCIPH_2$-treated-$CaCl_2$-TMF-2 (6), reconstitution of photoactivated membranes with 33 kD protein was necessary for expression of catalytic activity of the photoligated Mn. Those differences reflect the difference in membrane abundance of the 33 kD protein in NH_2OH- vs Tris-TMF-2 and $DCIPH_2$-treated-$CaCl_2$-TMF-2. In studies thus far, we find no dissimilarities in the Mn photoligation process in comparisons between NH_2OH- and Tris-TMF-2.

Fig. 1A shows typical time courses of photoactivation of NH_2OH-TMF-2 obtained in the presence of 100 µM DCIP/$MnCl_2$/$CaCl_2$ (circles), $MnCl_2$/$CaCl_2$ only (open triangles) and 200 µM atrazine/$MnCl_2$/$CaCl_2$ (closed triangles). These data show that the presence of DCIP (or other lipophilic PS2 e⁻ acceptors) enhance rate/yield of photoactivation only ~2-fold despite the limited PS2 e⁻ acceptor pool in these membranes (9). In the absence of an added e⁻ acceptor, atrazine was not inhibitory. In contrast, the addition of a lipophilic PS2 e⁻ acceptor conferred sensitivity (variable in different preps from 65 to 90%) of photoactivation to atrazine. Our data indicate: 1) Q_B is not absolutely essential in the mechanism of photoactivation and 2) turnover of reduced Q_A/Q_B by added lipophilic e⁻ acceptors enhances (~2 fold) yields of photoactivation. Fig. 1B shows the open circle data of Fig. 1A as a semilogarithimic plot of the population of inactive apo-S-state water oxidizing complex versus time of photoactivation. The observed kinetics and half-time ($t_{1/2}$~6 min) are similar to data previously reported with NH_2OH-extracted _Anacystis_ (10) and chloroplasts lacking functional Mn (4,11). Mechanistically, we apparently study with PS2 membranes the same process as studied with algae (1,10) or flash greened leaves (2). This conclusion is reinforced by the data of Fig. 2.

Here we show the time-courses of photoactivation induced by brief flashes

TIME COURSE OF PHOTOACTIVATION BY NH_2OH-TMF-2

TABLE 1. EFFECTS OF $MnCl_2$/$CaCl_2$/33 KDA PROTEIN ON PHOTO-ACTIVATION OF V_{O_2} AND PHOTOLIGATION G Mn BY Mn-DEPLETED NH_2OH-TMF-2

ADDITIONS	V_{O_2}		Mn/200 CHL	
	$-Ca^{2+}$	$+Ca^{2+}$	DARK	LIGHT
NONE	31.9	46.0	0.79	0.50
$MnCl_2$	42.5	197	1.22	3.82
$MnCl_2/CaCl_2$	288	363	0.88	3.33
$MnCl_2/CaCl_2$/33 KDA	326	346	0.89	3.46

separated by different dark intervals (t_d) between flashes (Fig. 2A).
These experiments, made with NH_2OH-TMF-2 in presence of DCIP, show that
essentially no photoactivation occurred with t_d
$\geq$5s, but with decreasing t_d the rate of photoactivation increased and reached a maximum at
t_d ~0.25-0.5s. Fig. 2B shows the relative quantum
efficiency of photoactivation vs the different t_d
values. The ascending portion of the curve
reflects a rate-limiting reaction ($t_{1/2}$ ~125 ms),
and the descending portion of the curve reflects
the decay ($t_{1/2}$ ~1-1.5 s) of an unstable
intermediate. Both of the half-times are very
similar to those reported in Refs (1,2,10).

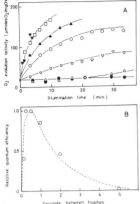

KINETIC ANALYSES OF
PHOTOACTIVATION BY
NH_2OH-TMF-2

A) Flash Spacings
were 0.12 (▼),
0.25 (□), 0.5 (▲),
1.0 (O), 2.0 (▽),
5.0 (△) and
10.0 s (●).
B) Relative Quantum
Efficiency From
Data of A).

Previously, we proposed that photoactivation
involved photooxidation of Mn^{2+} and photoligation
of Mn at valency states > +2 (12). Fig. 3 records
the inhibition by ascorbate of photoactivation of
NH_2OH-TMF-2 measured in absence of DCIP and at
different light regimes. Despite being a poor PS2
e donor, ascorbate proved an effective inhibitor.
As shown, the ascorbate concentration for 50%
inhibition diminished markedly to only 8 µM with
increase of t_d to ~2s. This inhibition is
attributed to the chemical reduction of the
$L_1/L_2/L_3$ states (Fig. 5) and not to Z^+ or the water
oxidizing tetra-Mn-S-state complex. Once photoactivated the tetra-Mn complex of NH_2OH-TMF-2 is
not inactivated by incubations with ascorbate or HQ
(13).

Fig. 4A and B show the effects of weak light (open
symbols) vs dark (closed symbols) preincubation of
NH_2OH-TMF-2 on photoactivation capacity (A) and
DCIP photoreduction (B) by DPC (circles) or Mn^{2+}

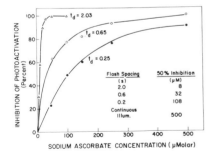

INHIBITION BY ASCORBATE

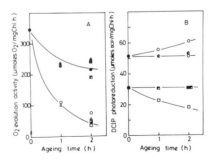

PREINCUBATION (WEAK LIGHT VS
DARK) ON PHOTOACTIVATION (A)
AND DCIP PHOTOREDUCTION (B)

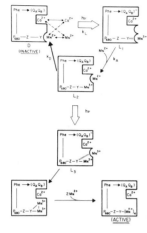

PHOTOACTIVATION MODEL

(squares). These data reveal that weak light pre-incubation in the absence but not the presence of Mn^{2+} (▰ symbol) promotes loss of photoactivation capacity and Mn^{2+}/DCIP photoreduction. In contrast, DPC (or TPB)/DCIP photoreduction was not diminished by light preincubation. The results suggest a causal relationship between loss of photoactivation capacity and loss of Mn^{2+} photooxidation. They also imply the sites of Mn^{2+} and DPC e⁻ donation to PS2 in NH_2OH-TMF-2 are not equivalent. We thus assign Mn^{2+} donation to Y in our working Model of the multiquantum photoactivation mechanism (Fig. 5). We postulate/assign k_A to be the ~125 ms rate-limiting step and k_D to be the ~1.0s decay of the unstable intermediate. According to this Model, two successive PS2 photoacts lead to an inactive Mn^{3+}-dimer-S-state complex which becomes catalytically active by ligation of two additional Mn^{2+} to form a mixed valency state Mn-tetramer. Other data indicate that neither Cl^- (14) nor $cyt.b_{559}$ are essential components in the photoactivation process.

REFERENCES

1. Cheniae, G. M. and Martin, I. F. (1971) Biochim. Biophys. Acta 253, 167–181.
2. Inoue, Y., Kobayashi, Y., Sakamoto, E. and Shibata, K. (1975) Plant Cell Physiol. 16, 327–336.
3. Yamashita, T., Inoue, Y., Kobayashi, Y. and Shibata, K. (1978) Plant Cell Physiol. 19, 895–900.
4. Yamashita, T. and Tomita, G. (1974) Plant Cell Physiol. 15, 69–82.
5. Pistorius, E. K. and Schmid, G. H. (1984) FEBS Lett. 171, 173–178.
6. Tamura, N. and Cheniae, G. (1986) FEBS Lett. 200, 231–236.
7. Becker, D. W., Callahan, F. E. and Cheniae, G. M. (1985) FEBS Lett. 192, 209–214.
8. Ono, T. A., Kajikawa, H. and Inoue, Y. (1986) Plant Physiol. 80, 85–90.
9. Radmer, R., Cammarata, K., Tamura, N., Ollinger, R. and Cheniae, G. (1986) Biochim. Biophys. Acta 850, 21–32.
10. Cheniae, G. M. and Martin, I. F. (1972) Plant Physiol. 50, 87–94.
11. Ono, T. and Inoue, Y. (1982) Plant Physiol. 69, 1418–1422.
12. Radmer, R. and Cheniae, G. (1971) Biochim. Biophys. Acta 253, 182–186.
13. Tamura, N. and Cheniae, G. (1985) Biochim. Biophys. Acta 809, 245–259.
14. Yamashita, T. and Ashizawa, A. (1985) Arch. Biochem. Biophys. 238, 549–557.

NUMBERS OF CALCIUM IONS ASSOCIATED WITH OXYGEN EVOLVING PHOTOSYSTEM II
PREPARATIONS WITH DIFFERENT AFFINITIES.

SAKAE KATOH, KAZUHIKO SATOH, TAKASHI OHNO, JIAN-REN CHEN AND YASUHIRO
KASINO
Department of Pure and Applied Sciences, University of Tokyo, Meguro-ku,
Tokyo 153, JAPAN

1. INTRODUCTION
    There is ample evidence suggesting that Ca^{2+} plays an important role
in PS II electron transport (for reviews see ref.(1-3)). Activation or
stimulation by Ca^{2+} of photosynthetic oxygen evolution has been
demonstrated in various plant materials with immature or impaired PS II.
The Ca^{2+} requirement for PS II has repeatedly been repoted in
cyanobacterial cells and membreane preparations depleted of Ca^{2+}. In
contrast, less is known about the abundance and binding affinity of Ca^{2+}
associated with PS II preparations. Generally, it is difficult to
determine accurately Ca^{2+} bound to biological membranes because chemicals
and glass wares are usually contaminated with significant amounts of Ca^{2+}
and proteins and lipids bind unspecifically the metal cations. Only two
and widely different values have been reported for Ca^{2+} contents in
oxygen evolvig membrane preparations from higher plants (4,5).
    In the present work, we have determined amounts of Ca^{2+} associated
with oxygen evolving preparations with low and high affinities. To this
end, a simple procedure to measure tightly bound Ca^{2+} by eliminating the
metal cations loosely associated with preparations and contaminating in
suspending media was developed (6,7).

2. MATERIALS AND METHODS
    Oxygen evolving PS II preparations were isolated from the thylakoid
membranes of spinach and a thermophilic cyanobacterium Synechococcus sp.
as in (8) and (9), respectively. Spinach preparations were suspended
in 40 mM Mes-NaOH (pH 6.5), 10 mM NaCl, 1mM $MgCl_2$ and 0.2 M sucrose and
Synechococcus preparations in the same medium except that pH was 5.7 and
sucrose concentration was 0.5 M. To removed contaminating Ca^{2+}, sample
suspensions were gently shaken with a chelating resin, chelex 100
(100-200 mesh, pH 6.5) for 1 min and, after standing still for 2 min to
sediment the resins but not PS II preparations, Ca^{2+} concentration of the
supernatant was determined with a Shimadzu atomic absorption
spectrophotometer equipped with a graphite furnace atomizer. Typically,
0.5 g of chelex 100 was added to 1 ml of the suspension containing 250 μg
chl. Cycle of the 1-min shaking and the two-min standing was repeated
and Ca^{2+} was measured after each cycle until a constant level of Ca^{2+}
concentration was attained.
    Oxygen evolution was determined with a Clark-type oxygen electrode as
in (8) for spinach and as in (9) for Synechococcus preparations.

Biggens, J. (ed.), Progress in Photosynthesis Research, Vol. I. ISBN 90 247 3450 9
© *1987 Martinus Nijhoff Publishers, Dordrecht. Printed in the Netherlands.*

3. RESULTS AND DISCUSSION

The total Ca^{2+} bound to spinach preparations was determined by measuring Ca^{2+} concentrations of suspensions containing various amounts of the sample without any treatment to eliminate contaminating Ca^{2+} (Fig. 1). A significant amount of Ca^{2+} was detected in the suspending medium used and, over this blank level, the Ca^{2+} conentration increased linearly with amounts of the PS II preparation added. Assuming that one PS II reaction center is present per 200 chl, the total Ca^{2+} bound to the preparation was estimated as about 40 Ca^{2+}/PS II. Synechococcus preparations gave a lower value of about 10 bound Ca^{2+}/PS II (6). Thus spinach preparations have larger binding capacity for Ca^{2+} than Synechococcus preparations, although the total bound Ca^{2+} thus determined for both preparations varied somewhat with experiments.

Fig. 2 shows that most of the bound Ca^{2+} are associated with low affinity sites, which are collectively called Site I. When Synechococcus preparations were subjected to repeated cycles of the chelex 100-treatment, the concentration of Ca^{2+} decreased to reach a low constant level. Because chelex 100 eliminated Ca^{2+} contaminating in suspending medium completely (curve a), the constant level of Ca^{2+} attained after a prolonged treatment with the resin is ascribed to the metal cations tightly associated with the preparation. In fact, amounts of the unextracted Ca^{2+} were proportional to the amounts of the preparation added, all giving ratios of about one Ca^{2+} to PS II (curves b-d). Oxygen evolving activity of the preparations was litte affected during 30 min of the treatment (not shown). Thus large amounts of loosely bound Ca^{2+} are not related to PS II electron transport.

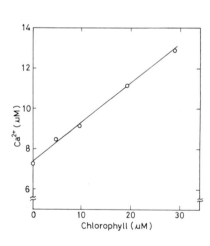

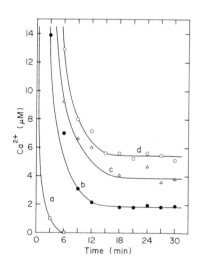

Fig. 1 Ca^{2+} concentration of suspensions containing various amounts of spinach PS II preparations.

Fig. 2 Effects of chelex 100-treatment on Ca^{2+} concentration of suspensions containing various amounts of Synechococcus PS II preparations. Curves a, b, c and d contained 0, 90, 177 and 253 μM chl, respectively.

Removal of loosely bound Ca^{2+} from spinach preparations had no
significant effect on oxygen evolution, either (Fig. 3). However,
constant level of Ca^{2+} attained afrer 30 min of the chelex 100-treatment
amounted to about two Ca^{2+}/PS II, which is twice larger than that of
Synechococcus preparations but agrees with the Ca2+ content reported for
wheat PS II preparation (5).

The amount of Ca^{2+} which can not be sequestered by the resin
increased to 13 Ca^{2+}/PS II when spinach preparations were incubated with
0.1 mM $CaCl_2$ for 50 min. The Ca^{2+}-binding was inhibited by Mg^{2+} and, in
the presence of 1 mM $MgCl_2$, the tightly bound Ca^{2+} decreased to two Ca^{2+}
/PS II. The results indicate that the spinach preparation has a
consideable number of high affinity binding sites for divalent metal
cations (Site II) and Mg^{2+} effectively competes with Ca^{2+} for these
sites. We routinely measured Ca^{2+} in the presence of 1 mM $MgCl_2$ to
block these unspecific sites.

The following experiments demonstrated that the two Ca^{2+} remained
unextracted by the chelex treatment are dissimilar from each other in
terms of binding affinity and environment surrounding the binding sites.
When the chelex 100-treatment was carried out in the presence of
digitonin, the ratio of Ca^{2+} to PS II decreased to one (Fig. 4).
Similarly, values around one Ca^{2+}/PS II were obtained in the presence of
A23187, an ionophore specific for divalent metal cations (not shown).
The results suggest that one of the two binding sites is a low affinity
site (Site III) which would be located in a hydrophobic domain of PS II
complexes. Thus only one Ca^{2+} binds to a high affinity site (Site IV).
Importantly, the chelex treatment in the presence of the detergent or the
ionophore did not appreciably affect the oxygen evolving activity. Thus
high rates of oxygen evolution occur in the presence of only one tightly

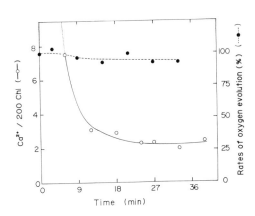

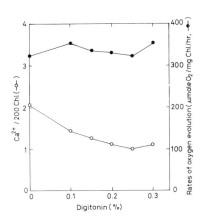

Fig. 3 Effects of chelex 100-
treatment on Ca^{2+} concentration
and oxygen evolving activity of
spinach PS II particle suspension.

Fig. 4 Effects of chelex 100-treatment
in the presence of various concentra-
tation of digitonin on Ca^{2+} concentration and oxygen evolving activity of
spinach PS II partivle suspension.

bound Ca^{2+} both in spinach and Synechococcus preparations.

We recently isolated highly purified PS II complexes from Synechococcus, which evolve oxygen at rates of 600 to 1300 μmoles O_2/mg chl.h [7]. The ratio of the tightly bound Ca^{2+} to PS II varied between 0.46 to 0.93 with preparations. Interestingly, the ratios above 0.90 were found in highly active complexes, whereas low Ca^{2+} contents were associated with relatively incompetent preparations, thereby suggesting that Ca^{2+} bound to Site IV is essential for PS II electron transport.

However, the functioning of loosely bound Ca^{2+} in oxygen evolving reaction has been suggested from observations that oxygen evolution is stimulated or activated on addition of relatively high concentrations (5–10 mM) of Ca^{2+} in various systems as described in Introduction. In particular, washing of PS II preparations with a high concentration of NaCl partially inactivates oxygen evolution concomitant with solubilization of the 24 and 18 kDa extrinsic proteins and the activity was considerably restored by the addition of 5–10 mM $CaCl_2$ [4,11]. Preliminary experiments showed that Ca^{2+} bound to Site IV was not affected by washing of spinach preparations with 1 M NaCl, although the activity was lowered and the Ca^{2+} demand was clearly manifested. Thus the stimulating effect of Ca^{2+} on oxygen evolution of NaCl-washed preparations cannot be ascribed to rebinding of Ca^{2+} to Site IV. The binding of Ca^{2+} to Site II or I or to sites created by NaCl-washing would be important for PS II electron transport to proceed maximally in NaCl-washed preparation. Experiments are in progress to relate the stimulation of oxygen evolution to binding of Ca^{2+} to specific sites in NaCl-washed preparation.

REFERENCES

1 Brand, J.J. and Becker, D.W. (1984) J. Bioenerg. Biomem. 16, 239–249
2 Govindjee, Kambara, T. and Coleman, W. (1985) Photochem. Photobiol. 42, 187–210
3 Ghanotakis, D.F. and Yocum, C.F. (1985) Photosyn. Res. 7, 97–114
4 Ghanotakis, D.F., Babcock, G.T. and Yocum, C.F. (1983) FEBS Lett. 167, 127–130
5 Tamura, N. and Cheniae, G. (1985) Biochim. Biophys. Acta 809, 245–259
6 Kashino, Y., Satoh, K. and Katoh, S. (1986) FEBS Lett. in press
7 Ohno, T., Satoh, K. and Katoh, S. (1986) Biochim. Biophys. Acta in press
8 Kuwabara, T. and Murata, N. (1982) Plant Cell Physiol. 23, 533–539
9 Satoh, K. and Katoh, S. (1985) Biochim. Biophys. Acta 806, 221–229
10 Satoh, K. and Katoh, S. (1985) FEBS Lett. 190, 199–203
11 Miyao, M. and Murata, N. (1984) FEBS Lett. 168, 118–120

INVOLVEMENT OF Ca^{2+} IN Cl^- BINDING TO THE OXYGEN EVOLVING COMPLEX
OF PHOTOSYSTEM II.

W.J. Coleman,[1] Govindjee,[1,2] and H.S. Gutowsky,[3] Departments of [1]Plant
Biology, [2]Physiology and Biophysics, and [3]Chemistry, University of Illinois,
Urbana, Illinois 61801 (U.S.A.)

INTRODUCTION

Although numerous kinetic studies have sought to explain the mechanism
of Cl^- activation of the oxygen-evolving complex (OEC) of Photosystem II
(PS II) (1,2), a thorough understanding of the mechanism has been hindered
by a lack of knowledge about Cl^- binding. The technique of ^{35}Cl-NMR was
previously used to examine Cl^- binding in thylakoids from halophytes (3,4),
but its application was restricted to plants requiring high Cl^-
concentrations because of low instrumental sensitivity. We have succeeded
in extending this NMR approach to measure Cl^- binding in spinach PS II
particles by using a specially designed probe, which has enabled us to
obtain ^{35}Cl-NMR spectra in the 0.1-10 mM range where oxygen evolution is
activated. Results obtained by this method provide additional insight into
the details of Cl^- binding in the OEC.

MATERIALS AND METHODS

PS II particles were prepared from market spinach by a modification of
the method of Berthold et al. (5), using only a single Triton X-100
treatment which was followed by several washes. Depletion of Cl^- was
achieved by incubating the particles in a Cl^--free buffer containing 50 mM
Na_2SO_4, and/or by a brief high-pH treatment (20s at pH 8.2). Salt-washed
particles (6) were prepared by incubation for 30 min in buffer containing
1.0 M NaCl on ice in the dark, followed by several washes in Cl^--free
buffer. The particles were stored at 77K at a concentration of 2.5 mg Chl
ml^{-1} in 400 mM sucrose, 20 mM MES at pH 6.0 until use.

The mean level of Cl^- depletion for the intact particles, as measured
by the Hill activity ±Cl^- ($H_2O \rightarrow$ferricyanide/2,6-dichloro-p-benzoquinone)
was 20%. The mean Hill activity at pH 6.0 in the presence of 50 mM NaCl was
402 μmol O_2 (mg Chl)$^{-1}$ hr^{-1}. For the NaCl-washed particles, the mean
activity (in μmol O_2 (mg Chl)$^{-1}$ hr^{-1}) was 28 in the presence of 50 mM NaCl,
102 in the presence of 50 mM NaCl/2.0 mM $CaSO_4$, and 198 in the presence of
25 mM $CaCl_2$.

The 8-ml NMR sample cell, 20 mm sideways-spinning probe, and 250 MHz
NMR spectrometer are described in ref. (4,7). The net line-broadening ($\Delta\nu_t$)
was calculated by measuring the full linewidth at half-maximum intensity
for a spectrum of an NaCl-containing buffer solution; this value was then
subtracted from the observed linewidth for each PS II particle suspension.

Biggens, J. (ed.), Progress in Photosynthesis Research, Vol. I. ISBN 90 247 3450 9

RESULTS AND DISCUSSION

A simple two-state model for Cl⁻ binding to proteins (8) or thylakoid membranes (4,9) predicts that a plot of $\Delta\nu_t$ vs. [Cl⁻] will be a smooth, descending hyperbola. This is not the case for Cl⁻ binding to spinach thylakoids (data not shown) or PS II particles (Fig. 1, top).

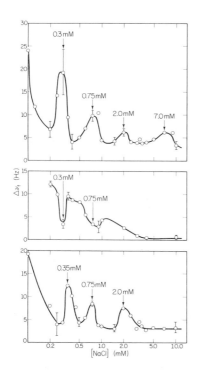

Figure 1. ³⁵Cl-NMR Binding Curves for Spinach PS II Particles. Top: Intact, Cl⁻-depleted particles. Middle: Particles washed with 1.0 M NaCl and then Cl⁻-depleted. Bottom: NaCl-washed particles with 2.0 mM CaSO₄ added to the NMR cell. Spectra were obtained at 24.51 MHz using a 33 μs 90° pulse and 360 ms recycle time. Signals were stored and averaged using a Nicolet 1180E computer. PS II particles were suspended in 400 mM sucrose, 20 mM MES at pH 6.0 and a Chl concentration of 0.5 mg Chl ml⁻¹. Error bars show the sample standard deviation for the mean value of $\Delta\nu_t$, with each point representing up to 7 different PS II preparations. The remainder of the error bars (which are approximately ±1.5 Hz) have been omitted for clarity. (After Coleman, Govindjee and Gutowsky, 1986; submitted for publication to Biochim. Biophys. Acta.)

In the latter case, the curve is interrupted by sharp increases in linewidth at 0.3 mM, 0.75-0.9 mM, and 2.0 mM, with an additional small broadening at 7.0 mM Cl⁻. The appearance of these maxima probably reflects the presence of 3-4 distinct Cl⁻ binding sites within the OEC. We propose that they arise because of Cl⁻ induced changes in the affinity of the OEC for Cl⁻ (i.e. cooperative binding and/or alterations in the exchange of bound and free Cl⁻).

In order to determine what proteins or cofactors might be involved in this phenomenon, we removed the extrinsic 18 and 24 kD polypeptides by washing with 1.0 M NaCl (6). As shown in Fig.1 (middle), NaCl-washing has two effects on the binding curve: 1) it lowers the overall curve, which probably reflects reduced affinity of the OEC for Cl⁻, and 2) it creates sharp decreases in linewidth (minima) at 0.3 mM and 0.75-0.9 mM Cl⁻. These binding effects are consistent with the decreased effectiveness of Cl⁻ as an activator of O₂-evolution in these particles.

When the assay mixture for the NaCl-washed particles was supplemented with 2.0 mM $CaSO_4$ in addition to 50 mM NaCl, the activity increased nearly four-fold. Likewise, addition of 2.0 mM $CaSO_4$ to the suspensions used in the Cl-NMR binding experiments also partially restored the linewidth maxima in the binding curve (Fig. 1, bottom).

The Ca^{2+}-dependent restoration of both O_2-evolution and Cl^- binding in NaCl-washed PS II particles strongly suggests that Ca^{2+} is required for Cl^- binding, either as a source of positive charge or as a stabilizer of protein conformation. Furthermore, since both the 18 and 24 kD polypeptides are absent, these results indicate that the 33 kD polypeptide or another intrinsic polypeptide is a functional location for Cl^- and Ca^{2+} binding.

Support for the 33 kD hypothesis can be found by examining the amino-acid sequence of the spinach 33kD polypeptide (10). In addition to potential Mn binding sites (10), this protein contains four regions that are rich in basic amino acids capable of binding Cl^-: 1) residues 1-20 (1 α-amino, 3 Lys, 1 Arg); 2) residues 41-80 (8 Lys, 1 Arg); 3) residues 101-161 (5 Lys, 3 Arg); and 4) residues 178-236 (7 Lys, 1 Arg).

There are also at least two regions which might function as Ca^{2+} binding sites (Fig. 2; see also ref. 11): residues 81-116 and 177-191. These potential Ca^{2+} binding sites overlap two of the Cl^- binding regions proposed above. The number of available Ca^{2+} ligands may not be optimal for tight Ca^{2+} binding, however, and this may explain why the 18 and 24 kD polypeptides appear to augment the ability of the depleted membrane to retain Ca^{2+} and Cl^- (see ref. 12 for a review). Furthermore, other evidence indicates that Ca^{2+} stimulates a low level of O_2-evolution at very high $[Cl^-]$ in the absence of the 33, 24, and 18 kD proteins (13). This suggests that intrinsic proteins such as D_2 may also be involved in Ca^{2+} and Cl^- binding.

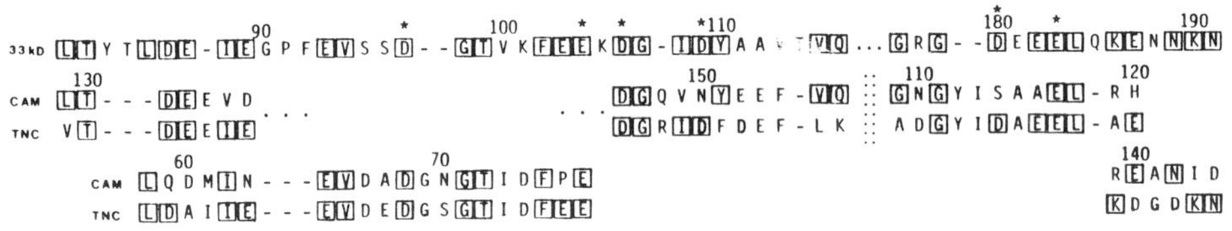

Figure 2. A Comparison of the Amino Acid Sequences for Two Regions of the 33 kD Polypeptide and Similar Sequences from Two Calcium Binding Proteins. The top line shows part of the sequence of the spinach 33 kD polypeptide ("33 kD"), as numbered in ref. 10 . The pairs of sequences arranged underneath it are those of mammalian calmodulin ("CAM") and troponin C from rabbit fast striated muscle ("TNC"). The numbering system for CAM and TNC is taken from ref. 11. Boxes over certain residues indicate homologies. Asterisks indicate potential Ca^{2+} ligands in the 33 kD protein, by analogy with similar binding sites in the two Ca^{2+}-binding proteins.

We have used information from our Cl-NMR binding studies, along with the published amino acid sequence, in order to show schematically how Cl⁻ and Ca²⁺ might interact with the 33 kD polypeptide (Fig. 3). As suggested by the close proximity in which we have placed the Cl⁻ and

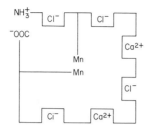

Figure 3. A Hypothetical Scheme for the Relative Positions of Cl⁻, Mn, and Ca²⁺ Binding Sites on the 33 kD Polypeptide from Spinach PS II. Similar Scheme may apply if any of the binding sites are on other intrinsic polypeptides (D_1 or D_2).

Ca^{2+} binding sites, we speculate that the binding of Ca^{2+} is necessary to increase substantially the affinity of the protein for Cl^-. The binding of Cl^- to specific sites in turn activates the water-splitting reactions by accelerating the extraction of protons from water, as described earlier (14).

REFERENCES

1. Kelley, P.M. and Izawa, S. (1978) Biochim. Biophys. Acta 502,198-210.
2. Homann, P.H. (1985) Biochim. Biophys. Acta 809, 311-319.
3. Critchley, C., Baianu, I.C., Govindjee and Gutowsky, H.S. (1982) Biochim. Biophys. Acta 682, 436-445.
4. Baianu, I.C., Critchley, C., Gutowsky, H.S. and Govindjee (1984) Proc. Natl. Acad. Sci. U.S.A. 81, 3713-3717.
5. Berthold, D.A., Babcock, G.T. and Yocum, C.F. (1981) FEBS Lett. 134, 231-236.
6. Murata, N., Miyao, M. and Kuwabara, T. (1983) in The Oxygen Evolving System of Photosynthesis (Inoue, Y.et al., eds.), pp.213-222, Academic Press, Tokyo.
7. Oldfield, E. and Meadows, M. (1978) J. Magn. Res. 31, 327-332.
8. Chiancone, E., Norne, J.E., Forsen, S., Antoniai, E. and Wyman, J. (1972) J. Mol. Biol. 70, 675-688.
9. Coleman, W.J. and Govindjee, in Biomembranes: Structure, Biogenesis, and Transport (Rajamanickam, C. and Packer, L., eds.), Today and Tomorrow Printers and Publishers, New Delhi, in press.
10. Oh-oka, H., Tanaka, S., Wada, K., Kuwabara, T. and Murata, N. (1986) FEBS Lett. 197, 63-66.
11. Grand, R.J.A. (1985) in Metalloproteins, Part 2: Metal Proteins with Non-Redox Roles (Harrison, P.M., ed.), Macmillan, London.
12. Ghanotakis, D.F. and Yocum, C.F. (1985) Photosynth. Res. 7, 97-114.
13. Kuwabara, T., Miyao, M., Murata, T. and Murata, N. (1985) Biochim. Biophys. Acta 806, 283–289.
14. Coleman,W. and Govindjee (1985) in Proc. of the 16th FEBS Congr., Moscow, pp. 21-28, VNU Science Press, BV, Utrecht.

INHIBITION AT THE CA^{2+} SENSITIVE SITE OF THE OXYGEN EVOLVING CENTER BY RUTHENIUM RED

Sylvie Lemieux and Robert Carpentier. Centre de recherche en photobiophysique, Université du Québec à Trois-Rivières, C.P. 500, Trois-Rivières, Québec, Canada, G9A 5H7.

1. INTRODUCTION

The requierement for Cl$^-$ and Ca^{2+} as cofactors for oxygen evolution is well documented. Althought there is no specific indication that the three extrinsic polypeptides (16, 24 and 33 kDa) associated with water splitting could bear a binding site for Cl$^-$ and/or Ca^{2+}, these proteins are believed to afford the high affinity of the oxygen-evolving complex for these ions (1).

The inhibition of oxygen evolution following addition of Cl$^-$ or Ca^{2+} channel blockers to photosystem II submembrane fractions suggested the presence of a protein with binding site(s) for Cl$^-$ and Ca^{2+} (2). Ruthenium red is used as an electron acceptor for photosystem I (3) but also as an inhibitor of Ca^{2+} transport across chloroplastic and mitrochondrial membranes (4, 5). We report here the inactivation of the water splitting complex by ruthenium red at the Ca^{2+} sensitive site and evidences that neither of the 16, 24 and 33 kDa proteins bears the inhibition site in a photosystem II submembrane fraction.

2. MATERIAL AND METHODS

Isolation of photosystem II submembrane fractions from stroma-free spinach thylakoids and alcali-salt treatment were performed as described in [6]. DCIP photoreduction was monitored at 580 nm with a Perkin-Elmer model 553 UV-VIS spectrophotometer in 20 mM Mes-NaOH pH 6.5 and 6.5 µg Chl ml^{-1} of photosystem II preparation. Oxygen evolution was monitored as in [7] at 22°C in 20 mM Mes-NaOH pH 6.5.

3. RESULTS AND DISCUSSION

Fig. 1 illustrates the inhibition of DCIP photoreduction and oxygen evolution by a photosystem II submembrane fraction following addition of ruthenium red. Complete inhibition was obtained at a ruthenium red concentration of 1.3 X 10^{-5} M. The inhibition site is located before or near the electron donation site of DPC since this donor could reactivate 25% of the DCIP photoreduction after full inactivation (Table 1). Higher percentages of reactivation were obtained at lower ruthenium red concentrations (not shown). However, H$_2$O$_2$ was inefficient in that respect.

Table 1. Restoration of DCIP photoreduction after inhibition by 10 µM RR.

Addition	Concentration	Percent of inhibition
None	---	94
H$_2$O$_2$	0.03 %	100
DPC	0.5 mM	73

Biggens, J. (ed.), Progress in Photosynthesis Research, Vol. 1. ISBN 90 247 3450 9
© *1987 Martinus Nijhoff Publishers, Dordrecht. Printed in the Netherlands.*

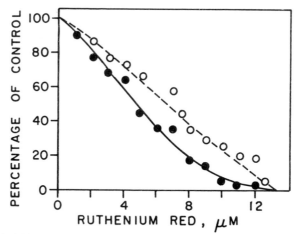

Figure 1. Inhibition of DCIP photoreduction ( ) and oxygen evolution ( ) following addition of ruthenium red to photosystem II submembrane fractions. Initial rates are 355 moles DCIP (mg Chl h)$^{-1}$ and 220 μmoles O_2 (mg Chl h)$^{-1}$.

The inhibition was also released by addition of different Cl$^-$ salts and by $Ca(NO_3)_2$ but not significantly by $NaNO_3$ or $Mg(NO_3)_2$ (Table 2). Complete reactivation can be obtained with 5 mN $CaCl_2$ after inhibition with 10 μM ruthenium red (not shown). Since ruthenium red is an inhibitor of Ca^{2+} membrane transport (4, 5), it is likely that this compound binds to a polypeptide with site(s) for Ca^{2+}. The specificity of Ca^{2+} and Cl$^-$ to compete with the inhibitor indicated that it interacts at the Ca^{2+} and/or Cl$^-$ site(s) in the oxygen evolving complex. Such a conclusion was also reached while studying Ca^{2+} and Cl$^-$ channel blockers (2).

Table 2. Restoration of oxygen evolution by the addition of salts after inhibition with 10 μM ruthenium red.

	None	$CaCl_2$	$MgCl_2$	$MnCl_2$	$NaCl_2$	$Ca(NO_3)_2$	$NaNO_3$	$Mg(NO_3)_2$
Concentration (mM)	--	2.5	2.5	2.5	5.0	2.5	5.0	2.5
Percent inhibition	84	30	29	30	34	43	77	74

The fact that the inhibition can be overcome by addition of $Ca(NO_3)_2$ is in good agreement with the idea of a Ca^{2+} modulation of the Cl$^-$ stimulation of oxygen evolution [6]. However, the restoration of oxygen evolution by $Ca(NO_3)_2$ without addition of Cl$^-$ can be explained only if Cl$^-$ ions are not displaced from their sites by ruthenium red. This conclusion also imply the independence of the binding sites for Cl$^-$ and Ca^{2+}. Reactivation of oxygen evolution by addition of salts not containing Ca^{2+} (Table 2) is explained if Na$^+$, Mg^{2+} and Mn^{2+} can play, to some extent, the role of Ca^{2+}. Another explanation would be that the presence of Ca^{2+} is not necessary when excess Cl$^-$ is added but this possibility is in opposition with the recent conclusion of Ikeuchi and Inoue [8].

Table 3. Prevention of DCIP photoreduction inhibition by 10 µM ruthenium red using alcali-salt treated photosystem II fractions with 0.03% H_2O_2 as electron donor.

Preaddition	Concentration (mM)	Percent inhibition
None	--	100
DPC	0.5	75
MnCl	3.5	57
MnCl	10.0	38
MnCl	17.5	24

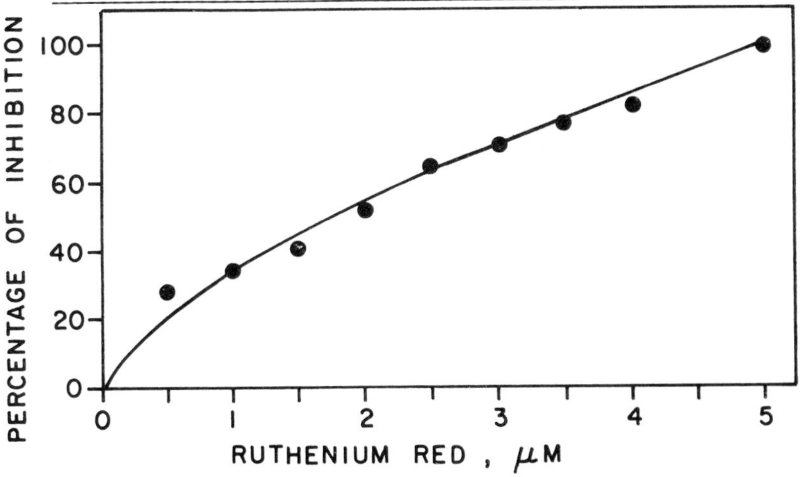

Figure 2. Inhibition of DCIP photoreduction in alcali-salt treated photosystem II fractions by ruthenium red. Initial rate was 55 µmoles O_2 (mg Chl h)$^{-1}$.

Ruthenium red was also causing inhibitory effect in the alcali-salt treated photosystem II submembrane fractions that are depleted from the polypeptides of 16, 24 and 32 kDa [6] (Fig. 2). Less than half the concentration needed for full inhibition in native photosystem II fractions was requested in treated samples. This could indicate that the inhibition site is more exposed after removal of the three extrinsic polypeptides. The inhibition in treated samples was overcome by 25% following addition of DPC as in native photosystem II fractions (Table 1) and it can be prevented by preaddition of $MnCl_2$ (Table 3) ($CaCl_2$ and $MgCl_2$ were inhibitory in these preparations). The $MnCl_2$ concentration needed for this effect (20 mM) was larger than the concentration used for electron donation to photosystem II by Mn^{2+} (0.1 - 0.3 mM) [9]. The initial activity with H_2O_2 as electron donor was enhanced by addition of 1 mM $MnCl_2$ (not shown), probably due to replacement of some lost Mn^{2+} at the oxygen evolving center, but a further increase in $MnCl_2$ concentration did not alter the rate of DCIP photoreduction without ruthenium red. However, in the presence of the inhibitor, the percent of inhibition was decreased from 57 to 24 when increasing the $MnCl_2$ concentration from 3.5 mM to 17.5 mM (Table 3). In respect to the above considerations, the inactivation of alcali-salt treated samples by ruthenium red thus behaves as in native photosystem II fractions if we except the fact that lower concentrations are needed.

I.5.**636**

The persistence of the inhibition after removal of the 16, 24 and 33 kDa polypeptides demonstrated that neither of these proteins could bear the bindind site for ruthenium red. We thus concluded that the Ca^{2+} and Cl^- binding sites are not located on these polypeptides. These sites are probably somewhere on the polypeptides (10, 34, 43 or 47 kDa) that are exposed after removal of the three extrinsic proteins (1).

REFERENCES

1. Ghanotakis, D.F. and Yocum, C.F. (1985) Photosynthesis Research 7, 917-114.

2. Carpentier, R. and Nakatani, H.Y. (1985) Biochim. Biophys. Acta 808, 288-292.

3. Barr, R., Crane, F.L. and Charke, M.J. (1982) Indiana Acad. Sci. Proc. 91, 114-119.

4. Kreiner, G., Melkonian, M. and Latzko, E. (1985) FEBS Lett. 180, 253-258.

5. Luthra, R. and Olson, M.S. (1977) FEBS Lett. 81, 142-146.

6. Nakatani, H.Y. (1984) Biochim. Biophys. Res. Comm. 120, 299-304.

7. Carpentier, R., Larue, B. and Leblanc, R.M. (1984) Arch. Biochm. Biophys. 228, 534-543.

8. Ikeuchi, M. and Inoue, Y. (1986) Arch. Biochem. Biophys. 247, 97-107.

9. Izawa, S. (1970) Biochim. Biophys. Acta 197, 328-331.

THERMOLUMINESCENCE STUDIES OF THE ABNORMAL S-STATES FORMED IN Cl^--DEPLETED OR 33 kDa EXTRINSIC PROTEIN-DEPLETED PSII

YORINAO INOUE

Solar Energy Research Group, The Institute of Physical and Chemical Research (RIKEN), Wako, Saitama 351-01, Japan

1. INTRODUCTION

Through recent development in biochemical manipulation of the oxygen evolving apparatus, it has been established that the overall process of water cleavage for molecular oxygen emission is catalyzed in a single multiprotein complex in which five hydrophobic membrane proteins and a single hydrophilic peripheral protein are associated with about 40 Chl $\underline{a}$, a few plastoquinone(s), 4 Mn and a few Fe and Ca atoms to form a functional integrity. One of the central problems under these circumstances is to understand the individual role of each particular component of the oxygen-evolving unit complex in executing the four step redox sequence. We have been approaching this problem by means, among others, of thermoluminescence measurements in combination with various depletion and repletion techniques. This communication describes briefly the general principle of thermoluminescence application to O_2 evolution studies and review the recent results obtained in our laboratory about the roles of the 33 kDa extrinsic protein and chloride anion in S-state turnovers.

2. PRINCIPLES OF THERMOLUMINESCENCE MEASUREMENTS

2.1 Assignment of charge pairs

Thermoluminescence is a burst of light emission during warming a frozen sample which had been excited by light during freezing or immediately before freezing. The energy for emission is supplied by the recombination between the positively and negatively charged pairs created by charge separation in photochemically reactive centers and stabilized by low temperature. Photosynthetic apparatus of higher plants usually show 6 or 7 glow peaks at different temperatures, most of which result from the photochemistry in PSII. Characterization studies of these glow peaks have made a notable progress in these years, and the charge pairs responsible for the emission of flash-excited glow peaks have been assigned (Table I). The negative charge carriers are either Q_B^- in the absence, or Q_A^- in the presence of DCMU, and the positive charge carriers are either S_2 (after 1 flash) or S_3 (after 2 flashes), hence there are four different types of charge pairs possible, and these recombination result in respective glow peaks having different peak emission temperatures [1,2].

Two types of information are available from the investigation of flash-excited thermoluminescence glow peaks: First, the intensity of a glow peak reflects the number of centers undergoing the charge recombination event, so that its oscillation under single turnover flashes provides us

Table 1 Assignment of TL bands

charge pair	TL-band	peak temp.
$S_2Q_B^-$	B2	$\approx 30°C$
$S_3Q_B^-$	B1	$\approx 25°C$
$S_2Q_A^-$	Q (D)	$\approx 10°C$
$S_3Q_A^-$	A	$\approx -15°C$

Biggens, J. (ed.), Progress in Photosynthesis Research, Vol. I. ISBN 90 247 3450 9
© *1987 Martinus Nijhoff Publishers, Dordrecht. Printed in the Netherlands.*

with information about the S state turnover. Second, the temperature for maximum emission reflects the amount of activation free energy required for recombination of the stabilized charge pairs, so
that it provides us with information
about the charge stabilization conditions: redox potential etc. In the following section some more details of the
two types of information are discussed.

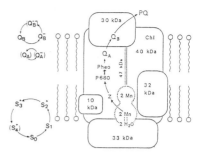

2.2 Mechanism of thermoluminescence
oscillation

Fig. 1. Redox cycles in donor
and acceptor sides of PSII.

Fig. 1 schematizes the electron
transport in the PSII multiprotein complex for water oxidation. When this
complex is illuminated with a series of
flashes at room temperature, the Mn catalyst on the donor side of the
reaction center undergoes a sequential S-state transition involving five S
states. Of these S-states, S_0, S_1, S_2 and S_3 are stable within the time
range of thermoluminescence measurement and can be stabilized by rapid
cooling after illumination, while S_4 cannot be stabilized because of its
short lifetime. Among the four stable S-states, S_2 and S_3 are associated
with one equivalent of positive charge, since the proton release occurs in
a [1,0,1,2] pattern during S_0 to S_4 transitions. On the acceptor side of
the reaction center, the negative charges are stabilized as electrons on Q_B
molecules, Q_B^-. Q_A^- is not stabilized because of its short lifetime unless
DCMU is present. The Q_B quinone operates as a two electron gate: the one
electron reduced form (Q_B^-) is stable, while the two electron reduced form
(Q_B^{2-}) is unstable. Thus Q_B and Q_B^- are the species can be stabilized
within the time range of thermoluminescence measurements.

When the center having an S_1Q_B redox pair in dark-adapted conditions
is illuminated with a series of flashes, the redox paris expected after
1,2,3 and 4 flashes are $S_2Q_B^-$, S_3Q_B, $S_0Q_B^-$ and S_1Q_B, respectively. Since S_2
and S_3 are the positively charged states, the probability of positive
charge stabilization on the donor side oscillates in a [1,1,0,0] pattern,
while the probability of negative charge stabilization on the acceptor side
oscillates in [1,0,1,0], a binary pattern, so that the probability of
stabilization of both positive and negative charges, the capability of
charge recombination for thermoluminescence oscillates in a [1,0,0,0] pattern.

In actual PSII preparations, this situation is perturbed by several
factors: (i) initial distribution of S_0:S_1, (ii) initial distribution of Q_B^-
:Q_B, (iii) misses and double hits probabilities and (iv) the ratio of
luminescence yield between $S_2Q_B^-$ and $S_3Q_B^-$ charge recombinations. Of these
four factors, the initial distribution of Q_B^-:Q_B varies much depending on
the sample and relaxation conditions, whereas other factors are rather
constant: S_0:S_1=25:75 in most cases, misses and double hits are around 10%
and 5%, respectively, and the luminescence yield of $S_2Q_B^-$:$S_3Q_B^-$=1:1.7,
whereas Q_B^-:Q_B is about 50:50 in well dark-relaxed intact chloroplasts and
algal cells or in broken thylakoids briefly illuminated with continuous
light followed by 10 min dark-relaxation, about 25:75 in well dark-relaxed
thylakoids or PSII particles. If we assume an even allocation of Q_B^- to the
centers in S_0 and S_1 states in dark-adapted condition, the transition of
the charge pairs in PSII preparation with different initial Q_B^-:Q_B distribution (but with the same initial S_0:S_1=25:75 distribution) can be described

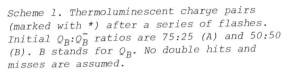
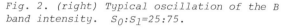

	(dark)	(1f)	(2f)	(3f)	(4f)	(5f)	(6f)	(7f)
	1SoB	1S1B⁻	1S2B	*1S3B⁻	1SoB	1S1B⁻	1S2B	*1S3B⁻
(A)	1SoB⁻ →	1S1B →	*1S2B⁻ →	1S3B →	1SoB⁻ →	1S1B →	*1S2B⁻ →	1S3B
	5S1B⁻	*5S2B⁻	5S3B	5SoB	5S1B⁻	*5S2B⁻	5S3B	5SoB
	1S1B⁻	1S2B	*1S3B⁻	1SoB	1S1B⁻	1S2B	*1S3B⁻	1SoB
	[0]	[5TL]	[2TL]	[1TL]	[0]	[5TL]	[2TL]	[1TL]
	1SoB	1S1B⁻	1S2B	*1S3B⁻	1SoB	1S1B⁻	1S2B	*1S3B⁻
(B)	1SoB⁻ →	1S1B →	*1S2B⁻ →	1S3B →	1SoB⁻ →	1S1B →	*1S2B⁻ →	1S3B
	3S1B⁻	*3S2B⁻	3S3B	3SoB	3S1B⁻	*3S2B⁻	3S3B	3SoB
	3S1B⁻	3S2B	*3S3B⁻	3SoB	3S1B⁻	3S2B	*3S3B⁻	3SoB
	[0]	[3TL]	[4TL]	[1TL]	[0]	[3TL]	[4TL]	[1TL]

Scheme 1. Thermoluminescent charge pairs (marked with *) after a series of flashes. Initial $Q_B:Q_B^-$ ratios are 75:25 (A) and 50:50 (B). B stands for Q_B. No double hits and misses are assumed.

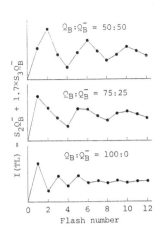

Fig. 2. (right) Typical oscillation of the B band intensity. $S_0:S_1=25:75$.

as in Scheme 1. The number of centers capable of recombination for thermoluminescence shows maxima after the 1st and 5th flashes if the initial $Q_B^-:Q_B=25:75$, or after the 2nd and 6th flash if the initial $Q_B^-:Q_B= 50:50$. By taking into account the other perturbation factors (misses, double hits and the ratio of luminescence yield), we can predict the oscillation pattern of thermoluminescence as shown in Fig. 2. These oscillation patterns agree well with observed patterns and support the above-described mechanism of oscillation of thermoluminescence intensity [2,3]. In Fig. 2, another type oscillation is shown. This binary pattern is observed when thylakoids are pretreated with ferricyanide or dark-adapted for extremely long period, and is explained by assuming the initial $Q_B^-:Q_B=0:100$.

2.3 Energetic parameters available from glow curves

The theoretical function for a glow peak of thermoluminescence can be described as follows according to Vass et al. [4]:

$$I(T) = cT \cdot \exp[-\Delta H^*/kT - SkT^3/B\Delta H^* \exp(-\Delta H^*/kT)]$$

I(T) is the luminescence intensity at temperature, T; C, proportionality constant; T, absolute temperature; ΔH^*, activation enthalpy (equivalent to activation energy, ΔE); k, Boltzmann's constant; S, frequency factor; B, heating rate.

By use of computer curve fitting technique, we can determine (or estimate the reasonable values) all these parameters: A glow curve recorded as a function of temperature provides us with I(T), T, B and k (constant) as input data. In actual experiments, the heating rate, B, is not always constant during measurement of a glow peak due to technical difficulties, so that B must be expressed as a function of T. Then, by varying the three parameters (c, ΔH^* and S), a set of values for these three parameters for the best fitting are determined. By using both measured and fitted parameters, we can calculate the activation free energy (ΔG^*) and half life ($t_{1/2}$) of the stored charge pairs:

$$\Delta G^* = \Delta H^* - kT\ln(hS/k) \qquad \text{(h, Planck's constant)}$$
$$t_{1/2} = \ln(2/ST) \cdot \exp(\Delta H^*/kT)$$

Of these two calculated parameters, the activation free energy, ΔG^*, provides us with important information: ΔG^* is the energy necessary to let the stored charge pairs recombine, and reflects the depth of stabilization, in turn the redox potentials of the charge pairs. Thus, if we find a change in ΔG^* value as a result of some treatment, we can conclude that the treatment affects the redox potential of the charge pair. However, it must be emphasized in this context that a change in ΔG^* value does not always provide us with information about the site of action by the treatment, whether it affects the redox potential of the positive charge or that of the negative charge, since the shift of redox potential on both acceptor and donor sides will result in a similar change in ΔG^*. As will be discussed later, we have to specify the site of action of the treatment by some other means in order to interpret the details of the effect induced by the treatment. A change in ΔG^* usually manifests, among others, as a shift in peak emission temperature, so that we can qualititatively follow the effect by the treatment simply from the shift in peak temperature of the glow curve. Another calculated parameter, $t_{1/2}$, is convenient in assessing the curve fitting: the coincidence between the predicted $t_{1/2}$ and the observed half decay time of the stored charge pair would verify a successful curve fitting. An example of such curve fitting and calculations is presented by Vass et al. in this book.

3. THERMOLUMINESCENCE FROM INACTIVATED O_2-EVOLVING CENTERS
3.1 Effect of Cl^- depletion

Cl^--depletion affects the thermoluminescence in two ways; upshift of the peak emission temperature and interruption of intensity oscillation [5,6]. As Fig. 3 shows, the glow curve from $S_2Q_B^-$ recombination in Cl^--depleted sample is shifted to higher temperature by about $10^\circ C$ accompanied by a marked sharpening of the band. Since we observe a similar upshift of the glow peak from $S_2Q_A^-$ recombination, we may assume that Cl^--depletion affects specifically the donor side but not the acceptor side, the Q_A and Q_B. Energetic parameters calculated for $S_2Q_B^-$ and $S_2Q_A^-$ recombination indicate that the redox potential of the S_2 state in the absence of Cl^- is lower by about 50 mV, indicative of deeper trapping of S_2, as compared with that in the presence of Cl^-. The calculation also predicts that the life time of $S_2Q_B^-$ charge pair in the absence of Cl^- is several times longer than that in the presence of Cl^-, and the predicted value agrees well with the observed life times. The long life-time and the lowered redox potential indicate that the S_2 state in the absence of Cl^- has some abnormal properties. (See details by Vass et al. in this book.)

One of the interesting properties of this abnormal S_2 state is its reversibility to form the normal S_2 state on readdition of Cl^-. As Fig. 3 shows, when Cl^- is added to the depleted sample immediately after 1 flash, a normal glow curve appears instead of the abnormal glow curve. Furthermore, when Cl^- is depleted by addition of SO_4^{2-} immediately after 1 flash illumination of Cl^--sufficient sample, an abnormal glow curve appears instead of the

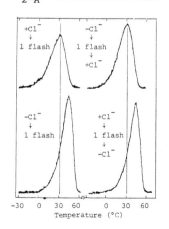

Fig. 3. Reversible formation of abnormal and normal S_2 states by Cl^--depletion and repletion (pH 7.5).

normal glow curve. This suggests that the normal and abnormal S_2 states are inter-convertible to each other depending on the activity of Cl^- anion [6]. By use of this interconvertibility, we can clearly see that Cl^- depletion affects the conformation of Mn atoms. When dark-adapted and then Cl^--depleted particles are illuminated with 1 flash, we do not see the low temperature multiline EPR signal which is associated with S_2 state. However, if Cl^- is added to the depleted sample immediately after the flash illumination, we can observe a forma-tion of the multiline EPR signal in dark-ness (Fig. 4) [7]. The lifetime of the multiline-less abnormal S_2 state can be measured by changing the time of Cl^- read-dition after the flash, which agrees well with the long lifetime of the abnormal S_2 state determined by thermoluminescence mea-surements. (See details by Ono et al. in this book.) The results indicate that the Cl^--depleted Mn catalyst is capable of storing one oxidizing equivalent even if it does not show the multiline EPR signal. Based on the view that the EPR signal origi-nates from the interaction between Mn(III) and Mn(IV) in binuclear or tetranuclear Mn cluster, we can inter-pret the above results as indicating that Cl^- depletion distorts the struc-ture of the Mn cluster to be incapable of this interaction and simultaneously loweres the redox potential of the S_2 state. It appears that the S state advancement as defined in terms of positive charge accumulation is possible to proceed at least by one digit without showing the multiline EPR signal.

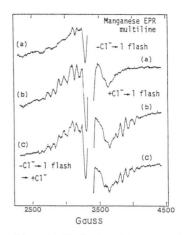

Fig. 4. *Dark development of the multiline EPR signal by Cl^--addition after flash.*

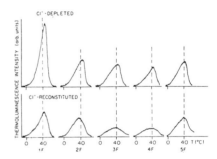

Fig. 5. *Interruption of TL-oscil-lation by Cl^--depletion (pH 6.0).*

Another effect of Cl^--depletion on the thermoluminescence is an in-hibition of the B band oscillation [5]. As shown in Fig. 5, the oscilla-tion of the B band is interrupted after the 2nd flash in Cl^--depleted particles. From this oscillation pattern, we once considered that the S state transition in the absence of Cl^- proceeds to S_3 but not beyond S_3. This interpretation, however, is inconsistent with the previous results obtained by fluorescence measurements that Cl^- depletion inhibits S_2 to S_3 transition [8,9]. We attacked this problem by curve fitting analysis. As shown in Fig. 6, the glow peak after 1 flash is composed mainly of the abnormal $S_2Q_B^-$ component peaking at $42^{\circ}C$, but this band is usually accompa-nied by an appreciable shoulder around $25^{\circ}C$, which is probably due to the centers irreversibly inactivated by Cl^--depletion. After the 2nd flash, the main $S_2Q_B^-$ band decreased to 1/3 both in Cl^--sufficient and Cl^--defi-cient samples, however the formation of $S_3Q_B^-$ component is clearly seen in Cl^--sufficient sample but not or very limited in Cl^--deficient sample. The above results show that S_2 to S_3 transition is inhibited in the absence of

Cl⁻, but do not clarify the reason why the abnormal $S_2Q_B^-$ charge pairs decreases on the 2nd flash. This question is partly answered by recent measurements of P680 rereduction. The P680 kinetics in the absence of Cl⁻ show a rapid rereduction after the 1st and 2nd flashes, but the rereduction after the 3rd flash is much slower [10]. This suggests that illumination by the 2nd flash of abnormal $S_2Q_B^-$ pair in the absence of Cl⁻ results in $S_2Z^+P680Q_B$ state (via $S_2Z^+P680Q_B^{2-}$), so that P680⁺ created by the 3rd flash recombines with Q_A^- to show a slower rereduction kinetics.

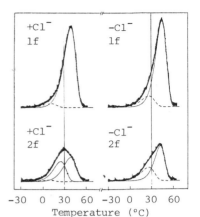

-30 0 30 60 -30 0 30 60
Temperature (°C)

Fig. 6. TL-component analysis in Cl⁻-depleted PSII.

3.2 Effect of removal of the 33kDa extrinsic protein

CaCl₂ (≥1 M)-wash of O₂-evolving PSII particles results in removal of all the three extrinsic proteins leaving most of the Mn atoms retained in the particles[11]. The washed particles do not evolve O₂ at all, unless the Cl⁻ concentration is raised over 150 mM, but are capable of exhibiting a normal thermoluminescence B band at a low concentration (10 mM) of Cl⁻.

Fig. 7 shows the effect of CaCl₂-wash on the B band oscillation. The unwashed particles exhibit an oscillation pattern having maxima at the 1st and 5th flashes, which can be simulated by assuming the following conditions: $S_0:S_1=Q_B^-:Q_B=25:75$; misses, 8%; double hits, 5%; luminescence yield ratio $S_2Q_B^-:S_3Q_B$ =1:2. The washed particles exhibit a very similar oscillation during the first 2 flashes, but no oscillatory behavior afterwards. This is interpreted as indicating that the Mn catalyst in the CaCl₂-washed particles is capable of undergoing the S-state transition from S_1 to S_3 via S_2, but cannot go beyond the S_3 state, and thereby

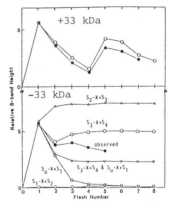

Fig. 7. Interruption of TL-oscillation by removal of 33 kDa extrinsic protein. (●) observed; other symbols, simulated.

incapable of O₂ evolution. Based on these, we proposed that the 33kDa protein is required for the Mn catalyst to undergo S_3 to S_0 transition [11].

There are two arguments against this view: (i) Removal of the 33kDa protein results in loss of the EPR multiline signal which is specifically associated with the S_2 state, hence neither S_2 nor S_3 can be formed in the absence of this protein [12,13]. (ii) The direct cause for the inhibition of O₂ evolution by CaCl₂-wash may not be the removal of the protein but will be the Cl⁻ deficiency due to an enhanced demand for Cl⁻ in the absence of the protein, since the O₂ evolution inactivated in the washed particles can be partially restored when the Cl⁻ concentration of the medium is elevated over 150 mM [13]. However, recent thermoluminescence studies have been revealing that both arguments may not be the case.

As shown in Fig. 8, the glow curves of CaCl₂-washed particles are simulated by a single component after the 1st flash but are deconvoluted into two components after the 2nd flash. The deconvoluted components and their relative intensities after the 2nd flash are more or less the same as those of untreated particles, indicating that $S_3Q_B^-$ component (low emission temperature) can be formed in the absence of the protein as well as in its presence. Similar deconvolution after the 3rd flash reveals that the relative intensity of the two components do not change any more in the washed particles, while $S_2Q_B^-$ further decreases and $S_3Q_B^-$ component increases in untreated particles. It is thus evident that both S_1 to S_2 and S_2 to S_3 transitions can take place in the absence of the 33 kDa extrinsic protein.

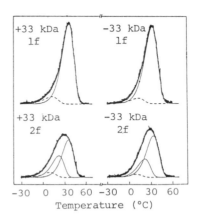

Fig. 8. TL-component analysis in 33 kDa protein-depleted PSII.

Based on the above deconvolution data and taking into account the information obtained from Cl^--depleted samples described in section 3.1, the arguments are reconciled with thermoluminescence observations: we consider that the Mn catalyst in the absence of the 33 kDa protein can accumulate one oxidizing equivalent in a form which is capable of charge recombination for thermoluminescence but incapable of multiline EPR signal. This situation is analogous to what found for Cl^--depleted samples. We also consider that the action site of the removal of the 33 kDa protein is different from that of Cl^--depletion: The former inhibits S_3 to S_0 transition while the latter inhibits S_2 to S_3 transition.

In view of the fact that the S_2 state in the absence of the 33 kDa protein does not show the multiline EPR signal, we have to consider this S_2 state is somewhat modified. However, as shown in Fig. 8, the $S_2Q_B^-$ and $S_3Q_B^-$ glow curves in the washed particles are very similar with respect to their shape to those of normal particles, indicating that the energetic parameters of the stabilized charges are not much affected by removal of the 33 kDa protein. It is worth while to note in this context that the $S_2Q_A^-$ glow curve is found to be affected by removal of the 33 kDa protein (See details by Vass et al. in this book.) : its peak temperature is shifted to higher temperature by about $20^{\circ}C$, and the shift is reversed by reconstitution with the 33 kDa protein. The shift of $S_2Q_A^-$ glow peak contrasts sharply with $S_2Q_B^-$ and $S_3Q_B^-$ glow curves, and suggests, at first glance, that removal of the protein affects the stabilization condition of Q_A^-. However, if we take into account the fact that the energetic parameters themselves do not specify the site of action of a treatment as we discussed in section 2.3, the phenomenon may be interpreted by following alternatives: (i) Removal of the protein affects the stabilization conditions of Q_A^- but not those of both Q_B^- and S_2. (ii) Removal of the protein affects the stabilization conditions of all of these components, but the effects on Q_B^- and S_2 compensate with each other to cancel fortuitously the changes in energetic parameters, whereas those on Q_A^- and S_2 are not compensated but strengthen with each other to enhance the changes in energetic parameters.

Table II compares the features of the two inhibitions of the S-state turnover. As we discussed in section 3.1, in Cl^--depleted particles, the

Table 2 Features of the inhibitions by Cl^--depletion and 33 kDa protein-depletion

	Cl^--depletion	33 kDa-depletion
Inhibition site	$S_2 \rightarrow S_3$	$S_3 \rightarrow S_0$
$S_2Q_B^-$	abnormal ($\approx 10°C$ upshift)	nearly normal ($\approx 3°C$ downshift)
$S_2Q_A^-$	abnormal ($\approx 10°C$ upshift)	abnormal ($\approx 20°C$ upshift)
EPR multiline	no S_2 signal	no S_2 signal
$S_3Q_B^-$	not formed	nearly normal ($\approx 3°C$ downshift)
$Em[S_2/S_1]$	lowered by 50 mV (due to abnormal S_2)	nearly no change (due to compensation ?)
Z	stable Z^+ is formed on illumination of abnormal $S_2Q_B^-$ state	$Z \rightarrow P680$ appears to be inactivated after formation of $S_3Q_B^-$ state

abnormal $S_2Q_B^-$ charge pair loses one reducing equivalent on the 2nd flash to yield $S_2Z^+Q_B^-$, a non luminescent pair, and the thermoluminescence intensity is markedly reduced. This behavior of Z is analogous to that reported for PSII reaction center particles, although the effect manifests one flash later in Cl^- deficiency. In $CaCl_2$-washed particles, however, such loss of negative charge by use of electrons on Z does not seem to occur, and thermoluminescence intensity keeps the amplitude attained after the second flash. A possible explanation is that in the absence of the protein especially after creation of the S_3 state, the properties of the reaction center, and possibly Z, are modified to cause rapid recombination between Q_A^- and $P680^+$.

REFERENCES
 1 Rutherford, A.W., Crofts, A.R. and Inoue, Y. (1982) Biochim. Biophys. Acta, 682, 457-465
 2 Rutherford, A.W., Renger, G., Koike, H. and Inoue, Y. (1985) Biochim. Biophys. Acta, 806, 25-34
 3 Inoue, Y. (1983) in "The Oxygen Evolving System of Photosynthesis" (Inoue, Y. et al. eds.) pp.103-112, Academic Press Japan, Tokyo
 4 Vass, I., Horvath, G., Herczeg, T. and Demeter, S. (1981) Biochim. Biophys. Acta, 634, 140-152
 5 Homann, P.H. and Inoue, Y. (1985) in "Proc. Intl. Workshop on Ion Interactions in Energy Transfer System" (Papageorgiou, G.C. et al. eds.) pp.279-290, Plenum Publishing Co., London
 6 Homann, P.H., Gleiter, G., Ono, T. and inoue, Y. (1986) Biochim. Biophys. Acta, 850, 10-20
 7 Ono, T., Zimmermann, J.L., Inoue, Y. and Rutherford, A.W. (1986) Biochim. Biophys. Acta, in press
 8 Itoh, S., Yerkes, C.T.., Koike, H., Robinson, H.H. and Crofts, A.R. (1984) Biochim. Biophys. Acta 766, 612-622
 9 Theg, S.M., Jursinic, P. and Homann, P.H. (1984) Biochim. Biophys. Acta, 766, 636-646
10 Ono, T., Conjeaud, H., Gleiter, H., Inoue, Y. and Mathis, P. (1986) FEBS Lett. in press
11 Ono, T. and Inoue, Y. (1985) Biochim. Biophys. Acta, 806, 331-340
12 Blough, N.V. and Sauer, K. (1984) Biochim. Biophys. Acta, 767, 377-381
13 Imaoka, A., Akabori, K., Yanagi, M., Izumi, K., Toyoshima, Y., Kawamori, A., Nakayama, H. and Sato, J. (1986) Biochim. Biophys. Acta, 848, 201

ACKNOWLEDGEMENTS: This paper is a summary of the work done in our laboratory by Drs. T. Ono, I. Vass in collaboration with Drs. P.H.Homann, P. Mathis, A.W. Rutherford and Mr. H. Gleiter. I thank to all these people.

TEMPERATURE DEPENDENCE OF THE S-STATE TRANSITION IN A THERMOPHILIC CYANO-
BACTERIUM MEASURED BY THERMOLUMINESCENCE

HIROYUKI KOIKE and YORINAO INOUE

Solar Energy Research Group, The Institute of Physical and Chemical
Research (RIKEN), Wako, Saitama 351-01, Japan

1. INTRODUCTION

Thermophilic cyanobacteria are unique photosynthetic organism. They
grow at temperatures as high as $60^{\circ}C$ and show maximal photosynthetic activities at high temperatures which cause heat inactivation of O_2 evolution
in mesophilic plants or algae [1,2]. The thermophilic properties including
the higher limiting temperature and heat stabilities of cell growth, O_2
evolution and electron transport reactions of this type of algae have been
extensively investigated [2,3,4]. However, very limited amount of information is so far available about the lower limiting temperature at which
the electron transport or O_2 evolution in these algae is blocked. It is
not established whether the functional temperature range is simply shifted
to higher temperature in these algae or the upper limiting temperature is
extended to higher temperatures without any changes in the lower limiting
temperature. In the present work, the temperature dependence of the S-
state transition in a thermophilic cyanobacterium, Synechococcus vulcanus
Copeland was studied by thermoluminescence in comparison with those of
spinach and a mesophilic cyanobacterium, Anabaena variabilis M3.

2. MATERIALS AND METHODS

Thylakoids of S. vulcanus and A. variabilis were prepared by an osmot-
ic shock as described in [4]. Spinach thylakoids were prepared as described in [5]. The samples were suspended in 25% (v/v) glycerol, 10 mM
$MgCl_2$, and 50 mM MES-NaOH (pH 6.5) at a chlorophyll concentration of 0.25
mg Chl/ml, and preilluminated with orange light (0.7 mW/cm^2) for 45 sec,
then kept in the dark for 5 min at $25^{\circ}C$ in the case of spinach and A.
variabilis thylakoids and 45 min in the case of S. vulcanus thylakoids.
Thermoluminescence measurements were done as in [5].

3. RESULTS AND DISCUSSION

Thermoluminescence B band originates from the recombination of $S_2Q_B^-$
and $S_3Q_B^-$ charge pairs [6]. In spinach thylakoids the B band is usually
emitted at $25-35^{\circ}C$, whereas in thermophilic cyanobacterial thylakoids the
band is emitted at $50-60^{\circ}C$, which is higher by $20-25^{\circ}C$ than the emission
temperature in mesophilic thylakoids [7]. The height of the B band depends
on both concentrations of S_2 (or S_3) and Q_B^-, so that its oscillation
pattern after single turnover flashes varies depending on the initial
conditions of both $S_0:S_1$ and $Q_B:Q_B^-$ [8]. However, if we control the $Q_B:Q_B^-$
ratio in the initial dark-adapted thylakoids to be 50:50, the B band shows
a period 4 oscillation having maxima at the 2nd and 6th flashes, which is
preferentially dependent on the S-state turnovers. By investigating such
oscillation patterns measured at various temperatures, we can estimate the
temperature dependence of the S-state turnover.

Fig. 1 depicts the oscillation pattern of the B band in thermophilic
cyanobacterial (S. vulcanus) thylakoids (A) in comparison with that in

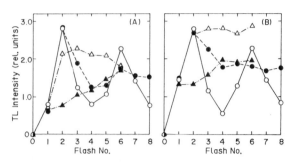

Fig. 1. Oscillation of thermoluminescence B band in S. vulcanus (A) and spinach (B) thylakoids after excitation with single turnover flashes at various low temperatures. A: S. vulcanus thylakoids were excited at 35°C (o), 0°C (•), -10°C (△) and -25°C (▲). B: Spinach thylakoids were excited at 10°C (o), -10°C (•), -20°C (△) and -45°C (▲).

spinach thylakoids (B). As shown by open circles, a normal period 4 oscillation is observed at T_{ex}=35°C (T_{ex}: excitation temperature) for S. vulcanus thylakoids and at T_{ex}=10°C for spinach thylakoids. On lowering the excitation temperature, a slight distortion of the oscillation pattern first takes place after the 3rd and 4th flashes, at around T_{ex}=0°C for S. vulcanus thylakoids and at T_{ex}=-10°C for spinach thylakoids (solid circles). When the excitation temperature is further lowered, the distortion is much enhanced to show almost flat or gradually decreasing pattern after the 2nd flash, indicative of complete blocking of photochemistry by the 3rd or more flashes (open triangles). Such complete blocking manifests at T_{ex}=-10°C for S. vulcanus thylakoids and at T_{ex}=-20°C for spinach thylakoids. At these temperatures, however, the B-band heights after the 1st and 2nd flashes are not much affected. When the excitation temperature is still further lowered, normal charging of the B band at a high efficiency can be observed only after the 1st flash, but the charging efficiency by the 2nd flash is seriously decreased (solid triangles). The critical temperature for this effect is at T_{ex}=-25°C for S. vulcanus thylakoids and at T_{ex}=-45°C for spinach thylakoids.

The above phenomena can be interpreted as reflecting the differences in temperature dependence of the S-state transitions between thermophilic cyanobacterial thylakoids and spinach thylakoids. In fact, the results for spinach thylakoids are in good agreement with the early report by Inoue and Shibata [9], and they roughly estimated the temperature dependence of each step of the S-state transitions based on the low temperature induced distortions in the oscillation pattern. However, this method is not always sufficient in view of the fact that there is a threshold temperature for the Q_A to Q_B electron transport at around -30°C. The 1st flash below -30°C will convert S_1 centers to $S_2Q_A^-$, but the 2nd flash will not effect any S-state conversion because of the presence of Q_A^-, so that it will result in an apparent inhibition of the S-state transition. Moreover, there is no available information at present about the threshold temperature for Q_A to Q_B electron transfer in thermophilic thylakoids. In order to avoid these artifacts due to the reduced equivalent on Q_A, we performed the present experiments according to the following protocol: The sample thylakoids were preilluminated with (n-1) flashes at high temperatures (35°C for S. vulcanus and 10°C for spinach thylakoids) which allow the normal S-state transitions and Q_A to Q_B electron transfer, then cooled to various low temperatures where the final (n)th flash was given, and the B-band height after the (n)th flash at low temperatures was compared with that after the same flash given at high temperatures. By this protocol, we could determine the temperature dependence of individual S-state transitions free from

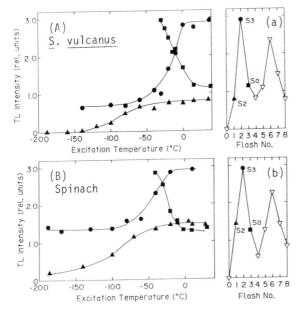

Fig. 2 Temperature dependences in S. vulcanus (A) and spinach (B) thylakoids of individual transitions in the S-state turnover. Thylakoids of S. vulcanus and spinach were excited by a single flash at indicated low temperatures after preillumination at 35°C and 10°C, respectively, with zero ($\blacktriangle$, $S_1 \rightarrow S_2$ transition), one ($\bullet$, $S_2 \rightarrow S_3$ transition) and two ($\blacksquare$, $S_3 \rightarrow S_0$ transition) flashes. (a) and (b) indicate the oscillation patterns excited at 35°C and 10°C for S. vulcanus and spinach, respectively.

the above mentioned artifacts.

Fig. 2A summarizes the results of such experiments done with S. vulcanus thylakoids. When a single flash was given at various low temperatures, the B-band height was not much affected above -40°C, then decreased gradually with lowering the temperature and almost totally abolished at -150°C (solid triangles). This indicates that S_1 to S_2 transition proceeds at a normal high quantum yield above -40°C, gradually inhibited below -50°C to be completely inhibited at -150°C. The half inhibition temperature was -90°C. Similar experiments were done with the sample preilluminated at 35°C with one flash. In this case the increase in B-band height induced by the 2nd flash at low temperatures corresponds to the quantum yield of S_2 to S_3 transition, since the recombination of $S_3Q_B^-$ lumineses more strongly than $S_2Q_B^-$ by a factor of 1.7 to 2.0 [ref.5]. As shown by solid circles, the conversion is sharply decreased below 0°C, and the 2nd flash illumination below -60°C practically did not change the B-band height which had been charged by one flash preillumination at 35°C. From these, the half inhibition temperature for S_2 to S_3 transition was determined to be -16°C.

The temperature dependence of S_3 to S_0 transition can be measured for 2 flash preilluminated samples. In this case the quantum yield of the transition can be indicated by the 3rd flash induced decrease in the B-band height which had been charged as $S_3Q_B^-$ pair by the preceeding two flashes. As shown by solid squares, the decrease in B-band height was gradually inhibited below 0°C and completely at -30°C showing a half inhibition temperature at -10°C.

Fig. 2B depicts the results of the same experiments done with spinach thylakoids. The half inhibition temperatures for S_1 to S_2, S_2 to S_3 and S_3 to S_0 transitions were determined to be -95, -45 and -23°C, respectively. These temperature dependencies are in agreement with those reported by Demeter et al. [10].

Table I summarizes these results together with those obtained with a mesophilic cyanobacterium, A. variabilis. In general, the transition from

TABLE 1. Half inhibition temperature for S-state transition in <u>S. vulcanus</u>, <u>A. variabilis</u> and spinach.

	Half inhibition temperature ($^{\circ}C$)		
	S. vulcanus	A. variabilis	Spinach
$S_1 \rightarrow S_2$	-90	-98	-95
$S_2 \rightarrow S_3$	-16	-30	-45
$S_3 \rightarrow S_0$	-10	-16	-23

a higher S state is more sensitive to low temperatures: There is no significant difference in the temperature dependence for S_1 to S_2 transition among the three species, whereas the S_2 to S_3 or S_3 to S_0 transitions in <u>S. vulcanus</u>, a thermophilic cyanobacterium, are more sensitive to low temperature, and their half inhibition temperatures are shifted by 29 and $13^{\circ}C$ toward high temperatures, respectively, as compared with those of spinach thylakoids. The temperature dependencies of these transitions in the mesophilic cyanobacterium are of intermediary characteristics. It may be of note that the S_1 to S_2 transition, among the three transitions, showes no difference between the three species. This might be related to the fact that S_2 to S_3 and S_3 to S_0 transitions involve the release of proton and oxygen plus proton, respectively, while S_1 to S_2 transition does none of them, i.e. the temperature dependence originates not in the electron abstraction but in the release process(es) of proton or oxygen. Apart from such speculations, it is interesting that the lower limiting temperature of S_2 to S_3 and S_3 to S_0 transitions in <u>S. vulcanus</u>, a thermophilic cyanobacterium, is shifted toward higher temperatures as well as the higher limiting temperatures, although there are no strict parallelism between the two limiting temperatures.

REFERENCES

1. Castenholz, R.W. (1973) in <u>Biology</u> <u>of</u> <u>Blue-green</u> <u>Algae</u> (N.G. Carr and B.A. Whitton eds.), pp.379-414, Blackwell Scientific Publications, Oxford.
2. Yamaoka,T., Satoh, K. and Katoh, S. (1978) Plant Cell Physiol. 19, 943-954.
3. Hirano, M., Satoh, K. and Katoh, S. (1981) Biochim. Biophys. Acta 635, 476-487.
4. Koike, H. and Inoue, Y. (1983) in <u>The</u> <u>Oxygen</u> <u>Evolving</u> <u>System</u> <u>of</u> <u>Photosynthesis</u> (Inoue et al. eds.) pp. 257-263, Academic Press Japan, Tokyo.
5. Rutherford, A.W., Renger, G., Koike, H. and Inoue, Y. (1984) Biochim. Biophys. Acta 767, 548-556.
6. Rutherford, A.W., Crofts, A.R. and Inoue, Y. (1982) Biochim. Biophys. Acta, 682, 457-465.
7. Govindjee, Koike, H. and Inoue, Y. (1985) Photochem. Photobiol. 42, 579-585.
8. Inoue, Y. (1983) in <u>The</u> <u>Oxygen</u> <u>Evolving</u> <u>System</u> <u>of</u> <u>Photosynthesis</u> (Inoue et al. eds.) pp.439-450, Academic Press Japan, Tokyo.
9. Inoue, Y. and Shibata, K. (1978) FEBS Lett. 85, 193-197.
10. Demeter, S., Rozsa, Zs., Vass, I. and Hideg, E. (1985) Biochim. Biophys. Acta 809. 379-387.

DEPLETION OF Cl^- OR 33 kDa EXTRINSIC PROTEIN MODIFIES THE STABILITY OF $S_2Q_A^-$ AND $S_2Q_B^-$ CHARGE SEPARATION STATES IN PS II

Imre Vass, Taka-aki Ono, Peter H. Homann[*], Hermann Gleiter and Yorinao Inoue
The Institute of Physical and Chemical Research (RIKEN), Wako, Saitama 351-01, Japan; [*]Institute of Molecular Biophysics, Florida State University, Tallahassee, Florida 32306, USA

1. INTRODUCTION

Removal of Cl^- [1] as well as 33 (and 17, 24) kDa extrinsic protein(s) [2,3] from the oxygen-evolving complex of PS II results in a reversible inhibition of O_2-evolution. The storage of positive charges in the water-oxidizing enzyme depleted of Cl^- [4] or 33 kDa protein [2] was previously shown by thermoluminescence measurements. Cl^--depletion has also been found to increase the stability of stored charges [1,4]. However, such stability increasing effect has not been reported yet for the removal of 33 kDa protein. Here we report that contrary to Cl^--depletion which increases the stability of both the $S_2Q_A^-$ and $S_2Q_B^-$ charge pairs, the removal of 33 kDa extrinsic protein preferentially increases the stability of $S_2Q_A^-$ leaving $S_2Q_B^-$ apparently uneffected or slightly destabilized.

2. MATERIALS AND METHODS

For all experiments dark adapted Triton PS II particles from spinach thylakoids were used . Cl^--depletion was attained either by alkaline shock in a Cl^--free medium or by SO_4^{2-} replacement [4]. The 33 (and 17,24) kDa protein(s) were released by washing with 1.2 M $CaCl_2$ [2] or by 2.7 M urea+0.2 M NaCl [3]. Thermoluminescence was measured as usual [4]. Digitized luminescence intensity and heating rate as functions of temperature were simultaneously obtained with a Nicolet Explorer IIIA digital oscilloscope and an HP 9825A microcomputer connected to the thermoluminescence apparatus. Deconvolution of measured glow curves and calculation of activation parameters was performed as in Ref. [5], with an HP 9826 microcomputer, taking into account the corrections needed due to the nonlinear heating of our apparatus.

3. RESULTS AND DISCUSSION

3.1. Glow curves of Cl^-- and 33 kDa protein-depleted PS II particles

Figure 1 shows the typical single flash-induced glow curves at pH 7.5. In dark adapted control samples (curve a) the main component is the so called B-band (solid line) at around 30 $^\circ$C which originates from the recombination of $S_2Q_B^-$ charge pair [6]. The curve resolution shows a small component as well at around 10 $^\circ$C which is probably due to the $S_2Q_A^-$ recombination in those centers in which Q_A and Q_B are disconnected. When the electron transfer between Q_A and Q_B is blocked by DCMU the main thermoluminescence component is the so called Q band at around 10 $^\circ$C (curve d) which originates from the $S_2Q_A^-$ recombination [6]. The small component at around 30 $^\circ$C may be due to the centers not affected by DCMU or having acceptors other than Q_A or Q_B. The blip on the thermoluminescence curves at around 0°C is the result of solid-liquid phase transition of the samples.

Single flash excitation of Cl^--depleted samples also results in the appearance of the B and Q bands (curves b and e) which, however, have higher intensity, higher peak temperature and narrower band shape than those found for the Cl^- sufficient samples. This suggests that Cl^--deple-

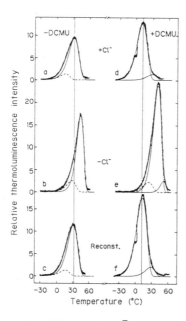 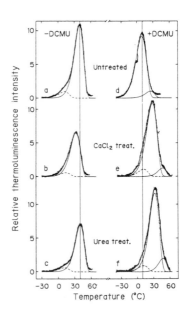

Figure 1. Effect of Cl⁻-depletion on single-flash induced thermoluminescence curves at pH 7.5. a(d), b(e) and c(f); control, Cl⁻ depleted and
reconstituted samples in the absence (or in the presence) of DCMU.

Figure 2. Effect of 33 kDa protein removal on single flash induced thermoluminescence curves at pH 5.5. a(d), b(e) and c(f); control, $CaCl_2$- and
urea treated samples in the absence (or in the presence) of DCMU.

tion induces some modification of $S_2Q_B^-$ and $S_2Q_A^-$ charge stabilization
states, as Homann et al. reported earlier [4]. The dashed lines in curves b
and e shows a small component with unexplained origin. Addition of 50 mM
NaCl to the Cl⁻-depleted samples completely reversed the changes in the
shape and peak position (curves c and f). The peak height, however, was not
decreased back to the control level. This is possibly due to the presence
of SO_4^{2-}. The cause for the Cl⁻ dependent reversible changes in the intensity is not known at present. It suggests formally that the absence of Cl⁻
(especially in DCMU poisoned samples) increases the net amount of stabilized charge pairs or the yield of recombination luminescence.

Figure 2 shows the typical single flash-induced glow curves measured
before and after the removal of the 33 (and 17, 24) kDa extrinsic pro-
tein(s). The suspension medium (pH 5.5) contained 20 mM NaCl in order to
suffice the enhanced Cl⁻ demand without reactivating the O_2-evolution.
Thermoluminescence characteristics are changed by the protein release: The
B band appearing at around 40 °C at pH 5.5 (curve a), is shifted slightly
to lower temperatures by $CaCl_2$-wash (curve b) and practically unaffected by
urea treatment (curve c). The intensity is decreased probably due a partial
Mn loss. As opposed to these slight changes in B-band properties the Q band
is markedly shifted from 10 °C (curve d) to 30 or 35 °C in $CaCl_2$- or urea
treated samples accompanied with a substantial increase in intensity
(curves e and f). These data indicate that differently from Cl⁻-depletion,
the removal of the three extrinsic proteins affects preferentially the $S_2Q_A^-$
charge stabilization state. This effect is specific for the absence of 33

kDa protein, because the release of 17 and 24 kDa proteins by NaCl-wash does not cause major changes in the characteristics of the Q band (data not shown), but not specific for the presence of DCMU, because a similar effect was observed with atrazine or ioxynil (not shown). It is also of note that contrary to the effect of Cl^--depletion, the changes in the shape and peak position of the B and Q bands induced by the removal of the 33 kDa protein are not reversed by the addition of 200 mM NaCl (data not shown) which can partially restore the O_2-evolving activity [3]. The slight differences in the thermoluminescence characteristics between $CaCl_2$- and urea-washed samples are probably due to different side effects of the two treatments.

3.2 Stability of $S_2Q_A^-$ and $S_2Q_B^-$ in the absence of Cl^- and 33 kDa protein

Based on the theoretical description of photosynthetic glow curves [5,7] the shape and peak position reflect the activation parameters of the charge recombination process: the activation enthalpy, ΔH^* (also called activation energy ΔE) and the preexponential factor which includes the activation entropy ΔS^*. The stability or lifetime of the charge separated state at a constant T temperature is best characterized by the activation free energy given by $\Delta G^* = \Delta H^* - T\Delta S^*$, which corresponds to the free energy loss during the charge stabilization process of forward electron transfer. The increase in activation free energy may indicate the same extent of decrease in free energy stored in the form of redox potential difference between the donor and acceptor components of PS II. (Free energy in meV corresponds redox potential difference in mV).

As Table I shows Cl^--depletion attained either by high pH shock (measured at pH 6.0) or by SO_4^{2-} replacement (measured at pH 7.5) results in the stability increase of both $S_2Q_B^-$ and $S_2Q_A^-$ charge pairs, ranging from 30 to 80 meV in terms of ΔG^*. The relatively small increase of ΔG^* is resulted from a much bigger and largely compensating increase of ΔH^* and ΔS^*, as a formal consequence of the narrow shape of Cl^--depleted thermoluminescence bands. Such activation entropy change, if not artefactual, would indicate a more rigid conformation of PS II in the absence of Cl^-. The data also reflect a more pronounced effect of SO_4^{2-} replacement than alkaline shock depletion. The ΔG^* increase obtained for B band suggest an at least 30-50 mV shift of $E_m(S_2/S_1)$ to more negative values in Cl^- depleted conditions. The higher increase of ΔG^* found in the presence of DCMU could be caused by the the shift of $E_m(Q_A/Q_A^-)$ to more positive values resulting in the extra stabilization of $S_2Q_A^-$ over $S_2Q_B^-$. It also cannot be excluded that a destabilizing shift of $E_m(Q_B/Q_B^-)$ to negative direction partly compensates the stabilization effect on $E_m(S_2/S_1)$, which in this case would be 50-80 mV. The reasonable agreement between the calculated and measured lifetimes at 25 °C supports the reliability of the obtained activation parameters as a measure of stability of charge separation.

The activation free energies as well as predicted (and measured) lifetimes confirm the different response of $S_2Q_A^-$ and $S_2Q_B^-$ charge pairs to the absence of 33 kDa protein: $S_2Q_B^-$ is slightly destabilized by $CaCl_2$-wash and practically unchanged by urea treatment but $S_2Q_A^-$ shows a 50-55 meV stability increase and about 10 times longer lifetime. Based on these data it can be concluded that the absence of 33 kDa protein results in the increased stability of the electron stored on Q_A^- e.g. by shifting $E_m(Q_A/Q_A^-)$ to a more positive value. An alternative explanation would be the stabilization of the positive charge on the S_2 state as revealed by the recombination with the unaffected Q_A^-, but hidden for the $S_2Q_B^-$ recombination because of a compensating destabilization of the electron on Q_B^-. Further studies are obviously needed to clarify this point together with the source of intensity increase in the presence of DCMU.

Our thermoluminescence data show that the removal of Cl^- and 33 kDa protein results in different modification or damage of PS II electron transfer processes. At 20 mM Cl^- concentration the absence of 33 kDa protein does not lead to Cl^--depletion in terms of thermoluminescence characteristics. This supports the earlier suggestion [2] that the role of high Cl^- concentration (200 mM) to replace some of the functions of 33 kDa protein in O_2-evolution is different from the role of Cl^- at the functional site influencing the redox- and catalytic properties of the Mn complex of the water oxidizing enzyme.

Charge pair		ΔH^* (meV)	$\Delta S^*/k$ (eV/K)	ΔG^* (meV)	ΔG^*_{stab} (meV)	$t_{1/2}$ calc. (s)	meas. (s)
Cl^- depletion (pH 6.0)							
$S_2Q_B^-$ (	$+Cl^-$	1048 ± 39	6.47 ± 1.7	882 ± 7		92 ± 17	
	$-Cl^-$	1247 ± 46	13.20 ± 1.9	908 ± 4	26 ± 4	250 ± 52	
$S_2Q_A^-$ (	$+Cl^-$	745 ± 15	-1.86 ± 0.3	793 ± 4		2.8 ± .5	
	$-Cl^-$	813 ± 42	-0.88 ± 1.1	836 ± 6	43 ± 5	16 ± 6	
Cl^- depletion (pH 7.5)							
$S_2Q_B^-$ (	$+Cl^-$	833 ± 11	-0.26 ± 0.5	840 ± 4		18 ± 5	(25)
	$-Cl^-$	1310 ± 63	16.19 ± 2.3	894 ± 6	54 ± 6	150 ± 33	(138)
$S_2Q_A^-$ (	$+Cl^-$	750 ± 11	-1.95 ± 0.4	800 ± 4		3.8 ± .4	(3)
	$-Cl^-$	1159 ± 58	10.93 ± 1.7	878 ± 5	78 ± 5	78 ± 15	(92)
33 kDa protein removal (pH 5.5)							
$S_2Q_B^-$ (	Non treat.	1014 ± 50	5.66 ± 1.6	869 ±10		56 ± 2	(63)
	$CaCl_2$ "	899 ± 12	1.70 ± 0.5	855 ± 7	-14 ± 8	32 ± 7	(43)
	Urea "	1042 ± 33	6.65 ± 1.2	871 ± 5	2 ± 7	59 ± 11	(60)
$S_2Q_A^-$ (	Non treat.	738 ± 16	-2.11 ± 0.6	792 ± 2		2.5 ± .3	(3)
	$CaCl_2$ "	841 ± 30	-0.05 ± 1.3	842 ± 6	50 ± 4	21 ± 4	(30)
	Urea "	901 ± 27	2.05 ± 1.0	848 ± 3	56 ± 3	25 ± 2	(30)

Table I. Effect of Cl^--depletion and 33 kDa protein removal on the energetics of $S_2Q_B^-$ and $S_2Q_A^-$ charge recombinations calculated from the curve fitting parameters of the B and Q thermoluminescence bands. The data represent mean values obtained from 3-5 independent measurements with the indicated standard deviations. The measured lifetimes, in parenthesis, are obtained from one series of decay measurements.

References
[1] Muallem, A., Farineau, J., Laine-Böszörmenyi, M. and Izawa, S. (1981) in Photosynthesis (Akoyunoglou G., ed.) Vol. II pp. 435-443, Balaban International Science Services, Philadelphia
[2] Ono, T. and Inoue, Y. (1985) Biochim. Biophys. Acta 806, 331-340
[3] Miyao, M. and Murata, N. (1985) FEBS Lett. 180, 303-308
[4] Homann, P.H., Gleiter, H., Ono, T. and Inoue, Y. (1986) Biochim. Biophys. Acta 850, 10-20
[5] Vass, I., Horváth, G., Herczeg, T. and Demeter, S. (1981) Biochim. Biophys. Acta 682, 140-152
[6] Rutherford, A.W., Crofts, A.R. and Inoue, Y. (1982) Biochim. Biophys. Acta 682, 457-465
[7] De Vault, D., Govindjee and Arnold, W. (1983) Proc. Natl. Acad. Sci. USA 80, 983-987

ABNORMAL S_2 STATE FORMED IN CHLORIDE DEPLETED PHOTOSYSTEM II AS REVEALED BY
MANGANESE EPR MULTILINE SIGNAL

T. Ono[2], J.L. Zimmermann[1], Y. Inoue[2] and A.W. Rutherford[1]

[1]Service de Biophysique, Département de Biologie, CEN-Saclay, 91191 Gif-sur-Yvette Cedex (France) and [2]Solar Energy Research Group, The Institute of Physical and Chemical Research (RIKEN), Wako, Saitama, 351-01 (Japan)

1. INTRODUCTION

Photosynthetic O_2 evolution requires Cl^- as an indispensable cofactor whose functional site has been shown to reside in S state system. The depletion of Cl^- leads to incomplete turnover of the S state, but some oxidized equivalents can be accumulated in the Cl^- free water oxidase [1-5]. There are, however, some ambiguities among the results as to the site of inhibition in the S state turnover or the properties of the formed S state(s) in the absence of Cl^-. The detection of the individual S state is possible by EPR spectroscopy and thermoluminescence(TL) measurement independent of O_2 evolution, since the multiline EPR signal arises from the magnetic interaction between Mn atoms in S_2 [6] and the TL B-band arises from the recombination between the negative charge on Q_B^- and the positive charge on S_2 and S_3 [7]. In the present study, the effects of Cl^- depletion on the S state system have been investigated by EPR and TL measurements. It is concluded from our measurements that an abnormal S_2 state is formed in the absence of Cl^-, and the further transition beyond this abnormal S_2 state is blocked.

2. MATERIALS AND METHODS

B.B.Y. type PSII membranes capable of O_2 evolution were prepared from spinach as in [8] and relaxed in the dark for 4 hr. Cl^- depletion was accomplished by SO_4^{2-} replacement by diluting the membranes with the medium containing 50 mM Na_2SO_4 at pH 7.5. For Cl^- repletion, NaCl was added to the depleted samples to a final concentration of 50 or 100 mM either before or after the membranes were illuminated. The samples were illuminated either with continuous light at 200 K or with a series of flashes at $0^{\circ}C$, and then cooled to 77 K. EPR spectra were recorded at liquid helium temperature as described [9]. TL glow curves were measured by recording the light emission during warming ($1^{\circ}C/s$) against the sample temperature as described [7].

3. RESULTS AND DISCUSSION

Fig. 1 shows the difference EPR spectra (light minus dark)(A) and the TL glow curves (B) induced by a continuous illumination at 200 K, which accumulates the S_2 state. The Cl^- sufficient control membranes showed a multiline EPR signal between 2500 and 4000 G and a broad EPR signal at g = 4.1 (A), and also showed a strong TL B-band around $30^{\circ}C$ (B). The formation of the multiline signal was markedly inhibited (<10 % of control) by Cl^- depletion which gave 95 % inhibition of O_2 evolution, while the g = 4.1 signal was much less affected by the depletion (80 % of control). In TL glow curves, the Cl^- depletion effected an increase in the emission temperature showing maximum around $40^{\circ}C$.

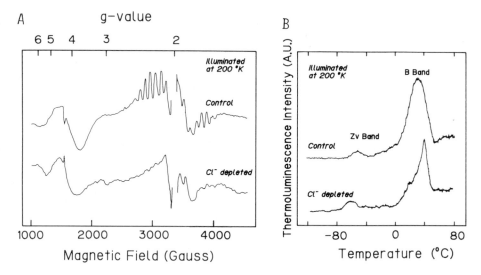

Fig. 1. Effect of Cl⁻ depletion on the EPR signals (A) and the thermoluminescence glow curves (B) induced by continuous illumination at 200 K.
Instrumental settings for EPR: temperature, 8 K; microwave power, 2 mW;
microwave frequency, 9.43 GHz; modulation amplitude, 32 G.

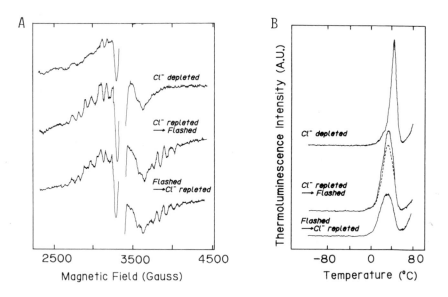

Fig. 2. Effect of Cl⁻ depletion and Cl⁻ repletion on the multiline EPR
signals (A) and the thermoluminescence glow curves (B) induced by a single
flash at 0°C. The repletion was performed by adding Cl⁻ to the depleted
samples before flash excitation or immediately after flash excitation
followed by dark incubation for 40 s at 0°C before being cooled rapidly to
77 K. Instrumental settings for EPR were the same as in Fig. 1.

Fig. 2 shows the effect of depletion and repletion of Cl^- on the multiline EPR signal (A) and the TL glow curves (B) by a single flash. The depletion of Cl^- inhibited the formation of the multiline signal and gave a modified TL band around $40^{\circ}C$, consistent with the result obtained by continuous illumination at 200 K (see Fig. 1). The repletion of Cl^- before a flash excitation restored the ability to form the multiline signal and the normal TL B-band, indicating that the loss of the multiline signal and the modification of the TL band is due to the depletion of Cl^-. Interestingly, the addition of Cl^- in the dark immediately after the flash excitation developed the multiline signal and the TL B-band which are almost identical to those of the control membranes. This indicates that an oxidizing equivalent can be stably stored on the O_2 evolving enzyme in the absence of Cl^- as an abnormal S_2 state although the capability to evolve O_2 is almost completely lost. The formation of the Cl^- free S_2 state would be also verified by the fact that the g = 4.1 signal was still largely induced in the absence of Cl^- since this EPR signal has recently reported to arise from the S_2 state [9]. The modified TL band probably implies some changes in redox properties of the Cl^- free S_2 state [4]

The loss of the multiline signal in the Cl^- free S_2 state indicates either that the oxidation state of Mn atoms in S_2 is not identical to that of the normal S_2 state or the interaction between Mn atoms in S_2 is modified in the absence of Cl^-. In either cases, it would be expected that the transition step beyond the modified S_2 would be somehow distorted. To check above consideration we measured the amplitude change of the multiline signal by a series of flashes (Fig. 3A), which gave a different signal amplitude depending on flash number with a maximum on the first flash followed by a sharp decline on the second and third flashes in the control and Cl^- repleted membranes (open circles and open squares in Fig.3 B). The multiline signal developed in Cl^- depleted membranes by the addition of Cl^- after a series of flashes, however, showed a constant amplitude after the second flash (closed circles in Fig. 3B). This result indicates the blocking of the Cl^- free S_2 to S_3 transition.

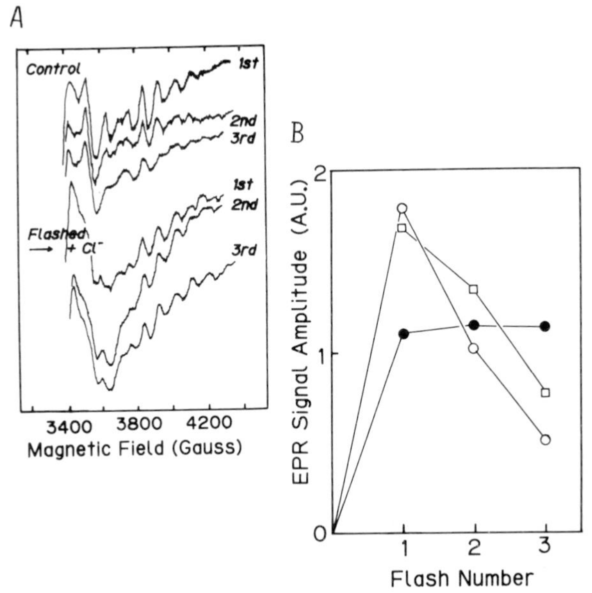

A

Fig. 3. Changes of the multiline EPR signal after a series of flashes at $0^{\circ}C$. A, multiline signal. B, amplitude changes of the multiline signal in control (O) and in chloride depleted membranes to which chloride was added to depleted membranes before (□) or immediately after (●) a series of flashes. 1 mM 2,5-dimethylquinone was added as an electron acceptor. Instrumental settings for EPR: temperature, 20 K; microwave power, 50 mW; microwave frequency, 9.42 GHz; modulation amplitude, 32 G.

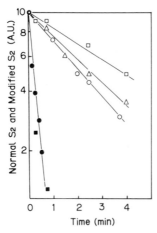

Fig. 4. Deactivation course of the normal S_2 state (closed symbols) and the modified S_2 state by chloride depletion (open symbols) at 20°C. Both S_2 states were estimated from multiline EPR signal ($\square\blacksquare$), thermoluminescence B-band (O●), and also 50 µs component of P-680$^+$ reduction course ($\triangle$) taken from Ref.5.

The abnormal property of the Cl$^-$ free S_2 state is also represented in its unusual stability. Fig. 4 compares the decay curves of the modified and normal S_2 states measured by various methods . Both the two S_2 states decayed monophasically during the dark incubation, but the decay kinetics of the modified S_2 ($t_{1/2}$ = 5-10 min) was about 10-20 times as slow as that of the normal S_2 ($t_{1/2}$ = 30 sec). This result indicates that the oxidized equivalent formed in the absence of Cl$^-$ is much more stable than S_2 in agreement with the qualitative results of Muallem et al. [2].

In conclusion, the present study has demonstrated that only one turnover from S_1 to S_2 is allowed in the Cl$^-$ depleted O_2 evolving center and the resulting Cl$^-$ free S_2 is abnormally stable in comparison with the normal S_2 state. This indicates that Cl$^-$ works to maintain the redox property of S_2 to guarantee the normal S state transition.

REFERENCES
1   Muallem, A., Farineau, J., Laine-Boszormenyi, M. and Izawa, S. (1981) in Photosynthesis (Akoyunoglou, G., ed.), Vol. II, pp. 435-443, Balaban International Science Service, Philadelphia
2   Theg, S.M., Jursinic, P. and Homann, P.H. (1984) Biochim. Biophys. Acta 766, 636-646
3   Itoh, S., Yerkes, C.T., Koike, H., Robinson, H.H. and Crofts, A.R. (1984) Biochim. Biophys. Acta 766, 612-622
4   Homann, P.H., Gleiter, H., Ono, T. and Inoue, Y. (1986) Biochim. Biophys. Acta 850, 10-20
5   Ono, T., Conjeaud, H., Gleiter, H., Inoue, Y. and Mathis, P. FEBS Lett. in the press
6   Dismukes, G.C. and Siderer, Y. (1980) FEBS Lett. 121, 78-80
7   Rutherford, A.W., Crofts, A.R. and Inoue, Y. (1982) Biochim. Biophys. Acta 682, 457-465
8   Ono, T. and Inoue, Y. (1985) Biochim. Biophys. Acta 806, 331-340
9   Zimmermann, J.L. and Rutherford, A.W. (1986) Biochemistry in the press

ACKNOWLEGEMENTS
    We gratefully acknowledge financial support from CEA, CNRS (France) and STA of Japan (grant for Solar Energy Conversion by Means of Photosynthesis at RIKEN). This collaboration was based on the Versailles Summit Cooporation Program on Photosynthesis and Photoconversion.

Cl⁻ DEPENDENT BINDING OF THE EXTRINSIC 23 kDa POLYPEPTIDE
AT THE WATER OXIDIZING SITE OF CHLOROPLAST PHOTOSYSTEM II

Peter H. Homann, Institute of Molecular Biophysics and Department of
Biological Science, Florida State University, Tallahassee, FL 32306-3015,
U.S.A.

1. INTRODUCTION

The cofactor role of Cl^- in photosynthetic water oxidation is still
very poorly understood even though it was discovered more than 40 years
ago (see review [1]). In previous studies, we have obtained evidence suggesting that Cl^- is associated with some protonated group(s) of a $pK_a \leq 6$
[2]. Surprisingly, a removal of the extrinsic 17 and 23 kDa polypeptides
did not appear to affect the estimated pK_a of the putative binding group
even though significantly higher Cl^- concentrations were required for
maximal activities under such a condition [3].

It is still obscure how the two polypeptides affect the Cl^- requirement of the water oxidizing system. In recognition of their effects on the
Cl^- requirement, Murata and Miyao have called them "Cl^- concentrators"
[4]. It is also known that the presence of the polypeptides allows the
water oxidizing site to retain the functional anion in Cl^- free media of
$pH < 7$ [1,2], and retards the acquisition of added Cl^- [3]. Recently,
Itoh and Iwaki [4] have looked into the cause of the alleged role of the
polypeptides as barriers to Cl^- diffusion from and to its binding site.
Studying the effect of various anions on the rate of Cl^- loss, they provided data which appear to reveal an overriding influence of simple
electrostatic surface charges on Cl^- binding and release.

2. RESULTS

With the above viewpoints in mind, we have continued our own studies
on Cl^- release and rebinding by PSII membranes prepared by triton-X treatment of chloroplast thylakoids (for exptl. conditions see [2]). We found
that several treatments purported to cause Cl^- deficiency, cause a
detachment of the extrinsic 23 kDa polypeptide from the water oxidizing
complex. In all instances, Cl^- protected against this loss, or even caused
its partial reversal. The relevant observations were:

1. After a brief Cl^- depleting exposure of the membranes to $pH \approx 10$ [2],
the Cl^- dependence of the restoration of activity occasionally yielded
biphasic double reciprocal plots suggesting a low and a high affinity
response. This was the case rather frequently with preparations from
tobacco (<u>Nicotiana tabacum</u> L. var. JWB) and spinach (<u>Spinacia oleracea</u>
L.), but never with preparations from pokeweed (<u>Phytolacca americana</u> L.)
(Fig. 1). PAGE analysis revealed that the biphasicity of the plots was
correlated with a partial loss of the 23 kDa polypeptide (Fig. 3).
2. The creation of a Cl^- deficiency syndrome at an elevated pH, and in
the presence of SO_4^{2-} [6,7] (but not of F^-) was similarly associated with
a loss of the 23 kDa polypeptide which, at neutral or slightly acid pH,
could be prevented by the presence of Cl^- (Figs. 2 and 3).

Biggens, J. (ed.), Progress in Photosynthesis Research, Vol. I. ISBN 90 247 3450 9
© *1987 Martinus Nijhoff Publishers, Dordrecht. Printed in the Netherlands.*

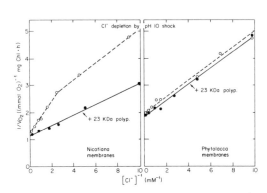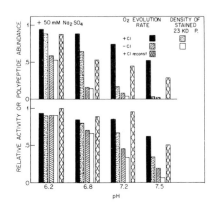

Fig. 1. Double reciprocal plots of O_2 evolution activity against concentration of added Cl⁻ after a 7 sec exposure to pH 9.8.(± Added 23 kDa pp.)
Fig. 2. Effect of a 5 min incubation of PSII preparations from <u>Nicotiana</u> at various pH in the presence or absence of 55 mM Na_2SO_4. NaCl, if added, at 25 mM; abundance of 23 kDa polypeptide given relative to untreated preparation as integrated optical density of Coomassie Blue stained band after SDS-PAGE separation.

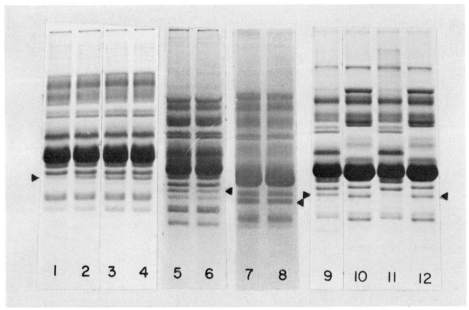

Fig. 3 SDS-PAGE analysis of PSII preparations after various 5 min incubations. Marking indicates position of 23 kDa polypeptide. 12 μg Chl/ml PSII membranes from <u>Nicotiana</u> (lanes 1-6, 9, 11) or <u>Phytolacca</u> (lanes 7, 8, 10, 12). Lanes 1 and 2: 2 mM $ZnSO_4$ at pH 6.0; lane 2, subsequent 2 min incubation with 25 mM NaCl. Lanes 3 and 4: 30 mM NaF at pH 6.3 minus or plus 25 mM NaCl. Lanes 5 through 8: 7 s at pH 9.8, then pH 6.5 and, for lanes 5 and 8, 2 min incubation with 25 mM NaCl. Lanes 9 and 10: 5 mM Na_2SO_4 at pH 6.35. Lanes 11 and 12: 5 mM $ZnSO_4$ at pH 6.35.

Table 1. Effect of Cl^- status on the response of O_2 evolution activity and binding of the 23 kDa polypeptide to 5 min incubations of PSII membranes in the presence of various cations at pH 6.35. Ions provided at 5 mM as Me-Mes except Zn as $ZnSO_4$ (values in parentheses for 2 mM $ZnSO_4$); presence of 10 mM Na_2SO_4 in Na-control had no effect. Activity in mmol O_2/mg Chl·h. Estimation of 23 kDa abundance as in Fig. 2.

| | $-Cl^-$ incubation | | $+Cl^-$ reconst. | $+Cl$ incubation | |
	activity	23 kDa pp.	activity	activity	23 kDa pp.
Nicotiana					
Na^+	355	.85	510	520	.90
Mg^{2+}	265	.40	465*	nd	.75
Ca^{2+}	180	.25	530	nd	.80
Zn^{2+}	35	.10	70	215	.80
	(70)		(110)	(390)	
Phytolacca					
Na^+	370		580	610	
Zn^{2+}	40		120	380	

*activity was raised to 510 after addition of 5 mM Ca^{2+}

3. The results under conditions of elevated pH or in the presence of SO_4^{2-} were quite analogous to those obtained in the presence of divalent cations like Mg^{2+}, Ca^{2+} or Zn^{2+} (Table 1). However, in the case of the latter, the rate of O_2 evolution could not be fully recovered by Cl^- addition after incubation, and remained the lower the higher the Zn^{2+} concentration and pH, presumably due to a loss of functional Mn from the water oxidase [8]. Again, thylakoid preparations from Phytolacca held on more tenaciously to their 23 kDa polypeptide than, for example, those from Nicotiana (Fig. 3). The Zn-effect on O_2 evolution, however, was comparable (Table 1).

It should be noted that the fate of the extrinsic 17 kDa polypeptide was not followed because ot its tendency to nonspecifically bind to the membranes.

3. DISCUSSION

While the influence of the 23 kDa polypeptide on Cl^- binding in PSII is well established, a reciprocal action of Cl^- has only rarely been recognized [6,9]. The results reported here demonstrate this role quite conclusively and suggest that the often described "protective" role of Cl^- against inactivations of the O_2 evolving mechanism [1] correlates with its apparent ability to maintain a configuration of the 23 kDa polypeptide conducive to a strong association with the lipids and proteins of the site of water oxidation. The electrostatic forces postulated by Itoh et al. [5] for Cl^- binding may in fact be the ones involved in the attachment of the 23 kDa polypeptide.

A Cl⁻ deficiency syndrome can be created by addition of F⁻ without losing the 23 kDa polypeptide, and PSII membranes from <u>Phytolacca</u> chloroplasts retained it also under other conditions. Thus, one cannot generalize that Cl⁻ release requires a detachment of the 23 kDa extrinsic polypeptides. However, F⁻ may not cause an actual Cl⁻ loss (we could not confirm the claim of a competitive interaction of F⁻ and Cl⁻), and the extrinsic polypeptide of <u>Phytolacca</u> homologous to the 23 kDa species of <u>Nicotiana</u> and <u>Spinacia</u> clearly is different chemically, displaying during SDS-PAGE a molecular mass somewhat lower. It is conceivable that even after the Cl⁻ sensitive binding sites of this particular polypeptide are severed, other regions remain attached at or close to the water oxidizing site. Revelations from our further studies of the molecular reasons for the different behavior of the various 23 kDa polypeptides may also bring us closer to an understanding of their physiological role.

Acknowledgments: These studies were supported by NSF grant DMB 8304416. Some initial data on this problem were collected by Dr. Diane Nash. The author thanks Ms. Michelle Currier and Ms. Valerie Pfister for skillful technical assistance, Ms. Pamela Lyons for typing and help with the layout, and Mr. B. Williams for the line drawings and photography.

References

1 Critchley, C. (1985) Biochim. Biophys. Acta 811, 33-46
2 Homann, P.H. (1985) Biochim. Biophys. Acta 809, 311-319
3 Homann, P.H. and Inoue, Y. (1985). In Ion Interactions in Energy Transfer Biomembranes (Papageorgiou, G.C., Barber, J. and Papa, S., eds.). Plenum Press, New York and London, pp. 279-290
4 Murata, N. and Miyao, M. (1985). Trends Biochem. Sci. 10, 122-124
5 Itoh, S and Iwaki, M. (1986) FEBS Lett. 195, 140-144
6 Homann, P.H. (1986) Photosynthesis Research, in the press
7 Sandusky, P.O. DeRoo, C.L.S., Hicks, D.B., Yocum, C.F., Ghanotakis, D.F. and Babcock, G.T. (1983) In The Oxygen Evolving System of Photosynthesis (Inoue, Y., Crofts, A.R., Govindjee, Murata, N., Renger, G. and Satoh, K., eds.), Academic Press, Tokyo, pp. 189-199
8 Miller, M. and Cox, R.P. (1983) FEBS Lett. 155, 331-333
9 Cammarata, K. and Cheniae, G.M. (1985) Plant Physiol. 77 suppl., 16
10 Sandusky, P.O. and Yocum, C.F. (1986) Biochim. Biophys. Acta 849, 85-93

EFFECTS OF CHLORIDE ON PARAMAGNETIC COUPLING OF MANGANESE IN CALCIUM
CHLORIDE-WASHED PHOTOSYSTEM II PREPARATIONS

Gopinath Mavankal, Douglas C. McCain, and Terry M. Bricker. Department
of Chemistry, University of Southern Mississippi, Hattiesburg, MS 39406-
5043

1. INTRODUCTION

The oxygen-evolving complex of Photosystem II requires both chloride
(1) and manganese (2) in order to oxidize water. In chloride-depleted
membranes, the oxygen-evolving complex loses its ability to proceed past
the Kok S_2 state (3,4). The functions of bound manganese and Cl^- appear
to be interrelated. In this communication, we have used Q-band ESR (35
GHz) to observe protein-bound but non-paramagnetically coupled manganese
associated with these PS II membranes. Our experiments have shown that
the degree of paramagnetic coupling of manganese depends on the chloride
concentration. We have also studied the effects of other anions on the
paramagnetic coupling of manganese.

2. MATERIALS AND METHODS

Oxygen-evolving PS II membrane preparations were prepared from market
spinach essentially according to the procedure of Ghanotakis and Babcock
(5) with the modifications noted in (6). The extrinsic polypeptides of
the oxygen-evolving complex were removed by washing the Photosystem II
membranes with resuspension buffer containing 1.0 M $CaCl_2$. These
membranes were depleted in Cl^- by washing with resuspension buffer (300
mM sucrose, 10 mM $MgCl_2$, 15 mM NaCl, 50 mM Mes-NaOH, pH 6.0). Non-
paramagnetically coupled manganese was removed from Cl^--depleted
membranes by washing with resuspension buffer containing 20 mM EDTA.
These membranes were then washed twice with resuspension buffer and
sampled for ESR spectroscopy. These samples were then heated to 100°C
for 5 min in the ESR tube and new spectra taken. The effect of anions on
the extent of paramagnetic coupling was studied by washing the
polypeptide-depleted membranes three times in chloride-free resuspension
buffer and then washing them in resuspension buffer with varying amounts
of the anion (0M to 2M). In each of these treatments the Photosystem II
membranes were resuspended to 1.0 mg/ml Chl when washed. Sample sizes
for ESR spectroscopy were 10 ul at a Chl concentration of 10 mg/ml Chl.
Q-band ESR spectra were obtained on a 35 GHz Varian E 109 Q spectrometer
at 90° K. Instrument conditions: microwave power 5mW, microwave
frequency 35 GHz, modulation amplitude 5.0 G, modulation frequency 100
kHz, scan rate 31.25 G/min, time constant 1 s.

Biggens, J. (ed.), Progress in Photosynthesis Research, Vol. I. ISBN 90 247 3450 9
© *1987 Martinus Nijhoff Publishers, Dordrecht. Printed in the Netherlands.*

RESULTS AND DISCUSSION

Paramagnetically uncoupled Mn(II)-protein complexes are easily detected with Q-band ESR, but not with X-band instruments (7). In aqueous solution Mn(II) ions exist in a highly symmetrical environment and their spectrum consists of six approximately equally spaced symmetrical lines (Fig. 1a). Protein-bound manganese, however, exists in an anisotropic crystal field, and exhibits "forbidden lines" in the spectrum (8) (Fig. 1b). In our studies we have exploited the advantages of Q-band ESR to investigate the effect of Cl⁻ and other anions on the paramagnetic coupling of PS II.

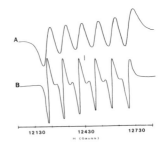

Fig. 1 Q-band ESR spectra of (A), 30 mM MnCl$_2$; and (B), the protein-bound manganese signal of concanavalin redrawn from (12).

Fig. 2 illustrates the results of a typical experiment in which the oxygen-evolving membranes (Fig. 2a) are first washed with 1M CaCl$_2$ (Fig. 2b) and then resuspended in low Cl⁻ media (resuspension buffer). A six-line signal which is indicative of protein-bound and paramagnetically uncoupled Mn(II) can be seen (Fig. 2c and 2e) when the membranes are in a low Cl⁻ environment. The intensity of the signal is greatly diminished (Fig 2d and 2f) when the membranes are resuspended in high Cl⁻ (2M NaCl in resuspension buffer). The membranes can be cycled repeatedly through these treatments (at least 4 times) without any apparent change in maximal signal intensity in the Cl⁻-depleted state.

Fig. 2 The effect of Cl⁻ depletion and re-addition on CaCl$_2$-washed oxygen-evolving PS II preparations. Q-band ESR spectra of: (A), untreated PS II membranes; (B), CaCl$_2$-washed PS II membranes; (C), sample (B) washed and suspended in resuspension buffer (35 mM Cl⁻); (D), sample (C) washed and suspended in 2.0 M NaCl; (E), sample (D) washed and suspended in resuspension buffer; (F), sample (E) washed and suspended in 2.0 M NaCl. Spectrometer sensitivity was essentially constant during all six runs. The small amount of manganese in (A) can be completely removed by treatment with EDTA with no effect on subsequent treatments.

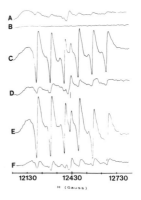

This strongly suggests that the decrease in signal intensity of the protein-bound Mn(II) is due to the induction of paramagnetic coupling within the Mn clusters of PS II. The results of Fig. 3 show that the uncoupled, protein-bound Mn which is seen after Cl⁻-depletion (Fig. 3b)

is accessible to the chelating agent EDTA (Fig. 3c). Protein-bound manganese which is paramagnetically coupled is not accessible to the chelating agent (not shown). If the membranes which have been depleted of Mn(II) by EDTA treatment are subsequently heated at 100°C for 5 min to denature the membrane proteins (after removal of EDTA), a second population of uncoupled protein-bound Mn appears (Fig. 3d). These data strongly suggest the existence of two pools of Mn associated with PS II.

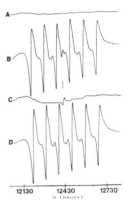

Fig 3. The identification of two pools of manganese in $CaCl_2$-washed PS II membrane preparations. Q-band ESR spectra of: (A), $CaCl_2$-washed PS II membranes; (B), (A) treated with 20 mM EDTA and removal of EDTA by washing and suspension in resuspension buffer; (D), (C) heated to 100°C for 5 min.

Fig. 4 shows the effects of various anions on the paramagnetic coupling of manganese in PS II. The order of effectivness is Cl^- > Br^- > NO_3^-. This order is the same as that observed for the reconstitution of oxygen evolution rates and in inverse order of their anion volumes (1). The overall profile of the curves shows two peaks, which are displaced to higher anion concentrations with increasing anion volume. We suggest that increasing anion concentration might be inducing conformational changes within or among the polypeptides which are associated with the bound manganese.

Fig. 4 The effects of different concentrations of (A), Cl^-; (B), Br^-; and, (C), NO_3^- on the intensity of paramagnetically uncoupled manganese. The manganese was uncoupled as in Fig. 2c and the membranes were resuspended in buffers containing anion concentrations ranging from 5.0 mM to 2.0 M. Error bars = 1.0 S.D., n=3 (3 replicates/n).

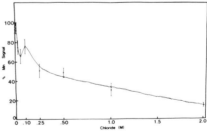

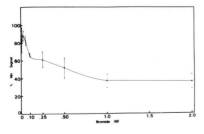

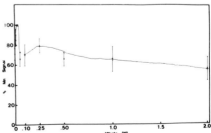

The model shown in Fig. 5 summarizes our observations. Two general
models have been proposed for Manganese-chloride interactions at the
oxygen-evolving site of PS II. Cl⁻ may either act as a bridging ligand
between Mn atoms (9) or may participate in a more general charge-
screening role at a site distant from the Mn (10). Our observations are
consistent with either model. It should be noted that the signals we
observe are those of protein-bound Mn(II). A number of models have been
proposed which suggest the presence of higher oxidation states of
manganese (11) at the oxygen-evolving site. We suggest that either
Mn(II) is present at the active site (more probable) or that the non-
paramagnetically coupled manganese is accessible to exogenous reductants
(less probable).

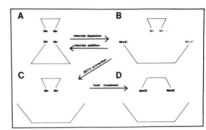

Fig. 5 Model illustrating the effects
of Cl⁻ depletion and addition, EDTA
treatment, and heating on the two
pools of manganese associated with PS
II.

4. REFERENCES

1. Critchley, C. (1985) Biochim. Biophys. Acta 811,33.
2. Dismukes, G.C., (1986) Photochem. Photobiol. 43,99.
3. Itoh, S., C.T. Yerkes, H. Koike, H.H. Robinson, and A.R. Crofts (1984)
   Biochim. Biophys. Acta 766,612.
4. Theg, S.M., P. Jursinic and P.H. Homann (1984) Biochim Biophys. Acta
   766,636.
5. Ghanotakis, D.F. and G.T. Babcock (1983) FEBS Lett. 153,231.
6. Ford, R.C. and M.C.W. Evans (1983) FEBS Lett. 160,159.
7. Mavankal, G., D.C. McCain, and T.M. Bricker (1986) FEBS Lett. 202,235.
8. Reed, G.H. and M. Cohn (1970) J. Biol. Chem. 245,662.
9. Sandusky, P.O. and C.F. Yocum (1983) FEBS Lett. 162,339.
10. Coleman, W. and Govindjee (1984) In: Proceedings of the 16th FEBS
    Congress (Moscow, USSR), UNV Science Press, Utrecht, The Netherlands.
11. Dismukes, G.C., K. Ferris, and P. Watnick (1982) Photobiochem.
    Photobiophys. 31,243.
12. Meirovitch, E. and R. Poupko (1978) J. Phys. Chem. 82,1920.

ACCESSIBILITY FOR, AND PRODUCTION OF H_2O_2 RELATED TO PS-II

Wolfgang P. SCHRÖDER and Hans-Erik ÅKERLUND

Dept. of biochemistry, University of LUND,
P.O. Box 124, S-221 00 LUND, SWEDEN.

1. INTRODUCTION

Several models for the oxygen evolving mechanism include some form of hydrogen peroxide as a bound intermediate. Although, this intermediate has never been shown to leak out, added hydrogen peroxide can interact with the water oxidizing system (1). Recently we found (2), from studies on inside-out thylakoids, that the hydrogen peroxide interaction with PS II was facilitated by removal of the extrinsic 23 and 16 kDa proteins associated with the oxygen evolving system. Also a light dependent hydrogen peroxide production was stimulated by the protein removal. This work has now been extended to also include detergent prepared PS II particles and inhibition studies. The results suggest that removal of the 23 and 16 kDa proteins (and possibly two low molecular weight polypeptides, Henrysson, et al., these proceedings) increase the PS-II accessibility for hydrogen peroxide, and that part of a light and PS II dependent hydrogen peroxide production originates from an exposed water oxidation site.

2. MATERIAL AND METHODS

Inside-out thylakoids were isolated, and subsequently washed with 250 mM NaCl as described earlier (3). The material was either used directly or stored in liquid nitrogen with 5% dimethylsulfoxide until use. PS-II-particles were isolated acording to (4), and salt-washed with 1 M NaCl and 50 mM MES pH 6.5.

Flash induced oxygen yields (Y_n) were measured with a rate electrode (2), with a flow medium containing 30 mM MES pH 6.5 and 2 mM KCl, and normalized to the average yield (Y_{ss}) on flash 3, 4, 5 and 6. To study the hydrogen peroxide production the samples, 0.5 mg Chl/ml in 30 mM MES pH 6.5 and 2 mM KCl were, illuminated with white light from two projectors.

The hydrogen peroxide production was followed either by the lactoperoxidase catalyzed reaction with ABTS as described earlier (2), or through the hydrogen peroxide dependent oxidation of methanol by catalase (5). In the latter case, 3.6 % methanol and 2000 U/ml catalase were present during the illumination. The formaldehyde formed were analyzed as described earlier (5).

Biggens, J. (ed.), Progress in Photosynthesis Research, Vol. I. ISBN 90 247 3450 9
© *1987 Martinus Nijhoff Publishers, Dordrecht. Printed in the Netherlands.*

3. RESULTS AND DISCUSSION

3.1. PS-II ACCESSIBILITY FOR HYDROGEN PEROXIDE

The oxygen yield pattern from salt-washed inside-out thylakoids in the presence of low concentrations of hydrogen peroxide (50 µM) is shown in Fig 1.a. A high yield was obtained already on the first flash and the normal oscillation with a period of four was almost completely lost. Similar, but less pronounced effects were seen at hydrogen peroxide concentrations down to 2 µM. Under the same conditions untreated inside-out thylakoids showed a normal pattern (Fig 1.b). However, at higher concentrations of hydrogen peroxide also the untreated inside-out thylakoids showed a high yield of oxygen on the first flash and a disappearance of the period of four oscillation (2). These results suggest that the salt-wash treatment, which releases the extrinsic 23 and 16 kDa proteins, open up the water oxidation site and thereby increase the accessibility for hydrogen peroxide to the donor side of photosystem II.
In the presence of hydrogen peroxide, the yield on the first flash was generally greater than that obtained on the second flash (Fig.1). One possible explanation to this is that hydrogen peroxide binds to the S_1-state in the dark and convert it into state S_3 which upon exitation produce oxygen.

3.2. HYDROGEN PEROXIDE PRODUCTION

The salt-washed inside-out thylakoids showed an oxygen yield on the first flash even if hydrogen peroxide was not added (2). Addition of catalase restored a normal oscillation pattern (Fig. 1.a), suggesting that hydrogen peroxide was present in the thylakoid material (2). Direct determination of hydrogen peroxide showed that both untreated and salt-washed inside-out thylakoids contained small amounts of hydrogen peroxide. This amount could be linearly increased by illumina-

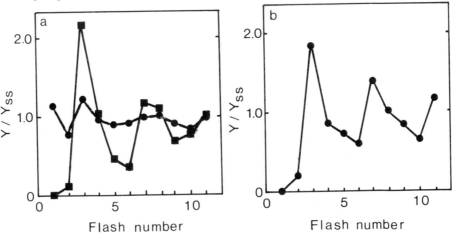

Fig. 1. Oxygen-yield pattern of a. salt-washed inside-out thylakoids (I.O.) in the presence of 50 µM H_2O_2 (●), or with catalase (■) and b. untreated inside-out in the presence of 50 µM H_2O_2.

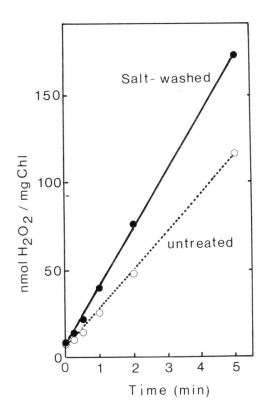

Fig. 2. H₂O₂ production
as a function of illumination
time for untreated (○) and
salt-washed (●) inside-out
thylakoids.

tion (Fig.2). Thus, the small amounts of hydrogen peroxide present in the
samples at time zero could be due to light absorbed during the isolation
and pretreatment of the samples.
Interestingly the hydrogen peroxide production was larger for the inside-
out material where the 16 and 23 kDa proteins had been removed, compared
to the untreated material. A question, that is not fully answered yet, is
how this hydrogen peroxide is produced.
To reduce the possible influence of hydrogen peroxide production from photosystem I, photosystem II particles isolated according to Berthold et al.
(4) were investigated. Although these particles are virtually free from
photosystem I, they still produce hydrogen peroxide upon illumination
(Fig. 3.a). The difference in hydrogen peroxide production seen between
the photosystem II particles and inside-out vesicles suggests that part of
the production found for inside-out thylakoids may be associated with the
low amount of photosystem I present in these thylakoids.
Also in the case of photosystem II particles was a higher light induced
hydrogen peroxide production seen after removal of the 23 and 16 kDa proteins, compared to untreated photosystem II particles.
Heat treatment (100° for 5 min.) prior to illumination inhibited a large
part of the hydrogen peroxide production in both untreated and salt-washed
photosystem II particles. Almost the same degree of inhibition was obtained on addition of DCMU. The remaining hydrogen peroxide production after
heat treatment or DCMU treatment was the same in both untreated and salt--
washed photosystem II particles and probably represents a photosensitized
nonenzymatic reaction. On the other hand, the heat and DCMU sensitive part

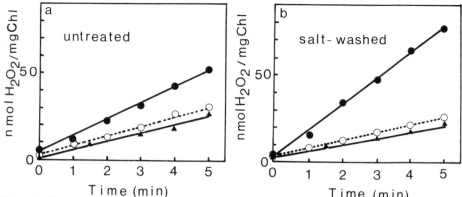

Fig. 3. H_2O_2 production as a function of illumination time for a. untreated, b. salt-washed PS-II particles, without additions (●), DCMU (○) and heat treated (▲).

of the hydrogen peroxide production is interpreted in terms of a photosystem II associated hydrogen peroxide production.

Light dependent hydrogen peroxide production associated with PS-II has been observed earlier (6). As autooxidation of hydroquinone produce hydrogen peroxide (7) and since no external electron acceptor has been added in these experiments, at least part of the hydrogen peroxide production seen in this work originate from the autooxidation of plastohydroquinones on the PS-II acceptor side.

This can not explaine the higher hydrogen peroxide production observed for salt-washed PS-II particles relative to untreated particles. However, if the presumed hydrogen peroxide intermediate, bound to the manganes cluster in the S_2 and S_3 states (see e.g. 8) could leak out when the 23 and 16 kDa proteins are removed by salt-washing, it would explain the observed difference. This possibility is supported by the results presented in section 3.2. , where the water splitting site was shown to be more accessible for hydrogen peroxide after removal of the 23 and 16 kDa proteins. If part of the hydrogen peroxide production really originate from the water splitting site it opens up new possibilities to study intermediate steps in the oxygen evolving reaction.

ACKNOWLEDGMENT
The authors appreciate the skillful technical assistance by Mrs. Sophie Bingsmark. This work was supported by the Swedish Natural Science Research Council and Magnus Bergvall foundation.

REFERENCES

1 Velthuys, B. and Kok, B. (1978) BBA: 502, 211-221.
2 Schröder, W.P. and Åkerlund, H.-E. (1986) BBA. 848, 359-363.
3 Åkerlund, H.-E., Jansson, C. and Andersson, B. (1982) BBA. 681, 1-10.
4 Berthold, D.A., Babcock, G.T. and Yocum, C.F. (1981) FEBS Lett.134,231-4
5 Hildebrandt, A.G., Roots, I. Tjoe, M. and Heinemeyer, G. in Methods in Enzymology (Fleischer, S., ed.), Vol. LII pp. 342-350, Academic press.
6 Elstner, E.F. and Frommaeyer, D. (1978) FEBS Letters 86, 143-146.
7 James, T.H. and Weissberger, A. (1938) J. Am. ch. soc. 60, 98-104.
8 Govindjee, Kamabara, T. and Coleman, W. (1985), Photochem. Photobiol. 42, 187-210.

REVERSIBLE INHIBITION OF PHOTOSYSTEM TWO ELECTRON TRANSFER REACTIONS
AND SPECIFIC REMOVAL OF THE EXTRINSIC 23 kDa POLYPEPTIDE BY ALKALINE pH

DAVID J. CHAPMAN AND JAMES BARBER

AFRC Photosynthesis Research Group, Department of Pure & Applied
Biology, Imperial College of Science & Technology, London, SW7 2BB, UK.

1.   INTRODUCTION
     A well characterised feature of photosystem two (PS2) electron
transfer activity is an acidic pH optimum with strong inhibition by
alkaline conditions (1-3). This is known to occur at a site which is
between water and the P680 component of the PS2 reaction centre and
physically located on the lumenal surface of the thylakoids (1-5).
However, little attention has been paid to the mechanism and
significance of this inhibition despite the fact that regulation by pH
is generally recognised as an important biochemical and physiological
control mechanism. One suggestion for the mode of inhibition is that
it is the result of dissociation from the membrane of the manganese and
extrinsic polypeptides which are thought to be involved in the water
oxidation complex. We have developed sensitive enzyme linked
immunosorbent assays (ELISA) for the 23 and 33 kDa extrinsic
polypeptides and used this quantitative approach to study the
relationships between polypeptide removal, manganese extraction and
loss of oxygen evolution activity.

2.   MATERIALS AND METHODS
     Membrane fractions enriched in PS2 (PS2-MB) were isolated by a
single solubilisation of Pisum sativum (pea) thylakoids in Triton
X-100 (6). PS2-MB samples were washed in a buffer medium containing
5 mM $MgCl_2$, 15 mM NaCl, 10% (w/v) glycerol and 20 mM bis-tris propane
(BTP) pH 6.5. The pelleted membranes (35,000 x g, 25 min, 4°C) were
resuspended at various pH values between 6.5 and 9.5 using buffer
medium with only 10 mM BTP. After incubation for 20 min on ice and in
the dark a second period of treatment was begun by addition of an equal
volume of buffer medium containing 50 mM BTP at the same pH as the
initial treatment; this ensured that the same incubation sequence was
followed as in the procedure which resulted in reconstitution.
Reconstitution was achieved by initiating the second incubation period
with a buffer medium containing 50 mM BTP, pH 6.5. After incubation,
membranes and supernatants were separated by centrifugation. Oxygen
evolution was determined with a Clark type oxygen electrode, at
saturating light intensity and with 20 ug chlorophyll of PS2-MB in the
buffer medium containing 20 mM BTP and additions to give 3 mM
phenyl-p-benzoquinone, 10 mM $K_3Fe(CN)_6$, 1 mM valinomycin and 1 mM
nigericin. Manganese was measured by flameless atomic absorption
spectrometry after digestion of samples in 0.2% nitric acid. The
quantities of specific proteins in samples were assessed with
antibodies raised in rabbits to extrinsic polypeptides which had been
isolated by selective washing of membranes and purified by SDS PAGE,
and also ion-exchange chromatography in the case of the 33 kDa
polypeptide. Different sample concentrations were sonicated in
0.1 g.l^{-1} SDS and applied to wells of ELISA plates. After blocking

Biggens, J. (ed.), Progress in Photosynthesis Research, Vol. I. ISBN 90 247 3450 9
© 1987 Martinus Nijhoff Publishers, Dordrecht. Printed in the Netherlands.

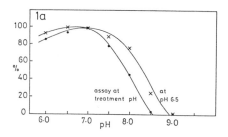
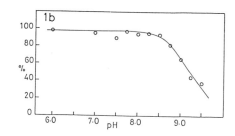

FIGURE 1. Effect of pH on oxygen evolution a) pH treatment, b) treatment and then reconstitution.
100%... 150 umoles. mg chlorophyll^{-1}.h^{-1}

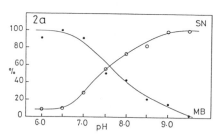
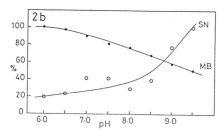

FIGURE 2. Effect of pH on relative amount of 23 kDa polypeptide associated with membrane (MB) and supernatant (SN) fractions. Specific assay by ELISA. a) pH treatment, b) treatment and then reconstitution.

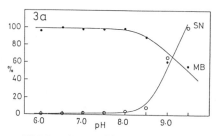
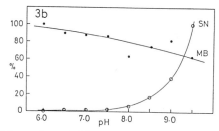

FIGURE 3. As figure 2, but for 33 kDa polypeptide.

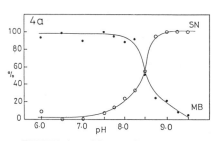
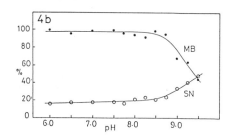

FIGURE 4. Effect of pH on manganese in membrane (MB) and supernatant (SN) fractions, 100%... 8.0 g.atoms Mn to 250 moles chlorophyll

with 2 g.l^{-1} casein, antibodies were added to all wells. The
concentrations of bound antibody were determined with anti-rabbit
peroxidase conjugate and spectrophotometric assay of enzyme activity to
give a linear response for sample dilutions over a range of absorbance
values up to 0.6.

3.   RESULTS AND DISCUSSION
       The optimum pH for the incubation and oxygen evolution assay was
between pH 6.5 and 7.0 with 50% inhibition at pH 7.9 and complete loss
of activity at pH 8.5 or above (Fig. 1a). This inhibition by alkaline
pH was partially reversible. A recovery of about 30% of the maximum
rate was achieved if, after alkaline pH treatment, the membranes were
returned to pH 6.5(after centrifugation). Even with this reactivation
treatment, initial incubation at pH 8.3 did inhibit activity (50%),
indicating an irreversible inhibition. These two distinct modes of
inhibition by alkaline pH have been characterised recently (4,5) and
the work of several groups (3,4,7,8,) points to the possibility that
removal of extrinsic polypeptides and/or manganese could cause the
apparently irreversible inactivation. To study these possibilities we
have used SDS-PAGE and developed a sensitive ELISA method to estimate
quantities of the 23 and 33 kDa polypeptides in membrane and
supernatant samples. The 17 kDa polypeptide was not detected in the
SDS-PAGE analysis of PS2-MB from pea although it was present, as found
by others, in preparations from spinach. Using the ELISA method we
found that treatment at pH 7.6 gave a 50% removal of the 23 kDa protein
from the membrane (Fig. 2a). This pH corresponds closely with the
point of half-maximal oxygen evolution (Fig. 1a). However, 50%
displacement of the 33 kDa polypeptide was at pH 9.0 (Fig. 3a), and a
similarly high pH was needed for extraction of manganese (Fig. 4a).
From these results it is clear that the loss of activity between pH 7.0
and 8.5 is related to removal of the 23 kDa polypeptide rather than
displacement of the 33 kDa polypeptide or manganese. It seems unlikely
that the presence of the 23 kDa polypeptide is an absolute requirement
for activity because other experiments have demonstrated high rates of
oxygen evolution in the absence of the protein (7,8,9). The reduction
in activity could also be the result of a pH induced change in a site
which is important for both water oxidation and the binding of the 23
kDa polypeptide. A similar explanation is that removal of this protein
by pH 7.0 to 8.5 exposes a site which is sensitive to incubation in
these mildly alkaline conditions. Indeed previous work by Ghanotakis
et al (10) has shown that removal of this protein can expose a pool of
functional manganese to exogenous reductants. One suggested role of
the 23 kDa polypeptide is that it reduces the chloride requirement of
water oxidation activity. However, an increased need for chloride
resulting from removal of the protein does not appear to be the cause
of the pH inhibition because 25 mM chloride was present in incubation
and assay media and further addition of chloride to inhibited samples
failed to stimulate activity.

       Recovery of high rates of water oxidation in the apparently
irreversibly inhibited samples was achieved by reconstitution
treatments in which the alkaline pH of the incubation media was lowered
to pH 6.5 by addition of a high concentration of buffer. This
procedure resulted in complete reversal of inhibition for all pH
treatments up to, and including, pH 8.5 (cf. Fig. 1a and b) but became
increasingly ineffective when initial treatments were extended to pH

9.5. The recovery of activity was matched closely by a rebinding of the 23 kDa polypeptide (Fig. 2b) and this further emphasised that this polypeptide, or its binding site, is intimately involved in water oxidation. Reconstitution had little or no effect on binding of the 33 kDa polypeptide (cf. Fig. 3a and b) but did alter the pH profile for manganese associated with the membrane (cf. Fig. 4a and b) with the 50% extraction point being moved from pH 8.5 to 9.5.

The manganese level in PS2-MB is often reported to be 4 g. atoms Mn/250 moles chlorophyll and loss from the membrane usually follows the removal of the 33 kDa polypeptide (7,8,9). Our estimate for total manganese is 8 g. atoms/250 moles chlorophyll and extraction did not appear to be closely correlated with removal of the 33 kDa polypeptide. One interpretation of our result is that there are two pools of manganese. One pool of 4 Mn/250 chlorophylls can be removed by treatments between pH 7.5 and 8.5, reducing the total from 8 to 4. The 33 kDa polypeptide was removed by incubations between pH 8.5 and 9.5 and this is the same pH range which accounted for reduction of the remaining pool from 4 to 0 Mn/250 chlorophylls. According to this two pool model it is possible to explain the change in overall pH profile for manganese due to reconstitution treatment by the reassociation of only one of the two manganese pools.

4. MAIN CONCLUSION
Inactivation of water oxidation by pH values between 7.0 and 8.5 is closely related to removal of the 23 kDa extrinsic polypeptide but not the loss of the 33 kDa polypeptide or manganese.

ACKNOWLEDGEMENTS
We thank Dr. P.A. Millner for help in the preparation of antisera and the AFRC (U.K.) for financial assistance.

REFERENCES
1   Harth, E., Reimer, S. and Trebst, A. (1974) FEBS Lett. 44,165-168
2   Cohn, D.E., Cohen, W.S. and Bertsch, W. (1975) Biochim. Biophys. Acta 376, 97-104
3   Briantais, J.M., Vernotte, C., Lavergne, J. and Arntzen, C. (1977) Biochim. Biophys. Acta 461, 61-74
4   Cole J., Boska, M., Blough, N.V. and Sauer, K. (1986) Biochim. Biophys. Acta 848, 41-47
5   Plijter, J.J., de Groot, A., van Dijk, M.A. and van Gorkom, H.J. (1986) FEBS Lett. 195, 313-318
6   Chapman, D.J., De-Felice, J. and Barber, J. (1985) Planta 166, 280-285
7   Murata, N. and Miyao, M. (1985) Trends Biochem. Sci. 10, 122-124
8   Ghanotakis, D.F. and Yocum, C.F. (1985) Photosynth. Res. 7, 97-114
9   Critchley, C. (1985) Biochim. Biophys. Acta 811, 33-46.
10   Ghanotakis, D.F., Babcock, G.T. and Yocum, C.F. (1984) Biochim. Biophys. Acta 765, 388-398

O$_2$ FLASH YIELD SEQUENCES OF PHOTOSYSTEM II MEMBRANES--SEQUENTIAL
EXTRACTION OF THE EXTRINSIC PROTEINS

MICHAEL SEIBERT, BRIGITTA MAISON-PETERI*, AND JEAN LAVOREL**.
SOLAR ENERGY RESEARCH INSTITUTE, GOLDEN, CO 80401 USA, *C.N.R.S., 91190
GIF-SUR-YVETTE, FRANCE, **A.R.B.S. CEN CARARACHE, 13108 ST-PAUL-LEZ-
DURANCE CEDEX, FRANCE

1. INTRODUCTION
    Three extrinsic proteins which bind to the lumenal surface of
appressed thylakoid membranes (see Staehelin, DeWit, and Seibert, these
Proceedings) have been associated with the water-oxidation process of
photosystem II (PSII). These proteins (17, 23, and 33 kDa) can be removed
sequentially from the surface of the membrane by exposure to various salt-
and Tris-washing treatments. We have examined both steady-state O$_2$
evolution and O$_2$ flash-yield sequence patterns in these membranes as the
three extrinsic proteins are removed. Several treatments give rise to an
anomalous amperometric signal on the first two flashes. It is suggested
that these anomalous signals are linked to the release of the extrinsic
proteins, but depend on functional Mn remaining present in the membrane.

2. PROCEDURE
2.1. Materials and Methods
    2.1.1. Experimental samples. O$_2$-evolving PSII (OESII) membranes were
        prepared by Triton X-100 treatment (1). The 17-kDa protein or
        the 17- and 23-kDa proteins were removed by 0.25 M and 1.0 M
        (or 2.0 M) NaCl treatment in the light, respectively (1). All
        three extrinsic proteins were removed by either alkaline-Tris
        treatment in the light (1), which also removes functional Mn
        required for O$_2$ evolution, or by CaCl$_2$-treatment in the dark,
        which does not remove functional Mn (2).
    2.1.2. Instrumentation. Steady-state O$_2$ evolution measurements were
        made using a Clark electrode with either 2,6-dimethylbenzo-
        quinone (DMBQ) or 2,6-dichlorobenzoquinone (DCBQ) as an
        acceptor. O$_2$ rates of OESII membranes were 250-300 µmol O$_2$·mg
        Chl^{-1}·hr^{-1} with DMBQ and 500 to 650 µmol·O$_2$·mg Chl^{-1}·hr^{-1} with
        DCBQ. O$_2$ rate electrode flash-yield sequences (2 µs Xenon
        flashes spaced 1 s apart) were obtained using a Joliot-type O$_2$
        rate electrode (3,4) with no added acceptor. The flow buffer
        contained 20 mM MES (pH 6.0), 15 mM NaCl, and 5 mM MgCl$_2$. Dark
        adapted samples were equilibrated on the electrode for 10 to 20
        min in the dark prior to flashing. A new sample was used for
        each flash sequence unless otherwise indicated.

3. RESULTS AND DISCUSSION
    Figure 1 shows that both the 17- and 23-kDa proteins have to be
removed for a significant decrease in O$_2$ evolution to be detected. A
direct relationship between the Y$_3$ yield and the steady-state O$_2$ yield is
observed for our different samples (Fig. 2). Treatment with 2 M NaCl

decreases the yield of O_2 evolution (the decrease is greater when samples
are treated in the light than in the dark) below that observed for 1 M
NaCl treatment (Fig. 2), but no loss of 33-kDa protein was observed. We
observe no anomalous O_2 production on the first flash with NaCl-treated
detergent-derived OESII membranes (Fig. 1), which is in contrast to the
results of Schroder and Åkerlund (5), who report anomalous amperometric
signals on the first flash with NaCl-treated inside-out thylakoid
vesicles. Both preparations lack the 17- and 23-kDa proteins. However,
we could detect a signal on the first flash in 1 M NaCl-treated OESII membranes after several treatments. In fresh preparations a signal on the
first flash was observed 10 min after pre-exposure of the sample (on the
rate electrode) to a single flash (data not shown). Aging the samples at
ice temperature (off the rate electrode) after having frozen and thawed
them several times also resulted in a signal on the first flash (Fig.
3). Furthermore, activating the rate electrode polarization (6) after
dark adapting a sample on the electrode and 1 min prior to running a flash
sequence (instead of dark adapting the sample on a polarized electrode)
elicited a similar response (data not shown). If the three extrinsic proteins are removed from OESII membranes by alkaline Tris treatment (which
also removes Mn), little O_2 emission was observed using either the rate or
Clark electrodes (Figs. 1 and 2). However, if they were removed by $CaCl_2$

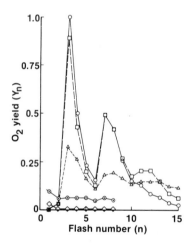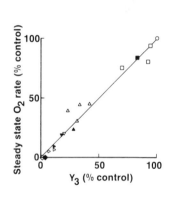

FIGURE 1. O_2 rate electrode flash sequences of control OESII membranes
(O), 0.25 M NaCl-treated OESII with the 17-kDa protein missing ($\square$), 1 M
NaCl-treated OESII with the 17- and 23-kDa proteins missing ($\triangle$), 1 M
$CaCl_2$-treated OESII with all three proteins missing ($\odot$), and Tris-treated
OESII with all three proteins missing ($\diamond$).

FIGURE 2. Relationship between the amount of steady-state O_2 evolution
and the yield of Y_3 (Fig. 1) in OESII membranes. Control (O), 0.25 M NaCl
treatment ($\square$), 1 M NaCl treatment ($\triangle$), 2 M NaCl treatment in the dark ($\triangledown$)
or in the light ($\triangleright$), and Tris treatment ($\diamond$). Steady-state measurements
used DMBQ (open symbols) or DCBQ (closed symbols) as an acceptor.

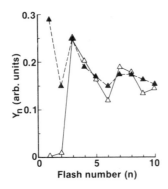

FIGURE 3. O_2 rate electrode flash sequences of 1 M NaCl-treated OESII membranes. Fresh (△); after several freeze-thaw cycles and aging at ice temperature (▲).

TABLE 1. The effect of $CaCl_2$ addition on steady-state O_2 and O_2 rate electrode yields in OESII membranes subjected to 2 M NaCl and 1 M $CaCl_2$ treatments.

Sample Treatment	$CaCl_2$ added during assay (mM)	Steady-state O_2 Yield (μmol $O_2 \cdot$mg $Chl^{-1} \cdot hr^{-1}$)	Y_3* (arb units)	Y_{25}* (arb units).
2 M NaCl	0	22	1.00	0.33
	15	158	0.83	0.21
1 M $CaCl_2$	0	4	0.26	0.16
	50	46	0.16	0.10

*Rate electrode results were obtained using 1.0 mg Chl/ml, and $CaCl_2$ was present in the flow buffer when the samples were added to the electrode.

treatment which does not remove functional Mn, O_2 emission including amperometric signals on the first two flashes were seen (Fig. 1). The addition of $CaCl_2$ to the Clark electrode assay medium reconstitutes O_2 evolution in both NaCl- and $CaCl_2$-treated PSII membranes (7,8). However, the presence of $CaCl_2$ in the flow buffer (prior to placing the sample on the electrode) did not lead to an increase in the amount of the rate electrode signal (Table 1).

These results indicate that anomalous O_2 production on the first two flashes in OESII membranes is related to the removal of the extrinsic proteins which protect the Mn complex required for normal O_2 evolution. The anomaly is seen in preparations lacking the 17- and 23-kDa proteins that in addition have been aged (Fig. 3), abused, or pre-exposed to light. It is also seen in $CaCl_2$-treated membranes lacking all three proteins (Fig. 1). Furthermore, the anomaly is related to the presence of functional Mn since it is seen in $CaCl_2$-treated membranes, but to a much lesser extent in Tris-treated material. For more information about this phenomenon see Berg and Seibert (these Proceedings). The relationship between this phenomenon and anomalous steady-state O_2 production seen in urea/NaCl-washed OESII membranes which were subsequently treated with NH_2OH (9) is not clear from the current data (but see Berg and Seibert, these Proceedings). However, these latter preparations were exposed to exogenous Mn, added H_2O_2, and an artificial electron acceptor (9) which was not the case in our experiments. Finally, there is a direct relationship between steady-state O_2 yields and Y_3 yields in preparations exposed to various salt treatments (Fig. 2). However, this relationship does not hold when $CaCl_2$ is added to the assay medium in NaCl- or CaCl-treated preparations under our conditions (Table 1). We suggest that the presence of non-physiological amounts of $CaCl_2$ in the samples can under some conditions interfere with contact between OESII preparations and the Pt electrode of the O_2 rate electrode system.

ACKNOWLEDGMENTS
      This work was supported by the Division of Biological Energy Research, U.S. Department of Energy contract FTP 006-86 (M.S.).

REFERENCES

1 Kuwabara, T. and Murata, M. (1983) Plant & Cell Physiol. 24, 741-747.
2 Ono, T.-A. and Inoue, Y. (1983) FEBS Lett. 164, 225-260.
3 Lavorel, L. and Seibert, M. (1982) FEBS Lett. 144, 101-103.
4 Seibert, M. and Lavorel, J. (1983) Biochim. Biophys. Acta 723, 160-168.
5 Schroder, W.P. and Akerlund, H.-E. (1986) Biochim. Biophys. Acta.
6 Akerlund, H.-E. (1984) in Adv. in Photosynthesis Res. (Sybesma, C., ed.), Vol. I, pp. 391-399, Martinus Nijhoff, Dordrecht.
7 Ghanotakis, D.F., Babcock, G.T. and Yocum, C.F. (1984) FEBS Lett. 167, 127-130.
8 Ono, T.-A. and Inoue, Y. (1984) FEBS Lett. 168, 281-286.
9 Boussac, A., Picaud, M. and Etienne, A.-L. (1986) Photobiochem. and Photobiophys. 10, 201-211.

COMPARATIVE STUDY OF PERIOD 4 OSCILLATIONS OF THE OXYGEN AND FLUORESCENCE
YIELD INDUCED BY A FLASH SERIES IN INSIDE OUT THYLAKOIDS

M. J. DELRIEU and F. ROSENGARD
Laboratoire de Photosynthese, C.N.R.S., BP 1,91190 Gif sur Yvette
(France)

1. INTRODUCTION
    The oxygen yield pattern induced by excitation of dark adapted
thylakoids with a train of short flashes reveals a characteristic
oscillation with a periodicity of 4. This can be explained by a 4 step
oxidation of water to oxygen. The flash induced advancement of the S_j
state "clock" (1) involves a certain probability of misses and double
hits (α and β respectively). For simplicity, it is generally accepted
that the probability of misses and double hits are the same after each
transition. Nevertheless, we have obtained evidence that this hypothesis
leads to several inconsistencies. For example, the least square fitting
yields a very high percentage of double hits in experimental oxygen yield
patterns induced by flashes of reduced energy (2). Based on the same
fitting method and other experimental facts, we consider that the misses
are different in different S_i states, and more precisely that there is a
predominant miss on the $S_2 \rightarrow S_3$ transition (2-4). However, in order to
compare the damping of our oxygen yield patterns with that found by other
authors, only equal miss and double hit values will be used in this
paper.
    A period 4 oscillation has been detected in the fluorescence yield
measured some ms after each of a series of flashes (5). In this paper, we
show that the particular structure of the oscillation pattern of
fluorescence looks like that of flash induced absorbance change in the
U.V. (6) or the red region (7), but is fundamentally different from that
of oxygen yield obtained under the same flash conditions. This difference
is independent of the respective weights of the S_i' states which could
contribute to the fluorescence oscillation.

2. MATERIALS AND METHODS
    Inside out thylakoids were prepared from pea chloroplasts according
to standard procedures (8). The suspension used was a medium containing
300mM Sorbitol, 10mM NaCl, 5mM MgCl$_2$, 40mM MES-NaOH buffered at pH 6.5.
    Flash excitation was provided by Stroboslave General Radio flash
lamps (half fall 3μs). The fluorescence yield of chlorophyll a was
measured 80ms after each flash of a series of 16 flashes as previously
described (3).
    The least square fitting method used in this work is based on a
general recurrence relation on Y_n (for oxygen yield) (2) or F_n (for
fluorescence). This method allows an analysis of the fluorescence
pattern, without any hypotheses concerning the fluorescence contribution
f_i of each state S_i' to the fluorescence oscillation $F = \sum_i f_i \, S_i'$.

3. RESULTS AND DISCUSSION
    Fig. 1 shows the oscillation pattern of the flash induced oxygen
yield and that of fluorescence measured 80ms after each flash of a series
under the same conditions: same inside out thylakoid preparation adapted

Biggens, J. (ed.), Progress in Photosynthesis Research, Vol. I. ISBN 90 247 3450 9
© *1987 Martinus Nijhoff Publishers, Dordrecht. Printed in the Netherlands.*

to darkness for 3 hours, same flashes, comparable flash energy and a flash spacing of 500ms. In order to avoid limitations of secondary electron transfer at the acceptor side, 1mM ferricyanide was added. In Fig. 1, the oxygen and fluorescence oscillations are shifted in order to be in phase at the first four flashes of the sequence, but progressively, the maxima of the oxygen pattern move 1 or 2 flashes to the right after the maxima of fluorescence yield which occurs strictly every four flashes. The feature of the fluorescence pattern is characterized by the same oscillation pattern repeated with a decreased size as a functon of flash number. This repeated structure as a function of flash number was also observed in absorbance changes even though the authors used short laser flashes (7). In contrast, the oxygen oscillation presents a change in phase with flash number, corresponding to the phase retardation associated with the damping coming from misses. The oxygen yield pattern of Fig. 1 is very typical. The numerical fitting yields α=0.12, β=0.03, values nearly similar to that obtained in chloroplasts by Vermaas et al (9) with a similar flash lamp (their data: α=0.13, β=0.04). The exact period 4 of oscillation in the oxygen yield pattern was observed only in the conditions of long flashes (100μs) (10), inducing double hits which advance the phase, counterbalancing the phase retardation by misses.

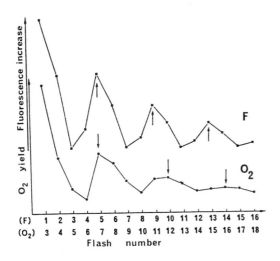

FIGURE 1. Oscillation patterns of the flash induced fluorescence measured 80ms after each flash of a series and that of oxygen yield under the same conditions (see text for details).

Figs. 2.A. and B. present the variation of the oscillation pattern of fluorescence as a function of flash energy (I=100%, 50%, 30%). The inside out thylakoid samples (pH 6.5) were all dark adapted for 3 hours. It is shown that in any cases, the period 4 of oscillation remains exact, with no phase change, and the fluorescence oscillation amplitude progressively decreases after many flashes. The initial amplitude of the oscillating part of the fluorescence yield does not saturate under our experimental conditions (Fig. 2. and 3.) , though the flash energy used (even for I=30%) was higher than necessary to saturate the S_i transitions of oxygen evolution (except for $S_2 \to S_3$ (4)). Furthermore, in Fig. 2.A, for I=100%, the large amplitude of the fluorescence oscillation after the first flash is associated with a fast decrease of the oscillation. Lower energy flashes (I=50%, 30%) give lower initial amplitude, but the oscillation appears more sustained over more flashes than with higher energy flashes. These results show that the fluorescence oscillation with period 4 is not directly related to the S_i states of the Kok cycle.

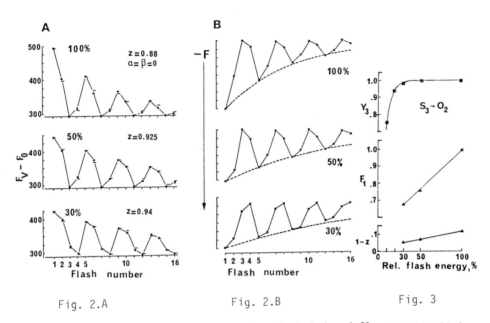

Fig. 2.A   Fig. 2.B   Fig. 3

FIGURE 2. Oscillation pattern of the flash induced fluorescence as a function of flash energy: I=100%, 50%, 30% (experimental conditions described in the text). A) Best least square fit of the experimental patterns : $\alpha =0$, $\beta =0$, z=0.88 for I=100%, z=0.925 for I=50%, z=0.94 for I=30%. - experimental values, . theoretical values. The flash induced fluorescence yield, F_V, was substracted from F_0, the fluorescence yield after darkness. B) Same results as in A) except that the origin of ordinate is reversed.

FIGURE 3. Initial amplitude of the oscillating part of the fluorescence yield F_1 , Proportion of fluorescent centers lost at each flash of the series 1-z, as a function of flash energy (from Fig. 2A), by comparison with the light saturation curve of the transition $S_3 \to O_2$ measured under the same conditions. The oxygen yield after the third flash Y_3 ,was measured as a function of the flash energy of the third flash I_3.

Thus, the fluorescence and oxygen oscillations have different properties. The oxygen flash yield pattern is damped due to misses (and double hits?) on some transitions, but the total amount $S_0 +S_1 +S_2 +S_3$ re   rather constant as a function of flash number. In Figs. 2.A, the ties of the fluorescence oscillations prove that they are related to other S'_i states, different from the S_i states producing oxygen. The S'_i states saturate at quite higher flash energy than that used to saturate the S_i states. One way to explain the decrease of the fluorescence oscillation with no phase change as a function of flash number is to assume that, in contrast to the oxygen yield oscillation, the total amount of S'_i states leading to fluorescence, decreases at each flash by a factor z (less than 1). The decrease of the oscillation amplitude as a function of flash number could be due to the consumption at each flash of a proportion 1-z of the total amount of S'_i states contributing to fluorescence. In Fig. 2.A.,the best fit of the fluorescence oscillations is obtained only with this new variable z, with no misses and no double hits (α =0, β =0): z=0.88 for I=100%, z=0.925 for I=50%, z=0.94 for I=30%. After a short dark adaptation period (10 min.), the best fitting yields a small percentage of misses which changes the features of the patterns. However, the percentage of misses generally remains smaller than in the oxygen yield pattern ($\leqslant$ 8%).

Thus, the fluorescence oscillations are controlled by the consumption of a product at each flash of the sequence, which is not renewed during the flash interval of 500ms (the half time of restoration in the dark is of order of 30 s. at pH 6.5, unpublished result). The quantity of product (or S'_i states) used at each flash is proportional to flash energy, as proved by the linear increase of the initial amplitude of the fluorescence oscillation as a function of flash energy and the decrease of the characteristic number of flashes 1/(1-z) necessary to exhaust the product, inversely proportional to flash energy as shown in fig. 3. The absence of saturation of this additional light reaction of consumption of the product controlling the S'_i states of fluorescence, at flash energy which saturates the O_2 flash yields (I around 20 t0 30 %), shows that this additional light reaction is different and less efficient than that leading to oxygen. A surprising result is that no misses are necessary to fit the fluorescence oscillations so that the S'_i states seem disconnected from the miss mechanism of oxygen evolution.

Further studies are necessary in order to characterize the four step reactions functioning in parallel to the oxygen evolving reactions and contributing to fluorescence.

REFERENCES
[1] Kok, B., Forbush, B. and McGloin, M. (1970) Photochem. Photobiol. 11, 457-475
[2] Delrieu, M.J. (1983) Z. Naturforsch. 38c, 247-258
[3] Delrieu, M.J. (1984) Biochim. Biophys. Acta 767, 304-313
[4] Delrieu, M.J., Phung Nhu Hung, S. and de Kouchkovsky, F. (1985) FEBS Lett. 187, 321-326
[5] Joliot, P. and Joliot, A. (1973) Biochim. Biophys. Acta 305, 302-316
[6] Dekker, J.P., Van Gorkom, H.J., Wensink, J. and Ouwehand, L. (1984) Biochim. Biophys. Acta 767, 1-9
[7] Saygin, O. and Witt H.T. (1985) Photobiochem. Photobiophys. 10, 71-82
[8] Akerlund, H.E. and Andersson, B. (1983) Biochim. Biophys. Acta 725, 34-40
[9] Vermaas, W., Renger, G. and Dohnt, G. (1984) Biochim. Biophys. Acta 764, 194-202
[10] Joliot, P., Joliot, A., Bouges, B. and Barbieri, G. (1971) Photochem. Photobiol. 14, 287-305

PURIFICATION OF AN OXYGEN EVOLVING PHOTOSYSTEM II REACTION
CENTER CORE PREPARATION

D.F. GHANOTAKIS, D.M. DEMETRIOU AND C.F. YOCUM

DEPARTMENT OF BIOLOGY, THE UNIVERSITY OF MICHIGAN
ANN ARBOR, MI 48109-1048, USA

INTRODUCTION
     The isolation of PS II membranes has advanced our
knowledge of the structural organization of PS II and has
also provided information about the roles of Mn, Cl^- and
Ca^{+2} as cofactors of oxygen evolution (for a review see ref.
1). Highly refined preparations of a PS II "core" complex
incapable of oxygen evolution activity have made it possible
to define the polypeptide composition of the PS II
hydrophobic domain and to identify the components of PS II
which are associated with this domain (2, 3). Recently, a
series of O_2-evolving PS II reaction center complexes
depleted of the LHCP have been isolated from spinach and
cyanobacteria (4-7). In this communication we report new
results on the isolation and characterization of a purified
oxygen evolving PS II RC core. We show here that one can,
by biochemical methods, define more closely the polypeptide
requirement for oxygen evolution activity.

MATERIALS AND METHODS
     Subchloroplast PS II membranes having high rates of
oxygen evolution were prepared as described in (7). The PS
II oxygen evolving reaction center complex was prepared by
the method described below. PS II membranes (2.5 mg
Chl/ml), resuspended in a solution containing 0.4 M Sucrose,
50 mM MES, pH 6.0, and 10 mM NaCl (Solution A), were mixed
with an equal volume of a solution containing 1.0 M Sucrose,
50 mM MES, pH 6.0, 0.8 M NaCl, 10 mM $CaCl_2$ and 70 mM OGP
(Solution B). After a 10 min incubation in the dark at 4 C
one part of the solubilized membranes was mixed with two
parts of a solution containing 1.0 M Sucrose, 50 mM MES, pH
6.0, 0.4 M NaCl, 5 mM $CaCl_2$ (Solution C); mixing was
followed by 5 min incubation and a 90 min centrifugation
step (40,000xg) to pellet the LHCP (see ref. 7). The
supernatant was desalted and the sucrose was removed by two
30 min dialysis steps against a solution containing 50 mM
MES, pH 6.0, 10 mM NaCl and 5 mM $CaCl_2$ (Solution D) (a
Spectrapor #6 dialysis tubing with 50,000 m.w. cutoff was
used), further diluted 50% with Solution D and subsequently

Biggens, J. (ed.), Progress in Photosynthesis Research, Vol. I. ISBN 90 247 3450 9
© 1987 Martinus Nijhoff Publishers, Dordrecht. Printed in the Netherlands.

centrifuged for 60 min at 40,000xg. The pellet which resulted from the last centrifugation step was resuspended in a solution containing 0.4 M Sucrose, 50 mM MES, pH 6.0, 10 mM NaCl and 5 mM $CaCl_2$ (Solution E), whereas the colorless supernatant was discarded. We should note that the increased amounts of sucrose in Solutions B and C result in recovery of the PS II reaction center complex in high yields (about 20% of the Chl in the starting PS II membranes was recovered in the PS II R.C complex preparation).

To isolate the purified PS II oxygen evolving core the PS II reaction center complex, prepared as described above, was solubilized with 0.5% of the detergent Dodecylmaltoside and was subsequently subjected to gel filtration chromatography on a Superose-12 Pharmacia column (attached to a Pharmacia FPLC system) which had been previously equilibrated with a solution containing 50 mM MES, pH 6.0, 20 mM NaCl and 0.05% Dodecylmaltoside (Solution F). The various fractions were eluted from the column with Solution F. Throughout the experiment precaution was taken to maintain the temperature below 10 C in order to avoid destruction of oxygen evolution activity.

RESULTS AND DISCUSSION

The PS II reaction center complex isolated by exposure of PS II membranes to the non-ionic detergent octylglucoside in the presence of high ionic strength is depleted of the water soluble 17 and 23 kDa polypeptides and thus requires the presence of non-physiological concentrations of Ca^{+2} and Cl^- for maximal oxygen evolution activity (7). To investigate the minimal structure required for O_2-evolution activity, we further purified the PS II RC complex by using gel filtration chromatography (Pharmacia FPLC system) (see "Materials and Methods" for details). When the PS II RC complex was solubilized with relatively high concentrations of Dodecylmaltoside and subsequently chromatographed on a gel filtration column, various protein complexes were resolved (Fig. 1), but none of these species retained oxygen evolution activity (data not shown). Solubilization of the system with a low concentration of Dodecylmaltoside (0.5%), however, resulted in isolation of two main fractions (Figs. 1 and 2). One of these, designated as Peak A, was a complex with a reduced number of polypeptides (Fig. 2) which retained high rates of oxygen evolution activity when ferricyanide was used as an artificial electron acceptor (Table I). As shown in Table I, the new preparation (designated as the PS II Reaction Center Core to distinguish it from the less highly resolved PS II Reaction Center Complex) ferricyanide is a more effective acceptor when compared to DCBQ; in addition, sensitivity of electron tranfer capacity to DCMU has also dissapeared in the new preparation.

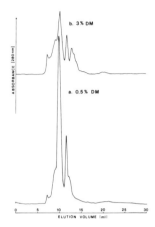

Figure 1: Elution profile of the PS II RC complex monitored by absorption at 280 nm. The complex was solubilized with the indicated concentration of the detergent Dodecylmaltoside and subsequently subjected to gel filtration chromatography. The running buffer contained 50 mM MES, pH 6.0, 20 mM NaCl and 0.05% Dodecylmaltoside and was maintained at 4 C.

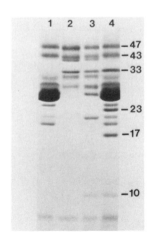

Figure 2: Gel electrophoresis patterns of 1) Tris treated PS II membranes; 2) Peak A (PS II RC core); 3) PS II RC complex; 4) intact PS II membranes (13.5 % acrylamide gel in the presence of 6.5 M urea).

An examination of the polypeptide content of the PS II RC core shows that the main components of this new complex are species with molecular weights of 47, 43, 33, 32, 30 and 9 kDa It is very interesting to note that the 10 kDa polypeptide which was previously implicated in manganese binding (8) is not present in this new oxygen evolving core preparation.

Table I: Oxygen-evolution activity of various PS II preparations

Additions	Activity (umol O_2/mg Chl per h)		
	[a]PSII membranes	[b]PSII RC complex	[c]PSII RC core
+DCBQ (500 uM)	680	1050	640
+DCBQ (500 uM)+ DCMU (5 uM)	60	340	520
+FeCN (2 mM)	160	620	1100
+FeCN (2 mM)+ DCMU (5 uM)	20	180	980

[a]The PSII membranes were assayed for oxygen evolution activity in a medium containing 0.4 M Sucrose, 50 mM MES, pH 6.4 and 10 mM NaCl (Solution G).
[b]The PSII RC complex was assayed in a medium containing 0.4 M Sucrose, 50 mM MES, pH 6.4, and 10 mM $CaCl_2$ (Solution H).
[c]The PS II RC core was assayed in Solution H in the presence of 0.01% Dodecylmaltoside which was present because of the isolation precedure followed.

Acknowledgements: This research was supported by grants to C.F.Y. from the National Science Foundation (PCM82-14240) and the Competitive Research Grants Office of USDA (G-82-1-1127).

References

1. Ghanotakis, D.F. and Yocum, C.F., (1985) Photosynthesis Research 7, 97-114.
2. Satoh, K., Nakatani, H.Y., Steinback, K.E., Watson, J. and Arntzen, C.J., (1983) Biochim. Biophys. Acta 724, 142-150.
3. Diner, B.A. and Wollman, F.A., (1980) Eur. J. Biochem. 110, 521-526.
4. Ikeuchi, M., Yuasa, M., and Inoue, Y., (1985) FEBS Lett. 185, 316-322.
5. Satoh, K., Ohno, T. and Katoh, S., (1985) FEBS Lett 180, 320-330.
6. Tang, X.S. and Satoh, K., (1985) FEBS Lett. 179, 60-64
7. Ghanotakis, D.F. and Yocum, C.F., (1986) FEBS Lett. 197, 244-248
8. Ljungberg, U., Akerlund, H.E., and Andersson, B., (1984) FEBS Lett. 175, 255-258.

SELECTIVE DEPLETION OF WATER-SOLUBLE POLYPEPTIDES ASSOCIATED
WITH PHOTOSYSTEM II

CHARLENE M. WAGGONER AND CHARLES F. YOCUM
DEPARTMENT OF BIOLOGY, THE UNIVERSITY OF MICHIGAN, ANN
ARBOR, MICHIGAN 48109-1048

1.   INTRODUCTION
     Research in several laboratories water-soluble polypep-
tides associated with PSII has provided evidence of both
structural and functional roles for the 17, 23, and 33 kDa
species (see 1 for review). There is, however, conflicting
evidence for the involvement of the 17 kDa polypeptide in
promoting maximal oxygen evolution activity at low chloride
concentrations (2,3,4). It is also not clear whether the 17
kDa species is an essential element of the structural shield
which protects the oxygen evolving complex from inactivation
by exogenous reductants (5). Andersson et al (2) have suggested that the chloride effect is not related to the presence of the 17 kDa polypeptide. To date, investigations of
the role of the 17 kDa polypeptide have been conducted on
systems that are missing both the 17 and 23 kDa polypeptides
as well as on systems that have been reconstituted with one
or both of these proteins. In both cases, the membranes are
perturbed and there is the potential for loss of calcium
from these systems as well as for incomplete reconstitution
of polypeptide function. In order to examine the role of
the 17 kDa polypeptide more closely, we have developed and
characterized a PSII preparation from which the 17 kDa
polypeptide has been selectively extracted.

2.   MATERIALS AND METHODS
     Photosystem II preparations were isolated using the
method of Berthold et al (6), and were exposed to salt at
the specified concentrations for one hour in the dark at 1.5
mg Chl/ml and washed once in 0.4M sucrose, 50mM MES and 15mM
NaCl (SMN) at pH 6.0. Oxygen evolution was assayed in a
Clark type O_2 electrode in SMN using 2,6-dichloro-p-benzo-
quinone as the accepter.

3.   RESULTS AND DISCUSSION
     Exposure of PSII membranes to non-physiological concen-
trations of divalent cations causes the sequential removal
of the 17 and 23 kDa polypeptides. Although 20 mM $MgCl_2$
produces membranes very similar in appearance to the controls (Fig 1C,1D), exposure to 40mM $MgCl_2$ begins to remove
the 17 kDa species (Fig 1B); 40mM $CaCl_2$ (data not shown) removes much more of the 17 kDa species than does 40mM $MgCl_2$.
Exposure of membranes to 60 mM $MgCl_2$ or 60 mM $CaCl_2$ removes

Biggens, J. (ed.), Progress in Photosynthesis Research, Vol. I. ISBN 90 247 3450 9
© *1987 Martinus Nijhoff Publishers, Dordrecht. Printed in the Netherlands.*

essentially all of the 17 kDa polypeptide while most of the
23 kDa is retained (Fig 1A); we cannot at this time quantify
the amount of 23 kDa remaining on the membranes.

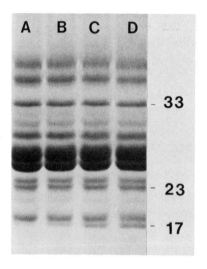

FIGURE 1: SDS-PAGE of salt washed PSII membranes.
a) 60 mM $MgCl_2$ b) 40 mM $MgCl_2$ c) 20 mM $MgCl_2$ d) control PSII
membranes.

TABLE 1: Effect of $CaCl_2$ and $MgCl_2$ washes on O_2 evolution

Wash treatment	Activity (%) (μmoles O_2 /hr/mg chl)	
	$-Ca^{+2}$	$+Ca^{+2}$
Control	540 (100)	542 (100)
40 mM $CaCl_2$	340 (63)	443 (82)
60 mM $CaCl_2$	351 (65)	481 (89)
40 mM $MgCl_2$	421 (78)	497 (92)
60 mM $MgCl_2$	356 (66)	470 (87)

In the presence of saturating chloride, PS II membranes
exposed to 60 mM divalent cations lose about 35% of their
oxygen evolving activity relative to the controls (Table 1).
Activity is almost completely restored upon the addition of
10 mM Ca^{+2}. At lower concentrations, $CaCl_2$ washing has a
greater effect on oxygen evolution than $MgCl_2$. Low concentrations of divalent cations (20mM) have little or no effect
on the activity of the membranes. In PSII membranes that
have been depleted of the 17 and 23 kDa polypeptides, oxygen
evolution drops to 15-20% of control levels and only returns
to 80% of control values upon the addition of calcium (1).

TABLE 2: Effect of Cl⁻ on oxygen evolution

mM NaCl	Activity (%) (µmoles O_2/hr/mg chl)		
	control PSII	60mM CaCl₂	washed
0	617 (100)	345	(56)
1	--	340	(55)
5	702 (114)	360	(63)
10	687 (111)	400	(65)
25	627 (102)	393	(64)

There is no chloride effect on our system which is depleted of the 17 kDa polypeptide. Controls and 60 mM CaCl₂
washed preparations both show a 10% increase in activity
when assayed in 5 mM NaCl (Table 2). At concentrations of
250 mM NaCl, which have been shown to remove 23 kDa polypeptide, oxygen activity declines (data not shown). Our results are contrary to Imaoka et al (3) who speculate that at
concentrations of NaCl less than 10 mM the 17 kDa is required for maximal activity; our data agree with those of
Andersson et al (2) who suggest that the 17 kDa species is
not involved in the chloride effect.

TABLE 3: Effect of hydroquinone on salt washed membranes

mM HQ	Activity (%) (µmoles O_2/hr/mg chl)		
	PSII	2M NaCl	60mM CaCl₂
0.00	542 (100)	411 (76)	477 (88)
0.10	527 (97)	78 (14)	363 (67)
1.00	579 (107)	35 (6)	342 (63)

All samples assayed in the presence of 10 mM CaCl₂.

Hydroquinone has been shown to reduce and extract manganese in PSII membrane preparations depleted of the 17 and
23 kDa polypeptides (5). Intact PSII membranes are not affected by the addition of hydroquinone, even at high concentrations (Table 3). PSII membranes washed in 2M NaCl that
are missing both the 17 and the 23 kDa polypeptides are totally inactivated by hydroquinone even at low concentrations, whereas samples washed in 60 mM CaCl₂ lose only 35%
of the control activity. Hydroquinone-treated samples have
the same activity as samples assayed without calcium in the
absence of hydroquinone, which indicates that about 65% of
the centers in the sample are functional although they are
missing the 17 kDa polypeptide. The remaining 35% of centers are missing the 17 and 23 kDa species, which gives rise
to the hydroquinone inactivation seen in Table 3. Centers
which do not require calcium for reactivation, and which are

insensitive to hydroquinone have retained tight calcium binding sites as well as the structural shield against reductant attack on manganese. From these observations, we would conclude that the 17 kDa polypeptide is not part of the shield.

4. SUMMARY

Exposure of PSII membranes to 60 mM divalent cations generates a preparation lacking the 17 kDa water-soluble polypeptide. These preparations retain 65% of control activity in the absence of calcium and regain 90% of control activity upon the addition of calcium to the assay medium. These salt-washed membranes are not chloride sensitive. Thus the 17 kDa polypeptide appears to have no role in the retention of chloride. Hydroquinone inhibition (35%) of PSII membranes washed with divalent cations suggests that a fraction of the PSII membranes have lost the 23 kDa polypeptide as well as the 17 kDa species. Because 65% of the centers are calcium insensitive and resistant to inactivation by hydroquinone, we would propose that the presence of the 33 and 23 kDa extrinsic polypeptides are sufficient not only for chloride retention, but also for protection of the oxygen evolving complex against inhibition byexogenous reductants.

5. ACKNOWLEDGMENT
The research was supported by grants from the National Science Foundation and the Competitive Research Grants Office of the United States Department of Agriculture to C.F.Y.

6. REFERENCES
1. Ghanotakis, D.F., and Yocum, C.F. (1985) Photosynthesis Res. 7, 97-114.
2. Andersson, B., Critchley,C., Ryrie,I.J., Jansson,C., Larsson,C., and Anderson,J.M. (1984) FEBS lett. 168, 113-117.
3. Imaoka, A., Akabori,K., Yangi,M., Izumi,K., Toyoshima,Y., Kawamori,A., Nakayama,H., and Sato,J. (1986) Biochimica et Biophpysica Acta 848, 201-211.
4. Miyao, M., and Murata, N. (1985) FEBS Lett. 180, 303-308.
5. Ghanotakis,D., Topper,J. and Yocum,C. (1984) Biochimica et Biophysica Acta 767, 524-531.
6. Berthold, D., Babcock,G., and Yocum,C. (1981) FEBS lett. 134, 231-234.

BINDING OF THE 17 AND 23 kDa WATER-SOLUBLE POLYPEPTIDES TO A HIGHLY-RESOLVED PSII REACTION CENTER COMPLEX

STEWART MERRITT, PATRIK ERNFORS, DEMETRIOS GHANOTAKIS AND CHARLES YOCUM
DEPARTMENT OF BIOLOGY, THE UNIVERSITY OF MICHIGAN, ANN ARBOR, MI   49109-1048

1. INTRODUCTION
    Recent research has focused on the components of PSII and how they interact to facilitate the oxidation of water. The role of water soluble polypeptides in O_2 evolution has been extensively investigated, and an extensive literature reviewed in (1) exists to document the involvement of 17 and 23 kDa polypeptides in calcium concentration, chloride retention and protection of the functional manganese pool of PSII against extraction by exogenously added reductants. In the absence of these polypeptides, O_2 evolution occurs only when Ca^{2+} is present at high, unphysiological concentrations (2,3); rebinding of the 17 and 23 kDa species by dialysis reconstitution restores high-affinity binding of Ca^{2+}. The recently-developed procedure (4,5; see also Ghanotakis, et al. in these Proceedings) in this laboratory for producing highly active oxygen-evolving particles depleted of LHCP, the 17 and 23 kDa species, and other polypeptides of unknown function raises the question of whether these more highly-resolved PSII preparations have lost the ability to rebind the 17 and 23 kDa polypeptides. Highly resolved PSII core particles lacking the 17 and 23 kDa polypeptides were utilized to determine if binding sites for the 17 and 23 kDa species had been retained on the proteins that remained in the core; as we show here, water-soluble polypeptides can be rebound to core particles, and the consequent disappearance of the Ca^{2+} requirement for O_2 evolution, as well as the appearance of resistance to inactivation by hydroquinone in the reconstituted material provides strong evidence for a functional reactivation of the roles of the 17 and 23 kDa species in the reconstituted cores.

2. MATERIALS AND METHODS
    Highly resolved PSII cores were prepared as described in (4,5). A mixture of cores and 17 and 23 kDa polypeptides was dialyzed in the dark (4 C, 90 min) in 0.4 M sucrose/50 mM Mes/5 mM $CaCl_2$/50 mM NaCl (pH 6.0) to allow the 23 kDa species to rebind. An additional 90 min dialysis in 0.4 M sucrose/50 mM Mes/5mM $CaCl_2$ (pH 6.0) promoted rebinding of the 17 kDa species. Core particles with the rebound 17 and 23 kDa polypeptides were recovered by centrifugation at 4 C for 10 min at 15,000 x g, resuspended in 0.4 M sucrose/50 mM Mes/15 mM NaCl (pH 6.0) to 0.5-1.5 mg chl/ml, and assayed for activity in the presence and absence of Ca^{2+}.

3. RESULTS AND DISCUSSION

Dialysis of PSII core particles in the presence of 17 and 23 kDa polypeptides led to partial restoration of Ca^{2+} independent O_2 evolution (Table I). Without Ca^{2+}, undialyzed cores initially evolved O_2 at only 20-30% of the control rate in the presence of Ca^{2+}, and after 1 min O_2 evolution had virtually ceased. Reconstituted cores, however, generated O_2 at 65-80% of the control rate in the presence of Ca^{2+} (Table 1). High rates of O_2 evolution continued well past the first minute. These results lead to the conclusion that the 17 and 23 kDa polypeptides have rebound to the core, since their presence is correlated with the high rates of O_2 evolution at low concentrations of Ca^{2+}.

TABLE 1. Effect of reconstitution of water-soluble polypeptides to cores on the Ca^{2+} requirement for O_2 evolution activity

Material assayed	Ca^{2+} present	Activity (μmoles O_2/hr·mg chl)
Cores	−	270
Cores	+	1170
Reconstituted cores	−	690
Reconstituted cores	+	1050

Dialysis of cores with 17 and 23 kDa polypeptides conferred resistance to inhibition of activity by hydroquinone (Table 2) which provides further evidence that a native rebinding of the polypeptides has occurred. Finally, gel electrophoresis shows the presence of 17 and 23 kDa species in the reconstituted core samples (Figure 1). Proteolysis of the 43 and 17 kDa species was frequently noted (Figure 1) in these experiments, even though the degree of proteolysis did not always correlate with the ability of dialyzed core samples to evolve O_2. This protease activity is of interest because no similar activity has been reported during reconstitution of salt-washed PSII particles and 17 and 23 kDa polypeptides. A new protease may have been exposed in the core preparation, which might be part of the core itself. Gel electrophoresis of concentrated 17 and 23 kDa polypeptides that did not bind to the core particles failed to show signs of proteolysis. The protease may be near the binding site for the 17 or 23 kDa polypeptides, and cleavage may occur after rebinding to produce a system containing bound, but proteolytically cleaved 17 kDa polypeptide.

TABLE 2. Effect of hydroquinone (HQ) on O_2 evolution activity of cores and reconstituted cores

Material assayed	[HQ] (mM)	% Activity
Cores	0.0	100
Cores	0.125	30
Cores	2.0	9
Reconstituted cores	0.0	100
Reconstituted cores	0.125	100
Reconstituted cores	2.0	92

Control activities (in μmoles O_2/hr·mg chl) were: core (10 mM $CaCl_2$ present) 790; reconstituted cores ($CaCl_2$ absent) 550. Samples were incubated in indicated concentrations of hydroquinone in the dark for 1 hr (4 C) before assaying.

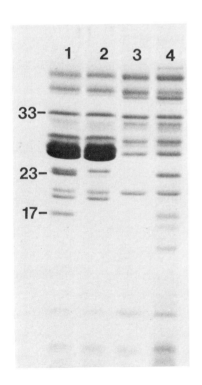

Figure 1. Gel electrophoresis showing rebinding of water soluble polypeptides to cores. (1) PSII; (2) 1 M NaCl wash; (3) core; (4) reconstituted core. Numbers to left are approximate molecular weights (x 1,000) of polypeptides.

4. SUMMARY

The highly functional rebinding of the 17 and 23 kDa polypeptides reported here is evidenced by Ca^{2+} independent O_2 evolution, insensitivity to hydroquinone inhibition of the reconstituted material, and gel electrophoresis. These results locate the binding sites for the 17 and 23 kDa species on the core particle with the OEC; binding sites are not associated with the polypeptides removed from PSII membranes. The ability to prepare even more highly-resolved oxygen evolving PSII reaction centers (5; Ghanotakis, et al., these Proceedings) has enabled us to initiate experiments to determine whether all functional binding sites for the 17 and 23 kDa species are restricted to the polypeptides of this reaction center preparation.

5. ACKNOWLEDGEMENT

The research was supported by grants from the National Science Foundation and the Competitive Research Grants Office of the United States Department of Agriculture to C.F.Y.

6. REFERENCES

1. Ghanotakis, D.F., and Yocum, C.F. (1985) Photosynthesis Res. 7, 97-114.
2. Ghanotakis, D.F., Babcock, G.T. and Yocum, C.F. (1983) FEBS Lett. 167, 127-130.
3. Ghanotakis, D.F., Topper, J.N., Babcock, G.T. and Yocum, C.F. (1984) FEBS Lett. 170, 169-173.
4. Ghanotakis, D.F. and Yocum, C.F. (1986) FEBS Lett. 197, 244-248.
5. Ghanotakis, D.F., Demetriou, D. and Yocum, C.F. (1986) Biochimica Biophysica Acta (submitted).
6. Ghanotakis, D.F., Topper, J.N. and Yocum, C.F. (1984) Biochimica Biophysica Acta 767, 244-248.

A MANGANESE CONTAINING PROTEIN COMPLEX ISOLATED FROM
PHOTOSYSTEM II PREPARATIONS OF SPINACH

NEIL R. BOWLBY AND WAYNE D. FRASCH, DEPARTMENT OF BIOLOGY,
THE UNIVERSITY OF MICHIGAN, ANN ARBOR, MICHIGAN   48109

1. INTRODUCTION
    Detailed knowledge about the structure of the oxygen
evolving complex (OEC) of Photosystem II (PSII) has been
hindered by the instability of the enzyme complex and the
lability of the Mn center when purified from thylakoid
membranes. However, two different approaches have been employed
to overcome these difficulties and have led to more highly
resolved preparations. The first approach utilizes sequential
solubilization of PSII membranes with non-ionic detergent and
allows the reaction center (RC) complex, including the OEC, of
PSII to be separated from the light harvesting complex (LHC)
(1-3). In the second approach, we employed chemical crosslinking to identify proteins which comprise the binding site for the
33 kDa extrinsic protein. This protein is known to be
associated closely with other protein(s) of the OEC and can be
reversibly removed from the membranes by high salt treatment
(5-7). The first approach gives information about the RC as a
whole, while the second approach provides a more direct probe of
the OEC.
    By using chemical crosslinking, we have shown that when
purified 33kDa protein, modified to contain adducts of the
heterobifunctional crosslinking reagent SADP, is rebound to PSII
membranes which lack the 33kDa protein but retain functional Mn,
oxygen evolving activity is reconstituted and a single major
crosslinked complex is formed. Two-dimensional SDS-PAGE
revealed that the complex was composed of 22, 24, 26, 28, 29 and
31 kDa proteins as well as the SADP-modified 33kDa, and the Mn
content was estimated to be 3-4 per crosslinked complex. Three
of the proteins were tentatively identified as LHC proteins. We
now report on the isolation of the crosslinked complex formed
either before or after isolation of the RC in order to take
advantage of the specificity of the ligand-receptor crosslinking
method and the resolution afforded by removal of the LHC using
non-ionic detergents. The crosslinked complex produced by
either method was essentially the same and consisted of 22, 29
and 34 kDa proteins crosslinked to the 33kDa.

2. MATERIALS AND METHODS
    Highly active PSII membranes were prepared (7) from which RC
complexes were isolated as described (3). Conditions for
crosslinking were as follows: 5 mg Chl (PSII) or 1.25 mg Chl
(RC) was reconstituted with 1 mg of SADP modified 33kDa in a
final volume of 20 ml containing the reconstitution buffer
described in (5). RC proteins from crosslinked PSII
preparations were purified by solubilization with

Biggens, J. (ed.), Progress in Photosynthesis Research, Vol. I. ISBN 90 247 3450 9
© *1987 Martinus Nijhoff Publishers, Dordrecht. Printed in the Netherlands.*

octyl-thio-glucoside (OTG) followed by ultrafiltration. The
bulk of the LHC was retained in the supernatant and the
crosslinked complex was recovered in the pellet following
centrifugation. Polyclonal antibodies to the 33kDa protein were
raised in rabbits and the whole serum was used to probe Western
blots of one- and two-dimensional SDS-PAGE gels.

3. RESULTS AND DISCUSSION
    The Mn content of the PSII membranes at various stages
during the preparation of the crosslinked complex was determined
by neutron activation analysis (Table I). More than 80% of the
Mn was retained by the membranes after extraction of the
extrinsic proteins with $CaCl_2$. After crosslinking, residual Mn
was removed with EDTA, and it was found that about 25% of the Mn
present in PSII was retained by the crosslinked membranes. This
is in close agreement with the level of oxygen evolution that
was reconstituted by the modified protein. This suggests that
the crosslinked photosystems that exhibited reconstitution with
SADP-33kDa protein contain about four Mn after EDTA washing.
The lack of oxygen evolving activity following UV irradiation to
crosslink the proteins was consistent with previous results (8)
which showed that UV irradiation inactivates PSII water
splitting activity.

Table I. Quantitation of Mn by Neutron Activation

SAMPLE	Manganese per 230 Chl (%)		% Oxygen Evolution
PS II Membranes	4.3	(100)	100
Calcium washed PS II Membranes	3.5	(>80)	10
Reconst. with SADP-33 kDa	---	---	25
X-Linked PS II after EDTA wash	1	(25)	0

    Crosslinked PSII membranes were solubilized with OTG to
purify PSII reaction centers. Figure 1 shows the constituents
of the OTG soluble fraction (A) and the fraction which was
insoluble in the detergent (B) that were separated by
2-D-SDS-PAGE. LHC proteins partitioned almost exclusively into
the OTG soluble fraction with the bulk of the LHC appearing as a
large spot on the diagonal. Some of the LHC appears below the
diagonal as a continuous band which is not due to crosslinking,
but probably results from incomplete dissociation during
electrophoresis in the first dimension. The crosslinked complex

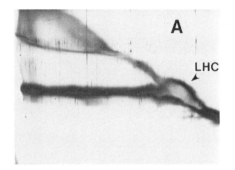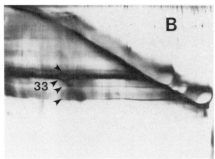

Figure 1. Two-dimensional SDS-PAGE of crosslinked PSII membranes
solubilized with OTG. (A) OTG soluble fraction; (B) OTG insoluble
fraction. Proteins were separated in the first and second dimensions by
10% acrylamide and 12.5% acrylamide with 2-mercaptoethanol, respectively.

co-purified with the RC proteins as shown by the set of
vertically aligned off-diagonal spots in Fig. 1B. The 33kDa
protein was identified in the crosslinked complex by Western
blot analysis using a polyclonal antibody made to the purified
protein. The apparent molecular masses of the proteins
crosslinked to the 33kDa protein are 22, 29 and 34 kDa.

Figure 2 shows the constituents of the crosslinked complex
formed when the 33kDa protein was bound and crosslinked to the
octyl-glucoside-purified PSII reaction center. This crosslinked
complex consisted of four off-diagonal polypeptides with similar
molecular masses to the protein constituents of the crosslinked
complex that was made from PSII membranes.

In our initial purification of the crosslinked complex (4)
we had observed a discrepancy between the sum of the molecular
masses of the constituents of the complex when dissociated and
the apparent molecular mass of the undissociated complex. We
now observe that when LHC is removed either before or after
crosslinking, the sum of the subunit molecular masses is more
consistent with the estimated molecular mass of the
undissociated complex. This would suggest that the LHC proteins
we had observed initially (4) as constituents of the crosslinked
complex were not actually crosslinked, but had co-purified with
the crosslinked complex in the 2-D SDS-PAGE.

We have combined the strengths of two different approaches
in order to probe the protein constituents of the OEC. The
specificity of the ligand-receptor crosslinking method and the
resolving power of detergent solubilization to purify the RC of
PSII have provided the means to identify the proteins which
comprise the binding site for the 33kDa protein. When the LHC
is removed either before or after rebinding and crosslinking,

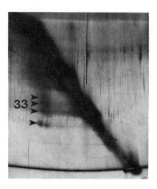

Figure 2. Constituents of the crosslinked complex formed by binding the 33kDa protein to OG-purified PSII reaction centers isolated as described in (3). Conditions for electrophoresis as in Fig. 1.

the isolated complex is shown to consist of 3 proteins crosslinked to the modified 33kDa protein. It is of note that the molecular masses of the subunits, estimated by SDS-PAGE, are similar to those of the L, M and H subunits of the photosynthetic bacterial reaction centers.

REFERENCES
1 Tang,X.-S. and Satoh,K.(1985) FEBS Lett. 179,60-64
2 Ikeuchi,M.,Yuasa,M. and Inoue,Y.(1985) FEBS Lett. 185,318-322
3 Ghanotakis,D. and Yocum,C.(1986) FEBS Lett. 197,244-248
4 Bowlby,N. and Frasch,W.(1986) Biochemistry 25,1402-1407
5 Ono,T. and Inoue,Y.(1983) FEBS Lett. 164,255-260
6 Imoaka,A.,Yanagi,M.,Akabori,K. and Toyoshima,Y.(1984)
  FEBS Lett. 176,341-345
7 Bertold,D.,Babcock,G. and Yocum,C.(1981) FEBS Lett. 134,231-234
8 Yamashita,T. and Butler,W.(1968) Plant Physiol. 43,2037-2040

ACKNOWLEDGEMENTS
This study was supported by NSF (DMB-8604118). The authors are indebted to the assistance of Matt Sanders.

PURIFICATION AND PROTEINCHEMICAL CHARACTERIZATION OF THE EXTRINSIC
MEMBRANE PROTEINS IN THE WATER SPLITTING SYSTEM OF SPINACH.

JOACHIM VATER, JOHANN SALNIKOW, RICCI ZEPMEUSEL AND CHRISTER JANSSON [*)]

INSTITUTE FOR BIOCHEMISTRY AND MOLECULAR BIOLOGY, TECHNICAL UNIVERSITY OF
BERLIN, FRANKLINSTRASSE 29, D-1000 BERLIN 10 and *) DEPARTMENT OF BIO-
CHEMISTRY, UNIVERSITY OF LUND, S-22100 LUND, SWEDEN.

1. INTRODUCTION
    The water splitting complex as a constituent of photosystem II is an
    integral part of the thylakoid membrane. Three peripheral membrane
    proteins with molecular masses of approx. 16, 23 and 33 kD have been
    characterized by several authors (1-4). These proteins show regula-
    tory functions in the water splitting system modulating the action
    of certain inorganic cofactors which are involved in this process.
    In this publication we present data on the primary and secondary
    structure of these proteins.

2. MATERIALS AND METHODS
    Column materials used for the purification of these proteins were
    Ultrogel AcA 54 from LKB (Gräfelfing, FRG) and Mono Q HR 5/5 as well
    as CM-Sepharose Cl-6 B from Pharmacia (Freiburg, FRG).
    For sequence determinations the proteins (ca. 2-4 nmoles) and peptide
    fragments were coupled to amino propyl glass via p-phenylenediiso-
    thiocyanate according to the method of Wachter et al. (5). Cyanogen-
    bromide peptides were also coupled by the homoserine lactone method.
    Automated solid phase Edman degradations were performed with 4-N,N-
    dimethylaminoazobenzene 4'-isothiocyanate/phenylisothiocyanate as
    described by Salnikow et al. (6). Analyses of the 4-N,N-dimethyl-
    aminoazobenzene 4'-thiohydantoins were performed by two dimensional
    micro thin layer chromatography on polyamide (7) and, in addition,
    by reversed phase HPLC using a methanol gradient (8).

3. RESULTS AND DISCUSSION
3.1. Protein purification.
    Photosystem II particles are prepared from spinach thylakoid membra-
    nes by treatment with Triton X-100. The extrinsic membrane proteins
    associated with the water splitting complex were extracted stepwise
    with solutions containing 0,5 M NaCl, 1 M NaCl and 1 M $CaCl_2$. The
    extracts were centrifuged in a Beckman Ti-60 rotor at approx. 200.000
    g for 5 h to remove pigment containing material of high molecular
    weight. Subsequently they were desalted by dialysis. The 16 and 23 kD
    proteins were obtained from the 0,5 or 1 M NaCl extract. They were
    first separated by anion exchange FPLC on Pharmacia Mono Q HR 5/5
    using a Pharmacia FPLC system. The 23 kD protein was eluted with a
    linear gradient ranging from 0-0,3 M NaCl either in 20 mM Tris- or
    20 mM NH_4HCO_3-buffer at pH = 8,0. It was finally purified by AcA 54
    gel filtration. The 16 kD protein was not retained by the Mono Q
    column under these conditions. The excluded volume was dialyzed
    against 20 mM MES-buffer, pH = 6,4 and loaded onto a CM-Sepharose
    column (2,5 x 20 cm). The protein was eluted with a linear gradient
    from 50 to 400 mM NaCl.
    The 33 kD protein was obtained in pure form by a one step procedure
    either by FPLC on Mono Q or by AcA 54 gel filtration from the lyophi-

lisate of the CaCl$_2$ extract.

3.2. Partial sequence of the 16 and 23 kD protein.
   The N-terminal sequences of the 16 and 23 kD protein have been
   determined by solid phase sequencing, as compiled in Table 1.

TABLE 1:   Partial sequences of the 16 and 23 kD protein in photosystem II
   of spinach

---

16 kD protein:

N-terminal sequence:

```
1           5                  10                 15
X-Ala-Arg-Pro-Ile-Val-Val-Gly-Pro-Pro-Pro-Pro-Leu-Ser-Gly-Gly-Leu-Pro-
     20              25
Gly-Thr-Glu-Asn-Ser-Asp-Gln-Ala-X-X-Gly...
```

23 kD protein:

N-terminal sequence:

```
1           5                  10                 15
X-Tyr-Gly-Glu-Ala-Ala-Asn-Val-Phe-Gly-X-Pro-X-X-Asn-Thr-Glu-Phe-Thr-
20
Pro-Tyr...
```

tryptic peptide:

```
X-Ile-Thr-Asp-Ile-Gly-Ser-Pro-Glu-Val-Phe-Leu-Ser-Gln-Val-Asp-Tyr-
Ile/Leu...
```

small BrCN fragment   (approx. 3 kD):

N-terminal segment:

```
1           5                  10                 15
X-Tyr-Gly-Glu-Ala-Ala-Asn-Val-Phe-Gly-Lys-Pro-Lys-Lys-Asn-Thr-Glu-Phe...
                                Gly-Lys-Pro-Lys-Lys-Asn-Thr-Glu-Phe-Thr..
```

1
X, the amino terminus could not be identified, because the amino group is
   coupled to the amino propyl glass matrix;

X, not identified (may be Lys or Cys for internal residues); Ile/Leu,
   Ile and Leu have not been differentiated.

The N-terminal sequence of the 16 kD protein shows an unusual proline rich
stretch of amino acids with obviously hydrophobic character (residues 5-14)
suggesting a distinct conformation for this part of the protein.
The N-terminal sequence of the 23 kD protein has been obtained for 21 residues. This protein possesses 1 Met residue. This is substantiated by
BrCN fragmentation. Two prominent BrCN peptides with molecular masses of
approx. 3 and 20 kD have been obtained by reversed phase HPLC on Pharmacia
Pro RPC using a linear acetonitrile gradient, as demonstrated in Fig. 1.

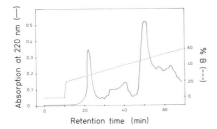

Retention time (min)

FIGURE 1   Separation of the BrCN fragments of the 23 kD protein
            by reversed phase HPLC on a Pharmacia Pro RPC HR 5/10
            column.
            1-3 mg of the freeze - dried 23 kD protein were dissol-
            ved in 1 ml 70 % trifluoroacetic acid and digested with
            1 g BrCN for 24 h at room temperature. The excess of
            BrCN was removed in vacuo. The residue was dissolved in
            2 ml 0,1 % trifluoroacetic acid, pH = 2,2 and injected
            into a LKB HPLC system.
            Gradient solutions: A) 0,1 % trifluoroacetic acid in
            water (v,v), pH = 2,2, B) acetonitrile with 20 % solu-
            tion A (v,v). A linear gradient ranging from 20 to 60 %
            B was used. The flow rate was 0,5 ml/min. The absorption
            was monitored at 220 nm.

Positions 11, 13 and 14 which could not be identified in the course
of the N-terminal sequencing of the whole protein coupled to the
amino propyl glass via amino groups were assigned as lysines by sequencing of the small N-terminal BrCN fragment which has been attached
to the matrix by the homoserine lactone method.
Apart from the presented main N-terminal sequence for some preparations of the 23 kD protein a more or less pronounced side sequence
starting with residue 10 (Gly) has been observed. Also this phenomenon
has been corroborated by the sequence of the 3 kD BrCN fragment which
during degradation proved to consist of two peptides implying partial
cleavage of the ^{9}Phe-^{10}Gly main chain. It has to be clarified, whether
this partial N-terminal processing of the 23 kD protein is due to a
thylakoid-specific protease solubilized by Triton X-100 or is induced
by other unspecific cytosolic enzymes.

3.3. Circular dichroism spectra and a secondary structure analysis of the
     16 and 23 kD protein.
     The circular dichroism spectra of these proteins were measured between
     195 and 240 nm in the region of the peptide backbone chromophores, as
     shown in Fig. 2. The proteins were dissolved in 20 mM NH_4HCO_3-buffer,
     pH = 8,0. Similar results have been obtained using 20 mM sodium
     phosphate buffer, pH = 7,5. A secondary structure analysis was per-
     formed using the method introduced by Provencher and Glöckner (10). In
     this procedure the CD-spectra of these proteins were analyzed as a
     linear combination of the CD-spectra of 16 proteins the secondary
     structure of which is known from X-ray crystallography. The secondary
     structure data obtained for both proteins are compiled in Table 2.

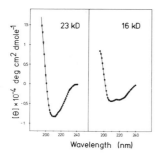

FIGURE 2   Circular dichroism spectra of the 16 and 23 kD proteins
of the water splitting complex of spinach in the range
of 195-240 nm.
Experimental data (x); fit to data (o). The spectra were
measured with a Jasco J-20 spectropolarimeter. Optical
path length: 1 mm; temperature: 25°C.

TABLE 2:   Secondary structure data of the 16 and 23 kD protein

	α-helix %	ß-structure %	remainder %
16 kD protein	6	51	43
23 kD protein	14	61	25

Obviously, both proteins in isolated form are rich in ß-structure,
while α-helical segments are less representative structural elements
of these polypeptides. Preliminary circular dichroism data indicate
a similar characteristic also for the 33 kD protein.

4. REFERENCES

1 Åkerlund, H.-E. and Jansson, C. (1981) FEBS Lett. 124, 229-232
2 Yamamoto, Y., Doi, M., Tomura, N. and Nishimura, M. (1981)
   FEBS Lett. 133, 265-268
3 Kuwabara, T. and Murata, N. (1982) Plant Cell Physiol. 23, 533-539
4 Murata, N. and Miyao, M. (1985) Trends Biochem. Sci. 10, 122-124
5 Wachter, E., Machleidt, W., Hofner, H. and Otto, J. (1973)
   FEBS Lett. 35, 97-102
6 Salnikow, J., Lehmann, A. and Wittmann-Liebold, B. (1981)
   Anal. Biochem. 117, 433-440
7 Chang, J.Y., Brauer, D. and Wittmann-Liebold, B. (1978)
   FEBS Lett. 93, 205-214
8 Chang, J.Y., Lehmann, A. and Wittmann-Liebold, B. (1980)
   Anal. Biochem. 102, 380-383
9 Jansson, C. (1984) in Proc. 6th Internat.Congr. on Photosynthesis
   (Sybesma, C., ed.) pp.1.3-375-1.3-378, M. Nijhoff / Dr. W. Junk,
   Publ., The Hague, Boston, Lancaster
10 Provencher, S.W. and Glöckner, J. (1981) Biochemistry 20, 33-37

PARTIAL AMINO ACID SEQUENCES OF THE PROTEINS OF PEA AND SPINACH
PHOTOSYSTEM II COMPLEX

N. Murata[1], H. Kajiura[1], Y. Fujimura[1], M. Miyao[1], T. Murata[2], A. Watanabe[3]
and K. Shinozaki[4]
[1]National Institute for Basic Biology, Myodaiji, Okazaki 444, [2]Department
of Chemistry, Faculty of Science, Toho University, Miyama, Funabashi
274, [3]Research Institute for Biological Regulation, School of Agriculture,
Nagoya University, Furocho, Chikusa, Nagoya 464 and [4] Center for Gene
Research, Nagoya University, Furocho, Chikusa, Nagoya 464, Japan

1. INTRODUCTION

The photosynthetic oxygen-evolving complex of higher plant chloroplasts
comprises several types of intrinsic and extrinsic proteins (1). Genes
of the intrinsic proteins of molecular masses 47 kDa, 43 kDa, 34 kDa,
32 kDa, 9 kDa and 4 kDa in the photosystem II (PSII) core complex have
been cloned, and their total amino acid sequences have been determined
(2). The PSII complex of spinach thylakoids contains some other proteins
of molecular masses 24 kDa, 22 kDa, 10 kDa and 5 kDa, but their functions
have not been elucidated. The 10-kDa phosphoprotein has been purified,
and its N-terminal sequence of nine amino acid residues determined (3).
The 22-kDa (4) and the 5-kDa protein (5) have been purified, but their
amino acid sequences have not been reported.

Three extrinsic proteins of 33 kDa, 23 kDa and 18 kDa are also involved
in the oxygen-evolving complex. The total amino acid sequence of the
33-kDa protein and a partial sequence of the 18-kDa protein both from
spinach have been determined (6,7).

In order to fully understand the molecular mechanism of the photochemical reaction and oxygen evolution in PSII, it is necessary to determine the amino acid sequences of all the proteins in the PSII complex.
In the present study, we determined partial sequences of the three extrinsic proteins from spinach and pea PSII, and also of the proteins of 22 kDa,
10 kDa and 5 kDa from spinach.

2. MATERIALS AND METHODS

Spinach (*Spinacia oleracea* L.) was purchased from a local market.
Peas (*Pisum sativum* L. Alaska) were grown from seed in a field for two
weeks. Thylakoids were isolated from spinach and pea leaves as described
previously (8). PSII particles were prepared from the thylakoids with
Triton X-100 as described previously (8). The isolated PSII particles
were suspended in 30% (v/v) ethylene glycol/10 mM NaCl/300 mM sucrose/25 mM
Mes-NaOH (pH 6.5) and kept in liquid nitrogen until use. After thawing,
the PSII particles were washed three times with 10 mM NaCl/300 mM sucrose/
25 mM Mes-NaOH (pH 6.5) by suspension and centrifugation. 0.0001% DNase
(Sigma, deoxyribonuclease 1, type IV, from bovine pancreas) was added
to the second washing medium, to facilitate the following chromatographic
procedures.

For purification of the 23-kDa and 18-kDa proteins, the washed PSII
particles were suspended in 1.0 M NaCl/300 mM sucrose/25 mM Mes-NaOH
(pH 6.5) at a chlorophyll (Chl) concentration of 2 mg ml^{-1} and incubated
at 0°C for 30 min (9). The suspension was centrifuged at 35,000 x g for

40 min. The supernatant was dialyzed at $4^{\circ}C$ against distilled water for
3 h and then against 2 mM $(NH_4)_2SO_4$/10 mM Mes-NaOH (pH 5.5) for 1.5 h.
After passage through a Millipore filter (0.45 μm), the dialyzate was
subjected to high performance liquid chromatography (HPLC) in a cation
exchange mode (Toyosoda, SP-650S), and the proteins were eluted with
a linear gradient of $(NH_4)_2SO_4$ from 2 mM to 100 mM in 10 mM Mes-NaOH
(pH 5.5).

For purification of the 33-kDa protein, the NaCl-washed PSII particles
were suspended in 1.0 M $CaCl_2$/25 mM Mes-NaOH (pH 6.5) at a Chl concentration of 2 mg ml^{-1} and incubated at $0^{\circ}C$ for 30 min under room light (10).
The suspension was centrifuged at 35,000 x $\underline{g}$ for 30 min and the supernatant
was dialyzed at $4^{\circ}C$ against distilled water for 2 h and then against
5 mM $(NH_4)_2SO_4$/10 mM Mes-NaOH (pH 6.5) for 5 h. After passage through
a Millipore filter, the dialyzate was subjected to anion exchange HPLC
(Toyosoda, DEAE-5PW), and the protein was eluted with a linear gradient
of $(NH_4)_2SO_4$ from 5 mM to 30 mM in 10 mM Mes-NaOH (pH 6.5).

The 5-kDa protein was purified essentially according to Ljungberg
et al. (5). The $CaCl_2$-treated PSII particles were suspended in 3% Zwitter-
gent 3-14/2% Triton X-100/8 mM Mes-NaOH (pH 6.5) at a Chl concentration
of 5 mg ml^{-1}, and incubated at $0^{\circ}C$ for 30 min. The suspension was centrifuged at 240,000 x $\underline{g}$ for 60 min. The supernatant was subjected to the
cation exchange HPLC, and eluted with a gradient of $(NH_4)_2SO_4$ from 2 mM
to 100 mM in 10 mM Mes-NaOH (pH 7.0).

The hydrophobic proteins of 10 kDa and 22 kDa were isolated by polyacrylamide gel electrophoresis in a preparative mode (Canal Co., Prep.
Disk). After washing with 10 mM NaCl/300 mM sucrose/25 mM Mes-NaOH
(pH 6.5), the $CaCl_2$-treated PSII particles were suspended in 90% acetone.
The resultant acetone powder was dissolved for 30 min at room temperature
in 2% SDS/5% mercaptoethanol/62.5 mM Tris-HCl (pH 6.8), and applied to
a 3 cm-high polyacrylamide gel column containing 4.6 M urea/0.1% SDS/15%
(for the 10-kDa protein) or 12% (for the 22-kDa protein) acrylamide.
Electrophoresis was carried out at a constant voltage of 500 V. Eluted
protein fractions were collected in 0.1% SDS/25 mM Tris-glycine buffer
(pH 8.8).

Purity of each protein preparation was examined by SDS-polyacrylamide
gel electrophoresis (11). The proteins were concentrated, if necessary.
The hydrophilic proteins were dialyzed against 10 mM NaCl for 8 h and
twice against 1 mM NaCl for 8 h each time, and the proteins were then
lyophilized. The hydrophobic proteins were dialyzed twice against 0.1%
SDS and lyophilized. The amino acid sequences of the proteins at the
N-terminal side were determined by a gas-phase protein sequence analyzer
(Applied Biosystems, 470A).

3. RESULTS AND DISCUSSION
Three extrinsic proteins were purified from pea PSII particles in
a way similar to that used for the corresponding proteins from spinach
PSII particles. Their molecular masses estimated by SDS-polyacrylamide
gel electrophoresis were 33 kDa, 20 kDa and 17 kDa. Although the molecular
masses of the latter two were smaller than those of the corresponding
proteins from spinach, they were similar to those of spinach in their
chromatographic behavior. Therefore, we have tentatively named them pea
23-kDa protein and pea 18-kDa protein. The 33-kDa protein from pea appeared
at the same migration range as the 33-kDa protein from spinach. However,
it should be noted that the molecular mass of spinach 33-kDa protein
determined from the total amino acid sequence is actually 27 kDa (6).

TABLE 1. Partial amino acid sequences at the N-terminus of the proteins from PSII particles

No.	33 kDa Pea	33 kDa Spinach	23 kDa Pea	23 kDa Spinach	18 kDa Pea	18 kDa Spinach	22 kDa	10 kDa	5 kDa
1	Glu	Glu	Ala	Ala	Glu	Glu	Ala	Gly	Glu
2	Gly	Gly	Tyr	Tyr	Ala	Ala	Ala	Gly	Glu
3	Ala	Gly	Gly	Gly	Ile	Arg	Ala	Val	Pro
4	Pro	---	Glu	Glu	Pro	Pro	Ala	Lys	Lys
5	Lys	Lys	Ala	Ala	Ile	Ile	Pro	Lys	Arg
6	Arg	Arg	Ala	Ala	Lys	Val	Lys	Ile	Gly
7	Leu	Leu	Asn	Asn	Val	Val	Lys	Lys	Thr
8	(Thr)	Thr	Val	Val	Gly	Gly	Ser	Val	Pro
9	Phe	Tyr	Phe	Phe	Gly	Pro	(Trp)	X	Glu
10	Asp	Asp	Gly	Gly	Pro	Pro	Ile	Lys	Ala
11	Glu	Glu	Lys	Lys	Pro	Pro	Pro	Pro	Lys
12	Ile	Ile	Ala	Pro	Pro	Pro	Ala	Leu	Lys
13	Gln	Gln	Lys	Lys	Leu	Leu	Val	(Gly)	Lys
14	Ser	Ser	Thr	Lys	Ser	Ser	Lys	Ile	Tyr
15	Lys	Lys	Asn	Asn	X	Gly	Gly	(Gly)	Ala
16	(Thr)	Thr	(Thr)	Thr	Gly	Gly	Gly		Pro
17	Tyr	Tyr	Asp	Glu	Leu	Leu	Gly		Val
18	Leu	Leu	Tyr	Phe	Pro	Pro	Asn		X
19	Glu	Glu	Leu	Met	Gly	Gly	Phe		Val
20	Val	Val	Pro	Pro	Thr	Thr	Leu		Thr
21	Lys	Lys	Tyr	Tyr	Leu	Glu	Asp		Met
22	Gly	Gly	Asn	Asn	Asn	Asn	Pro		Pro
23	X	Thr*	(Gly)	Gly	Ser	Ser	(Glu)		Ser
24	X	Gly*	Asp	Asp	Asp	Asp	(Trp)		Ala
25	X	Thr*	(Gly)	Gly	Glu	Gln	Leu		Arg
26	Ala	Ala*	Phe	Phe	Ala	Ala**			Ile
27	Asn	Asn*	Lys	Lys	Arg	Arg			X
28	Gln	Gln*	Leu	Leu	Asp	Asp			Tyr
29	X	Cys*	Leu	Leu	Leu	Gly			Lys
30	Pro	Pro*	Val	Val	Lys	Thr**			
31		Thr*	Pro	Pro	Leu	Leu			
32		Val*	Ala	Ser	Pro	Pro**			
33		Glu*	Lys	Lys		Tyr**			
34		Gly*	(Lys)	Trp		Thr**			
35		Gly*	Asn	Asn		Lys**			
36		Val*	Pro	Pro		Asp**			
37		Asp*				Arg**			
38		Ser*				Phe**			
39		Phe*				Tyr**			
40		Ala*				Leu**			
41		Phe*				Gln**			
42		Lys*				Pro**			
43		Pro*				Leu**			
44		Gly*				Pro**			

*Taken from the total sequence of the protein (6). **Taken from the sequence of the 17-kDa fragment (7).

Table 1 shows the N-terminal amino acid sequences of the extrinsic proteins of 33 kDa, 23 kDa and 18 kDa from pea and spinach and also of the proteins of 22 kDa, 10 kDa and 5 kDa from spinach. When the 33-kDa proteins from pea and spinach are compared, the most striking difference is an insertion of proline at the 4th residue of the pea protein. The amino acid substitutions occur at the 3rd and 9th residues from the N-terminus in the 33-kDa protein, at the 12th, 14th, 17th, 18th, 19th and 32nd residues in the 23-kDa protein, and at the 3rd, 6th, 9th, 21st, 25th, 29th and 30th residues in the 18-kDa protein.

The 22-kDa protein from spinach has a unique sequence at the N-terminus, i.e., four alanine residues in series. The sequence of the 10-kDa protein differs from those of the phosphoprotein (3) and the large polypeptide of cytochrome b-559 (12) which have similar molecular masses. This 10-kDa protein is assumed to correspond to the one purified by Ljungberg et al (4). Two-thirds of the total sequence of the 5-kDa protein was determined. Within this partial sequence, there was no sign of a hydrophobic region capable of spanning the membrane.

The amino acid sequences of the extrinsic 33-kDa, 23-kDa, 18-kDa and 5-kDa proteins and the hydrophobic 22-kDa and 10-kDa proteins from spinach PSII particles were compared with the amino acid sequence derived from the complete nucleotide sequence of the tobacco chloroplast genome (13). No homology higher than 50% was observed between them, suggesting that the genes of these proteins are not encoded on the chloroplast DNA, but probably on the nuclear DNA.

REFERENCES

1. Murata, N. and Miyao, M. (1985) Trends Biochem. Sci. 10, 122-124
2. Cramer, W.A., Widger, W.R., Herrmann, R.G. and Trebst, A. (1985) Trends Biochem. Sci. 10, 125-129
3. Farchaus, J. and Dilley, R.A. (1986) Arch. Biochem. Biophys. 244, 94-101
4. Ljungberg, U., Åkerlund, H.-E. and Andersson, B. (1986) Eur. J. Biochem., in press
5. Ljungberg, U., Henrysson, T., Rochester, C.P., Åkerlund, H.-E. and Andersson, B. (1986) Biochim. Biophys. Acta 849, 112-120
6. Oh-oka, H., Tanaka, S., Wada, K., Kuwabara, T. and Murata, N. (1986) FEBS Lett. 197, 63-66
7. Kuwabara, T., Murata, T., Miyao, M. and Murata, N. (1986) Biochim. Biophys. Acta 850, 354-361
8. Kuwabara, T. and Murata, N. (1982) Plant Cell Physiol. 23, 533-539
9. Miyao, M. and Murata, N. (1983) Biochim. Biophys. Acta 725, 87-93
10. Kuwabara, T., Miyao, M., Murata, T. and Murata, N. (1985) Biochim. Biophys. Acta 806, 283-289
11. Laemmli, U.K. (1970) Nature 227, 680-685
12. Widger, W.R., Cramer, W.A., Hermodson, M., Meyer, D. and Gullifor, M. (1984) J. Biol. Chem. 259, 3970-3876
13. Shinozaki, K., Ohme, M., Tanaka, M., Wakasagi, T., Hayashida, N., Matsubayashi, T., Zaita, N., Chunwongse, J., Obokata, J., Yamaguchi-Shinozaki, K., Ohto, C., Torazawa, K., Meng, B.Y., Sugita, M., Deno, H., Kamogashira, T., Yamada, K., Kusuda, J., Takaiwa, F., Kato, A., Tohdoh, N., Shimada, H. and Sugiura, M. (1986) EMBO J., in press.

PROLINE-RICH STRUCTURE AT AMINO-TERMINAL REGION OF THE 18-kDa
PROTEIN OF PHOTOSYNTHETIC OXYGEN-EVOLVING COMPLEX

TOMOHIKO KUWABARA, TERUYO MURATA, MITSUE MIYAO[*], AND NORIO
MURATA[*], DEPARTMENT OF CHEMISTRY, FACULTY OF SCIENCE, TOHO
UNIVERSITY, MIYAMA, FUNABASHI 274, JAPAN AND [*]NATIONAL INSTITUTE
FOR BASIC BIOLOGY, MYODAIJI, OKAZAKI 444, JAPAN

1.   INTRODUCTION
    The machinery for photosynthetic oxygen evolution in higher plants
contains three extrinsic proteins of 33 kDa, 24 kDa, and 18 kDa (1). The 18-
kDa protein is known to be a regulatory factor which functions in retaining
Cl^-, an essential factor for the oxygen evolution, at the functional site of the
oxygen-evolving complex (2,3). The functioning of the 18-kDa protein requires
the 24-kDa protein, since the binding site for the 18-kDa protein on the
thylakoids is afforded by the 24-kDa protein (4). Molecular characteristics of
the 18-kDa protein have been partially studied (5,6). However, this protein is
easily degraded to a polypeptide of 17 kDa during purification, and this
degradation has retarded the study on the structure and function of the
protein. We have partially characterized the degradation, and have purified the
18-kDa protein and the 17-kDa fragment. In this report, we compare the
molecular characteristics of the protein and the fragment, and discuss the
structure of the 18-kDa protein in light of its N-terminal sequence.

2.   MATERIALS AND METHODS
    Photosystem (PS) II particles were prepared from spinach chloroplasts as
in (7) except that the chloroplasts were suspended in 100 mol m^{-3} NaCl/300 mol
m^{-3} sucrose/50 mol m^{-3} Na-K phosphate (pH 6.4) for Triton X-100 treatment.
The modifications in NaCl concentration and pH value provided a higher yield
for the PS II particles.
    The N-terminal amino acid sequence of protein was determined by the
Edman degradation method where amino acid phenylthiohydantoins were
identified (8) with protein sequence analyzers. The M_r of the protein was
estimated by SDS-PAGE in the buffer system of Laemmli (9). The effective
molecular size of protein in the absence of SDS was estimated by gel filtration
chromatography as in (10).

3.   RESULTS
Characterization of protein degradation.   Treatment of PS II particles with 1
kmol m^{-3} NaCl (pH 6.5) released the 18-kDa and 24-kDa proteins with trace
amounts of the 33-kDa protein (10). When the extract was dialyzed against a
low-salt medium, the 18-kDa protein was degraded to a polypeptide of 17 kDa.
Fig. 1 shows a typical time course of the degradation; the 17-kDa band was
detected after 3 h and increased with further dialysis, whereas the 18-kDa
band became faint with dialysis and diminished at 48 h. The 24-kDa protein
also seemed to be degraded to a polypeptide of 22 kDa. The degradation of
these proteins did not occur when the dialysis medium contained 500 mol m^{-3}
NaCl, or when the dialysis was performed at pH 9.0 instead of pH 6.6. These
dependences suggested the involvement of a protease in the degradation.
Several protease inhibitors were tested to see if they could elicit an inhibitory
effect on the degradation. The following reagents were ineffective;

Biggens, J. (ed.), Progress in Photosynthesis Research, Vol. I. ISBN 90 247 3450 9
© *1987 Martinus Nijhoff Publishers, Dordrecht. Printed in the Netherlands.*

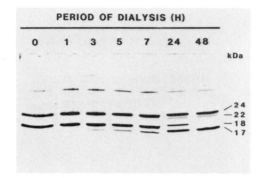

PERIOD OF DIALYSIS (H)

0 1 3 5 7 24 48

kDa

24
22
18
17

FIGURE 1. Change in the polypeptide composition of NaCl-extract from PS II particles during dialysis. The NaCl-extract was dialyzed against 20 mol m^{-3} Na phosphate (pH 6.6) at 7°C for a designated period.

0.1 mol m^{-3} p-chloromercuribenzoate, 1 mol m^{-3} N-ethylmaleimide, 10 mol m^{-3} Na iodoacetate, 1 mol m^{-3} DTT, 2 mol m^{-3} PMSF, 5 mol m^{-3} benzamidine-HCl, 2.5 mol m^{-3} 1,10-phenanthroline, 10 mol m^{-3} EDTA, 5 mol m^{-3} 6-amino-n-caproic acid, 1 mol m^{-3} α-N-benzoyl-L-arginine. However, once the extract was dialyzed against 1 kmol m^{-3} CaCl$_2$ (pH 6.5) or 1 kmol m^{-3} Tris-HCl (pH 9.3), the proteins were not degraded during the subsequent dialysis in a low-salt medium at pH 6.5, indicating that these treatments can inactivate the protease.

Characterization of the 18-kDa protein and the 17-kDa fragment   The 18-kDa protein and the 17-kDa fragment were purified by column chromatography and their molecular characteristics were compared. The $\underline{M}_r$ of the 18-kDa protein was estimated to be 18 kDa by SDS-PAGE, but the effective molecular size estimated by gel filtration chromatography in the absence of SDS was 23 kDa (10). This observation suggests that the molecular shape of the 18-kDa protein is not spherical, which would make its effective size much larger than expected from the $\underline{M}_r$. In contrast, the 17-kDa fragment showed similar values for the $\underline{M}_r$ and the effective molecular size in the absence of SDS, 17 kDa and 18 kDa, respectively. This finding suggests that the molecular shape of the 18-kDa protein was considerably altered along with the conversion to the 17-kDa fragment. The isoelectric points of the 18-kDa protein and the 17-kDa fragment were almost the same, 9.5 and 9.6, respectively. No significant difference could be detected between their absorption spectra (data not shown).

Figure 2 shows the N-terminal amino acid sequence of the 18-kDa protein. The most striking feature was the occurrence of sequential 4 proline residues, Pro-9 to Pro-12. These residues may form a rigid structure similar to the type II trans helix of poly-L-proline. The secondary structures of other parts in the N-terminal region were predicted according to Garnier et al (11). The prediction suggested that the rigid structure is preceded by a β-sheet, and followed by a β-turn (Fig. 2). No part of the region was predicted to be an α-helix.

The N-terminal sequence of the 17-kDa fragment was the same as that from Leu-13 of the 18-kDa protein, indicating that the protease of PS II particles cleaved the peptide bond at the carboxy end of the sequential proline residues, between Pro-12 and Leu-13 (Fig. 2). The molecular weight of the removed 12 amino acid residues, about 1200, explained the observed $\underline{M}_r$

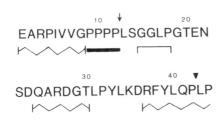

FIGURE 2. The N-terminal amino acid sequence of 18-kDa protein. Common secondary structures, β-sheet (ʍ) and β-turn (⌐⌐), and a rigid structure by sequential proline residues (▬), are shown below the sequence. An arrow indicates the Pro-Leu bond cleaved by the protease of PS II particles. A triangle shows another Pro-Leu sequence found in the N-terminal region.

difference between the 18-kDa protein and 17-kDa fragment. The C-terminal region of the protein might not be cut by the protease, since the difference in total amino acid compositions of the 18-kDa protein and the 17-kDa fragment agreed well with the composition of the 12 amino acids (data not shown).

4. DISCUSSION

It was shown that the oxygen-evolving PS II particle preparation (7) contained an extrinsic protease which could be liberated with 1 kmol m^{-3} NaCl. This finding suggests that the protease is endogenously associated with thylakoids by ionic interaction. According to the cleavage site in the 18-kDa protein, between Pro-12 and Leu-13, the protease may be classified into a category of prolyl endopeptidase (E.C. 3.4.21.26), which is the only enzyme species known that can cleave the peptide bond at the carboxy end of proline residue within a peptide chain (12).

Another Pro-Leu sequence was found in the N-terminal region, Pro-42 — Leu-43 (Fig. 2). However, this site did not seem to be a good substrate for the protease; the accumulation of the 17-kDa fragment in the course of the degradation (Fig. 1) suggested that the 17-kDa fragment was not effectively degraded by the protease. This Pro-Leu sequence may be located at a site to which the protease cannot gain access; this site is probably buried in the protein folding. Alternatively, the protease may recognize a sequence of more than two amino acid residues.

The N-terminal structure of the 18-kDa protein seems to present a structural basis for the large difference between its M_r and effective molecular size in the absence of SDS. The sequential 4 proline residues should form a rigid structure, and thus prevent the polypeptide part of Glu-1 to Gly-8 from being folded. Since the secondary structure of that part was predicted to be a β-sheet, it is likely to have an extended conformation without significant bending. This situation may explain the larger effective molecular size of the 18-kDa protein. This idea was supported by the finding that the 17-kDa fragment, which lacks the N-terminal 12 amino acid residues, showed similar values for the M_r and the effective molecular size. Based on the above considerations, we propose an N-terminal protrusion model for the 18-kDa protein (Fig. 3). Since the 17-kDa fragment cannot bind to NaCl-treated and 24-kDa protein-supplemented PS II particles with high affinity (13), the

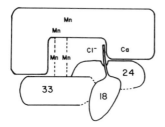

FIGURE 3. A tentative model for the binding of the 18-kDa protein to the oxygen-evolving complex. The intrinsic protein part is shown without defining individual proteins.

polypeptide part of Glu-1 to Pro-12 should be responsible for the binding of the protein to the oxygen-evolving complex. It is of note that the sequential proline residues may be relatively negative and have a potential to form intermolecular hydrogen bonds with their carbonyl group as hydrogen acceptors (14). It is very likely that the hydrogen bonds at the non-charged amino acid residues as well as the ionic interactions at the charged ones are involved in the binding of the protein.

ACKNOWLEDGEMENT
We thank Professor L.A. Sherman, University of Missouri, for the use of a computer program for the prediction of secondary structures of proteins.

REFERENCES
1   Murata, N. and Miyao, M. (1985) Trends Biochem. Sci. 10, 122-124
2   Akabori, K., Imaoka, A. and Toyoshima, Y. (1984) FEBS Lett. 173, 36-40
3   Miyao, M. and Murata, N. (1985) FEBS Lett. 180, 303-308
4   Miyao, M. and Murata, N. (1983) Biochim. Biophys. Acta 725, 87-93
5   Kuwabara, T. and Murata, N. (1983) in Advances in Photosynthesis Research (Sybesma, C., ed.), Vol. 1, pp. 371-374, Martinus Nijhoff/Dr. W. Junk Publishers, The Hague
6   Jansson, C. (1983) in Advances in Photosynthesis Research (Sybesma, C., ed.), Vol. 1, pp. 375-378, Martinus Nijhoff/Dr. W. Junk Publishers, The Hague
7   Kuwabara, T. and Murata, N. (1982) Plant Cell Physiol. 23, 533-539
8   Zimmerman, C.L., Appella, E. and Pisano, J.J. (1977) Analyt. Biochem. 77, 569-573
9   Laemmli, U.K. (1970) Nature 227, 680-685
10   Kuwabara, T. and Murata, N. (1983) Plant Cell Physiol. 24, 741-747
11   Garnier, J., Osguthorpe, D.J. and Robson, B. (1978) J. Mol. Biol. 120, 97-120
12   Wilk, S. (1983) Life Sciences 33, 2149-2157
13   Kuwabara, T., Murata, T., Miyao, M. and Murata, N. (1986) Biochim. Biophys. Acta 850, 146-155
14   Vies, A. and Nawrot, C.F. (1970) J. Am. Chem. Soc. 92, 3910-3914

TOPOGRAPHICAL STUDIES ON SUBUNIT POLYPEPTIDES OF OXYGEN-EVOLVING PHOTOSYSTEM II PREPARATIONS BY REVERSIBLE CROSSLINKING: FUNCTIONS OF TWO CHLOROPHYLL-CARRYING SUBUNITS

ISAO ENAMI[a], TAKESHI MIYAOKA[a], SAHOKO IGARASHI[a],
KAZUHIKO SATOH[b] and SAKAE KATOH[b]
[a]Department of Biology, Faculty of Science, Science University of Tokyo, Kagurazaka, Shinjuku-ku, Tokyo 162; [b]Department of Pure and Applied Sciences, College of Arts and Sciences, University of Tokyo, Komaba, Meguro-ku, Tokyo 153, (Japan)

INTRODUCTION

The PS II reaction center complex contains two chlorophyll-carrying subunit polypeptides of about 47 and 43 kDa, together with several smaller intrinsic polypeptides (1). The primary photochemistry takes place on the 47 kDa polypeptide (2), whereas the 43 kDa polypeptide serves as antenna (3). Of the three extrinsic proteins of 33, 24 and 18 kDa involved in water oxidation, the 33 kDa protein was recently shown to be directly associated with the PS II reaction center complex (4-6). The PS II reaction center complex is also linked to large numbers of antenna pigment protein complexes, light-harvesting chlorophyll a/b proteins (LHCP) in higher plants and green algae, and phycobiliproteins in red algae and cyanobacteria (7,8).

Here we report the nearest-neighbor relationships among subunit polypeptides of two oxygen-evolving preparations examined with cleavable bifunctional crosslinking reagents (dithiobis-succinimidyl esters). The results suggest that the 47 kDa polypeptide carries the Mn-binding sites, whereas the 40 kDa chlorophyll protein of cyanobacterial PS II complex accepts excitation energy from phycobilisomes.

MATERIALS and METHODS

Oxygen-evolving PS II membrane preparations were isolated from spinach with Triton X-100 as in Ref.(9) and from the thermophilic cyanobacterium Synechococcus sp. with β-octyl glucoside as in Ref.(10). Spinach preparations were crosslinked by adding 4 μl of freshly prepared dimethyl-sulfoxide solution of dithiobis(succinimidyl propionate) (DSP, 10 mg/ml, Pierce) successively 6 times with 10 min intervals to 1 ml of sample suspension containing 1 mg chlorophyll, 20 mM Mes-NaOH (pH 6.5), 0.4 M sucrose, 10 mM NaCl and 5 mM $MgCl_2$. To analyze crosslinked products, samples were treated with 8 M urea and 5% SDS for 30 min at 45°C, and their polypeptide compositions were examined by electrophoresis with a 9.5-11.5% acrylamide gradient gel containing 6 M urea. A 10 mm wide strip of the gel was then incubated for 1 h at 30°C in 125 mM Tris-HCl (pH 8), 10% 2-mercaptoethanol, 8 M urea and 5% SDS, and applied to the second dimensional electrophoresis on a 11.5% acrylamide gel containing 6 M urea.

Synechococcus oxygen-evolving preparations (0.5 mg Chl./ml) were crosslinked with 0.1% dithiobis(succinimidyl glycolate) (DSG) in a medium containing 50 mM Hepes (pH 7), 1 M sucrose, 10 mM NaCl and 5 mM $CaCl_2$ for 1 h at room temperature. Electrophoretic conditions were described in Text.

Biggens, J. (ed.), Progress in Photosynthesis Research, Vol. I. ISBN 90 247 3450 9
© *1987 Martinus Nijhoff Publishers, Dordrecht. Printed in the Netherlands.*

RESULTS and DISCUSSION

1) <u>Crosslinking of spinach oxygen-evolving membrane preparations</u>

Fig. 1-Ia shows polypeptide profile of untreated spinach oxygen-evolving PS II preparations examined by one dimensional electrophoresis. The 47 kDa reaction center chlorophyll-binding polypeptide, the 43 kDa antenna chlorophyll-binding polypeptide, 29-26 kDa subunits of LHCP, and three extrinsic proteins of 33, 24 and 18 kDa associated with water oxidation were resolved together with other polypeptides. Essentially, all polypeptides migrated along a diagonal path on the second dimensional electrophoresis (Fig. 1-Ib).

DSP is a crosslinking reagent which specifically reacts with amino groups with a crosslinking span of 1.2 nm (11). Because the crosslinker has a disulfide bond in the middle of the symmetrical molecule, proteins crosslinked can be identified by gel electrophoresis after cleavage of the disulfide bond. Treatment of the PS II preparations with DSP produced three crosslinked product bands labelled A, B and C on the one dimensional electrophoresis (Fig. 1-Ic). Second dimensional electrophoresis after cleavage of the disulfide bond with 2-mercaptoethanol revealed many off-diagonal spots originated from A, B and C (Fig. 1-Id). Three spots of 24, 20 and 18 kDa were resolved from A. Crosslinking of NaCl-washed preparations which lack the 24 and 18 kDa proteins produced no band A but the 20 kDa spot was still present in its second dimensional electrophoretogram (Fig. 1-IId). This indicates that A ia a crosslinked product of the

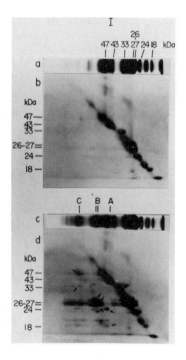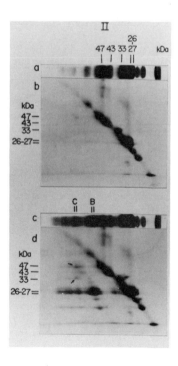

Fig. 1   One- and two-dimensional electrophoreses of spinach oxygen-evolving preparations crosslinked with DSP.   I. unwashed preparations; II. preparations washed with 1 M NaCl.

24 and 18 kDa proteins and the 20 kDa spot is a protein comigrated with band A.

Band B was identified as a crosslinked product of 29-26 kDa apoproteins of LHCP. Its long tailing may be ascribed to oligomeric forms of the products. Band B itself was often split into two bands.

The diagonal electrophoresis showed several off-diagonal spots including the 47 kDa chlorophyll-carrying protein and the 33 kDa extrinsic protein at the positions corresponding to C (Figs. 1-Id and 1-IId). Note that the pattern of the 33 kDa spot with a faint tailing was very similar to that of the 47 kDa protein but considerably different from those of other spots of 43, 29-26 and 20 kDa. The 29-26 and 20 kDa spots are considered to be contamination due to their long tailings. Polypeptides of the 43 kDa and at the gel front were separated from C by reelectrophoresis (not shown). In addition, C had an apparent molecular mass of 80 kDa which well corresponds to the sum of molecular masses of the 47 kDa and 33 kDa proteins. It is therefore concluded that the 33 kDa protein is specifically crosslinked with the 47 kDa protein. Because the 33 kDa protein is considered to stabilize the Mn center (12), the result suggests that the 47 kDa protein carries the Mn-binding sites.

2) <u>Crosslinking of Synechococcus oxygen-evolving preparations</u>

The oxygen-evolving preparations isolated from <u>Synechococcus</u> do not contain intact phycobilisomes but still are associated with a considerable amount of allophycocyanin (10). The bound allophycocyanin molecules were completely solubilized by washing with 2 M NaBr. However, when the preparations had been crosslinked with dithiobis(succinimidyl glycolate) (DSG), another bifunctional crosslinking reagent with a cleavable disulfide bond, repeated NaBr-washings failed to liberate a significant fraction of phycobiliprotins, solubilization of which required further treatment with 2-mercaptoethanol to cleave the disulfide bond of the crosslinker.

Fig. 2 shows absorption and fluorescence spectra of the phycobiliprotein which was crosslinked and then solubilized with 2-mercaptoethanol. The

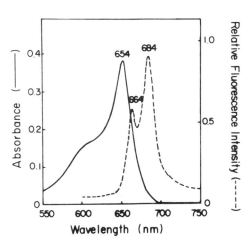

Fig. 2 Absorption and fluorescence spectra of phycobiliproteins crosslinked with <u>Synechococcus</u> oxygen-evolving preparations.

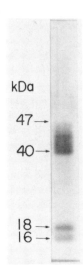

Fig. 3 Polypeptides resolved from a crosslinked product b.

absorption spectrum measured after removal of 2-mercaptoethanol corresponds to that of allophycocyanin trimers. Interestingly, the fluorescence emission spectrum determined at liquid nitrogen temperature shows, besides an emission band of allophycocyanin at 664 nm, a prominent band at 684 nm which is ascribed to a long wavelength emitter of phycobilisomes. The results suggest that a terminal energy trap of phycobilisome is crosslinked with the PS II complex.

To identify a protein(s) of the PS II preparation crosslinked with allophycocyanin, the DSG-treated preparations were incubated with 4% SDS for 30 min at room temperature, then applied to electrophoresis on a 9.2% acrylamide gels with 0.05% SDS present only in reservoir buffer. Two green bands (called a and b) of crosslinked products were resolved under the mild electrophoresis conditions (not shown). Polypeptide compositions of bands a and b were examined by gel electrophoresis after treatment with 10% 2-mercaptoethanol, 10% SDS and 8 M urea for 2 h at room temperature. Band a was identified to be the 47 kDa protein crosslinked with the 31 kDa intrinsic protein which may correspond to the D-2 protein (not shown). The 40 kDa antenna chlorophyll-binding polypeptide and 18-16 kDa subunits of allophycocyanin were resolved from band b (Fig. 3). This strongly suggests that allophycocyanin B which emits a long wavelength fluorescence at 77K (8) is crosslinked with the 40 kDa subunit. However, Fig. 3 also shows a diffuse band of about 43 kDa. The 43 kDa polypeptide may be a proteolytic product of the anchor protein, which also emits the 684 nm band (8), because digestion of the anchor protein with endogenous proteases yielded a fragment of a comparable size (not shown). At any event, the results indicate that the terminal energy trap(s) of phycobilisome is crosslinked with the 40 kDa protein. It is concluded that the 40 kDa antenna chlorophyll-carrying protein plays a key role in excitation energy transfer from phycobilisomes to the PS II reaction center.

REFERENCSES
1 Satoh, Ki. (1985) Photochem. Photobiol. 42, 845-853
2 Satoh, Ka. FEBS Lett. in press
3 Yamagishi, A. and Katoh, S. (1985) Biochim. Biophys. Acta 807, 74-80
4 Tang, X.-S. and Satoh, Ki. (1985) FEBS Lett. 179, 60-64
5 Satoh, Ka., Ohno, T. and Katoh, S. (1985) FEBS Lett. 180, 326-330
6 Ikeuchi, M., Yuasa, M. and Inoue, Y. (1985) FEBS Lett. 185, 316-322
7 Zuber, H. (1985) Photochem. Photobiol. 42, 821-844
8 Glazer, A.N. (1984) Biochim. Biophys. Acta 768, 29-51
9 Kuwabara, T. and Murata, N. (1982) Plant Cell Physiol. 23, 533-539
10 Satoh, Ka. and Katoh, S. (1985) Biochim. Biophys. Acta 806, 221-229
11 Middaugh, C.R., Vanin, E.F. and Ji, T.H. (1983) Mol. Cell Biochem. 50, 115-141
12 Ono, T. and Inoue, Y. (1984) FEBS Lett. 168, 281-286

TENACIOUS ASSOCIATION OF THE 33kDa EXTRINSIC POLYPEPTIDE (WATER SPLITTING) WITH PS II PARTICLES

EDITH L. CAMM and BEVERLEY R. GREEN, Department of Botany, University of British Columbia, Vancouver, B.C., Canada, V6T 2B1.

1. INTRODUCTION

Recent reports suggest that it is possible to prepare PS II particles with a very limited number of polypeptides which carry out both water-splitting and core functions (1, 2, 3). These preparations are very depleted in the polypeptides of the antennal complexes, but retain the 33kDa polypeptide associated with the water-splitting complex. This polypeptide has been shown to be hydrophilic (4). Its presence in PS II core preparations suggests that a so-called extrinsic component of PS II is more tightly linked to the core complex than are the intrinsic membrane components of the antennal complexes. In this report, we demonstrate directly that a fraction of the population of the 33kDa extrinsic protein (33kDa EP) appears to be very tightly bound to the membrane, remaining even after the application of published methods of removal. We suggest that an hydrophobic section of an otherwise hydrophilic molecule might act as an anchor to the PS II core proteins to facilitate integration of water-splitting and core functions.

2. PROCEDURE

2.1. Preparation of washed particles

PS II particles were prepared by the method of Berthold, Babcock and Yocum as outlined by (5). The particles were stored frozen at -80° in the B4 medium of (5) (15 mM NaCl, 5 mM $MgCl_2$, 20 mM HEPES pH 7.5 and 400 mM sucrose). Aliquots of the frozen PS II particles were used for various washing procedures.

2.1.1. 1 M $CaCl_2$-cholate wash. The method of (6) as modified by (7) was followed with the single substitution of HEPES for MOPS during the three initial washes of the thawed PS II particles.

2.1.2. NaCl/urea wash. The method of (8) employing washes of 1M NaCl and of 2.3 M urea in the light was followed.

2.1.3. Tris wash. PS II particles were resuspended in 16 mM tris-HCl pH 8.3, 10 mM NaCl at 250 ug/ml and were shaken on ice in the dark for 15 min (9). The particles were pelleted and then resuspended for a second cycle.

In the case of each wash, the final pellet was resuspended in B_4 medium prior to electrophoresis or activity measurement.

Biggens, J. (ed.), Progress in Photosynthesis Research, Vol. I. ISBN 90 247 3450 9
© *1987 Martinus Nijhoff Publishers, Dordrecht. Printed in the Netherlands.*

2.2. Octyl glucoside gradient

PS II particles were prepared from 6-day old barley plants as in part 2.1. Particles were washed with $CaCl_2$ as above. The washed pellets were suspended in 300 mM octyl glucoside at a detergent: chl ratio of 10:1, and at once diluted to 30 mM octyl glucoside. The solutions were layered on sucrose gradients (30 mM octyl glucoside) (10) and were centrifuged for 4 hrs at 65,000 xg in an 80 Ti rotor (Beckman; fixed angle). Five fractions and the pellet at the bottom of the tube were collected.

2.3. Electrophoresis and Immunoblotting

Samples were electrophoresed using the methods in (10) on 10% acrylamide gels. Gels were blotted onto nitrocellulose in 50 mM acetate buffer, pH7. The blots were reacted with antibody at a dilution of 200:1. Anti-33kDa EP and anti-D1 were the generous gifts of Dr. David Allred, (University of Colorado, Boulder, CO) and Dr. Lee MacIntosh (Michigan State University, East Lansing, MI) respectively. Antigen-antibody complexes were visualized using goat anti-rabbit-linked alkaline phosphatase (Kirkegard and Perry Labs) by the method of White and Green (unpublished).

The relative amounts of each antigen were determined by densitometry of the nitrocellulose blots. The value for the intensity of each spot was multiplied by a factor representing the width of the spot. Linear relationships were obtained between the volume of test preparations of 33kDa EP-containing material applied to a gel, and the corrected intensity of the resulting spots (correlation co-efficients equal to or greater than 0.98).

3. RESULTS AND DISCUSSION

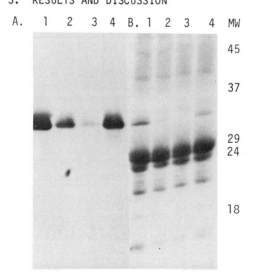

FIGURE 1. The removal of 33kDa EP from spinach PS II particles by various washing procedures.
A. immunoblots from a gel with 0.25 ug chl/lane.
B. Commassie stained gel, 2.5 ug chl/lane.
Lane 1. control membranes.
Lane 2. Ca++/cholate-washed membranes.
Lane 3. NaCl/urea-washed membranes.
Lane 4. 16 mM tris-washed membranes.

Figure 1 shows Coomassie-stained lanes of control and washed PS II membranes. The 33kDa EP was identified in each case by immunoblotting (Figure 1, left side). The three washing procedures differ in their effectiveness in removing the 33kDa EP. Washing with 16 mM tris is least effective and the polypeptide is easily recognized both in stained gels and immunoblots. The Ca++/cholate wash is more effective in removal, although a considerable amount of 33kDa EP still remains. Most effective under our conditions is the two-step wash with NaCl and urea. The 33kDa EP is not visible by Coomassie staining, although immunoblotting (Figure 1, Table 1) suggests that about 10% of the population still remains attached. Table 1 shows that there is no correlation between PS II activity (water-splitting or core activity) and the residual 33kDa EP.

TABLE 1. Residual 33kDa EP (quantified from immunoblots) and PS II activity after various washing procedures.

Treatment	33kDa EP	percent of control remaining PS II activity	
		water-splitting $H_2O \rightarrow$ DCPIP	core DPC $\rightarrow$ DCPIP
Ca++/cholate wash	46.8 ± 5.0%	33 ± 13%	96 ± 29%
NaCl/urea wash	9.8 ± 2.0%	below detection	81 ± 13%
16 mM tris wash	82.2 ± 17.9%	below detection	96%

Figure 2 shows the distribution of 33kDa EP along a detergent gradient separation of Ca++/cholate washed PS II particles from barley. The distribution of this so-called extrinsic protein is compared with that of the intrinsic protein D1 (the Q_B-binding protein). Since these polypeptides have similar molecular masses, discrimination between them is difficult except by antibodies.

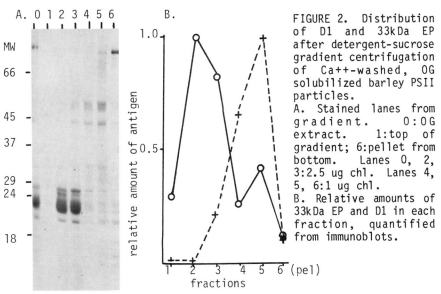

FIGURE 2. Distribution of D1 and 33kDa EP after detergent-sucrose gradient centrifugation of Ca++-washed, OG solubilized barley PSII particles.
A. Stained lanes from gradient. 0:OG extract. 1:top of gradient; 6:pellet from bottom. Lanes 0, 2, 3:2.5 ug chl. Lanes 4, 5, 6:1 ug chl.
B. Relative amounts of 33kDa EP and D1 in each fraction, quantified from immunoblots.

The uppermost fraction (containing solubilized material that did not migrate into the gradient) has a very low content of stainable bands, yet a moderately high content of 33kDa EP, indicating that some was solubilized by the detergent treatment. However, even after the initial Ca++-wash and the subsequent detergent treatment, there is still a substantial proportion of the molecules migrating into the gradient fractions occupied by D1 and the polypeptides of the reaction centre (see stained lanes 4 and 5).

There are a number of conclusions to be drawn. First, it is difficult to assess the degree of depletion of the 33kDa EP on the basis of Coomassie stain alone. Immunoblotting is much more sensitive and furthermore is specific to the protein in question. Second, not all washing procedures are equally effective. While other workers claim perfect depletion by a Ca++-wash (7, 11), in our hands this was not so, and the NaCl/urea wash (8) was the only resonably effective one. This persistence of the 33kDa EP could be of special importance in experiments involving the characteristics of a washed membrane (e.g. Mn++ content).

Some more general conclusions can also be drawn. It appears that a fraction of the population of molecules of 33kDa EP is very tightly bound to the thylakoid membrane. This is in agreement with work showing a high affinity binding site and a low affinity binding site for 33kDa EP on washed PS II membranes (12). We suggest that while the 33kDa molecule as a whole exhibits hydrophilic and extrinsic properties, there could be a localized hydrophobic domain. This domain, whether a consequence of primary or secondary structure, could be the part of the molecule binding to high affinity sites resulting in the intimate association with the membrane. The rest of the population could occupy the more hydrophilic sites.

4. REFERENCES

1. Tang, X.-S. and Satoh, K. (1985). FEBS Lett. 179, 60-64.
2. Ikeuchi, M., Yuasa, M. and Inoue, Y. (1985). FEBS Lett. 185, 316-22.
3. Ikeuchi, M. and Inoue, Y. (1986). Arch. Biochem. Biophys. 247, 97-107.
4. Bricker, T.M. and Sherman, L.A. (1982). FEBS Lett. 148, 197-202.
5. Dunahay, T.G., Staehelin, L.A., Seibert, M., Ogilvie, P.D. and Berg, S.P. (1984). Biochim. Biophys. Acta 764, 179-93.
6. Ono, T.-A. and Inoue, Y. (1983). FEBS Lett. 164, 255-60.
7. Imaoka, A., Akabori, K., Yanagi, M, Izumi, K., Toyoshima, Y., Kawamori, S., Nakayama, H. and Sato, J. (1986). Biochim. Biophys. Acta 848, 201-11.
8. Miyao, M. and Murata, N. (1984). Biochim. Biophys. Acta 765, 253-57.
9. Kuwabara, T. and Murata, N. (1982). Plant Cell Physiol. 23, 533-39.
10. Camm, E. and Green, B.R. (1983). Journ. Cell. Biochem. 23, 171-79.
11. Chapman, D.J. and de Felice, J. (1986). Biochem. Soc. Trans. 14, 38-39.
12. Bowlby, N.R. and Frasch, W.D. (1986). Biochemistry 25, 1402-07.

Thermodynamic Constraints to Photosynthetic Water Oxidation

Lee Spencer and Donald T. Sawyer*
Department of Chemistry
Texas A&M University
College Station, Texas 77843

Andrew N. Webber and Robert L. Heath
Department of Botany and Plant Sciences
University of California
Riverside, California 92521

Most reviewers now argue that photosynthetic oxygen evolution results from a sequential four-step electron transfer process in which oxidizing equivalents from chl $\underline{a}^{+\cdot}$ are accumulated in a "charge-storing" complex to accomplish the concerted four-electron oxidation of two H_2O molecules to one O_2 molecule. The photooxidant (chl $\underline{a}^{+\cdot}$) and reductant (pheo$^{-\cdot}$) are one-electron transfer agents and the matrix is the lipoprotein thylakoid membrane. Hence, evaluation and consideration of the one-electron redox potentials for PSII components within a lipoprotein matrix are necessary in order to assess the thermodynamic feasibility of proposed mechanistic sequences.On the basis of recent membrane models[1] dipolar aprotic solvents such as dimethylformamide (with a peptide bond), dimethylsulfoxide, acetonitrile, and butyronitrile (with their low proton availability and polar hydrophilic character) provide environments that closely parallel that of a lipid bilayer with embedded protein. This concept has prompted the evaluation of the reduction potentials for some of the PSII components in dipolar aprotic media. Several model systems have been chosen which, on the basis of recent evidence, closely approximate the structure and redox properties of the in-vitro system.

1. The Primary Charge Separation.

The primary oxidizing component of PSII is chl $\underline{a}^{+\cdot}$, which is the product of the primary photoact and must be the most electropositive species in the system. Measurements of transmembrane potentials for chl $\underline{a}$ and [Mg^{II}(OEP)] in simulated lipid bilayers yield derived potentials of approximately +0.7 V vs. NHE,[2] which are close to the measured potential for the formation of [Mg^{II}(OEP)]$^{+\cdot}$ in dimethylformamide (+0.77 V vs. NHE). This result is consistent with the evaluations by other workers for a wide variety of model compounds and solvents (+0.76 ± 0.10 V). Thus, any chemical transformation at the active site requiring a redox potential in excess of +0.77 V is not a viable reaction pathway.

2. Oxidation of Z.

Measurement of the redox potentials[3] for 1,4-hydroquinone (H_2Q) in dipolar aprotic media has demonstrated that oxidation to its corresponding cation radical ($H_2Q^{+\cdot}$), which could be the mechanism for the operation of Z, only occurs within base-free media. For such conditions the oxidation potential is greater than +1.0 V vs NHE. Consideration of the one-electron oxidation potential of H_2Q in acetonitrile (+ 1.42V vs. NHE) with that for the HQ$^-$ anion:(-0.13V vs. NHE) provides an estimate of + 0.58V vs. NHE for the one-electron oxidation potential for H_2Q under neutral acid-base conditions [picolinic acid (HA)/picolinate

Biggens, J. (ed.), Progress in Photosynthesis Research, Vol. I. ISBN 90 247 3450 9
© *1987 Martinus Nijhoff Publishers, Dordrecht. Printed in the Netherlands.*

TABLE 1. Redox Thermodynamics for PSII Components in Acetonitrile.

A. Redox Potentials of Oxidizing Side of PSII.

Couple	$E^{o'}$, V vs NHE
Chl $\underline{a}^{+\cdot} + e^- \rightarrow$ Chl $\underline{a}$	+0.77
$PQH\cdot + PAH + e^- \rightarrow PQH_2 + PA^-$	+0.58
$\rightarrow 1/2PQ + 1/2PQH_2$	

Complex	Mn(III)/(II)	Mn(IV)/(III)
$[Mn^{III}(DPA)(PA)(H_2O)]$	+0.58	+1.60
$[Mn^{III}(PA)_3]$	+0.62	+1.61
$[Mn^{III}(PA)_2(acac)]$	+0.41	+1.45
$[Mn^{III}(8\text{-}Q)_3]$	+0.16	+1.18
$[Mn^{III}(acac)_3]$	+0.11	+1.21

B. Redox Potentials of Water Oxidation to Dioxygen.

Couple	$E^{o'}$, V vs NHE
$O_2 + e^- \rightarrow O_2^-\cdot$	-0.6
$HO_2\cdot + e^- + H^+ \rightarrow H_2O_2$	+1.0
$H_2O_2 + e^- \rightarrow OH\cdot + OH^-$	+0.4
$OH\cdot + e^- + H^+ \rightarrow H_2O$	+1.4
$O_2 + 2e^- + 2H^+ \rightarrow H_2O_2$	+0.2
$H_2O_2 + 2e^- + 2H^+ \rightarrow 2H_2O$	+0.9
$O_2 + 4e^- + 4H^+ \rightarrow 2H_2O$	+0.6

(A$^-$) buffer]. These values should shift only slightly when corrected for the inductive effect of the substituents that occur in plastoquinone.

3. Manganese Complexes

The redox potentials for several mononuclear tris chelates of Mn(III) that are formed by the bidentate ligands 8-quinolinate (8-Q), picolinate (PA), and acetylacetonate (acac), and the tridentate pyridine-2,6-dicarboxylate (DPA) have been measured.[4] These complexes constitute a series of uncharged Mn(III) complexes that have oxo and pyridyl nitrogen donor groups analogous to the histidine, glutamic acid, aspartic acid and tyrosine residues that act as ligands in proteins. As such, their redox potentials should be representative of those required to generate similar manganese oxidation states at the active PSII site (see Table I).

4. Oxygen

Selected values for the reduction of O_2 and of its reduced species (in various states of protonation) in acetonitrile with a picolinic acid/picolinate buffer are summarized in Table I. The assumptions that (a) there are no major solvent effects for neutral species and that (b) the picolinic acid/picolinate buffer system in acetonitrile provides an effective acidity equivalent to pH 9 lead to estimated two-electron oxidation potentials of +0.9 V and +0.2 V vs. NHE for the H_2O/H_2O_2 and H_2O_2/O_2 couples, respectively (Table I).

(4) Conclusions

Combination of the redox potentials for the PSII components (and their appropriate model systems) serves to emphasize that on the basis of the thermodynamic constraints described:

(a) The most highly oxidizing species in PSII is chl $\underline{a}^{+\cdot}$, which has a redox potential of +0.77 V vs. NHE.

(b) The energy available for chemical transformation at the active site of water oxidation is governed by the redox potential of Z for which the $PQH_2/PQH\cdot$ couple, with an estimated potential of +0.58 V vs. NHE, is the only reasonable candidate.

(c) This Z_{ox}/Z_{red} redox manifold (+0.58 V) precludes the formation of any known Mn(IV) intermediates ($\sim$ +1.3V) at the active site.

(d) The electron-transfer oxidation of water to peroxide at the active site via one- (+1.4 V) or two-electron (+0.9 V) mechanisms that are driven by the +0.58-V manifold are precluded on thermodynamic grounds.

5. Proposed Mechanisms

A sequence of reactions that is consistent with the observed redox potentials of PSII components would involve the initial oxidation of plastoquinol. The resulting quinone, PQ, makes possible a water-oxidation scheme that involves an initial concerted removal of the protons from the water molecules of a binuclear manganese(II) center $[Mn^{II}(OH_2)]_2$ by a two-electron oxidant (PQ) to yield hydroxide ions bound to Mn(III):

$$4\ Chl\ \underline{a}^{+\cdot} + 4\ PQH_2 \rightarrow 4\ Chl\ \underline{a} + [4HPQ\cdot] + 4\ H^+$$
$$\hookrightarrow 2\ PQ + 2PQH_2$$

$$[Mn^{II}(OH_2)]_2 + PQ \rightarrow PQH_2 + [Mn^{III}(OH)]_2$$

The oxidation of such bound hydroxide then can be facilitated by (a) stabilization of the hydroxyl radical by the four partially filled d-orbitals of Mn(III), (b) the removal of protons by the product of the oxidant, and (c) a concerted two-centered process with the formation of a peroxo bridge between the metal centers

$$PQ + [Mn^{III}(OH)]_2 \rightarrow PQH_2 + [Mn^{III}(O\cdot)]_2$$
$$\hookrightarrow [Mn^{III}\text{-O-O-}Mn^{III}]$$

The resulting peroxide would then spontaneously decompose to dioxygen and regenerate the catalyst.

$$[Mn^{III}\text{-O-O-}Mn^{III}] \xrightarrow{\ 2H_2O\ } [Mn^{II}(OH_2)]_2 + O_2$$

A recently proposed model redox system for PSII[5] provides a pathway for water oxidation that also is compatible with the thermodynamic limits summarized in Table 1. Features of this model include (a) formation of a peroxide linkage prior to coupling and electron transfer at the manganese active site, which obviates the need for a highly oxidizing and thermodynamically unfavorable manganese center; (b) the predicted pattern of proton release on going from $S_0 \rightarrow [S_4] \rightarrow S_0$ of 1:0:1:2, which is in accord with recent kinetic experiments;[6] and (c) a proposed role for cytochrome b_{559} [base-induced reduction of PQ and formation of quinone peroxide intermediate, $PQ_2(H_2O_2)$], which is consistent with its observed reactivity in PSII.

I. Photoact of Reaction Center

$4[\text{Chl } \underline{a} - \text{Pheo}]_{PSII} + 4h\upsilon \rightarrow 4 \text{ Chl } \underline{a}^{+\cdot}_{PSII} + 4 \text{ Pheo}^{-\cdot}$

II. Reducing side

A. $4 \text{ Pheo}^{-\cdot} + 4Q_A \rightarrow 4 \text{ Pheo} + 4Q_A^{-\cdot}$

B. $2 Q_A^{-\cdot} + 2 \text{ Cyt}^{III}b_{559}(OH_2) \rightarrow 2 \text{ Cyt}^{II}b_{559}(^-OH) + 2H^+ + 2Q_A$

C. $2Q_A^{-\cdot} + PQ + 2H^+ \rightarrow 2Q_A + PQH_2$ (To PSI system)

III. Oxidizing side

A. $2 \text{ Chl } \underline{a}^{+\cdot} + (Mn^{II})_2 \overset{a}{\rightarrow} 2 \text{ Chl } \underline{a} + (Mn^{III})_2$

B. $2 \text{ Chl } \underline{a}^{+\cdot} + 2 \text{ Cyt}^{II}b_{559} + 2 H_2O \rightarrow 2 \text{ Chl } \underline{a} + 2 \text{ Cyt}^{III}b_{559}(OH_2)$

C. $2 \text{ Cyt}^{II}b_{559} (^-OH) + 2Q_w \rightarrow 2Q_w(^-OH) + 2 \text{ Cyt}^{II}b_{559}$

D. $2Q_w(^-OH) + 2Q_A \rightarrow 2Q_A^{-\cdot} + 2Q_w(\cdot OH)$

E. $2Q_A^{-\cdot} + 2H^+ + PQ \rightarrow 2Q_A + PQH_2$ (To PSI system).

F. $2Q_w(\cdot OH) \rightarrow Q_w\text{-O-O-}Q_w$

G. $Q_w\text{-O-O-}Q_w + (Mn^{III})_2 \rightarrow 2Q_w + (Mn^{II})_2 + O_2$

[a]Via the intermediacy of the Z redox couple.

Acknowledgement
The work was supported by the National Institutes of Health under grant No. GM-36289 to D.T.S

References and Notes
1. K. Sauer, Acc. Chem. Res. 11, 257 (1978).
2. A. Ilani and D. Mauzerall, Biophys J. 35, 79 (198
3. M. D. Stallings, M. M. Morrison and D. T. Sawyer, Inorg. Chem., 20, 2655 (1981)
4. K.S. Yamaguchi and D.T. Sawyer, Inorg. Chem, 24, 971 (1985).
5. A.N. Webber, L. Spencer, D.T. Sawyer and R.L. Heath, FEBS. Lett. 189, 258 (1985)
6. W. A. Cramer, J. Whitmarsh and W. Widger, in "Photosynthesis Electron Transport and Phosphorylation" G. Ahogunoglou, Ed. (Balaban International Science Services Philadelphia, 1981) p. 509.

BINUCLEAR AND TETRANUCLEAR MANGANESE COMPLEXES: AS MODELS FOR THE SITE FOR PHOTOSYNTHETIC WATER OXIDATION[1]

J.E. Sheats[a], B.C. UnniNair[b], V. Petrouleas[c], S. Artandi[b]
R. S. Czernuszewicz and G. C. Dismukes[b]
[a]Department of Chemistry, Rider College, Lawrenceville, NJ 08648
[b]Department of Chemistry, Princeton University, Princeton,NJ 08544
[c]Department of Physics, NRC "Demokritos", Athens, Greece.

The catalytic site for photosynthetic water oxidation has been shown by spectroscopic and biochemical studies to contain two pairs of manganese ions electronically coupled to one another.[1-3] It is known that chloride ion is required for photo-oxidation of this system. Therefore, we have attempted to prepare and study the chemistry of polynuclear manganese complexes containing halide ions.

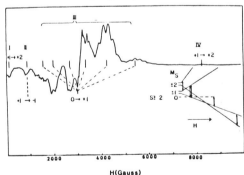

I a X=Br
b X=ClO$_4$

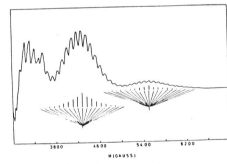

We find that the planar tridentate 2,6-diacetylpyridine dioxime, H_2dapd, forms only mono-nuclear 2:1 complexes $(H_2$dapd$)_2$MnX$_2$ Ia,b in the presence of Cl⁻. The EPR spectra of Ia,b in Ch$_3$CN agree closely with those of other Mn(II) complexes having D_{3h} symmetry or lower. By contrast, the solution EPR (in DMF) of $(H_2$dapd$)$MnCl$_2$ shows zero-field splitting and resolved hyperfine structure characteristic of an even spin cluster S≥ 2 (Fig. 1).

Fig. 1 EPR SPECTRUM OF $(H_2$dapd$)$MnCl$_2$

Fig. 2 EXPANSION OF FIG. 1

The symmetric hyperfine pattern (Fig. 2) on each zero field peak consists of 21 lines separated by 86 gauss. The intensities and spacing are consistent with electronic coupling between four ^{55}Mn(II) (I=5/2).

[1]Supported by Research Grant CHE-82-17920 from the NSF (Princeton)
The Research Corporation Cottrell Grant (Rider) and a NATO Travel Grant #183.

Biggens, J. (ed.), Progress in Photosynthesis Research, Vol. I. ISBN 90 247 3450 9
© *1987 Martinus Nijhoff Publishers, Dordrecht. Printed in the Netherlands.*

The magnetic susceptibilty changes little from 294K to 65K (6.02-6.34 BM/Mn), but rises sharply to 8.40 Bm/Mn at 4.2K (16.8 BM/Mn$_4$) and appears to be approaching a limiting value of 21.0 Bm/Mn$_4$, consistent with an S=10 cluster comprised of four ferromagnetically coupled Mn(II) ions. The exchange interaction is weak with J=0.2 cm^{-1}. The spectroscopic equivalence of the Mn(II) ions and infrared spectral evidence for both bridging and terminal chloride ions has led us to propose the cyclic structure II (Unni Nair, Petrouleas, Dismukes, submitted). Attempts to obtain an X-ray crystal structure of II are in progress.

The unique ability of Cl$^-$ to assemble a tetrameric Mn complex can be explained by the stronger ionic bonding afforded by the smaller Cl^{-1} ion. This suggests the possibility that part of the Cl$^-$ requirement in photosynthesis may arise from the need for the initial assembly of a tetrameric unit from very dilute Mn(II) solutions.

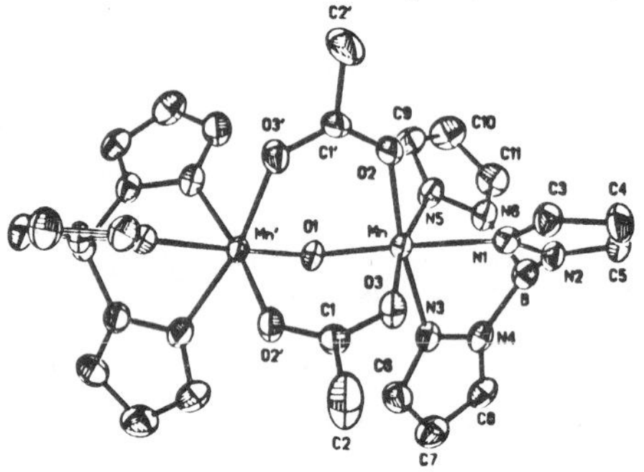

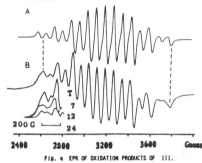

Fig. 3 X-RAY CRYSTAL STRUCTURE OF III

Fig. 4 EPR OF OXIDATION PRODUCTS OF III.
A, KMnO$_4$; B, I$_2$
INSET, TEMPERATURE DEPENDENCE OF B

An understanding of the biological Mn site must also encompass the coordination chemistry of H$_2$O, OH$^-$ and O$^=$. This led us to prepare the dimanganese analog[6], III, of the binuclear iron complex HBPz$_3$FeO(CH$_3$CO$_2$)$_2$FePz$_3$BH, a model for the oxygen-binding site in hemerythrin.[7] The X-ray crystal structure of III is shown in Fig. 3. Variable temperature magnetic susceptibility of III shows that the two Mn(III) ions are magnetically uncoupled, in contrast to the strong antiferromagnetic coupling the Fe(III) analog. The solution structure of III was confirmed by proton NMR spectroscopy. All eight protons were assigned and the distance of several of these from the Mn ions can be determined from the dipolar line broadening. These results make it very clear that high resolution proton NMR spectroscopy can be a powerful structural tool for understanding Mn(III) sites in chemistry and biology.

Scheme I summarizes the reactions of III. Chemical exchange of both the oxo-bridge and carboxylate bridges occurs readily in contact with aqueous buffers and is catalyzed by H$^+$. Below pH 1 rapid loss of the carboxylate bridges occurs and a symmetrical complex forms, tentatively assigned structure IV, it shows only three contact-shifted resonances in the proton NMR, indicating that all three pyrazoles are equivalent. Cyclic voltammetry shows a partially reversible oxidation to a Mn$_2$(III,IV) dimer at 1.06 V (NHE) and an irreversible oxidation to a putative Mn$_2$(IV,IV) species at 1.76V. Different mixed valence states can be formed chemically depending on the choice of oxidants and solvent. The oxidation product with aqueous KMnO$_4$ is an electrically neutral complex soluble in nonpolar solvents. Resonance Raman spectroscopy reveals four peaks in the region characteristic of a four-membered Mn(III) (O)$_2$Mn(IV) ring. The

Scheme I

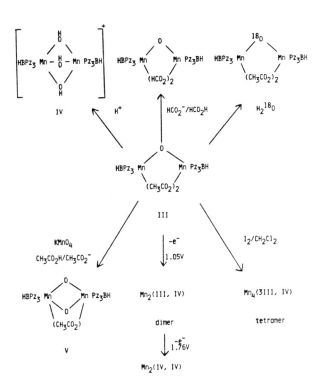

III

expected isotopic shifts were observed on substitution with ^{18}O(Table I).

Table I Raman Frequencies (cm^{-1}) for V and its ^{18}O Analogue.

^{16}O	705	665	634	603	570	515	430	391
^{18}O	676	665(sh)	610	584	570(sh)	494	413	391

These data, as well as IR data and mass spectral data for apparent mass of m/e =641 are consistent with the neutral di-μ-oxo structure (V). Thus, oxidation to the mixed valence state, which is formally equivalent to the presumed oxidation state of Mn in the S_2 state of the water oxidizing enzyme[1], leads to the displacement of carboxylate ligands by water and the subsequent hydrolysis to form the di-μ-oxo structure. These data establish a chemical model for the binding of water in what is equivalent to the S_2 state of the water oxidizing complex.

The EPR spectrum of V at 6-200K yields a 16-line multiplet (Fig. 4), with a width of 1180 gauss, analogous to that found for $(bpy)_2 Mn^{III}(O)_2 Mn^{IV}(bpy)_2^{+3}$.[8] Good agreement with simulated spectra is obtained using hyperfine parameters of 77 gauss and 159 gauss for the two Mn ions. The temperature dependence of the EPR signal is consistent with that of a simple dimer, having a ground state spin S=1/2 (Fig. 5). A similar behavior is observed for the S_2 multiline EPR signal in spinach PS-II particles when formed by continuous illumination at 300K.

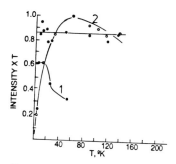

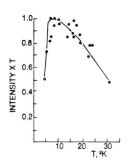

Fig. 5 TEMPERATURE DEPENDENCE OF THE EPR INTENSITY FOR:
(●) KMnO₄ OXIDATION PRODUCT: (O) I₂ OXIDATION PRODUCT
PEAKS LABELLED 1 AND 2 (SEE FIG. 6).

Fig. 6 TEMPERATURE DEPENDENCE OF THE EPR INTENSITY OF
THE S₂ MULTILINE SIGNAL PRODUCED BY 200K ILLUMINATION

Another mixed valent species produced by oxidation of III with I_2 in the absence of water shows a 16-line pattern of width 1130 gauss at 6K, which gradually changes into a different 16-line pattern of width 1200 gauss at 60K. At intermediate temperatures more than 16 lines are observed. Behavior of this type has not been observed for any antiferromagnetically coupled dimer but could arise either from a ferromagnetically coupled dimer or from a Mn_4(3 III,IV) tetramer,[8] which could be formed by reaction of an initially formed dimer with another molecule of III. Temperature dependence of this type has also been observed for the 19-line multiline signal for the S_2 state formed by illumination at 200K (Fig. 6).

In summary, we can mimic both types of thermal behavior observed for the S_2 state by chemical models derived from III. The unusual change in the temperature dependence which occurs upon going from the 200K form of the S_2 state to the 300K form (multiple turnover) is duplicated by the model system simply by allowing the mixed-valence state, derived from oxidation of III, to react with water. This constitutes the first experimental evidence that the room temperature form of the S_2 state differs from the low temperature form of the S_2 state by the reaction with water, the substrate for the enzyme.

References

1. Dismukes, G. C. (1986) *Photochem. Photobiol.*, 43, 99.
2. DePaula, J. C.; Beck, W. F.; Brudvig, G. W. (1986) *J. Am. Chem. Soc.*, 108, 4002.
3. Yachandra, Z. K.; Guiles, R. D.; McDermott, A.; Britt, R. D.; Dexheimer, S. L.; Sauer, K.; Klein, M. P. (1986) *Biochimica et Biophysica Acta*, 850, 324.
4. Critchely, C. (1985) *Biochimica Biophysica Acta* 811, 33.
5. Damoder, R.; Klimov, V. V.; Dismukes, G. C. (1986) *Biochimica et Biophysica Acta*, 848, 378.
6. Sheats, J.E.; Czernuszewicz, R.S.; Dismukes, G.C.; Rheingold, A.; Petrouleas, V.; Stubbe, J.; Armstrong, W. H.; Beere, R. H.; Lippard, S.J. (1986) *J. Am. Chem. Soc.* (submitted).
7. Armstrong, W. H.; Lippard, S. J. (1983) *J. Am. Chem. Soc.* 105, 4837.
8. Dismukes, G. C.; Ferris, K.; Watnick, P. (1982) *Photobiochem. Photobiophys.* 3, 343.

MODELS FOR MANGANESE CENTERS IN METALLOENZYMES.

Vincent L. Pecoraro,* Dimitris P. Kessissoglou, Xinhua Li and
William M. Butler. Contribution from the Department of Chemistry,
University of Michigan, Ann Arbor, MI 48109 USA.

Interest in the chemistry of manganese has blossomed as the awareness
of this element's role in photosynthetic oxygen evolution has increased.
Reports of photochemically assisted oxygen production by small molecule
manganese complexes [1], structurally characterized multinuclear manganese
complexes [2,3] and mononuclear Mn(IV) compounds [4,5] have appeared.
These studies are developing a foundation for the understanding of the
structure and reactivity of manganese centers in biological systems. We
report herein the structural, spectroscopic and magnetic properties of a
series of mono- and multinuclear manganese coordination compounds. The
first of these are mononuclear Mn(IV) species. The second are bi- and
trinuclear Mn(III) complexes. The third are mono- and binuclear Mn(II)
compounds containing Mn-S (disulfide) bonds.

Mononuclear Mn(IV) Complexes

A red-brown solid, $Mn(SALADHP)_2$, is obtained when Mn(II) or Mn(III)
Acetate is reacted, in methanol, with SALADHP (below) and KOH (2
equivalents). The ORTEP diagram of $Mn(SALADHP)_2$ illustrates that the

SALADHP   **SALAHP**

Mn(IV) ion is six-coordinate, forming a bis(tridentate) complex. All of
the bond angles about the Mn(IV) are nearly $90°$. The Mn(IV)-O distances
are 1.889 Å (alkoxide) and 1.906 Å (phenolate) and the Mn(IV)-N distance is
2.005 Å. The cathodic potential in DMF is -0.52 V (vs. SCE) indicating
that the Mn(IV) oxidation state is stabilized dramatically by the alkoxide
oxygen atoms. The epr spectrum shown below is typical of a Mn(IV) ($S=3/2$
system) in an axial environment. The feature at $g=2$ exhibits both [55]Mn
nuclear hyperfine and nitrogen superhyperfine coupling. Another broad
signal is observed at lower field ($g=5.2$). Four other mononuclear Mn(IV)
complexes have been synthesized and show these axial type EPR signals;
however, a rhombic signal with well resolved g_x, g_y and g_z is obtained
from $Mn(SALAHP)_2$. The materials which display axial signals form one 5
and one 6-membered chelate ring with the Mn(IV), presumably as the
meridional isomer. In contrast, SALAHP forms two six-membered rings. The
increased flexibility imparted by the larger chelate ring may allow the
SALAHP ligand to form a facial isomer giving rise to the rhombic signal.

Biggens, J. (ed.), Progress in Photosynthesis Research, Vol. I. ISBN 90 247 3450 9
© *1987 Martinus Nijhoff Publishers, Dordrecht. Printed in the Netherlands.*

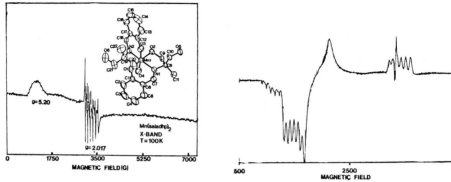

Binuclear and Trinuclear Mn(III) Complexes

When SALADHP is reacted with Mn(II) Acetate in methanol in the absence of a base such as KOH a red-brown solid which has a trinuclear composition is obtained. The structure of $Mn_3(SALADHP)_2(Acetate)_4$ (methanol)$_2$, a Mn(III)$_2$/Mn(II) mixed valence complex, is shown below. The

$Mn_3(SALADHP)_2(Acetate)_4(methanol)_2$ $[Mn(SALAHP)(Acetate)]_2$

complex has Mn(II) ion sitting on a center of symmetry. This Mn(II) ion satisfies octahedral coordination by being bridged to two terminal Mn(III) ions through two acetate groups and an alkoxide oxygen atom of SALADHP. Each Mn(III) ion forms a distorted octahedron by filling the sixth site with a methanol oxygen atom. The solid state room temperature magnetic moment is consistent with this oxidation state assignment. The Mn ions do not appear to interact strongly, even at liquid helium temperatures. Selected bond lengths and angles are given in Table 1.

Table 1. Bond Lengths (Å) and Angles (°) for $Mn_3(SALADHP)_2$ $(OAc)_4(MeOH)_2$.

Mn1-Mn2	3.551	Mn1-O2	1.885	Mn2-O2	2.144
Mn1-O6	1.962	Mn2-O7	2.207	Mn1-O8	2.346
Mn1-O2-Mn2	121.1	Mn1-Mn2-Mn1	180	O2-Mn1-O6	98.3

While SALADHP readily forms this trinuclear material, SALAHP has been reported to form a binuclear Mn(III/III) acetate bridged dimer schematically illustrated at right above. We have been unable to synthesize this acetate dimer with SALADHP; however, we can isolate the dimeric benzoate analogue. This material is green and has a magnetic moment slightly reduced from the spin only value for Mn(III) suggesting that the manganese ions show small magnetic coupling.

Mn(II)-Disulfide Complexes

Although containing lone pair electrons, disulfide sulfur atoms are often considered to be unable to form complexes with metal ions, especially those which are relatively strong Lewis acids. We have prepared Mn(II) complexes of the pentadentate chelating agent SALPS. This ligand uses two phenolate oxygen, a disulfide sulfur and two imine nitrogen atoms to bind to the metal. The other disulfide sulfur does not bind to the Mn(II). As shown in the ORTEP diagram of the mononuclear material, the sixth coordination site is filled by solvent. When the monomeric material is dissolved in non-donor solvents the methanol is

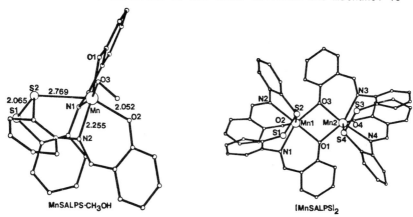

MnSALPS-CH$_3$OH   [MnSALPS]$_2$

displaced and a new dimeric material crystallizes. An ORTEP of the dimer, [Mn(SALPS)]$_2$, is also shown above. In both complexes, the Mn(II)-S bond is very long (Table 2). In addition, the Mn(II) octahedron is highly distorted with angles ranging between 75 and 105°. Bond lengths for the monomer are very similar to those of the dimer reported in Table 2.

Table 2. Bond Distances for [Mn(II)SALPS]$_2$ (Å).

Mn1-Mn2	3.300	Mn1-O1	2.140	Mn1-N1	2.266
Mn1-S2	2.757*	Mn1-S1	3.929	S1-S2	2.064
Mn2-S4	2.707*	Mn2-S3	3.887	S2-S4	2.052

*Bonded to Mn(II). S3 and S4 are not coordinated to the metal.

The complex epr spectrum of the monomer (90 K) in DMF illustrates the highly assymmetric electronic environment of the complex. At first, we were surprised to observe an epr spectrum of the dimer (in toluene solution, 90 K). That this spectrum does arise from the dimer is supported by two pieces of evidence. First, the number of hyperfine lines for a binuclear system is predicted to be $2(I_{Mn1} + I_{Mn2}) + 1$. The ^{55}Mn nuclear hyperfine of the dimer can be resolved into groups of eleven, not six. Second, the nuclear hyperfine coupling in the dimer should be 1/2 that of the monomer. In this case, the monomer has a 93 Gauss coupling while the dimer is 47 Gauss. The complex exhibits a magnetic moment of 5.9 BM at room temperature. Variable temperature magentic data indicate that the Mn(II) ions are weakly antiferromagnetically coupled with

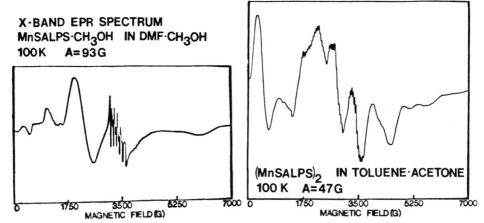

X-BAND EPR SPECTRUM
MnSALPS·CH₃OH   IN DMF·CH₃OH
100K   A=93G

(MnSALPS)₂   IN TOLUENE·ACETONE
100 K   A=47G

MAGNETIC FIELD (G)   MAGNETIC FIELD (G)

$J <-2$ cm^{-1}. Thus, the simple observation of an epr spectrum alone is not sufficient evidence to assign a material as a mixed valence complex. Additional information such as detailed magentic data should be obtained to confirm this assignment.

Conclusions

We have presented in this report data useful in the evaluation of models for the manganese center in the oxygen evolving center of photosystem II. Strong Lewis bases such as alkoxide oxygen atoms stabilize the Mn(IV) oxidation state. In contrast, poorer bases such as carboxylates favor the Mn(III) oxidation state [4]. Depending on the geometry of the metal complex, one can observe either axial or rhombic epr spectra for mononuclear Mn(IV) complexes. Multinuclear Mn(II) and Mn(III) complexes which are bridged by acetate, phenolate or alkoxide oxygen atoms exhibit weak magnetic interaction. In no case reported herein are values of $J >-15$ cm^{-1}. The epr spectrum for the [MnSALPS]₂ illustrates that simply observing a complex epr signal for a binuclear center is not sufficient evidence for the assignment of the material as a mixed valence complex. Additional information such as detailed magnetic data are required. Future studies aimed at defining the reactivity of these and other Mn compounds should be useful in defining structurally and mechanistically the involvement of manganese centers in biological systems.

References
1. Ashmawy, F.M.; McAuliffe, C.A. and Parish, R.V. (1984) J. Chem. Soc., Chem. Comm., 14-15.
2. Kirby, J.A.; Robertson, A.S.; Smith, J.P.; Thompson, A.C.; Cooper, S.R. and Klein, M.P. (1981) J. Amer. Chem. Soc., 103, 5529-5536.
3. Lynch, M.W.; Hendrickson, D.N.; Fitzgerald, B.J. and Pierpont, C.G. (1984) J. Amer. Chem. Soc., 106, 2041-2049.
4. Pavacik, P.S.; Huffman, J.C. and Christou, G. (1986) J. Chem. Soc., Chem. Comm., 43-44.
5. Kessissoglou, D.P.; Butler, W.M. and Pecoraro, V.L. (1986) J. Chem. Soc., Chem. Comm., in press.

MOLECULAR ORBITAL STUDY (IV) ON THE "MICROSURFACE" MODEL OF CATALYTIC
BINUCLEAR MANGANESE COMPLEX IN PHOTOSYNTHETIC WATER-SPLITTING AND OXYGEN-
EVOLVING REACTION

MASAMI KUSUNOKI, Fuculty of Engineering, Meiji Univ., Kawasaki, 214, Japan

1. INTRODUCTION
    The complete oxidation of water molecules in photosystem II(PSII) is un-
derstood to be realized by an enzymatic organization of a special Mn cluster
in a protein architecture maintained at appropriate solvent ionic condition
(for a review see [1]). The most fundamental questions are (1) why this
water-splitting enzyme(WSE) undergoes a cyclic change of redox states; known
as S_i(i=0 to 4), with the minimum period of 4 in the abstracted electron
number(S_4 is short-lived(<1 ms) and returns to S_0 releasing O_2)[2], (2) why
S_1 is most stable into which S_0 decays with the life-time of order of 30 min
and the S_2 and S_3 states do with that of 1 min in normal conditions in the
dark[3]. We attempt to explain this minimum periodicity of four electron
oxidation by a "microsurface" model for catalytic two Mn ions[4,5], schematically illustrated in Figure 1. The idear that only two among four Mn ions
directly interact with a water molecule each seems to be supported by a number of biochemical findings of heterogeneity in the Mn pool associated with
PSII(see review in [6])as well as the observation of hyperfine broadening
of the multiline EPR signal from the antiferromagnetically coupled Mn(III)
and Mn(IV) ions[7] in the presence of ^{17}O-enriched water[8]. Further, EXAFS
(Extended X-ray Absorption Fine Structure) data on PSII particles show a Mn-
Mn distance of approx. 2.7 Å[9] in compatible with our microsurface model,
favoring neither "a pair of separated site" model exemplified by oxo-bridged
binuclear Ru(III) complex[10]nor "micropocket" model analogous to a binuclear Fe--Cu site in cytochrome c oxidase[11]. There is, however, only a poor
experimental information on the molecular nature of the catalytic Mn cluster
and its interaction mode with water as substrate. Therefore, we present a
molecular orbital(MO) study on an improved model enzyme-substrate complex
with more probable ligands than those invetigated[4,5].

2. DEPROTONATION ENERGETICS AND ELECTRON TRANSFER IN MODEL SYSTEMS
    The active site of WSE consists of two Mn ions which are designated by
Mn(V',σ';V",σ") with indication of the valency and spin states: V', σ' for
a right Mn' ion and V", σ" for a left Mn" ion, two bridge-ligands(L_b', $L_b"$),
two sets of in-plane ligands(L_p', $L_p"$), two fifth ligands(L_5', $L_5"$) and a
fractional species of decomposed water dimer(W). Since Mn' and Mn" appear
biochemically equivalent[1], we consider a near-symmetric case of $L_p'=L_p"$
and $L_5'=L_5"$. Further, L_b' and $L_b"$ are assumed to be OH⁻ because OH⁻ ions are
ubiquitous in aqueous solvent and Cl⁻ ions would result in much larger Mn-Mn
distance than the observed value of 2.69 Å[9] and other possible O^{2-} ions
would be difficult to be synthesized from O_2. We require that L_5' and $L_5"$
are equally acidic amino-acid residues in order to suppress the ET's from
either bound (HO'HO"H)⁻ or (HO'O"H)⁻ anion to a high-valent Mn(III,IV)
complex. Hence, we use a model ligand, OH⁻, both for L_5' and $L_5"$. A consideration on the net charge of the active site favors neutral in-plane ligands
so that one may adopt in-plane ligands of $L_p'=L_p"=(OH_2)_2$. A set of ligands,
L=($L_b'L_p'L_5'$)($L_b"L_p"L_5"$), are assumed to undergo neither oxidation nor
reduction reaction accompanied with any S-state transition.

Biggens, J. (ed.), Progress in Photosynthesis Research, Vol. I. ISBN 90 247 3450 9
ⓒ *1987 Martinus Nijhoff Publishers, Dordrecht. Printed in the Netherlands.*

TABLE 1. Proton transfer(PT) energies as a theoretical index of proton releasibility in each S-state defined in Figure 2.

S-states	Relations to model system PT energies		ΔF_{pt} (kcal/mol)	
	right-up	left-up	right-up	left-up
S_0	$= C_{II,3}$	$= C_{III,2}$	29.4	0.9
S_1^*	$= C_{III,3}$	the same	-0.7	the same
S_1	$= D_{III,3}$	the same	24.1	the same
S_2	$= D_{III,4}$	$= D_{IV,3}$	22.5	7.5
S_3^*	$\simeq C_{III,4}$	$\simeq C_{IV,3}$	$\simeq -3.9$	$\simeq -17.3$
S_3	$> D_{III,4}$	$> D_{IV,3}$	> 22.5	> 7.5
S_4^*	$\simeq C_{III,4}$	$\simeq C_{IV,3}$	$\simeq -3.9$	$\simeq -17.3$
S_4^{3*}	$\simeq C_{III,3}$	none	$\simeq -0.7$	none

The ab initio MO method to calculate the "proton transfer energy"(ΔE_{pt}), defined by the total energy difference between the post- and pre- proton transferred(PT) states, was described[4,5,12]. In the deprotonation energetics, a "noncentral" Mn ion together with the surrounding in-plane and 5th ligands were replaced by point charges, and the second and third solvation shells as a proton-accepter(i.e. $-O(-H-OH_2)_2$) were also took into account. These model Mn-complexes and their PT energies are denoted by the same symbols: $C_{V,v}$ for $W=H_2O-O^*H'(\equiv W_0)$ and $D_{V,v}$ for $W=H-(O-H-O^*)^--H'(\equiv W_1)$, where the suffices; V and v, indicate the valencies of the central and noncentral Mn ions respectively, O^* represents an oxygen atom coordinated to the central Mn ion and H' a proton releasing along the right-up or left-up direction in the x-z plane(the Mn'-Mn" axis is chosen as an x-axis and a z-axis is normal to the Mn(V',V") microsurface). The results of calculations were used in Table 1 to estimate the proton releasibility from each S-state schematically defined in Fig.2. Here, all ΔE_{pt}'s are added a constant shift of ca. 5.5 kcal/mol expected when Gaussian-type orbitals of double-zeta quality are used also for ligand atoms. The other two "noncatalytic" Mn atoms, polar amino acids, bound Ca^{2+} ions and bound Cl^- ions are also expected to cause another constant or S-dependent shift in the total sum to the ΔE_{pt}.

3. GROUND STATES OF S_0 AND S_1 STATES: GEOMETRY OPTIMIZATION

In order to compare the Mn-ligand and Mn'-Mn" distances in our model S_0 and S_1 states with reported EXAFS data[9], we optimized the geometries of model binuclear Mn complexes, L:Mn(II,↑;III,↑):W_0 and L:Mn(III,↑;III,↓):W_1, performing ab initio MO calculations with use of STO-3G basis set in the unrestricted Hartree-Hock approximation(we used the IMSPACK program system). In these calculations, the Mn-O in L_p, Mn-O in L_5 and Mn-O in W bond-lengths were optimized in mononuclear Mn complexes, but the bridging parts including the Mn':L_b'L_b":Mn" moiety and especially W_1 in S_1 were required to be optimized in the whole systems. The ground of S_0 was found to have approximately a total energy of $E_0=-3042.7010$ a.u. at the Mn-Mn distance: R(Mn-Mn)=2.74 Å. In this calculation, two water molecules in W_0 were taken to have their molecular plane parallel to the x'-z plane(which bisects the x-z and y-z planes). Significantly, the ground state of S_1 was found to be realized when the water bridge of $W_1=H'-(O'-H-O")^--H"$ takes a C_2-symmetric geometry

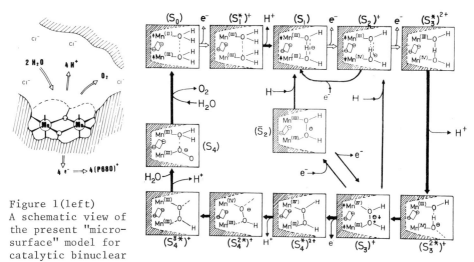

Figure 1(left)
A schematic view of
the present "micro-
surface" model for
catalytic binuclear
Mn site in the oxygen-evolving complex in photosynthesis; The catalytic
microsurface forms a bottom of the cavity into which just a water dimer is
absorbed to be oxidized by the cation radical, P680$^+$. The optimum size of
the cavity might be maintained at some biochemical conditions.

Figure 2(right) A theoretical model for the molecular mechanism of photosynthetic water decomposition; The catalytic binuclear Mn complex is assumed to
be nearly C_2-symmetric having OH$^-$ ion as two bridge-ligands(θ‿θ), O- or N-
as the other in-plane ligands and O$^-$ as the 5th ligands. Each symbol represents: →e$^-$, an electron abstraction; →H$^+$, a proton release; θ, an excess
electron; •, an unpaired electron;↑ or ↓, an electronic spin state of each
Mn atom. Relative charges of the defined Kok's S_i or transient $S_i^{m\cdot\cdot}$ states
relative to the S_0 state are written down outside each parenthesis.

at R(O'-O")=2.48 Å and R(O'-H)=R(O"-H)=1.24 Å as shown in Fig.2. Then, it
has a total energy of E_1=-3048.2864 a.u.(<<E_0) at R(Mn-Mn)=2.70 Å. This
remarkable stabilization(note that E_1<< E_0) is noteworthy, because, if W_1
is asymmetric, keeping the middle proton at R(O'-H)=0.957 Å, S_0 will take
the lowest energy of E_1'=-3042.0848 a.u.(> E_0) around R(Mn-Mn)=2.77 Å.
    Since the net charges of S_0 and S_1 are equally +|e$^-$|, one may conclude
that, if the interaction between the active site and the surroundings does
not compensate with the energy difference between E_1 and E_0, the S_1 state
could be more stable than the S_0 state in accord with experimental results
[3]. Further, a good agreement between the experimental and theoretical
values appears to strongly support the present theoretical model.

4. MOLECULAR MECHANISM
    An improved model for the molecular mechanism of cyclic water oxidation
is schematically given in Figure 2. This is the result of new model ligands,
L=(OH$^-$.OH$_2$.OH$_2$.OH$^-$)$_2$. Especially, the anionic fifth ligand(OH$^-$) reduces an
effective valency of Mn, which in turn increases the PT energy by an order
of 7 kcal/mol as compared with the case of a neutral fifth ligand(NH$_3$).
Another effect of the anionic fifth ligand is to make the 3d($2z^2-x^2-y^2$)
orbital vacant in both Mn(III) and Mn(IV) ions. Further, in the Mn(IV) ion,
another vacant orbital becomes one of three dε orbitals, i.e. 3d(x^2-y^2) or
3d(x'y') in standard coordinates, because its exchange repulsion to electrons
in occupied 3d(yz) and 3d(xz) orbitals is strongly enhanced by a large

asymmetry of crystal field. These effects seem to be quite effective in suppresing the ET's from W to the high-valent Mn(III,IV) complex in Fig.2.

Significantly, the presumed S_0 state can become stable against the deprotonation processes at least for such ligands adopted in this paper. Hence, we may not necessitate to assume a rapid electron-shuttle between two sites[5]. However, another assumption of ferromagnetic exchange coupling may be still necessary, since the S_0 state is known to display no EPR-sinal. Unfortunately, a little is known about the effective exchange interaction(EI) in di-μ-hydroxo-bimanganese complexes as compared with di-μ-oxo bridged ones. Quantum mechanical calculations show that the Mn-OH$^-$ distance is in general larger than the Mn-O^{2-} distance. This indicates that the kinetic EI between Mn spins via the bridge-ligands should be less in the former than in the latter. In fact the observed antiferromagnetic EI between high-spins of Mn(III) and Mn(IV) ions is much less in S_2 ($2J_2^*=-19\pm4$ cm^{-1}[8]) than in di-μ-oxo-tetrakis(2,2'-bipyridine) dimanganese(III,IV) perchlorate($2J=-300\pm14$ cm^{-1}[7]), in spite of the S_2 state involving an excess antiferromagnetic EI($\equiv J_{W1}$) through MO's of the $-(O'-H-O'')^-$ bridge. Hence, if the Mn-Mn distance is kept short by an external force of surrounding protein and is best fit in the most stable S_1 state, the sum of four ferromagnetic EI's in S_0 is likely to overcome such weak antiferromagnetic EI's. In S_1, the effective EI is approx. given by $J_1^*\simeq[12J_2^*+J_D(x^2-y^2)+2J_K(xy,x^2-y^2)]/16$, where J_D (>0) is a direct EI between the $3d(x^2-y^2)$ electrons and J_K an antiferromagnetic EI between the $3d(xy)$ and $3d(x^2-y^2)$ electrons via a filled $2p(x-y)$ orbital of OH$^-$. Further, the effective EI in S_3 is found to be $J_3^*\simeq J_2^*+(|J_{W1}|+J_{W2})/2$ with J_{W2} being a ferromagnetic EI between the $3d(xz)$ electrons via a half-filled MO of the $-(O'-O'')^-$ bridge. J_1^* and J_3^* are expected to be antiferromagnetic and ferromagnetic respectively.

REFERENCES
1 Renger, G. and Govindjee(1985) Photosynth; Res. 6, 33-55.
2 Kok, B., Forbush, B., and McGloin, M.(1970) Photochem. Photobiol. 11, 457 -475.
3 Vermaas, W.F.J., Renger, G. and Dohnt, G.(1984) Biochim; Biophys. Acta 704, 194-202.
4 Kusunoki, M.(1983) in The Oxygen Evolving System of Photosynthesis(Inoue, Y. et al. eds) pp.165-173, Academic Press, Tokyo.
5 Kusunoki, M.(1984) in Advances in Photosynthesis Research(Sybesma, C. ed.) Vol.1, pp.275-278, Nijhoff/Junk, Hague.
6 Mavankal, G., McCain, D.C. and Bricker, T.M.(1986) FEBS Lett. 202, 235-239.
7 Dismukes, G.C. and Siderer, Y.(1981) Proc. Natl. Acad. Sci. USA 78,274-278.
8 Hansson, Ö., Andreasson, L.-E. and Vänngård (1986) FEBS Lett. 195, 151-154.
9 Yachandra, V.K., Guiles, R.D., McDermott, A., Britt, R.D., Dexheimer, S.L. Sauer, K. and Klein, M.P.(1986) Biochim; Biophys; Acta 850, 324-332.
10 Gersten, S.W., Samuels, G.J. and Meyer, T.J.(1982) J. Am. Chem. Soc. 104, 4029-30.
11 Kusunoki, M., Kitaura, K., Morokuma, K. and Nagata, C.(1980) FEBS Lett. 117, 179-182.

ACHNOWLEDGEMENTS
 This work is supported by the Grant-in-Aid for Scientific Research from the Ministry of Education, Science and Culture. Calculations were done at the computer centers of Univ. of Tokyo, Institute for Molecular Science and Meiji Univ..

DYNAMIC LINEARITY OF THE BARE PLATINUM ELECTRODE FOR OXYGEN EXCHANGE
MEASUREMENTS IN MARINE ALGAE

S.I. SWENSON, C.P. MEUNIER, K. COLBOW

Department of Physics, Simon Fraser University, Burnaby, B.C., Canada
V5A 1S6

1.   INTRODUCTION

The bare platinum and Ag/AgCl electrode system (1) provides a current
due to O_2 reduction at the cathode and is linear over a wide range of
ambient O_2 concentrations during steady-state measurements. Dynamic
measurements of O_2 evolution and/or uptake in biological systems are
limited by the response time of the electrode system. Oxygen exchange
measurements in marine algae due to short, saturating light flashes (5 µs
at 3.3 Hz) result in a "pile-up" of individual oxygen pulses because the
relaxation time of the system to a steady-state value is much longer than
the time between flashes. A minimum flash frequency of 1 Hz is needed to
minimize decay of the S_2 and S_3 states during the dark interval between
flashes (2,3); thus the pile-up of pulses is unavoidable. If the pile-up
of pulses affects the linearity of the system, then quantitative calculations of oxygen evolution are inaccurate.
The O_2 reduction current is independent of the applied potential in
the region of -0.7 V (Pt vs Ag/AgCl) (4), and thus the current is
diffusion-controlled. Since the diffusion equations are linear, the system
should be linear. Linearity means that the total output function due to a
series of excitations is simply the sum of all individual responses (5)
(e.g. the O_2 current pulses). Thus, it should be possible to describe the
waveform of the entire flash sequence as a superposition of the O_2 signals
from each flash in the sequence.

2.   MATERIALS AND METHODS

Oxygen exchange (evolution and uptake) measurements on Ulva sp. were
made using a bare platinum electrode (6,7) in direct contact with the
sample, and a silver/silver chloride counter electrode. The Pt electrode
was biased to -0.700 ± 0.001 V with respect to the Ag/AgCl electrode by a
potentiostat. The excitation source was an EG&G FX 249 xenon flash lamp
operated at 1.0 kV with a 10 µF discharge capacitor resulting in a pulse
width at half height of 5 µs. The flash frequency was set to 3.3 Hz by a
Hewlett Packard 3301B function generator which also triggered a Tracor
Northern Model TN 1710 signal averager simultaneously with the first flash.
The energy output per pulse of the xenon flash at 3.3 Hz was determined to
be 3.0 ± 0.3 mJ using a Scientech 361 power meter with detector model VPH-2
(Newport Research Corporation). The signal produced by O_2 reduction at the
Pt cathode was measured across a 40 kΩ resistor inside the potentiostat and
input to the signal averager.
Total chlorophyll content in Ulva was measured in 80% acetone using
the method of Arnon (8).

Biggens, J. (ed.), Progress in Photosynthesis Research, Vol. I. ISBN 90 247 3450 9
© *1987 Martinus Nijhoff Publishers, Dordrecht. Printed in the Netherlands.*

3.   RESULTS AND DISCUSSION

To correct for the pile-up of pulses, the data were fitted with the Simplex algorithm (9,10).   The characteristic electrode response for O_2 evolution in <u>Ulva</u> was determined by giving three 5 µs saturating light flashes to dark-adapted samples, and allowing the signal to decay back to the original baseline, as shown in Fig. 1.   The O_2 exchange curve was

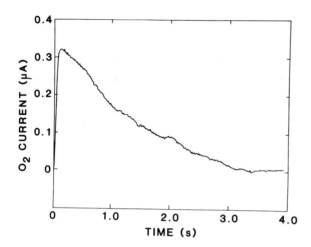

Figure 1.   The O_2 reduction current pulse due to the third 5 µs light flash at 3.3 Hz flash frequency represents the characteristic (reference) response of the system to O_2 produced by <u>Ulva</u>.

generated by adding pulses with the same shape as the third pulse, shifted along the time axis by the appropriate time interval between pulses.   The amplitude of each pulse was determined to give the best fit to the experimental O_2 exchange data (Fig. 2).   This fitting procedure assumes that the

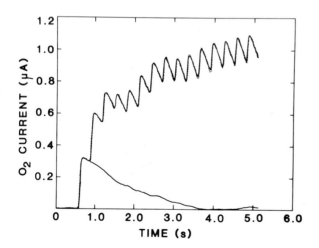

Figure 2. O_2 exchange in Ulva (————) using a bare Pt electrode is shown to be a sum of multiples (-----) of the characteristic curve (Fig. 1) (————).   As can be noted from the figure, the experimental and fitted data are nearly indistinguishable.

time course of the O_2 current from a single flash delivered at different times changes in scale, but not in shape.   This is a reasonable assumption since the evolution of O_2 follows four-step kinetics which only change in magnitude as a function of flash number (2), the half-time for O_2 production is independent of flash number (11,12), and O_2 is most likely produced at the same locations within the algae. In previous experiments, the pile-up of O_2 due to several light flashes (as seen in Fig. 2) was corrected for by extrapolating the decay of the O_2 signal from the previous pulse to account for residual O_2 (7). The use of the Simplex algorithm to fit the experimental data not only corrects for the pile-up of O_2 current pulses, but supports the initial hypothesis that the system is linear under dynamic measurements.

During the computer fitting, the amplitude of each O_2 pulse relative to the reference response (Fig. 1) was determined. Plotting the relative amplitudes as a function of flash number gives the flash yield sequence for the computer-generated oxygen exchange curve. This is compared to the flash yield sequence for the previous method (7) of determining the amplitudes of the O_2 current pulses in Fig. 3.

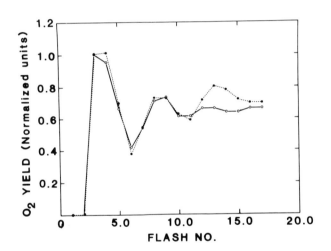

Figure 3. The O_2 yield as a function of flash number is plotted for the O_2 exchange curve of Fig. 2. The flash yield sequence for the computer fit (•------•) is compared to the method of extrapolating the decay of the previous pulse (o———o) (7). The extrapolation was carried out until it intersected a vertical line drawn from the maxima of the given pulse.

The first two oscillations in yield are similar for both flash yield sequences in Fig. 3. However, the third oscillation in yield for the computer-fitted curve is larger than was previously obtained by extrapolating the decay of the O_2 current pulses. This could be due to an increased error in the fitting after the twelfth flash, or may indicate that the O_2 yield for <u>Ulva</u> may not be damped to a steady-state value as quickly as was previously determined (7).

The net amount of O_2 evolved during the third flash (Fig. 1) can be determined by integration of the O_2 reduction current. The net charge produced by reduction of O_2 at the cathode is computed to be 4.15×10^{-7} Coulombs. With four electrons removed at the Pt electrode per O_2 molecule reduced, the amount of O_2 produced by Ulva during the reference response (Fig. 1), was calculated to be 6.63×10^{11} molecules. The net amount of O_2 produced per flash can then be determined by multiplying this value by the relative amplitude determined during the computer fitting for each of the pulses in the sequence.

This information can be used to determine the photosynthetic unit size for Ulva sp. The relative amplitudes of O_2 produced on the first four flashes (one cycle of the oxygen-evolving complexes) were 0.003, 0.006, 0.984, and 0.992, corresponding to 1.3×10^{12} O_2 molecules. Each sample was determined to contain 2.8×10^{15} chlorophyll molecules (4.2 µg chl). This is equivalent to a photosynthetic unit size of 2290 ± 150 chlorophyll molecules, close to the value of 2360 ± 160 chl/O_2 calculated for Ulva lactuca by Mishkind and Mauzerall (13).

4. CONCLUSIONS

The behavior of a bare platinum electrode to O_2 exchange in marine algae is linear. The response of the electrode/algae system to the third saturating light flash for dark-adapted algae is characteristic of the electrode system. The experimental O_2 exchange curve from series of saturating flashes can be obtained by adding time-axis shifted multiples of the response to the third flash, indicating dynamic linearity of the system. This allows quantification of the net amount of oxygen produced per flash and determination of a photosynthetic unit size of 2290 ± 150 chl/O_2 for Ulva sp.

5. REFERENCES

1 Haxo, F.T., Blinks, L.R. (1950) J. of Gen. Physiol. 33, 389-422
2 Kok, B., Forbush, B. and McGloin, M. (1971) Photochem. Photobiol. 11, 457-475
3 Joliot, P., Joliot, A., Bouges, B. and Barbieri, G. (1971) Photochem. Photobiol. 14, 287-305
4 Olson, R.A., Brackett, F.S. and Crickard, R.G. (1949) J. of Gen. Physiol. 32, 681-703
5 Oppenheim, A.V., Willsky, A.S. and Young, I.T. (1983) Signals and Systems, pp. 43-45, Prentice Hall, Englewood Cliffs, N.J.
6 Chandler, M.T. and Vidaver, W.E. (1971) Rev. Sci. Instrum. 42, 143-146
7 Swenson, S.I., Colbow, K. and Vidaver, W.E. (1986) Plant Physiol. 80, 346-349
8 Arnon, D.I. (1949) Plant Physiol. 24, 1-15
9 Nelder, J.A. and Mead, R. (1965) Computer J. 7, 308-314
10 Caceci, M.S. and Cacheris, W.P. (1984) Byte, 340-362
11 Joliot, P., Hofnung, M. and Chabaud, R. (1966) J. Chim. Phys. 63, 1423-1441
12 Dekker, J.P., Plijter, J.J. and Ouwehand, L. (1984) Biochim. Biophys. Acta 767, 1-9
13 Mishkind, M. and Mauzerall, D. (1977) Biol. Bull. 153, 440.

A DYNAMIC MODEL FOR THE BARE PLATINUM ELECTRODE

C.P. MEUNIER, S.I. SWENSON, K. COLBOW

Department of Physics, Simon Fraser University, Burnaby, B.C., Canada
V5A 1S6

1. INTRODUCTION
    Measurements always perturb the system on which they are performed.
Knowledge of the measuring apparatus and the system to be measured generally permits correction and/or prediction of distortions in the system. The
objective of this paper is to determine the perturbations in marine algae
caused by measurements of oxygen exchange with a bare Pt electrode.

2. PROCEDURE
2.1. Materials and methods
    Algae (Ulva sp.) was held tightly on a circular (0.6 mm diam.) Pt
electrode by a dialysis membrane. The membrane was soaked in sea water and
stretched to cover a Ag/AgCl counter electrode. The Pt electrode was
biased to -0.700 ± 0.001 V vs Ag/AgCl by a potentiostat. The output of the
potentiostat was fed to a Tracor-Northern TN-1710 signal averager interfaced to a microcomputer. The algae were illuminated by 5 µs white light
flashes from an EG&G FX249 xenon flash lamp with a 10 µF capacitor. The
light intensity was determined to be saturating when the flash lamp was
operated at 1kV. For experiments the lamp was operated at 1.5kV as a
safety margin. The bare platinum electrode, the sample holder and experimental procedure have been previously described [1]. The thickness of Ulva
was determined to be 70 microns with a micrometer. Care was taken not to
pressurize the algae. The thickness of unstretched dialysis membrane was
measured to be 20 microns.

2.2. Data processing
    The experimental data was fitted by multivariable regression. The
coefficients which could not be put into linear form were slowly varied by
a master procedure using the Simplex algorithm. The coefficient of determination (r^2) is the objective function maximized by the Simplex algorithm.
This coefficient is the fraction of initial variance removed by the fit
[2]. The regressions were performed in an imaginary space with N+1 dimensions and no time axis, where N is the number of independent variables.
The dependent variable is the amplitude of the data. The first 30 ms after
each flash were not considered by the regression procedure in order to
avoid the influence of noise. In Fig. 1, the corresponding regions were
deleted and replaced by straight lines.

3. THEORY
3.1. Diffusion of oxygen
    The observed current from the oxygen reduction reaction does not
depend on variations of the voltage. Therefore, it is assumed that the
oxygen concentration at the Pt electrode is close to zero; the current is
then diffusion-controlled [3]. At the surface of the dialysis membrane,

Biggens, J. (ed.), Progress in Photosynthesis Research, Vol. I. ISBN 90 247 3450 9
© 1987 Martinus Nijhoff Publishers, Dordrecht. Printed in the Netherlands.

oxygen is exchanged with air, and the exchange rate can be approximated to be K times the difference between the actual concentration and the equilibrium concentration, where K is a constant.

The general solution to the diffusion equations can be decomposed in two parts. The steady-state solution is a straight line going from zero concentration at the Pt electrode to some finite concentration at the air-water interface. The transient solution is:

$$[O_2]_{tr}(x,t) = \sum_{n=1}^{\infty} B_n * SIN(P_n * x) * EXP(-P_n^2 * D * t) \tag{1}$$

where x is the distance from the Pt electrode, D is the diffusion constant, and the Fourier development coefficients B_n are given by:

$$B_n = \int_0^d f(x) * SIN(P_n * x) \, dx; \tag{2}$$

where f(x) is the difference between the initial distribution of oxygen and the steady-state distribution. P_n can be found from the transcendental equation:

$$-D * P_n / K = TAN(P_n d - n\pi) \tag{3}$$

where n is an integer greater or equal to one, d is the distance from the Pt electrode to the air-water interface, and K is the proportionality constant mentioned before.

Four electrons are donated to O_2 by the Pt electrode during the cathodic reaction [4]. Thus the current at the electrode is four times the oxygen flux towards the Pt electrode.

3.2. System analysis

The response of a linear system to a delta function input characterizes completely that system [5]. An input with a duration shorter than the fastest time constant of importance in that system can be considered a delta function input. The Pt and Ag/AgCl electrode system is linear (S.I. Swenson, C.P. Meunier, K. Colbow, unpublished). The input to that system is given by the oxygen sources in the electrolyte (the algae). For that input to be considered a delta function, the time scale of oxygen production must be such that the corresponding diffusion length of oxygen is shorter than the distance from the sources to the Pt electrode. The slowest step of oxygen production is supposed to be the $S_3-(S_4)-S_0$ transition, with a time scale of 1 ms [3,6]. The diffusion length of oxygen during this period is approximately 1 μm. Therefore, if there are no sources closer than 1 micron from the Pt electrode, the oxygen produced on the third flash can be considered a delta function input to the electrode system, and the oxygen reduction current obtained characterizes completely that system.

4. RESULTS AND DISCUSSION

The oxygen produced on the third flash causes a rise in concentration which can be considered to be the function f(x) in Eq. 2. The characteristic response (e.g. Fig. 1) can be fitted to the theoretical expression of the current derived from Eq. 1 with a determination coefficient (r^2) usually around 0.9995; the fitted curves (not shown) are indistinguishable from the data. The coefficients of the space-Fourier development of the initial distribution of oxygen, B_n, are determined by the computer fit. The reconstructed distribution is shown in Fig. 2.

The distribution of oxygen after the third flash (Fig. 2) shows two maxima corresponding to the two layers of chloroplasts known to exist in Ulva. On the average, the first layer seems to be 10 µm away from the Pt electrode, and therefore the curve in Fig. 1 is the characteristic response of the electrode system. The algae extends for about 70 µm, as measured before, and the dialysis membrane has only a part of its 20 µm represented. The layer closer to the Pt electrode seems to have produced less oxygen, although both chloroplasts received saturating light. It is known that anaerobic conditions inhibit O_2 production [7,8]. It is therefore reasonable to assume that before the third flash, the layer closer to the Pt electrode suffered from low O_2 concentration.

Because of the Pt electrode, the algae does not behave like a linear system. The amount of O_2 evolved is dependent on the ambient concentration of oxygen, which in turn depends on previous amounts of evolved oxygen. In the frame of the S-state theory, this introduces a dependence of the 'misses' on O_2 concentration. Care should be taken to minimize this dependence or correct data to account for it.

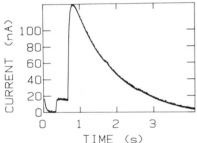

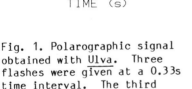

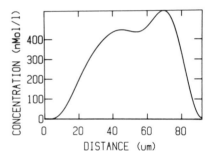

Fig. 1. Polarographic signal obtained with Ulva. Three flashes were given at a 0.33s time interval. The third flash occurs at 0.67s.

Fig. 2. The initial O_2 distribution after the third flash, computed from Fig. 1.

5. CONCLUSIONS

The theoretical expressions resulting from the solution to the diffusion equation were fitted to data with $r^2 = 0.9995$. The oxygen distribution after the third flash can be computed from the coefficients of the fit. This initial distribution shows that less oxygen is produced close to the Pt electrode, most likely due to inhibition by low oxygen concentration. The amount and localization of evolved oxygen would vary from one flash to another due to changing ambient O_2 concentration, as well as due to the S-state transitions. The algae does not behave like a linear system under these conditions. Therefore the data should be corrected, or a different polarization method used which consumes less oxygen.

REFERENCES

1 Swenson, S.I., Colbow, K. and Vidaver, W.E. (1986) Plant Physiol. 80, 346-349
2 Kleinbaum, D.G., Kupper, L.L. (1978) in Applied Regression Analysis and other Multivariable Methods (Service to Publishers Inc., ed.), pp. 158-162, Duxbury Press, Belmont, California
3 Joliot, P., Hofning, M. and Chabaud, R. (1966) J. Chim. Phys., 63, 1423-1441
4 Hoare, J.P. (1985) J. Electrochem. Soc., 132, 301-305
5 Gille, J-Ch., Decaulne, P. and Pelegrin, M. (1981) in Dynamique de la commande lineaire (Dunod, ed.), 6th ed., pp. 275-284, Bordas, Paris
6 Dekker, J.P., Plijter, J.J., Ouwehand, L. and Van Gorkom, H.J. (1984) Biochim. Biophys. Acta, 767, 176-179
7 Diner, B.A. (1974) in Proceedings of the IIIrd International Congress on Photosynthesis (Avron, M., ed.), Vol. 1, pp. 589-601, Elsevier, Amsterdam
8 Diner, B.A. (1977) Biochim. Biophys. Acta 260, 247-258

INDEX OF NAMES